MATERIALS SCIENCE

MATERIALS SCIENCE

Professor Arthur L. Ruoff

Department of Materials Science and Engineering
College of Engineering
Cornell University

Prentice-Hall, Inc., Englewood Cliffs, New Jersey

© 1973 by
Prentice-Hall, Inc.
Englewood Cliffs, N.J.

10 9 8 7 6 5 4 3 2 1

ISBN: 0–13–560805–8

Library of Congress Catalog Card Number: 72-3875

Printed in the United States of America

PRENTICE-HALL INTERNATIONAL, INC., *London*
PRENTICE-HALL OF AUSTRALIA, PTY., *Sydney*
PRENTICE-HALL OF CANADA, LTD., *Toronto*
PRENTICE-HALL OF INDIA PRIVATE LIMITED, *New Delhi*
PRENTICE-HALL OF JAPAN, INC., *Tokyo*

To My Parents:
Who Believe in Freedom

CONTENTS

PREFACE   xix

FOREWORD TO THE STUDENT   xxv

1 **AN INTRODUCTION TO THE SUBJECT OF
   MATERIALS SCIENCE**   1
   REFERENCES   3
   PROBLEMS   4

2 **MECHANICAL PROPERTIES**   7

   2.1   Young's Modulus and Tensile Yield Strength
         as Design Parameters   7
   2.2   The Shear Stiffness Constant   17
   2.3   Definition of Stress, Strain, and Tensile Modulus
         and Shear Modulus   21
   2.4   Mechanical Damping   28
   2.5   Plastic Deformation   35
   2.6   Fatigue, Creep and Notch Fracture   44
   2.7   Stiffness to Weight and Strength
         to Weight Ratios   49
   2.8   The Flow of Fluids   50
   2.9   Friction   55
         REFERENCES   56
         PROBLEMS   56

3 PHENOMENOLOGY OF ELECTRICAL BEHAVIOR   **65**

3.1   Resistivity as a Design Parameter   65
3.2   Metals   68
3.3   Semiconductors   72
3.4   Insulators   75
3.5   Ferroelectrics   81
3.6   Piezoelectric Crystals   83
       REFERENCES   85
       PROBLEMS   85

**4 SOME PHENOMENOLOGICAL ASPECTS OF
    MAGNETIC MATERIALS**   **89**

4.1   A Solenoid   89
4.2   Ferromagnetic Materials   91
4.3   Soft Magnetic Materials   95
4.4   Hard Magnetic Materials   99
4.5   Special Magnetic Materials   101
4.6   Magnetostriction   101
       REFERENCES   102
       PROBLEMS   102

5 SUPERCONDUCTIVITY   **107**

5.1   Superconducting Magnet   107
5.2   Superconductors   108
5.3   Hard Superconductors   110
5.4   Synthetic Hard Superconductors   112
5.5   Applications   112
       REFERENCES   113
       PROBLEMS   113

6 PHENOMENOLOGY OF TEMPERATURE FIELDS   **117**

6.1   Space Rockets and Thermal Expansion Coefficient   117
6.2   Thermal Conductivity   119
6.3   Heat Capacity   123
6.4   Equation of Continuity and the
       Parabolic Heat Equation   125
6.5   Energy Changes During Phase Transformation   127
6.6   Temperature Effects on Properties   128
6.7   Heat Treatment   129
6.8   Electrical-Thermal Coupling   131
       REFERENCES   133
       PROBLEMS   133

7 LINEAR BEHAVIOR   137

7.1   Linear Relations   137
7.2   More Detailed Discussion of Stress   139
7.3   Transformation of Axes   141
7.4   Other Second-Rank Tensors   144
       REFERENCES   145
       PROBLEMS   146

8 ELECTRONS AND ATOMS   151

8.1   Wave Mechanics   151
8.2   The Hydrogen Atom   155
8.3   The Hydrogen-like Atom and the Periodic Table   158
8.4   Magnitudes of Binding Energies and
       Particle Sizes   163
       Appendix 8A Electronic Structures of the Atoms   164
       REFERENCES   166
       PROBLEMS   167

9 BINDING OF ATOMS   173

9.1   Introduction   173
9.2   The Ionic Bond   173
9.3   Covalent Bonding   176
9.4   The Metallic Bond   179
9.5   Secondary Bonds   182
       REFERENCES   186
       PROBLEMS   187

10 CRYSTALS   195

10.1   Introduction   195
10.2   Periodicity and Lattices   196
10.3   Symmetry Properties of Crystals   198
10.4   Some Simple Crystals   203
10.5   Density of Crystals   205
10.6   Crystallographic Directions and Planes   206
10.7   Piezoelectricity and Symmetry   209
10.8   Anisotropy and Crystal Structure   211
10.9   More About Symmetry   214
        REFERENCES   218
        PROBLEMS   218

11 DIFFRACTION BY CRYSTALS   223

11.1 Bragg's Law   223
11.2 Bragg's Law and Simple Cubic Crystals   224
11.3 Bragg's Law and Cubic Crystals in General   228
11.4 The Effect of Atomic Vibration   229
11.5 Neutrons and Electrons   230
      REFERENCES   232
      PROBLEMS   232

12 PACKING   237

12.1 Introduction   237
12.2 Closest Packing of Spheres   237
12.3 Voids in Closest-Packed Structures   239
12.4 The Body-Centered Cubic Structure   242
12.5 The Structure of Elemental Metals   243
12.6 Structure of Simple Alloys   244
12.7 Structure of Simple Ionic Compounds   248
12.8 The Structure of Simple Interstitial Compounds   252
12.9 The Structure of $BaTiO_3$ 253
12.10 Structure of Silicates   254
12.11 Glasses   258
      Appendix 12A   Metallic Valence and Metallic   261
      Radii of the Elements in Angstroms for $CN = 12$
      Appendix 12B   Radii of Covalent Elements in Angstroms   262
      Appendix 12C   Radii of Ions in Angstroms   262
      REFERENCES   263
      PROBLEMS   263

13 CRYSTALLINE IMPERFECTIONS   267

13.1 Imperfections and Crystal Properties   267
13.2 Point Imperfections   268
13.3 Line Imperfections   274
13.4 Planar Imperfections   282
13.5 Spatial Imperfection   292
      REFERENCES   292
      PROBLEMS   292

14 POLYMERS   299

14.1 Introduction   299
14.2 An Idealized Random Chain   301

14.3   Degree of Polymerization   303
14.4   Vinyl Polymers   304
14.5   Other Addition Polymers   306
14.6   Copolymers   308
14.7   Condensation Polymers   309
14.8   Network Polymers   312
14.9   Thermoplastics and Thermosetting Resins   313
14.10   Crystallinity in Polymers   313
14.11   Macromolecules in Living Matter   314
        REFERENCES   317
        PROBLEMS   318

15   MICRO- AND MACROSTRUCTURE OF MATERIALS   321

15.1   Introduction   321
15.2   The Reflection Microscope   321
15.3   Polycrystalline Materials   324
15.4   Polyphase Materials   330
15.5   Composite Materials   331
15.6   Quantitative Microscopy   336
        REFERENCES   337
        PROBLEMS   338

16   PHASE DIAGRAMS   341

16.1   The Phase Rule   341
16.2   Pure Substances (Elements or Compounds)   342
16.3   Variable-Composition Systems   345
16.4   Nonequilibrium Transformations   355
16.5   Two Important Nonequilibrium
        Transformations   360
16.6   Segregation in Binary Alloys
        During Solidification   369
16.7   Zone Refining and Leveling   372
16.8   Application of Phase Diagrams   374
16.9   Determination of Phase Diagrams   376
16.10   Ternary Diagrams   377
        REFERENCES   380
        PROBLEMS   381

17   A DISCUSSION OF STATISTICAL THERMODYNAMICS   385

17.1   Introduction   385
17.2   Atom Motion and Temperature   386

17.3   Random Distribution   387
17.4   Velocity Distribution in Ideal Gases   392
17.5   Internal Energy   395
17.6   Randomness and Entropy   401
17.7   Equilibrium at Constant Temperature
       and Pressure   409
17.8   Applications of Thermodynamics   413
17.9   Phase Equilibria in Two-Component Systems   416
17.10 Calculation of Internal Energy   422
17.11 Vibrational Energy   425
17.12 Vibrational Entropy   430
17.13 Thermal Expansion   432
17.14 Thermal Conductivity   433
       REFERENCES   435
       PROBLEMS   435

18   KINETICS   441

18.1   Introduction to Kinetics   441
18.2   Introduction to Diffusion   445
18.3   Thin Layer of Tracer Atoms on a Thick Slab   447
18.4   The Microscopic Viewpoint of Diffusion   451
18.5   Interstitial Impurity Diffusion   454
18.6   The Diffusion of Vacancies   457
18.7   Self-Diffusion and Tracer Diffusion   460
18.8   Ionic Crystals   461
18.9   Diffusion at Surfaces, in Grain Boundaries,
       and Along Dislocations   465
18.10 Chemical Diffusion   466
18.11 Diffusion in Polymeric Materials   467
18.12 Illustration of Applications   470
18.13 Generalized Diffusion Theory   474
18.14 Nucleation   476
18.15 Spinodal Decomposition   484
       REFERENCES   485
       PROBLEMS   485

19   EXTREMES OF MECHANICAL BEHAVIOR   493

19.1   Introduction   493
19.2   Elastic Behavior of Crystals
       Under Hydrostatic Pressure   494

19.3   Elastic Constants of Crystals   496

19.4   Rubber-like Elasticity   501

19.5   Viscosity of Gases   509

19.6   Viscosity of Liquids   511

19.7   Viscosity of Crystals   515

     References   518

     Problems   519

20   ANELASTICITY AND VISCOELASTICITY   **525**

20.1   Introduction   525

20.2   The Two-Element Model   525

20.3   Loss Factor in Rubber   528

20.4   Generalized Viscoelasticity   530

20.5   Internal Friction in Single Crystals Owing to Motion
       of Interstitial Atoms   533

20.6   Other Damping Mechanisms in Crystals   535

20.7   Viscoelasticity of Polymeric Materials   537

     References   539

     Problems   540

21   PLASTIC FLOW OF CRYSTALS   **545**

21.1   Introduction   545

21.2   Slip   546

21.3   The Stress Required for Slip   550

21.4   The Structure-Sensitive Nature of Slip
       in Crystals   553

21.5   Fracture of Crystals   556

21.6   Deformation Twinning   559

     References   561

     Problems   561

22   PROPERTIES OF DISLOCATIONS   **565**

22.1   Introduction and History   565

22.2   The Stress Fields of Dislocations   568

22.3   Free Energy of Formation of Dislocation   573

22.4   Dislocation Reactions   575

22.5   The Force on a Dislocation Line   579

22.6   Dislocation Climb   581

22.7   The Peierls-Nabarro Stress   583

22.8   Dislocation Intersections   584
22.9   Multiplication   586
22.10   Dislocation Velocity   589
22.11   The Forces Between Dislocations   590
22.12   Arrays of Dislocations   593
22.13   Interaction of Impurity Atoms with
        Dislocations   596
        REFERENCES   597
        PROBLEMS   597

23 ON UNDERSTANDING DEFORMATION AND
    ANNEALING IN CRYSTALLINE MATERIALS   603

23.1   Introduction   603
23.2   Yield Points   604
23.3   Flow Stress and Dislocation Densities   607
23.4   Annealing of Cold-Worked Materials   609
23.5   Recrystallization   612
23.6   Grain Growth   614
23.7   Yield Strength vs Temperature   616
23.8   High-Temperature Creep   618
23.9   Fracture of Materials   624
23.10   Fatigue   634
        REFERENCES   636
        PROBLEMS   636

24 PLASTICITY OF POLYCRYSTALLINE AGGREGATES   641

24.1   Yielding   641
24.2   The Tension Test   648
24.3   Cold-Working Processes   650
24.4   The Hardness Test   651
24.5   Anisotropic Plasticity   653
        REFERENCES   653
        PROBLEMS   654

25 STRENGTHENING MECHANISMS   659

25.1   Introduction   659
25.2   Grain Size   660
25.3   Precipitation Hardening   661
25.4   Dispersion-Hardened Alloys   666
25.5   Periodic Structures   667

25.6 Martensite Formation 669
25.7 Strengthening at High Temperatures 678
25.8 Strengthening Mechanisms in Polymers 678
25.9 Composites 679
25.10 Materials Selection for
      Mechanical Strengths 685
25.11 A Review of Mechanical Behavior 686
      REFERENCES 687
      PROBLEMS 688

26 ELECTROCHEMICAL PROPERTIES   691

26.1 Introduction 691
26.2 Electrode Potentials 694
26.3 Concentration Cells 698
26.4 Electrode Potential and Nernst Equation 699
26.5 Transference 701
26.6 Corrosion 702
26.7 Protecting Against Corrosion 705
26.8 Polarization and Overvoltage 708
      REFERENCES 712
      PROBLEMS 712

27 ELECTRONS IN CONDENSED PHASES   717

27.1 The Electron Gas 717
27.2 The Quantized Electron Gas 721
27.3 Fermi-Dirac Distribution 726
27.4 Electrons in a Periodic Potential 728
27.5 Brillouin Zones 731
27.6 Constant Energy Curves 733
27.7 Number of States in an Energy Band and
      Effective Mass 736
27.8 Real Crystals and Conductivity 737
27.9 The Origin of Resistivity in Metals 742
27.10 Electron Emission 746
      REFERENCES 754
      PROBLEMS 754

28 SEMICONDUCTORS   761

28.1 Introduction 761
28.2 Intrinsic Semiconductors 762

28.3 Mobilities 765
28.4 Extrinsic Semiconductors 766
28.5 Recombination of Excess Carriers 770
28.6 The p-n Junction 772
28.7 The Junction Transistor 778
28.8 Semiconductor Devices 782
       REFERENCES 782
       PROBLEMS 783

29 SUPERCONDUCTIVITY   789

29.1 Introduction 789
29.2 Superconducting Phase and the Normal Phase 794
29.3 The Penetration Depth 797
29.4 Phase Separation 799
29.5 Hard Superconductors 803
       REFERENCES 805
       PROBLEMS 805

30 DIELECTRICS   809

30.1 Introduction 809
30.2 Macroscopic and Microscopic
       Meaning of Polarization 809
30.3 Molecular Mechanisms of Polarization 811
30.4 Polarizability 813
30.5 Approximate Evaluation of Internal Fields 820
30.6 Behavior of Dielectrics in Varying Fields 822
30.7 The Classical Theory of Electronic Polarization and
       Resonance Absorption 827
30.8 Dielectric Strength 828
30.9 Other Dielectric Properties 829
       REFERENCES 832
       PROBLEMS 833

31 OPTICAL PROPERTIES   837

31.1 Introduction 837
31.2 Refraction 837
31.3 Absorption 843
31.4 Reflection 848
31.5 The Photographic Process 849
31.6 Liquid Crystals 849

31.7   Some Electro-Optic Phenomena   851
31.8   Excitons   851
31.9   Luminescence   853
31.10 Lasers   855
       REFERENCES   859
       PROBLEMS   859

32 MAGNETIC MATERIALS   **863**

32.1   Introduction   863
32.2   Diamagnetism   865
32.3   Paramagnetism   865
32.4   Spontaneous Magnetization   872
32.5   The Origin of Ferromagnetism   874
32.6   Antiferromagnetism   876
32.7   Ferrimagnetism   876
32.8   Domains   878
32.9   Magnetization Processes According to Domain
       Theory   883
       REFERENCES   887
       PROBLEMS   887

APPENDIX   **891**

INDEX   **895**

PREFACE

PURPOSE OF THE BOOK

This book is intended to introduce students of engineering and science to those fields of materials science which, from the modern point of view, seem to be the most important in connection with practical problems [all things (excepting ideas, plans, etc.) are made from materials]. Topics are selected according to the expected frequency of occurrence in applications. New concepts and methods of modern materials science instruction as expressed in various recent symposia on engineering education were taken into account. The book should suit those institutions that have offered extended materials science courses for a long time as well as those that intend to follow the general trend of broadening their programs of instruction in materials science.*

A course in calculus with analytic geometry, and elementary chemistry and physics are the prerequisites.†

The material included in the book has formed the basis of various one and two semester courses given to sophomores and juniors.

CONTENT AND ARRANGEMENT

A student of materials science has to be familiar with material properties (*phenomenology*), with the electronic and atomic structure and microstructure and macrostructure of materials (*structure*) and with the fundamental theory at the atomic and molecular level which predicts the behavior (*fundamental theory*). He then has the tools to combine these areas in what von Hippel (MIT Professor and author of several advanced books) calls **molecular engineer-**

---

* The present and future dynamic growth of Materials Science has not been overlooked by the economists. Drucker [see Peter F. Drucker, *The Age of Discontinuity*, Harper and Row (1969)] mentions this area as one of four areas likely to undergo profound change and growth in the remainder of this century. It is therefore reasonable that its presence in the curriculum should be broadened.

† At Cornell the sophomores take the second half of their four semester University physics course concurrently with the materials science course.

ing; this is the synthesis which achieves improved properties [higher yield strengths] and even new behavior [amplification by n-p-n junction transistor].

The arrangement of the subject matter into various chapters can be seen from the diagram.

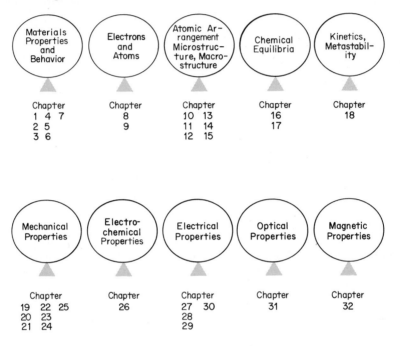

The chapters are divided into relatively short subsections.

Historical notes, references [our purpose is to give enough introduction so that the students can continue learning materials science on their own] and more than 450 figures and 350 worked examples are included in the text. So are occasional statements about what is *not* known, i.e., what are some active areas of current research.

EXAMPLES

The text contains over 350 examples. These are given as questions. The student should always make a conscientious effort to solve these and should write down his answer before reading the answer provided. These examples have several purposes:

To illustrate theory with a *concrete example*.
To clarify a concept.

To reinforce learning.

To help develop a feeling for the order of magnitude of the properties involved.

To help develop the skill of making quick estimates.

To relate materials technology to socioeconomic factors.

To make the student use his thinking capabilities and his creative talents.

PROBLEMS

There are 856 problems in this book. They are divided into three classes:

Problems.
More Involved Problems.
Sophisticated Problems.

The Problems involve *informational knowledge* such as terminology, facts, classifications sequences, methodology, etc. and *intellectual skills* such as comprehension, translation, interpretation and extrapolation.

The More Involved Problems involve *intellectual skills* at a deep level, *application abilities*, and at a modest level the abilities of *analysis, synthesis* and *evaluation*.

The Sophisticated Problems involve *analysis, synthesis* or *evaluation* at a deep level.

REFERENCES

Lists of books and articles for further study are given at the end of each chapter.

SUGGESTIONS FOR COURSES

A one semester course on Mechanical Properties can be based on Chapters 1, 2, 6, 7 and selected sections of Chapters 10–18, and Chapters 19–25. Chapter 26 might also be covered.

A one semester course on Electrical Properties can be based on Chapters 1, 3–6, selected sections of Chapters 8–16, 18, 27–30. Chapter 26 might also be covered.

A two semester course can cover all the topics in the book but not in depth.

The book also can be used for a longer in-depth course (3–4 semesters) in which a more quantitative coverage is desired, where some outside reading is required and where the student is expected to do the More Involved Problems and some of the Sophisticated Problems.

SELECTION OF TOPICS

The subject of materials science is so diverse that it is impossible to give due justice to all its many facets. Choices must be made. The topics are selected by the author with one major thought in mind: Eventually these materials are going to perform a function in a device! The student must therefore develop a feeling for how materials behave. That is why the book starts with *how* materials behave. In the general area of phenomenology there are a number of laws of behavior (such as Hooke's law for elasticity or Ohm's law for conductivity) which should be part of an engineer's knowledge. The voltage-current characteristics of a p-n junction rectifier also fall into this class, but this is a relatively new (post World War II) behavior. It should be stated here that there is a very large number of effects (responses of materials to stimuli). Those which are introduced in this book are selected on the basis of the author's research experience, teaching experience, consulting experience, etc.

FEEDBACK

A strong effort was made to produce a clear and well-illustrated text directed to the students' needs. Criticism by colleagues, research associates, graduate students and most importantly, feedback from the undergraduates who used the text was solicited, evaluated and used to hopefully achieve these ends.

ACKNOWLEDGMENTS

It is a pleasure indeed to acknowledge the help of the many persons who cooperated in the production of this book. I am indebted to my mother, Mrs. L. A. Ruoff, who foresaw the emergence of materials science during World War II and to my father, L. A. Ruoff, who taught me that worthwhile contributions require work. The cooperation of Dr. Seid Ghafelehbashi is particularly appreciated as he read and reread and unabashedly and ruthlessly criticized some parts and even sometimes made kindly comments on others. Mr. Thomas Schmid, Stephen Danyluk, Warren Yohe and John Paul Day also read the manuscript in full and made worthwhile contributions. Many of my colleagues who have taught these courses over the years deserve mention for their help. These include J. M. Blakely, E. J. Kramer and H. H. Johnson in particular, as well as T. N. Rhodin, H. W. Weart, J. P. Howe, A. Taylor and M. S. Burton. I am also indebted to others for profoundly affecting my attitude of materials science: These include H. Eyring, R. L. Sproull, D. R. Corson, H. S. Sack, F. Seitz, J. A. Krumhansl, and R. W. Balluffi.

The very able typing of Mrs. Andrea Lucente is much appreciated. The help with the drawings by William Van Duzer, Mrs. Lucente and George Chevalier is acknowledged as is the willing help of my wife, Enid, with the proofreading.

Lastly, I am indebted to the many former students for stimulating ideas and discussion which provided the incentive for writing this book.

FOREWORD TO THE STUDENT

There are some 10^{17}–10^{23} objects on the earth which are components of structures, systems, or devices which were designed and constructed by man to serve the purpose of his survival, his enjoyment, and his destruction. These objects (which are made of materials) exhibit 10^3 or so types of properties (which are reactions to external stimuli). An organized structure of thought is needed to study such a diversity of materials and properties. We call this subject Materials Science.

The properties which materials exhibit determine their utility. For a given application, we might require an electrical insulator or an electrical conductor. Electrical conductivity of ordinary materials varies over a range of 10^{26} or so. Thus both good conductors and good insulators are available under ordinary conditions. Moreover superconducting materials exist in which electrical currents (if they are not too big) will persist indefinitely in current loops (which could be naively stated to correspond to zero resistance); so far such properties have been found to exist only below 20°K. Even so these are extremely practical materials. It would be interesting and rewarding to find materials in which larger currents could persist at higher temperatures.

We might require a material which is soft (not stiff) and which absorbs much energy on impact or we might require a material which is as stiff as possible and as strong as possible. Imagine our transportation systems without rubber-like materials for tires or imagine our energy conversion systems with nothing stronger or stiffer than rubber or lead. Thus in the case of stiffness and strength we find materials which have a wide range of properties.

We might require a material which has high fluidity when a shear stress is applied or we might require materials which flow hardly at all when a shear stress is applied. Almost an infinite range of such properties exist. It would be interesting and rewarding to find materials which would support larger stresses for longer times and higher temperatures than the best materials currently available.

It is good to look closely at one's surroundings and to ask: What is this stuff of which things are made? One does not have to think of unusual applications such as superconducting magnets at 15°K generating magnetic fields of

100 kilogauss or turbine blades highly stressed by body forces at 1700°F and operating for long periods of time to be intrigued by the study of materials. Look at the world around you. Look at a pair of spectacles. More likely than not they have plastic frames made of polymers. How are the atoms arranged in the molecules? How are the molecules arranged in the solids? How did they get themselves in such an arrangement and why? Why use plastic? The lenses are made of a glass, more likely than not a silica based glass although it could be an organic polymer. What kinds of atoms are there in a silica glass anyhow? How are they arranged? Where are the outer shell or valence electrons? Are they running about wildly throughout the glass or are they localized? Why is the glass transparent? Is this related to the behavior of the valence electrons? Why is it brittle? Is there a way of making it less prone to fracture? The hinges are metallic. Often they are made of stainless steel. What kinds of atoms are there in this material? Why is it a relatively good conductor of electricity? Heat? Why are the atoms arranged in a periodic array? Do they remain more or less in such a periodic array (called crystals) if we beat on a piece of the stainless steel with a hammer? Why? And what really is going on within the material when we brutally deform it? Do its macroscopic properties as well as its shape change because of this deformation? What makes a material strong, i.e., what makes it resist permanent deformation? How does it react to dipping in solutions of NaCl (table salt)? What are its properties in general? Is it good for anything else? What?

You can continue the above process on and on by considering the objects which surround you right now. They may be simple objects which you have taken for granted up to now. Perhaps you wish to consider an incandescent bulb with its threaded brass base, solder contact, phenolic cement connecting the glass to the thread, the tungsten filament, and the glass shell. The solder is polycrystalline (lots of little crystals which you can see by properly preparing the surface and looking in a microscope, the whole examination process being called metallography). So is the tungsten, but its properties are very much different than the solder. The solder is an alloy. What's an alloy? What is a phenolic resin or cement? What is brass? Is it an alloy? Why use brass? Couldn't the cheaper material steel be used? Why is the glass shell not crystalline?

Why is there a magnetic core (soft iron) used in large transformers? Would you be surprised to learn that these transformers would have to be larger by factors of 10^3 to 10^6 if we did not have such magnetic materials? Is soft iron the best of all possible materials or is it used because it is cheap and readily formed? Why is the iron magnetic? Why is mineral oil an insulator?

You might consider the paper from which you are now reading. Have you ever asked questions such as: Just what is this paper? What are the atoms, the molecules, the fibers that make up this paper? Why does the ink wet it and how? And what is the ink?

There are many materials problems involved in the design of a nuclear reactor. It is often necessary to encapsulate the fissionable material. The capsule material is then subjected to intense bombardment by neutrons and other particles causing damage. Moreover the fission reaction produces gases as products (e.g., argon) which coalesce into bubbles within the material and build up pressure. Why does radiation damage occur? What are ways to avoid it? How do the bubbles form? One of the coolants which is considered for the breeder reactor is a liquid alloy of sodium and potassium (called NaK) which melts below room temperature even though both Na and K melt above room temperature. Why? Why is NaK used instead of water? NaK tends to penetrate the grain boundaries of the metals with which it comes into contact. What is a grain boundary? Does such penetration tend to destroy the tubing?

You might look at a concrete floor or highway. This material is a composite of rocks, sand, and Portland cement. The cement itself is a mixture of hydrates of different calcium aluminum silicates; it forms on its own account the basis for an extensive area of research. It is interesting to ask if there are other composites which are useful materials; indeed there are such materials as glass fiber reinforced plastics (GRP), tungsten wire reinforced ceramics and boron fiber reinforced epoxy. It is of interest to see how two or more materials can be put together in composites to give materials with excellent properties, not exhibited to the same extent by any of the individual materials.

If we continue this process just with the objects around us we begin to notice the enormous number of materials that exist. We then clearly see the need for a systematic method of studying them.

Materials are collections of nuclei and electrons and the manner in which these interact and become arranged determine the property of the material. We find that binding can be classified in four groups; this is a very useful classification to help us in understanding and categorizing the 10^{20} objects of construction. We also find that materials may be crystalline or noncrystalline. This too helps us categorize and understand them. We find that there are certain properties which do not change much (say a few percent) when small changes are made in the arrangement of the atoms; we call these structure insensitive properties. However certain properties change drastically (by factors of thousands) when only minor changes are made in the structure; we call these structure sensitive properties. The distinction between these two types of properties will help us also. As you progress further in your studies you will see further grouping and subdivisions which provide a logical basis of knowledge of materials and their properties. You will see that the study of materials includes a vast display of atomic splendor, that here as in music, art, and literature there is cultural value for here you can find enlightenment on what is the nature of things. And the study of materials has a practical side as well, for the physical processes of the real world involve things which are made of materials. You can be sure that when you understand materials you will be able to make better materials and utilize those already existing with even higher efficiency. And along the way, as our understanding deepens, we will make new materials and discover new material properties. The proper frame of mind for beginning this study is stated nicely by Rudyard Kipling in the Just-So stories, "I keep six honest serving men (they taught me all I knew); Their names are What and Why and When and How and Where and Who".

"*I do not know what I may appear to the world, but to myself I seem to have been only like a boy playing on the seashore, diverting myself in now and then finding a smoother pebble or a prettier shell than ordinary, whilst the great ocean of truth lay all undiscovered before me.*"

SIR ISAAC NEWTON

"*Was there an awareness on the part of many researchers, particularly the younger graduates, as to how big was the beach and how many pebbles were there, how many pebbles had been picked up, and how many were worth picking up?*"

C. P. SNOW

1

AN INTRODUCTION TO THE SUBJECT
OF MATERIALS SCIENCE

The study of materials can be conveniently divided into three areas, as illustrated in Figure 1.1. The student in this field must develop a feeling for all three areas. On the assumption that students who are studying materials are, at the beginning at least, interested in materials from the viewpoint of the applicability in devices, we shall begin the course with a study of **phenomenology**, i.e., a study of macroscopic behavior. In this way the student will become aware of some of the important properties of materials and will develop a feeling for the magnitude of these properties. Moreover, he will learn that there is room for improvement because the actual magnitude of a given property of a certain material may be far below the potential magnitude of this property. Let us illustrate briefly. (The illustration which follows involves a mechanical property, the yield strength. We could just as well have chosen a magnetic property, such as saturation induction, or a property of electrical insulators, such as dielectric strength. These properties will be studied later in the text.)

Suppose we consider the maximum load that a rod of 1-in.^{2} cross-sectional area can lift without causing permanent deformation of the rod (i.e., it returns to its original length when the load is removed). This quantity is called the **tensile yield strength**. It has units of force per area; engineers often use pounds per square inch (psi), while scientists usually prefer dynes per square centimeter. An ordinary aluminum rod might have a tensile yield strength of 10,000 psi, and a specially heat-treated aluminum alloy rod, 60,000 psi; while a truly perfect single-crystal rod (which has yet to be made in the form of a 1-in.-diameter rod) would have a tensile yield strength of about 1,200,000 psi. So what?

This is of profound commercial importance. Airplanes use a great deal of aluminum in their construction (usually an alloy with a yield strength in the

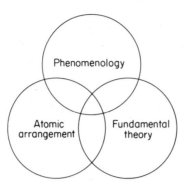

FIGURE 1.1. Three areas of study
of materials science.

neighborhood of 60,000 psi). If, instead, only the more ordinary 10,000-psi material were available, the weight of aluminum in the aircraft would increase, roughly, by a factor of 6 (more about this later on in the book). Each extra pound of material in the airplane's construction displaces a pound of cargo, and during the lifetime of the plane this extra pound will cost about $10. Since a pound of ordinary aluminum in ingot form costs $0.285 and a pound of the alloy costs about $0.32, the economics is obvious. One does not have to use dollars as a measure. The waste of natural resources (both aluminum and fuel) is an equally good measure. Now suppose that the material with a tensile yield strength of 1,200,000 psi were available in large sections. What a profound effect this would have on aircraft economy! It would also have a similar effect on design criteria and production techniques (we shall bring this up later in the book). The important point to consider here is that the yield strength is capable of improvement by a *factor* of nearly 20. Fantastic! True!

It is interesting to know what tensile yield strength is and how large it is for a given material if you are interested in designing a machine. But if you are to improve the properties of materials (and the production of the 60,000-psi tensile-yield-strength aluminum has to be considered such an improvement), you have to understand the detailed **atomic structure** of the material and how it can be varied. You will also be interested in how such variations affect properties. Therefore, you need **fundamental theories** which correctly describe (or predict) the magnitude of certain properties. You have already seen an example of the importance of such theories, namely, the prediction that a perfect crystal of aluminum would have a tensile yield strength of about 1,200,000 psi.

These three areas of study—phenomenology of material properties, the nature of atomic arrangement, and fundamental theory—together make up the subject of materials science. (The subject of materials engineering deals with the selection and processing of materials to produce a product having the required physical and mechanical properties at the *lowest cost*.) Materials science is a subject of tremendous practical importance because material properties set the limit on the design of all things which we make, as shown in Figure 1.2. It is also a subject of high esthetic value, for the essence of living surely includes an attempt at understanding the materials in our environment and beyond.

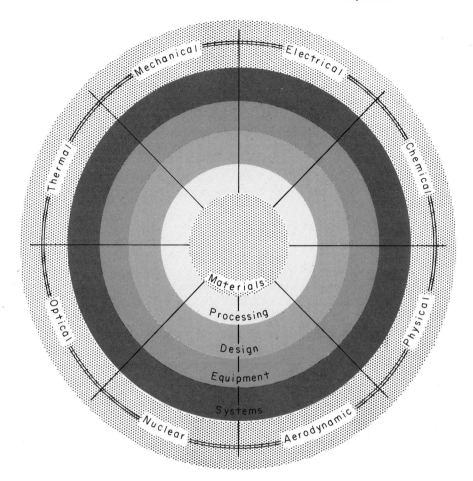

FIGURE 1.2. Materials limited technologies.

REFERENCES*

Alexander, W. O., "The Competition of Materials," *Scientific American* (Sept. 1967) p. 225. The economics of materials, particularly with respect to their use as mechanical components, is discussed. The new systems approach to materials selection is emphasized.

Holden, A., *The Nature of Solids*, Columbia University Press, New York (1965).

Smith, C. S., "Materials," *Scientific American* (Sept. 1967) p. 69. The entire Sept. 1967 issue of *Scientific American* is devoted to materials. The purchase of this issue by students of materials is recommended.

---

* References were put in order of increasing difficulty, especially in later chapters, as well as grouped by common material.

PROBLEMS

1.1. Into what three areas can materials science be divided?

1.2. Suppose that later on (because you have read this book) you invent a cheap way to raise the yield strength of aluminum from 60,000 to 180,000 psi. Assume that you are willing to share equally the fruits of your invention with the aircraft industry and the consumer. Show that your profit from the following project would be of the order of $1 billion if it resulted in the reduction of the structural weight of the aircraft by 67%. Assume that 2000 giant transports are to be built with a loaded weight of 500 tons each, of which 30% is now structure (before your invention).

1.3. Give several reasons for studying materials. Include some reasons not already mentioned in the text.

More Involved Problem

1.4. Find ten journals which are concerned with various aspects of materials science. Ask your instructor, if necessary, to help you to get started. Include at least one journal for each of the following specialized areas: polymers, ceramics, metals, semiconductors, glasses, composites.

Sophisticated Problem

1.5. Tiny crystals, called whiskers, have been grown which have tremendous strengths, often approaching that expected of a perfect crystal. Write an essay on the current status of whisker growth.

Prologue

For every properly designed component which must carry a load, the design engineer should be able to specify an allowable load and an expected lifetime. To accomplish this he must understand the mechanical behavior of matter. The production engineer often will use mechanical force to shape a component during manufacturing. Hence he must understand the mechanical behavior of matter. In this chapter we shall find that mechanical force and displacement of material can be described conveniently in terms of *stress* and *strain*. Stiffness (in stretching or compressing) is described in terms of a material property known as *Young's modulus*, while torsional stiffness is described in terms of a material property known as the *shear modulus*. The student should begin to develop a feeling for the *order of magnitude* of these quantities for different materials.

The difference between elastic and plastic deformation is discussed, as is the fact that the quantity called *yield strength* is an important design parameter. It is pointed out that elastic moduli are *structure-insensitive properties*, while yield strength is a *structure-sensitive property*. The importance of *elastic* and *plastic instabilities* in design is noted. The phenomena of *mechanical damping* (internal friction), *mechanical loss angle*, *creep* (continued deformation under fixed load), *cyclic fatigue* (fracture at reduced load owing to repeated loading and unloading), *stress concentration*, and *brittle fracture* are briefly described.

Temperature is shown to be a vital parameter in mechanical behavior, especially in creep and brittle fracture phenomena. In addition to the deformation of solids, the flow of liquids is studied, and the concept of a *viscosity coefficient* is introduced. Finally, *sliding friction* is briefly discussed.

2

MECHANICAL PROPERTIES

2.1 YOUNG'S MODULUS AND TENSILE YIELD STRENGTH AS DESIGN PARAMETERS

The bending of a beam is used to illustrate the importance of two material properties, Young's modulus and yield strength.

UNIFORMLY LOADED BEAM. Consider a cantilever beam mounted in a rigid support at one end and having a hollow rectangular cross section as shown in Figure 2.1.1(a). A load w per unit length of beam is uniformly applied to

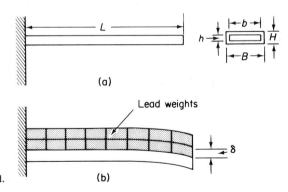

(a)

Lead weights

FIGURE 2.1.1. Cantilever beam
(a) without load
and (b) with load.

(b)

the beam as shown in Figure 2.1.1(b) and causes a deflection δ of the end of the beam. If the deflection is not too large (which means very small relative to the length), the deflection will be related linearly to the total downward force $F = wL$ as shown in Figure 2.1.2. The linear deflection-force relation can be written

7

as

$$\delta = sF = s(wL) \tag{2.1.1}$$

or

$$F = c\delta, \tag{2.1.2}$$

where s, the compliance constant, or c, the stiffness constant, of the beam is related according to $c = s^{-1}$. The student of elasticity could readily show that

$$c = \left(\frac{2}{3} \frac{BH^3 - bh^3}{L^3} \right) E. \tag{2.1.3}$$

The stiffness constant of the beam depends on the dimensions of the beam and a material property called E or *Young's modulus*, which therefore must be an important parameter in design criteria.

$F = c\delta$

FIGURE 2.1.2. Total downward force causing bending of beam versus deflection of end of beam.

If too large a load is placed on the beam it will be permanently deformed (meaning that it will not recover its original shape when the load is removed). The design engineer knows that yielding occurs when the load per unit length reaches

$$w_{max} = \left(\frac{BH^3 - bh^3}{3L^2 H} \right) \sigma_0, \tag{2.1.4}$$

where σ_0 is the *tensile yield strength* already discussed briefly in Chapter 1. Design formulas such as (2.1.3) and (2.1.4) are derived in textbooks of elasticity. See H. D. Conway, *Mechanics of Materials*, Prentice-Hall, Englewood Cliffs, N. J. (1950).

[Students have an undesirable habit of wanting to memorize every equation. Equations (2.1.3) and (2.1.4) are placed here *only* for the purpose of illustrating the fact that E and σ_0 are important design parameters. When the student studies strength of materials or elasticity he will probably derive these equations. In the meanwhile we shall proceed to use these equations so that the student can develop a feeling for the order of magnitude of these mechanical properties.]

Although a very crude approximation, the beam of Figure 2.1.1 may be considered to be a wing of an aircraft (in an actual wing, B and H would decrease as the tip is approached; moreover, the shape would not be rectangular; in addition there would be a superstructure within the wing which would give support but which we have ignored). We shall assume that the wing is made of a heat-treated titanium alloy in which $E = 18 \times 10^6$ psi and $\sigma_0 = 160,000$ psi.

The wings have $L = 20$ ft, $B = 48$ in., $b = 47.75$ in., $H = 8$ in., and $h = 7.75$ in. The required load capacity is 225 lb/ft² = 900 lb/ft (since $B = 4$ ft) = 75 lb/in.

EXAMPLE 2.1.1

a. Calculate the deflection of the wing tips under maximum static loading.

Answer. We use the expressions (2.1.1) and (2.1.3):

$$\delta = 8.8 \text{ in.}$$

b. Calculate the maximum loading allowable without getting permanent deformation.

Answer. The maximum loading (up to yielding) per unit area is from (2.1.4), given by

$$\frac{w_{max}}{B} = 816 \text{ lb/ft}^2.$$

c. What is the *safety factor*?

Answer. On the basis of allowable stress levels it is 3.6—i.e., the maximum tensile stress is only $\sigma_0/3.6$ and could be increased by a factor of 3.6—but we must remember that these calculations are based on static loading.

EXAMPLE 2.1.2

It is conceivable that titanium will eventually be made with a σ_0 value as large as 1.8×10^6 psi.

Suppose that the dimensions now are $L = 20$ ft, $B = 48$ in., $b = 47.975$ in., $H = 8$ in., and $h = 7.975$ in. but that the loading is the same. It should be noted that the skin is very thin and might buckle readily; we shall ignore this problem for now. The student could readily show using (2.1.4) that if this new light wing were made with titanium now in use ($\sigma_0 = 160,000$ psi) it would fail by yielding for a loading of only 84 lb/ft², which is far below the required value of 225 lb/ft². However, he can now show that if the new light wing were made of the futuristic titanium ($\sigma_0 = 1,800,000$ psi) it would not yield.

a. Calculate the maximum loading without getting yielding.

Answer. 947 lb/ft².

b. Calculate the safety factor.

Answer. 4.2.

c. Calculate the deflection with the actual loading of 225 lb/ft².

Answer. If we use Equation (2.1.1) and (2.1.3) we see that $\delta = 86$ in.; of course the formula used would no longer apply since it applies only to cases where the deflection is very small compared to the length.

d. Is this tolerable?

> *Answer.* No. The deflection is intolerable even though the stress is tolerable.

We see that stiffness is an extremely important aspect of design and that improper attention to it, as illustrated in Figure 2.1.3, can lead to failure. The quantity E, a material property, is thus an important design parameter.

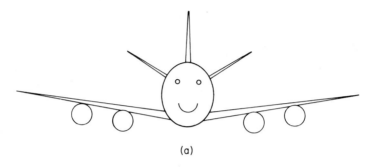

(a)

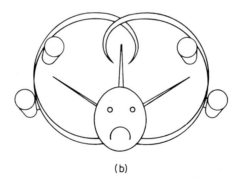

(b)

FIGURE 2.1.3. (a) Front view of airplane before takeoff. (b) Front view of airplane at an instant in flight. It was designed so that solid wings will not yield. Elastic stiffness and deflection criteria were not used. Note that the wings are twisting as well as bending.

Likewise we have already seen that tensile yield strength is an important design parameter.

END-LOADED BEAM. We now consider a similar problem where both E and σ_0 also enter as important parameters in design. This problem involves the process of mechanical machining in a lathe as shown in Figure 2.1.4.

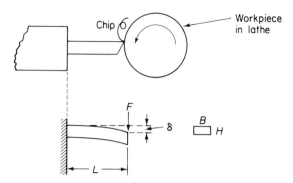

FIGURE 2.1.4. Cutting tool on a lathe and its idealization as
a rigidly supported cantilevered beam.

The deflection of the cutter tip, if small, is given by

$$\delta = \left(\frac{4L^3}{BH^3}\right)\frac{1}{E}F,$$ (2.1.5)

and the elastic energy stored in the cutter is

$$W = \tfrac{1}{2}F\delta.$$ (2.1.6)

Why?

EXAMPLE 2.1.3

Calculate δ for a steel cutter with $L = 1$ in., $B = \tfrac{1}{4}$ in., and $H = \tfrac{1}{4}$ in.
subject to a load $F = 300$ lb. Young's modulus is 30×10^6 psi.

Answer. 0.01 in.

The load F_{max} which may be applied without permanently deforming
the cutter is

$$F_{max} = \frac{BH^2}{6L}\sigma_0.$$ (2.1.7)

Again, a student who has taken a course in strength of materials or mechanics
of materials could readily derive (2.1.5) and (2.1.7). See H. D. Conway, *Mechanics of Materials*, Prentice-Hall, Englewood Cliffs, N. J. (1950).

EXAMPLE 2.1.4

Calculate F_{max} for the cutter of Example 2.1.3 assuming $\sigma_0 = 240{,}000$
psi.

Answer. 625 lb.

EXAMPLE 2.1.5

A cutter is made of a futuristic steel with $E = 30 \times 10^6$ psi and $\sigma_0 =$ 2.4 $\times$ 10^6 psi. $L = 1$ in., $B = \frac{1}{4}$ in., and $H = \frac{1}{16}$ in. It is subjected to a load of $F = 100$ lb. Calculate δ and $F_{\max}$. Discuss.

Answer. $\delta = 0.22$ in. and $F_{\max} = 390$ lb. Insofar as the yield strength is concerned there is a safety factor of 3.9, but the displacement of the cutter tip is clearly intolerable.

A COLUMN.   A straight circular column of radius R is loaded until it buckles, as shown in Figure 2.1.5. Assume that the column is rigidly supported

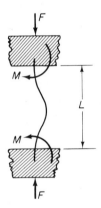

FIGURE 2.1.5. Column with clamped ends.

in the end pieces, which are constrained by the moment M to move only in the vertical direction.

1. Where on an airplane might such a component be found?
2. Where on a steel bridge might such a component be found?
3. Where on a building might such a column be found?

The load, $F_{\max}$, which just causes buckling is

$$F_{\max} = \frac{\pi^3 R^4}{16 L^2} E, \qquad (2.1.8)$$

where R is the radius of the circular column, L is the length, and E is Young's modulus. It should be noted that until this load is reached the column remains straight and that when this load is reached it suddenly buckles.

The phenomena of buckling is an example of an **elastic instability** (gross, nearly instantaneous, deflection of a member without plastic deformation). This is another example of failure with which the designer must concern himself. Note that the maximum force depends on the material property E, and not the yield strength σ_0 unlike the situation of (2.1.7).

We have seen (1) how important both stiffness (determined by E) and the onset of tensile plastic yielding (determined by σ_0) are in design, (2) the magnitude of these parameters for two important materials, (3) the fact that σ_0 might in the future be increased by about an order of magnitude for both of these materials, and (4) the fact that E will not change appreciably even if σ_0 does change by a factor of 10. In future chapters we shall be concerned with the origin of E and σ_0. We conclude this section with some tables of E and σ_0 so that the reader can develop a feeling for the range of each of these properties.

Table 2.1.1 shows the values of E for various materials. We give E in units of pound per square inch (or psi) and dynes per square centimeter. Another unit for stress is the **bar** (1 bar $= 10^6$ dynes/cm² $= 14.5$ psi). Sometimes the atmosphere is used as a unit of stress (1 atm $= 14.7$ psi). In the mks system of units, stress is measured in Newtons/m² $= 10$ dynes/cm². The hydrostatic pressure at the deepest spot in the ocean is about 1 kbar. (Tables such as Table 2.1.1 can be found in the appropriate books listed under REFERENCES at the end of this chapter.) Note that the range of Young's modulus, E, is very large (and considered as a ratio, it can be infinite). Gases and liquids have no tensile stiffness. Rubber has a very low stiffness, which is one of the properties

Table 2.1.1. YOUNG'S MODULUS E (TENSILE STIFFNESS MODULUS) FOR VARIOUS BULK MATERIALS

Material	10^6 psi	10^{12} dynes/cm²
Gas	0	0
Liquid	0	0
Rubber (gum)	0.00015	0.00001
Rubber (hard)	0.3	0.02
Nylon	0.3	0.02
Concrete	2.0	0.14
Lead	2.6	0.18
Magnesium	6.3	0.43
Limestone	8.3	0.57
Glass	10	0.69
Aluminum (99.45% annealed)	10	0.69
Aluminum (99.45% cold rolled)	10	0.69
Aluminum (2024 heat-treated alloy)	10.6	0.72
Copper	16	1.10
Composite boron fibers 50% volume in a matrix of 6061 aluminum (along the direction of the fibers; the modulus perpendicular to the fiber is considerably less)	30	2.06
Steel	30	2.06
Beryllium	37	2.5
Boron fibers	55–60	3.8–4.1
Tungsten	60	4.1
Sintered carbide	100	6.9
Diamond	162	11.0

that makes tires behave as they do (we shall study some of the other properties in Section 2.4) and hence makes the present-day automobile possible. Plastics (and nylon is a typical example) are less stiff than common structural metals by an order of magnitude ($\times 10^1$) or even two orders of magnitude ($\times 10^2$).

Of the common materials shown, most of them have the same Young's modulus in every direction. That is, if we were to take a large block of commercial aluminum and cut out rods of material with the axis of each rod being a different direction in the original block and were to measure E for these rods, we would find that E does not vary with direction. The material is said to be **isotropic with respect to** E. Many materials are **anisotropic with respect to** E. An example is a glass-reinforced plastic fishing rod where very stiff fibers of glass run longitudinally along the rod and are embedded in soft plastic (see Table 2.1.1). Here E measured along the rod would be very different from E measured normal to the rod. Moreover, all single crystals are anisotropic with respect to E. As a rule a large copper rod contains billions of tiny crystals and yet the copper rod may be isotropic with respect to E (if these crystals are more or less randomly oriented) or anisotropic with respect to E (if the crystals tend to be oriented in a specific way, as is often the case after cold rolling, drawing, or even casting). Likewise, wood, plywood, many composites such as the glass-reinforced plastic already mentioned, and honeycomb structures are extremely anisotropic with respect to E.

The development of anisotropic materials which have superior properties means that the design engineer will have to become particularly concerned with developing and using the design criteria of such materials just as in the past he has been concerned with isotropic materials. We, therefore, shall study the way in which E varies with direction in Chapter 19. We shall also study the reason for the enormous range of E shown in Table 2.1.1.

Table 2.1.2 lists some values of tensile yield stress for several materials. It also includes (in some cases) another property which sometimes might be used in an equivalent fashion to σ_0 by designers. This property is the **tensile fracture strength**. This is defined in the same fashion as the tensile yield strength except that the fracture strength is the maximum static uniaxial tension force that a rod initially of 1-in^2 cross section carries.

Several things become clear after an inspection of Table 2.1.2. First we have to be specific. Nylon is not just called nylon. It is a specific material, nylon 66. But that is not enough. It may be bulk nylon 66 or fibers of nylon 66. Copper is not just copper. Even though the chemical content is unchanged, σ_0 varies by a factor of 4 owing to the fact that in one case the copper is annealed (held at a sufficiently high temperature for a sufficient period of time—more about that later) while in the other case the copper is cold drawn (deformed permanently by drawing a copper rod through a die to produce a smaller rod). We also see various possibilities for aluminum. There is nearly pure annealed aluminum with a yield strength of 4000 psi, though researchers study single crystals of 99.9999% purity, which have much lower yield strength and which

Table 2.1.2. TENSILE YIELD STRENGTH, σ_0*

Material	psi	kbars (10^9 dynes/cm^2)
Gas	0	0
Liquid	0	0
Nylon 66 bulk	10,000 (tensile strength)	0.7
Nylon 66 fiber	100,000 (tensile strength at 65% relative humidity)	7.0
OFHC 99.95% annealed copper†	10,000	0.7
OFHC 99.95% cold-drawn copper	40,000	2.8
99.45% annealed aluminum	4,000	0.28
99.45% cold-drawn aluminum	24,000	1.7
2024 (3.8–4.9% Cu, 0.3–0.9% Mn, 1.2–1.8% Mg) heat-treated aluminum alloy	50,000	3.4
Molded phenolic (similar materials are called bakelite)	5,000 (tensile strength)	0.34
Bulk soda lime glass	10,000 (this is a typical tensile strength or fracture strength)	0.7
Freshly drawn (in vacuum) glass fiber	2,000,000	140
Silicone-coated commercial glass fiber for use in glass-reinforced plastics	300,000 (fracture strength)	20.6
Boron filament	400,000 (fracture strength)	28
Graphite filament	300,000 (fracture strength)	20.6
Sapphire whisker	10,000,000 (fracture strength)	700
Gray cast iron (#20)	20,000 (fracture strength)	1.4
Pearlitic malleable cast iron	45,000	3.1
AISI 1020 steel**	35,000– 40,000	2.4–2.8
AISI 1095 steel (hardened)	100,000–188,000	7–13
AISI 4340 annealed alloy steel	65,000– 70,000	4.5–4.8
AISI 4340 fully hardened alloy steel	130,000–228,000	9–16
Maraging (300) steel	290,000	20
Piano wire	350,000–500,000	24–34
Iron whisker	1,900,000	131
Concrete	500 (tensile strength; varies considerably with nature of concrete and in fact we are usually more concerned with compression strength)	0.03

* At room temperature and at rates of loading typical of the usual static loading test (perhaps an hour to apply the load would be typical; more about this later). σ_0 is defined, except where noted, by 0.5% permanent extension. Certain of the quantities are tensile fracture strengths, as noted, rather than tensile yield strength.

† OFHC means oxygen-free high-conductivity.

** AISI is the abbreviation for the American Iron and Steel Institute. The 1020 is a code developed by them to specify the alloy content of the steel. See Table 25.6.1.

15

may eventually have yield strengths of 1 psi. What is interesting here is that when the annealed 99.45% aluminum is cold drawn, the yield strength goes up by a factor of 6, and it is entirely conceivable that if we were (in the future) very clever about the way we do this cold drawing, the yield strength could be raised by a few orders of magnitude instead. We see in addition that a chunk of aluminum may not be just aluminum but may contain 5% or so of **alloying elements** (elements which are purposely added) which make it possible, after going through an appropriate temperature-time history (**heat treatment**), to obtain a yield strength of over 12 times that of ordinary commercially pure aluminum. This is rather interesting, but what you may find truly intriguing is the fact that if we really knew how to do this heat treatment, we could do much better than that, perhaps by a factor of 20.

We note that fine fibers or filaments of glass, boron, and graphite as well as whiskers of iron and sapphire can have enormous strengths and provide potential for future materials. Some of the more recent developments using such materials involve boron-epoxy composites made by United Aircraft and used for various parts of airplanes and graphite filament composites used by Rolls Royce.

The range of properties attainable with cast irons and steel (and here we refer primarily to σ_0) is fantastic. Cast iron and steel are alloys of iron with carbon (and frequently other elements). The number of such alloys is vast, and many properties in addition to σ_0 are involved in the design criteria for a specific application.

> Gold is for the mistress—silver for the maid—
> Copper for the craftsman, cunning at his trade.
> "Good," said the Baron, sitting in his hall,
> "But iron—cold iron—is master of them all."*

However, the volume of plastics produced per year now exceeds the volume of steel, and the diversity of plastics is also as great if not greater. Perhaps, when the reader has learned more about plastics, he will provide us with appropriate poetry about the plastics age.

We shall study the reason for the enormous variation of σ_0 not only between dissimilar materials but within similar materials such as the three different aluminum materials or the several different irons and steels. Properties which show large changes such as σ_0 for the three different aluminum materials, with only a modest change in atomic structure, are called **structure-sensitive properties**. Young's modulus is an example of a **structure-insensitive property** (note the tiny variation of E among the three different aluminum materials of Table 2.1.1). This dichotomy exists not only for mechanical properties but for various other properties, which will be discussed in later chapters. Both structure-insensitive and structure-sensitive mechanical properties are important

* J. G. Parr, *Man, Metals and Modern Magic,* Collier Books, New York (1958).

in design and must be thoroughly understood to design a vehicle such as the Star II shown in Figure 2.1.6. The student needs only to read the article by W. Bascom, "Technology and the Ocean," *Scientific American* (Sept. 1969), p. 199, to develop an appreciation for the need for understanding mechanical properties.

FIGURE 2.1.6. Star II undersea research vehicle. This two-man vehicle has a 5-ft-diameter pressure hull of $\frac{5}{8}$-in.-thick HY-80 steel and weighs 10,000 lb. (*Courtesy of General Dynamics.*)

2.2 THE SHEAR STIFFNESS CONSTANT

THE CIRCULAR SHAFT. The circular shaft has many industrial applications. It transmits the torque from a turbine to an electric power generator, and it transmits the torque from a ship's engines to its propellers. The torsion wire in a deflection galvanometer is another example of the application of the circular shaft (Figure 2.2.1).

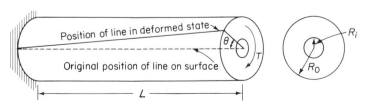

FIGURE 2.2.1. Hollow shaft twisted through an angle θ by a torque T.

If the angle of twist, θ, is not too large (so that the shaft does not become permanently twisted), one observes that the torque T is

$$T = K\theta, \tag{2.2.1}$$

where K is the torsion constant of the shaft.

EXAMPLE 2.2.1

Show that the energy stored in such a twisted shaft is $K\theta^2/2$.

Answer. The work done in twisting the shaft is given by

$$\int T\,d\theta = \int K\theta\,d\theta = \frac{K\theta^2}{2}.$$

A person knowledgeable in the elasticity of homogeneous, elastically isotropic materials could readily show [see H. D. Conway, *Mechanics of Materials*, Prentice-Hall, Englewood Cliffs, N.J., (1950)] that

$$K = \left[\frac{\pi(R_0^4 - R_i^4)}{2L}\right]G, \tag{2.2.2}$$

where R_i is the inner radius, R_0 is the outer radius, and L is the length of the shaft.

Thus the torsion constant depends on the geometry of the shaft (the term in the brackets) and on the quantity G, which is a *material property* called the *shear modulus*.

Materials which are elastically anisotropic have different shear moduli in different directions, e.g., wood; this will be discussed further in Section 19.3. In Example 2.2.2, we note that the engineer can increase the stiffness of a shaft of given weight not only by choosing a material with a higher G but by changes in geometry as well, which amounts to putting the material where it will do the most good.

EXAMPLE 2.2.2

Show that a hollow shaft is stiffer, weight for weight, than a solid one.

Answer. If the solid shaft has an outer radius b, then K for the solid shaft is

$$K_s = \frac{\pi b^4}{2L}\,G,$$

and the volume of material is

$$\pi b^2 L.$$

Let $R_i = aR_0$, where $0 < a < 1$ for the hollow shaft. Then, if the volume of material in the two shafts is the same,

$$\pi b^2 = \pi R_0^2(1 - a^2),$$

and K for the hollow shaft is therefore

$$K_h = \frac{(1 + a^2)}{1 - a^2} K_s > K_s.$$

The implication is that a shaft should be made "paper-thin" with a large radius. What new problems might this cause?

There are thus three ways to increase the torsion stiffness K for a given weight of a circular shaft of given length: (1) Make it hollow; (2) choose a material with a larger G; (3) choose a material with a smaller density ρ.

There is a limit to the torque that can be applied to a shaft without causing permanent deformation. This is given by

$$T_{max} = \frac{\pi(R_0^4 - R_i^4)}{2R_0} \tau_0. \tag{2.2.3}$$

Here τ_0 is a material property known as the *shear yield strength*. It will be shown in Chapter 24 that

$$\frac{\sigma_0}{2} < \tau_0 < \frac{\sigma_0}{\sqrt{3}}, \tag{2.2.4}$$

where σ_0 is the *tensile yield strength*, discussed in Section 2.1.

THE HELICAL SPRING. Consider the close-coiled helical spring represented by Figure 2.2.2.

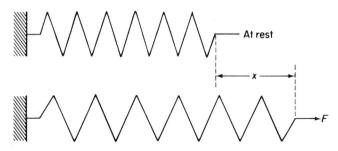

FIGURE 2.2.2. Upper part shows coil spring at rest and lower part shows spring extended by x under force F.

Experimentation shows that the force (for slow extension) is directly proportional to the deflection (if the latter is not too large), i.e.,

$$F = kx, \tag{2.2.5}$$

where k is the spring constant. This is another example of a **linear response**: the displacement (the response) varies linearly with the force,

$$x = k^{-1}F. \tag{2.2.6}$$

A plot of F vs. x is shown in Figure 2.2.3.

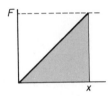

FIGURE 2.2.3. Response curve for a spring. The shaded area represents the energy stored in the spring.

The work done in stretching the spring is

$$W = \int F \, dx = \frac{Fx}{2} = \frac{F^2}{2k} = \frac{kx^2}{2}.$$ (2.2.7)

This work is stored in the spring as elastic energy. Note the quadratic dependence of this stored energy on x.

The student of elasticity can readily show that, if the spring is made of a homogeneous, elastically isotropic material,

$$k = \left(\frac{d^4}{8D^3 n}\right)G.$$ (2.2.8)

The term in parentheses involves geometric aspects of the spring: d is the diameter of the spring wire, D is the diameter of the coil, and n is the number of coils in the spring.

The maximum force which can be applied to the spring without causing permanent stretching of the spring (which is equivalent to permanent twisting of the wire) is

$$F_{max} = \left(\frac{\pi d^3}{8D}\right)\tau_0.$$ (2.2.9)

If, however, the spring is made of a material, such as glass, which fractures without any permanent twisting,

$$F_{max} = \left(\frac{\pi d^3}{8D}\right)\sigma_f,$$ (2.2.10)

where σ_f is the **tensile fracture strength** (stress at fracture).

Table 2.2.1 shows some typical shear moduli for elastically isotropic

Table 2.2.1. SHEAR STIFFNESS (MODULI) FOR ELASTICALLY ISOTROPIC MATERIALS

Material	10^6 psi	10^{12} dynes/cm^2
Gas	0	0
Liquid	0	0
Gum rubber	0.00005	0.000003
Nylon	0.1	0.007
Aluminum	4	0.28
Copper	6	0.41
Steel	12	0.83
Hematite (α-Fe$_2$O$_3$)	13.7	0.94

materials. Comparison with Table 2.1.1 shows that G is approximately one half to one third of E. In fact, for elastically isotropic materials, it can be shown that

$$\frac{E}{3} \leq G \leq \frac{E}{2}.$$
(2.2.11)

G, like E, often varies with the direction in materials. Such materials are called elastically anisotropic materials.

It should be noted that the student is not expected to know or remember the details of Equations (2.2.2), (2.2.3), (2.2.8), and (2.2.9). They are used here solely for the purpose of showing a few examples from the real world where G and τ_0 enter as parameters.

2.3   DEFINITION OF STRESS, STRAIN, AND TENSILE MODULUS AND SHEAR MODULUS

TENSILE BEHAVIOR.   To understand the tensile stiffness modulus E, we must first understand tensile stress and strain. We consider stretching a specimen as shown in Figure 2.3.1.

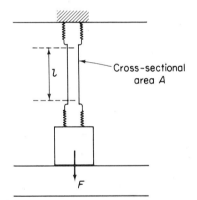

Cross-sectional area A

FIGURE 2.3.1. Tensile test specimen threaded into fixed holder and moving platen which applies the force F.

The initial length between the gauge markers on the test specimen is l_0 before the force F is applied. The change of length, divided by the initial length, is called the **nominal** or **engineering tensile strain**,

$$\epsilon = \frac{l - l_0}{l_0}.$$
(2.3.1)

Likewise, the **nominal tensile stress** is

$$\sigma = \frac{F}{A_0},$$
(2.3.2)

where A_0 is the initial cross-sectional area.

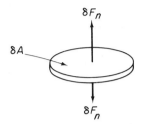

FIGURE 2.3.2. Forces causing a tensile stress.

We shall use the convention that if σ is positive, the stress is tensile, and if negative, it is compressive. More generally, we define such a stress according to Figure 2.3.2. The definition is

$$\sigma \equiv \lim_{\delta A \to 0} \frac{\delta F_n}{\delta A}. \qquad (2.3.3)$$

Hence, σ is usually called a **normal stress**. We note that it is due to a force acting normal to the area under consideration.

If careful measurements are performed, it will be found that there is a linear relationship between normal stress and strain [if the stresses (or strains) are not too large; more about that later],

$$\sigma = E\epsilon. \qquad (2.3.4)$$

The quantity E, a constant, is called **Young's modulus** or the **tensile modulus**. Equation (2.3.4) holds in both tension and compression. Equation (2.3.4) is an example of **Hooke's law**. Note that the elastic modulus E has the same units as stress. Unlike the equations given in earlier sections, the equations in this section are not just for the purpose of illustration. Rather they help to define fundamental fields (stress fields and strain fields) or they define important properties; hence the student must become familiar with the equations of this section.

In most ordinary materials, the tensile stress will reach some critical value σ_0, called the **tensile yield strength**, such that when this stress is removed, a certain small permanent strain remains (say 0.1 %, although 0.01 %, 0.2 %, etc., are often used). The corresponding strain will be quite small, and below this stress Equation (2.3.4) will be satisfactory; this is called the **linear elastic range**.

EXAMPLE 2.3.1

For special, heat-treated alloy steel, $\sigma_0 = 300,000$ psi (ultrastrength steel) and $E = 30 \times 10^6$ psi. What is the elastic strain at yielding for a permanent strain of 0.01 %?

Answer. The elastic strain is

$$\epsilon = \frac{\sigma}{E} = \frac{30 \times 10^4}{30 \times 10^6} = 0.01$$

when the tensile yield strength is reached.

The materials which are exceptions to this linear expression in the elastic range are rubber, which can often be reversibly strained by several hundred percent, and many plastics. Also, the elastic tensile strain in perfect crystals often exceeds 0.05, and Equation (2.3.4) is not sufficient to describe the relationship between stress and strain (except for small strains) for these materials.

SHEAR BEHAVIOR.   A simple shear strain is shown in Figure 2.3.3.

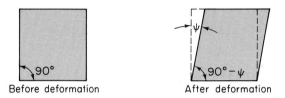

Before deformation   After deformation

FIGURE 2.3.3. Shear strain.

The shear strain is a measure of the change in angle between two lines which, prior to deformation, were normal to each other. We define **shear strain** quantitatively as

$$\gamma = \tan \psi. \tag{2.3.5}$$

If ψ is very small,

$$\gamma \doteq \psi. \tag{2.3.6}$$

Shear deformations are caused by shear stresses. We define such a stress according to Figure 2.3.4.

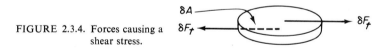

FIGURE 2.3.4. Forces causing a shear stress.

The **shear stress** is defined as

$$\tau = \lim_{\delta A \to 0} \frac{\delta F_t}{\delta A}. \tag{2.3.7}$$

Note the distinct difference between shear stress and tensile or normal stress.

To specify a stress, we have to specify not only the direction of the force, but the orientation of the area on which this force is acting. The usual way to specify the orientation of an area is to specify the direction of the normal to the area.

It is worth noting that there are indeed shear stresses in the tensile specimen of Figure 2.3.1; these shear stresses vary with direction and have their maximum on planes at 45 deg to the tensile axis. See Figure 2.3.5.

Show that the maximum shear stress in a tensile test, in which the tensile stress is σ, is $\sigma/2$.

Answer. The tensile stress on the horizontal plane in Figure 2.3.5 is $\sigma = F/A_0$. The shear stress on the inclined plane is

$$\tau = \frac{F \sin \theta}{A_0/\cos \theta} = \frac{F}{A_0} \sin \theta \cos \theta.$$

τ has a maximum when $\theta = 45$ deg. If this is not obvious, find the value of θ for which $d\tau/d\theta = 0$.

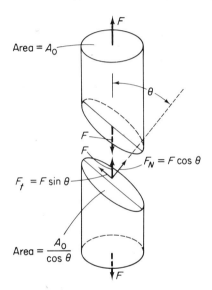

FIGURE 2.3.5. State of stress in a tensile specimen.

Another point is that in general the normal stresses and shear stresses are interrelated. As an example, consider the two states of stress shown in Figure 2.3.6. These are *identical* states of stress if $\tau = \sigma$ (in magnitude) with directions as shown (proof of this comes later). What the student should be aware of is this: Figure 2.3.6(a) appears to show only normal stresses present in the specimen. The actual situation is far from this. Only on the planes shown in (a) are there no shear stresses. On any other plane inclined at some angle, there are shear stresses, and, if we choose planes at 45 deg to the original, these planes have only shear stresses and no tensile stresses.

The state of stress described in Figure 2.3.6 is two-dimensional. The general three-dimensional state of stress is discussed in Section 7.2 (and could be studied now).

When experiments are performed involving shear stress and shear strain, it is found that for small strains ($\gamma \gtrsim 0.01$), reversible behavior exists

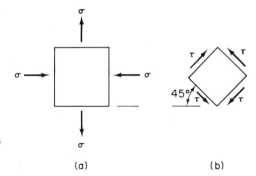

FIGURE 2.3.6. Equivalent states
of stress if $|\sigma| = |\tau|$.

(a)   (b)

such that the shear stress is related to the shear strain according to

$$\tau = G\gamma \qquad (2.3.8)$$

and G is the **shear modulus** or the **shear stiffness constant**. Equation (2.3.8) even holds for large strains in rubberlike materials. Equation (2.3.8) is another example of **Hooke's law** (named after Robert Hooke who discovered it in 1660).

ELASTIC ISOTROPY AND ANISOTROPY.   Consider a huge block of zinc, such as shown in Figure 2.3.7.

If tensile specimens cut from regions 1, 2, 3, and 4 (note the same orientations) all have the same E, the material is said to be **macroscopically homogeneous** (i.e., the same everywhere on a macroscopic scale, measured, e.g., by the tensile test). Let us assume that the body is homogeneous. Now if tensile specimens

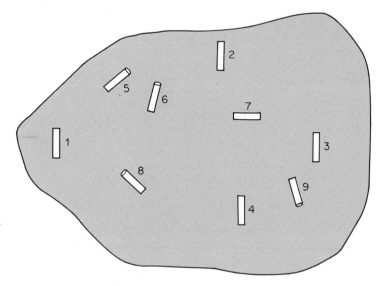

FIGURE 2.3.7. Block of material from which tensile test specimens are cut.

cut from regions 1, 5, 6, 7, 8, and 9 (note the *different* orientation of the resultant tensile axes) all have the same E, the material is said to be **macroscopically elastically isotropic.** If the body is homogeneous but the E's are different for specimens 1, 5, 6, 7, 8, and 9, the material is said to be **elastically anisotropic.**

The two elastic constants E and G are sufficient to describe the elastic behavior (for small strain) of elastically isotropic homogeneous bodies. All *single* crystals are elastically *anisotropic* so that more than two elastic constants are needed to describe their properties.

ELASTIC ENERGY PER UNIT VOLUME.   The work which is done to slowly and elastically stretch the uniform portion of the rod of Figure 2.3.1 is

$$W = \int_0^l F \, dl = A_0 l_0 \int_0^\epsilon \sigma \, d\epsilon. \tag{2.3.9}$$

This is stored in the specimen as elastic energy. The elastic energy stored per unit volume is obtained by dividing by the volume, $A_0 l_0$, and is

$$w = \int_0^\epsilon \sigma \, d\epsilon = \int_0^\epsilon E\epsilon \, d\epsilon = \frac{E\epsilon^2}{2} = \frac{\sigma^2}{2E}. \tag{2.3.10}$$

This is known as the **elastic energy per unit volume** or the **strain energy density.**

If the specimen is deformed elastically by shear only, as in Figure 2.3.3, then

$$w = \int_0^\gamma \tau \, d\gamma = \frac{G\gamma^2}{2} = \frac{\tau^2}{2G}. \tag{2.3.11}$$

MORE ABOUT ELASTICALLY HOMOGENEOUS ISOTROPIC MATERIALS.   During the tensile test (see Figure 2.3.1), we obtained a longitudinal strain ϵ and, even though we do not apply lateral forces, we note a lateral strain. We define the **Poisson ratio** ν as

$$\nu = -\frac{\text{lateral normal strain in tension test}}{\text{longitudinal normal strain in tension test}}. \tag{2.3.12}$$

The longitudinal strain has already been defined in Equation (2.3.1). The lateral strain is defined in the same manner, namely by

$$\text{lateral strain} = \frac{r - r_0}{r_0}, \tag{2.3.13}$$

where r_0 is the initial radius of the test rod and r is the final radius.

It can be shown [see H. D. Conway, *Mechanics of Materials*, Prentice-Hall, Englewood Cliffs, N. J. (1950), p. 24] that

$$\nu = \frac{E}{2G} - 1 \tag{2.3.14}$$

so that ν is not an independent elastic constant. ν has values between 0 and $\frac{1}{2}$ as shown in Table 2.3.1.

Table 2.3.1. SOME VALUES OF POISSON'S RATIO FOR ELASTICALLY ISOTROPIC SOLIDS

Material	v
Beryllium	0.05
Glass	0.20–0.25
Rock	0.25
Steel	0.28
Copper	0.33
Nylon	0.48
Rubber	$\infty\,0.50$

If a hydrostatic pressure is applied to a material, its volume decreases according to the expression

$$dP = -K\frac{dV}{V},\tag{2.3.15}$$

where dV is the change in volume, V is the original volume, dP is the change in the magnitude of the pressure, and K is the **bulk modulus**. If the volume changes are small ($dV/V \sim 0.01$), it can be assumed that the bulk modulus is a constant (is not a function of pressure). The bulk modulus varies rapidly with pressure for a gas, slowly for liquids, and very slowly for most solids.

For homogeneous isotropic elastic solids, it can be shown that

$$K = \frac{E}{3(1-2v)}.\tag{2.3.16}$$

Values of K for some materials are given in Table 2.3.2.

Table 2.3.2. SOME VALUES OF BULK MODULI AT 1 ATM AND 25°C

Material	10^6 psi	10^{12} dynes/cm^2
Ideal gas (at 1 atm)	0.0000147	0.000001
H_2O	0.34	0.023
Sodium	0.91	0.063
Quartz	3.54	0.24
Sodium chloride	5.42	0.37
Aluminum	11.1	0.76
Hematite (α-Fe_2O_3)	29.4	2.0

We should compare the data of Table 2.3.2 with those of Tables 2.1.1 and 2.2.1. Note that K is nonzero for gases and liquids, while both E and G are zero for these materials.

A convenient way to measure K (and G if nonzero) is to measure the velocity of sound in the materials. It can be shown that

$$K + \tfrac{4}{3}G = \rho v_l^2, \qquad (2.3.17)$$

where ρ is the density of the material and v_l is the longitudinal wave velocity. See J. C. Jaeger, *Elasticity, Fracture and Flow*, Wiley, New York (1956).

It can also be shown (again see Jaeger) that

$$G = \rho v_s^2, \qquad (2.3.18)$$

where v_s is the shear wave velocity.

Equations (2.3.17) and (2.3.18) provide a basis for a convenient way of measuring K and G (and hence E and v). They also provide a method for finding out whether the interior of the earth is liquid or solid. How? The sending of such sound waves into the earth and recording the results is known as seismology. It plays a vital role in prospecting for oil. See A. F. Fox, *The World of Oil*, Pergamon, New York (1964).

It should be noted that there is a difference between K measured in a dynamic experiment according to (2.3.17) and K measured in a static experiment according to (2.3.15). The dynamic experiment is adiabatic (the material is compressed so rapidly that heat cannot flow out—all materials heat up during compression), while the static experiment is isothermal (the material is compressed so slowly that the heat generated during the compression can flow out of the specimen, which remains at constant temperature). This difference is significant for gases, small for liquids, and quite small (a few percent or less) for solids.

2.4   MECHANICAL DAMPING

FREE VIBRATIONS.   Consider the spring-mass system shown in Figure 2.4.1. [In this section we shall assume that the pertinent aspects of mechanical damping are obtained by an experimentalist without the use of mathematical modeling, i.e., the use of models involving springs and dashpots (shock absorbers) which lead to differential equations. In Chapters 19 and 20 we shall consider the fundamental atomic nature of the springs and the dashpots, the physical models, and the mathematics of modeling.] We imagine that the system is placed in a vacuum so that there is no damping owing to motion through the viscous media air (though clearly not so viscous as honey). We shall assume that the spring material is a strong steel. We shall also assume that the vacuum chamber is isolated from external vibrations. The experiment is set up so that the mass can be displaced by x_0 from its rest position and can be released from this new position. The experimentalist also has set up devices for measuring the displacement of mass versus time, i.e., $x(t)$, and for measuring the force on the spring versus time, i.e., $F(t)$. The experimentalist performs an experiment and

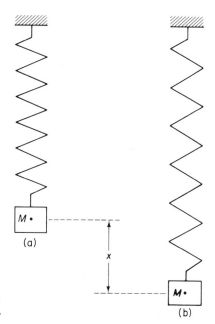

FIGURE 2.4.1. Spring-mass system (a) at rest and (b) in motion.

finds that $x(t)$ is very closely fitted by the cosine function

$$x = x_0 \cos \omega_0 t. \tag{2.4.1}$$

x_0 is the amplitude of the oscillation.

Further, using several springs with different stiffnesses k and several masses M, he is able to show that

$$\omega_0 = \sqrt{\frac{k}{M}}. \tag{2.4.2}$$

The frequency ω_0 is called the **natural frequency** of the spring-mass system. This is the frequency at which the spring-mass system will naturally vibrate in the absence of externally applied force. The experimentalist also notes that the force on the spring is

$$F = F_0 \cos \omega_0 t \tag{2.4.3}$$

and he readily finds that the amplitude of this cosine function is

$$F_0 = kx_0. \tag{2.4.4}$$

The experimentalist also has a setup whereby the instantaneous value of F and x can be shown as a spot on an oscilloscope screen, with F being the vertical input and x being the horizontal input. He finds in the present case that the spot moves back and forth over a line as shown in Figure 2.4.2. This illustrates the behavior of a system with little damping; the experimentalist did no-

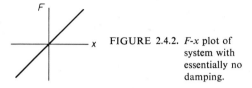

FIGURE 2.4.2. *F-x* plot of
system with
essentially no
damping.

tice, however, that the length of the line of Figure 2.4.2 appeared to shorten
very slowly in time.

EXAMPLE 2.4.1

Show that the equation of the motion in Figure 2.4.1 is $M(d^2x/dt^2) +$
$kx = 0$, that the initial conditions are $x = x_0$ at $t = 0$ and $dx/dt = 0$ at $t = 0$,
and that $x = x_0 \cos \sqrt{(k/M)}\, t$ is a solution.

Answer. Using Newton's law $F = Ma$, we note that $a = d^2x/dt^2$
and that the force acting on this mass is $-kx$. Hence $M(d^2x/dt^2) + kx$
$= 0$. It is stated that the spring is released from a stretched position x
$= x_0$ at $t = 0$. Likewise the velocity is zero at $t = 0$.

To prove that

$$x = x_0 \cos \sqrt{\frac{k}{M}}\, t$$

is a solution, simply substitute it into the equation of motion and show
that it satisfies the identity. Also show that the alleged solution satisfies the
initial conditions.

Suppose we now replace the steel coil spring by a rubber band and make
the same set of measurements as before. This time, for $F(t)$ we would find the
curve shown in Figure 2.4.3 and for $F(x)$ the curve shown in Figure 2.4.4.

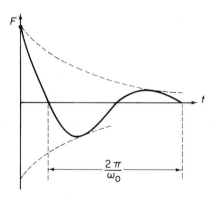

FIGURE 2.4.3. Free damped
oscillations.

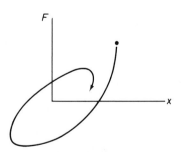

FIGURE 2.4.4. $F(x)$ for free
damped
oscillations.

The curve-fitting experimentalist would notice that the envelope of the $F(t)$ curves (the upper dashed line of Figure 2.4.3) would be closely fitted by

$$F_0(t) = kx_0 e^{-t/\tau}, \tag{2.4.5}$$

where τ is an experimental parameter called the **relaxation time**. He would note that the actual curves are well fitted by

$$F = kx_0 e^{-t/\tau} \cos \omega_0 t. \tag{2.4.6}$$

However, he would also note that x reached its peak in general at a later time than F and that

$$x = x_0 e^{-t/\tau} \cos (\omega_0 t - \delta); \tag{2.4.7}$$

i.e., there is a phase difference, δ, between F and x. Measurement of the quantity δ would show that τ and δ (which are the quantities which did not appear in the undamped situation) are related by

$$\tau\omega_0 = \frac{2}{\delta}, \tag{2.4.8}$$

so the damping is really described by one parameter δ. This angle δ is called the **angle of loss** or the **loss angle**.

The stored elastic energy, which was $kx_0^2/2$ initially [see Equation (2.2.7)], decreases with each full cycle, and it can be shown that the fractional decrease per cycle is

$$\frac{\Delta u}{u} = 2\pi\delta. \tag{2.4.9}$$

The mechanical energy which is lost is converted into heat, i.e., the temperature of the spring rises. If the damping is small, the $F(x)$ curve approximates an ellipse.

Some values of δ are given in Table 2.4.1.

Table 2.4.1. ANGLE OF LOSS FOR DIFFERENT
MATERIALS

Material	Angle of Loss, rad
Quartz	0.0002
Steel	0.0004
Brick	0.003
Concrete	0.007
Wood	0.02
Cork	0.07
Rubber	(hard) 0.01–0.1 (soft)

The loss angle is a structure-sensitive property and so may depend to a large degree (by orders of magnitude) on small changes in composition, mechanical deformation, and thermal treatment. Later in the book, we shall describe the origin of δ in terms of atomic and molecular behavior.

EXAMPLE 2.4.2

A rubberlike material has $\delta = 0.032$. What is the fractional change in amplitude during each cycle?

Answer. The amplitude of the oscillatory wave is the coefficient of $\cos(\omega_0 t - \delta)$ in Equation (2.4.7); i.e.,

$$x_0 e^{-t/\tau} = x_0 e^{-(\omega_0 \delta t/2)}.$$

The period T of the cosine function is a time $2\pi/\omega_0$, so the fractional decrease in amplitude between the n and $n + 1$ period is

$$\frac{x_0 e^{-(\omega_0 \delta n T/2)} - x_0 e^{-[\omega_0 \delta(n+1)T/2]}}{x_0 e^{-(\omega_0 \delta n T/2)}} = 1 - e^{-(\omega_0 \delta T/2)} = 1 - e^{-\pi \delta}$$

$$= 1 - e^{-0.1} \approx 0.1,$$

where we have used $e^{-y} \approx 1 - y$ if $y \ll 1$.

EXAMPLE 2.4.3

Assume that there is a viscous drag force proportional to the velocity of stretching of the rubber band with the proportionality constant equal to μ. Show that the equation of motion is

$$M\frac{d^2 x}{dt^2} + \mu\frac{dx}{dt} + kx = 0.$$

Answer. Proceeding as in Example 2.4.1, we note that the force F acting in the positive x direction is $-kx - \mu(dx/dt)$ and this force F equals $M(d^2 x/dt^2)$.

The student who has studied differential equations may wish to obtain (as most elementary differential equation books do) the solution of this equation with the associated side conditions $x = x_0$ at $t = 0$ and $dx/dt = 0$ at $t = 0$. See Problems 2.39–2.42.

FORCED VIBRATIONS.   In many real physical situations, a periodic external force

$$F = F_0 \cos \omega t \tag{2.4.10}$$

is applied to a spring-mass system and it is then found that

$$x = x_0(\omega) \cos(\omega t - \beta), \tag{2.4.11}$$

where the amplitude of the displacement depends on ω, as shown for a typical case in Figure 2.4.5. Analysis of this system in terms of a mathematical model is given in Problems 2.43–2.45. The equations which follow in this section could be obtained experimentally for small damping or they could be derived from the mathematical model for the case of small damping.

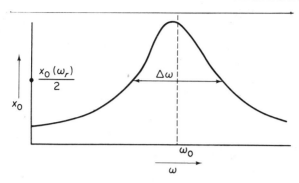

FIGURE 2.4.5. Amplitude-frequency curve for forced vibration.

The phase difference β is a different quantity here than δ in the preceding discussion, except for small ω. Of particular interest here is the maximum deflection obtained under the vibrational force of amplitude F_0 [Equation (2.4.10)] compared to the deflection obtained by a static force F_0. For a material with small damping ($\delta \approx 0.01$) this maximum amplitude occurs at $\omega = \omega_r$ (the resonance frequency), which is very close (but not equal) to ω_0, where $\omega_0 = \sqrt{k/M}$. The experimentalist would find that

$$\frac{x_0(\omega_r)}{x_0(0)} = \frac{1}{\delta} \tag{2.4.12}$$

and later on we shall derive this theoretically as well. This ratio is called the **resonance amplification**. We note that for steel it has a value of 2500.

The **Q factor** or **quality factor** of the system is given by $Q = \omega_r/(\omega_1 - \omega_2)$, where ω_1 and ω_2 are the frequencies where the power dissipation falls to half of that at resonance; it is experimentally found to be

$$Q = \frac{1}{\delta} \tag{2.4.13}$$

The peak width at half the resonance amplitude is given by

$$\Delta\omega = \omega_0\sqrt{3}\,\delta. \tag{2.4.14}$$

EXAMPLE 2.4.4

A bar of steel of 1-in.^{2} cross-sectional area is stretched in tension by hanging on a weight of 2000 lb. It is therefore stressed to 2000 psi. If an additional cyclic force with an amplitude of only 200 lb were applied with a frequency equal to the natural frequency ω_0 of the system, what additional stress would be present?

Answer. An additional stress of 500,000 psi would be present since the resonant amplification is 2500. Even a very strong steel would be near its breaking point.

EXAMPLE 2.4.5

Find the stiffness constant of a steel rod having a 1-in.² cross-sectional area and a length of 100 ft.

Answer. $F = kx$. But $F = \sigma A$ and $kx = kL(x/L) = kL\epsilon$ so $\sigma A = kL\epsilon$. But $\sigma = E\epsilon$ so

$$k = \frac{EA}{L} = \frac{30 \times 10^6 \text{ lb}}{100 \text{ ft}} = 3 \times 10^5 \text{ lb/ft},$$

where E is Young's modulus (see Table 2.1.1).

EXAMPLE 2.4.6

What would ω_0 be for the rod of Example 2.4.5, which carries a mass of 2000 lb.

Answer.

$$\omega_0 = \sqrt{\frac{k}{M}} = \sqrt{\frac{(3 \times 10^5 \text{ lb/ft})(32 \text{ ft/sec}^2)}{2000 \text{ lb}}} \simeq 70/\text{sec}.$$

The factor of 32 ft/sec² is introduced to convert k to the units of lb/sec² which is consistent with the unit of mass (lb) used here.

The resonance amplification factor (and hence the loss angle) is a very important quantity in design, not only in design of dynamic equipment, such as lathes, airplane wings, engines, tires, etc., but for what might appear to be static structures, such as bridges. The student has probably already read of marching soldiers breaking cadence as they cross a bridge, of the story of a tiny dog causing near failure of a foot bridge as it trotted across, and of the

FIGURE 2.4.6. Resonance failure of the Tacoma Narrows Bridge.

story of the infamous bridge, "Galloping Gertie" (see "Bridges" in *Encyclopaedia Britannica* and note the discussion of the Tacoma Narrows Bridge). The resonance failure of Galloping Gertie in which the wind provided the driving force is illustrated in Figure 2.4.6.

There are applications in which a low value of δ are desirable, e.g., in bells. However, in most applications, such as tires, a large damping (measured by a large loss angle) is desirable. (Imagine cars with steel wheels roaring down our concrete highways.) Rubber has two properties which make it highly desirable for tires: (1) It can be stretched elastically to high strains, and (2) it has a very high loss factor.

2.5   PLASTIC DEFORMATION

If a copper rod is twisted, it will at first behave elastically. But if twisted too much, it deforms permanently. We call this permanent deformation **plastic deformation**. One can get a shear stress-shear strain curve by measuring the torque versus the angular twist. The exact analysis does not concern us here. See A. Nadai, *Theory of Flow and Fracture of Solids*, McGraw-Hill, New York (1950), p. 347.

The data obtained at room temperature at a strain rate of $\dot{\gamma} = 10^{-3}$ sec^{-1} appear in Figure 2.5.1.

The values of τ and γ are the values at the surface of the specimen. We note that it is necessary to specify *as exactly as possible* the nature of the copper; copper with a slightly different composition or a different thermal history or copper which has been mechanically deformed would behave in a different

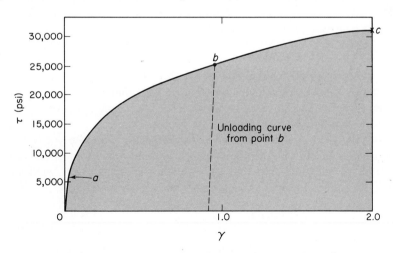

FIGURE 2.5.1. Shear stress-strain curve for a copper rod annealed 1 hr at 800°C.

fashion. It is also necessary to specify the test conditions (temperature and strain rate) for a given copper. For γ preceding point a, the elastic behavior of Equation (2.3.8) holds and the slope is G. Plastic deformation then begins if the stress is increased. Such plastic deformation would continue until the rod fractures at point c. The fracture would appear almost the same as if the rod were sawed off perpendicular to the axis. (The author had this happen to the drive shaft of his automobile late one night outside Fort Bridger, Wyoming. Such occurrences help one to remember the nature of the fracture and, for that matter, the nature of that isolated bus stop.)

The original rod is capable of a large amount of plastic deformation; this is called **ductility**. γ_{max}, which equals 2 in the present case, is a measure of the ductility. We note that during this deformation a large amount of work per unit volume is done, namely,

$$w_{max} = \int_0^{\gamma_{max}} \tau \, d\gamma. \tag{2.5.1}$$

Just prior to fracture, a tiny portion of this is stored as elastic energy,

$$w_{elastic \, max} = \frac{\tau_{max}^2}{2G}, \tag{2.5.2}$$

so the maximum plastic work done is

$$w_{plastic \, max} = \int_0^{\gamma_{max}} \tau \, d\gamma - \frac{\tau_{max}^2}{2G}. \tag{2.5.3}$$

The amount of work done on the specimen material per unit volume at fracture is called the **toughness** of the material.

Suppose the original material had been stressed only to the point b and the stress were removed. The unloading curve (except perhaps near point b) would be a straight line with slope G. On reloading, it would follow the unloading curve to point b. Both the unloading and reloading would be examples of elastic behavior. Note that the shear yield strength would be increased from τ_a to τ_b, i.e., from 6000 to 25,000 psi. It should be noted that during plastic deformation the volume remains constant (at least that is a very good approximation, except in special cases, as we shall see later). The dimension of the test specimen does not change much. The process whereby plastic deformation increases the yield strength is called **strain hardening**. Suppose you were given a rod which had been plastically deformed to point b and told to determine its properties. You would find that your rod has a higher shear yield strength (25,000 psi) and a lower toughness than the original material.

EXAMPLE 2.5.1

a. Compute the work done per unit of volume in deforming a copper specimen until yielding just begins at $\tau_0 = 6000$ psi.

Answer.

$$w_{\text{elastic}} = \frac{\tau_0^2}{2G} = \frac{(6000)^2}{2(6,000,000)} = 3\frac{\text{in.-lb}}{\text{in.}^3}.$$

b. Compute the work done in deforming (by torsion) the specimen described in Figure 2.5.1 to fracture.

Answer.

$$w_{\text{max}} = \int_0^{\tau_{\text{max}}} \tau \, d\gamma.$$

By numerical integration, the area under the τ-γ curve is 45,000 in.-lb/in.³

Before we continue with our discussion of plastic deformation, let us consider a material such as ordinary window glass, which has a shear stress-strain curve as shown in Figure 2.5.2.

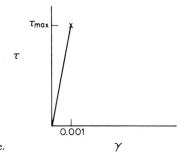

FIGURE 2.5.2. Shear stress-strain curve for window glass at room temperature.

In this case, the behavior is entirely elastic; there is no plastic deformation. Consequently, the energy stored per unit volume at fracture is very small. The crack will also be on a helicoidal surface inclined at 45 deg to the specimen axis. (Why? See Figure 2.3.6.) (The student can readily check this on a piece of chalk which shows similar **brittle behavior** compared to the ductile behavior shown by the copper in Figure 2.5.1.) If a tension test were performed on this glass, it would fracture at a tensile stress equal to τ_{max} and would behave elastically up to that point.

Let us now consider carrying out a tensile test on a tensile specimen made of the same copper for which the shear data of Figure 2.5.1 were obtained. The tensile data were obtained at room temperature at a strain rate of 10^{-3} sec^{-1}. The stress plotted in Figure 2.5.3 is the nominal tensile stress F/A_0 and the strain is the nominal tensile strain $(l - l_0)/l_0$.

The specimen yielded when the tensile stress reached σ_0 and this was found to be nearly equal to $\sqrt{3}\,\tau_0$. The specimen then deformed plastically with essentially no volume change. Since the length increases, the cross-sec-

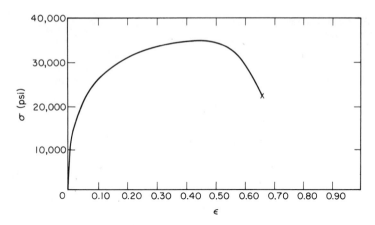

FIGURE 2.5.3. Tensile stress-strain curve for a copper rod
annealed at 800°C for 1 hr.

tional area decreases according to

$$\frac{A_0}{A} = \frac{l}{l_0}.$$
(2.5.4)

(Here we ignore the tiny elastic strains present for which the assumption
of constant volume does not hold unless Poisson's ratio $v = \frac{1}{2}$. For copper,
$v = \frac{1}{3}$.)

Remember that there were no such changes in shape in the torsion
specimen. As the plastic deformation proceeds, the material also strain-hardens.
The load that can be carried depends on two things: (1) how fast the cross-
sectional area decreases as the length increases and (2) how fast the material
strain-hardens as the strain increases. Initially, (2) wins out and the load increases. However, eventually, (1) wins out, and the load that can be supported
then decreases. There is, therefore, a maximum load or a maximum in the
nominal stress-strain curve which occurs long before the specimen fractures.
Before this maximum load $F_{\max}$ is reached, the cross-sectional area decreases
uniformly along the gauge length. What happens as the strain increases? It
would be easy to find out if you had a tension machine handy. An alternative
way is to try to predict what happens.

It is more convenient to work with real stress and real strain when dealing
with large strains. **Real stress** is defined by

$$\sigma_r = \frac{F}{A}$$
(2.5.5)

and **real strain** by

$$\epsilon_r = \ln \frac{l}{l_0}$$
(2.5.6)

from

$$d\epsilon_r = \frac{dl}{l}.$$
(2.5.7)

EXAMPLE 2.5.2

Explain the origin of (2.5.6) and (2.5.7).

Answer. Since $\epsilon = (l - l_0)/l_0$ is a satisfactory definition of strain when dealing with small strains, we note that by differentiation we have

$$d\epsilon = \frac{dl}{l_0}$$

if the change in length dl occurs when the initial length is l_0. If the change in length dl occurs when the length is l, we would correspondingly write

$$d\epsilon_r = \frac{dl}{l}$$

and

$$\epsilon_r = \int_{l_0}^{l} \frac{dl}{l} = \ln\frac{l}{l_0}.$$

The expression

$$\sigma_r = K\epsilon_r^n \tag{2.5.8}$$

can be used to represent the plastic portion of the data for ductile metals. Here K and n are constants. n is the **strain-hardening exponent**. Equation (2.5.8) is an empirical expression for work hardening.

EXAMPLE 2.5.3

Show that σ has a maximum when $\epsilon_r = n$. A maximum in σ corresponds to

$$\frac{\partial \sigma}{\partial \epsilon} = 0.$$

Answer. Noting that $\sigma = \sigma_r(A/A_0) = \sigma_r(l_0/l) = \sigma_r e^{-\epsilon_r}$ and that $d\epsilon = d\epsilon_r(l/l_0)$, we have

$$\frac{\partial \sigma}{\partial \epsilon} = \frac{l_0}{l}\frac{\partial(\sigma_r e^{-\epsilon_r})}{\partial \epsilon_r} = 0.$$

Substitution of $\sigma_r = K\epsilon_r^n$ gives $nK\epsilon_r^{n-1}e^{-\epsilon_r} - K\epsilon_r^n e^{-\epsilon_r} = 0$ or

$$\epsilon_r = n \quad \text{when} \quad \sigma = \sigma_{\max}.$$

EXAMPLE 2.5.4

Given $K = 73{,}000$ psi and $n = 0.4$ for the copper rod of Figure 2.5.3, plot σ vs. ϵ *assuming that the entire gauge length continues to decrease uniformly even after $\sigma = \sigma_{\max}$ is reached.*

Answer. We use $\sigma = \sigma_r e^{-\epsilon_r}$ from Example 2.5.3 and $\sigma_r = K\epsilon_r^n$ from Equation (2.5.8) and $\epsilon_r = \ln(1 + \epsilon)$ from Equations (2.3.1) and (2.5.6) to obtain

$$\sigma = \frac{K[\ln(1 + \epsilon)]^n}{1 + \epsilon},$$

which is then plotted in Figure 2.5.4 as the dashed line.

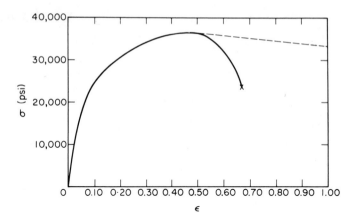

FIGURE 2.5.4. Calculated σ vs. ϵ curve shown by dashed line.

Why do the curves in Example 2.5.4 not agree? Perhaps the italicized assumption in Example 2.5.4 is wrong. Why? Up to the maximum, all is well. Now suppose that slightly beyond the maximum one section of the rod is *slightly* weaker than the rest (it is too much to expect perfect homogeneity). Besides, the rod heats when deformed and the point midway between the ends gets hottest since heat is carried away fastest at the ends; materials get weaker at higher temperatures. Hence we should not expect homogeneity. The weakest part deforms more than the rest, decreasing its area more than the rest. Since we have passed the maximum, the load that this area can now carry will be even less than before and it will therefore deform further with a *smaller* load than before. The specimen will appear before fracture to have a neck, as shown in Figure 2.5.5. The phenomenon is called **necking**. It is an example of a **plastic**

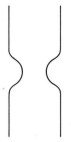

FIGURE 2.5.5. Necked tensile specimen.

instability. (Remember the elastic instability associated with the buckling of the column in Figure 2.1.5.)

Ductility is the ability to be plastically strained. The **maximum real strain**

$$\epsilon_r^{max} = \ln\frac{A_0}{A_f} \qquad (2.5.9)$$

is one measure of ductility. The **percent reduction in area** at the neck

$$\frac{A_0 - A_f}{A_0} \times 100\% \qquad (2.5.10)$$

is another measure of ductility. In very ductile materials, the specimen will neck down almost to a line and the materials will slide apart. The ductility of a material plays a major role in design since it determines the actual safety factor in many cases and since it plays a major role in determining the formability of components (i.e., how easy a given part can be made), as we shall see later in this section. Figure 2.5.6 illustrates brittle and ductile fractures.

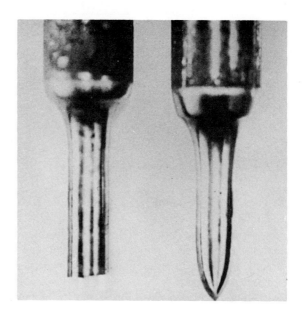

FIGURE 2.5.6. Brittle fracture and ductile fracture.

A **brittle material** has a stress-strain curve similar to Figure 2.5.2 while a ductile material has a stress-strain curve similar to Figure 2.5.3. As we shall see later, various factors, such as temperature, affect the ductility of materials.

Certain materials, including some metals, can be stretched 1000 % without fracture. This behavior is called **superplasticity**.

TEMPERATURE AND RATE OF STRAINING.   In general, as the temperature is increased, smaller stresses will be needed to reach the same strain. As the rate of straining is increased, larger stresses will be needed to reach the same strain.

ANNEALING.   After the copper rod shown in Figure 2.5.1 is cold-worked to $\gamma = 1$, it has a shear yield strength of 25,000 psi. If this cold-worked copper bar is unloaded and annealed by heating to a high temperature (say 800°C for 1 hr), it will lose the strength gained during cold working, and the shear yield strength at room temperature will drop to 6000 psi. The annealed material will behave the same as the annealed starting material described by Figure 2.5.1.

STRESS CONCENTRATION.   Consider a plate with a circular hole as shown in Figure 2.5.7. The tensile stress is σ. However, next to the hole the stress is σ', where $\sigma' > \sigma$. We write

$$\sigma' = K\sigma. \tag{2.5.11}$$

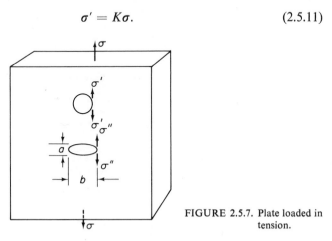

FIGURE 2.5.7. Plate loaded in tension.

K is known as the **stress concentration factor**. In the present case, $K = 3$, as can be shown theoretically by elasticians [J.C. Jaeger, *Elasticity, Fracture and Flow*, Wiley, New York (1962) p. 187.] or experimentally by experimental stress analysts using either photoelasticity techniques, which measure the stresses directly, or electrical strain gauges, which measure elastic strain which is then converted to stress.

The stress concentration at the tip of the elliptical hole is much larger and, in fact, for very sharp cracks (large b/a ratio of the ellipse) can take on values of the order of 1000. Such sharp cracks should be avoided by designers, and even the stress concentrations of a circular hole are undesirable. There are some situations where very sharp cracks occur quite naturally, for example, on the surface of glass (*glass* here means simply a material, such as window glass, although later the word *glasses* will refer to a group of materials with special atomic structure). The presence of these cracks makes window glass fracture at a stress of about 10,000 psi with a stress-strain curve as shown in Figure 2.5.2. In the absence of such cracks, glass has a strength in excess of 1,000,000 psi.

Imagine now an airplane wing with the choice of materials being glass or an aluminum alloy. Each of these materials has about the same stiffness modulus. Their density is about the same. Let us suppose that the glass has a tensile fracture stress of 50,000 psi and that the aluminum has a tensile yield strength of 50,000 psi. If a crack is present in a glass component and begins to grow, the component will fracture. However, assuming that the aluminum has considerable ductility and toughness (like the copper of Figure 2.5.3), then the presence of a crack might not cause much concern, even though the tensile stress across the crack has reached 50,000 psi. The reason for this is shown in Figure 2.5.8. The sharp crack may have a stress concentration factor $K = 1000$ but if plastic deformation is possible at the tip of the crack, the crack tip will become rounded with a fairly large radius of curvature and the stress concentration factor will be of the order of one. [See: References (at end of chapter), Jaeger, p. 167.] Hence, we see why ductility and toughness are such important properties to the design engineer.

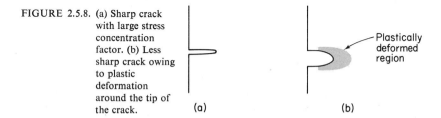

FIGURE 2.5.8. (a) Sharp crack with large stress concentration factor. (b) Less sharp crack owing to plastic deformation around the tip of the crack. (a)   (b)

FORMING OF MATERIALS. The ease of forming a given material into a component (along with its eventual properties) is vitally important in its effective utilization. Thus a given material may have exactly the properties which the designer wants, but the cost of achieving the desired shape may be so high that this material will not be used.

Early in our lives, most of us have had the opportunity to form materials: We molded clays and we whittled sticks. In so doing, we have performed two of the most important processes of obtaining a desired shape—shaping by material movement and shaping by material removal. Both of these processes are often performed by mechanical forces. Cutting machines, such as the lathe, the miller, the shaper, and the grinder shape by material removal by mechanical force. Removal of material (from electrical conductors) may also be achieved by electric machining processes, such as electrical discharge machining, plasma arc cutting, and electrochemical machining. Laser beams can also be used to remove material, e.g., to drill holes in diamonds to be used for wire drawing. When cutting machines are used, particularly those in which chips are formed (lathe, miller, shaper), the material being formed must be ductile, and the details of its stress-strain behavior determine its machinability.

When a rod is drawn through a die to make wire (it is actually drawn through several dies and annealed in between), shaping is achieved by material movement. This process has the advantage that there is no material waste. The forming of sheet by rolling, of beams by extrusion, and of gears by forging and the use of explosive forming are other examples of forming by material movement. All of these processes depend strongly on the details of the stress-strain curve of the material being worked, and these, in turn, depend on the temperature and velocity at which ·the processing is performed. They also depend strongly on the hydrostatic pressure superimposed on the stresses which cause deformation. For example, limestone, which is very brittle in an ordinary tension test, behaves as a ductile material when tested in tension while also under a hydrostatic pressure of 150,000 psi.

Frank Fuchs and his colleagues at the Western Electric Company use this concept for wire drawing. Using this method, they are able to cause a reduction in area by a factor of 10,000 in a single pass. (They do not have to draw the wire through numerous dies with intermittent annealing to achieve this.) Recently, they have developed techniques for doing this continuously.

2.6   FATIGUE, CREEP, AND NOTCH FRACTURE

CYCLIC STRESS FATIGUE.   In studying the phenomenology of the mechanical behavior of solids, three other important aspects are worthy of brief mention in an introduction. When a component is repeatedly stressed in tension and compression, such as

$$\sigma = S \cos \omega t, \tag{2.6.1}$$

we find that it fractures at a tensile stress considerably below even the tensile yield strength, as shown in Figure 2.6.1. We say that the material is subject to **cyclic fatigue**. An example of such stressing would be the repeated bending back and forth of a ruler. (Any devoted paper clip bender could tell you about fatigue.) One very important attribute of fatigue behavior is the wide statistical spread of the results. Whereas the yield strength for each of 1000 specimens may fall within 2% of the mean, the fatigue strength of one of the specimens at a given number of cycles may be 50% below the average.

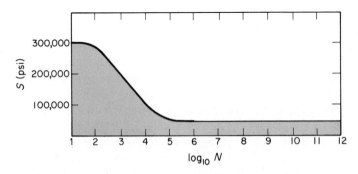

FIGURE 2.6.1. Plot of average fatigue strength versus the logarithm of the number of cycles of reversed loading $\sigma = S \cos \omega t$ for a high-strength steel. The curve is called an S-N curve.

Fatigue is of considerable importance in machinery involving oscillations, e.g., airplane wings, which are buffeted by air turbulence, fuselage stresses owing to air compression and decompression, landing struts which are flexed during landing, etc. Pressure cycles in the British Comet airplane caused catastrophic failures of the planes owing to fatigue stressing at the corners of the rectangular window frames. See T. Bishop, "Fatigue and the Comet Disasters," *Metal Progress*, **67**, 77 (May 1955). Cyclic fatigue causes the designer a number of problems; for example, he can build a plane which is 99.99% safe, say, but not one which is 100% safe.

Some materials show an **endurance limit**; i.e., after a certain large number of cycles, S stops decreasing and reaches a limiting value. For many materials, however, S simply continues to decrease. A rough rule is that for 10^8 cycles, $\sigma_0/4 < S < \sigma_0/3$.

Many books have been written about fatigue. See G. Sines and J. L. Waisman, *Metal Fatigue*, McGraw-Hill, New York (1959). The experimental methods of studying fatigue are described in such books. The student should try to invent methods of studying fatigue at 30 Hz and at 10^5 Hz.

Why do materials fatigue and what can be done about it? Before we delve into such questions, we shall have to learn more about the atomic structure of materials.

STATIC FATIGUE. Some materials, such as glass, fracture at a given stress without any apparent plastic deformation only after stress has been present for a long time. Such behavior is known as **static fatigue**.

CREEP. An engineer tries to run an engine at the highest possible temperature in order to improve its efficiency. The maximum possible efficiency of a heat engine (as any student of thermodynamics can show) is

$$\text{maximum efficiency} = \frac{T - T_0}{T}. \tag{2.6.2}$$

Here T is the internal temperature and T_0 is the exhaust temperature (which tends to be fixed by the environment, say room temperature or river temperature, etc.). The components of an engine, e.g., the turbine blades, are also under stress at these high temperatures. Under such conditions, the material continually deforms under fixed stress (unlike the situation at low temperatures where the stress has to be increased to get additional deformation, as in Figure 2.5.1). This is called **creep**.

Typical behavior at constant stress is shown in Figure 2.6.2. Creep depends very strongly upon temperature. A rough rule is that it becomes quite important above $T_m/3$, and critical above $T_m/2$ where T_m is the melting temperature. Creep failure can occur in two ways: either the magnitude of deformation becomes intolerable or the component ruptures. An empirical expression for the rate of steady-state creep (region where $d\epsilon/dt$ is a constant in Figure

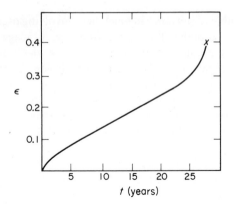

FIGURE 2.6.2. Creep curve strain ϵ vs. time t.

2.6.2) is

$$\dot{\epsilon} = A\sigma^n e^{-Q_c/RT}. \tag{2.6.3}$$

Here A is a constant, σ is the tensile stress, n is a constant usually equal to about five, R is the gas constant, T is the absolute temperature, and Q_c, called the **activation energy for creep**, has a characteristic value for each material.

EXAMPLE 2.6.1

A nickel sample shows a steady-state creep rate of

$$\dot{\epsilon} = 10^{-12} \text{ sec}^{-1}$$

at

$$T = 862°\text{K}$$

(which is half the melting point) at a tensile stress of only 1000 psi. If $Q_c = 69,000$ cal/mole, estimate the creep rate at $1293°\text{K}$ ($0.75T_m$) if the stress is the same.

Answer.

$$\frac{\dot{\epsilon}(1293°\text{K})}{\dot{\epsilon}(862°\text{K})} = \frac{A\sigma^n e^{-Q_c/R \times 1293}}{A\sigma^n e^{-Q_c/R \times 862}}.$$

Hence, since $R = 1.987$ cal/mole-°K,

$$\dot{\epsilon}(1293°\text{K}) = 10^{-12} e^{-(69,000/1.987)[(1/1293)-(1/862)]}.$$

Remembering that $e^{-x} = 10^{-x/2.3}$, we have

$$\dot{\epsilon} = 10^{-12} \times 10^{5.8} \text{ sec}^{-1}.$$

The creep rate at the higher temperature is faster by nearly 1,000,000 times. If the creep rate at $0.5T_m$ is tolerable for only 30 years, at $0.75T_m$ the lifetime is only about 0.3 hr.

For more about creep, see I. Finnie and W. R. Heller, *Creep of Engineering Materials*, McGraw-Hill, New York (1959).

It should be noted that if the stresses are very high (near σ_0), creep even occurs at low temperatures.

The question of why creep occurs and why it is so strongly dependent on temperature requires an understanding of the atomic structure of solids and of the motion of atoms in solids.

NOTCH FRACTURE.   Earl A. Parker begins his book *Brittle Fracture of Engineering Structures*, Wiley, New York (1957), as follows:

> On January 16, 1943, a T-2 tanker lying quietly at her fitting-out pier at Portland, Oregon, suddenly cracked in a brittle manner . . . "without warning and with a report that was heard for at least a mile . . . ." The sea was calm, the weather mild, her computed deck stress was only 9,900 psi. There seemed to be no reason why she should have broken in two, but she did.

FIGURE 2.6.3. Brittle fracture of a T-2 tanker.

See Figure 2.6.3.

There were cracks owing to faulty welding and these cracks (remember the stress concentration factor K of Section 2.5) opened slowly at first (note the possibility of creep at high stresses even though the temperature is only about $T_m/6$). We have already noted the role which ductility and toughness play in affecting the propagation of a crack.

Many steels (and also many other materials) show a special behavior when notched specimens are impacted at high velocity and the energy absorbed is plotted versus temperature (Figure 2.6.4). The **Charpy V-notch impact test** is a specific type of test which refers to the sample size, notch size, and testing procedure. (The student should sketch a design of an impact apparatus which will measure the total energy absorbed in bending and breaking the notched specimen. The specimen should be simply supported near its ends and should be struck by the anvil at the middle.)

At high temperature, the specimen absorbs lots of energy because of the plastic deformation surrounding the crack (see Figure 2.5.8), but as the temperature drops to around room temperature, there is a sudden sharp drop

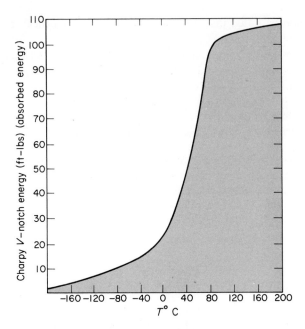

FIGURE 2.6.4. Impact test. Charpy V-notch energy versus temperature for a steel with $\sigma_0 = 60,000$ psi.

in the energy absorbed. Thus at the higher temperatures, the steel is very ductile, while at the lower temperatures, it is brittle (or only slightly ductile). This peculiar behavior was responsible for the demise of many welded ships (and of several welded bridges in Canada and Belgium) during the winter of 1943 (which was especially cold).

Several materials, such as austenitic 18-8 stainless steel, do not show such behavior and remain ductile to temperatures approaching absolute zero. Why do some materials show the behavior of Figure 2.6.4 whereas others do not? We shall study these problems later. Engineers often specify a minimum Charpy V-notch energy of 15 ft-lbs.

EXAMPLE 2.6.2

A tank car is being built to transport liquid nitrogen (boiling point, 78°K). Would the steel of Figure 2.6.4 be used as the inner dewar?

Answer. This temperature corresponds to −195°C. The steel has a Charpy V-notch impact energy of only 2–3% of its high-temperature value. A small crack or notch in the vessel would be critical (because K would effectively be very high).

2.7 STIFFNESS TO WEIGHT AND STRENGTH TO WEIGHT RATIOS

The rocket or aircraft designer wants a structure which has the desired stiffness and the desired strength with the least weight (and cost). His desire to reduce the weight may be simply one of improving efficiency (so that cargo will replace the dead weight in a commercial aircraft) or one of feasibility (so that a certain space voyage can be made). Consequently, when designing columns, he is not only interested in E [as we were when we discussed Equation (2.1.8)] but in the **density of the material** ρ. He is actually interested in the ratio of the stiffness of a component to its weight. In terms of materials properties he is interested in the ratio E/ρ, i.e., tensile stiffness modulus/density; this is often called the **specific stiffness**. Similarly when designing a component so that it will not fracture in tension, he is not only interested in knowing that the member is strong enough (and hence in knowing tensile strength) but in the strength to weight ratio, which in terms of material properties means the ratio $\sigma_{\text{T.s.}}/\rho$; this is often called the **specific tensile strength**.

These two ratios are very important not only in the design of aircraft, rockets, and deep submergence vessels but in a great variety of dynamic machinery. Table 2.7.1 gives some typical values. We note that the ratio $\sigma_{\text{T.s.}}/\rho$ is simply the maximum length of a rod of material which can hang freely without fracturing.

Table 2.7.1. DENSITY, STIFFNESS/DENSITY, AND TENSILE STRENGTH/DENSITY FOR SOME MATERIALS

Material	ρ		E/ρ		$\sigma_{\text{T.s.}}/\rho$	
	lb/in.^{3}	g/cm^3	10^8 in.	10^8 cm	10^6 in.	10^6 cm
Aluminum (2024-T3)	0.100	2.76	1.06	2.7	1	2.5
Steel (maraging)	0.283	7.84	1	2.5	1.5	3.8
Boron-reinforced epoxy (boron fibers parallel to rod)	0.068	1.88	5	13	4	10

It should be obvious that the allowable fatigue stress/density ratio and the allowable creep strength/density ratio are also very important parameters in design.

Strength to weight ratios are important for space craft and ocean craft; however, this property has been important in the construction of bridges for a long time. Figure 2.7.1 shows a large suspension bridge. We note that the suspension cables must support, in addition to their own weight, the bridge floor and the traffic.

FIGURE 2.7.1. Mackinac Bridge, Michigan. After D. B. Steinman and S. R. Watson, *Bridges and Their Builders*, Dover New York (1957).

The importance of the technology of bridge building was described in a speech by Franklin D. Roosevelt in 1931: "There can be little doubt that in many ways the story of bridgebuilding is the story of civilization. By it we can readily measure an important part of a peoples' progress." An excellent book on the subject of bridges is D. B. Steinman and S. R. Watson, *Bridges and Their Builders*, Dover, New York (1957).

2.8 THE FLOW OF FLUIDS

We have noted in the previous sections that gases and liquids do not support shear stresses in the static case, i.e., the shear modulus $G = 0$. However, liquids and gases are more or less viscous media in that they resist flow. For example, in a long transmission pipeline there are periodic pumping stations which build up the pressure and provide the driving force for moving the fluid to the next pumping station. Pressure measuring devices could be installed at a distance L apart (with no pumping station in between); say the measured pressure drop is $\Delta P = P_1 - P_2$. Then the flow rate Q (volume/time) passing a particular cross section is given by

$$Q = \frac{dV}{dt} = \frac{1}{\eta}\left(\frac{\pi R^4}{8L}\right)\Delta P \qquad (2.8.1)$$

where R is the pipe radius. Thus dV/dt is found to be directly proportional to the pressure drop, to a geometric factor, and to the inverse of η, where η is a *material property* known as the viscosity coefficient. Equation (2.8.1) holds for incompressible fluids and also for ideal gases if the volume is measured at the average pressure, i.e., $(P_1 + P_2)/2$. Equation (2.8.1) is known as Poiseuille's equation (1844). If one were to study experimentally the velocity of the liquid as it passes

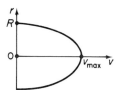

FIGURE 2.8.1. Velocity versus radius profile.

down the tube as a function of radial position, one would find a velocity profile shown in Figure 2.8.1 and given by

$$v = v_{max}\left(1 - \frac{r^2}{R^2}\right). \qquad (2.8.2)$$

If we consider two concentric cylindrical sheets of fluid as shown in Figure 2.8.2, they will move along at different velocities. Newton proposed that there

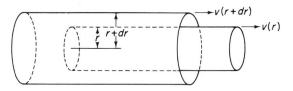

FIGURE 2.8.2. Fluid flow of adjacent layers.

was a viscous drag proportional to the velocity gradient normal to the layers which were flowing over each other. Thus the shear stress τ is related to dv/dr according to

$$\tau = -\eta\frac{dv}{dr} \qquad (2.8.3)$$

where η is a constant, characteristic of the material, called the **viscosity coefficient**. Materials which behave in this fashion are called **Newtonian fluids**. Equation (2.8.3) defines the viscosity coefficient.

EXAMPLE 2.8.1

Show that Equation (2.8.2) follows from Equation (2.8.3).

Answer. The following sketch shows the forces acting on the inner cylinder of Figure 2.8.2. A balance of forces must equal zero for steady

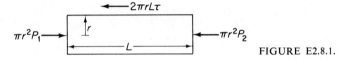

FIGURE E2.8.1.

flow:

$$-2\pi r L \eta \frac{dv}{dr} = \pi r^2 (P_1 - P_2).$$

Using the condition that $v = 0$ when $r = R$ as we experimentally observed (see Figure 2.8.1), we can show by integration that

$$v = \frac{P_1 - P_2}{4L\eta}(R^2 - r^2)$$

and hence

$$v_{\max} = \frac{P_1 - P_2}{4L\eta} R^2.$$

Hence

$$v = v_{\max}\left(1 - \frac{r^2}{R^2}\right).$$

Assuming that you have the results of this example, how would you proceed to evaluate the total flow rate through the pipe?

Table 2.8.1 shows the viscosity coefficient of some fluids. η has units of force-time/area. A system commonly used by hydrodynamicists is the centimeter-gram-second (cgs) system and the unit for η is then the **poise**: 1 P = 1 dyne-sec/cm^2.

Hydrodynamicists often use the **fluidity coefficient** ϕ, where

$$\phi = \frac{1}{\eta}. \tag{2.8.4}$$

Table 2.8.1. VISCOSITY COEFFICIENTS AT 20°C

Material	η, mP
Hydrogen (1 atm)	0.084
Ethylene (1 atm)	0.093
Oxygen (1 atm)	0.167
Nitrogen (1 atm)	0.192
Hexane	3.26
Octane	5.42
Carbon tetrachloride	9.09
Water	10.09
Mercury	15.47
Light machine oil	~1,000
Heavy machine oil	~5,000
Glycerol (glycerine)	10,690
Shoe wax	~5 × 10^9
Pitch (for roads)	~10^{13}

The viscosity coefficient for ideal gases is independent of pressure and increases slowly with absolute temperature ($\propto T^{3/4}$). However, the viscosity coefficient of liquids varies strongly with the temperature and pressure. The viscosity of simple liquids decreases rapidly with temperature according to

$$\eta = \eta_0 e^{Q_{vis}/RT},\tag{2.8.5}$$

where η_0 and Q_{vis} are characteristics of each material. In terms of fluidity this would be

$$\phi = \phi_0 e^{-Q_{vis}/RT}.\tag{2.8.6}$$

(Compare this with the temperature dependence of steady-state creep, i.e., the viscous flow of solids, given in Section 2.6.) For a number of applications it would be desirable to have liquids for which η does not vary with T, e.g., lubricants for engines.

Viscosity is both foe and friend of the design engineer. Because of viscosity, work is required to send gas or oil through a pipeline. However, it is also true that because of viscosity, hydrodynamic bearings are possible so that a shaft rotates in a sleeve without rubbing against the sleeve. As the shaft rotates a pressure is built up in the fluid, and if the shaft tends to be nonconcentric, the fluid pressure builds up even higher at the point of smallest clearance. The shaft therefore rides on this pressurized layer of fluid. The pressure buildup often exceeds 10 kbars (we use this pressure unit here to compare with typical values of the yield strength σ_0 of Table 2.1.2). The viscosity of air makes possible the air bearing. Pumps supply a continuous flow of oil to the region between the flat oil pads at the bottom of the carriage of the Hale telescope and the flat foundation surface on which they rest; this makes possible the easy rotation of this enormous structure, which weighs hundreds of tons, using only a small motor since the structure is actually riding on oil.

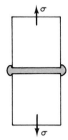

FIGURE 2.8.3. Butt joint with liquid adhesive.

Many adhesives are simply high-viscosity fluids. Figure 2.8.3 shows two solids with matching optical flat surfaces connected by an adhesive which wets the solid surfaces. It can be shown that if the initial adhesive thickness is h_0 and the value at the end of time t is h, then

$$t = \frac{3\eta R^2}{4\sigma}\left(\frac{1}{h_0^2} - \frac{1}{h^2}\right),\tag{2.8.7}$$

where R is the radius of the rods and σ is the tensile stress. The assumption is made that there is sufficient excess adhesive present to flow into the joint to prevent necking.

EXAMPLE 2.8.2

Calculate the lifetime of a butt joint between two rods whose radii are 2 cm, assuming that glycerine is used at 20°C, that the initial separation is 10^{-5} cm, that the final separation is 10^{-2} cm, and that $\sigma = 10$ bars (1 bar = 10^6 dynes/cm^2 = 14.5 psi).

Answer.

$$t = \frac{3 \times 10 \times 2^2}{4 \times 10^7}\left(\frac{1}{10^{-10}} - \frac{1}{10^{-4}}\right)$$

$$= 3 \times 10^4 \text{ sec}$$

$$\sim 8 \text{ hr.}$$

Note: The appropriate units are poise for viscosity, dynes per square centimeter for stress, and centimeters for length.

POROUS MEDIA.   Note that the flow through a single tube given by Poiseuille's equation (2.8.1) can be written as

$$\frac{1}{\pi R^2}\frac{dV}{dt} = \frac{1}{\eta}\frac{R^2}{8}\frac{\Delta P}{L}. \tag{2.8.8}$$

The term on the left is simply flux, J; flux is the flow rate across unit area. If we replace L by x and consider P to be a function of x, then we have over a length of pipe dx,

$$\frac{\Delta P}{L} = \frac{P(x) - P(x + dx)}{dx} = \frac{-dP}{dx} \qquad \text{as } dx \longrightarrow 0$$

and hence we have

$$J = -\frac{R^2}{8\eta}\frac{dP}{dx}. \tag{2.8.9}$$

Since a porous media such as soil consists of a lot of interconnected tubes, we might expect that such a material would have a relationship in which flux is proportional to the pressure gradient:

$$J = -k_P\frac{dP}{dx}. \tag{2.8.10}$$

This is known as **Darcy's equation** and k_P is known as the **permeability coefficient**. More generally for flow in an isotropic porous media,

$$\mathbf{J} = -k_P \text{ grad } P. \tag{2.8.11}$$

Here, $\mathbf{J}$ is a vector representing the flux and grad P is the gradient of P, also a vector.

2.9 FRICTION

Figure 2.9.1 shows a block being pulled along a surface. It is found experimentally that

$$F_S = \mu_f F_N, \tag{2.9.1}$$

where μ_f is called the **coefficient of friction**. For solids sliding on solids the area could be doubled or halved or changed in nearly any way, but if F_N remained the same, so would F_S; i.e., μ_f does not depend on the area of contact. Moreover, μ_f does not depend on the sliding velocity. These three statements [Equation (2.9.1) plus the two conditions on μ_f] are known as **Amonton's laws** (1699).

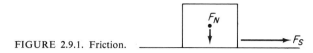

FIGURE 2.9.1. Friction.

The coefficient of friction is very important in rotating and reciprocating machinery. It (along with viscosity) is important in the handling of automobiles, particularly in the wintery months, and in sports such as skiing and skating. In fact a major contribution to these sports could be made by someone who invented a replacement for snow and ice so that the appropriate surfaces could be available in all seasons. Some values for the coefficient of friction are given in Table 2.9.1.

Table 2.9.1. COEFFICIENTS OF FRICTION

Material	μ_f
Steel on polytetrafluoroethylene (PTFE or Teflon)	0.01–0.03
Steel separated from steel by powdered MoS_2	0.06
Steel coated with 10^{-4}-cm indium on steel	0.05
Steel on steel	0.34
Clean Ni on clean glass (in vacuum)	0.5
Clean Fe on clean Fe (in vacuum)	0.8
Clean Ni on clean Ni (in vacuum)	1.2

We note that "clean" metals in vacuum have a coefficient of friction between themselves of about 1.

Certain solid lubricants give much smaller values. Moreover, lubrication by liquids can be much more efficient and the coefficient of friction for roller bearings can often be $10^{-(3-4)}$, for ball bearings 10^{-5}, and for hydrostatically pumped oil pads 10^{-7}. A major engineering problem today is lubrication at very high temperatures.

REFERENCES

Jaeger, J. C., *Elasticity, Fracture and Flow*, John Wiley and Sons, Inc., New York (1962). This little monograph describes stress in simple but thorough fashion, pp. 1–20. Strain is similarly discussed, pp. (20–49). Few undergraduates (even seniors) in any of the disciplines understand stress or strain.

Timoshenko, S., and Young, D. H., *Elements of Strength of Materials*, Van Nostrand, Reinhold, New York (1968). This book analyzes the stresses in and the deflections of simple mechanical components such as the tension rod, torsion bar, beam, column, pressure vessel, coil spring, etc.

Conway, H. D., *Mechanics of Materials*, Prentice-Hall, Englewood Cliffs, N. J. (1950). This is similar to above.

Wulff, J., et. al., *Structure and Properties of Materials*, Vol. III, *Mechanical Behavior*, John Wiley & Sons, Inc., New York (1965). Chapter 1 gives a summary of the mechanical tests which are performed to study the mechanical properties of solids.

"Materials Selector Issue" of "Materials Engineering". This annual issue contains extensive tables of properties; in particular, mechanical properties.

Handbook of Chemistry and Physics, ed. by R. C. Weast, Chemical Rubber Co., Cleveland (1969). Extensive lists of material properties, particularly those which are *not* structure sensitive.

Parker, E. R., *Materials Data Book for Engineers and Scientists*, McGraw-Hill Book Co., Inc. New York (1967).

Sines, G., and Waisman, J. L., *Metal Fatigue*, McGraw-Hill Book Co., New York (1959).

Finnie, I. and Heller, W. R., *Creep of Engineering Materials*, McGraw-Hill Book Co., New York (1959).

Parker, E. R., *Brittle Fracture of Engineering Structures*, John Wiley and Sons, Inc., New York (1957).

Tetelman A. S., and McEvily A. J., Jr., *Fracture of Structural Materials*, John Wiley & Sons, Inc., New York (1967).

PROBLEMS

2.1. Discuss why E and σ_0 are important parameters in the design of beams.

2.2. Why is flexural rigidity (or the stiffness constant of a beam) important in the design of a cutting tool for a lathe?

2.3. Give some reasons why both torsional and flexural rigidity are important in the design of an airplane wing.

2.4. Design a diving board out of a glass-reinforced epoxy composite.

The glass fibers run parallel to the length of the board. Assume $E = 300$ kbar and $\sigma_0 = 10$ kbar for the composite.

2.5. Make a table with three columns—Material, σ_0, and E—with the materials being 99.45% annealed aluminum, 99.45% cold-drawn aluminum, and 2024 heat-treated aluminum alloy. What is significant about your table?

2.6. The columns in an ordinary airplane wing are made of aluminum.
a. If their buckling load is given by

$$F_{max} = \frac{\pi^2 EI}{4L^2},$$

should they be hollow or solid from an equal weight basis to maximize F_{max} if

$$I = \frac{\pi(R_0^4 - R_i^4)}{4}?$$

Here, R_0 is the outer radius and R_i is the inner radius of a hollow pipe.
b. In addition to configurational changes (hollow versus solid), what other possibilities does the design engineer have for increasing the buckling load without increasing the weight of a column?

2.7. Assume the wooden beams under the floor of a house are uniformly loaded, with a weight w per unit length. Assume also that they simply rest on the foundation at their ends. The maximum deflection is given by

$$\delta = \frac{5wL^4}{384EI},$$

where I is the cross-sectional moment of inertia of the beam, E is the stiffness modulus, and L is the length of the beam. Assume $E = 1.2 \times 10^6$ psi. For a rectangular beam whose width is B and whose height is H (in the direction of loading),

$$I = \frac{BH^3}{12}.$$

How many times stiffer is a 2×12 in. wooden beam with the 12 in. vertical than one with the 2-in. dimension vertical?

2.8. Discuss the range of σ_0 for different materials.

2.9. a. If there is so much variation of σ_0 among different irons and steels, why is not the strongest steel always used in applications?
b. Suggest some reasons why there might be such variation in σ_0.

2.10. a. Define tensile stress; define shear stress. Suppose the test specimen of Figure 2.3.1 has a cross-sectional area of 1 in.² and that it carries a load of 20,000 lb.
b. What is the tensile stress on an area whose normal is along the tensile axis?
c. What is the shear stress on an area whose normal is along the tensile axis?

 d. What is the tensile stress acting on an area whose normal is 45 deg to the tensile axis?

 e. What is the shear stress acting on an area whose normal is 45 deg to the tensile axis?

 f. Which, if any, of the following is the correct description of the state of stress?

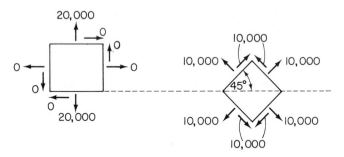

FIGURE P2.10f.

2.11. a. How many elastic constants are needed to describe the elastic behavior, for small strain, of elastically homogeneous and isotropic bodies?

  b. Of single crystals?

2.12. Would wood show isotropic tensile strength behavior?

2.13. Give some situations in which the value of Poisson's ratio may be important?

2.14. Describe briefly how the elastic constants of solids can be measured using sound waves.

2.15. Describe an experience in your life where mechanical resonance in a forced vibration either caused failure or concern.

2.16. In driving an automobile down a concrete thruway at 65 mph, a student notices a discomfiting subsonic vibration at about 3 Hz. Discuss the possible origin of this.

2.17. Describe the meaning of the term *loss angle*. How is it related to relaxation time? fractional energy loss per cycle? Q factor? fractional change in amplitude?

2.18. Suppose all the components of an airplane could be readily and cheaply welded together. Why might you still want to use rivets or other fasteners?

2.19. Suppose we ignored the mechanical difficulty associated with using a rigid rod of material to lift an elevator. Give two good reasons (from the viewpoint of materials properties) why a cable would be preferred to the rod.

2.20. A solid circular shaft is twisted carefully (no bending) to failure. Sketch the form of the break if

    a. The material is ductile, e.g., copper.
    b. The material is brittle, e.g., chalk.

2.21. What is work hardening? What practical value does it have?

2.22. a. Explain why necking occurs in a tensile specimen.
    b. Necking is an example of what phenomenon?

2.23. Give two different measures of ductility in a tension test.

2.24. When a copper rod is drawn into wire, it repeatedly is passed through a die where it is cold drawn and a heating chamber where it is annealed. Explain why.

2.25. a. What is the stress concentration factor?
    b. How large is it at a rivet hole on an airplane skin?
    c. How large is it in minute cracks on the surface of glass?
    d. Why are cracks more dangerous in brittle materials than in ductile materials?

2.26. Name four applications in which cyclic fatigue is likely to be very important.

2.27. Name four applications in which creep is likely to be one of the main design-limiting properties.

2.28. Why are the results of an impact energy test useful to the design engineer who is building bridges, ships, etc.?

2.29. Viscosity data for CCl_4 are shown below:

T°C	0	20	40	60	80
η, mP	13.47	9.09	7.38	5.84	4.68

Show that these data obey Equation (2.8.5), where T in (2.8.5) is the absolute temperature, and evaluate Q_{vis}.

2.30. The following Table 20 from M. D. Hersey's book *Theory of Lubrication*, Chapman & Hall, London (1936), shows the relation between the logarithm of the reduced viscosity coefficients and pressure for an oil:

Pressure, atm	ln η/η_1
1	0
500	0.32
1000	0.63
2000	1.24
3000	1.84
4000	2.45
5000	2.94
6000	Solid

What simple relationship exists between viscosity and pressure?

2.31. Derive Equation (2.8.1) using $v = (\Delta P/4L\eta)(R^2 - r^2)$.

2.32. Fatigue cracks appear on the portion of a shaft which rotates within a heavy-duty hydrodynamic bearing. Why?

2.33. How long can a uniform steel rod be if it is supported at the top end only and is not to yield? Assume that $\sigma_0 = 300,000$ psi and that the specific gravity is 7.80.

More Involved Problems

2.34. Show how each of the mechanical properties discussed in the present chapter is involved in the design of
a. A motorcycle.
b. An automobile.
c. A motorboat.
d. An airplane.

2.35. Describe seismology and its applicability in prospecting for oil and in studying the structure of the earth.

2.36. Calculate the velocity of shear waves in polycrystalline quartz at 1 atm pressure if quartz has a density of 2.648 g/cm^3 and a shear modulus of 377 kbars (1 kbar $= 10^9$ dynes/cm^2).

2.37. The deepest point in the ocean is about 35,000 ft.
a. Calculate the hydrostatic pressure there in pounds per square inch and in kilobars.
b. Would such a pressure have much effect on the volume of an aluminum bar? H_2O? an ideal gas?

2.38. Show that the lamellar (nonturbulent) flow rate Q through a rectangular pipe of width w, height h (where $w \gg h$), and length L is given by

$$Q = \frac{wh^3}{12\eta L}P,$$

where P is the pressure drop and η is the viscosity coefficient.

Sophisticated Problems

Problems 2.39–2.42 are a learning sequence concerned with internal damping. It is necessary to know something about linear second-order differential equations to solve Problem 2.39.

2.39. Given that the differential equation of motion of a damped spring-mass system is

$$M\ddot{x} + \mu\dot{x} + kx = 0$$

and that

$$x(0) = x_0, \qquad \dot{x}(0) = 0,$$

find $x(t)$ for the case where $\mu^2 < 4Mk$.

$$\textit{Answer. } x = x_0 e^{-(\mu/2M)t}\left(\cos \omega t + \frac{\mu}{2M\omega}\sin \omega t\right),$$

where

$$\omega = \sqrt{\frac{k}{M} - \frac{\mu^2}{4M^2}}.$$

This is the natural damped frequency.

2.40. Find the force versus time relationship $F(t)$ for the force on the spring. *Answer.* The force on the mass is given by $-kx - \mu\dot{x}$ and on the spring by $kx + \mu\dot{x}$. Define $F(t) = kx + \mu\dot{x}$. We can obtain $x(t)$ and $\dot{x}(t)$ from the answer to Problem 2.39 and hence $F(t)$:

$$F(t) = kx_0 e^{-(\mu/2M)t}\left[\cos \omega t - \frac{\mu}{2M\omega} \sin \omega t\right].$$

2.41. Define $\delta/2 = \tan^{-1} \mu/2M\omega$.

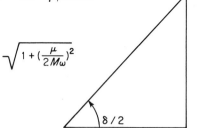

FIGURE P2.41

Show that the expression for x in Problem 2.39 is

$$x = x_0 e^{-(\mu/2M)t}\sqrt{1 + \left(\frac{\mu}{2M\omega}\right)^2} \cos\left(\omega t - \frac{\delta}{2}\right)$$

and for F in Problem 2.40 is

$$F = kx_0 e^{-(\mu/2M)t}\sqrt{1 + \left(\frac{\mu}{2M\omega}\right)^2} \cos\left(\omega t + \frac{\delta}{2}\right).$$

Note the phase difference δ.

2.42. If the damping is very small ($\mu/2M\omega \ll 1$),

$$\sqrt{1 + \left(\frac{\mu}{2M\omega}\right)^2} \doteq 1$$

and

$$\tan^{-1} \frac{\mu}{2M\omega} \doteq \frac{\mu}{2M\omega}.$$

Therefore $\delta \doteq \mu/M\omega$. What is an approximate expression for the natural damped frequency of this system? *Answer.* $\sqrt{k/M}$, which is the natural frequency ω_0. What is the relaxation time? *Answer.* $2/\delta\omega_0$.

2.43. Given a damped spring-mass system subject to an oscillatory force $F_0 \cos \omega t$, the differential equation is

$$M\ddot{x} + \mu\dot{x} + kx = F_0 \cos \omega t.$$

Find the particular solution of this equation in the form

$$x = A \cos \omega t + B \sin \omega t.$$

Answer.

$$x = F_0 \frac{(k - M\omega^2)}{(k - M\omega^2)^2 + (\omega\mu)^2} \cos \omega t + \frac{F_0\omega\mu}{(k - M\omega^2)^2 + (\omega\mu)^2} \sin \omega t.$$

2.44. Define $\beta = \tan^{-1}[\omega\mu/(k - M\omega^2)]$. Hence, show that the solution in Problem 2.43 can be written as

$$x = \frac{F_0}{\sqrt{(k - M\omega^2)^2 + (\mu\omega)^2}} \cos (\omega t - \beta).$$

2.45. In Problem 2.44, for what value of ω would the amplitude reach a maximum? We call this the resonance amplitude and the corresponding frequency the resonance frequency. What is an approximation for the resonance frequency if the damping is very small? What is the resonance amplitude for the case of very small damping? *Answer.* $F_0/\mu\omega_0$, where $\omega_0 = \sqrt{k/M}$. How is the resonance amplitude for very small damping related to the loss angle (see Problem 2.42 for the value of δ)? What is the resonant amplification?

2.46. One of the oil pads of the Hale telescope is shown below. The pad is a 28-in. square with 7×7 in. recesses located symmetrically and 6 in. apart. The pad supports 164,000 lb.

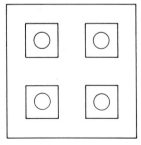

FIGURE P2.46

a. What pressure is necessary to lift the load?

b. If a pressure of 275 psi is actually used, calculate the pumping rate per pad for an SAE 20 oil at 60°F. The viscosity is about 2.7 P.

c. If the pad moves on a radius of 23 ft, making one revolution per day, what is the friction coefficient?

Note: You may assume that the flow is outward through a rectangular slit and use the results of Problem 2.38. The pressure in the small square areas is actually 275 psi.

Prologue

The most important electrical property is the *electrical resistivity*. Based on resistivity, materials are classified as *metals, semiconductors, insulators*, or *superconductors*. The resistivity of metals decreases rapidly as the temperature is *lowered* at cryogenic temperatures but varies only slowly at room temperature and above. Semiconductors behave differently; in semiconductors the resistivity at *high* temperature decreases rapidly as the temperature is *raised*. The concepts of *charge carriers* and the *Hall coefficient* are introduced. Several important properties of insulators or dielectrics in addition to resistivity are described: *permittivity, dielectric constant, dielectric strength*, and *dielectric loss angle*. In addition, some dielectric materials which exhibit special properties, namely, *ferroelectricity* and *piezoelectricity*, are studied. Ferroelectric materials always lose their ferroelectric property at high temperatures.

It is emphasized again that the reader should begin to develop a feeling for the order of magnitude of these properties for different materials. Moreover, it is helpful to think of how these properties can be effectively utilized in constructing apparatus, machinery, devices, etc.

3

PHENOMENOLOGY OF
ELECTRICAL BEHAVIOR

3.1 RESISTIVITY AS A DESIGN PARAMETER

The importance of resistivity as a material property will be illustrated by examples in Sections 3.1 and 3.2.

SPIRAL HEATER.   Consider the spiral of nichrome wire, shown in Figure 3.1.1, which is found in many toasters. (Nichrome contains nickel, chromium,

FIGURE 3.1.1. Heating element of a toaster.

and possibly some iron.) The power is given by

$$P = Ri^2 = \frac{V^2}{R},$$ (3.1.1)

where P is measured in watts, R is measured is ohms, i is the current measured in amperes, and V is the voltage in volts. This electrical energy is converted into heat. Assuming that the heater is connected across a standard 115-V, 60-Hz ac line, the voltage V is understood to be the root mean square (rms) voltage, which is 0.707 of the peak voltage. The resistance of a wire of constant cross-sectional area A is given by

$$R = \frac{\rho l}{A},$$ (3.1.2)

so that

$$P = \frac{1}{\rho}\left(\frac{A}{l}\right)V^2.$$ (3.1.3)

65

Here l is the length of the wire (so A/l is a size parameter) and ρ is a material property called the **resistivity**. ρ has the dimensions ohm-meters [in the rationalized meter-kilogram-second (mks) system].

A toaster might have several heater elements which provide the heat needed. To calculate the power required, it would be necessary to know how fast heat is lost to the surroundings, how hot the toast should be, and how long it should take to make the toast. Let us suppose that all these factors are known and proceed to consider the factors which determine what wire we are to use. Since the wire will get rather hot and since such spirals usually hang vertically and support their own weight, the material must have sufficient hot strength. The wire must also have high oxidation resistance. It must also have a value of ρ which leads to an element of reasonable dimensions. If ρ is exceedingly small, then A must be very small and l very large. The spiral, viewed from a mechanical viewpoint, is simply a helical coil spring as discussed in Section 2.2; we note that the stiffness constant is described by Equation (2.2.8) and the maximum allowable tensile force by Equation (2.2.9). As is obvious from these expressions, if A is made small, the spring will be too flexible and too weak. The designer, therefore, has to choose a material which satisfies mechanical and chemical as well as electrical requirements. It is for these reasons that he chooses nichrome wire for the heating element.

Table 3.1.1 shows some values of resistivity. The first column shows values for **metals** which are very good electrical conductors and the second column shows two different groups: **semiconductors**, which have intermediate values of ρ, and **insulators**, which have very high values of ρ or are very poor electrical conductors. Note the enormous range of properties; the data of Table 3.1.1 suggest

$$\frac{\rho_{\text{insulator}}}{\rho_{\text{metals}}} \sim 10^{24}.$$

In addition, there are materials called **superconductors** which carry electrical current without any detectable loss. Not all materials have superconducting properties, but those which do are superconductors only at very low temperatures (cryogenic temperatures), below about 20°K. Superconductors are discussed briefly in Chapter 5. Table 3.1.1 gives resistivity values only near room temperature; some of the materials in Table 3.1.1, such as tin, become superconductors at low temperatures.

ANISOTROPY.   Suppose test samples are cut from a block of material as in Figure 2.3.7 and their resistance is measured. If samples 1, 2, 3, and 4 give the same resistivity, the material is said to be homogeneous. If the block of material is homogeneous and if samples 1, 5, 6, 7, 8, and 9 all have the same value of ρ, the material is said to be isotropic with respect to resistivity; if not, it is anisotropic with respect to resistivity. Some materials are highly anisotrop-

Table 3.1.1. ELECTRICAL RESISTIVITIES IN MKS UNITS OF METALS
AND NONMETALS AT 20°C*

Metals	Resistivity, 10^{-8} ohm-m†	Nonmetals	Resistivity, ohm-m†
Silver	1.6	Semiconductors	
Copper	1.67	⎰Silicon	0.00085
Gold	2.3	⎱Germanium	0.009
Aluminum	2.69	Insulators	
Magnesium	4.4	⎧Diamond	10^{10}–10^{11}
Sodium	4.61	⎪Quartz	1.2×10^{12}
Tungsten	5.5	⎪Ebonite	2×10^{13}
Zinc	5.92	⎨Sulfur	4×10^{13}
Cobalt	6.24	⎪Mica	9×10^{13}
Nickel	6.84	⎪Selenium	2×10^{14}
Cadmium	7.4	⎩Paraffin wax	3×10^{16}
Iron	9.71		
Tin	12.8		
Lead	20.6		
Uranium	29		
Zirconium	41		
Manganin	44		
Titanium	55		
Lanthanum	59		
96% Iron–4% Si	62		
Cerium	78		
Nichrome	100		

* From *American Institute of Physics Handbook*, Dwight E. Gray, ed. McGraw-Hill, New York (1963), pp. 4–90; 9–38.
† Note the different units in the two columns.

ic; a single crystal of graphite, which has a layer-like structure, is a metal along the layers and a semiconductor normal to the layers. If the material is isotropic,

$$\rho = \frac{1}{\sigma}, \qquad (3.1.4)$$

where σ is the **electrical conductivity**. In general, electrical conductivity is defined in terms of the **current density** $\mathbf{J}$ (charge flowing across the unit area in unit time) and the **electrical field** $\boldsymbol{\xi}$. For isotropic materials in fields which are not extremely high, **Ohm's law** holds:

$$\mathbf{J} = \sigma \boldsymbol{\xi}. \qquad (3.1.5)$$

The boldface font, as used for $\mathbf{J}$ and $\boldsymbol{\xi}$, is used to designate a vector quantity. When dealing with simple situations in which the vector has only one nonzero component, an ordinary font, J and ξ, is often used. In such a case the magnitude of $\mathbf{J}$ is simply J (some authors use $|\mathbf{J}|$). (The case for anisotropic materials will be treated later.)

EXAMPLE 3.1.1

Derive the expression $V = iR$ for the voltage drop in a wire.

Answer. Starting with (3.1.5) we note that $\xi = (1/\sigma)J$ (with both the electrical field and current density along the wires). Since $\xi = V/l$, $1/\sigma = \rho$, and $JA = i$, we have

$$\frac{V}{l} = \frac{\rho i}{A} \qquad \text{or} \qquad V = \frac{\rho l}{A} i = Ri.$$

3.2   METALS

Good electrical conductors play an important role in our society. At least part of the reason that we are not all lowly peasants sweating behind a plow (or at least 99% of us) is the existence of electrical generation and transmission equipment. Conductors are vital components of many communications systems as well. The fact that they are not perfect conductors means that energy is lost (converted into heat). In some cases this means only a small fraction of energy is wasted. However, in others the heat created causes problems of its own; e.g., if the heat is not carried away, the machine will destroy itself. Consequently, the manner of carrying away the heat may become very important; this is discussed in Chapter 6. An example of such a situation is a high field magnet. A high field magnet is an example where heating is important; such a magnet will be needed to contain the nuclear fusion reaction. The harnessing of this reaction to create cheap electric energy without pollution and with great safety (relative to the fission reactor) will be the great technological development of the twentieth century.

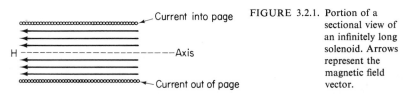

FIGURE 3.2.1. Portion of a sectional view of an infinitely long solenoid. Arrows represent the magnetic field vector.

Let us consider the role of resistivity in the behavior of a solenoid as shown in Figure 3.2.1. The magnetic field intensity is

$$H = ni, \tag{3.2.1}$$

where n is the number of turns per unit length in one layer (as a rule the solenoid would consist of many layers m of the type shown). The magnetic induction is

$$B\left(\frac{\text{webers}}{\text{meter}^2}\right) = 4\pi \times 10^{-7} \, mn\left(\frac{\text{turns}}{\text{meter}}\right) i(\text{amps}), \tag{3.2.2}$$

where we have assumed that the current is the same in each turn in each layer.

The power dissipated per meter of length of the coil is

$$P_1 = \frac{\rho}{4\pi^2 \times 10^{-14}} \left(D + \frac{t}{2} \right) \frac{B^2}{t}, \tag{3.2.3}$$

where D is the inside diameter of the solenoid and t is the total thickness of the winding as shown in Figure 3.2.2. We have assumed that the insulation is of negligible thickness and that the windings are close and tight. P_1 will have units of watts per meter if ρ has units of ohm-meters, B has units of webers per square meter, and the length is in meters.

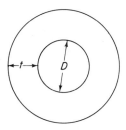

FIGURE 3.2.2. End view of solenoid.

EXAMPLE 3.2.1

Suppose the inner diameter is to be 0.1 m and suppose we wish to have $B = 10$ Wb/m² (for those people who like to think of B in units called gauss, this would be 100,000 G or 100 kG). Suppose the wire is copper and $t = 0.1$ m. Calculate the heat created per unit length.

Answer. From Table 3.1.1 we note that $\rho = 1.67 \times 10^{-8}$ Ω-m. Hence

$$P_1 = 6.3 \times 10^6 \frac{\text{watts}}{\text{meter}} = 6300 \frac{\text{kilowatts}}{\text{meter}}.$$

The ordinary iron used by the housewife uses about 1 kW of power when heating. The standard furnace for heating a three-bedroom house uses about 10 kW at maximum capacity. Needless to say, a magnet which produces heat at the above rate would be destroyed by overheating in a very short time. To prevent this destruction a *major* cooling system would have to be installed.

Example 3.2.1 illustrated the heating problems in a high field solenoid magnet.

To solve these problems the designer might:

1. Cool by passing through large amounts of water. This is a major thermal engineering problem. This is precisely the procedure used at the National Magnet Laboratories (Figure 3.2.3).
2. Use a superconductor (more about that later).
3. Use a material with a smaller ρ.

FIGURE 3.2.3. 250-kG (25 Wb/m²) magnet at National Magnet Laboratories. Inner bore is 2 in. in diameter. Magnet contains 3 tons of copper (hollow tubing), has a power consumption up to 16 MW, and requires 2000 gal of cooling water per minute. (*Courtesy of Francis Bitter National Magnetic Laboratory, Cambridge, Mass.*)

A rapid check of Table 3.1.1 shows that silver is only slightly better. However, the resistivity of metals drops considerably as temperature is decreased as illustrated in Figure 3.2.4. Potassium has a melting temperature of $333°K$; note that the resistivity varies slowly with temperatures above $T_m/5$. This temperature variation is characteristic of metals.

Very high-purity aluminum has been made for which the resistivity at $4.2°K$ (the boiling point of liquid He) is only 2×10^{-5} times the room-temperature (20°C) value. The **resistivity ratio** is $\rho_{293.2°K}/\rho_{4.2°K}$; for high purity aluminum this is 50,000. The higher the purity, the higher the resistivity ratio. The room-temperature resistivity is only slightly affected by impurities, but the low-temperature resistivity is greatly affected.

EXAMPLE 3.2.2

Consider a solenoid which will operate at $4.2°K$. It is made of Al with a resistivity ratio of 50,000 and made to give $B = 1$ Wb/m² in a cavity of 1-cm diameter with an overall length of 10 cm. Calculate the power dissipated.

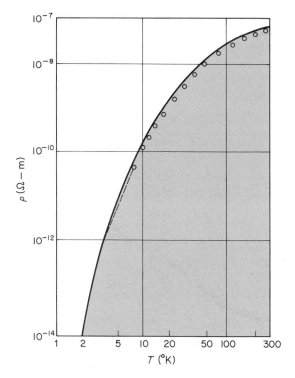

FIGURE 3.2.4. Log-log plot of bulk resistivity versus absolute
temperature for 99.99% potassium. Circles and
dashed line are experimental. Solid line is computed. [From G. T. Meaden, *Electrical
Resistance of Metals*, Plenum Press, New York
(1965).]

Answer. Using $\rho_{293°K} = 2.69 \times 10^{-8}$ Ω-m we have $\rho_{4.2°K} =$
5.38×10^{-13} Ω-m. We shall assume $t = 2$ cm. Hence the power is $P =$
0.14 W. (This would boil off 0.18 liters of liquid helium in 1 hr, which would
cost about \$1.)

The ultimate resistivity ratio of bulk metals is much higher than 50,000
but there is a size effect on resistivity which will make it difficult to utilize this
low bulk resistivity at low temperature; namely, the measured resistivity increases as the wire diameter decreases when the wire diameter is quite small
and when the resistivity is low. (See Problem 3.20.)

It should be noted that at very high frequencies the currents move
toward the surface of the conductor (**skin effect**) so another size effect arises.
(See Problem 3.21.)

3.3   SEMICONDUCTORS

RESISTIVITY.   Pure metals are very good conductors at very low temperature (see Figure 3.2.4) but their resistivity *increases* as temperature increases. Pure semiconductors (such as pure germanium) are insulators at very low temperatures but their resistivity *decreases very rapidly* as temperature increases, although they are still poor conductors relative to metals at high temperatures.

This behavior is shown in Figure 3.3.1 for a gallium-doped germanium crystal. Note that the behavior from 100 to 900°C is of the form

$$\rho = \rho_0 e^{\Delta e/2kT}, \tag{3.3.1}$$

where ρ_0 and Δe are constants for a given material, k is Boltzmann's constant,

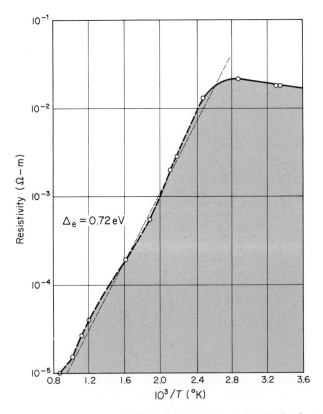

FIGURE 3.3.1. Log resistivity versus $1/T$ for single crystals of gallium-doped germanium in the temperature range 25–900°C. [From W. C., Dunlap, *An Introduction to Semiconductors*, Wiley, New York (1957).]

and T is the absolute temperature. For gallium-doped germanium, $\Delta e = 0.72$ eV and $\rho_0 = 2.62 \times 10^{-7}$ Ω-m. The quantity Δe is called the **energy band gap**. It is determined by the electronic properties of the material as discussed in Chapters 27 and 28. (Note that the plot for semiconductors in Figure 3.3.1 involves $\log_{10} \rho$ vs $1/T$, while the plot for metals in Figure 3.2.4 involves $\log_{10} \rho$ vs. $\log_{10} T$. The variation with temperature is very different in the two cases!)

Note that while the temperature increases from 400 to 1200°K for the semiconductor, the resistivity changes by a factor of 10^{-3}. Because of this very strong change with temperature, semiconductors are useful as temperature measuring devices called thermistors. The point where deviations from the straight line occur (at low temperatures) depends strongly on the amount of gallium added, and small fractions of 1% cause profound changes. Hence the resistivity of semiconductors is a structure-sensitive property at low temperatures.

EXAMPLE 3.3.1

Show that the argument of the exponential in Equation (3.3.1) is $4180/T$ for gallium-doped germanium.

Answer. We note that

$$0.72 \text{ eV} = 1.60 \times 10^{-19} \frac{\text{coulombs}}{\text{electron}} \times 0.72 \text{ electron volts}$$

$$= 1.15 \times 10^{-19} \text{ joules}$$

since

$$1 \text{ J} = 1 \text{ coulomb-volt.}$$

The universal physical constant k, or **Boltzmann's constant**, is given by

$$k = 1.380 \times 10^{-23} \text{ joules/°K.}$$

Hence

$$\frac{\Delta e}{2kT} = \frac{4180}{T}.$$

HALL EFFECT. There is another property which illustrates the structure-sensitive nature of electrical conduction in semiconductors. A long thin slab carrying current in the x direction in the presence of a magnetic induction in the z direction develops an electric field (and hence potential or voltage) in the y direction; this is the **Hall effect** (1888), Figure 3.3.2. The ratio of electric field to the product of current density and magnetic field is the **Hall constant** R_H; thus in the present case

$$R_H = \frac{\xi_y}{J_x B_z}. \tag{3.3.2}$$

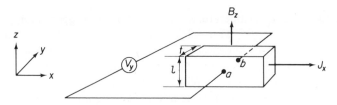

FIGURE 3.3.2. Hall effect experiment.

If the moving charge carriers are considered to be free (i.e., independent of their environment), it can be shown (see Problem 3.19) that

$$R_H = \frac{1}{Nq}, \tag{3.3.3}$$

where N is the number of charge carriers per unit volume and q is their charge. [The Hall coefficient R_H is often represented in various units. In the mks system the units are volt-cubic meters per ampere-weber or cubic meters per coulomb. (This makes use of the following identity: $1\ C = 1\ \text{A-Wb/V}$.) In the mks system we have ξ_y (volts per meter), J_x (amperes per square meter), and B_z (webers per square meter).] It is found that germanium (which is tetravalent) doped with a trivalent impurity such as gallium has a positive Hall coefficient (so that the charge carrier appears to be *positive*); this is called **p-type** germanium. If it is doped with a pentavalent impurity such as arsenic, it has a negative Hall coefficient and negative charge carriers and is called **n-type** germanium. As we shall see later, the Hall effect measurements combined with resistivity measurements are a powerful combination for studying the details of conductivity from a fundamental viewpoint. The Hall effect is also observable in metals (where the coefficient may be positive or negative) and in insulators.

EXAMPLE 3.3.2

Design a device which will continuously multiply two variables together. Consider that variable A is in the form of a current output from one device and variable B is in the form of a current output from a second device.

Answer. One possible solution is to wind a solenoid coil of n turns and feed one current i_A through it which generates a field $H_z = ni_A$ or a magnetic induction $B_z = 4\pi \times 10^{-7}ni_A$ parallel to its axis. Place a semiconductor slab normal to this axis and feed the other current i_B through it in the x direction as shown in Figure 3.3.2; hence $J_x = i_B/lt$. The potential drop across the slab in the y direction is $V_y = \xi_y t$. This voltage is [from Equation (3.3.2)]

$$V_y = R_H J_x B_z t = R_H 4\pi \times 10^{-7}\frac{n}{l}i_A i_B.$$

The product of i_A and i_B is the measured voltage V_y divided by the constant

$R_H n 4\pi \times 10^{-7}/l$. This output may be read continuously off a calibrated potential measuring device, such as an oscilloscope or potential recorder or intermittently by a digital voltmeter.

JUNCTIONS.   Suppose germanium is p type on one side of an interface (excess gallium) and n type on the other side (excess arsenic). This couple is known as a **p-n junction**. This couple has the ability to rectify, i.e., conduct current readily in one direction but allow only a very tiny flow in the reverse direction. If we make a crystal having two junctions, e.g., *p-n-p* (or *n-p-n*), we have made a **junction transistor**, a device capable of amplifying a signal. The invention of the point contact transistor by J. Bardeen and W. H. Brattain in 1947 and of the junction transistor shortly later by William Shockley revolutionized the electronics industry and had a major part to play in the recent development of reliable high-speed digital computers. To make high-quality transistors it is necessary to work with materials, such as silicon, of exceptional high purity and to have essentially perfect crystals. Achieving both of these ends has been a major development in materials science. Later on we shall see why junctions operate and why ultrapurity and perfect crystallinity are needed in some cases and how they are obtained.

There are a number of other important semiconductor devices including the varistor, electristor, and the Esaki diode which also can be more conveniently discussed after the electronic nature of solids is studied. The Esaki diode can switch in 10^{-11}–10^{-12} sec.

3.4   INSULATORS

A CONDENSER.   Insulators, in addition to having a resistivity which differs by a factor of about 10^{24} from metals and 10^{10} from semiconductors, have several other properties which are important. Consider a simple parallel plate condenser with plates of area A separated by a distance d as in Figure 3.4.1.

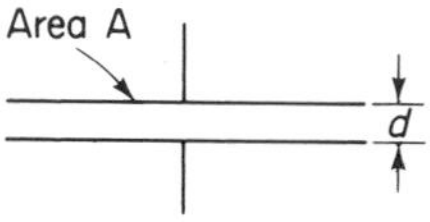

FIGURE 3.4.1. Parallel plate condenser.

When a voltage is applied across the plates, one of the plates attains a charge of $+Q$ while the other attains a charge of $-Q$; i.e., there is a *displacement* of charge from one plate to the other. The charge Q is found to be directly proportional to the applied potential V,

$$Q = C_0 V. \tag{3.4.1}$$

Here C is a proportionality constant called the **capacitance**. We use the subscript zero to designate that the space between the plates is a vacuum. [Note the analogy between Equations (3.4.1) and (2.2.6), which describes the displacement of a spring under an applied force.]

The student who remembers his electricity studies knows that the capacitance of the parallel plate condenser is

$$C_0 = \frac{\epsilon_0 A}{d} \qquad (3.4.2)$$

when the space is vacuum and the lateral plate dimensions are very large with respect to d. Here ϵ_0 is called the **permittivity** (with the subscript zero denoting a vacuum) and is, in rationalized mks units, given by $\epsilon_0 = 8.854 \times 10^{-12}$ F (farads)/m. If a homogeneous isotropic insulating material fills the space between the plates, the proportionality constant in (3.4.1) will now be C (where $C > C_0$) and we write

$$\kappa = \frac{C}{C_0} = \frac{\epsilon}{\epsilon_0}. \qquad (3.4.3)$$

The quantity κ is called the **dielectric constant**. Typical values are given in Table 3.4.1; note the necessity of specifying frequency.

Table 3.4.1. VALUES OF DIELECTRIC CONSTANTS AT ROOM TEMPERATURE

Material	Dielectric Constant	Frequency Hz
Air (1 atm, 0°C)	1.00059	0
Mineral oil	2.2	0
Polytetrafluoroethylene (PTFE)	2.1	$0-10^{10}$
NaCl	2.25	10^{10}
NaCl	5.62	0
Polyvinyl chloride (PVC)	2.6	10^{10}
Polyvinyl chloride (PVC)	6.5	10^2
Porcelain	8	0
Mica	65–85	0
Water	80	0
Titanates	~ 1000	0

We note that the energy stored in a capacitor is given by

$$U = \int V \, dQ = \frac{Q^2}{2C} = \frac{CV^2}{2} = \frac{\kappa C_0 V^2}{2}. \qquad (3.4.4)$$

Thus the energy stored in a capacitor of a given volume at a given voltage is increased by the factor κ by the presence of the insulating or **dielectric material**. Hence extremely small condensers can be built using materials such as the titanates.

EXAMPLE 3.4.1

Explain why water is such a good solvent for polar compounds such as sodium chloride.

Answer. Since the force between charges is inversely proportional to the permittivity and hence the dielectric constant, the attractive force between a Na$^+$ ion and a Cl$^-$ ion is only 1/80 as much as in a vacuum.

The analogy with the spring of Section 2.2 can be carried further. Although the quantity G appeared in the equation for the stiffness constant of the spring, G was not defined in terms of the force and deflection of the spring. Rather G was defined in terms of the fundamental field quantities, stress and strain. A similar situation applies in the definition of permittivity, which is defined as the proportionality coefficient ϵ (for homogeneous isotropic behavior) between the **electric displacement D** and the electrical field ξ:

$$\mathbf{D} = \epsilon\xi. \tag{3.4.5}$$

D is a property of electrical fields which depends only on the distribution and strength of the charge sources producing the field. Thus the integral of the normal component of **D** taken over a closed surface equals q, the total charge enclosed by the surface:

$$\int_{\substack{\text{closed} \\ \text{surface}}} D_n \, dS = q. \tag{3.4.6}$$

This is **Gauss's law**.

In most types of crystals, but not all, the permittivity varies with direction, and hence these materials are anisotropic with respect to permittivity. In such cases a more detailed mathematical framework is needed to describe the permittivity.

The electrical energy per unit volume stored in a dielectric is given by

$$u = \int \xi \, dD. \tag{3.4.7}$$

This is effectively the integral of force times the differential of displacement.

EXAMPLE 3.4.2

Use Equation (3.4.7) to derive the expression for the energy stored in a parallel plate capacitor $U = \kappa C_0 V^2/2$.

Answer. From Equation (3.4.5), $dD = \epsilon \, d\xi$ and

$$u = \int \epsilon\xi \, d\xi = \frac{\epsilon\xi^2}{2} = \frac{\epsilon V^2}{2d^2}.$$

Since $C = \epsilon A/d = \kappa C_0$,

$$U = Ad\,u = \frac{\epsilon A}{2d}\,V^2 = \frac{\kappa C_0 V^2}{2}.$$

The origin of the dielectric constant will be studied later. Measurement of dielectric constants provides valuable data to the scientist attempting to understand molecular structure.

DIELECTRIC STRENGTH. **Dielectric strength** or **breakdown strength** is the dielectric analog of the tensile fracture strength in plastic flow. It is defined as

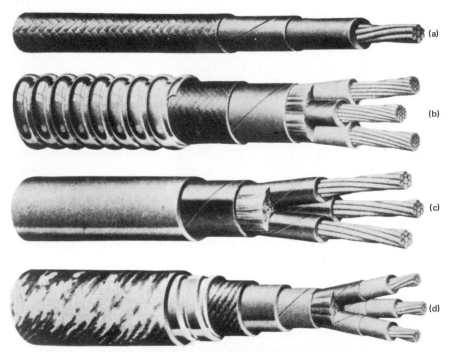

FIGURE 3.4.2. Cable construction (*polymeric insulators*). (a) Rubber-insulated cable with weatherproof braided finish (single-conductor stranded). (b) Varnished-cambric-insulated cable with filler, belt, braided finish, and interlocked spiral steel armor (three-conductor stranded). (c) Varnished-cambric-insulated with filler, belt, and lead sheath. (d) Parkway cable with rubber-insulated conductors belted, lead covered, protected by jute, interlocked armor, and jute overall. [From J. F. Young, *Materials and Processes*, Wiley, New York (1954).]

the voltage per unit distance of plate separation (voltage gradient or electric field) at which failure occurs. Failure is characterized by a rapid increase in current and may be associated with melting, burning, and vaporization of the dielectric. It is irreversible. Some values of dielectric strength are shown in Table 3.4.2. Examples of solid insulators in use are shown in Figures 3.4.2 and 3.4.3.

Table 3.4.2. DIELECTRIC STRENGTH

Material	kV/cm
Air (1 atm)	30
Dry paper	45
SF_6 (1 atm)	80
Porcelain	15–120
Mineral oil	150
Varnish	80–350
SF_6 (7 atm)	400
Air (7 atm)	480
Mica	1200–2000

Pressurized gases are often used as insulating dielectrics in very high-voltage X-ray machines which are used to make radiographic studies of thick

FIGURE 3.4.3. Five-stage Cockcroft-Walton multiplier circuit 1200-kV dc generator for testing high-voltage cables. Note *ceramic* insulator. (*Courtesy of Ferranti Ltd.*)

structural sections or for 1-MeV Van de Graaff generators whose electron or ion output may be used in such phases of materials processing as paint drying on automobiles, polymerization of polymers, and ion implantation to make *p-n* junctions. Liquids such as mineral oil may be used in heavy transformers and regulators where heavy heat transfer losses necessitate the excellent heat transfer brought about by convection. In such cases the presence of a small concentration of water is highly detrimental since the dielectric strength is greatly reduced.

DIELECTRIC LOSS.　Since

$$i = \frac{dQ}{dt} = C\frac{dV}{dt}, \tag{3.4.8}$$

we note that in the *ideal* capacitor in the presence of a sinusoidal voltage variation of amplitude V_0 the current and the voltage are out of phase by 90 deg; the charge and voltage are in phase. The maximum stored energy is

$$U_{max} = \frac{CV_0^2}{2} = \frac{\kappa C_0 V_0^2}{2}. \tag{3.4.9}$$

The work done in a single cycle is

$$\Delta U = \int_0^T Vi\, dt = 0, \tag{3.4.10}$$

where T is the period.

The maximum stored energy per unit volume is

$$u_{max} = \frac{\epsilon}{2}\xi^2 \tag{3.4.11}$$

(where $\xi = V/d$ for the parallel plate case) and the work done per unit volume per cycle is $\Delta u = 0$.

However, a real insulating material is not a perfect insulator, just as a spring is not a purely elastic spring. In a real insulating material there are resistive losses,

$$\frac{\Delta u\,(\text{dielectric})}{u_{max}(\text{vacuum})} = 2\pi\kappa\delta_{el}, \tag{3.4.12}$$

and the voltage leads the charge by δ_{el}, the **dielectric loss angle**. (Note the complete analogy to the mechanical case.) The product $\kappa\delta_{el}$ is called the **dielectric loss factor**. The electric loss angle, like its mechanical counterpart, is a structure-sensitive property. At 10^{10} Hz, pure SiO_2 has $\delta = 5 \times 10^{-4}$, while SiO_2 with 2% K_2O has $\delta = 7.5 \times 10^{-3}$, a change of a factor of 15 for a 0.02 fractional composition change. As in the mechanical case the energy dissipated is converted into heat. Note that heat can cause degradation of the material, which leads to eventual breakdown.

The dielectric loss angles are *approximately* frequency-independent. Some values are shown in Table 3.4.3.

Table 3.4.3.   ELECTRIC LOSS ANGLE OF
DIELECTRIC MATERIALS

Material	Loss Angles at 60 Hz
Pure silica	0.0001
Polystyrene	0.0002
Paraffin	0.002
Porcelain	0.01–0.02
Hard rubber	0.01
Dry paper	0.01
Lucite	0.03
Bakelite	0.05

One interesting aspect of the electric loss angle and the mechanical loss angle is shown in Figure 3.4.4. The nearly one-to-one correspondence suggests the same molecular origin of the loss angle for the two processes.

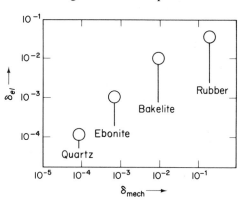

FIGURE 3.4.4. Comparison of mechanical and electrical angle of loss. [From C. Zwikker, *Physical Properties of Solid Materials*, Pergamon Press, New York (1955).]

An elementary discussion of many of the practical aspects of dielectrics is given in Chapters 10 and 13 of J. F. Young, *Materials and Processes*, Wiley, New York (1954).

3.5   FERROELECTRICS

It is useful to have a variable which compares $\mathbf{D}$ for the case in which a dielectric is present to the case in which it is absent; in the latter case $\mathbf{D}_0 = \epsilon_0 \boldsymbol{\xi}$. This variable is defined as

$$\mathbf{P} = \mathbf{D} - \mathbf{D}_0 = \mathbf{D} - \epsilon_0 \boldsymbol{\xi} \tag{3.5.1}$$

and is called the **polarization**. We can also write

$$\mathbf{P} = \epsilon_0 \chi \xi, \qquad (3.5.2)$$

where χ, called the **dielectric susceptibility**, is given by

$$\chi = \kappa - 1. \qquad (3.5.3)$$

The polarization is the net dipole moment per unit volume. A **dipole moment** is a vector quantity; it arises owing to the separation of a charge q and $-q$ by a distance r. It is equal in magnitude to qr, and the direction of the vector is from the $-q$ charge to the $+q$ charge. A simple dielectric contains either no dipoles or many dipoles with random orientation and hence no net dipole moment for $\xi = 0$. An electrical field tends to create dipoles in the former material and to preferentially orient those in the latter.

For the simple dielectric materials which we have discussed thus far, χ does not vary with the magnitude or sign of ξ; as the electric field in such a solid increases, the material becomes more polarized in direct proportion to the field, and as the field decreases it retraces the original path. Equation (3.5.2) states that polarization is directly proportional to the electrical field for these simple dielectrics. This is another example of a *linear response*. The presence of the additional coefficient ϵ_0 in (3.5.2) arises from the manner in which electrical units are defined.

There is an important class of materials for which χ not only varies with ξ but is not even a single-valued function of ξ. These are the **ferroelectrics** and their behavior is illustrated by the P-ξ curve of Figure 3.5.1.

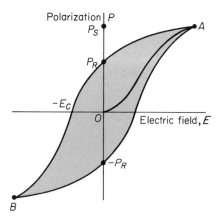

FIGURE 3.5.1. *P*-ξ curve for ferroelectric behavior.

The path followed as ξ is increased is OA. At the point A the polarization closely approaches its maximum possible value called the **saturation polarization**. Upon cycling ξ, the path subsequently traced is [A to P_R to E_C to B to $(-P_R)$] to [A to P_R] to E_c to B to $(-P_R)$] $\cdots$. P_R is called the **remanent polarization** and E_C the **coercive field**. Note that when ξ goes to O, the polarization is either P_R or $-P_R$; i.e., the solid remains polarized even in the absence of a field. There is actually a decay of polarization with time after the ξ field is switched off so that

$$P_R = P_R(0)e^{-t/\tau}, \tag{3.5.4}$$

where τ is the **relaxation time** for the process. τ depends strongly on temperature; as temperature increases, τ decreases.

An increase in temperature tends to destroy ferroelectric behavior. One of the most common ferroelectrics is $BaTiO_3$ (barium titanate). Above 120°C it loses its ferroelectric characteristics; this temperature is called the **ferroelectric Curie point,** T_C.

Some values of P_S and T_C are shown in Table 3.5.1.

Table 3.5.1. FERROELECTRIC CRYSTAL DATA

Material	Formula	$T_C(°K)$	P_S Coulombs/m²	(at T°K)
Rochelle salt	NaK $(C_4H_4O_6)\cdot 4H_2O$	297	0.0024	278
KDP	KH_2PO_4	123	0.053	96
Perovskite	$BaTiO_3$	393	0.260	296
TGS	Tri-glycine sulphate	322	0.028	293

*See F. Jona and G. Shirane, *Ferroelectric Crystals*, Pergamon Press, New York (1962).

3.6   PIEZOELECTRIC CRYSTALS

Certain crystals when stressed mechanically in the elastic range develop a polarization while stressed. Quartz and Rochelle salt are examples of this (also all ferroelectric crystals show this effect). This is known as the **piezoelectric effect**. Alternately, if the same crystals are placed in an electrical field, they will be strained (and if constrained will develop stresses). This is known as the **inverse piezoelectric effect**.

These effects have numerous applications in industry, e.g., as an electromechanical transducer in sonar (to send and receive sound pulses, which are really stress waves, through the water). Near the end of Section 2.3 we noted that one way to measure the elastic constants of crystals was to send a stress pulse into the crystal and to measure the sound velocity. To do this a small quartz crystal is glued to the sample and is pulsed electrically. The inverse effect sends a stress pulse into the sample, which reflects off the other parallel end of the sample back to the transducer where it causes a direct effect and thus a second voltage pulse. Since the path traveled is known and the time between the two electrical pulses is measured, the velocity is obtained.

Hundreds of millions of tiny quartz crystals are used as frequency references. A quartz crystal will have a resonance frequency for mechanical oscillations determined by its elastic constants, density, and geometry. Quartz crystals have been grown synthetically with quality factor $Q > 10^6$ [see Eq. (2.4.13)] so the resonance frequency will be very sharp since the peak width is given by

$$\Delta\omega_0/\omega_0 = \frac{\sqrt{3}}{Q}. \tag{3.6.1}$$

Since the mechanical oscillations are coupled to the electrical oscillations, this provides a fixed electrical frequency reference. Piezoelectric behavior is usually described by linear relationships between polarization and strains and between strains and electric field. The coefficients which describe this relationship are called the piezoelectric coefficients. A mathematical framework using Cartesian tensors is needed to describe this behavior.

A field of 10^5 V/m (100 V across a 1-mm-thick sample) along a certain crystallographic direction of quartz called the diad axis would produce a strain of -2.3×10^{-7}. A compressive mechanical stress of 1.8×10^5 dynes/cm^2 (2.6 psi) would yield the same strain.

Figure 3.6.1 shows a small piezoelectric motor capable of delivering 15 hp (to and fro motion, not rotary) at its tip.

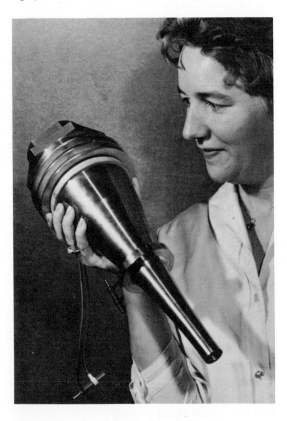

FIGURE 3.6.1. 15-hp piezoelectric motor weighing 22 lb. (*Courtesy of Department of Welding Engineering, Ohio State University.*)

REFERENCES

Young, J. F., *Materials and Processes*, Wiley, New York (1954). Pages 179–200 contain a description of conductor contacts, resistor materials, while pp. 439–485 are concerned with dielectric materials.

Von Hippel, A. R., ed., *Dielectric Materials and Applications*, Wiley, New York (1954). Von Hippel states in the preface, ". . . . we hope to establish alliances between research worker, development engineer, manufacturer, field engineer and actual user of 'nonmetals.' " He succeeds! The student might well read this book over lightly just to develop a feeling for the many applications of dielectrics.

Handbook of Chemistry and Physics (R. C. Weast, ed.), The Chemical Rubber Co., Cleveland (1969).

"Materials Selector Issue," annual issue of *Materials Engineering*.

Heath, F. G., "Large Scale Integration," *Scientific American* (Feb. 1970) p. 22. This is an excellent discussion of how the materials scientist has devised ways of placing 10^5 electronic components per square inch of semiconductor area.

Duckworth, H. F., *Electricity and Magnetism*, Holt, Rinehart and Winston, New York (1961). Systematically develops Maxwell's equations and applies theory to a number of simple electronic and magnetic components and devices.

Pugh, E. M., and Pugh, E. W., *Principles of Electricity and Magnetism*, Addison-Wesley, Reading, Mass. (1960). Has an excellent discussion of units and conversion tables for units.

Jona F. and Shirane, G., *Ferroelectric Crystals*, Pergamon Press, New York (1962). Has excellent crystal data on ferroelectric and piezoelectric behavior.

PROBLEMS

3.1. What are the three important classifications of materials according to
  a. State of aggregation?
  b. Electrical conductivity?
  Define each of the six classifications carefully.

3.2. The copper wire used for overhead electrical transmission is usually heavily cold-worked. Does this affect the economics of electrical transmission greatly? Would it if the transmission were at $4.2°K$ rather than at around room temperature?

3.3. Sodium metal is a very good conductor. Moreover, NaCl is much more available than CuS. Is it ever likely that sodium will be used extensively to transmit power?

3.4. a. Define Ohm's law in fundamental terms.
   b. When does this lead to

$$R = \rho \frac{l}{A}?$$

   c. Under which circumstances do we not get $R = \rho(l/A)$?

3.5. Name some important electrical properties of
   a. Metals (one property).
   b. Semiconductors (two properties).
   c. Insulators (three properties).

3.6. What are typical room-temperature resistivity values for
   a. Metals?
   b. Semiconductors?
   c. Insulators?

3.7. a. What is the typical behavior of $\rho(T)$ for metals?
   b. What is meant by the resistivity ratio?
   c. How is $\rho(4.2°K)$ affected by impurities? by cold working?

3.8. Assuming that there is only one type of charge carrier present, what quantities does the Hall coefficient measurement yield?

3.9. a. What is the typical behavior of $\rho(T)$ for semiconductors?
   b. How is this function affected by impurities?

3.10. Give the analog for dielectric behavior of the following aspects of mechanical behavior:
   a. Hooke's law.
   b. Tensile fracture strength.
   c. Internal damping.

3.11. Is there any relationship between δ_{mech} and δ_{elect}?

3.12. What is the general expression for energy per unit volume stored in a dielectric?

3.13. Why are dielectric materials used in capacitors?

3.14. Why are liquids, such as mineral oils, used as the dielectric material in heavy transformers?

3.15. Sketch the P-ξ curve for
   a. Ideal dielectric.
   b. Real dielectric.
   c. Ferroelectric.

3.16. A ferroelectric, such as $BaTiO_3$, is polarized and the electric field is removed. The polarization is then P_R. Is this a permanent polarization?

3.17. Give a possible application of a ferroelectric material.

3.18. Define the piezoelectric effect. What possible applications does it have?

More Involved Problems

3.19. A charge q, moving through space at a velocity $\mathbf{v}$ in a region where the magnetic induction is $\mathbf{B}$ and the electrical field is $\boldsymbol{\xi}$, is acted upon by a force $\mathbf{F}$ given by

$$\mathbf{F} = q(\boldsymbol{\xi} + \mathbf{v} \times \mathbf{B}).$$

Referring to Figure 3.3.2,

$$\mathbf{B} = [0, 0, B_z], \qquad \mathbf{v} = [v_x, 0, 0].$$

Why is $J_x = v_x N q$? The potential measuring device has a very high impedance, so we can assume no current flows across the specimen in the y direction. This means that the total force in the y direction is zero. Use this result and $R_H = \xi_y / J_x B_z$ to obtain $R_H = 1/Nq$.

Sophisticated Problems

3.20. At very low temperatures, there is a size effect on resistivity.
  a. Explain what is meant by this. This is discussed from a theoretical viewpoint by R. B. Dingle, *Proceedings of the Royal Society (London)*, **A201**, 545 (1950), and E. H. Sondheimer, *Advances in Physics*, **1**, 1 (1952), and from an experimental viewpoint by K. Forsvoll and I. Holweck, *Journal of Applied Physics*, **34**, 2230 (1963), and F. Dworschak, H. Schuster, H. Wollenbergen, and J. Wurm, *Physica Status Solidi*, **21**, 741 (1967).
  b. Calculate the actual resistance of a 0.003-cm-diameter wire of copper (whose bulk resistivity ratio is 50,000) at 4.2°K.
  c. Calculate the apparent resistivity at 4.2°K of a thin foil of copper (1000 Å) whose bulk resistivity ratio is 50,000.

3.21. At very high frequencies, electrical current moves to the surface of a conductor. The skin depth (layer in which most of the current passes) is readily derivable from Maxwell's equation. See J. C. Slater and N. H. Frank, *Electromagnetism*, McGraw-Hill, New York (1947) p. 126. What is the skin depth for copper at room temperature at 1 Mhz? for copper at 4.2°K with a resistivity ratio of 100,000 at 10 Mhz?

Prologue

Magnetic materials are classified according to whether they attract or expel magnetic flux. Ordinary *diamagnetic materials* expel flux slightly, *paramagnetic materials* attract flux somewhat, while *ferromagnetic materials* attract flux strongly. From an engineering viewpoint the last are the most important materials. The behavior of magnetic materials is described in terms of their *magnetic induction* versus *magnetic field* curves, i.e., their *B-H curves*. In general the *B-H* curve for a material will be a *hysteresis loop*. Important properties of magnetic materials are the *permeability coefficient*, the *saturation induction*, the *remanent induction*, and the *coercive force*. There are two important classes of ferromagnetic materials: *soft* and *hard*.

We shall study the difference between these two types of materials and why soft magnetic materials are used for transformer coils, while hard magnetic materials are used for permanent magnets. It will be pointed out that increased temperature causes ferromagnetism to disappear in a characteristic way. Magnetic materials become strained in the presence of magnetic fields, an effect called *magnetostriction*. The student should begin to see why different types of magnetic materials have present and potential commercial importance.

4

SOME PHENOMENOLOGICAL ASPECTS
OF MAGNETIC MATERIALS

4.1   A SOLENOID

In Chapter 3, a simple solenoid was discussed briefly from the viewpoint of using it as a magnet (see Figure 3.2.1). The voltage drop across such a coil in a case in which the current is changing is

$$V = L\frac{di}{dt},$$ (4.1.1)

if we assume the i^2R losses are negligible because the winding is large-diameter wire of high conductivity. The quantity L is called the **self-inductance** of the coil, i is the current, and t is the time. If the coil is in a vacuum we use a subscript zero, L_0. Every student of electromagnetism should be able to show that

$$L_0 = \mu_0 n^2(Al).$$ (4.1.2)

Here μ_0 is the **permeability coefficient of a vacuum** [$4\pi \times 10^{-7}$ H (henry)/m in rationalized mks units], n is the number of turns per unit length, A is the cross-sectional area of the coil, and l is the length of the coil. We assume l is very large relative to the coil diameter.

EXAMPLE 4.1.1

Find the energy stored in the above coil.

Answer. The power is Vi so that the magnetic energy is

$$U_{\text{mag}} = \int Vi\, dt = \tfrac{1}{2}Li^2.$$ (4.1.3)

EXAMPLE 4.1.2

Compare the spring-mass (kM) system of Figure 2.4.1 with the coil-condenser (LC) system shown in Figure 4.1.1.

FIGURE 4.1.1. Simple LC circuit with switch.

Answer. From previous discussions (see Example 2.4.1) the equation of motion was shown to be

$$M\frac{d^2x}{dt^2} + kx = 0.$$

In the present case the sum of the potential drops around the closed circuit must be zero and so

$$L\frac{di}{dt} + \frac{Q}{C} = 0.$$

Since $i = dQ/dt$, this can be written as

$$L\frac{d^2Q}{dt^2} + \frac{Q}{C} = 0. \tag{4.1.4}$$

The differential equations are clearly analogs. $\frac{1}{2}Li^2$ is the analog of the kinetic energy $\frac{1}{2}Mv^2$ and $Q^2/2C$ is the analog of the potential energy $kx^2/2$.

If a rod of homogeneous isotropic material (from the magnetic viewpoint) fills the space within the coil, then the self-inductance is

$$L = \mu n^2(Al), \tag{4.1.5}$$

where μ is the permeability coefficient of the substance. It is useful to introduce the term **relative permeability coefficient** for the ratio μ/μ_0. Note that this gives the ratio of the self-inductance of the coil with magnetic material in the coil and without, i.e., L/L_0; it also gives the ratio of the energy stored in these two cases.

From a more fundamental viewpoint the permeability relates the **magnetic induction B** to the **magnetic field intensity H** according to

$$\mathbf{B} = \mu\mathbf{H}. \tag{4.1.6}$$

The permeability coefficient or the permeability ratio (unlike the conductivity coefficient σ, or the electric permittivity coefficient ϵ, or the dielectric coefficient κ) is often not a constant so that Equation (4.1.6) does not always have the general linear nature; this point will be discussed later.

The energy per unit volume in a magnetic field is given by

$$u_{\text{mag}} = \int H\,dB. \tag{4.1.7}$$

EXAMPLE 4.1.3

Using Equation (4.1.7), obtain an expression for the magnetic energy per unit volume for the case in which μ does not vary with H.

Answer.

$$\int H\,dB = \mu \int H\,dH = \frac{\mu H^2}{2} = \frac{B^2}{2\mu} \qquad (\mu = \text{constant}).$$

Materials can be classified from the magnetic viewpoint according to the magnitude of μ/μ_0 as follows:

$$\mu \lessdot \mu_0, \quad \textbf{diamagnetic;}$$

$$\mu \gtrdot \mu_0, \quad \textbf{paramagnetic;}$$

$$\mu \gg \mu_0, \quad \textbf{ferromagnetic.}$$

For diamagnetic materials the magnitude of the fractional variation of μ from μ_0 [i.e., $(\mu - \mu_0)/\mu_0$, which is also called the **magnetic susceptibility**] is about -10^{-5}, while for paramagnetic materials it is about 10^{-4}. For such materials μ is a constant (except for unusually intense magnetic fields). For many applications such materials can be considered as nonmagnetic. However, the actual study of their magnetism provides, in many cases, detailed knowledge of their electronic structure.

4.2   FERROMAGNETIC MATERIALS

From a technological viewpoint the ferromagnetic materials are extremely important; note that μ is not generally a constant for these materials. In fact, the *B-H* relationship is not even reversible; its behavior is thus not at all like the *F-x* relation for an ideal spring (see Figure 2.4.2) but is more like that of the rubber band which has strong damping (i.e., a large loss angle).

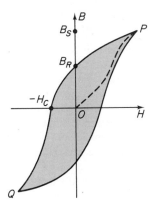

FIGURE 4.2.1. Typical *B-H* curve.

However, it is not exactly analogous. A typical *B-H* curve is shown in Figure 4.2.1. The first *B-H* curves were measured by Ewing in 1885; he called them **hysteresis curves**. The path followed is O to P to (B_R to H_C to Q to P) to (B_R to H_C to Q to P). We have the following important definitions:

B_S—**saturation induction** (the limiting value of $B - \mu_0 H$ for large H);

B_R—**remanent induction** (magnitude of B when $H = 0$);

H_C—**coercive force** (magnitude of H when $B = 0$);

μ_i—$\lim\limits_{H \to 0} (dB/dH)$ along line $\overline{OP}$—**initial permeability**;

Maximum of dB/dH along line $\overline{OP}$—**maximum permeability**;

$$M\text{—\textbf{magnetization}} = \frac{B}{\mu_0} - H.$$

EXAMPLE 4.2.1

What is the reason for using the iron core in the transformer shown in Figure 4.2.2?

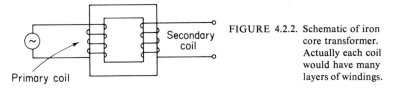

Primary coil

FIGURE 4.2.2. Schematic of iron core transformer. Actually each coil would have many layers of windings.

Answer. The emf produced in the second coil is given by the product of the mutual inductance M and the rate of change of current in the primary. The mutual inductance is given by

$$M = k\sqrt{L_1 L_2}, \tag{4.2.1}$$

where k is the coupling constant (if the entire magnetic flux of one coil passes through the other, $k = 1$; in the absence of the iron core k would be small in the above case). The iron core serves the dual purpose of enhancing the magnetic flux (i.e., increasing both L_1 and L_2 by the factor μ/μ_0) and of leading the flux from the primary to the secondary coil (i.e., increasing k). We note in Section 4.3 that μ/μ_0 can have values of several hundred or more. If the magnetic core were not present, the number of windings would have to be increased greatly. Hence compare the size of a transformer with and without the use of a ferromagnetic core! The size would have to increase by 10^3 to 10^6 if the magnetic core were not used. Figure 4.2.3 shows a transformer station. Imagine the cost of the copper and the size of the station if magnetic core material were not available!

FIGURE 4.2.3. Power transformer stations. (*Courtesy of Pacific Gas and Electric Company.*)

These quantities define the important aspects of the initial *B-H* behavior and the *hysteresis loop* obtained by periodic cycling of *H*. An important feature of magnetism is the saturation induction, B_S; beyond the point *P*, $dB/dH = \mu_0$; i.e., the ferromagnetic material makes zero contribution to the cause of increasing *B* as *H* increases. B_S has a characteristic value for a given material. B_S is relatively independent of impurities, mechanical deformation history, thermal history, etc., and is hence called a structure-insensitive property. The **saturation magnetization** M_S is the value of *M* when $B = B_S$. The *M-H* curve hence has the shape shown in Figure 4.2.4.

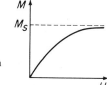

FIGURE 4.2.4. Magnetization curve.

The **magnetic susceptibility** χ_m is defined by

$$\mathbf{M} = \chi_m \mathbf{H}. \qquad (4.2.2)$$

B, **H**, and **M** are related by

$$\mathbf{B} = \mu_0 \mathbf{H} + \mu_0 \mathbf{M} \qquad (4.2.3)$$

in the rationalized mks system. Magnetization is the magnetic analog of polarization. Magnetization could be defined as the net magnetic dipole moment per unit volume. Note that $\chi_m \approx -10^{-5}$ for diamagnetic materials. Thus diamagnetic materials tend to repel the lines of flux. Note also that $\chi_m \approx 10^{-4}$ for paramagnetic materials. Paramagnetic materials are weakly attracted by a magnet. For example, liquid oxygen clings to a magnet. Note that $\chi_m \gg 0$ for ferromagnetic materials at magnetization much less than M_S. However, ferromagnetic materials become saturated at high fields, so that the response owing to further increase in the field is the same as for a vacuum.

Temperature has a strong effect on ferromagnetism; it tends to destroy it. See Figure 4.2.5.

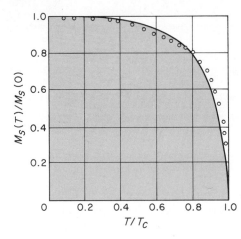

FIGURE 4.2.5. Saturation magnetization versus temperature for nickel. [Experimental values are from P. Weiss and R. Forrer, *Annales de Physique*, **5**, 153 (1926).] M_S at $T = 0$ is 4.8×10^5 A/m and $T_C = 631°$K.

All ferromagnetism vanishes for $T > T_C$. T_C is the **Curie temperature**. It could also be called the **ferromagnetic transition temperature**. Table 4.2.1 shows some values of T_C and M_S for strongly magnetic materials.

Table 4.2.1. CURIE POINTS AND SATURATION CONSTANTS

Material	$T_C°$K	M_S 10^5 A/m
Fe	1043	17.1
Co	1400	14
Ni	631	4.85
Gd	289	19.8
MnBi	630	6.75
$3Y_2O_3-5Fe_2O_3$*	560	2.0

* Also called YIG for yttrium iron garnet.

B_R and H_C (and hence the shape of the hysteresis loop) are structure-sensitive properties. We note that there is a dissipation of magnetic energy per unit volume per one hysteresis loop given by

$$\Delta u = \int_{\substack{\text{closed} \\ \text{loop}}} H\,dB = \oint H\,dB, \qquad (4.2.4)$$

which is converted into heat. Ferromagnetic materials can be divided into two classes: soft magnetic materials and hard magnetic materials. We shall discuss each class in more detail in the next two sections.

4.3 SOFT MAGNETIC MATERIALS

SOFT FERROMAGNETICS.   A material for which B is nearly a single-valued function of H is a **soft magnetic material**. (They are usually mechanically soft also; i.e., they have a low yield strength.) They are characterized by a high initial permeability μ_i and a small hysteresis loop, i.e., small coercive force H_C.

Some examples are iron, 3.25% silicon–96.75% iron, 45 Permalloy (45% Ni–55% Fe), Mumetal (75% Ni–5% Cu–2% Cr–18% Fe), and Super-malloy (80% Ni–15% Fe–5% Mo).

Silicon-iron has important application in power equipment involving ac currents (such as the 60 Hz of our utilities). Such equipment includes transformers, generators, motors, controllers, and meters. In 1968 about 670,000 tons of such materials were produced. Philip Sporn in the fascinating book *Research in Electric Power*, Pergamon Press, New York (1966) predicts an increase by a factor of 6.6 in power generation by the year 2000. The critical parameter to the design engineer is the energy loss per cycle. The two important contributions to this loss are (1) the hysteresis loss given by Equation (4.2.4), and (2) the eddy current losses. Eddy current losses arise as follows. The changing magnetic field induces a flow of electrical current (eddy current) in the magnetic material. This leads to ordinary i^2R losses, i.e., ordinary electrical heating as in the spiral heating element of Section 3.1.

EXAMPLE 4.3.1

Given in the year 2000 that 4.4×10^6 tons of silicon iron will be used showing a hysteresis loss per cycle of 35 W-sec/m^3 in 60-Hz utility equipment, estimate the annual kilowatt-hour dissipation owing to $\oint H\,dB$ losses. How much would be saved if the dissipation could be reduced by 50% (assume 1 kWh = \$0.02).

Answer. The 4.4×10^6 tons of steel correspond to 4×10^{12} g and, assuming a density of 8 g/cm^3 (actually 7.8), gives a volume of 5×10^{11} cm^3 or 5×10^5 m^3. There are about 10^5 sec/day so the energy dissipated per year is $35 \times 5 \times 10^5 \times 60 \times 10^5 \times 365 \times (1/1000) \times (1/3600)$ kWh, which is about 10^{10} kWh. The money saved would be about 10^8

per year, which would run a decent sized university. *Note*: 1 J = 1 W-sec. (Materials such as Supermalloy are not used because the initial cost is too high.)

There are two ways that the design engineer can eliminate eddy currents losses: Choose a material with a higher resistivity ρ and use laminated cores. (Note the analogy with the problem of the design engineer who wanted to increase the stiffness of a torsion shaft and had the choices of increasing G or changing geometry, in that case using a hollow shaft, as discussed in Example 2.2.2.)

Table 4.3.1 shows some important parameters of some common soft ferromagnetics. For those students familiar with units of gauss for B, 10 kG = 1 Wb/m².

Table 4.3.1. IMPORTANT PROPERTIES OF SOME SOFT FERROMAGNETICS*

Material	μ_i/μ_0	$\oint H \, dB$, J/m³	B_S, Wb/m²
Commercial iron ingot	250	500	2.16
4% Si–96% Fe (random texture)	500	50–150	1.95
3.25% Si–96.75% Fe (oriented)	15,000	35–140	2.0
45 Permalloy	2,700	120	1.6
Mumetal	30,000	20	0.8
Supermalloy	100,000	2	0.79

* From J. Wulff et al., *The Structure and Properties of Materials*, Wiley, New York (1966).

Because of their small hysteresis loss, the nickel-iron alloys can be used at higher frequencies, e.g., in communication equipment. [The power lose is the product of the energy per cycle times the frequency.] Just as some materials are made mechanically soft by proper thermal treatment (compare the yield strength of annealed copper in Table 2.1.2 with the cold-worked copper) so they often can be made magnetically soft by proper thermal treatment. Figure 4.3.1 shows the effect of annealing (holding at a high temperature) Permalloy in the presence and in the absence of a magnetic field.

Impurities can have an enormous effect on the relative permeability ratio, as shown in Table 4.3.2. Concurrent with the increase in permeability as the carbon is decreased, there is a huge drop in the mechanical yield strength.

Table 4.3.2. RELATIVE PERMEABILITY AND
CARBON CONTENT OF IRON

Material	Relative Permeability
Ultrapure iron	1,500,000
Fe with 0.004% C	16,000
Fe with 0.01% C	8,000

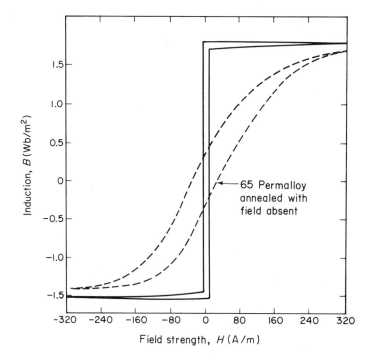

FIGURE 4.3.1. Magnetic annealing effects on 65 Permalloy (65% Ni–35% Fe). Solid line, annealed with magnetic field present. Dashed line, annealed without field. [From R. M. Bozorth, *Ferromagnetism*, Van Nostrand, Princeton, N.J., (1951).]

The author remembers the consternation of a graduate student working with him who decided he needed a higher μ for the core of a linear variable differential transformer so he proceeded to thoroughly anneal his core at high temperatures in the presence of hydrogen (which removes the carbon and forms CH_4). The iron, of course, became very soft magnetically, but to his chagrin the core, which was about a 1-in.-long cylinder of 0.1-in. diameter, was also so mechanically soft that it deformed plastically under its own weight when held horizontally from one end. The student should at this stage have a deep curiosity concerning the origin of the simultaneous magnetic and mechanical softening of these ferromagnetic materials.

Figure 4.3.2 shows a profound difference in the *B-H* curves of sheet silicon-iron produced by rolling in rolling mills.

MAGNETICALLY SOFT CERAMICS. Magnetically soft ceramics have the same general *B-H* characteristic as the soft ferromagnetic materials. The most common materials are ferrites and garnets. They are made by mixing powders of metallic oxides which are then formed into a solid block of material by a

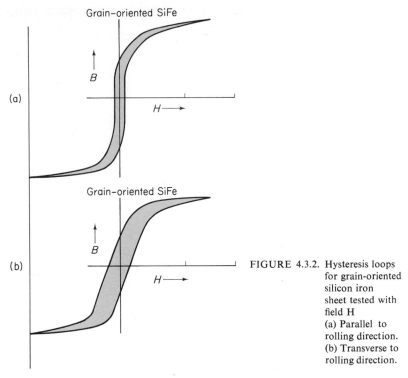

(a)

(b)

FIGURE 4.3.2. Hysteresis loops for grain-oriented silicon iron sheet tested with field H (a) Parallel to rolling direction. (b) Transverse to rolling direction.

process known as sintering (which we shall study later after we have studied the surfaces of solids and the motion of atoms in solids). A typical composition is Ferroxcube A (48% MnO–Fe_2O_3, 52% ZnO–Fe_2O_3). It is characterized by a **"square" hysteresis loop** (actually rectangular), as shown in Figure 4.3.3. Its electrical resistivity is large (0.5×10^6 Ω-m) compared to metals such as iron-silicon (see Table 3.1.1 for typical values for metals) and hence its eddy

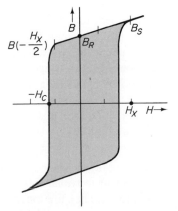

FIGURE 4.3.3. Nearly "square" loop.

current losses are small compared to the silicon-irons. For this reason it is used in transformers in high-frequency (~ 1 MHz, i.e., 10^6 cycles/sec) communication equipment. The squareness ratio is defined in terms of Figure 4.3.3. Let H_x be the value of H for which B first reaches B_S. The **squareness ratio** is then defined as $B(-H_x/2)/B_S$, i.e., by B evaluated at $-H_x/2$ divided by B_S. Because of the squareness of its hysteresis loop it is useful in computers since there is a very abrupt change in the direction of magnetization when H decreases from $-H_x/2$ to $-H_x$. Because of the sharpness of this switching, a magnetic core of this material maintains the strongest binary signal.

An important garnet, yttrium iron garnet (YIG), $3Y_2O_3$–$5Fe_2O_3$, has hysteresis losses sufficiently small that it can be used at microwave frequencies.

4.4   HARD MAGNETIC MATERIALS

Materials characterized by huge hysteresis losses, large B_R, and large H_C are called **hard magnetic materials**. These are permanent magnet materials. In rough terms the designer of a permanent magnet wants B_R to be large so that the greater part of the magnetization will remain when the magnetizing field is removed and he wants H_C to be large in order that the magnet will not be easily demagnetized.

The coercive force is perhaps the ferromagnetic property most sensitive to control through changes in composition, heat treatment, and mechanical deformation; further, it is a most important property to consider in the selection of ferromagnetic materials for practical applications. Coercive force values in commercial materials range from 50,000 A/m in a loudspeaker magnet (Alnico V) and 200,000 A/m in a special high-stability magnet (Fe–Pt) to 20 A/m in an iron-silicon transformer core and 0.3 A/m in a Supermalloy pulse transformer [for students familiar with other units, 1 A/m $= 4\pi \times 10^{-3}$ Oe (oersted)]. Thus the coercive force may be varied over a range of approximately 5×10^6. Coercive force values are closely related to the saturation hysteresis loss at low frequencies since the area of the hysteresis loop is approximately the product of the coercive force and the saturation magnetic induction.

A close correlation exists between the initial relative permeability and the coercive force, as shown in Figure 4.4.1, which spans the scale from soft to hard magnetic materials.

The permanent magnet design engineer is particularly interested in the maximum value of the product BH along the demagnetization curve (along line $\overline{B_R H_C Q}$ in Figure 4.2.1). This is called the **maximum energy product** and some typical values are shown in Table 4.4.1.

The maximum energy product is the largest rectangle which can be inscribed in the second quadrant of the B-H curve.

Research in the last two decades into the nature of the rare earths and

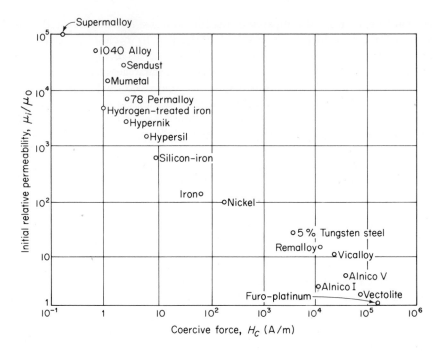

FIGURE 4.4.1. Correlation between initial permeability and coercive force. Note the log-log plot. (*Courtesy of Charles Kittel.*)

into their chemistry (they essentially all have the same valence and hence are very difficult to separate by standard means) has paid off handsomely in the case of the samarium-cobalt alloy, which costs about one eightieth of its nearest competitor, Pt–Co, and is better!

Table 4.4.1. **MAXIMUM ENERGY PRODUCT FOR HARD MAGNETIC MATERIALS**

Material	Max. Energy Product, J/m³
Samarium-cobalt	120,000
Platinum-cobalt	70,000
Alnico V*	36,000
Ferroxdur†	12,000
Fe–Co sintered powder	7,000
Carbon steel	1,450

* Alnico V has a composition 51 % Fe–24 % Co–14 % Ni–8 % Al–3 % Cu.
† Ferroxdur is a hard ferrite, $BaFe_{12}O_{19}$.

4.5   SPECIAL MAGNETIC MATERIALS

Special methods of fabrication make possible a number of different kinds of magnetic materials. **Thin magnetic films** usually of Fe–Ni which can be formed in a thickness of ~ 3000 Å by vacuum deposition of the vapor on a nonmagnetic substrate can have switching times of 10^{-8} sec or better and hence have applications in computers. The fact that it should take some time for the magnetic core material to respond to the magnetization force has not been mentioned here before. After we have studied the atomic and electronic structure of solids we shall be in a position to understand the reason for this switching time.

Fine particle magnets (sufficiently small particles of ferromagnetic materials always exist as magnetized particles when the material is in the ferromagnetic state) are also particularly interesting. Tiny elongated particles of iron having a diameter of only a few hundred Ångstroms are placed in a nonmagnetic molten binder such as lead (or epoxy resin) which is then solidified or hardened in the presence of a magnetic field. Such a composite material can have a very high maximum energy product (10^4 J/m) compared to the value for bulk carbon steel. (See Table 4.4.1.) Why should such fine particles be such hard magnets?

The magnetic tape of our common tape recorders is also a composite consisting of oriented elongated fine particles of γ-Fe_2O_3 on a plastic film.

Recently it has been possible to generate cylindrical domains in **ortho-ferrites**, materials having the composition $RFeO_3$, where R is a rare earth or yttrium. These domains (sometimes referred to as **magnetic bubbles**) can be propagated to perform memory and logic functions. They have a potential density of 10^6/in.[2] and have many potential applications in devices. See A. H. Bobeck, et. al. *IEEE Transactions on Magnetics* **5**, 544 (1969).

4.6   MAGNETOSTRICTION

The magnetization of a ferromagnetic material is accompanied by a small change in length; this effect is called **magnetostriction**. This is the origin of another energy loss phenomenon during cyclic magnetization. Magneto-strictive effects set up strong unwanted mechanical oscillations in many cases and so represent another factor of concern to the design engineer. The sample may either shrink in the direction of **H** (cobalt) or stretch in the direction of **H** (as iron does for small fields, although it shrinks for large fields).

Magnetostriction provides one mechanism of converting electrical energy into mechanical energy (via an intermediate, magnetic energy). The resultant device is called an electromechanical transducer.

REFERENCES

Young, J. F., *Materials and Processes*, Wiley, New York (1954) pp. 200–240. This contains a good discussion of the phenomenological aspects of magnetic behavior.

"Magnetic Materials," *Metals Handbook*, American Society for Metals, Cleveland (1948) pp. 587–600. Contains a good description of the structure-sensitive nature of many magnetic properties.

Handbook of Chemistry and Physics (R. C. Weast, ed.), The Chemical Rubber Co., Cleveland (1969).

Bardell, P., *Magnetic Materials in the Electrical Industry*, Macdonald, London (1960). The student should look over this book (or the book by Say, below) just to become more familiar with the many applications of magnetic materials.

Say, M. G., ed., *Magnetic Alloys and Ferrites*, G. Newnes, London (1954).

Jacobs, I. S., "The Role of Magnetism in Technology," *Report No. 68-C-432*, General Electric Research and Development Center, Schenectady, N.Y. (Nov. 1968). Also: *Journal of Applied Physics* **40**, 917(1969). This paper nicely describes the developments taking place in magnetic technology. The interested student should obtain a copy of this paper.

PROBLEMS

4.1. Sketch a *B-H* curve for (a) a magnetically soft material and (b) a hard magnetic material.

4.2. For a 60-Hz transformer which operates at high fields, which is the more important design parameter:
a. The maximum relative permeability?
b. The initial relative permeability?

4.3. Why are soft ferromagnetic cores used in transformers?

4.4. Plot *B* vs. *H* for line *OP* and beyond of Figure 4.2.1.

4.5. Which of the quantities are essentially the same for magnetically hard and soft iron: μ_i, B_S, B_R, H_C?

4.6. a. Explain why the tiny elongated residue particles which arise from the finish grinding of a nonmagnetized steel rod are magnetized.
b. Discuss why Si–Fe is used for utility equipment, Permalloy at audio frequencies, ferrites at higher frequencies, and garnets at very high frequencies.

4.7. By placing a ferromagnetic core in a coil, the energy stored for a fixed current is increased, while, if a dielectric slab is placed between the plates of a condenser, the energy stored for a fixed potential is increased.

What effect does the use of these materials have on the size of a resonant *LC* circuit?

4.8. Describe the advantage of using a ferromagnetic core in an electromagnet operating to $B = 1\text{Wb/m}^2$. Would the same be true if $B = 15 \text{ Wb/m}^2$?

4.9. Sketch the geometry of a transformer and discuss with equations, if possible, the role of the magnetic material in changing the self-inductances of the primary and the secondary and in changing the mutual inductance.

4.10. What are the most important material properties of a soft magnetic material for transformer applications?

4.11. What is one of the most important properties of a magnetic core in a digital computer?

4.12. What is the most important property of a hard magnetic material insofar as the design engineer is concerned?

4.13. In some scientific experiments, it is necessary to have a region that is free of magnetic field. How would you achieve this?

4.14. The accompanying figure shows an electromagnet. The gap l_g is very small, say 1/100th of the average magnetic flux path $(2\pi R = l_m)$. l_g is also small with respect to the pole face diameter, so that there is very little fringing of the flux. In this case the magnetic induction at any point in the magnetic circuit can be written

$$B \doteq \frac{Ni}{[(2\pi R - l_g)/\mu] + (l_g/\mu_0)}.$$

Here N is the total number of turns.
a. The effect of the presence of a core with $\mu/\mu_0 = 2000$ is to increase B by what factor over the case of no core?
b. What would the factor be if there were no gap with $\mu/\mu_0 = 2000$ (of course, this would be useless as an electromagnet)?

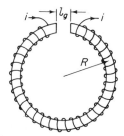

FIGURE P4.14

4.15. The accompanying table lists corresponding values of H and B for a specimen of commercial hot-rolled silicon steel, a material widely used in transformer cores.
a. Construct graphs of B and μ as functions of H, in the range from $H = 0$ to $H = 1000$ A-turn/m.

b. What is the maximum permeability?

c. What is the initial permeability ($H = 0$)?

d. What is the permeability when $H = 800{,}000$ A-turns/m?

Magnetic field intensity H, A-turns/m	Flux density B, Wb/m^2
0	0
10	0.050
20	0.15
40	0.43
50	0.54
60	0.62
80	0.74
100	0.83
150	0.98
200	1.07
500	1.27
1,000	1.34
10,000	1.65
100,000	2.02
800,000	2.92

4.16. How large a magnetic induction can one obtain in a magnet whose design is based on a magnetic core?

4.17. Given that energy loss per cycle is $\oint H \, dB$, what would be the expression for energy loss per cycle in terms of H and M?

More Involved Problems

4.18. The accompanying circuit contains the three simple elements L, C, and R, which have been discussed in Chapters 3 and 4, plus an ac voltage supply. Show that it has the differential equation

$$L\frac{d^2q}{dt^2} + R\frac{dq}{dt} + \frac{q}{C} = E_0 \cos \omega t.$$

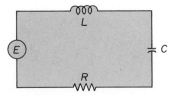

FIGURE P4.18

4.19. a. Show by direct substitution that the function

$$q(t) = \frac{E_0}{\sqrt{[(1/C) - \omega^2 L]^2 + (\omega R)^2}} \cos(\omega t - \beta)$$

is a solution of the equation given in Problem 4.18, where

$$\beta = \tan^{-1} \frac{\omega R}{(1/C) - \omega^2 L} \cdot$$

b. Show that the value of ω for which the amplitude is a maximum is

$$\omega_r = \left(\frac{1}{LC} - \frac{R^2}{2L^2}\right)^{1/2}.$$

When $\omega = \omega_r$, the system is in *resonance*. When the damping is small, $\omega_r \doteq (LC)^{-1/2} = \omega_0$, which equals the natural frequency (of the undamped system).

c. Hence, show that the amplitude of $q(t)$ for the resonant circuit for the case of small damping is

$$\sqrt{LC} \frac{E_0}{R} \cdot$$

d. Use the result of Problem 4.19c to show that the resonance amplification of this circuit is

$$\frac{1}{R}\sqrt{\frac{L}{C}}.$$

e. How does ω_0 depend on μ (the permeability of the inductor core) and ϵ (the permittivity of the capacitor material)?

f. What is the quality (Q factor) of the circuit?

g. What is the width of the resonance peak at half-amplitude?

4.20. Silicon-iron is used in a transformer built for continuous 30-year 60-Hz service. If instead you used a material for which $\oint H \, dB$ was one sixtieth as much as for the silicon-iron, how many times as much could you pay for this material with equal economy. Assume that the silicon-iron costs $1.00/kg. Assume that 1 kWh = $0.02 and that the density of the two materials is 8 g/cm^3.

Sophisticated Problem

4.21. Write an essay on the current state of development of magnetic tapes, including their production.

Prologue

Superconductors can carry dc currents with no energy losses. Superconductors exist only at low temperature (about 20°K or below). Above a *critical temperature* (characteristic of each material) they become normal conductors. A *soft superconductor* completely excludes magnetic flux up to a certain critical field; beyond that field it is completely penetrated and behaves as a normal conductor. The *critical magnetic field* needed to destroy superconductivity is a characteristic function of temperature.

A hard *superconductor* completely expels magnetic flux for small fields but eventually allows some flux penetration; this flux penetration increases as the field increases. Certain hard superconductors are capable of carrying enormous dc currents without loss. Such materials have great commercial potential. They can be used to create large magnetic fields such as might be needed to contain a plasma in a fusion reactor. They can be used to transmit electrical power from a large reactor across long distances with no I^2R losses. They can be used to generate magnetic fields in generators and motors, thus making possible much more powerful units without increasing the size.

Because current in a superconducting loop will persist indefinitely unless the material is made normal, superconductors can be used for computer memory storage.

5

SUPERCONDUCTIVITY

5.1   SUPERCONDUCTING MAGNET

Figure 5.1.1 shows the cross section of a projected design of a controlled fusion reactor. We note that superconducting magnet coils are on the outside. These are likely to be wound from Nb_3Sn ribbon. Nb_3Sn is a **hard superconductor** at low temperatures (several degrees Kelvin): Such materials will carry very high dc-current densities with essentially no loss; coils which generate very high magnetic fields can be made from it. However, there are losses with ac currents. The advantages over a normal metallic conductor in making a high field coil magnet are immediately obvious; see Section 3.2, where the heat generation in a magnet made of normal conductors is discussed. Even at 4.2°K, one of the best normal conductors would generate about 2000 W/m of length for a 10-cm

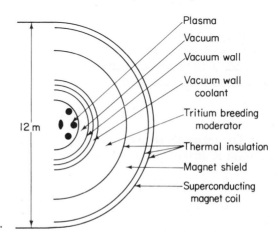

FIGURE 5.1.1. Proposed 5000 Mw steady-state fusion reactor. (Oak Ridge National Laboratory, Oak Ridge, Tennessee).

Plasma
Vacuum
Vacuum wall
Vacuum wall coolant
Tritium breeding moderator
Thermal insulation
Magnet shield
Superconducting magnet coil

12 m

bore in creating a magnetic induction of 10Wb/m² or a magnetic field of ~8 × 10^6 A/m (or 100 kG and 100 kOe, respectively).

Hard superconductors will also be used in the field coils of the next generation of electric generators; this will eliminate the need for heavy iron cores in the armature. In addition superconductors may serve for long distance power transmission lines.

5.2   SUPERCONDUCTORS

In 1911 Kamerlingh Onnes discovered that the resistivity of mercury vanished at 4.15°K. This temperature is called the **superconducting critical temperature**, T_c. Since that time T_c has been measured for a number of elements, and the results for a few of these are shown in Table 5.2.1. It has also been noted that above a certain magnetic field the superconducting property is destroyed. We call this the **superconducting critical field** H_c. It varies with temperature; the value at absolute zero is $H_c(0)$. Some values of $H_c(0)$ are given in Table 5.2.1. Materials are known as **soft superconductors** if $H_c \gtrsim 10^6$ A/m. (For many materials the quantity H_c is an effective quantity defined by a thermodynamic relation; this will be discussed later.)

Table 5.2.1. T_c AND $H_c(0)$ VALUES FOR LOW FIELD
SUPERCONDUCTORS

Type	Material	T_c, °K	$H_c(0)$, A/m
I	Tin	3.72	2×10^4
I	Lead	7.19	7×10^4
II	Vanadium	5.03	9×10^4
II	Niobium	9.1	20×10^4

Figure 5.2.1 shows the temperature dependence of the critical field. We note that both magnetic and temperature fields have a strong effect on superconductivity; they tend to destroy it. Note the similarity with the effect of temperature on saturation magnetization (see Figure 4.2.5). Roughly,

$$H_c(T) = H_c(0)\left(1 - \frac{T^2}{T_c^2}\right). \tag{5.2.1}$$

The materials shown in Table 5.2.1 would not be of much use in the bulk form in making high field magnets ($H \sim 10^7$ A/m) since $H_c(0)$ is much less than this.

TYPE I SUPERCONDUCTORS.   Sn and Pb are examples of Type I superconductors. A long cylinder of Type I material parallel to a magnetic field will

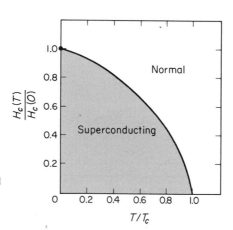

FIGURE 5.2.1. Reduced critical field versus reduced temperature.

completely exclude magnetic induction ($B = 0$) if the field is below H_c, in which case $M = -H$; above H_c it is completely penetrated. This behavior, known as the **Meissner effect**, is illustrated in Figure 5.2.2. The path is retraced as H decreases; the process is reversible. When magnetic flux is completely excluded, the material is said to be perfectly diamagnetic: This corresponds to a magnetic susceptibility of $\chi_m = -1$ and a magnetic permeability $\mu = 0$. When the same material is in the normal state, the flux readily penetrates (since $\mu \doteq \mu_0$). No

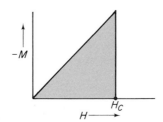

FIGURE 5.2.2. Magnetization curve of Type I superconductor.

material which becomes a superconductor is strongly magnetic ($\mu \gg \mu_0$) in the normal state; at least none has been found. F. B. Silsbee (1916) showed that the superconductivity in a long circular wire of radius a of a Type I superconductor is destroyed when the current i in the wire exceeds the value

$$H_c = \frac{2i}{a}. \tag{5.2.2}$$

(Note that the field generated by current in a wire is given by $H_c = 2i/r, r \gtrless a$.) This is the **Silsbee effect**. **Type I superconductors** show a Meissner effect and a Silsbee effect.

TYPE II SUPERCONDUCTORS. **Type II superconductors** show a Meissner effect only for small H and then deviations occur. See Figure 5.2.3. Below H_{c_1} the material is superconducting, above H_{c_2} it is normal, and in between it is

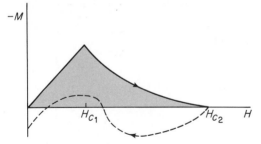

FIGURE 5.2.3. Magnetization of a Type II superconductor. Dashed line shows behavior of nonideal Type II as H decreases.

mixed. For such materials the quantity H_c is defined by

$$\frac{H_c^2}{2} = -\int_0^{H_{c_2}} M dH. \tag{5.2.3}$$

The origin of this expression is discussed in Chapter 29. Values of H_{c_2} as large as 3×10^7 A/m have been found.

The ideal Type II superconductors will retrace the *M-H* path when H is decreased. Most real Type II superconductors, however, are irreversible and follow a path such as that shown by the dashed line in Figure 5.2.3.

5.3   HARD SUPERCONDUCTORS

In 1961 J. E. Kunzler showed that Nb_3Sn had an $H_c(0)$ value in excess of 10^7 A/m. The critical current density versus field intensity for a Nb_3Sn wire at 4.2°K (after a specific heat treatment) is shown in Figure 5.3.1. Nb_3Sn

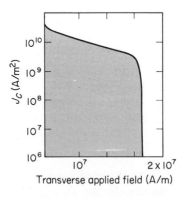

FIGURE 5.3.1. Critical current density versus transverse magnetic field for Nb_3Sn at 4.2°K.

has a critical temperature of 18.5°K. The highest critical temperature known is slightly above 20°K. (A high field superconductor with a transition temperature above room temperature is still a magnet-builder's dream.) Typical high critical temperature hard superconductors are either primarily Nb- or V-based compounds or alloys. They are Type II superconductors.

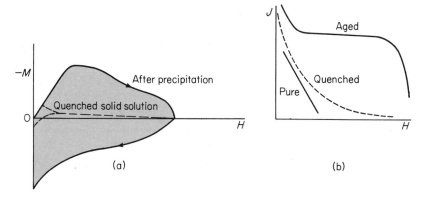

FIGURE 5.3.2. (a) Magnetization curves at 4.2°K of a Pb–8.3 at. % Na alloy as quenched from solution temperature and after aging 4 hr at room temperature to allow precipitation. (b) Sketch of corresponding critical current curves.

The J_c-H curves of superconductors (like the B-H curves of magnetic materials or the σ-ϵ curves of mechanical behavior) are highly dependent on impurity content, thermal history, and mechanical working. This is illustrated in Figure 5.3.2. Superconductors which can carry large currents at large fields (such as illustrated in Figure 5.3.1) are known as **hard superconductors**.

Remember that pure lead is a Type I superconductor. The addition of 8.3 atomic % Na followed by quenching gives a nearly ideal Type II superconductor while the age precipitation heat treatment drastically alters this behavior. Cold working would introduce similar nonideal Type II behavior, i.e., make

FIGURE 5.3.3. 15-Wb/m² superconducting magnet. Compare with normal coil magnet of Figure 3.2.3. (*Courtesy of RCA, Laboratories, Princeton, N.J.*)

the superconductor harder. Cold working makes the material mechanically harder (stronger) and the age precipitation heat treatment behavior also increases the mechanical strength.

EXAMPLE 5.3.1

How large is the allowable J in a $\#18$ copper wire exposed to room-temperature air?

Answer. Such wire has a diameter of 10^{-3} m and can carry a current up to 5 A (handbook value). Hence $J = i/A \sim 6 \times 10^6$ A/m². Compare with $J_c \sim 10^{10}$ A/m² values for superconducting Nb_3Sn at 4.2°K.

The high current-carrying capacity with no resistance heating makes possible the construction of high-field superconducting magnets which are midgets compared to normal conducting magnets. Compare the magnet of Figure 5.3.3 with that of Figure 3.2.3.

5.4   SYNTHETIC HARD SUPERCONDUCTORS

By forcing molten lead into porous glass (pore size, 100 Å and less) under high pressures (several kilobars; the deepest place in the ocean is at a pressure of about 1 kbar), Bean and his colleagues (1962) were able to produce high field, hard superconductors [C. P. Bean, M. V. Doyle, and A. G. Pincus, *Physical Review Letters*, **9**, 93 (1962)]. It should be noted that the starting materials are glass, which is an insulator, and lead, which is a soft superconductor. Materials with critical fields in excess of 10^7 A/m have been made, and, potentially, fields in excess of 10^8 A/m can probably be reached.

5.5   APPLICATIONS

Superconductors have a wide range of applications and potential applications. These include:

1. Electromagnets
2. Motors and generators
3. Transmission cables
4. Suspension systems (such as high-speed trains)
5. Cryotron (switch)
6. Persistent current cells for binary storage
7. Heat valves
8. Bolometers.

For further study of these devices see V. L. Newhouse, *Applied Superconductivity*, John Wiley and Sons, Inc., New York (1964). Superconducting magnets are discussed in W. B. Sampson, P. P. Craig and M. Strongin, "Advances in Superconducting Magnets," *Scientific American* (March 1967). The possible uses of the cryotron and persistent current cells in digital computers is discussed in J. A. Rajchman, "Integrated Computer Memories," *Scientific American* (July 1967) p. 18.

The heat valve is based on the fact that in a normal conductor most of the heat is carried by the electrons. In the superconducting state the electron pairs cannot interact with the lattice (which is why there is no scattering, hence no electrical resistivity). The thermal coupling between these electron pairs and the warm or cold regions outside of the superconductor has disappeared. Hence a metal such as tantalum has a thermal conductivity in the superconducting state of only 10^{-2} that of the normal state at about $1°K$.

REFERENCES

Handbook of Chemistry and Physics (R. C. Weast, ed.), The Chemical Rubber Co., Cleveland (1969). Has tables of properties.

Roberts, B. W., "Superconducting Materials," *Progress in Cryogenics*, **4**, 161 (1964). Has tables of properties.

Cohen, Morrel H., ed., *Superconductivity in Science and Technology*, University of Chicago Press, Chicago (1968).

Newhouse, V. L., *Applied Superconductivity*, Wiley, New York (1964). The student should at first use this book (or the one by Bremer, below) to become familiar with the many applications of superconductors.

Bremer, J. W., *Superconductive Devices*, McGraw-Hill, New York (1962).

Rajchman, J. A., "Integrated Computer Memories," *Scientific American* (July 1967) p. 18. Includes a discussion of superconducting memory cores.

Sampson, W. B., et al., "Advances in Superconducting Magnets," *Scientific American* (March 1967) p. 114.

Fishlock, David, Editor, *A Guide to Superconductivity*, American Elsevier, Inc., New York (1969). This book is concerned with the technology involving superconductors.

PROBLEMS

5.1.   a. What is the highest value of T_c currently available?

      b. Why would superconductors be more interesting if $T_c > 77°K$? if $T_c > 300°K$?

5.2. Why are bulk Type I superconductors not useful for high field magnets?

5.3. a. Describe the Meissner effect.
b. Describe the Silsbee effect.

5.4. Is the *M-H* curve of Type II hard superconductors structure-sensitive? Discuss.

5.5. What is meant by a synthetic hard superconductor?

5.6. A metal, such as Cu, is often used as the covering material on a super-conducting wire or ribbon which is wound into a magnet. Discuss.

More Involved Problems

5.7. Discuss the possible advantage of a composite coil magnet for reaching ultrahigh static magnetic fields (2.5×10^7A/m); make the inner core of ultrahigh-purity aluminum and the outer core of superconducting wire.

5.8. Discuss the possibility of making a motor from hard superconductors —all magnetic fields are generated, no permanent ones. See A. D. Appleton, "The Superconducting Motor," *Industrial Research* (Sept. 1969) p. 72.

5.9. Superconducting solenoid magnets provide a convenient source of large magnetic fields. However, their inductance is large and they are subject to instabilities that limit the voltage that can be imposed across them. These limitations dictate some special characteristics for their power supplies. A simple series *RL* circuitc onnected to a variable dc voltage source $E(t)$ gives a good representation of the problem in which the magnet provides the pure inductance L and the lead resistance is R.
a. Find the source $E(t)$ that will increase the current linearly with time at the rate of $dI/dt = \alpha$ until a current I_m is reached and then hold the current steady at I_m. Plot it. *Hint:* The differential equation is

$$L\frac{dI}{dt} + IR = E(t) \qquad \text{or} \qquad L\frac{d^2q}{dt^2} + R\frac{dq}{dt} = E(t).$$

b. If the maximum voltage that can be imposed across the magnet (L alone) is 2 V, $L = 10$ H, and $R = 0.01$ Ω, find the maximum value of α and the time required to reach $I = 20$ A.

Sophisticated Problems

5.10. Discuss the possibility of using a superconducting Nb_3Sn wire for transmission of dc current across a country. Since the Nb_3Sn will have to be cooled to near absolute zero to be superconducting, one important economic consideration is heat transfer (see Chapter 6).

5.11. Union Carbide has announced a development program for using very pure niobium wire in the superconduction state for transmission of ac current across a country. Describe the current status of this project and the economic feasibility relative to present transmission methods.

Prologue

The effects of temperature can be characterized as those owing to a change in temperature such as *thermal expansion*, those owing to a gradient of temperature such as *thermal conductivity*, and those in which the absolute temperature itself plays a major role in determining the existence or magnitude of a given property. Thermal expansion is shown to be an important parameter to the design engineer. *Thermal conductivity* and *thermal diffusivity* are important material properties in describing heat conduction in solids. Thermal conductivity is a structure-sensitive property of solids at low temperature. We have previously observed that ferroelectricity, magnetism, and superconductivity disappear at high temperatures. So does the solid state; the *melting temperature* of a solid is an important parameter. A number of material properties are described in terms of coefficients which depend exponentially on the negative reciprocal of temperature. Certain materials, when subjected to an appropriate temperature cycle (*heat treatment*), exhibit drastically altered properties. Certain *thermoelectric properties*, which form the basis for the thermoelectric refrigerator and the thermoelectric generator, are introduced. Temperature fields clearly have a profound effect on the behavior of materials.

6

PHENOMENOLOGY OF TEMPERATURE FIELDS

6.1   SPACE ROCKETS AND THERMAL EXPANSION COEFFICIENT

A space rocket with a pyroceram nose cone plummets into the atmosphere whereupon its surface temperature is quickly raised thousands of degrees. We say the material is subjected to **thermal shock**. To understand the significance of this to the design engineer we consider a more mundane problem, namely, rapidly placing a hot glass circular rod at temperature T_1 into water at temperature T_2. The outside surface readily approaches the temperature of the water (and attempts to shrink), while the rest of the glass is still hot (and hence does not shrink and tends to keep the outer layer from shrinking). The outer layer is therefore in tension and the maximum circumferential tensile stress (which occurs at the beginning of the ideal quench) is given by

$$\sigma = \frac{E\alpha}{(1 - v)}\Delta T, \qquad (6.1.1)$$

where E is Young's modulus, v is Poisson's ratio, α is the thermal expansion coefficient, and $\Delta T = T_1 - T_2$.

EXAMPLE 6.1.1

The thermal expansion coefficient for a glass is $5 \times 10^{-6}/°\mathrm{K}$. Assume $E = 10^7$ psi and $v = \frac{1}{4}$. What is the maximum tensile stress when a rod is quenched from 770°C into water at room temperature?

117

Answer.

$$\sigma = \frac{10^7}{0.75} \times 5 \times 10^{-6} \times 750$$

$$= 50{,}000 \text{ psi.}$$

Table 2.1.2 suggests a fracture stress of 10,000 psi for glass.

Since ΔT is fixed and E does not vary much with temperature, we must choose α small to minimize the stresses owing to thermal shock; in fact the ideal choice would be to choose a material for which $\alpha = 0$ over the temperature range of concern (which is why pyroceram was used for the nose cone).

The **linear thermal expansion coefficient** is defined as the fractional change,

$$\alpha = \frac{1}{l}\frac{dl}{dT} = \frac{d\epsilon_r}{dT}, \tag{6.1.2}$$

of length with temperature. Here ϵ_r is the real strain defined by Equation (2.5.7). For small temperature ranges α may be considered a constant and so there exists a *linear* strain response to the temperature change

$$\epsilon_r = \alpha\,\Delta T. \tag{6.1.3}$$

Some materials behave isotropically; i.e., the strain in all directions is the same when the temperature of the entire body is changed from T_2 to T_1, i.e., by ΔT. However, many materials show anisotropic behavior; i.e., their linear thermal expansion coefficient varies with direction.

The **volume thermal expansion coefficient** is

$$\beta = \frac{1}{V}\left(\frac{dV}{dT}\right). \tag{6.1.4}$$

For isotropic materials, $\beta = 3\alpha$.

There are some applications in which a value of $\alpha \approx 0$ is necessary, e.g., the thermal shock situation, tapes for surveying, and gyroscopes. There

Table 6.1.1. LINEAR THERMAL EXPANSION
COEFFICIENT AT 20°C*

Material	α, (°K^{-1})
Pyroceram	0.4×10^{-6}
Invar	0.9×10^{-6}
Porcelain	$(1\text{–}4) \times 10^{-6}$
Glass (crown)	9×10^{-6}
Steel	11×10^{-6}
Ice (at 0° C)	50×10^{-6}
Sodium	60×10^{-6}

* *Handbook of Chemistry and Physics*, (R.C. Weast, ed.), The Chemical Rubber Co., Cleveland (1969).

are other applications in which $\alpha = $ constant $\neq 0$ is desired, e.g., liquid in thermometers and metals for bimetallic strip temperature indicators. There are many applications in which a value of $\alpha = 0$ would be desirable but for which it is not possible. Examples are housing materials, bridge materials, and highway materials. Some values of the thermal expansion coefficient are given in Table 6.1.1. At **cryogenic temperatures** (near absolute zero) the thermal expansion coefficient is quite small; it is roughly proportional to T^3 and goes to zero as T goes to zero.

6.2 THERMAL CONDUCTIVITY

Just because the pyroceram nose cone of the previous section withstood the thermal shock does not mean the astronauts' problems are over. The surface of the nose cone is now at a very high temperature and this heat will be conducted into the interior. Assuming that the temperature on the outside is now fixed, the time in which the inside temperature changes by a small fraction f of the outside temperature change is given by

$$t_L \approx \frac{fL^2}{4D_T}, \tag{6.2.1}$$

where L is the thickness of the heat barrier and D_T is the **thermal diffusivity**, which is given by

$$D_T = \frac{k_T}{\rho C_p}, \tag{6.2.2}$$

Here k_T is the thermal conductivity, ρ is the density, and C_p is the specific heat. It should be noted that D_T is more nearly independent of temperature than are the other variables.

EXAMPLE 6.2.1

Assuming that $D_T = 8 \times 10^{-3}$ cm²/sec, that the initial temperature of the heat shield was 0°C, and that the outside temperature change was 1000°C (and remained 1000°C), calculate the time for the inside wall temperature to reach 40°C if $L = 2$ cm.

Answer.

$$f = \frac{40 - 0}{1000 - 0} = \frac{1}{25}.$$

Hence from (6.2.1) we have $t = 5$ sec. Actually the nose cone would be subject to a much higher external stagnation temperature than 1000°C.

We are concerned here with the thermal conductivity. It was proposed by Jean Fourier (1824) that the fundamental law regarding the conduction of heat in one dimension, say the x direction, is

$$J_x = -k_T \frac{\partial T}{\partial x}. \tag{6.2.3}$$

Here J_x is the quantity of heat passing a unit area whose normal is in the x direction and $\partial T/\partial x$ is the gradient of the temperature. Here k_T is called the **thermal conductivity coefficient**; it is often found that over a moderate temperature range k_T does not vary with temperature and can be considered a constant. The negative sign accounts for the fact that heat flows from hot to cold. J_x has units of heat (energy) per area-time. Hence k_T has units of

$$k_T = \frac{\text{energy}}{\text{area-time}} \frac{\text{length}}{\text{temperature}}$$
$$= \frac{\text{energy}}{\text{length-time-temperature}} \tag{6.2.4}$$

A common unit is joules per centimeter-second-degrees Kelvin or watts per centimeter-degrees Kelvin.

However, different length, temperature, and energy scales are often used. We note

$$4.186 \text{ J} = 1 \text{ cal},$$
$$252 \text{ cal} = 1 \text{ Btu}.$$

Table 6.2.1 shows some values of the coefficient k_T for solids. Equation (6.2.3) is known as the **Fourier heat conduction law**.

There is an interesting relationship for metals, called the **Wiedemann-Franz ratio**, which states that the ratio of thermal conductivity to electrical conductivity at a fixed temperature is the same for all metals, or

$$\frac{k_T}{\sigma} = k_T \rho = \text{constant} \times T(^\circ \text{K}), \tag{6.2.5}$$

where k_T values are given in Table 6.2.1 and ρ values in Table 3.1.1. This suggests a common origin for these two conductivities in metals.

The thermal conductivity of solids at low temperatures is a highly structure-sensitive property, as illustrated for copper of different purity in Figure 6.2.1. It should be noted that for all of the copper types k_T will drop rapidly as $T \to 0$. It should, however, be noted that k_T approaches the same value (about 4 W/cm-°K) near room temperature and is fairly constant thereafter.

Figure 6.2.2 shows the thermal conductivity versus temperature behavior of different SiO_2 compounds: quartz, quartz which has displaced atoms, and fused quartz (quartz glass). Although the composition is the same in all cases, the thermal conductivity varies by a factor of 10^4.

Because of its high thermal conductivity and its low viscosity (which aids heat transport by convection), hydrogen is used as a coolant fluid in elec-

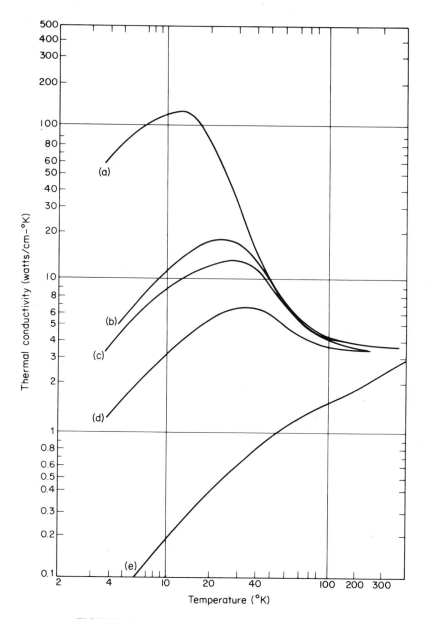

FIGURE 6.2.1. Low-temperature thermal conductivities of different copper. (a) High-purity copper. (b) Coalesced copper. (c) Copper, electrolytic tough pitch. (d) Free-machining tellurium copper. (e) Copper, phosphorous-deoxidized. [From R. B. Scott, *Cryogenic Engineering*, Van Nostrand, Princeton, N.J. (1959).]

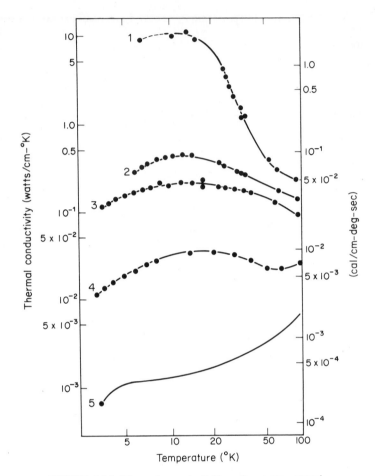

FIGURE 6.2.2. Thermal conductivity of quartz crystal, irradiated quartz crystal, and quartz glass. (1) Quartz crystal perpendicular to axis 5-mm² cross section. (2) Concentration of displaced atoms 1×10^{-4}. (3) 2×10^{-4}. (4) 2×10^{-3}. (5) Quartz glass. [From C. Kittel, *Introduction to Solid State Physics*, Wiley, New York (1956).]

trical generators. It is also a reducing agent so that it helps prevent the degradation by oxidation of insulating plastics and the like in the generator.

The most effective solid insulators involve porous solids (preferably without interconnecting pores) since these can be filled with gas, which has a low conductivity (except for He and H_2 gas) compared to solids. A dilemma which aquanauts face is the need for good thermal insulating suits since the ordinary insulators are crushed by the pressure.

Table 6.2.1. THERMAL CONDUCTIVITIES OF METALS AND NONMETALS
          AND GASES AT 20°C*

Element	Thermal Conductivity, cal/cm-sec-°K	Material	Thermal Conductivity, cal/cm-sec-°K
Silver	1.00	Quartz	0.012
Copper	0.94	Gas carbon	0.010
Gold	0.70	Marble	0.007
Aluminum	0.57	Ice (at 0°C)	0.005
Magnesium	0.40	Mica	0.002
Tungsten	0.39	Sulfur	0.0006
Zinc	0.27	Paraffin wax	0.0006
Cobalt	0.165	Ebonite	0.0004
Nickel	0.21	Sand	0.0001
Cadmium	0.20	Hydrogen (1 atm)	0.000406
Iron	0.17	Oxygen (1 atm)	0.000059
Tin	0.155	Nitrogen (1 atm)	0.000058
Lead	0.08	Argon (1 atm)	0.000039
Uranium	0.06	Ethylene (1 atm)	0.000041
Zirconium	0.04		
Titanium	0.036	Vacuum (perfect)	(zero)

* From *American Institute of Physics Handbook*, Dwight Gray, ed. McGraw-Hill, New York (1963) pp. 4–90, 9–38.

6.3   HEAT CAPACITY

Let us call ΔU the quantity of heat which we add to 1 mole of a system whose volume is kept constant. Assuming there are no phase changes, such as melting or vaporization, and no chemical reactions, the temperature increases and we have a change of **internal energy**

$$\Delta U = \int C_v \, dT, \qquad (6.3.1)$$

where C_v is called the **molar heat capacity at constant volume**. It is useful to define C_v because there are fairly wide temperature ranges over which it is often nearly a constant, and hence we can write

$$\Delta U \approx C_v \, \Delta T. \qquad (6.3.2)$$

It is more usual to work with a system to which heat is added while the pressure remains constant; we call the heat added at constant pressure ΔH, or **enthalpy**,

$$\Delta H = \int C_p \, dT, \qquad (6.3.3)$$

where C_p is called the **molar heat capacity at constant pressure**. It is the amount of heat which must be added to increase the temperature of 1 mole of the material by 1 deg at constant pressure.

IDEAL GASES.   For an ideal gas (i.e., one which obeys the law $PV = nRT$),

$$C_p = C_v + R. \qquad (6.3.4)$$

For an ideal monotonic gas such as argon,

$$C_v = \tfrac{3}{2}R. \qquad (6.3.5)$$

For diatomic and other polyatomic gases, C_v is larger than this. We shall not discuss this further here (the student can look in a good book on statistical thermodynamics).

SOLIDS.   For solids at high temperature, for each mole of atoms,

$$C_v \doteq 3R. \qquad (6.3.6)$$

This is the **"law" of Dulong and Petit**; by high temperature we mean a temperature greater than the Debye temperature Θ_D, which is a characteristic temperature for every solid (it is roughly $\Theta_D \approx 0.2T_M$, where T_M is the melting temperature). Figure 6.3.1 shows how C_v varies with temperature for several solids (note that it goes to zero at absolute zero). At very low temperatures C_v is proportional to T^3.

Thermodynamicists can show that

$$C_p - C_v = \frac{\beta^2 TV}{K}, \qquad (6.3.7)$$

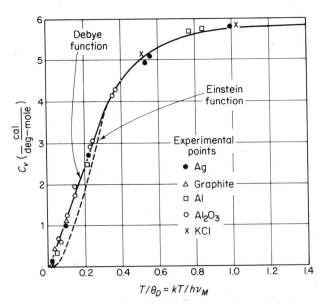

FIGURE 6.3.1. Molar heat capacity of solids. [After F. Seitz, *The Modern Theory of Solids*, McGraw-Hill, New York (1940).]

a difference which is quite small for solids (a few percent). Here β is the volume coefficient of thermal expansion [Equation (6.1.4)], V is the molar volume, and K is the bulk modulus [Equation (2.3.15)]. Handbooks are likely to put specific heat in units of calories per gram-degrees Kelvin rather than calories per mole-degrees Kelvin.

The unit of heat, the calorie, was originally defined as the heat required to raise the temperature of 1 g of water 1°C. Since the specific heat of water varies slightly with temperature, the calorie was later defined to be the heat needed to raise the temperature of 1 g of water from 14.5 to 15.5°C. Presently the calorie is defined in terms of the joule: 4.186 J = 1 cal.

6.4 EQUATION OF CONTINUITY AND THE PARABOLIC HEAT EQUATION

In certain situations, heat flow takes place essentially in one direction. Suppose a plate of hot steel is rapidly lowered into a quenching medium. Then (except near the edges) the heat flow takes place normal to the two large areas of the plate. We consider the general analysis of such problems now.

FIGURE 6.4.1. One-dimensional heat flow through a rod of uniform cross-sectional area A. We assume there is a perfect insulator at the cylindrical surface of this rod.

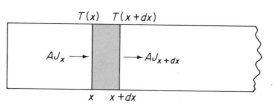

Consider the one-dimensional heat flow problem of Figure 6.4.1. We shall do a heat flow rate balance on the element located between x and $x + dx$:

heat flow rate in − heat flow rate out = rate of heat accumulation.   (6.4.1)

This can be written as

$$AJ_x - AJ_{x+dx} = A\left(C_p \rho \frac{\partial T}{\partial t}\right)_{x+c\,dx}, \qquad 0 < c < 1, \qquad (6.4.2)$$

where we assume the heat capacity is in units of energy per mass, ρ is the density of the material, and t is the time.

Since the Taylor series expansion for J_{x+dx} about x is

$$J_{x+dx} = J_x + \frac{\partial J_x}{\partial x} dx + \frac{\partial^2 J_x}{\partial x^2} \frac{dx^2}{2} + \cdots, \qquad (6.4.3)$$

we have in the limit as dx approaches zero

$$-\frac{\partial J_x}{\partial x} = C_p \rho \frac{\partial T}{\partial t}. \tag{6.4.4}$$

This is the **equation of continuity**.

If we use Fourier's heat conduction law assuming that k_T does not depend on T (and hence on x), we have

$$\frac{k_T}{C_p \rho} \frac{\partial^2 T}{\partial x^2} = \frac{\partial T}{\partial t}. \tag{6.4.5}$$

This is called the **parabolic heat equation**.

Note that the coefficient of the second derivative in Equation (6.4.5) is the thermal diffusivity D_T, defined by (6.2.2). At high temperature ($T \geq \Theta_D$), it is often found that D_T varies negligibly with temperature. Under these conditions Equation (6.4.5) becomes a *linear* partial differential equation. We consider heat flow in such a material now. Let us now suppose that we have a slab of thickness L whose temperature initially is $T = 0°C$ everywhere and whose end at $x = 0$ is *very* rapidly brought to T_0 at $t = 0$. We shall also assume that there is no heat flux through the wall at L. Using concepts from an advanced engineering mathematics course, we could readily show that the solution is given in terms of a Fourier series,

$$T = T_0 - T_0 \sum_{n=1,3,5,\ldots} \frac{4}{n\pi} e^{-(D_T n^2 \pi^2 t / 4L^2)} \sin\frac{n\pi x}{2L}. \tag{6.4.6}$$

Under certain conditions the leading term ($n = 1$) dominates this expression. Then the temperature at a given value of x has a time dependence governed by

$$e^{-D_T (\pi^2 t / 4L^2)};$$

the thermal response time is very similar to that given in Equation (6.2.1). In fact we define the **thermal response time** by

$$\tau = \frac{L^2}{4D_T}. \tag{6.4.7}$$

The problem just considered involves **time-dependent heat flow**.

Another important set of problems arises in the following way. Suppose the inside of a wall is at 25°C and the outside of the wall is at −20°C. Assume that this situation has existed for a very long time: It would then be found that the temperature in the wall does not change with time (so $\partial T/\partial t = 0$ at each x). This is called **time-independent heat flow**. Note that there is still heat flow. It is called **steady-state heat flow**.

EXAMPLE 6.4.1

What is the difference between the steady state and the equilibrium state?

Answer. In the **steady state** there is a net flow which does not vary with time. In the **equilibrium state** there is *no net* flow. In the later case the temperature everywhere in the system is a constant; i.e., there are no temperature gradients.

EXAMPLE 6.4.2

Find the steady-state temperature in a wall made of a material for which k_T does not vary with x for the case where $T = 25°C$ at $x = 0$ and $T = -20°C$ at $x = L$.

Answer. Equation (6.4.6) reduces to

$$\frac{d^2T}{dx^2} = 0,$$

which has solutions of the form

$$T = ax + b.$$

Applying the given boundary conditions we have

$$T = 25 - 45\frac{x}{L}.$$

How would you proceed to calculate the heat flux J_x through the wall? Note that this involves k_T only and not C_p and p. Why?

6.5 ENERGY CHANGES DURING PHASE TRANSFORMATION

When materials undergo simple phase transformations such as melting, vaporization, or sublimation the process is associated with a finite quantity of heat being absorbed over an infinitesimal temperature range at constant pressure. We define:

ΔH_m = heat absorbed at constant pressure per mole when solid melts; it is also called the **heat of melting.**

ΔH_v = heat absorbed at constant pressure per mole when liquid vaporizes; it is also called the **heat of vaporization.**

ΔH_s = heat absorbed at constant pressure per mole when solid vaporizes; it is also called the **heat of sublimation.**

We saw in Example 6.2.1 that the inside wall of the astronauts' nose cone heated up rapidly in a few seconds. One way to prevent this would be to coat the nose cone with an ablative compound which would absorb considerable heat in *decomposing* and *vaporizing*; this heat would simply be carried away with the

Table 6.5.1. HEATS OF MELTING AND VAPORIZATION

Material	ΔH_m, kcal/mole	ΔH_v, kcal/mole
H_2	0.028	0.22
H_2O	1.43	11.3
Na	0.63	24.6
Al	2.55	67.6
Fe	3.56	96.5
NaCl	7.22	183
KCl	6.41	165

air rushing past. We shall consider the energies associated with the decomposition of molecules in Chapter 9. In general, ΔH_m is much less than ΔH_v and ΔH_s. A few typical examples are shown in Table 6.5.1 in units of kilocalories per mole.

6.6   TEMPERATURE EFFECTS ON PROPERTIES

So far we have discussed the effects of change in temperature fields and of gradients in temperature fields. We now want to discuss some of the direct effects of the temperature field itself, i.e., T (and not just ΔT or grad T).

EFFECTS DESTROYED BY TEMPERATURE.   We have already noted that temperature tends to destroy:

1. Magnetism.
2. Superconductivity.
3. Ferroelectricity.
4. Shear rigidity (solids melt or vaporize).
5. Togetherness (liquids which occupy a fixed volume vaporize to gases which occupy as large a volume as they are given).

Temperature has a profound effect on the mechanical yield strength, on the rate of creep [see Equation (2.6.3)], on the fluidity of liquids [see Equation (2.8.6)], on the electrical conductivity (or resistivity) of metals (see Figure 3.2.4) and semiconductors [see Equation (3.3.1)], and on the thermal conductivity of solids (see Figure 6.2.1).

As temperature increases, the rate of degradation of rubber increases, the rate of oxidation of metals increases, and the rate of chemical reactions in general increases.

Because temperature so strongly affects most of the properties of materials, be they gases, liquids, or solids, it is very important to always specify the temperature field. Along with the temperature field, other fields such as magnetic, electrical, stress, and gravitational fields may have more or less importance. The detailed study of temperature and its significance is called thermodynamics and statistical mechanics.

One of the important general properties of a material is its melting temperature. As we shall see in Chapter 9, solids in which the atoms or molecules are weakly bonded to each other have low melting points, while the strongly bonded solids have high melting points (and high boiling points). Examples of melting temperatures are shown in Table 6.6.1. Materials which have high melting points ($T_M > 1500°C$) are called **refractory materials**.

There is an approximate empirical relation called **Trouton's rule** which relates the heat of vaporization (which is a measure of the strength of bonding)

Table 6.6.1. MELTING TEMPERATURES

Material	°C	Material	°C
		Al	660
H_2	−259.14	NaCl	801
A	−189.2	Fe	1535
H_2O	0	α-Al_2O_3	2015
Na	97.8	Mo	2610
		MgO	2800
Polyethylene*	140	W	3380
Nylon 610	220	Graphite†	~3400
Nylon 66	260		
Pb	327.3	HfC	3890

* Infinite chain extrapolation.
† Sublimes.

to the boiling temperature:

$$T_B(K°) = \frac{\Delta H_v(\text{cal/mole})}{22}. \qquad (6.6.1)$$

EXPONENTIAL TEMPERATURE EFFECTS. A number of material properties are macroscopically defined in terms of material coefficients which depend exponentially on the negative reciprocal of absolute temperature:

$$\text{material coefficient} \propto e^{-Q/RT}. \qquad (6.6.2)$$

Examples are creep rate [Equation (2.6.3)], fluidity [Equation (2.8.6)], conductivity of semiconductors [Equation (3.3.1)], diffusion coefficients, chemical reaction rates, rate of electron emission from a filament (hot wire), and vapor pressure above a liquid (where Q means the heat of vaporization) or solid (where Q means the heat of sublimation). We particularly emphasize the exponential temperature dependence because the properties vary so strongly with temperature in this case. For example, the self-diffusion coefficient for copper is 10^{-8} cm^2/sec just below T_M; 10^{-16}cm^2/sec at $T_M/2$ and 10^{-32} cm^2/sec at $T_M/4$. The diffusion of atoms in solids is important in reactions which take place in solids. Diffusion, as we shall see later, controls the rate of high-temperature creep. Solid-state transistors are made by diffusing atoms into silicon. Diffusion will be studied in Chapter 18.

6.7   HEAT TREATMENT

The metallurgist, the ceramicist, the glassmaker, and the polymer chemist often subject a material to a certain temperature cycle to achieve desired properties. In the simplest case a solid is melted and the lower-viscosity fluid can then be cast into a desired shape. We have already mentioned annealing of cold-worked materials in which the result of the annealing process is to decrease the yield strength. There are two other very important heat treatment processes which we shall discuss briefly here.

HEAT TREATMENT OF STEEL.   Steel is an alloy of iron and carbon (up to 2% by weight) and possibly other elements. When certain of these alloys are held near 850°C for, say, 1 hr (this is called **austenitizing**) and then **quenched** rapidly in water (or oil) at room temperature, there is a large increase in strength and hardness. The tensile yield strength may increase from 80,000 to 240,000 psi. Whereas the material could be readily machined before the heat treatment, it can be machined only by grinding after the heat treatment. Because it is too hard and brittle after this process, it is often **tempered** (held at an intermediate temperature for a while) to increase its toughness. In later chapters we shall study from the atomic viewpoint how and why this three-step heat treatment works.

AGE PRECIPITATION HARDENING.   Certain alloys such as aluminum with 4% Cu and perhaps some other alloying elements can have their yield strength increased in the following way. They are first held at 550°C for 1 hr (**solution heat treatment**) and then quenched in water to room temperature. Unlike the steels mentioned earlier, they are now very soft. If the temperature is increased to say 100°C and held there for several hours (**aging**) and then cooled to room temperature, the alloy will be hard and have a high yield strength. The reason this three-step heat treatment works will also be studied later from the atomic viewpoint.

EXAMPLE 6.7.1

You want to waste as little energy flattening rivet heads on an airplane construction job as possible. Suggest a way of using an appropriate age-precipitation-hardenable alloy for this job.

Answer. Choose an alloy which ages at room temperature. Solution-heat-treat the rivets, quench the rivets, and store them cold until used. After installation they will age at room temperature. *Note:* Certain aircraft have over 100,000 rivets.

OTHER PROPERTIES.   Heat treatments not only have profound effects on certain mechanical properties but also on thermal, magnetic, electric, super-conducting, optical, and other properties. We have already seen examples of this and we shall see more examples and also the reasons these effects occur in later chapters.

6.8 ELECTRICAL-THERMAL COUPLING

We have seen previously how a mechanical stress creates a voltage across certain materials (piezoelectric effect). A temperature change causes a strain (thermal expansion), and if the material is constrained, it causes a stress. A tourmaline crystal (an electrical insulator) whose temperature is changed by 1°C becomes polarized; this is the **pyroelectric effect** (a voltage of 740 V across a 1-cm-thick crystal would produce the same polarization).

Temperature and electrical behavior are also coupled in interesting and useful ways in conductors.

THERMOCOUPLES.   Figure 6.8.1 shows two dissimilar electrical conductors with junctions at different temperatures T_1 and T_2. There is a potential

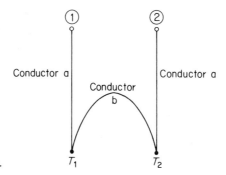

FIGURE 6.8.1. Thermocouple.

Table 6.8.1. THERMAL EMF'S OF METALS RELATIVE TO PLATINUM
(MILLIVOLTS; COLD JUNCTION AT 0°C)*

Temp. of Hot Junction, °C	Copper	Iron	Constantan	90% Pt, 10% Rh	Chromel	Alumel
100	0.76	1.98	−3.51	0.643	2.81	−1.29
200	1.83	3.69	−7.45	1.436	5.96	−2.17
300	3.15	5.03	−11.71	2.316	9.32	−2.89
400	4.68	6.08	−16.19	3.251	12.75	−3.64
500	6.41	7.00	−20.79	4.221	16.21	−4.43
600	8.34	8.02	−25.47	5.224	19.62	−5.28
700	10.49	9.34	−30.18	6.260	22.96	−6.18
800	12.84	11.09	−34.86	7.329	26.23	−7.08
900	15.41	13.10	−39.45	8.432	29.41	−7.95
1000	18.20	14.64	−43.92	9.570	32.52	−8.79
1100	—	—	—	10.741	35.56	−9.58
1200	—	—	—	11.935	38.51	−10.34
1300	—	—	—	13.138	41.35	−11.06
1400	—	—	—	14.337	44.04	−11.77

* From *American Institute of Physics Handbook*, McGraw-Hill, New York (1957).

difference between point 1 and point 2. This potential difference depends on the conductors a and b and on the temperatures T_1 and T_2. Assuming T_1 is fixed at the ice point, the potential will be directly related to T_2. With suitable calibration this device can be used to measure temperature potentiometrically. It is known as a **thermocouple**. The fact that thermal emf's (voltages) are generated is known as the **Seebeck effect**. Typical thermocouple materials are shown in Table 6.8.1. The thermal emf for copper-constantan with the hot junction at 800°C and cold junction at 0°C would be $12.84 - (-34.86) = 47.70$ mV.

EXAMPLE 6.8.1

Calculate the average thermal emf per degree between 700 and 800°C for copper-constantan.

Answer. At 700 the thermal emf is $10.49 - (-30.18) = 40.67$ mV and hence

$$\frac{47.70 - 40.67}{100} = \frac{0.0703 \text{ mV}}{\text{deg}}.$$

The emf per degree is called the **thermoelectric power**.

HEAT PUMP.   Consider now the thermocouple of Figure 6.8.2 in which we pass a current i. In this case heat is generated at one of the junctions and absorbed at the other according to

$$\dot{Q}_1 = -\pi_{ab}J \tag{6.8.1}$$

and

$$\dot{Q}_2 = +\pi_{ba}J. \tag{6.8.2}$$

Here $\dot{Q}$ is the rate of heat production at a junction and J is the electrical current density. π_{ab} and π_{ba} are known as the **Peltier coefficients**. The effect is known as the **Peltier effect**. The device is a thermoelectric heat pump, or a thermoelectric refrigerator, since one of the junctions will be cooled and the other will be heated. The efficiency of such a device depends on i^2R losses in the circuit (and hence on electrical conductivities) and on losses owing to heat flow down the temperature gradients (and hence on the thermal conductivities) and on radiation.

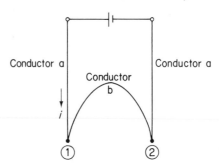

FIGURE 6.8.2. Thermoelectric
heat pump.

A junction of p-type versus n-type Bi_2Te_3 (semiconductor) has a thermoelectric power of about 0.25 mV/°C.

Using an alloy $(Bi_2Se_3-Bi_2Te_3)$, which is a semiconductor, with platinum as the other conductor, a maximum cooling to $-37°C$ can be obtained from $0°C$. Operated in the reverse fashion (one junction kept cold, the other heated), the thermocouple becomes a thermoelectric generator. Thermoelectric devices have been described by S. I. Freedman [see G. W. Sutton, ed., *Direct Energy Conversion*, McGraw-Hill, New York (1966) p. 105].

REFERENCES

Handbook of Chemistry and Physics (C. R. Weast, ed.), The Chemical Rubber Co., Cleveland (1969). Contains extensive data on melting points, specific heats, heats of vaporization, thermal conductivities, etc.

Scott, R. B., *Cryogenic Engineering*, Reinhold Van Nostrand, New York (1959). Contains considerable data on thermal properties, particularly in the low-temperature range where these properties may vary rapidly with temperature. Also contains a convenient table for conversion of units.

Gebhart, B. I., *Heat Transfer*, McGraw-Hill, New York (1961).

Carslaw, H. S., and Jaeger, J. C., *Conduction of Heat in Solids*, Oxford, New York (1959). A large number of problems involving heat conduction through differently shaped objects and methods for solving them are discussed.

Dye, R. P., ed., *Thermal Conductivity*, Academic Press, New York (1969). This two-volume work is a critical review of the methods available for making measurements of thermal conductivity.

PROBLEMS

6.1. a. Explain why many glasses fracture when taken from scalding water and placed in cold water.
b. How would you solve this problem?

6.2. In addition to the examples in the text, give three cases where a thermal expansion coefficient equal to zero is desired.

6.3. Design two devices which depend on a difference in thermal expansion coefficient.

6.4. Design a device which depends on a difference in coefficient of linear compressibility.

6.5. Ice cubes at $-10°C$ fracture when dropped into toluene at $-35°C$. Estimate the maximum tensile strength of the ice. Assume $v = \frac{1}{3}$ and $E = 3 \times 10^5$ psi.

6.6. Why is the thermal diffusivity the important parameter in time-dependent heat flow?

6.7. Why do ρ and C_p not enter into steady-state heat flow problems?

6.8. What is the Wiedemann-Franz ratio and what does it suggest?

6.9. To what do you attribute the different $k_T(T)$ curves for copper in Figure 6.2.1?

6.10. Is there any difference in k_T of crystalline quartz (SiO_2) and noncrystalline silica (SiO_2)?

6.11. Clothing, the thermal insulation in the walls of houses, etc., have a k_T approximately equal to what?

6.12. What is the value of the molar heat capacity of crystals at high temperatures, say, $T_m/2$?

6.13. Derive the equation of continuity.

6.14. Name five important properties which are destroyed by temperature.

6.15. Name five properties which depend exponentially on temperature (actually on the reciprocal of temperature).

6.16. a. Illustrate the Seebeck and Peltier effects.
b. What possible application does each have?

6.17. What three factors determine the efficiency of a Peltier refrigeration system?

More Involved Problems

6.18. In Example 3.2.1 we considered the design of a 10-Wb/m² solenoid magnet. This magnet generated considerable heat. Assuming that water is used to cool it and that the water temperature is changed by 20°C, how many gallons are needed per hour?

6.19. Show for materials in which the linear thermal expansion coefficient is isotropic that $\beta = 3\alpha$.

6.20. Describe three temperature measuring devices where the temperature is measured electrically.

Sophisticated Problem

6.21. Derive Equation (6.4.6).

Prologue

Material behavior is due to the presence of such fields as temperature, pressure, force, electric, magnetic, thermal gradient, chemical gradient, and stress. Mathematically, such fields are described in various cases by *scalars*, *vectors*, and second-rank *tensors*. The response of materials to such fields is also a scalar, a vector, or a tensor. There is often a *linear relation* between the cause and the effect. It should be noted that the property itself may be described by a scalar, a vector, or a tensor. Thus density is a scalar, the pyroelectric effect is a vector, and the thermal conductivity is a second-rank tensor in general. *Stress*, which is usually a symmetric second-rank tensor, is studied in detail.

7

LINEAR BEHAVIOR

7.1   LINEAR RELATIONS

We have seen a number of examples of linear conservative behavior and linear dissipative behavior. A **dissipative process** is one in which the work done goes to produce heat; a **conservative process** is one in which the work done is stored as potential energy. Examples of components which respond with a general displacement to some generalized force are shown in Table 7.1.1; these are examples of conservative behavior. We have

generalized force = (component constant) (generalized displacement).   (7.1.1)

Table 7.1.1. COMPONENTS SHOWING LINEAR BEHAVIOR

Component	Generalized Force	Generalized Displacement	General Energy	Linear Behavior Energy	Energy of Linear System in Terms of Generalized Displacement
Cantilever beam with loaded end	F	δ	$\int F\,d\delta$	$\dfrac{F\delta}{2}$	$\dfrac{c\delta^2}{2}$
Coil spring	F	x	$\int F\,dx$	$\dfrac{Fx}{2}$	$\dfrac{kx^2}{2}$
Torsion rod	T	θ	$\int T\,d\theta$	$\dfrac{T\theta}{2}$	$\dfrac{K\theta^2}{2}$
Condenser	V	Q	$\int V\,dQ$	$\dfrac{VQ}{2}$	$\dfrac{Q^2}{2C}$
Inductor	V	i	$\int Vi\,dt$		$\dfrac{Li^2}{2}$

137

The component constants c, k, K, $1/C$, or L are determined by some material property along with the dimensions of the component. Table 7.1.2 gives linear *material property* relations for conservative behavior. A relation similar to (7.1.1) applies but with a material coefficient now replacing the component constant.

Table 7.1.2. LINEAR MATERIAL PROPERTIES FOR CONSERVATIVE BEHAVIOR

Property	Generalized Force	Material Constant	Generalized Displacement	General Energy Per Unit Volume	Energy for Linear Behavior	Energy for Linear Behavior in Terms of Displacement
Elastic (tensile)	σ	E	ϵ	$\int \sigma \, d\epsilon$	$\dfrac{\sigma\epsilon}{2}$	$\dfrac{E\epsilon^2}{2}$
Elastic (shear)	τ	G	γ	$\int \tau \, d\gamma$	$\dfrac{\tau\gamma}{2}$	$\dfrac{G\gamma^2}{2}$
Dielectric	ξ	$\dfrac{1}{\epsilon}$	D	$\int \xi \, dD$	$\dfrac{D\xi}{2}$	$\dfrac{D^2}{2\epsilon}$
Magnetic—for diamagnetic, paramagnetic, and soft magnetic materials	H	$\dfrac{1}{\mu}$	B	$\int H \, dB$	$\dfrac{HB}{2}$	$\dfrac{B^2}{2\mu}$

There are also a number of dissipative processes which we have discussed. Several of these take the form of a flux equal to a material coefficient times a negative gradient. Examples are shown in Table 7.1.3.

We note that $\xi = -\mathrm{grad}\ V$ so that Ohm's law, which was discussed

Table 7.1.3. DISSIPATIVE FLUX PROCESSES

Flux of	Flux	Material Coefficient	Negative Gradient	Name of Law
Charge	$\mathbf{J}_{\text{elec}}$	σ (conductivity coefficient)	$-\mathrm{grad}\ V$	Ohm
Heat	$\mathbf{J}_{\text{therm}}$	k_T (thermal conductivity coefficient)	$-\mathrm{grad}\ T$	Fourier
Atoms or matter	$\mathbf{J}_{\text{matter}}$	D (diffusion coefficient)	$-\mathrm{grad}\ C$ (C is concentration)	Fick
Fluid in porous media	$\mathbf{J}_{\text{fluid}}$	k_P (permeability coefficient)	$-\mathrm{grad}\ P$ (P is pressure)	Darcy

earlier as $\mathbf{J} = \sigma\boldsymbol{\xi}$, does indeed take this form. The gradient of V (grad V) is a vector quantity which in a Cartesian coordinate system takes the form

$$\left(\frac{\partial V}{\partial x_1}, \frac{\partial V}{\partial x_2}, \frac{\partial V}{\partial x_3}\right).$$

The flux law governing diffusion will be discussed in detail later. Diffusion is one of the most important processes in materials science. It is involved in the production of microelectronic circuits, in radiation damage in nuclear reactors, in the heat treatment of steel and aluminum alloys, in creep and in many other phenomena. Darcy's equation is used in discussing flow through porous media, such as through soils, by civil and agricultural engineers and through percolating towers, beds, etc., by chemical engineers. In the discussions which follow we shall discuss mechanical stress in more mathematical detail. We shall find that it is a second-rank tensor. In addition we shall look in more detail at the linear behavior of materials which exhibit anisotropic behavior and we shall find that this behavior is also describable by tensors.

7.2   MORE DETAILED DISCUSSION OF STRESS

Previously we have used two symbols to represent components of stress, σ for a normal stress component and τ for a shear stress component. It is more useful to use a single symbol with two subscripts, σ_{ij}. Consider a right-hand Cartesian coordinate system as shown in Figure 7.2.1. Let σ_{ij} be a stress com-

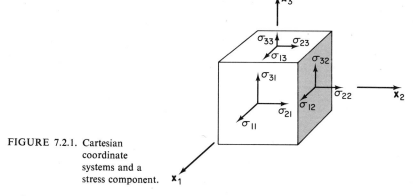

FIGURE 7.2.1. Cartesian coordinate systems and a stress component.

ponent owing to a force in the x_i direction acting on a plane whose normal is in the x_j direction. Note the definition of the normal stress component in the discussion associated with Figure 2.3.2 and of the shear stress components in the discussion associated with Figure 2.3.4. We note that σ_{ij} is a normal stress component if $i = j$ and that σ_{ij} is a shear stress component if $i \neq j$.

There are nine stress components in all:

$$\begin{matrix} \sigma_{11} & \sigma_{12} & \sigma_{13} \\ \sigma_{21} & \sigma_{22} & \sigma_{23} \\ \sigma_{31} & \sigma_{32} & \sigma_{33}. \end{matrix} \qquad (7.2.1)$$

These are illustrated in Figure 7.2.2.

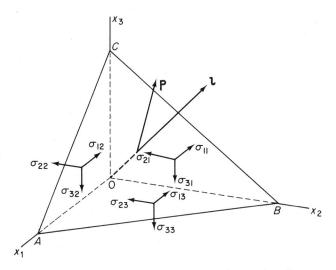

FIGURE 7.2.2. Nine stress components acting on three faces
of a tetrahedron are balanced at equilibrium by
the force p acting on the face ABC.

p is some force acting on area ABC and **l** is the unit vector normal to
the plane ABC. *Note:* l_1, l_2, and l_3 are the direction cosines of the unit length
vector; hence l_1 is the cosine of the angle between **l** and x_1, l_2 is the cosine of the
angle between **l** and x_2, etc. We assume the face ABC has unit area. Then, since
l is the unit normal vector to the area ABC, we have

$$\text{area } BOC = l_1$$
$$\text{area } AOC = l_2$$
$$\text{area } AOB = l_3.$$

Consider now a balance of forces in the x_1 direction. We have

$$p_1 = \sigma_{11}l_1 + \sigma_{12}l_2 + \sigma_{13}l_3. \qquad (7.2.2)$$

Note: Of the nine stress components in (7.2.1) only the three with the first coefficient equal to 1 arise as the result of a force in the x_1 direction. In general we
can write

$$p_i = \sum_{j=1}^{3} \sigma_{ij}l_j, \qquad i = 1, 2, \text{ or } 3. \qquad (7.2.3)$$

The reader should write out the expression for p_2 from (7.2.3) and should derive it by a balance of forces from Figure 7.2.2.

    p and **l** are both Cartesian vectors, and a quantity which relates a Cartesian vector of one kind (e.g., force) to a Cartesian vector of another kind (e.g., direction) in a linear fashion as in (7.2.3) is a **second-rank Cartesian tensor**. (The vectors themselves are also called **first-rank Cartesian tensors**.) The **stress tensor** is the array σ_{ij} represented by (7.2.1).

    There is a shorthand notation used to describe Equation (7.2.3). In the **Einstein convention** the summation sign is omitted but it is assumed that a repeated index means "sum." Instead of (7.2.3), we write

$$p_i = \sigma_{ij}l_j, \qquad j = 1, 2, 3. \tag{7.2.4}$$

Since we already know that j takes on values 1, 2, and 3, we usually do not write this down either.

EXAMPLE 7.2.1

    Find the stress normal to the plane of *ABC* in Figure 7.2.2.

    Answer. All we need to do is to find the component of **p** along **l** since *ABC* has unit area. This is given by the projection of **p** on **l**. Hence

$$\sigma = \text{projection of } \mathbf{p} \text{ on } \mathbf{l},$$
$$= \mathbf{p} \cdot \mathbf{l}$$
$$= p_i l_i = p_1 l_1 + p_2 l_2 + p_3 l_3.$$

Using Equation (7.2.3) we have

$$\sigma(\text{along } \mathbf{l}) = \sigma_{ij} l_i l_j. \tag{7.2.5}$$

The right-hand side of this equation is the sum of *nine* terms.

7.3   TRANSFORMATION OF AXES

    Figure 7.3.1 shows two sets of Cartesian axes. We assume that all of the angles (and hence their cosines) between all of the lines are known. We define

$$a_{ij} = \text{cosine of the angle between } x_i' \text{ and } x_j. \tag{7.3.1}$$

The vector from the origin to point P, $\overline{OP}$, can be represented by the set $[x_1, x_2, x_3]$ referred to the unprimed axes or to the set $[x_1', x_2', x_3']$ referred to the primed axes. The components of these two sets are related to each other in the simple fashion

$$x_i' = a_{ij}x_j. \tag{7.3.2}$$

[The proof of (7.3.2) will not be given here. The two-dimensional case involving a simple rotation of axes is treated in most elementary calculus books. Students

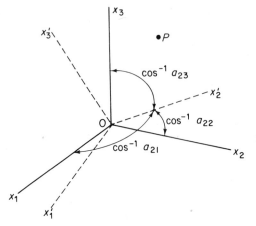

FIGURE 7.3.1. Two sets of orthogonal axes. Unprimed set shown by solid lines. Primed set shown by dashed lines.

who have studied orthogonal matrix transformations will know all about it and students who have not should have the motivation to do so.] It is also true that

$$x_i = a_{ji}x'_j. \tag{7.3.3}$$

Consider now that the vector l of Figure 7.2.2 is one of the primed axes, such as x'_1. Then in this new notation

$$p'_1 = a_{1j}p_j,$$

and more generally

$$p'_i = a_{ij}p_j. \tag{7.3.4}$$

Thus the components of the vector $\mathbf{p}$ in the new coordinate system (primed) are related to the components of the vector $\mathbf{p}$ in the old (unprimed) system. The vector $\mathbf{p}$ itself is unchanged; only the way in which we represent it mathematically has changed.

In this new notation Equation (7.2.5) becomes

$$\sigma'_{11} = a_{1k}a_{1l}\sigma_{kl}. \tag{7.3.5}$$

Since the axis x'_1 is now along l, σ along $\mathbf{l}$ would now be called σ'_{11}. Also, l_1, which is the cosine of the angle between $\mathbf{l}$ and x_1, would now be called a_{11} since a_{11} is the cosine of the angle between x'_1 and x_1. Similarly, $l_2 = a_{12}$ and $l_3 = a_{13}$.

More generally it is true that stresses in the primed coordinate system are given by

$$\sigma'_{ij} = a_{ik}a_{jl}\sigma_{kl}. \tag{7.3.6}$$

EXAMPLE 7.3.1

Evaluate the a_{ij} if the new axes (primed axes) are obtained from the old by rotation about x_1 (taking x_3 toward x_2) by 30 deg.

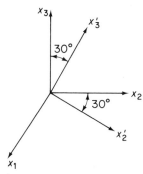

FIGURE E7.3.1

Answer. Using the definition of a_{ij} in (7.3.1) we have $a_{11} = \cos 0 = 1$, $a_{12} = \cos 90° = 0$, $a_{13} = \cos 90° = 0$; $a_{21} = \cos 90° = 0$, $a_{22} = \cos 30° = \sqrt{3}/2$, $a_{23} = \cos 120° = -\frac{1}{2}$; $a_{31} = \cos 90° = 0$, $a_{32} = \cos 60° = \frac{1}{2}$, $a_{33} = \cos 30° = \sqrt{3}/2$.

The array of coefficients

$$(a_{ij}) = \begin{pmatrix} a_{11} & a_{12} & a_{13} \\ a_{21} & a_{22} & a_{23} \\ a_{31} & a_{32} & a_{33} \end{pmatrix} = \begin{pmatrix} 1 & 0 & 0 \\ 0 & \sqrt{3}/2 & -\frac{1}{2} \\ 0 & \frac{1}{2} & \sqrt{3}/2 \end{pmatrix} \tag{7.3.7}$$

is called the **rotation matrix**.

EXAMPLE 7.3.2

A rod has a tensile stress σ along its axis which is considered to be the x_3 direction. The x_1 and x_2 directions are perpendicular to x_3. The stress tensor referred to this set of axes is therefore the array

$$\begin{bmatrix} 0 & 0 & 0 \\ 0 & 0 & 0 \\ 0 & 0 & \sigma \end{bmatrix};$$

i.e., $\sigma_{33} = \sigma$ and all the other stress components are zero. Find the stress components acting on the plane whose normal is x'_3 of Example 7.3.1.

Answer. From (7.3.6),

$$\sigma'_{33} = a_{33}a_{33}\sigma_{33} \text{ plus eight terms which are zero} = \tfrac{3}{4}\sigma$$
$$\sigma'_{13} = 0$$
$$\sigma'_{23} = a_{23}a_{33}\sigma_{33} = -\frac{\sqrt{3}}{4}\sigma.$$

In most of the applications which the student will see,

$$\sigma_{ij} = \sigma_{ji}; \tag{7.3.8}$$

i.e., the stress tensor is **symmetric**. [A couple would have to act at the same point as the force which produces the stresses acts for (7.3.8) not to be true.] It can be shown that given a symmetric stress, a set of axes x_1^p, x_2^p, x_3^p can be chosen so that $\sigma_{ij}^p = 0$ for $i \neq j$ and only the terms σ_{11}^p, σ_{22}^p, and σ_{33}^p are nonzero. Physically this means that with respect to these axes there are only normal stresses and all the shear stresses are zero. These axes are called the **principal axes** and the stresses are called the **principal stresses**. This is a general property not only of the stress tensor but of any symmetric second-rank Cartesian tensor. The student familiar with matrix theory knows that a real symmetric matrix can be diagonalized. The eigenvalues of such a matrix would in this case be the principal stresses and the eigenvectors would be the principal axes of stress. Thus, if we possess a knowledge of matrix algebra, the concept of stress is much easier to understand. Suppose that $\sigma_{11}^p > \sigma_{22}^p > \sigma_{33}^p$ algebraically. Then the maximum shear stress in the sample is

$$\tau_{max} = \frac{\sigma_{11}^p - \sigma_{33}^p}{2} \tag{7.3.9}$$

and acts on a plane which bisects the planes on which these two principal stresses are acting.

7.4   OTHER SECOND-RANK TENSORS

The material properties coefficients given in Tables 7.1.2 and 7.1.3 were for isotropic materials. More generalized relations are needed in dealing with anisotropic materials. Examples are

$$D_i = \epsilon_{ij}\xi_j \tag{7.4.1}$$

$$B_i = \mu_{ij}H_j \tag{7.4.2}$$

$$J_i = \sigma_{ij}\xi_j = -\sigma_{ij}\frac{\partial V}{\partial x_j} \tag{7.4.3}$$

$$J_i = -k_{ij}^T\frac{\partial T}{\partial x_j}. \tag{7.4.4}$$

Each of these expressions relates a vector to another vector in the same linear fashion as in (7.2.4). The array of coefficients for each is therefore a second-rank Cartesian tensor. Here ϵ_{ij} represents the permittivity, μ_{ij} the permeability, σ_{ij} the electrical conductivity, and k_{ij}^T the thermal conductivity. It is also true, for physical reasons, that each is a symmetric tensor. Thus to describe properly the electrical conductivity of an anisotropic material, one must have knowledge of six coefficients. The above tensors are called second-rank **matter tensors** since they describe properties of matter. The stress tensor is called a second-rank **field tensor** since it describes a field, just as electrical field would be called a first-rank field tensor. Strain is also a second-rank field tensor; we shall not discuss it further here. See J. F. Nye, *Physical Properties of Crystals*, Oxford, New York (1960) Chapter VI.

Thermal expansion coefficient is also a second-rank matter tensor. In this case it relates a second-rank field tensor, strain, to a zero-rank field tensor, temperature change:

$$\epsilon_{ij} = \alpha_{ij} \Delta T. \tag{7.4.5}$$

We use the symbol ϵ_{ij} for a strain component and σ_{ij} for a stress component [although these were used for certain matter tensors in Equations (7.4.1) and (7.4.3)] because these symbols are in common use in the literature. A strain component for which $i = j$ is a normal strain and one for which $i \neq j$ is a shear strain. The tensor strain component for shear strain has a magnitude only half that of the engineering strain γ. Small strains are defined by

$$\epsilon_{ij} = \frac{1}{2}\left(\frac{\partial u_i}{\partial x_j} + \frac{\partial u_j}{\partial x_i}\right), \tag{7.4.6}$$

where the vector $\mathbf{u} = [u_1, u_2, u_3]$ represents the displacement of a particle from its initial position at $\mathbf{x} = [x_1, x_2, x_3]$. The piezoelectric coefficient d_{ijk} is a third-rank Cartesian tensor relating polarization (a vector or first-rank tensor) to stress (a second-rank tensor) according to

$$P_i = d_{ijk}\sigma_{jk}. \tag{7.4.7}$$

Elasticity is described in terms of a fourth-rank Cartesian tensor c_{ijkl}, where

$$\sigma_{ij} = c_{ijkl}\epsilon_{kl}. \tag{7.4.8}$$

Anisotropy is common among single crystals, although not all crystals are anisotropic toward every property. We shall discuss this further after we have studied crystals.

The energy per unit volume stored in an elastically strained body is

$$u_{\text{elast}} = \int \sigma_{ij}\, d\epsilon_{ij}. \tag{7.4.9}$$

The energy per unit volume stored in a dielectric is

$$u_{\text{diel}} = \int \xi_i\, dD_i. \tag{7.4.10}$$

The energy per unit volume in a magnetic material is

$$u_{\text{mag}} = \int H_i\, dB_i. \tag{7.4.11}$$

The Einstein summation convention is being used in all cases.

Because anisotropic materials are being used more and more by engineers, the study of the Cartesian tensors is more important than ever.

REFERENCES

Nye, J. F., *Physical Properties of Crystals*, Oxford, New York (1960). The linear properties of solids are discussed in mathematical detail in terms of linear algebra (vectors, matrices, and Cartesian tensors). Some knowledge of vectors and matrices is a prerequisite to reading this book. Tensor theory is developed as needed.

Zwikker, C., *Physical Properties of Solid Materials*, Pergamon Press, New York (1955).

Mason, W. P., *Crystal Physics of Interaction Processes*, Academic Press, New York (1966). See comments on Nye, above.

PROBLEMS

7.1. In building mechanical systems, an engineer often uses simple components.
   a. Give examples of several of these components.
   b. Several of these components behave in a linear fashion. Describe three which show linear behavior.

7.2. In building electrical systems, an engineer often uses simple components.
   a. Give examples of several of these.
   b. Several of these components behave in a linear fashion. Describe three which show linear behavior.

7.3. Several material properties are described by linear relations. Give examples of four of these which are associated with reversible storage of energy.

7.4. Several dissipative processes have fluxes proportional to negative gradients.
   a. Describe these processes.
   b. Tell where each is important to the engineer.

7.5. Consider the two-dimensional state of stress:

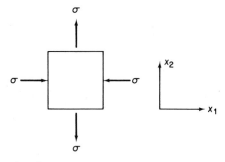

FIGURE P7.5

The two-dimensional stress tensor

$$\begin{bmatrix} \sigma_{11} & \sigma_{12} \\ \sigma_{21} & \sigma_{22} \end{bmatrix}$$

is then

$$\begin{bmatrix} -\sigma & 0 \\ 0 & \sigma \end{bmatrix}$$

Show that this is so.

7.6. Given:

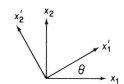

FIGURE P7.6a

a. Show that

$$x'_1 = (\cos\theta)x_1 + (\sin\theta)x_2$$
$$x'_2 = (-\sin\theta)x_1 + (\cos\theta)x_2.$$

b. Hence, show, using Equations (7.3.2) and (7.3.6), that the state of stress in Problem 7.5 can be represented in the primed system for which $\theta = 45°$ by

$$\begin{bmatrix} 0 & \sigma \\ \sigma & 0 \end{bmatrix}$$

and physically by

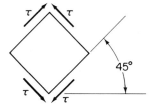

FIGURE P7.6b

where the magnitude of τ is the magnitude of σ.

7.7. a. Show using algebra and the given relationship in Problem 7.6a that

$$x_1 = (\cos\theta)x'_1 - (\sin\theta)x'_2$$
$$x_2 = (\sin\theta)x'_1 + (\cos\theta)x'_2.$$

b. Show this also using Equation (7.3.3).
c. Show it by simple physical reasoning, using Equation (7.3.2).

7.8. Consider tension as a two-dimensional stress

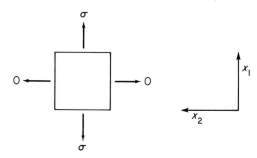

FIGURE P7.8

where the stress tensor is

$$\begin{bmatrix} \sigma & 0 \\ 0 & 0 \end{bmatrix}$$

a. Show that the maximum shear stress is at 45 deg to the tensile axis.
b. What is the magnitude of this shear stress?

7.9. Write down the stress tensor for a state of hydrostatic pressure of magnitude P.

More Involved Problems

7.10. It can be shown theoretically, as well as experimentally, that the permittivity tensor for a single crystal of sodium chloride has the form

$$\begin{bmatrix} \epsilon & 0 & 0 \\ 0 & \epsilon & 0 \\ 0 & 0 & \epsilon \end{bmatrix}$$

What does this mean physically?

7.11. Equation (7.2.4) could also be written as

$$p_r = \sigma_{rs} l_s.$$

a. Why?
b. Why is it also true that

$$p'_k = \sigma'_{kl} l'_l,$$

where the primes refer to some primed Cartesian coordinate system?

Sophisticated Problems

7.12. Derive Equation (7.3.6).

7.13. Using the fact that second-rank Cartesian tensors transform according to Equation (7.3.6) when we change from one orthogonal coordinate system to another, show that the elastic constant tensor transforms according to

$$C'_{ijkl} = a_{ip} a_{jq} a_{kr} a_{ls} C_{pqrs}.$$

Prologue

Quantum mechanics describes the behavior of electrons in atoms, in molecules, and in solids. A few special applications of quantum mechanics are discussed in this chapter. One of these, namely, the behavior of a *particle in a box*, is often a good approximation for the valence electrons in a metal. In another example the allowed *energy levels* in the *hydrogen atom* are described. This is followed by a study of the *hydrogen-like atom*. With this elementary background it is possible to estimate the energy of X-ray emission lines. Also, the origin of the atomic table can be understood in part from a theoretical viewpoint.

8

ELECTRONS AND ATOMS

8.1 WAVE MECHANICS

Classic Newtonian mechanics provides an adequate mathematical description of large slow-moving objects. However, when the relative velocity of interacting objects approaches the speed of light, classic mechanics does not hold and we must use relativistic mechanics. Moreover, if the interacting objects are tiny and close together, e.g., electrons, protons, and nuclei, we must use quantum mechanics to describe their motion. The quantum mechanical description takes cognizance of the wave-particle duality of matter. Thus the behavior of electrons may sometimes be described as particles with a mass, momentum, etc., and sometimes as waves with a wavelength, frequency, etc. The recognition of this by L. de Broglie in 1923 led to the development of quantum mechanics by E. Schrödinger, W. Heisenberg, P. M. Dirac, and others. de Broglie notes in his Nobel Prize speech why he was led to the wave-particle duality of matter.

> When I began to consider these difficulties [of contemporary physics] I was chiefly struck by two facts. On the one hand the quantum theory of light cannot be considered satisfactory, since it defines the energy of a light corpuscle by the equation $E = h\nu$, containing the frequency ν. Now a purely corpuscular theory contains nothing that enables us to define a frequency; for this reason alone, therefore, we are compelled, in the case of light, to introduce the idea of a corpuscle and that of periodicity simultaneously.
>
> On the other hand, determination of the stable motion of electrons in the atom introduces integers; and up to this point the only phenomena involving integers in Physics were those of interference and of normal modes of vibration. This fact suggested to me the idea that electrons too could not be regarded simply as corpuscles, but that periodicity must be assigned to them also.

See L. de Broglie, *Matter or Light*, Dover, New York (1946).

The behavior of electrons in atoms, in molecules, and in condensed phases such as liquids and solids is described by quantum mechanics. The simplest case involves the motion of one electron moving in a potential V which in general varies with position. The behavior is then described by **Schrödinger's wave equation**; i.e., the electron behaves as a wave. In Cartesian coordinates this takes the form

$$\frac{-h^2}{8\pi^2 m}\left(\frac{\partial^2\psi}{\partial x_1^2} + \frac{\partial^2\psi}{\partial x_2^2} + \frac{\partial^2\psi}{\partial x_3^2}\right) + [V(x_1, x_2, x_3) - E]\psi = 0. \qquad (8.1.1)$$

Here h is **Planck's constant**, m is the mass of the electron, and E is the unknown energy of the electron. ψ is called the **wave function**. $|\psi^2|\,dv$, where dv is an element of volume, is the probability of finding the electron in the volume element dv. We call $|\psi^2|$ the **probability density** since it is the probability of finding the electron in a unit volume. Note that this is similar to the definition of light intensity, which is proportional to the square of the amplitude.

We shall not discuss the background of the Schrödinger equation here. This is done in detail in courses in atomic physics and quantum mechanics. We only note in passing that the Schrödinger equation is the wave mechanical expression for "kinetic energy plus potential energy equals total energy E"; i.e., K.E. $+ V = E$. This can also be written in terms of the momentum p as $p^2/2m + (V - E) = 0$. The first term corresponds to the first term of (8.1.1) and the second term to the second term of (8.1.1).

In quantum mechanics the component p_1 is represented by the operator

$$p_1 = \frac{h}{2\pi i}\frac{\partial}{\partial x_1} \qquad (8.1.2)$$

and

$$\frac{p^2}{2m} + V = E \qquad (8.1.3)$$

becomes in quantum mechanics

$$\left(\frac{p^2}{2m} + V\right)\psi = E\psi. \qquad (8.1.4)$$

(A more general statement is that the Hamiltonian operating on ψ gives the energy E times ψ.)

Note how Equation (8.1.1) follows from $H\psi = E\psi$ for the case where $H = p^2/2m + V$.

A simple metal such as sodium can be considered, to a first approximation, to consist of ions of Na^+ at fixed points while a corresponding number of free electrons move *freely* through the crystal but cannot escape from it. To move freely through the crystal means that V is a constant; since we can arbitrarily choose the reference point for potential energy, we set $V = 0$. To be unable to escape from the crystal means that $\psi \rightarrow 0$ as the boundaries are approached. We are therefore interested in solving

$$\frac{h^2}{8\pi^2 m}\left(\frac{\partial^2\psi}{\partial x_1^2} + \frac{\partial^2\psi}{\partial x_2^2} + \frac{\partial^2\psi}{\partial x_3^2}\right) + E\psi = 0 \tag{8.1.5}$$

subject to the conditions that $\psi = 0$ at the boundaries.

Because of its simplicity, the one-dimensional case for motion restricted to $0 \leq x \leq a$ will be treated instead:

$$\frac{d^2\psi}{dx_1^2} + \frac{8\pi^2 mE}{h^2}\psi = 0 \tag{8.1.6}$$

$$\psi = 0 \qquad \text{at } x = 0 \tag{8.1.7}$$

$$\psi = 0 \qquad \text{at } x = L. \tag{8.1.8}$$

Physically, this means that the electron is restricted to the infinite potential, well shown in Figure 8.1.1. The problem to which we seek a solution is often

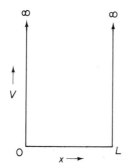

FIGURE 8.1.1. Infinite potential well.

called the one-dimensional **particle-in-a-box problem**. We shall find that a solution exists only for *discrete* values of E. Mathematically, the problem is referred to as a characteristic value problem, an eigenwert problem (German), or most commonly as an eigenvalue problem. We must solve a linear second-order differential equation with constant coefficients subject to boundary value conditions. Even if you have not studied differential equations you can show that

$$\psi = A\sin\alpha x + B\cos\alpha x \tag{8.1.9}$$

is a solution of Equation (8.1.6), where

$$\alpha^2 = \frac{8\pi^2 mE}{h^2}. \tag{8.1.10}$$

Here A and B are constants. Let us now try to satisfy the boundary conditions. From Equations (8.1.7) and (8.1.9),

$$\psi(0) = 0 = A\cdot 0 + B\cdot 1$$

or

$$B = 0.$$

Hence

$$\psi = A \sin \alpha x. \tag{8.1.11}$$

To satisfy Equation (8.1.8) we have

$$\psi(L) = 0 = A \sin \alpha L.$$

Either $A = 0$ (in which case ψ is identically zero everywhere and the electron does not exist) or $\sin \alpha L = 0$, which can be true if and only if

$$\alpha L = n_1 \pi, \tag{8.1.12}$$

where $n_1 = 1, 2, 3, \ldots$, i.e., an integer. Using Equations (8.1.10) and (8.1.12), we obtain

$$E = \frac{h^2 n_1^2}{8mL^2}. \tag{8.1.13}$$

Thus the energy of the electron is quantized; i.e., it can have discrete energies only. The integer n_1 is called a **quantum number.**

To evaluate A of the wave function in Equation (8.1.11) we note that the electron must be somewhere on the interval $0 \leq x \leq a$; hence

$$\int_0^L \psi^2 \, dx = 1. \tag{8.1.14}$$

Substitution of (8.1.11) into (8.1.14) allows us to solve for A.

The three-dimensional case is more mathematically involved (the two-dimensional case is discussed in the programmed learning sequence of Problems 8.15–8.18). We assume that the box is a parallelopiped with dimensions L_1, L_2, and L_3. The result is

$$E = \frac{h^2}{8m} \left(\frac{n_1^2}{L_1^2} + \frac{n_2^2}{L_2^2} + \frac{n_3^2}{L_3^2} \right), \tag{8.1.15}$$

where n_1, n_2, and n_3 are integers. The three-dimensional case therefore involves three **quantum numbers.** Actually, there is a fourth quantum number, the **electron spin,** which can have the two values, $m_s = \frac{1}{2}$ and $m_s = -\frac{1}{2}$. The state of the electron is therefore described by four quantum numbers. By this we mean that given n_1, n_2, n_3, and m_s, the wave function and energy are defined. For the case considered, the energy is independent of m_s so that the states with the same n but different m_s ($\frac{1}{2}$ or $-\frac{1}{2}$) have the same energy and are thus said to be **degenerate states**.

The **Pauli exclusion principle** states that in any atom, molecule, solid, etc., there cannot be two electrons in the same quantum state, i.e., with the same four quantum numbers.

Thus each of the free electrons in sodium (the so-called electron gas) has its own specific set of four quantum numbers which are not shared by any other free electron. In the case of sodium metal at absolute zero temperature, the various free electrons have energies from nearly 0 up to 3.12 eV as will be shown in Chapter 27.

EXAMPLE 8.1.1

An electron has a kinetic energy of 3 eV. Calculate its wavelength using the **de Broglie relation**

$$\lambda = \frac{h}{p}. \tag{8.1.16}$$

Answer : Since

$$K.E. = \frac{p^2}{2m},$$

we have

$$\lambda = \frac{h}{\sqrt{2m(K.E.)}} = 7 \text{ Å}.$$

8.2   THE HYDROGEN ATOM

In the case of the hydrogen atom we write Schrödinger's equation in spherical coordinates (this seems reasonable since in earlier concepts of the atom the electron traveled around the nucleus):

$$\frac{1}{r^2}\frac{\partial}{\partial r}\left(r^2\frac{\partial\psi}{\partial r}\right) + \frac{1}{r^2\sin^2\theta}\frac{\partial^2\psi}{\partial\phi^2} + \frac{1}{r^2\sin\theta}\frac{\partial}{\partial\theta}\left(\sin\theta\frac{\partial\psi}{\partial\theta}\right) + \frac{8\pi^2m}{h^2}\left(E + \frac{e^2}{r}\right)\psi = 0. \tag{8.2.1}$$

Here, r, θ, and ϕ have their usual significance, as shown in Figure 8.2.1.

We also note that we have replaced V by

$$V = -\frac{e^2}{r}, \tag{8.2.2}$$

which is the ordinary coulombic potential between the negatively charged electron and the positively charged proton. Physicists and chemists often use the electrostatic system of units. In the rationalized mks system a factor of $1/4\pi\epsilon_0$, where ϵ_0 is the permittivity of free space, would be introduced on the right-hand side of this equation.

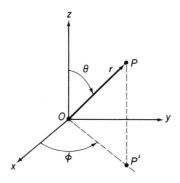

FIGURE 8.2.1. Spherical coordinate system.

The boundary conditions in the present case are

$$\psi(r, \theta, \phi) = \psi(r, \theta, \phi + 2\pi) \tag{8.2.3}$$

$$\psi(r, \theta, \phi) = \psi(r, \theta + 2\pi, \phi) \tag{8.2.4}$$

$$\lim_{r \to \infty} \psi(r, \theta, \phi) = 0 \tag{8.2.5}$$

$$\int_{r=0}^{\infty} \int_{\theta=0}^{\pi} \int_{\phi=0}^{2\pi} |\psi|^2 \, dv = 1. \tag{8.2.6}$$

The first two conditions state that ψ must be a periodic function of ϕ and θ, respectively, with a period of 2π in each case. The third condition states that the wave function of the electron approaches zero for very large distances from the proton and hence the probability, $\psi^2 d\tau$, of finding the electron in a small volume element $d\tau$ approaches zero as the distance approaches infinity. The fourth condition states that the probability of finding the electron anywhere is one. We shall not go through the mathematical details here. See, e.g., L. Pauling and E. B. Wilson, *Introduction to Quantum Mechanics*, McGraw-Hill, New York (1935).

The general solution for ψ will have the form

$$\psi(r, \theta, \phi) = R(r)\Theta(\theta)\Phi(\phi). \tag{8.2.7}$$

From our previous experience, we might expect that these wave functions would be characterized by three quantum numbers. In the present case these are called n, l, and m_l, where $n = 1, 2, 3, \ldots$, and $l = n - 1, \ldots, 0$, while $m_l = -l, \ldots, 0, \ldots, l$.

The energy is found to be

$$E = -\frac{2\pi^2 me^4}{h^2} \frac{1}{n^2}. \tag{8.2.8}$$

The lowest energy state of the atom is the $n = 1$ state, called the **ground state**. This corresponds to an energy $E = -13.6$ eV. This means that the energy of the electron and proton in the hydrogen atom in its most stable configuration is 13.6 eV less than the energy of a static proton and a static electron which are infinitely separated. The energy to strip off the outer electron of an atom, i.e., to take it from its ground state to infinity, is called the **ionization potential**; for hydrogen it is experimentally found to be 13.60 eV.

We note that the hydrogen atom can exist in discrete excited electronic states. When it is in an excited state such as $n = 2$, it can drop to the state $n = 1$. In the process it emits a photon whose energy is given by the **Einstein relation**

$$\Delta E = h\nu = E_2 - E_1, \tag{8.2.9}$$

where ν is the frequency of the photon. More generally the emission spectra are given by

$$\Delta E = h\nu = E_n - E_m = \frac{2\pi^2 me^4}{h^2} \left(\frac{1}{m^2} - \frac{1}{n^2} \right), \tag{8.2.10}$$

where $m < n$.

Table 8.2.1. BALMER SPECTRAL SERIES*

Name of Line	Value of n	Calculated Wavelength, Å	Measured Wavelength, Å
H_α	3	6562.793	6562.8473
			6562.7110
H_β	4	4861.327	4861.3578
			4861.2800
H_γ	5	4340.466	4340.497
			4340.429
H_δ	6	4101.738	4101.7346
H_ϵ	7	3970.075	3970.0740
	8	3889.052	3889.0575
	9	3835.387	3835.397
	10	3797.900	3797.910
	11	3770.633	3770.634
	12	3750.154	3750.152
	13	3734.371	3734.372
	14	3721.948	3721.948
	15	3711.973	3711.980

*J. E. Goldman, ed. *The Science of Engineering Materials,* John Wiley & Sons. Inc., New York (1957) p. 28.

Also, a photon of the correct frequency can be absorbed. Equation (8.2.10) also describes this process. Table 8.2.1 gives calculated and experimental values for the case where $m = 2$. [The fine structure (note the two values of measured wavelength) is studied in courses in quantum mechanics and modern physics]. Wavelength λ and frequency v of light are related by

$$\lambda v = c, \tag{8.2.11}$$

where c is the speed of light. The wavelength in Ångstroms is related to the energy in electron volts by

$$\lambda = \frac{12{,}400}{\Delta E}. \tag{8.2.12}$$

It is of particular interest to study the wave functions for the hydrogen atom. To do so we must introduce some additional notation. The l states are specified by either a number or a letter (spectroscopy notation) as follows:

$$l = 0 \quad 1 \quad 2 \quad 3 \quad 4 \quad 5 \quad 6 \quad 7 \quad 8$$
$$s \quad p \quad d \quad f \quad g \quad h \quad i \quad k \quad l.$$

Thus an electron for which $n = 2$, $l = 0$ is called a $2s$ electron. The ground state electron for which $n = 1$, $l = 0$, $m_l = 0$ is a $1s$ electron and it has a wave function

$$\psi_{1s} = \frac{1}{\sqrt{\pi}} \frac{1}{a_0^{3/2}} e^{-r/a_0}, \tag{8.2.13}$$

where

$$a_0 = \frac{h^2}{4\pi^2 me^2} = 0.529 \text{ Å}. \tag{8.2.14}$$

Note that ψ_{1s} does not vary with θ and ϕ. The probability of finding the electron in the volume element between r and $r + dr$ is

$$\psi_{1s}^2 4\pi r^2 \, dr. \tag{8.2.15}$$

EXAMPLE 8.2.1

At what radius is the $1s$ electron most likely to be found?

Answer. This is the radius for which $\psi_{1s}^2 4\pi r^2$ (which is called the **radial probability density**) is a maximum. A maximum occurs for that value of r for which

$$\frac{d}{dr} [\psi_{1s}^2 4\pi r^2] = 0.$$

By substituting from Equation (8.2.13) for ψ_{1s} and carrying out the differentiation, the student can show that the maximum occurs for $r = a_0$. This is called the **Bohr radius**; it equals 0.529 Å.

The ground state of the hydrogen atom therefore consists of a proton surrounded by a spherically symmetric electron cloud. A system of excited hydrogen atoms such as an incandescent gas will emit radiation spontaneously, unless external excitation is continually provided, and settle eventually into the ground state. The description of this process involves the use of time-dependent quantum mechanics. For example, the analysis of such optical devices as the laser would involve this subject.

8.3   THE HYDROGEN-LIKE ATOM AND THE PERIODIC TABLE

A **hydrogen-like atom** has a nucleus of charge Ze with a single electron, e.g., H, He$^+$, Li^{2+}, Be^{3+}, ... It is described by Schrödinger's equation but with e^2/r replaced by Ze^2/r. Its energy is given by

$$E = -\frac{2\pi^2 mZ^2 e^4}{h^2} \frac{1}{n^2}, \tag{8.3.1}$$

i.e., by

$$E = -13.6 \frac{Z^2}{n^2} \text{ (eV)}. \tag{8.3.2}$$

Table 8.3.1 gives a few of the hydrogen-like wave functions. The quantum number n is described by either a number or a letter (spectroscopy notation) as follows:

$$n = 1 \quad 2 \quad 3$$
$$K \quad L \quad M.$$

Table 8.3.1. THE HYDROGEN-LIKE WAVE FUNCTION

K Shell

$n = 1$, $l = 0$, $m_l = 0$:

$$\psi_{1s} = \frac{1}{\sqrt{\pi}} \left(\frac{Z}{a_0}\right)^{3/2} e^{-Zr/a_0}$$

L Shell

$n = 2$, $l = 0$, $m_l = 0$:

$$\psi_{2s} = \frac{1}{4\sqrt{2\pi}} \left(\frac{Z}{a_0}\right)^{3/2} \left(2 - \frac{Zr}{a_0}\right) e^{-Zr/2a_0}$$

$n = 2$, $l = 1$, $m_l = 0$:

$$\psi_{2p_z} = \frac{1}{4\sqrt{2\pi}} \left(\frac{Z}{a_0}\right)^{3/2} \frac{Zr}{a_0} e^{-Zr/2a_0} \cos\theta$$

$n = 2$, $l = 1$, $m_l = \pm 1$:

$$\psi_{2p_x} = \frac{1}{4\sqrt{2\pi}} \left(\frac{Z}{a_0}\right)^{3/2} \frac{Zr}{a_0} e^{-Zr/2a_0} \sin\theta \cos\phi$$

$$\psi_{2p_y} = \frac{1}{4\sqrt{2\pi}} \left(\frac{Z}{a_0}\right)^{3/2} \frac{Zr}{a_0} e^{-Zr/2a_0} \sin\theta \sin\phi$$

The $1s$, $2s$, $3s$, and all s-state electrons have spherical symmetry. As n increases, the maximum of the radial distribution function would move to larger r values. See Figure 8.3.1. None of the other states have radial symmetry. The p states all have a dumbbell shape, e.g., $|\psi_{2p_z}|^2 = 0$ at the origin, and a maximum along the z axis at a distance of $\pm(2a_0/Z)$.

The main axis of the dumbbell is the z axis for the $2p_z$ state; i.e., it has rotational symmetry around the z axis. Likewise, the y axis is the main axis of the dumbbell for the $2p_y$ state. The general nature of the electron distribution for different orbitals is shown in Figure 8.3.2.

EXAMPLE 8.3.1

How many $2p$ states are there?

Answer. We have $n = 2$, $l = 1$. Hence m_l can take on the values $-1, 0, 1$, and the electron spin can be $m_s = -\frac{1}{2}$ or $\frac{1}{2}$. Hence there are six $2p$ electrons as follows:

n	l	m_l	m_s
2	1	0	$\frac{1}{2}$
2	1	0	$-\frac{1}{2}$
2	1	1	$\frac{1}{2}$
2	1	1	$-\frac{1}{2}$
2	1	-1	$\frac{1}{2}$
2	1	-1	$-\frac{1}{2}$.

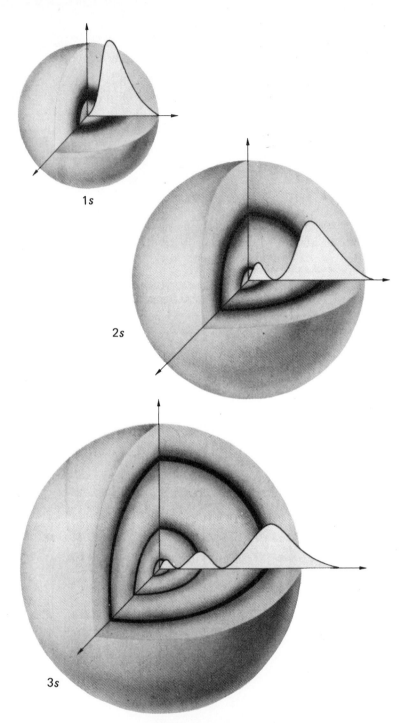

1s

2s

3s

FIGURE 8.3.1. Electron densities for different states in
hydrogen-like atoms. [From S. M. Edelglas,
Engineering Materials Science, Ronald Press,
New York (1966).]

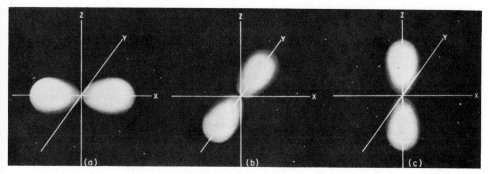

FIGURE 8.3.2. Polar representation of angular dependence of
ψ_{2p}^2. The magnitude of a vector from the origin
to a point on the surface represents the value
of the angular dependence of ψ_{2p}^2. (a) ψ_{2p_x}.
(b) ψ_{2p_y}. (c) ψ_{2p_z}.

The first approximation of an atom is that all of its electrons have the
general form of **hydrogen-like wave functions.** Thus the electron configuration
of the light elements is as shown in Table 8.3.2. At least for the lighter atoms,

Table 8.3.2. ELECTRON CONFIGURATIONS OF THE LIGHT ELEMENTS

Atom	Atomic Number	Electron Configuration
Hydrogen	1	$1s^1$
Helium	2	$1s^2$
Lithium	3	$1s^2 2s^1$
Beryllium	4	$1s^2 2s^2$
Boron	5	$1s^2 2s^2 2p^1$
Carbon	6	$1s^2 2s^2 2p^2$
Nitrogen	7	$1s^2 2s^2 2p^3$
Oxygen	8	$1s^2 2s^2 2p^4$
Fluorine	9	$1s^2 2s^2 2p^5$
Neon	10	$1s^2 2s^2 2p^6$
Sodium	11	$1s^2 2s^2 2p^6 3s^1$
Magnesium	12	$1s^2 2s^2 2p^6 3s^2$

successive atoms are seen to have the electron configuration of the preceding
atom plus one more electron. Because of the Pauli exclusion principle, each
electron has a different wave function, i.e., a different set of four quantum
numbers.

EXAMPLE 8.3.2

Suppose the $2p$ states are all filled as in neon. Show that the sum of the
ψ^2 for all of these p electrons is spherically symmetric.

Answer. We have

$$\psi^2 = \psi^2_{2p_x} + \psi^2_{2p_y} + \psi^2_{2p_z}$$

for each spin state. Each of these functions (see Table 8.3.1) has the same radial factor, which we call R_{2p}. Hence

$$\psi^2 = R^2_{2p}(\cos^2 \theta + \sin^2 \theta \cos^2 \phi + \sin^2 \theta \sin^2 \phi)$$
$$= R^2_{2p}.$$

More generally it is true that a full "shell" of electrons (same n) has a spherically symmetric distribution. For example, He and Ne have this structure which is characteristic of chemical stability. Thus, for example, sodium tends to ionize to Na^+, which has the Ne configuration, and fluorine readily becomes F^-, which has the Ne electron configuration when the ionic compound NaF is formed.

It is also true that a full complement of d electrons (10 in all) or f electrons (14 in all) gives a spherically symmetric distribution.

It should be noted that our model of the atom is only approximate since it ignores the interactions of electrons with each other. This approximation could be expected to be the best for innermost electrons.

EXAMPLE 8.3.3

Using the present approximation, calculate the energy and wavelength when an electron drops from an L shell to a K shell in copper. (*Note:* This is the source of the K_α line for copper, which is X-ray radiation widely used in diffraction studies. An electron is initially knocked out of the K shell by bombarding the copper with high-energy electrons. Then an electron from an L shell spontaneously falls into the lower energy state with the emission of the X-ray radiation.)

Answer. We use Equation (8.3.2) to describe the energy levels with $Z = 29$, the atomic number of copper. Then

$$h\nu = \Delta E = 13.6(29)^2 \left(\frac{1}{1^2} - \frac{1}{2^2}\right)$$

$$\Delta E = 8580 \text{ eV}.$$

The wavelength can be obtained from Equation (8.2.12), which gives 1.45 Å. The experimental value is $\lambda = 1.54$ Å. The present approximation would give poor results for transitions involving outer state electrons. The reason for this is that such outer electrons would be screened by the other electrons from the charge Ze so that the effective nuclear charge would be very much less than this.

Appendix 8A gives the electron configuration for 98 atoms as obtained by spectroscopists. We note that the $4s$ electrons add before the $3d$, the $5s$ before the $4d$, the $6s$ and $5d$ before the $4f$, etc.

8.4   MAGNITUDES OF BINDING ENERGIES AND PARTICLE SIZES

How large are the various binding energies in nature? The nuclear binding energy per large particle only (proton or neutron) is in the range of 1–2 MeV. Inner electrons (K electrons) in medium atomic number elements have a binding energy around 20 keV. The outer electron of an atom is bonded to the ion with binding energy of about 5–20 eV.

Table 8.4.1 gives the binding energy of outer electrons for the light elements. Finally, atoms are bound together to make molecules or solids with binding energies near 1–10 eV/atom.

Table 8.4.1. IONIZATION POTENTIALS (eV)

Element	First Ionization Potential	Second Ionization Potential
H	13.60	
He	24.58	54.41
Li	5.39	75.62
Be	9.32	18.21
B	8.30	25.12
C	11.27	24.38
N	14.55	29.61
O	13.62	35.08
F	17.42	34.98
Ne	21.56	40.96

EXAMPLE 8.4.1

Why is the second ionization potential of the helium atom four times the first ionization potential of the hydrogen atom? We are comparing exact hydrogen-like states of H and He$^+$, i.e., nuclei with single electrons.

Answer. For H, $Z = 1$, and for He$^+$, $Z = 2$. In each case the ground state is $n = 1$ and the fully ionized state is $n = \infty$. The ionization energy is proportional to Z^2 [Equation (8.3.2)]; hence the ionization potential of He$^+$ is four times that of the H atom.

Note the increase in ionization potential as a given state (n) is filled. Note how the tendency to form a positive ion increases as the ionization potential decreases.

Electrons and nuclei have radii of $\sim 10^{-13}$ cm and atoms have radii of $\sim 10^{-8}$ cm (1 Å); large molecules such as a nylon polymer chain have "lengths" of $\sim 10^{-4}$ cm, and crystals in steel often have "diameters" of $\sim 10^{-3}$ cm.

APPENDIX 8A ELECTRONIC STRUCTURES OF THE ATOMS

Element and
Atomic Number   Principal and Secondary Quantum Numbers

		$n=1$	2		3			4				Ground State Notation
		$l=0$	0	1	0	1	2	0	1	2	3	
1	H	1										$1s$
2	He	2										$1s^2$
3	Li	2	1									[He]$2s$
4	Be	2	2									— $2s^2$
5	B	2	2	1								— $2s^22p$
6	C	2	2	2								— $2s^22p^2$
7	N	2	2	3								— $2s^22p^3$
8	O	2	2	4								— $2s^22p^4$
9	F	2	2	5								— $2s^22p^5$
10	Ne	2	2	6								— $2s^22p^6$
11	Na	2	2	6	1							[Ne]$3s$
12	Mg	2	2	6	2							— $3s^2$
13	Al	2	2	6	2	1						— $3s^23p$
14	Si	2	2	6	2	2						— $3s^23p^2$
15	P	2	2	6	2	3						— $3s^23p^3$
16	S	2	2	6	2	4						— $3s^23p^4$
17	Cl	2	2	6	2	5						— $3s^23p^5$
18	Ar	2	2	6	2	6						— $3s^23p^6$
19	K	2	2	6	2	6		1				[Ar]$4s$
20	Ca	2	2	6	2	6		2				— $4s^2$
21	Sc	2	2	6	2	6	1	2				— $3d4s^2$
22	Ti	2	2	6	2	6	2	2				— $3d^24s^2$
23	V	2	2	6	2	6	3	2				— $3d^34s^2$
24	Cr	2	2	6	2	6	5	1				— $3d^54s$
25	Mn	2	2	6	2	6	5	2				— $3d^54s^2$
26	Fe	2	2	6	2	6	6	2				— $3d^64s^2$
27	Co	2	2	6	2	6	7	2				— $3d^74s^2$
28	Ni	2	2	6	2	6	8	2				— $3d^84s^2$

		$n=$ 1	2		3			4				5			Ground State Notation
		$l=$ 0	0	1	0	1	2	0	1	2	3	0	1	2	
29	Cu	2	2	6	2	6	10	1							— $3d^{10}4s$
30	Zn	2	2	6	2	6	10	2							— $3d^{10}4s^2$
31	Ga	2	2	6	2	6	10	2	1						— $3d^{10}4s^24p$
32	Ge	2	2	6	2	6	10	2	2						— $3d^{10}4s^24p^2$
33	As	2	2	6	2	6	10	2	3						— $3d^{10}4s^24p^3$
34	Se	2	2	6	2	6	10	2	4						— $3d^{10}4s^24p^4$
35	Br	2	2	6	2	6	10	2	5						— $3d^{10}4s^24p^5$
36	Kr	2	2	6	2	6	10	2	6						— $3d^{10}4s^24p^6$

Element and
Atomic Number　　　　　Principal and Secondary Quantum Numbers

	$n=$ 1	2	3	4, $l=0$	1	2	3	5, $l=0$	1	2	
	$l=$ —	—	—	0	1	2		0	1	2	
37 Rb	2	8	18	2	6			1			[Kr]$5s$
38 Sr	2	8	18	2	6			2			— $5s^2$
39 Y	2	8	18	2	6	1		2			— $4d5s^2$
40 Zr	2	8	18	2	6	2		2			— $4d^2 5s^2$
41 Nb	2	8	18	2	6	4		1			— $4d^4 5s$
42 Mo	2	8	18	2	6	5		1			— $4d^5 5s$
43 Tc	2	8	18	2	6	6		1			— $4d^6 5s$
44 Ru	2	8	18	2	6	7		1			— $4d^7 5s$
45 Rh	2	8	18	2	6	8		1			— $4d^8 5s$
46 Pd	2	8	18	2	6	10		—			— $4d^{10}$
47 Ag	2	8	18	2	6	10		1			— $4d^{10} 5s$
48 Cd	2	8	18	2	6	10		2			— $4d^{10} 5s^2$
49 In	2	8	18	2	6	10		2	1		— $4d^{10} 5s^2 5p$
50 Sn	2	8	18	2	6	10		2	2		— $4d^{10} 5s^2 5p^2$
51 Sb	2	8	18	2	6	10		2	3		— $4d^{10} 5s^2 5p^3$
52 Te	2	8	18	2	6	10		2	4		— $4d^{10} 5s^2 5p^4$
53 I	2	8	18	2	6	10		2	5		— $4d^{10} 5s^2 5p^5$
54 Xe	2	8	18	2	6	10		2	6		— $4d^{10} 5s^2 5p^6$

	$n=$ 1	2	3	4, $l=0$	1	2	3	5, $l=0$	1	2	6, $l=0$	Ground State Notation
	$l=$ —	—	—	0	1	2	3	0	1	2	0	
55 Cs	2	8	18	2	6	10		2	6		1	[Xe]$6s$
56 Ba	2	8	18	2	6	10		2	6		2	— $6s^2$
57 La	2	8	18	2	6	10		2	6	1	2	— $5d6s^2$
58 Ce	2	8	18	2	6	10	2	2	6		2	— $4f^2 6s^2$
59 Pr	2	8	18	2	6	10	3	2	6		2	— $4f^3 6s^2$
60 Nd	2	8	18	2	6	10	4	2	6		2	— $4f^4 6s^2$
61 Pm	2	8	18	2	6	10	5	2	6		2	— $4f^5 6s^2$
62 Sm	2	8	18	2	6	10	6	2	6		2	— $4f^6 6s^2$
63 Eu	2	8	18	2	6	10	7	2	6		2	— $4f^7 6s^2$
64 Gd	2	8	18	2	6	10	7	2	6	1	2	— $4f^7 5d6s^2$
65 Tb	2	8	18	2	6	10	8	2	6	1	2	— $4f^8 5d6s^2$
66 Dy	2	8	18	2	6	10	10	2	6		2	— $4f^{10} 6s^2$
67 Ho	2	8	18	2	6	10	11	2	6		2	— $4f^{11} 6s^2$
68 Er	2	8	18	2	6	10	12	2	6		2	— $4f^{12} 6s^2$
69 Tm	2	8	18	2	6	10	13	2	6		2	— $4f^{13} 6s^2$
70 Yb	2	8	18	2	6	10	14	2	6		2	— $4f^{14} 6s^2$
71 Lu	2	8	18	2	6	10	14	2	6	1	2	— $4f^{14} 5d6s^2$
72 Hf	2	8	18	2	6	10	14	2	6	2	2	— $4f^{14} 5d^2 6s^2$

	$n=$ 1	2	3	4	5, $l=0$	1	2	6, $l=0$	
	$l=$ —	—	—	—	0	1	2	0	
73 Ta	2	8	18	32	2	6	3	2	— $4f^{14} 5d^3 6s^2$
74 W	2	8	18	32	2	6	4	2	— $4f^{14} 5d^4 6s^2$

APPENDIX 8A ELECTRONIC STRUCTURES OF THE ATOMS (Contd.)

Element and
Atomic Number   Principal and Secondary Quantum Numbers

	$n=$ 1	2	3	4	5			6		
	$l=$ —	—	—	—	0	1	2	0		
75 Re	2	8	18	32	2	6	5	2		— $4f^{14}5d^56s^2$
76 Os	2	8	18	32	2	6	6	2		— $4f^{14}5d^66s^2$
77 Ir	2	8	18	32	2	6	7	2		— $4f^{14}5d^76s^2$
78 Pt	2	8	18	32	2	6	8	2		— $4f^{14}5d^86s^2$

	$n=$ 1	2	3	4	5				6			7	Ground State
	$l=$ —	—	—	—	0	1	2	3	0	1	2	0	Notation
79 Au	2	8	18	32	2	6	10		1				— $6s$
80 Hg	2	8	18	32	2	6	10		2				— $6s^2$
81 Tl	2	8	18	32	2	6	10		2	1			— $6s^26p$
82 Pb	2	8	18	32	2	6	10		2	2			— $6s^26p^2$
83 Bi	2	8	18	32	2	6	10		2	3			— $6s^26p^3$
84 Po	2	8	18	32	2	6	10		2	4			— $6s^26p^4$
85 At	2	8	18	32	2	6	10		2	5			— $6s^26p^5$
86 Rn	2	8	18	32	2	6	10		2	6			— $6s^26p^6$
87 Fr	2	8	18	32	2	6	10		2	6		1	[Rn]$7s$
88 Ra	2	8	18	32	2	6	10		2	6		2	— $7s^2$
89 Ac	2	8	18	32	2	6	10		2	6	1	2	— $6d7s^2$
90 Th	2	8	18	32	2	6	10		2	6	2	2	— $6d^27s^2$
91 Pa	2	8	18	32	2	6	10		2	6	3	2	— $6d^37s^2$
92 U	2	8	18	32	2	6	10	3	2	6	1	2	— $5f^36d7s^2$
93 Np	2	8	18	32	2	6	10	5	2	6		2	— $5f^57s^2$
94 Pu	2	8	18	32	2	6	10	6	2	6		2	— $5f^67s^2$
95 Am	2	8	18	32	2	6	10	7	2	6		2	— $5f^77s^2$
96 Cm	2	8	18	32	2	6	10	7	2	6	1	2	— $5f^76d^17s^2$
97 Bk	2	8	18	32	2	6	10	7	2	6	2	2	— $5f^76d^27s^2$
98 Cf	2	8	18	32	2	6	10	9	2	6	1	2	— $5f^96d^17s^2$

REFERENCES

Leighton, R. B., *Principles of Modern Physics*, McGraw-Hill, New York (1959).

Sproull, R. L., *Modern Physics*, Wiley, New York (1963). The reader will learn to appreciate Sproull's ability to make modern physics clear and exciting.

PROBLEMS

8.1. Give the binding energy of
  a. Nuclear particles.
  b. Inner electrons of the atom to the nucleus.
  c. Outer (valence) electrons to the ion.
  d. Chemical bonds.
  The materials scientist is primarily concerned with which of the above binding energies?

8.2. Give in Ångstroms the approximate size of
  a. Protons.
  b. Atoms.
  c. The wavelength of optical light.
  d. The length of a nylon polymer molecule.
  e. The size of a crystal in steel.

8.3. Calculate the electron speed in the Bohr orbit. Compare this with the speed of light in free space.

8.4. Calculate the binding energy of the electron to the proton in the ground state of the hydrogen atom.

8.5. X-rays are generated by bombarding a target material, such as molybdenum, with electrons which have been accelerated by an electric field. Approximately what potential (volts) would be necessary to knock a K electron (i.e., a $1s$ electron) from molybdenum?

8.6. a. Discuss the assumption being made when it is stated that the electron configuration of carbon is $1s^2 2s^2 2p^2$.
  b. Describe the nature of the electron distribution for each of these orbitals.

8.7. An approximation for a metal, such as silver, is the following: Each silver atom loses one electron. The resultant ions are localized at some position in the crystal, but the lost electrons wander freely throughout the crystal. It is assumed that these electrons do not interact with each other nor with the ions and that they cannot leave the metal.
  a. Given a cube of silver 1 cm on the side, what is the energy of the ground state?
  b. What does the Pauli exclusion principle say about the possible values of n_1, n_2, n_3, and m_s of these "free" electrons?

8.8. Give three distinctly different physical situations in which eigenvalues are involved.

8.9. What is a spectrometer? Discuss the design of a simple spectrometer. Discuss its basis as a tool in analytical chemistry.

8.10. Transition metals often exhibit more than one valence. Explain how iron (see Appendix 8A) can have a valence of 3 as well as 2.

8.11. a. What is the physical significance of the function ψ?

b. What is the meaning of $\psi_{1s}^2 4\pi r^2 \, dr$, where ψ_{1s} is given in Table 8.3.1, when $Z = 1$?

c. In the hydrogen atom, calculate the most probable distance of an electron from the proton.

More Involved Problems

8.12. a. Show that for a wave function with spherical symmetry Equation (8.2.6) becomes

$$\int_0^\infty |\psi(r)|^2 4\pi r^2 \, dr = 1.$$

b. Show for the wave function given in Table 8.3.1 for the 1s state of the hydrogen atom that the above equation does in fact hold.

8.13. Write an essay on the productions of characteristic X-rays. Among the many references which you might use are B. D. Cullity, *Elements of X-ray Diffraction*, Addison-Wesley, Reading, Mass. (1956).

8.14. Early chemists made predictions about the existence of new elements on the basis of an assumed periodic behavior. Discuss this statement from an historical viewpoint. More recent studies of this type are given by G. Schneider, *Solid State Physics*, Vol. 16 [F. Seitz and D. Turnbull, eds.] Academic Press, New York (1965) p. 276.

Note: Problems 8.15–8.18 are a programmed learning sequence for solving the motion of an electron in a two-dimensional box with infinite walls.

8.15. Suppose an electron moves in a two-dimensional area in a square region with edge a in which $V = 0$ within the region and $V = \infty$ at the edges. The wave equation representing the electron in the region where $V = 0$ is

$$\frac{\partial^2 \psi}{\partial x^2} + \frac{\partial^2 \psi}{\partial y^2} + \frac{2m}{\hbar^2} E\psi = 0,$$

where

$$\hbar = \frac{h}{2\pi}.$$

If a solution in the form $\psi(x, y) = X(x)\, Y(y)$ exists, show that

$$\frac{1}{X}\frac{d^2 X}{dx^2} = -\left\{ \frac{1}{Y}\frac{d^2 Y}{dy^2} + \frac{2mE}{\hbar^2} \right\}.$$

The variables are now said to be separated.

8.16. In Problem 8.15, the expression on the left-hand side is clearly a function of x only, while the expression on the right-hand side does not vary with x. Give a function of $f(x)$ which does not vary with x.

8.17. From Problems 8.15 and 8.16, we can write

$$\frac{1}{X}\frac{d^2 X}{dx^2} = -\left\{ \frac{1}{Y}\frac{d^2 Y}{dy^2} + \frac{2mE}{\hbar^2} \right\} = \pm k^2,$$

where k^2 is a constant, possibly zero. Show that the solutions of

$$\frac{d^2X}{dx^2} \pm k^2X = 0$$

for the three cases where the sign is negative, where the sign is positive, or where k is zero are

$$X = Ae^{kx} + Be^{-kx},$$

$$X = C \cos kx + D \sin kx,$$

$$X = Ex + F.$$

8.18. In Problems 8.15–8.17, the fact that $V = \infty$ at the edges means

$$\psi = 0 \text{ at } x = 0, \qquad 0 \le y \le a,$$

$$\psi = 0 \text{ at } x = a, \qquad 0 \le y \le a.$$

Physically, this means that the probability of having an electron in a region where the potential is infinite is zero.

a. Show that this excludes the first solution of Problem 8.17.

b. Show that the second solution is acceptable if $C = 0$ and $ka = n_1\pi$, where $n_1 = 1, 2, 3, \ldots$.

Thus we have

$$X(x) = D \sin \frac{n_1\pi x}{a}.$$

c. The equation

$$\frac{d^2Y}{dy^2} + \left(\frac{2mE}{\hbar^2} - k^2\right) Y = 0$$

can be rewritten as

$$\frac{d^2Y}{dy^2} + \alpha^2 Y = 0,$$

where

$$\alpha^2 = \frac{2mE}{\hbar^2} - k^2.$$

Show that the acceptable solution is

$$Y = G \sin \alpha y,$$

where G is a constant and $\alpha a = n_2\pi$, where $n_2 = 1, 2, 3, \ldots$.

d. The solution for $\psi = XY$ is therefore of the form

$$\psi = N \sin \frac{n_1\pi x}{a} \sin \frac{n_2\pi y}{a},$$

where N is a constant equal to the product DG. Using the fact that $|\psi|^2 \, dx \, dy$ represents the probability of finding the electron in the area $dx \, dy$ and the fact that the electron is in the square region, find N.

e. Since α and k are now known, find the expression for the allowable energy levels E. Compare with the three-dimensional case of Equation (8.1.15).

8.19. a. Write the Schrödinger equation in spherical coordinates for the hydrogen atom.

    b. Complete the solution for the hydrogen atom. This will require either a prior knowledge of Legendre and Laguerre differential equations or some study of them. See H. Eyring, J. Walter, and G. E. Kimball, *Quantum Chemistry*, Wiley, New York (1944) p. 48.

8.20. A cubic centimeter of metallic silver has about 6×10^{22} atoms and hence about 6×10^{22} free electrons (Ag has a valence of 1). Hence about 3×10^{22} have an electron spin $m_s = \frac{1}{2}$. Assuming that Equation (8.1.15) correctly gives the energy for each electron and that each level is filled from the lowest state to the highest which has energy E_F, obtain an expression for E_F. Evaluate E_F for this case.

Prologue

The interaction and binding between atoms is described by quantum mechanics. However, the exact calculation of this binding is impossible except in a few special cases. Hence approximate models are used to describe the bonding. These models involve the *ionic bond*, the *covalent bond*, *metallic bonding*, and the *secondary bond*. The purpose of this chapter is to study these models briefly and to become familiar with the types of bonds and the order of magnitude of the strength of these bonds.

9

BINDING OF ATOMS

9.1 INTRODUCTION

Atoms are bonded together to form molecules or condensed phases by several different types of bonds: (1) ionic bonding, (2) covalent or homopolar bonding, (3) metallic bonding, and (4) secondary-type bonds.

It is only in a few special cases that the actual bonding is exactly of one type of another; more often it is a combination. In the sections which follow, each of these bond types will be taken up individually. The student should attempt to develop a feeling for the strength of these bonds and for how the behavior of the material depends on the nature of the bonds. The first three types of bonds have strengths of about 100 kcal/mole, while the secondary-type bonds have strengths of only 1–10 kcal/mole. Note that

$$1 \text{ eV/atom} = 23.0 \text{ kcal/mole}.$$

9.2 THE IONIC BOND

Certain atoms readily lose or gain an electron to form an ion, with a resultant electron configuration of an inert gas atom but with a net negative or positive charge. Thus sodium readily loses an electron to form the sodium ion (with electron configuration $1s^2 2s^2 2p^6$) while chlorine readily gains an electron to form the chloride ion (with electron configuration $1s^2 2s^2 2p^6 3s^2 3p^6$). The calculation of the energy change in each case is a detailed quantum mechanical problem; alternatively these quantities can be measured. We shall assume that we know these quantities and ask: Can these Na^+ and Cl^- ions now be put together to form a stable molecule?

A true NaCl molecule exists in the vapor in which the binding is due mainly to the electrostatic attraction between one Na^+ and one Cl^- ion.

The attractive force between two ions with charges Q_1 and Q_2 can be represented at moderate distances of separation r_{12} by the coulombic force $Q_1 Q_2/r_{12}^2$ or by the net potential

$$U_{12} = \frac{Q_1 Q_2}{r_{12}} + \frac{b}{r_{12}^n}, \qquad (9.2.1)$$

where the second term is called the **Born repulsive potential;** the ions repel each other at very close distances because the electron clouds of the ions overlap. Here b and n are positive constants which are characteristics of the particular

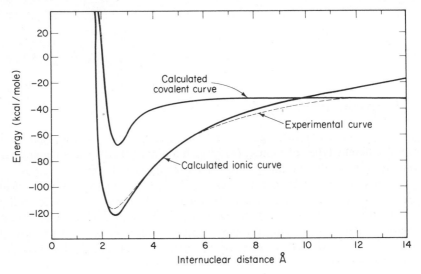

FIGURE 9.2.1. Potential energy of one sodium-chlorine pair as a function of internuclear separation. [Fròm W. J. Moore, *Physical Chemistry*, Prentice-Hall, Englewood Cliffs, N. J. (1955).]

ions. Usually, $n \approx 10$. The function is plotted in Figure 9.2.1 for one NaCl molecule; the minimum in the curve represents the stable internuclear separation for the molecule held together by an **ionic bond.** Also shown is the result for a NaCl molecule calculated on the basis of covalent bonding, which is discussed in the next section. Note that at large separations Na + Cl is a more stable system than $Na^+ + Cl^-$, and thus we find that the molecule NaCl dissociates into atoms.

We next ask: Is it possible to form a stable solid from the Na^+ and Cl^- ions? The solid which is formed is shown in Figure 9.2.2. It consists of a periodic regularly alternating array of positive and negative ions. The particular array of ions of which a portion is shown in Figure 9.2.2 is known as the **sodium chloride structure.** All of the lithium, sodium, potassium, and rubidium halides have this

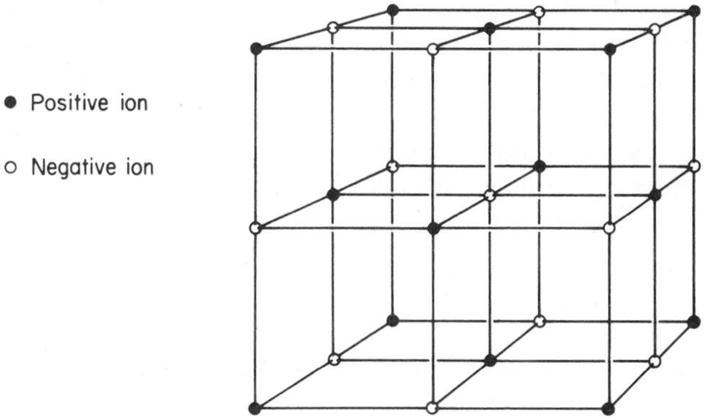

FIGURE 9.2.2. Portion of the sodium chloride structure.

crystal structure at room temperature and pressure. So does the important refractory material MgO. Remember that the distance between the atoms is of the order of a few Ångstroms.

The ions are electrostatically attracted to each other. The lattice does not collapse because when the electronic charge clouds begin to overlap, they repel each other. As a first approximation we can consider the ions as *rigid* spheres. We can define the radius of these spheres in such a way that the sum of the cation radius and anion radius equals the interionic distance. These radii are called **ionic radii**. We can consider that the electrostatic attraction brings the ions together until they just touch; since the spheres are rigid, they can get no closer. Such a solid would be incompressible and would have a binding energy which is about 10% too large. Thus we must remember that the idea of a rigid sphere ionic radius is a fiction—but a useful fiction.

We assume that a potential of the type

$$u_{ij} = \frac{Q_i Q_j}{r_{ij}} + \frac{b}{r_{ij}^n} \tag{9.2.2}$$

exists between each pair of ions; then, by summing over all ions in the infinite crystal, we can obtain an expression for the binding energy per mole of NaCl.

For N ions of Na^+ and N ions of Cl^-, the equilibrium value of the lattice energy $U(r_0)$ is

$$U(r_0) = -1.74755 \frac{Ne^2}{r_0}\left(1 - \frac{1}{n}\right). \tag{9.2.3}$$

Here e is the electron charge and r_0 is the separation between the nearest neighbor Na^+ and Cl^- ions at equilibrium. The separation distance r_0 can be measured by X-ray diffraction. One can also show from theoretical considerations that n is related to the bulk modulus K by

$$n = 1 + \frac{18 r_0^4 K}{1.74755 e^2}. \tag{9.2.4}$$

Using experimental values of r_0 and K, we find that $n \approx 10$ for many ionic solids. The first term in Equation (9.2.3) represents the coulombic attraction, while the second (which is about a 10% contribution) represents the Born repulsion. Note that the repulsive term $(1/r^{10})$ is a **short-range interaction**. Its contribution is negligible unless the ions are very close together. Table 9.2.1 gives some values of the lattice energy $U(r_0)$ obtained from Equations (9.2.3) and (9.2.4). Thus the binding energies are in the neighborhood of 10 eV.

EXAMPLE 9.2.1

Which has the greater binding energy per mole of NaCl: molecules of NaCl or a crystal of NaCl?

Answer. From Figure 9.2.1 the binding energy for molecules is about 120 kcal/mole, while for the crystal (see Table 9.2.1) it is about 183 kcal/mole.

Table 9.2.1. LATTICE ENERGY (KCAL/MOLE)

Crystal	Theory (Mayer)	Experiment
NaCl	183.1	182.8
KCl	165.4	164.4
AgCl	203	205.7
TlCl	167	170.1
CuCl	216	221.9

Note: 23 kcal/mole = 1 eV/atom pair.

We note from Equation (9.2.4) that if r_0 and n were known, the bulk modulus, which is an elastic stiffness constant, could be calculated. In general all of the elastic constants of a crystal can be calculated from a detailed theory of binding.

A detailed discussion of the Born (also called Born-Mayer) model for ionic bonding is given in the learning sequence of Problems 9.8–9.15.

9.3   COVALENT BONDING

Consider two hydrogen atoms which when brought closely together combine to form a hydrogen molecule. Why? In previous chemistry courses the student was told that this was due to the formation of an **electron pair bond**, designated by H: H. The general picture is that the electrons are no longer localized around their respective protons but that each is in the field of both protons. The detailed mathematical analysis of this is an involved quantum mechanical problem which is discussed in quantum chemistry books. A simplified mathematical discussion is considered in Example 9.3.1.

EXAMPLE 9.3.1

The kinetic energy of the electron in the Bohr orbit is $E_K = 13.6\,\text{eV}$. Let us then consider this electron as a particle in a cubical box moving in a potential $V = 0$. The edge of the cube, a, is chosen so that the kinetic energy of the electron in the box is also 13.6 eV. Its energy in the lowest state is

$$E_K = \frac{h^2}{8ma^2}(1^2 + 1^2 + 1^2).$$

See Equation (8.1.15). Consider the interaction of two such (imaginary) atoms to form a molecule, i.e., a box of dimensions a, a, $2a$. Calculate the binding energy for this model of the hydrogen molecule and atoms.

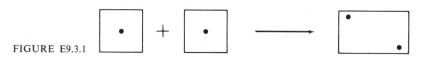

FIGURE E9.3.1

Answer. The energy of each electron in the large box, E_L, is

$$E_L = \frac{h^2}{8m}\left[\frac{1^2}{a^2} + \frac{1^2}{a^2} + \frac{1^2}{(2a)^2}\right].$$

Since two electrons are involved, the decrease in kinetic energy when the bond is formed is

$$\Delta E_K = 2(E_K - E_L) = \frac{1}{2}\left(\frac{h^2 3}{8ma^2}\right) = \frac{E_K}{2} = 6.8\,\text{eV}.$$

The experimental value is 4.7 eV. It is not surprising that we do not get the correct answer in this oversimplified model, which completely ignores the potential or coulombic interaction of an electron with each of the protons and with the other electron.

It is pointed out here that the bonding energy would be only 0.6 eV if the electrons had not been delocalized. The energy of the system is lowered because the electrons become delocalized. Actually the average charge is concentrated more between the nuclei when the bond is formed and the two protons are pulled closer together; thus the electron pair concept of G. N. Lewis has substantiation from detailed quantum mechanical calculations. This is the **covalent bond**.

The outer electron configuration of sulfur is $3s^2 3p^4$. It thus needs two electrons to achieve the inert gas configuration of argon. In the compound H_2S, two p bonds of sulfur are primarily involved to give a molecule with a bond angle near 90 deg, as expected (actually 92°20′). Recall from Figure 8.3.2 that p orbitals are at 90 deg to each other. Values of bond angles in similar compounds are shown in Table 9.3.1.

The covalent bond formed with carbon in saturated organic compounds and in diamond is of special interest. Here four bonds extend outward from the

Table 9.3.1. OBSERVED VALUES OF BOND ANGLES*

Substance	Method	Bond Angle	Experimental Value, deg
H_2O(g)	Sp.	H–O–H	105
OF_2(g)	E.D.	F–O–F	100
Cl_2O(g)	E.D.	Cl–O–Cl	115
$(CH_2)_2O$(g)	E.D.	C–O–C	111
H_2S(g)	Sp.	H–S–H	92°20'
S_8(c)	C.S.	S–S–S	106
S_8(g)	E.D.	S–S–S	100
BaS_3(c)	C.S.	S–S–S	103
$K_2S_3O_6$(c)	C.S.	S–S–S	103
SCl_2	E.D.	Cl–S–Cl	103
Se(c)	C.S.	Se–Se–Se	105
Te(c)	C.S.	Te–Te–Te	102

* g = gas, c = crystal, Sp. = spectroscope, E.D. = electron diffraction, and C.S. = crystal structure. (From L. Pauling, *Nature of Chemical Bond*, Cornell University Press, Ithaca, N.Y. (1940) p. 79.)

carbon atom at 109°28'. Carbon atoms, in the gas state, have the $1s^2 \ 2s^2 \ 2p^2$ electron configuration. Thus there are four electrons in the valence shell. Suppose that we excite a $2s$ electron to a $2p$ state. We then have a valence shell configuration of $2s2p^3$. These combine to form four equivalent **hybridized bonds** called sp^3 **bonds** directed toward the corners of a regular tetrahedron. (Four classic electrons placed on the surface of a sphere and allowed to move on the surface of the sphere would position themselves in the same manner.) Such bonds are sometimes called **tetrahedral bonds**. The simplest of compounds based on such bonds is methane, CH_4. However, the carbons in all of the compounds

$$H[-\underset{\underset{\displaystyle H}{|}}{\overset{\overset{\displaystyle H}{|}}{C}}-]_n H$$

also show such tetrahedral bonding. Such compounds include from small molecules such as methane and ethane,

$$H-\underset{\underset{\displaystyle H}{|}}{\overset{\overset{\displaystyle H}{|}}{C}}-\underset{\underset{\displaystyle H}{|}}{\overset{\overset{\displaystyle H}{|}}{C}}-H$$

all the way to huge molecules such as polyethylene, where *n* is about 1000; these are called polymeric molecules. Figures 9.3.1 and 9.3.2 illustrate how sp^3 orbitals form tetrahedral bonds in methane and ethane, respectively.

Covalent bonding in *solids* is the same as in *molecules*. The attractive forces arise from the concentration of electronic charge along the bonding

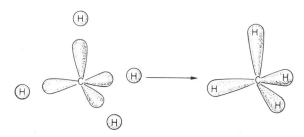

FIGURE 9.3.1. Formation of methane from carbon with four
sp^3 hybrid orbitals and four hydrogens, each
with one $1s$ orbital.

directions joining nuclei. The distribution of electrons in a crystal structure can
be determined relatively precisely by the degree to which they scatter X-rays.
This has been used to define effectively the location of highly directed bonds
in specific covalent crystals of the diamond type. Typical examples of crystals
where bonding is essentially all covalent are diamond, silicon, germanium,
and silicon carbide. Diamond, silicon, germanium, and gray tin all have the
tetrahedral structure associated with sp^3 hybridized bonding found in simple
hydrocarbon molecules such as methane. In diamond each carbon is tetrahedrally bonded to four other carbon atoms. This arrangement in solids is referred
to as the **diamond structure** and will be discussed in Chapter 10.

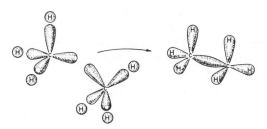

FIGURE 9.3.2. Formation of the
ethane molecule.

The strengths of these covalent bonds are about 93 kcal/mole (4.0
eV/bond) for a C—H bond to 83 kcal/mole (3.6 eV/bond) for a C—C bond.
Examples of bond lengths and dissociation energies for bonds ordinarily found
in polymeric materials are given in Table 9.3.2 on page 180.

9.4　THE METALLIC BOND

The crystal structure and physical properties of the large class of metals
and alloys are associated with another unique type of binding called the **metallic
bond**. The most striking characteristic of this type of bonding is the formation
of closely packed structures associated with high cohesive energies and high
mobility of electrons at all temperatures.

It is significant that a great number of metals and alloys can be described

Table 9.3.2. BOND LENGTHS AND ENERGIES*

Bond	Bond Length	Dissociation energy, kcal/mole
O—O	1.32	35
Si—Si	2.35	42.5
S—S	1.9–2.1	64
C—N	1.47	73
C—Cl	1.77	81
C—C	1.54	83
C—O	1.46	86
N—H	1.01	93
C—H	1.10	99
C—F	1.32–1.39	103–123
O—H	0.96	111
C=C	1.34	146
C=O	1.21	179
C≡N	1.15	213

*From J. A. Brydson, *Plastic Materials*, D. Van Nostrand Reinhold, New York (1966).

effectively in terms of three simple structures: face-centered closest-packed (fcc), hexagonal closest-packed (hcp), and body-centered cubic (bcc), as indicated in Figure 9.4.1. The two important structural features characteristic of metals in this example are the close packing of the atoms and the fact that each atom is coordinated equally with many others, i.e., for hcp and fcc the **coordination number** is 12 (an atom has 12 nearest neighbors) and for bcc it is 8. This indicates strong binding in the crystal as a whole but lack of any localized bonding interaction characteristic of, for example, covalent bonds in molecules. Compare these high coordination numbers for metals with the coordination

(a)

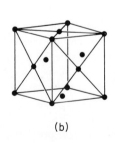

(b)

(c)

FIGURE 9.4.1. Three most important crystal structures for metals. (a) Body-centered cubic. (b) Face-centered closest-packed. (c) Hexagonal closest-packed.

number of ionic sodium chloride, which is 6, and the coordination number of 4 in the covalently bonded diamond crystal. The concept of metallic binding, therefore, is identified with the strong attractive interaction among a relatively great number of atoms and electrons. Bond strengths vary greatly in metals. A tightly bound crystal has a large heat of vaporization ΔH_v. This quantity is a good indication of the strength of binding in liquids and solids since ΔH_v is the energy required to separate the atoms from nearby positions in the liquid to widely separated positions in the vapor. Consider the range in cohesiveness in Table 9.4.1, for example, from sodium to tungsten.

Table 9.4.1. COMPARISON OF COHESIVENESS IN METALS

	ΔH_v, kcal/mole	
	Molecule	Crystal
Sodium	17.6	26.2
Copper	—	81.7
Iron	—	96.5
Tungsten	—	203.0

Unusually high electrical and thermal conductivity is the second striking characteristic of the metallic bond. This characteristic is due to the presence of a large number of loosely tied electrons which move with relative freedom through the lattice under the influence of external electrical or thermal gradients. A satisfactory theory of metallic binding must therefore explain how this "sea" of electrons at the same time results in the substantial cohesion and high electrical conductivity characteristic of the metals.

There are many simple metals which are accurately described by this picture, such as the alkali metals, the alkaline earth metals, and magnesium and aluminum. Let us consider, for example, a typical metal, lithium. It crystallizes in the bcc structure, separated from each of its eight nearest neighbors by 3.03 Å. If these are considered as spheres which just touch, then the radius of these spheres is 1.51 Å, and we call this the **metallic radius** of the atom. The difference in size is particularly striking when one considers that the volume of the atom is approximately 12 times that of the ion core without the valence electron. *Note*: The concept of metallic radii and metallic valence is useful in describing the structure and bonding in metals. Both parameters are typical of each metal. The electronic structure of the lithium atom, $1s^2 2s$, is characterized by the $1s^2$ electron pair, which is tightly held, as indicated by the small ion radius for Li^+ of 0.60 Å, and by the $2s$ valence electron, which is freely shared with all its neighbors and easily displaced in an electrical field. Thus lithium and many other simple metals can be described to a first approximation on the basis of the *free electron theory of metals*.

Assuming that the electrons are classical free electrons (meaning describable by classical mechanics, not quantum mechanics), we can predict the form of Ohm's law, and we can predict within about 30% the Hall coefficient for simple monovalent metals using Equation (3.3.3) assuming that there is one free electron contributed by each atom. However, we see in Table 9.4.2 that although the agreement is fair for the monovalent metals (the ratio should equal 1), the prediction fails miserably (even in sign) for the divalent metals.

Table 9.4.2. RATIO OF EXPERIMENTAL HALL
COEFFICIENT TO CLASSICAL
THEORETICAL VALUE

Metal	$R_H(\text{measured})/(1/-Ne)$
Li	1.3
Na	0.9
K	0.9
Rb	1.0
Cs	1.1
Cu	0.8
Ag	0.8
Au	0.7
Be	−5.0
Cd	−0.5

A *detailed* theoretical analysis in which the electrons are treated quantum mechanically removes these discrepanicies. There is another problem with classical electron theory: If the electrons are really a classical gas (such as argon gas), they should have an electronic specific heat of $C_v^{\text{elect}} = \frac{3}{2}R$. In fact, the electron specific heat is only about $\frac{1}{100}$ of this. Again the electron should be treated quantum mechanically. The essential reason for the existence of metallic bonding and hence of metals is that the solid has lower energy if the valence electrons form an electron gas and become delocalized.

9.5   SECONDARY BONDS

There is a large group of solids called **molecular solids** which are held together primarily by forces between molecules. These forces were originally described by van der Waals to explain how molecular gases condense to liquids. Here the bonding is relatively weak and so the bonds are called **secondary bonds**. These forces actually arise from several different interactions: *permanent dipole* forces, *induced dipole* forces, and *dispersion* forces. Although the magnitude of these effects is relatively small, in many cases their contribution is often critical in defining the stabiiity of solids. For example, they contribute to the condensation of the noble gases, to the equilibrium distance in ionic crystals, and to the stability of many solids, including proteins. The **orientation effect** between permanent dipoles was suggested first by W. H. Keesom in 1912. If the centroid of the positive charges does not coincide with the centroid of the negative charges, the molecule is said to possess a **permanent dipole**.

The **dipole moment** is defined as

$$\mathbf{p} = q\mathbf{a},\tag{9.5.1}$$

where $\mathbf{a}$ describes the charge separation as shown in Figure 9.5.1.

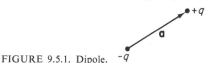

FIGURE 9.5.1. Dipole.

For example, in the hydrogen chloride gas molecule the hydrogen end of the molecule has a permanent positive charge δ, while the chlorine end has a permanent negative charge, $-\delta$. Here δ is a fraction of the unit electron charge magnitude. At low temperature (or in large electrical fields) such dipoles would align and they would attract each other. However, at high temperatures (and low electrical fields) they tend to be randomly oriented. (This is one of the important effects of temperature, namely, to cause randomization or disorientation.) However, there will still be some attraction at high temperature. The potential interaction (in electrostatic units) between two dipoles (see Problems 9.20–9.27) can be shown to be (by statistical mechanics)

$$U_0 = -\frac{2}{3}\frac{p_1^2 p_2^2}{r^6}\frac{1}{kT}.\tag{9.5.2}$$

Here r is the distance between the molecules, k is Boltzmann's constant, and T is the absolute temperature.

The **induction effect** was pointed out by Debye, who noted that orientation alone cannot account for van der Waals cohesion since there is an interaction term which is independent of temperature. The dipole is not rigid, but rather the charge distribution of one dipole changes the field of a neighboring one in such a manner as to result in an additional attraction between the two dipoles. Again the potential interaction U_I varies as $1/r^6$.

To explain the van der Waals interaction observed in gases which possess no permanent dipole moment such as argon, nitrogen, and methane, the existence of **dispersion forces** was postulated and proved to exist by F. London in 1930. These are transient or fluctuating electric moments induced between molecules having no permanent dipole moment which result in a significant attractive interaction. The dispersion energy U_D also varies as $1/r^6$.

Table 9.5.1 gives the sublimation energies for solids in which the molecules are attracted to each other by one or more of the **van der Waals bonds**.

It should be noted that the energies of molecular crystals are of the order

Table 9.5.1. COMPARISON OF THEORETICAL AND
EXPERIMENTAL HEATS OF SUBLIMATION
(KCAL/MOLE)*

Substance	Theoretical Values	Experimental Values
Ne	0.47	0.59
Ar	1.92	2.03
Kr	3.27	2.80
N_2	1.64	1.86
O_2	1.69	2.06
CO	1.86	2.09
CH_4	2.42	2.70
HCl	3.94	5.05
HBr	4.45	5.52
HI	6.65	6.21

*F. London, *Transactions of the Faraday Society*, **33**, 8 (1937).

of 1–5 kcal/mole as contrasted to values of about 200 kcal/mole for ionic solids and even higher values for pure covalent solids. The large polymer molecules of cotton and silk are bound to each other by secondary bonds. The contribution of the van der Waals binding will be considered again in many instances when we consider crystal structures in more detail.

Mention should be made of **hydrogen bonding** which has a contribution similar to that of a van der Waals bond but is completely different in origin. We know that hydrogen can form only one normal covalent electron pair bond since it has only the single $1s$ orbital available for bond formation. Since the proton is extremely small, its electrostatic field is intense, and bonding can occur owing to the attraction of the positive proton for the electrons on the bonded atom. In some situations, therefore, a hydrogen atom can form a bond to two other atoms instead of to only one. It is an example of a particularly strong dipole-dipole interaction. It occurs in general between hydrogen and the electronegative elements N, O, and F of small atomic volumes. Hydrogen bonding, like the van der Waals bonding, is a secondary bonding effect and has many important applications.

The **hydrogen bond**, while not very strong, usually has a dissociation energy of about 5–10 kcal/mole (which is a strong secondary bond), but it is important in many structures such as water and organic compounds. The structure of ice is shown in Figure 9.5.2. The high freezing and boiling points of water are due to the hydrogen bonds.

It is interesting that hydrogen bonding also makes a significant contribution to the structure of proteins, such as hair and muscle, and of complex amino acids, such as DNA (deoxyribonucleic acid). It is likewise very important in

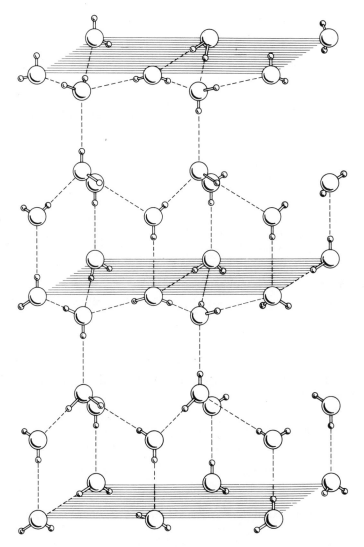

FIGURE 9.5.2. Arrangement of H_2O molecules in the ice
crystal. The large spheres are the oxygens.
Note that the orientation of the H_2O molecule
as indicated is arbitrary. It is significant that
every hydrogen atom is on a line between two
oxygen atoms and is closer to one than to the
other.

determining the structure of such synthetic polymers as nylon and plays a major role in the strength of nylon fibers. The structure of such polymeric materials will be discussed later.

EXAMPLE 9.5.1

The bonding energy per mole of argon in the solid phase is 2.03 kcal/mole, according to Table 9.5.1. Here the distance of closest approach of the atoms is 3.87 Å. Estimate the bonding energy per mole of gas at 1 atm and 0°C.

Answer. A mole of ideal argon gas under these conditions has a volume of 22.4 liters. It contains 6.02×10^{23} argon atoms. Hence

$$\text{volume/atom} = \frac{22.4 \times 10^3 \times 10^{24} \text{ Å}^3}{6.02 \times 10^{23}}$$
$$= 37 \times 10^3 \text{ Å}^3.$$

The distance between atoms is roughly the cube root of this or 33.4 Å. The binding energy is approximately determined by the attractive term $U \approx -(a/r^6)$. Then

$$\frac{U_{\text{gas}}}{U_{\text{solid}}} \approx \frac{r^6_{\text{solid}}}{r^6_{\text{gas}}} \approx \left(\frac{3.87}{33.4}\right)^6 \approx 10^{-6}.$$

Hence in the present case $U_{\text{gas}} \approx 2 \times 10^{-6}$ kcal/mole or 2×10^{-3} cal/mole.

EXAMPLE 9.5.2

The kinetic energy per mole of an ideal monatomic gas is $\frac{3}{2}RT$. Compare the potential energy of argon gas (see Example 9.5.1) with its kinetic energy at 0°C.

Answer. Since $R = 1.987$ cal/mole-°K, we have

$$U_K \doteq \tfrac{3}{2} \times 2 \times 273.2 \doteq 820 \text{ cal.}$$

The potential energy is thus only a tiny fraction $(2 \times 10^{-3}/820 \approx 2 \times 10^{-6})$ of the kinetic energy.

In a truly perfect gas the potential interactions between the molecules are exactly zero. Argon at 1 atm and 0°C comes very close to this.

REFERENCES

Sproull, R. L., *Modern Physics*, Wiley, New York (1963).

Hill, T. L., *Matter and Equilibrium*, Benjamin, New York (1966). Chapter 3 on intermolecular forces is a good simple introduction.

Speakman, J. C., *Molecules*, McGraw-Hill, New York (1966). The structure of molecules is discussed from an elementary viewpoint.

Ryschkewitsch, G. E., *Chemical Bonding and the Geometry of Molecules*, Reinhold, New York (1963). An elementary paperback which would make nice reading while waiting for a plane.

Coulson, C. A., *Valence*, Oxford, New York (1952).

Moore, W. J., *Physical Chemistry*, Prentice-Hall, Englewood Cliffs, N.J. (1964).

PROBLEMS

9.1. a. Give in kilocalories per mole the binding energies for
1. Ionic bonding of NaCl.
2. Metallic bonding of Fe.
3. Covalent bonding of diamond.

b. Give in kilocalories per mole the binding energies for secondary bonding of
1. Argon by dispersion forces.
2. H_2O molecules in ice by hydrogen bonding.

9.2. What is the essential feature of
a. The ionic bond in MgO?
b. The covalent bond in H_2?
c. The metallic bond in copper?
d. The van der Waals bond in solid argon?
e. The hydrogen bond in water?

9.3. Define ionic radius. If the distance between the Na^+ center and the Cl^- center in solid NaCl is 2.79 Å and the Cl^- has a radius of 1.82 Å, what is the Na^+ radius?

9.4. Describe the nature of the covalent bond in diamond.

9.5. Why is it that water vapor in equilibrium with water at room temperature behaves as an ideal gas?

9.6. At room temperature, iron has a body-centered cubic crystal structure. See Figure 9.4.1. The length of the edge of the cube is 2.8664 Å. Assume that the atoms touch along the cube diagonal. What is the atomic radius of the metal iron?

9.7. Describe the origin of the Keesom, Debye, and London secondary bonds which are often grouped together and called the van der Waals bonds.

More Involved Problems

Note: Problems 9.8–9.15 form a learning sequence associated with the Born-Mayer model.

9.8. Consider a one-dimensional prototype crystal of NaCl with the nearest neighbor separation equal to r.
a. Show why the coulomb attraction term of a given ion to all the other

ions can be written as

$$U_c = -\frac{2e^2}{r} + \frac{2e^2}{2r} - \frac{2e^2}{3r} + \frac{2e^2}{4r} - \cdots$$

or

$$-\frac{2e^2}{r}\left(1 - \frac{1}{2} + \frac{1}{3} - \frac{1}{4} + \cdots\right).$$

b. Show that the term in brackets is simply the Taylor series expansion of $\ln(1 + x)$ about $x = 0$, which is then evaluated for the case $x = 1$.

c. Hence, show that

$$U_c = -\frac{1.38e^2}{r} = -\frac{Ae^2}{r}.$$

The coefficient A is the **Madelung constant**.

9.9.   a. Show that the total coulombic attraction of the $2N$ ions in Problem 9.8 (N cations and N anions) is

$$U_c = -\frac{1.38Ne^2}{r}.$$

b. Assuming that we write the total repulsive potential in the form

$$U_R = \frac{B}{r^n},$$

we have

$$U = U_c + U_R = -\frac{1.38Ne^2}{r} + \frac{B}{r^n}.$$

What conditions must apply to U at equilibrium? At equilibrium r has the value r_0.

c. Show that

$$U_0 = U(r_0) = -\frac{1.38Ne^2}{r_0}\left(1 - \frac{1}{n}\right)$$

since

$$\frac{1.38Ne^2}{r_0^2} - \frac{nB}{r_0^{n+1}} = 0.$$

9.10. Suppose an experimentalist has available apparatus for applying a tensile force F to the one-dimensional crystal of Problem 9.9 and can measure

$$\lim_{r \to r_0} \frac{\partial F}{\partial r} = k.$$

Show that if k and r_0 are known, then B and n can be obtained.

9.11. There is an alternative way of obtaining Madelung's constant. This is very useful in three dimensions, but we shall illustrate it for the case of

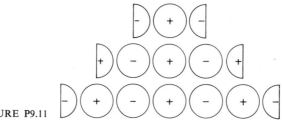

FIGURE P9.11

one dimension. In this technique, successively larger electrically neutral sections of the crystal are considered:

$$U_1 = -\frac{e^2}{r} \quad (1)$$

$$U_2 = -\frac{2e^2}{r} + \frac{e^2}{2r} = -\frac{e^2}{r}(1.5)$$

$$U_3 = -\frac{2e^2}{r} + \frac{2e^2}{2r} - \frac{e^2}{3r} = -\frac{e^2}{r}(1.33)$$

a. Show that $U_4 = -(e^2/r)(1.417)$ and $U_5 = -(e^2/r)(1.367)$. In this method, called Evjen's method, the numbers in parentheses are approximations for Madelung's constant. Note that the U_5 approximation is within 1% of the answer.

b. Compare this with the answer obtained by truncating the series expression in Problem 9.8a at five terms.

c. Give physical reasons why Evjen's method converged much more rapidly than the series of Problem 9.8a.

9.12. Use Evjen's method to evaluate the Madelung constant for a square alternating charge array.

9.13. Use Evjen's method to estimate the Madelung constant for NaCl.

9.14. The energy of the $2N$ ions of a NaCl crystal is given by

$$U = -\frac{NAe^2}{r} + \frac{B}{r^n}.$$

The pressure is related to U by

$$P = -\frac{\partial U}{\partial V}.$$

The volume of this crystal is

$$V = 2Nr^3.$$

What is $P(r)$? Use the equilibrium condition result, i.e.,

$$\frac{nB}{r_0^n} = \frac{NAe^2}{r_0}.$$

9.15. The bulk modulus is defined by

$$K = -V\frac{\partial P}{\partial V}.$$

What is $K(r)$? The bulk modulus at $P = 0$, i.e., $r = r_0$, can be measured. Show the relation of K_0 to n.

Sophisticated Problems

9.16. Show that

$$K'_0 = \lim_{r \to r_0 \text{ or } P \to 0} \frac{dK}{dP} = \frac{n+7}{3}$$

for the Born model of NaCl.

9.17. Given the polyene molecule ion

$$\begin{array}{c} \text{H} \quad \text{H} \quad \text{H} \quad \text{H} \quad \text{H} \\ | \quad\ | \quad\ | \quad\ | \quad\ | \\ \text{H}-\text{N}-\text{C}=\text{C}-\text{C}=\text{C}-\text{C}=\text{N}^+-\text{H} \\ | \qquad\qquad\qquad\qquad | \\ \text{H} \qquad\qquad\qquad\qquad \text{H} \end{array}$$

a. Draw a resonance form of this molecule.
b. The six π electrons (the resonating electrons) may be considered as free electrons in a one-dimensional box of length

$$L = 7 \times 1.39 \text{ Å},$$

where we assume the ends of the box are one-half bond; the bond distance is taken as 1.39Å. Remembering that the electrons have two possible spins, show that the energy of the highest occupied π electron state is

$$E = \frac{h^2 n^2}{8mL^2}, \qquad \text{where} \qquad n = 3.$$

c. The first excited state corresponds to exciting a π electron to $n = 4$ from $n = 3$. Calculate the change in energy for this transition.
d. Calculate the wavelength of the absorbed light for this transition and compare with the experimental value of 4250 Å. For a further discussion of calculating the color of dye molecules, see W. Kauzman, *Quantum Chemistry*, Academic Press, New York (1957) p. 675.

9.18. a. Given the interaction potential

$$U = -\frac{C}{r^6} + \frac{D}{r^{12}},$$

which is a fair description for solid xenon, carry out a Born-Mayer analysis to evaluate $U(r_0)$, assuming r_0 and K_0 are known.
b. Using the data in T. H. K. Barron and M. L. Klein, *Proceedings of the Physical Society (London)*, **85**, 533 (1965) evaluate $U(r_0)$.

9.19. In the usual application of the Born calculation, we start with Na$^+$ and Cl$^-$ ions. However, if an actual chemical reaction were occurring, we would have

$$\text{Na(metal)} + \tfrac{1}{2}\text{Cl}_2(\text{gas}) \longrightarrow \text{NaCl(ionic crystal)}.$$

a. Define all the terms which enter when we attempt experimentally to calculate the energy of the reaction

$$Na^+ + Cl^- \longrightarrow NaCl.$$

The collection of all the terms which add up to the energy of the reaction is known as the **Born-Haber cycle**. See, e.g., W. J. Moore, *Physical Chemistry*, Prentice-Hall, Englewood Cliffs, N.J. (1964).

b. Discuss how each of the terms in the Born-Haber cycle is measured.

Note: Problems 9.20–9.27 are a learning sequence concerned with charge-dipole and dipole-dipole interactions.

9.20. Show that the interaction energy between a positive charge and an aligned dipole $p = qa$ which are separated by r (where $r \gg a$) is given by

$$U = -\frac{qp}{r^2}.$$

We are using electrostatic units in which the interaction energy between two charges is $U = q_1 q_2 / r$, the force is $F = q_1 q_2 / r^2$, and the field owing to q_1 is $\xi = q_1 / r^2$. (In mks units each of the expressions would have a factor of $4\pi\epsilon_0$ added to the denominator of the right side.)

9.21. It can be shown (Chapter 30) that the average dipole moment $\bar{p}$ in a small electrical field ξ and a high temperature field T in which the dipoles are free to orient is given by

$$\bar{p} = \frac{p^2}{3kT} \xi,$$

where p is the moment qa of an individual dipole. Hence show that the charge-dipole random interaction is given by

$$U = -\frac{q^2 p^2}{3kTr^4}.$$

9.22. Show that the field ξ at a point $\mathbf{r}$ (where $\mathbf{r}$ is colinear to a dipole $\mathbf{p}$) is

$$\xi = \frac{2\mathbf{p}}{r^3}.$$

9.23. Show that the interaction energy between two parallel dipoles is

$$U = \frac{-2p_1 p_2}{r^3}.$$

9.24. A molecule (without a dipole) is placed in an electrical field ξ and becomes polarized with a dipole moment $p = \alpha\xi$, where α is called the polarizability of the molecule. Show that the work done on the molecule is $\alpha\xi^2/2$. *Hint:* $dW = F\,dx = q\xi\,dx = \xi\,dp$.

9.25. Show that a charge-induced dipole interaction is given in electrostatic units by

$$U(r) = -\frac{q^2\alpha}{2r^4}.$$

Here α is called the **polarizability** of the molecule (or atom); i.e., a dipole moment **p** is created in the molecule owing to the presence of the electric field ξ according to $\mathbf{p} = \alpha\xi$. You must include the energy needed to form the dipole as well as the attractive energy between the charge and the dipole.

9.26. Show that a dipole such as $CHCl_3$ creates a dipole in CH_4 and that the resultant interaction, averaged over all orientations, is

$$U(r) = -\frac{p^2\alpha}{r^6}.$$

9.27. Derive Equation (9.5.2).

Prologue

Many important materials are crystalline solids. In this chapter certain geometric aspects of perfect crystals are introduced: *lattice, unit cell, basis,* and *lattice parameter.* Crystals exhibit certain *symmetry* which can be described in terms of *symmetry operations. Crystal systems* and *Bravais lattices* are described, as are *lattice coordinates.* Four crystal structures are studied in detail because of their simplicity and frequency of occurrence in important engineering materials. The calculation of the density of a perfect crystal from a knowledge of the crystal structure and the lattice parameter is illustrated. The conventions used to describe planes and directions in crystals (*Miller indices* and *direction indices*) are explained.

It is noted that the symmetry of a crystal determines the particular anisotropy of a given property. For example, for a cubic crystal, the thermal expansion coefficient is isotropic (does not vary with direction) while for hexagonal crystals two coefficients are required to describe the thermal expansion behavior. Finally, there is a brief introduction to show how *group theory* and *matrix theory* can be used to describe the symmetry of crystals.

10

CRYSTALS

10.1 INTRODUCTION

In crystalline solids the atoms or molecules are arranged in a highly periodic manner in three dimensions over distances that are very great relative to the distances between them, as was proposed by Abbé Haüy in 1784. This proposal was based on the fact that interfacial angles of various crystals of the same materials, e.g., quartz, were always found to be equal, as first shown by Niels Stenson (Steno) in 1669, even though the overall shapes of the various crystals were radically different. This periodicity of the atoms affects almost all

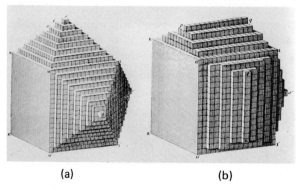

(a)   (b)

FIGURE 10.1.1. Relation of the external form of crystals to the form of the building blocks. Identical building blocks were used in (a) and (b) but different crystal faces, and hence interfacial angles are developed. (From A. Haüy, atlas to the 1822 edition of his *Traite de cristallographie*.)

FIGURE 10.1.2. Synthetically grown quartz crystal. Note the angles between the faces. (*Courtesy of Sawyer Products Co.*)

of the properties of crystals; hence it is important to be able to describe it quantitatively and in a manner which expresses effectively the significant relationships involved. Figure 10.1.1 shows why one might expect the interfacial angle relationship to hold. Figure 10.1.2 is an example of a quartz crystal grown hydrothermally.

10.2   PERIODICITY AND LATTICES

A **lattice** is a collection of points in a periodic arrangement. A **plane lattice** is defined by two noncolinear *translations* and a **space lattice** by three noncoplanar *translations*. A line joining any two points is a translation, as indicated in Figure 10.2.1 in which a plane lattice is depicted. These translations are *vectors*. It is seen that for a plane lattice any two of these (noncolinear) vectors with a common origin define a **unit cell**. The unit cell is so called because the entire lattice can be derived by repeating this cell as a unit by means of the translations that serve as the unit cells edges. The unit cell is a **primitive cell** if all the lattice points in it are at vertices; or, alternatively, if the primitive cell is displaced slightly only one lattice point falls within its area. In Figure 10.2.1 the cell t_5, t_6 is a nonprimitive unit cell; all of the others shown are primitive cells. The choice of a unit cell is made on the basis of that which best represents

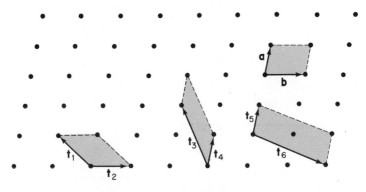

FIGURE 10.2.1. A (portion of a) plane lattice.

the *symmetry* of the lattice. It is assumed at this point that the student has a feeling for what is meant by symmetry; Figure 10.2.2 will help clarify this feeling. The unit cell designated by t_1 and t_2 represents the greater symmetry (clearly a square is more symmetric than a parallelogram).

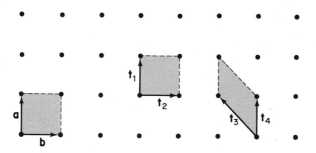

FIGURE 10.2.2. A (portion of a) plane lattice. Here $a = b$.

In the three-dimensional case we can consider the fundamental translation vectors **a**, **b**, and **c**. If the position of one of the lattice points is designated by **r**, then any other lattice point **r′** is designated by

$$\mathbf{r}' = \mathbf{r} + n_1\mathbf{a} + n_2\mathbf{b} + n_3\mathbf{c}, \tag{10.2.1}$$

where

$$n_1, n_2, n_3 = 0, \pm1, \pm2, \pm3, \ldots. \tag{10.2.2}$$

We can also write

$$\mathbf{r}' = \mathbf{r} + \mathbf{T}, \tag{10.2.3}$$

where

$$\mathbf{T} = n_1\mathbf{a} + n_2\mathbf{b} + n_3\mathbf{c}. \tag{10.2.4}$$

If every lattice point is translated according to **T**, we say that a **translation operation** has been carried out. A unit cell is shown in Figure 10.2.3. The quantities **a**, **b**, and **c** are called **lattice vectors**, while a, b, c, α, β, and γ are called **lattice parameters**. We can consider the lattice as built up from building blocks having the shape of the unit cell.

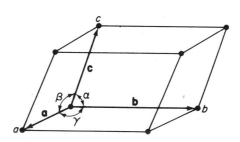

FIGURE 10.2.3. Unit cell.

10.3   SYMMETRY PROPERTIES OF CRYSTALS

A **crystal structure** is formed by the addition of a basis to every lattice point of the space lattice. By a **basis** we mean an assembly of atoms located at a lattice point which has the same composition, arrangement, and orientation as the assembly of atoms located at every other point. A two-dimensional illustration is shown in Figure 10.3.1. Here the lattice points are located at the intersection of the lines and the dumbbell-shaped assemblies of atoms (the basis) are as shown.

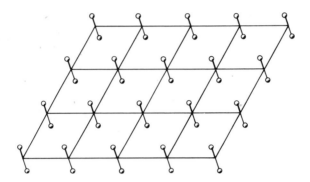

FIGURE 10.3.1. Planar crystal structure.

Suppose that we consider a physical property of a crystal such as tensile stiffness along a direction **t**. We have the following definition:

tensile stiffness along **t**

$$= \frac{\text{tensile stress along } \mathbf{t} \text{ (with no other stresses present)}}{\text{tensile strain along } \mathbf{t}}. \qquad (10.3.1)$$

Consider the crystal being made of metallic atoms located on the lattice points of the plane lattice of Figure 10.2.1 (this is an imaginary two-dimensional metallic crystal). If we were to measure the tensile stiffness along $\mathbf{t}_1$ and then along $\mathbf{t}_2$ for such a crystal, we would find in general that the tensile stiffness along $\mathbf{t}_1$ does not equal the tensile stiffness along $\mathbf{t}_2$. This is an example of an-isotropic behavior. However, if we were to make a crystal from the square lattice of Figure 10.2.2 by placing metallic atoms at each of the lattice points, then because of symmetry we see that the tensile stiffness along $\mathbf{t}_2$ equals the tensile stiffness along $\mathbf{t}_1$. Thus a rotation of the square about its center by 90 or 180 deg brings it into a position such that it superimposes on the original square. This is not true of a parallelogram when it is rotated 90 deg. It is clear that the manner in which many physical properties vary with direction in a crystal depends on the symmetry of the crystal.

Symmetry is described in terms of **symmetry operations** which transform the crystal into itself. In the example just discussed (Figure 10.2.2) we have noted that a rotation about a lattice point by 90 deg accomplished just that; *this symmetry operation is called a fourfold rotation*. An **n-fold rotation** is a rotation by $360°/n$. In crystals the only values of n which give allowable symmetry operations are $n = 1, 2, 3, 4, 6$. See Problem 10.23. The square lattice of Figure 10.2.2 clearly contains symmetry properties that the oblique lattice of Figure 10.2.1 does not. When a certain operation such as rotation brings a body into superposition with the original, we say that the body possesses this **symmetry element**.

EXAMPLE 10.3.1

Other than the ± 90-deg rotation and the 180- and 360-deg rotation, what symmetry operations can be performed on the square; i.e., what symmetry elements does the square have?

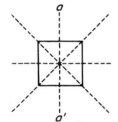

FIGURE E10.3.1

Answer. It contains **reflection planes** (or mirror planes) as shown by the dashed lines in the accompanying figure. Thus for the plane aa', if all the points on the left are reflected through the plane aa' to the right and vice versa, a new square will be obtained which superposes on the original. The parallelogram of Figure 10.2.2 does not have such symmetry.

Figure 10.3.2 shows a snowflake which exhibits symmetry characteristics of its crystal structure. Nakaya's book (see figure caption) shows over 1400 plates of snowflakes and ice crystals.

EXAMPLE 10.3.2

Consider the snowflake shown in Figure 10.3.2 to be a two-dimensional object. What symmetry elements does it possess?

Answer. Rotation by ± 60, ± 120, 180, 360 deg, and six mirror planes, three of which pass through opposite vertices and three of which pass through the midpoints of opposite edges.

We note that **a** and **b** in the square lattice are related by

$$a = b \quad \text{and} \quad \mathbf{a} \cdot \mathbf{b} = 0. \qquad (10.3.2)$$

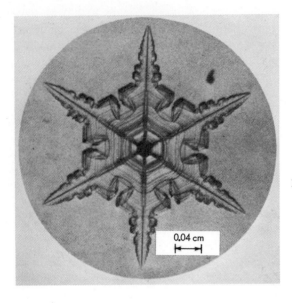

FIGURE 10.3.2. Snowflake. [From U. Nakaya, *Snow Crystals*, Harvard University Press, Cambridge, Mass. (1954).]

We might ask, *Are there other relations between* **a** *and* **b** *which yield other lattices having symmetry properties that the oblique lattice does not possess?* There are in fact five two-dimensional lattices. These five lattices are shown in Figure 10.3.3. How would you proceed to prove that there are five and only five two-dimensional lattices? The five lattices are grouped into four **crystal systems**, as shown in Table 10.3.1. A crystal system is defined on the basis of minimal symmetry. See C. Kittel, *Introduction to Solid State Physics*, Wiley, New York (1968).

Table 10.3.1. TWO-DIMENSIONAL BRAVAIS LATTICES

Lattice	Relation Between **a** and **b**	Crystal System	Lattice Points, in Conventional Cell	Minimal Symmetry
Oblique	$a \neq b$, angle arbitrary	Oblique	1	Twofold rotation axis
Square	$a = b$, $\gamma = 90°$	Square	1	Fourfold rotation axis
Hexagonal	$a = b$, $\gamma = 120°$	Hexagonal	1	Threefold rotation axis
Rectangular	$a \neq b$, $\gamma = 90°$	Rectangular	1	Mirror plane
Centered— rectangular	$a \neq b$, $\gamma = 90°$	Rectangular	2	Mirror plane

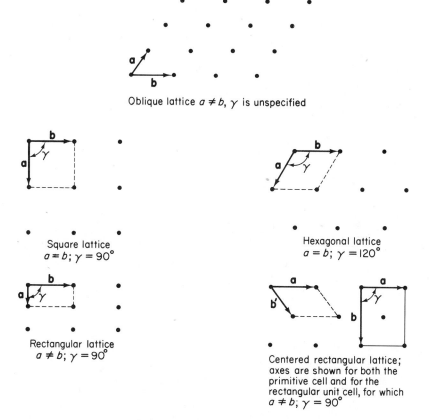

FIGURE 10.3.3. The five two-dimensional lattices. A point in one type of lattice has a distinctly different set of surroundings than a point in a different type of lattice.

There are 14 lattices in three dimensions. They are called the **Bravais lattices.** (M. L. Frankenheim had deduced in 1842 that there were 15 lattices; however, in 1848 A. Bravais showed that 2 of those 15 were the same.) Their conventional unit cells are shown in Figure 10.3.4. (The numbers 1–14, used there, have no general significance.) Notice that not all the cells chosen are primitive (although they could have been). The 14 lattices are grouped into seven crystal systems, as noted in Table 10.3.2. In a first course such as this we shall be primarily concerned with crystals based on the cubic lattices and the hexagonal lattice (there will only be a few occasions where other lattices are discussed). The unit cells of Figure 10.3.4 contain four types of symmetry elements: rotation, reflection, inversion, and rotoinversion (or rotoreflection). **Inversion** means taking each point located at any $\mathbf{r}$ to $-\mathbf{r}$. **Rotoinversion** is a combined but single operation of inversion followed by rotation.

Table 10.3.2. THE BRAVAIS LATTICES

Number in Figure 10.3.4	Space Lattice	Relationship Between a, b, and c		Corresponding Crystal Systems	Lattice Points in the Conventional Unit Cells	Minimal Symmetry
		Axes	Angles			
1	Triclinic	$a \neq b \neq c$	$\alpha \neq \beta \neq \gamma$	Triclinic	1	Onefold axis
2	Simple monoclinic	$a \neq b \neq c$	$\alpha = \gamma = 90° \neq \beta$	Monoclinic	1	One twofold axis
3	Base-centered monoclinic				2	
4	Simple orthorhombic	$a \neq b \neq c$	$\alpha = \beta = \gamma = 90°$	Orthorhombic	1	Three orthogonal twofold axes
5	Base-centered orthorhombic				2	
6	Face-centered orthorhombic				4	
7	Body-centered orthorhombic				2	
8	Hexagonal	$a = b \neq c$	$\alpha = \beta = 90°$; $\gamma = 120°$	Hexagonal	1	One sixfold axis
9	Rhombohedral	$a = b = c$	$\alpha = \beta = \gamma \neq 90°$, $< 120°$	Trigonal	1	One threefold axis
10	Simple tetragonal	$a = b \neq c$	$\alpha = \beta = \gamma = 90°$	Tetragonal	1	One fourfold axis
11	Body-centered tetragonal				2	
12	Simple cubic	$a = b = c$	$\alpha = \beta = \gamma = 90°$	Cubic	1	Four threefold axes
13	Body-centered cubic				2	
14	Face-centered cubic				4	

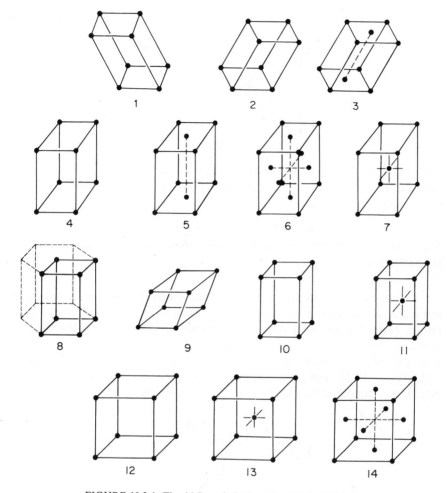

FIGURE 10.3.4. The 14 Bravais lattices. See Table 10.3.2 for
the definitions of 1-14.

10.4   SOME SIMPLE CRYSTALS

This short section has one purpose: to familiarize the student with four
crystal structures containing only one kind of atom.

BODY-CENTERED CUBIC (BCC).   The **body-centered cubic** cell is shown in
Figure 10.4.1. This is not a primitive cell (although one can be chosen). Two
atoms are contained *within* this cell (the cube center atom and one eighth of
each of the eight corner atoms). **Lattice coordinates** are used to describe the
position of a point relative to the cell origin. The position is measured in units

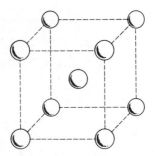

FIGURE 10.4.1. The bcc unit cell (enclosed by the dashed line).

of a along **a**, of b along **b**, and of c along **c**. Thus the position of the atom in the body center of the cube is $\frac{1}{2}\frac{1}{2}\frac{1}{2}$, or in general, uvw, so a vector r from the origin to the point uvw is given by

$$\mathbf{r} = u\mathbf{a} + v\mathbf{b} + w\mathbf{c}. \tag{10.4.1}$$

Molybdenum, tungsten, iron, and sodium have this crystal structure at room temperature and pressure.

FACE-CENTERED CUBIC (FCC). The **face-centered cubic** cell is illustrated in Figure 10.4.2. Here an atom is located at each point of a face-centered cubic lattice. Note that an atom is centered at each of the six faces. The unit cell contains four atoms. Note that we would get the same structure by starting with a simple cubic lattice and adding a basis of four atoms at positions 000, $\frac{1}{2}\frac{1}{2}0$, $\frac{1}{2}0\frac{1}{2}$, and $0\frac{1}{2}\frac{1}{2}$ (remember that the basis is added to *each* lattice point). Aluminum, copper, and nickel have this crystal structure.

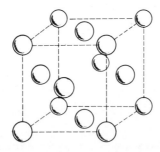

FIGURE 10.4.2. The fcc unit cell.

DIAMOND CUBIC. The **diamond cubic** cell is illustrated in Figure 10.4.3. Here the unit cell contains eight atoms. It can be considered as two interlocking fcc crystals related by a translation of $\frac{1}{4}\frac{1}{4}\frac{1}{4}$. Thus the diamond structure is a face-centered cubic lattice with a basis of two atoms, 000 and $\frac{1}{4}\frac{1}{4}\frac{1}{4}$. There are eight atoms per unit cell. Diamond, silicon, and germanium have this crystal structure.

HEXAGONAL CLOSEST-PACKED (HCP). The **hcp crystal structure** is formed from the simple hexagonal space lattice [Bravais lattice, see Figure 10.3.4 ($\#$8)]

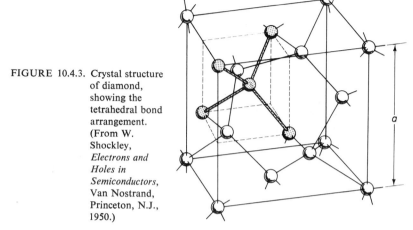

FIGURE 10.4.3. Crystal structure of diamond, showing the tetrahedral bond arrangement. (From W. Shockley, *Electrons and Holes in Semiconductors*, Van Nostrand, Princeton, N.J., 1950.)

with a basis of two atoms, 000 and $\frac{2}{3}\frac{1}{3}\frac{1}{2}$. This is illustrated in Figure 10.4.4. Beryllium, zinc, and magnesium have this crystal structure.

Of the elemental metals, more than two thirds commonly crystallize in the bcc, fcc, and hcp crystal structures which we have just described.

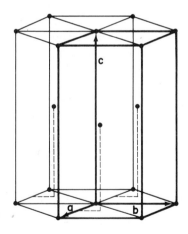

FIGURE 10.4.4. The hcp structure.

10.5 DENSITY OF CRYSTALS

We illustrate for the case of cubic crystals containing only one kind of atom how the density is related to the lattice parameter:

$$\text{density of cell} = \frac{\text{mass of cell}}{\text{volume of cell}} = \rho \qquad (10.5.1)$$

$$= \frac{m}{v} = \frac{n m_a}{v}, \qquad (10.5.2)$$

where m_a is the atomic mass and n is the number of atoms per unit cell. If Avogadro's number of atoms has a mass M, then

$$m_a = \frac{M}{N_0} \qquad (10.5.3)$$

and

$$\rho = \frac{nM}{N_0 v} = \frac{nM}{N_0 a^3}. \qquad (10.5.4)$$

From previous sections we know that $n = 2$ for bcc, $n = 4$ for fcc, and $n = 8$ for diamond cubic crystals. You have studied how M and N_0 are determined in *general chemistry*. Thus the density of an ideal cubic crystal is determined if, *first*, the crystal structure is given and, *second*, the lattice parameter a is known. The crystal structure and the lattice parameter are determined by X-ray diffraction.

10.6   CRYSTALLOGRAPHIC DIRECTIONS AND PLANES

DIRECTIONS.   A **crystallographic direction** in a crystal is designated by the vector $u\mathbf{a} + v\mathbf{b} + w\mathbf{c}$, where u, v, and w are whole numbers. Since we have already decided upon our choice of $\mathbf{a}$, $\mathbf{b}$, and $\mathbf{c}$ for the unit cell, we need only specify this direction by the ordered set enclosed in brackets $[uvw]$. Thus if the cell is simple cubic and the direction is from the origin along the body diagonal of the cube toward positive values of $\mathbf{a}$, $\mathbf{b}$, and $\mathbf{c}$, we specify the direction by [111]. All lines parallel to it are also [111] directions.

To specify a negative component a bar is placed above the number, e.g., [11$\bar{2}$]. There are many times when we wish to talk about a certain type of direction such as any body diagonal in a cube rather than specifying a specific one. We designate such a family by carets, $\langle 111 \rangle$; all the directions outward from the origin along cube body diagonals are specified by [111], [$\bar{1}$11], [1$\bar{1}$1], [11$\bar{1}$], [$\bar{1}\bar{1}\bar{1}$], [1$\bar{1}\bar{1}$], [$\bar{1}$1$\bar{1}$], and [$\bar{1}\bar{1}$1]; the last four are antiparallel, respectively, to the first four.

MILLER INDICES.   A plane which passes through a lattice point is called a **crystallographic plane**. Each crystallographic plane has an infinite number of planes parallel to it. It is, however, possible to designate the orientation of a crystallographic plane in a simple way by a set of three integers, called the **Miller indices** and designated by (hkl). This is illustrated with the help of Figure 10.6.1.

The procedure is as follows:

1. Determine the intercepts of the plane on the three crystal axes in units of a, b, and c.
2. Take the reciprocals of these numbers.
3. Clear fractions.

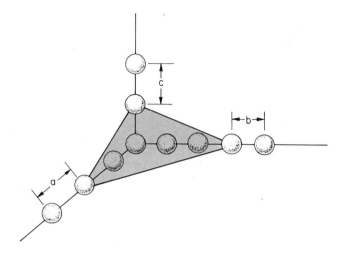

FIGURE 10.6.1. Intercepts of a plane of atoms on the three
crystal axes. Only a few of the lattice points
are shown.

The intercepts along the **a**, **b**, and **c** axes in the present case are 2, 3, and 1, respectively. The reciprocals are $\frac{1}{2}$, $\frac{1}{3}$, and 1, respectively. These have a common denominator of 6; hence we multiply by 6 to obtain 3, 2, and 6, respectively.

The Miller indices 3, 2, and 6 are expressed by the ordered set enclosed in parentheses (326), or, in general form, (*hkl*). There will be a plane parallel to this plane which passes through the origin. It should be noted that the set of Miller indices specifies not merely a single plane but the whole array of planes parallel to it.

If a plane has a negative intercept such as is the case for the intercepts −3, 2 and 1, we indicate it as ($\bar{2}$36) with an overbar on the negative intercept. Braces, {*hkl*}, signify all the planes in a crystal which are equivalent; e.g., {100} means, for a cubic crystal, all the cube faces. It should be noted that the distance *d* between two neighboring parallel planes can be determined and general mathematic expressions written out; this distance will depend on the lattice constants and the Miller indices. They are given, for example, in C. S. Barrett and T. B. Massalski, *Structure of Metals*, McGraw-Hill, New York (1966). The cubic case is particularly simple since the crystal axes form a Cartesian coordinate system. From his background in analytical geometry the student can show that for a *simple cubic* crystal *only*,

$$d = \frac{a}{\sqrt{h^2 + k^2 + l^2}}. \tag{10.6.1}$$

Another interesting aspect of *cubic* crystals (only) is that the crystallographic direction [*hkl*] is normal to a plane whose Miller indices are (*hkl*). Because cubic crystals are particularly simple (and important), we often use them as illustrations. The Miller indices of some important planes in a cubic crystal are shown in Figure 10.6.2.

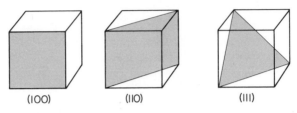

(100)   (110)   (111)

FIGURE 10.6.2. Important planes in a cubic crystal.

In the hexagonal crystal we may choose a unit cell as shown by the heavy lines in Figure 10.6.3. Often, however, it is convenient to use a cell which is three times as large and is defined by four lattice vectors: a_1, a_2, a_3, and c. Here three of the crystal axes lie in the base of the hexagon (the **basal plane**). The intercepts of a crystal plane on these four axes lead to the four Miller-Bravais indices (*hkil*). If the (*hkl*) planes are the Miller indices, then the (*hkil*) planes are the **Miller-Bravais indices**, where we must have $i = -(h + k)$.

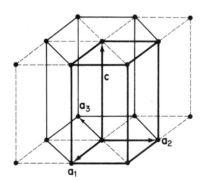

FIGURE 10.6.3. Hexagonal system.

EXAMPLE 10.6.1

Which planes are the farthest apart in the simple cubic crystal?

Answer. The {100} planes are a distance a apart, where a is the lattice parameter of the cubic crystal. Note that to specify the lattice parameters of a cubic crystal we need to specify only the length of an edge.

EXAMPLE 10.6.2

Which planes are the farthest apart in a bcc crystal?

Answer. We note that Equation (10.6.1) applies to simple cubic crystals only. The answer in this case is the {110} planes which are a distance $a/\sqrt{2}$ apart. Consider Figure 10.4.1 to be infinite in extent. Note that the {100}-type planes are only $a/2$ apart. Because the body-centered cell is not a primitive cell, many of the planes {hkl} are closer than given by Equation (10.6.1).

The cell dimensions of a few elements are shown in Table 10.6.1.

Table 10.6.1. CELL DIMENSIONS OF SOME ELEMENTS AT ROOM TEMPERATURE (UNLESS NOTED)

Element	Structure	a, Å	c, Å
Aluminum	fcc	4.05	
Argon	fcc	5.43 (20°K)	
Beryllium	hcp	2.28	3.58
Cadmium	hcp	2.98	5.62
Carbon	Diamond	3.57	
Copper	fcc	3.61	
Germanium	Diamond	5.66	
Iron	bcc	2.87	
Magnesium	hcp	3.21	5.21
Molybdenum	bcc	3.15	
Nickel	fcc	3.52	
Niobium	bcc	3.30	
Silicon	Diamond	5.43	
Titanium	hcp	2.95	4.68
Tungsten	bcc	3.16	
Zinc	hcp	2.66	4.95

For further information regarding specific crystal structures, see R. W. G. Wycoff, *Crystal Structures*, Wiley-Interscience, New York (1963) or the *International Tables for X-Ray Crystallography*, Kynock Press, Birmingham (1952–1962).

10.7   PIEZOELECTRICITY AND SYMMETRY

Because three-dimensional crystals are more complicated than two-dimensional crystals (which are imaginary), it is often helpful to consider the behavior of a two-dimensional crystal. Figure 10.7.1 shows a two-dimensional ionic crystal. This crystal does not have an inversion center, and such crystals are called **noncentrosymmetric crystals**. Recall from Section 10.3 that the inversion operation means taking every point **r** to $-$**r**. If this leaves the crystal unchanged, it possesses an inversion center. Consider the center of one of the negative atoms in Figure 10.7.1 to be an origin from which **r** is measured. Performing the inversion operation moves positive charges to positions previously

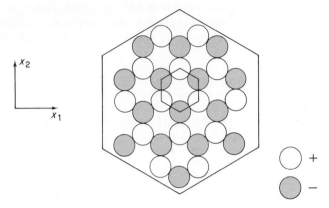

FIGURE 10.7.1. Two-dimensional ionic crystals.

not occupied by any atoms. Hence, this crystal does not possess an inversion center. Let us consider the ions to be rigid spheres which are just touching and let us apply a compressive stress in the x_2 direction. See Figure 10.7.2. After the compressive stress is applied, ions 5 and 6 move inward, 1 and 2 move to the left, and 3 and 4 move to the right. From the diagram it can be seen that the centroids of the two triangles do not coincide with each other (as they did before the deformation), so that an electric dipole moment **p** of magnitude qd exists, as is clear from Figure 10.7.2. Since a similar dipole occurs in each cell, the crystal has become polarized (the polarization **P** is the sum of all the dipole moments per unit volume). The development of a polarization as a result of application of a stress is called the **piezoelectric effect**.

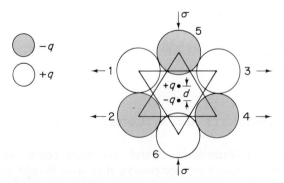

FIGURE 10.7.2. Stress applied to ionic crystal.

If a crystal has a center of symmetry, the charge centroids would coincide when the crystal deforms and there would be no piezoelectric effect. You can illustrate this by considering the two-dimensional crystal formed by a single (100) plane of cubic rocksalt (see Figure 9.2.2).

10.8 ANISOTROPY AND CRYSTAL STRUCTURE

In Section 10.2 we noted that the symmetry of a crystal determines its degree of anisotropy. As an example, linear thermal expansion behavior is described in general by a tensor having six

$$
\begin{matrix}
\alpha_{11} & \alpha_{12} & \alpha_{13} \\
\alpha_{21} & \alpha_{22} & \alpha_{23} \\
\alpha_{31} & \alpha_{32} & \alpha_{33}
\end{matrix}
\tag{10.8.1}
$$

independent components since $\alpha_{ij} = \alpha_{ji}$. See Equation (7.4.5). Specifying this behavior for triclinic crystals would require exactly six coefficients. However, for a hexagonal closest-packed crystal such as zinc (see Figure 10.4.4) only *two* independent coefficients would be required. The tensor would be

$$
\begin{matrix}
\alpha_a & 0 & 0 \\
0 & \alpha_a & 0 \\
0 & 0 & \alpha_c
\end{matrix}
\tag{10.8.2}
$$

where $x_1 \perp \mathbf{c}$ and $x_3 \,||\, \mathbf{c}$. Here x_1, x_2, and x_3 are the Cartesian axes to which the tensor properties are referred and $\mathbf{a}$, $\mathbf{b}$, and $\mathbf{c}$ are the crystallographic axes as given in Figure 10.4.4. The tensor simplifies in this fashion because of the symmetry properties of the crystal. (There are four types of symmetry operations which affect macroscopic behavior: rotation, reflection, inversion, and rotation-inversion.)

It can be shown that the thermal expansion coefficient along a line inclined at θ to the c axis of the hcp crystal is

$$
\alpha = \alpha_a \sin^2 \theta + \alpha_c \cos^2 \theta.
\tag{10.8.3}
$$

Table 10.8.1 gives some other second-rank tensor properties for hexagonal crystals. (The strain ϵ_{ij} produced by the presence of a hydrostatic pressure P is related to P by

$$
\epsilon_{ij} = -\beta_{ij}P,
$$

where β_{ij} represents the **linear compressibility**.)

Using the equations developed for tensors in Chapter 7, we could show that only one thermal expansion coefficient has to be specified for cubic crystals and that the thermal expansion coefficient is the same in all directions (the behavior is therefore isotropic). The tensor is then

$$
\begin{matrix}
\alpha & 0 & 0 \\
0 & \alpha & 0 \\
0 & 0 & \alpha
\end{matrix}
$$

However, for other properties, such as elastic properties which involve fourth-rank tensors, cubic crystals behave anisotropically.

Table 10.8.1. EXAMPLES OF ANISOTROPIC PROPERTIES OF
HEXAGONAL CRYSTALS*

Crystal	Structure	Linear Compressibility, 10^{-13} cm²/dyne, Parallel		Coeff. of Thermal Exp., $10^{-6}/°C$, Parallel		Spec. Elec. Resistivity, 10^{-6} Ω-cm, Parallel		Thermal Conductivity, W/cm-°C, Parallel	
		c	*a*	*c*	*a*	*c*	*a*	*c*	*a*
Mg	Hcp	9.9	9.5	27.0	25.4	3.78	4.53		
Zn	Hcp	13.2	1.75	63.9	14.1	6.05	5.83	1.24	1.24
Cd	Hcp	16.9	1.5	52.6	21.4	8.36	6.87	0.83	1.04
Hg	Hex.	3.0	14.0	47.0	37.5	5.87	7.78	0.399	0.290

* From W. Boas and J. K. Mackenzie, *Progress in Metal Physics*, Vol. II, Pergamon Press, New York (1949) p. 90.

EXAMPLE 10.8.1

An aluminum bar consists of an enormous number of randomly oriented crystals. If the temperature of the bar is changed uniformly, are internal stresses built up?

Answer. No, because thermal expansion is isotropic in cubic crystals. However, this would not be the case for zinc which is hcp or uranium which is orthorhombic.

EXAMPLE 10.8.2

Prove for a tetragonal crystal that the thermal expansion coefficient along **a** is the same as the thermal expansion coefficient along **b**. See Table 10.3.2.

Answer. We note that a 90-deg rotation about **c** brings the crystal back into itself. Hence the properties are the same in directions differing by 90 deg.

EXAMPLE 10.8.3

Show for a tetragonal crystal that the thermal expansion tensor also reduces to (10.8.2). Assume $x_1 \| \mathbf{a}$, $x_2 \| \mathbf{b}$, and $x_3 \| \mathbf{c}$.

Answer. Use Equations (7.3.2) and (7.3.6), except that (7.3.6) now takes the form

$$\alpha'_{ij} = a_{ik} a_{jl} \alpha_{kl}. \qquad (10.8.4)$$

We note that a 90-deg rotation about x_3 is a symmetry operation. Let us therefore rotate the axes by 90 deg, as shown.

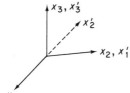

FIGURE E10.8.3 x_1

Then the a_{ij}'s of Equation (7.3.2) are given by

$$\begin{pmatrix} a_{11} & a_{12} & a_{13} \\ a_{21} & a_{22} & a_{23} \\ a_{31} & a_{32} & a_{33} \end{pmatrix} = \begin{pmatrix} 0 & 1 & 0 \\ -1 & 0 & 0 \\ 0 & 0 & 1 \end{pmatrix}. \tag{10.8.5}$$

Hence from (10.8.4) and (10.8.5),

$$\begin{aligned}
\alpha'_{11} &= \alpha_{22} \\
\alpha'_{12} &= -\alpha_{21} \\
\alpha'_{13} &= \alpha_{23} \\
\alpha'_{22} &= \alpha_{11} \\
\alpha'_{21} &= \alpha'_{12} \quad \text{since } \alpha_{ij} = \alpha_{ji} \\
\alpha'_{23} &= -\alpha_{13} \\
\alpha'_{33} &= \alpha_{33} \\
\alpha'_{31} &= \alpha'_{13} \quad \text{since } \alpha_{ij} = \alpha_{ji} \\
\alpha'_{32} &= \alpha'_{23} \quad \text{since } \alpha_{ij} = \alpha_{ji}.
\end{aligned} \tag{10.8.6}$$

However, because of the symmetry property we note that the primed tensor equals the unprimed tensor for a 90-deg rotation. Hence,

$$\begin{aligned}
\alpha'_{11} &= \alpha_{11} \\
\alpha'_{12} &= \alpha_{12} \\
\alpha'_{13} &= \alpha_{13} \\
\alpha'_{22} &= \alpha_{22} \\
\alpha'_{21} &= \alpha_{21} \\
\alpha'_{23} &= \alpha_{23} \\
\alpha'_{33} &= \alpha_{33} \\
\alpha'_{31} &= \alpha_{31} \\
\alpha'_{32} &= \alpha_{32}.
\end{aligned} \tag{10.8.7}$$

From (10.8.6) and (10.8.7) we have

$$\begin{aligned}
\alpha_{22} &= \alpha_{11} \\
-\alpha_{21} &= \alpha_{12} = 0 \\
\alpha_{23} &= \alpha_{13} \\
-\alpha_{13} &= \alpha_{23} \quad \text{(and since } \alpha_{13} = \alpha_{23}, \, \alpha_{13} = \alpha_{23} = 0).
\end{aligned}$$

Hence the thermal expansion tensor of Equation (10.8.1) is reduced by this 90-deg rotational symmetry operation to

$$\begin{matrix} \alpha_{11} & 0 & 0 \\ 0 & \alpha_{11} & 0 \\ 0 & 0 & \alpha_{33}. \end{matrix} \qquad (10.8.8)$$

The tetragonal crystal also has other possible symmetry operations; e.g., a 180-deg rotation about x_2 brings the crystal into itself. It is left as an exercise to the student to show that (10.8.8) does not reduce further under this 180-deg symmetry operation.

All the symmetry operations discussed thus far (rotations, reflections, inversion, and rotoinversions) can be represented mathematically by orthogonal matrices.

EXAMPLE 10.8.4

A Cartesian set of axes x_i is set in a cube with the axes along the edges. We consider a new set of axes x'_j generated from the old by a 120-deg rotation about the [111] direction (cube diagonal) such that x_1 moves toward x_2, etc. How are the components of the two sets of axes related?

Answer. Clearly, x'_1 and x_2 coincide, x'_2 and x_3 coincide, and x'_3 and x_1 coincide. This is represented by a matrix transformation

$$\begin{pmatrix} x'_1 \\ x'_2 \\ x'_3 \end{pmatrix} = \begin{pmatrix} 0 & 1 & 0 \\ 0 & 0 & 1 \\ 1 & 0 & 0 \end{pmatrix} \begin{pmatrix} x_1 \\ x_2 \\ x_3 \end{pmatrix}.$$

Since the rotation described is a symmetry property of the cube (either the axes or the crystal could have been rotated in the manner given), the matrix represents one of the symmetry operations of the cube. Rotation in the opposite direction by 120 deg would also be a symmetry operation. Find its matrix representation.

10.9   MORE ABOUT SYMMETRY

To discuss symmetry further, it is useful to define and illustrate mathematical entities called groups. A **group** is a distinct set of elements $A, B, C, \ldots$ satisfying the following postulates:

1. A composition rule called **product** is defined such that AB is also an element of the group.
2. The set contains an identity element I such that RI (or IR) $= R$ for all elements R in the group.
3. The set contains for every member R an inverse designated as R^{-1}, such that $R^{-1}R = RR^{-1} = I$ for all R in the group.
4. The associative law for a product $A(BC) = (AB)C$ holds.

A simple example of a group is the set of elements 1, -1, i, and $-i$ (where $i = \sqrt{-1}$), where the composition rule is ordinary multiplication. The easiest way to prove that this is indeed a group is to make a multiplication table as in Table 10.9.1. Clearly this satisfies all the group requirements.

Table 10.9.1. MULTIPLICATION TABLE

	1	-1	i	$-i$
1	1	-1	i	$-i$
-1	-1	1	$-i$	i
i	i	$-i$	-1	1
$-i$	$-i$	i	1	-1

The set of all symmetry operations of a finite figure such as a cube form a point group. In this case the product AB means the operation B followed by the operation A. The product CC will be presented by C^2. The identity operator in this case is the onefold rotation.

Consider now the *two-dimensional* symbol of Figure 10.9.1.

FIGURE 10.9.1. Two-dimensional symbol.

EXAMPLE 10.9.1

What symmetry elements does the symbol of Figure 10.9.1 contain?

Answer.

A 90-deg (counterclockwise) rotation. Call this operation C.

A 180-deg rotation. This could be called C^2.

A -90-deg rotation. This is C^{-1}.

A 360-deg rotation. This is I.

There are no other symmetry elements. The group is $\{C\ C^2\ C^{-1}\ I\}$.

EXAMPLE 10.9.2

Prove that the elements of Example 10.9.1 do indeed form a group.

Answer. To do this form the multiplication table.

Actually each of these rotation operations is representable by a matrix transformation

$$\begin{pmatrix} x_1' \\ x_2' \end{pmatrix} = \begin{pmatrix} \cos\theta & \sin\theta \\ -\sin\theta & \cos\theta \end{pmatrix} \begin{pmatrix} x_1 \\ x_2 \end{pmatrix}.$$

Thus

$$(C) = \begin{pmatrix} 0 & 1 \\ -1 & 0 \end{pmatrix}.$$

We now have the necessary background to discuss the **crystallographic point groups** in two dimensions where the only possible operators are the n-fold rotations, where $n = 1, 2, 3, 4,$ and 6, and the mirror plane m. These point groups are illustrated in Figure 10.9.2.

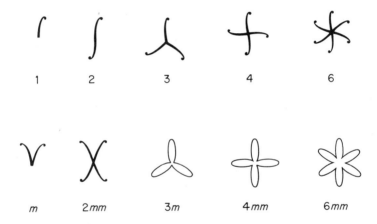

FIGURE 10.9.2. Top row represents the five rotation groups. Bottom row represents the five vertical mirror groups.

Could you figure out a scheme to prove that these are the only possible 2-d point groups?

We note that there are four 2-d crystallographic point groups associated with the 2-d hexagonal crystal system: 3, 6, 3 m and 6 mm. Why are point groups important? Because the point group to which a crystal belongs determines the nature of its anisotropic macroscopic behavior, as already discussed (in part) in Section 10.8. (The reader may have been left with the erroneous impression that it was the crystal system which determines this in the general case.)

In three dimensions, there are 32 point groups based on the four types of operations: rotation, reflection, inversion, and rotoinversion (or rotoreflection).

We might ask how we would generate all the different 2-d crystal symmetries. We would do this by associating the symbols of Figure 10.9.2 with the lattice points of the five 2-*d* lattices. This would give 13 distinct patterns. To obtain the remaining 4, we would have to use an additional operator, the **glide-mirror** operation. This is a combined operation of glide of the symbol followed by reflection. (However, from the viewpoint of a macroscopic property such as elastic coefficients, this has the same effect as a reflection.)

Thus from the viewpoint of structure there are 17 patterns called plane groups. (In 3-*d* there are 230 space groups. These are based on the six operations rotation, reflection, inversion, rotoinversion, glide-mirror, and screw axes.

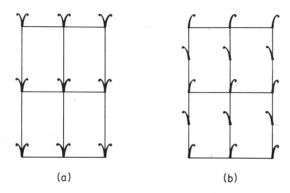

FIGURE 10.9.3. Two plane groups.

The last is a combined rotation-translation operation.) Figure 10.9.3 illustrates the formation of two plane groups. The pattern (crystal) in Figure 10.9.3(a) could be constructed by starting with the basic symbol ʃ, reflecting to form ⌐, and then combining the two to form Υ, which then undergoes the translation operation of the simple rectangular lattice. The crystal in Figure 10.9.3(b) could be constructed by starting with the basic symbol and proceeding as follows:

$$ʃ + [\ ʃ \text{ reflection} \rightarrow\ ⌐ + \text{glide} \rightarrow\ ⌐\ ] =\ ⌐ ʃ.$$

This new symbol then undergoes the translation pattern of the simple rectangular lattice.

To summarize: Point groups are important because they determine the anisotropic macroscopic nature of crystals. (All cubic point groups have an isotropic thermal expansion coefficient; symmetry cannot tell us the magnitude of α.) The external features of a crystal, except for the extent of the development of the faces, are also determined by the point group symmetry. To completely determine the structure of a crystal, the space group must be obtained. For further study: Pictures of the 32 point groups are given in Holden and Singer (see References), where the elements of the groups are described verbally. The elements of the 32 point groups are described mathematically by Jaswon (see References).

REFERENCES

Holden, Alan, and Singer, Phyllis, *Crystals and Crystal Growing*, Doubleday, New York (1960). This is an elementary paperback book. If the student has not already read it, he should. The student interested in the type of symmetry exhibited by a crystal about a point should refer to p. 284.

Weyl, Hermann, *Symmetry*, Princeton University Press, Princeton, N.J. (1952). Illustrates symmetry in art, architecture, and nature as well as in crystals. This is a great book.

Jaswon, M. A., *Mathematical Crystallography*, Longmans, London (1965). Discusses the 32 point groups and the space groups. Requires an appreciation of matrix algebra. What Holden and Singer do pictorially is done here with matrix theory. The student could progress from Jaswon to a book utilizing formal group theory.

Mott, N., "The Solid State," *Scientific American* (Sept. 1967) p. 80. Sir Nevill Mott gives a clear introduction to the crystalline state plus an introduction to solid state physics.

Brice, J. C., *The Growth of Crystals from the Melt*, Wiley, New York (1965).

Gilman J. J., ed., *The Art and Science of Growing Crystals*, Wiley, New York (1963).

Bragg, Sir William, *Concerning the Nature of Things*, G. Bell, London (1948), reprinted by Dover, New York (1954).

Wycoff, R. W. G., *Crystal Structures*, Wiley-Interscience, New York (1963). Look in this or in the International Tables for specific crystal structures. *International Tables for X-ray Crystallography*, Kynock Press, Birmingham (1952–1962.)

PROBLEMS

10.1. What is the volume of a unit cell defined by **a**, **b**, and **c**?

10.2. There are three cubic lattices. Draw the unit cell for each. Imagine the cells to be moved slightly along a body diagonal. How many lattice points are then completely within the cell for each of three lattices?

10.3. Comment on the following: The bcc crystal structure of molybdenum is a body-centered cubic lattice with a basis of one molybdenum atom at the origin or the bcc crystal structure of molybdenum is a simple cubic lattice with a basis of two atoms: Mo at 000 and Mo at $\frac{1}{2}\frac{1}{2}\frac{1}{2}$.

10.4. The unit cell of the fcc crystal structure of aluminum contains four atoms. Prove this.

10.5. While we ordinarily think of the fcc crystal structure of aluminum as a fcc lattice with a basis of one aluminum atom centered at each lattice point, we could also form this crystal structure from a simple cubic lattice with a basis of four atoms: 000, $\frac{1}{2}\frac{1}{2}0$, $\frac{1}{2}0\frac{1}{2}$, $0\frac{1}{2}\frac{1}{2}$. Prove this.

10.6. A NaCl crystal is a fcc lattice with a basis of Na$^+$ at 000 and Cl$^-$ at $\frac{1}{2}00$. When such a crystal grows from a water solution, cubes are formed. What can you say about the relative growth rates of the (100), (110), and (111) planes?

10.7. A sodium chloride crystal grows as a cube from pure water-solution, but the growth habit is changed when borax is added to the solution and regular tetrahedrons are then formed.
a. Describe the rate of growth of specific planes.
b. What is a possible explanation of the effect of the borax?

10.8. Copper has a fcc crystal structure with a lattice parameter of 3.62 Å. If the atoms are rigid spheres in contact along the ⟨110⟩ directions, calculate the sphere diameters.

10.9. Lead has a fcc crystal structure with a lattice parameter of 4.95 Å and an atomic weight of 207.19. Calculate the density.

10.10. Copper has a fcc structure with $a = 3.615$ Å. Calculate its density.

10.11. Molybdenum has a bcc structure with $a = 3.15$ Å. Calculate its density.

10.12. Zinc has a hcp structure with $c = 4.95$ Å and $a = 2.66$ Å. Calculate its density.

10.13. In the simple cubic crystal unit cell shown here, give the direction indices of

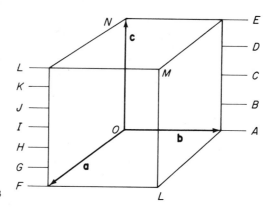

FIGURE P10.13

a. G to D.
b. O to L.
c. F to M.
Give the Miller indices of the planes
d. FAN.
e. KFA.
f. HAN.

10.14. Define (a) isotropic and (b) anisotropic. Give some examples of anisotropic properties of crystals.

10.15. What is the maximum thermal expansion coefficient (depending on direction) in zinc?

10.16. Because of their symmetry, some crystals do not exhibit certain behavior. Illustrate.

More Involved Problems

10.17. Draw a fcc lattice (several cells) and pick out a primitive cell which could be used instead.

10.18. Clearly sketch the (112) plane of a cubic crystal. Does a $\langle 110 \rangle$ direction lie in this plane? Which one? Prove that for *cubic* crystals a direction $[uvw]$ lies in a plane (hkl) if $uh + vk + wl = 0$.

10.19. Define Miller indices. Then show that for a simple cubic crystal the distance from the origin to the plane with Miller indices hkl is given by Equation (10.6.1).

Sophisticated Problems

10.20. Complete Example 10.8.3.

10.21. Given that the thermal expansion tensor for a tetragonal crystal is

$$\begin{matrix} \alpha_{11} & 0 & 0 \\ 0 & \alpha_{11} & 0 \\ 0 & 0 & \alpha_{33}, \end{matrix}$$

show that the thermal expansion for a cubic crystal is

$$\begin{matrix} \alpha_{11} & 0 & 0 \\ 0 & \alpha_{11} & 0 \\ 0 & 0 & \alpha_{11}. \end{matrix}$$

10.22. Derive Equation (10.8.3) for a tetragonal crystal.

10.23. Show that the only rotations consistent with the translational symmetry of the two-dimensional lattice are n-fold rotations, where $n = 1, 2, 3, 4, 6$. *Hint:* Start with two parallel rows of atoms as shown.

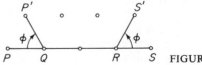

FIGURE P10.23

10.24. Show that there are five and only five two-dimensional lattices.

10.25. a. After doing outside reading, write down all the symmetry operators associated with a cube.

b. Could each of your symmetry operators be represented by an orthogonal matrix?

c. Illustrate question b.

10.26. a. What are the symmetry elements of a regular tetrahedron?

b. Does it contain a center of symmetry?

c. Do diamond crystals have a center of symmetry?

Prologue

Perfect crystals are periodic arrays of atoms. Such periodic arrays lead to diffraction of waves whose wavelength is the same order of magnitude as the periodicity of the lattice. Such diffraction leads to *Bragg's law*. The origin of *diffraction lines* or *diffraction peaks* of simple cubic crystals is discussed. We also show how to use a set of diffraction data to determine whether a cubic crystal has a simple cubic crystal structure, a body-centered cubic crystal structure, or a face-centered cubic crystal structure and to obtain its lattice parameter. We observe that the intensity of a diffraction peak decreases exponentially as the temperature increases. Finally, we note that electrons and neutrons can also be diffracted by crystals and that *electron diffraction* and *neutron diffraction* can often provide important information about the structure of crystals which cannot be obtained by *X-ray diffraction*.

11

DIFFRACTION BY CRYSTALS

11.1 BRAGG'S LAW

It has been known for a long time that a ruled grating causes diffraction. In 1912 Max von Laue, a German scientist, suggested that if crystals were composed of regularly spaced atoms, and if X-rays were electromagnetic waves with a wavelength of the same order as the atom size, then crystals should diffract X-rays. Experiments carried out under his direction on copper sulfate verified both of these assumptions. In 1913 W. L. Bragg obtained the crystal structure of NaCl. This was the first determination of a crystal structure.

Let us suppose that we have a parallel beam of copper K_α X-rays (wavelength $\lambda = 1.54$ Å). (We have discussed the origin of such X-rays in Example 8.3.3.) Figure 11.1.1 shows a set of parallel planes (normal to the paper) of a crystal represented by the lines AA and BB. The line LL_1 represents the crest

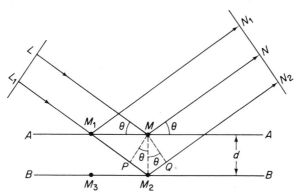

FIGURE 11.1.1. Bragg's law.

of an incoming wave [actually this wave will "strike" huge numbers of scattering centers (atoms) but only three are shown scattering here]. *LM* was chosen to be an integral number of wavelengths and *MN* is also an integral number of wavelengths. We consider now the three path lengths: LMN, $L_1M_1N_1$, and $L_1M_2N_2$. The line N_1NN_2 has been chosen to be the crest of the outgoing wave. We now proceed to prove that there is such a crest under certain conditions. Clearly,

$$LMN = L_1M_1N_1 \tag{11.1.1}$$

if the angle of incidence equals the angle of reflection (and we have chosen that case). (The point M_1 could be *any* atom on the *AA* plane which scatters the incoming wave.) We note that

$$L_1M_2N_2 - LMN = PM_2Q. \tag{11.1.2}$$

For N_2 to be at a crest the extra path difference must equal an integral number of wavelengths, or

$$n\lambda = 2d \sin \theta, \tag{11.1.3}$$

where n is an integer. A similar relation would apply to an incoming wave scattered from M_3. It would, moreover, apply between any neighboring pair of planes, e.g., *BB* and *CC*. Equation (11.1.3) is therefore the condition for mutual reinforcement of the scattered rays. It is called **Bragg's law** or the **Bragg equation**. It relates the **order of the reflection** n, the wavelength of the radiation λ, the distance between the planes d, and the angle of incidence θ. θ is also called the **diffraction angle**.

11.2   BRAGG'S LAW AND SIMPLE CUBIC CRYSTALS

Although simple cubic crystals are not common (polonium has this structure at room temperature), we can easily analyze their diffraction pattern.

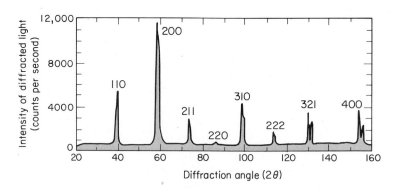

FIGURE 11.2.1. Record of diffraction angles for a tungsten sample obtained by use of a diffractometer with characteristic copper radiation.

If we combine Equation (11.1.3) with Equation (10.6.1), we have

$$n^2(h^2 + k^2 + l^2) = \frac{4a^2}{\lambda^2} \sin^2 \theta. \tag{11.2.1}$$

We choose to write this in the form

$$(H^2 + K^2 + L^2) = \frac{4a^2}{\lambda^2} \sin^2 \theta, \tag{11.2.2}$$

where $H = nh$, etc. We shall write the set HKL without the parentheses used to describe Miller indices (hkl); thus 200 is possible, while (200) is not since Miller indices cannot have a common factor. The experimentalist measures a set of θ's. An example is shown in Figure 11.2.1, in which case the radiation is monochromatic K_α radiation of copper and the sample is a powder.

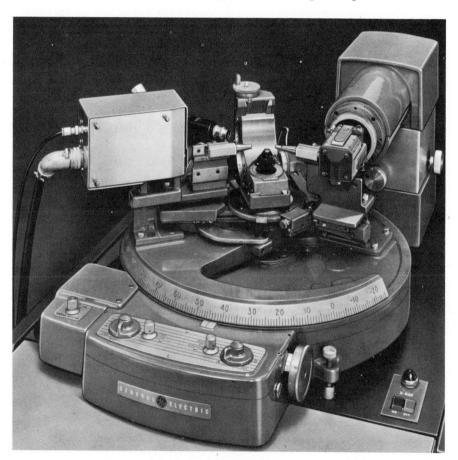

FIGURE 11.2.2. General Electric Diffractometer. (*Courtesy of General Electric Company.*)

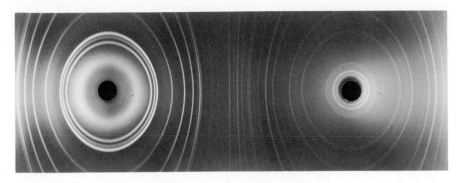

FIGURE 11.2.3. X-ray diffraction lines.

The wavelengths of characteristic X-ray lines are related to the voltage V by

$$\lambda \text{ (Ångstroms)} = \frac{12,400}{V \text{ (volts)}}. \qquad (11.2.3)$$

See Example 8.3.3. The various peaks are called diffraction peaks, e.g., the 110 **diffraction peak**. The X-ray diffractometer used to obtain this data is shown in Figure 11.2.2. The peaks are measured using a scintillation counter.

The data shown in Figure 11.2.1 were obtained from a very fine powder of tungsten containing a huge number of randomly oriented crystals. This is called the **Debye-Scherrer method**; in the original powder method the diffraction data were recorded on film. When the data are recorded on film each peak appears as a line, e.g., the 110 **diffraction line**. An example of such a film is shown in Figure 11.2.3.

The experimental arrangement for the film technique is shown in Figure 11.2.4.

Suppose that the experimentalist obtains $\sin^2 \theta$ values as follows: 0.100,

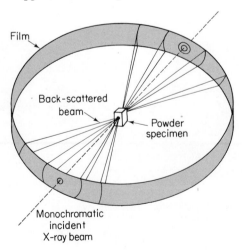

Film

Back-scattered beam

Powder specimen

Monochromatic incident X-ray beam

FIGURE 11.2.4. Sketch of Debye-Scherrer powder experiment using film.

0.201, 0.299, 0.401, 0.502, 0.599, 0.801, and 0.900. We now show that this is consistent with a simple cubic crystal structure. We do not know the lattice parameter a. We *assume* $4a^2/\lambda^2 = 10$ in the present case. This is a guess! We then make a table:

$\sin^2 \theta$	$\dfrac{4a^2}{\lambda^2} \sin^2 \theta$	$H^2 + K^2 + L^2$	HKL
0.100	1.00	1	100
0.201	2.01	2	110
0.299	2.99	3	111
0.401	4.01	4	200
0.502	5.02	5	210
0.599	5.99	6	211
—	—	—	—
0.801	8.01	8	220
0.900	9.00	9	300

The data are consistent with Equation (11.2.2), except for small experimental errors which are to be expected. Assuming that the wavelength used was $\lambda = 2.00$ Å, we have $a^2 = 10$ Å^2 and $a = \sqrt{10}$ Å.

It should be noted that the diffraction peaks in Figure 11.2.1 have finite widths. The Bragg equation, however, suggests that for monochromatic radiation, the diffraction peak for a perfect crystal would be perfectly sharp. What do you think causes a finite peak width? We note that at large θ a peak appears to split into two peaks. Could this be due to the fact that the characteristic emission spectra of copper contains two nearby lines?

Suppose that instead of bombarding a huge number of randomly oriented single crystals with *monochromatic* radiation, we bombard a single crystal with *polychromatic* radiation. The result we obtain (on a flat film) is shown in Figure 11.2.5. This is known as the **Laue method**. Note the symmetry. This method provides a means of orienting single crystals. It is necessary to have oriented crystals not only for research but in many applications such as when quartz is used for a delay line or as a piezoelectric transducer.

These X-ray techniques are described in the book by Cullity (see References at the end of the chapter).

11.3   BRAGG'S LAW AND CUBIC CRYSTALS IN GENERAL

All HKL are possible for simple cubic crystals. However, for bcc crystals various diffraction lines HKL are missing, i.e., have zero intensity. For example, 100 is missing. The reason for such behavior is that the bcc crystal is based on a nonprimitive cell and the distance between (100) planes is actually only $a/2$, and not a (see Example 10.6.2). If the crystal is set up for diffraction from planes a apart, as shown in Figure 11.3.1, so PMQ is one wavelength, then

FIGURE 11.2.5. Back reflection Laue patterns of tungsten. Incoming X-ray normal to (111) plane.

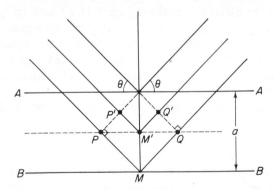

FIGURE 11.3.1. Diffraction from (100) planes of a bcc crystal.

$P'M'Q'$ is exactly one-half wavelength and the scattered wave is then exactly out of phase. Since waves from alternating planes are out of phase, the total intensity is zero. It can be shown that the intensity of radiation is proportional to $[\mathcal{S}(HKL)]^2$, where

$$\mathcal{S}(HKL) = \sum_j f_j e^{2\pi i(Hu_j + Kv_j + Lw_j)}. \qquad (11.3.1)$$

$\mathcal{S}(HKL)$ is called the **geometric scattering factor**. The summation is over all the atoms in the basis that is added to the primitive cell. Here $i = \sqrt{-1}$ and f_j is the **atomic scattering factor** of the jth atom (at $\theta = 0$ it equals the atomic number Z but it decreases as θ increases). Equation (11.3.1) is derived in C. Kittel, *Introduction to Solid State Physics*, Wiley, New York (1968) p. 63.

For a bcc crystal it can be shown that $S = 0$ if $H + K + L =$ odd and $S \neq 0$ otherwise. For a fcc crystal it can be shown that $S \neq 0$ if all HKL are even (e.g., 422) or if all of HKL are odd (e.g., 311), but $S = 0$ otherwise.

EXAMPLE 11.3.1

Obtain S for a bcc crystal structure such as iron.

Answer. The basis is an iron atom at 000 and an iron atom at $\frac{1}{2}\frac{1}{2}\frac{1}{2}$. Hence

$$S = f_{Fe}[1 + e^{\pi i(H+K+L)}].$$

Since $e^{m\pi i} = 1$ if m is even and -1 if m is odd, we have $S = 2f_{Fe}$ if $H + K + L$ is even and $S = 0$ if $H + K + L$ is odd.

EXAMPLE 11.3.2

Give the values of $H^2 + K^2 + L^2$ for which there is a diffraction peak for fcc nickel.

Answer. From the stated **selection rules**, namely, that $S \neq 0$ for HKL all even or HKL all odd, we find

HKL	$H^2 + K^2 + L^2$
111	3
200	4
220	8
311	11
222	12
400	16
331	19
420	20
.	.
.	.
.	.

Note that $H^2 + K^2 + L^2$ never equals 7 or 13, etc.

11.4   THE EFFECT OF ATOMIC VIBRATION

In the derivation of Bragg's law the assumption was made that the atoms were located exactly at their lattice site. It was known even at the time of von Laue's first studies (from the work of Einstein and others) that atoms are vibrating so that nearest-neighbor distances are changing by 10–20% in a nearly random fashion. One's first thought is that these vibrations will destroy the diffraction effect completely. This is not the case. What actually happens is that the *intensity* of a given peak decreases exponentially as the temperature increases. Surprisingly, this does not cause broadening of the peak.

11.5   NEUTRONS AND ELECTRONS

The de Broglie relation,

$$\lambda = \frac{h}{p},$$
(11.5.1)

is the key expression in the dual wave-particle nature of matter.

This is one of the most fundamental equations in physics. It relates the wavelength λ to the momentum p; here h is Planck's constant.

An electron accelerated through a potential V has kinetic energy equal to eV. If the velocities are nonrelativistic, then the kinetic energy is $\frac{1}{2}mv^2$ ($=eV$) and the momentum is mv, where m is the rest mass. Hence it can be shown since

$$\lambda = \frac{h}{mv}$$
(11.5.2)

that

$$\lambda \, (\text{Ångstroms}) = \sqrt{\frac{150}{V\,(\text{volts})}};$$
(11.5.3)

at low voltages ($\approx 10^2$ V) this expression is satisfactory, while at high voltages ($\approx 10^5$ V) it is necessary to use relativistic expressions.

We would expect electron waves to be scattered from crystals ($\lambda = 1$ Å with $V = 150$ V, which is in the neighborhood of lattice plane spacings) and to lead to diffraction phenomena. There are some interesting applications of electron diffraction which are complementary to X-ray studies. Two examples are given here. If very low voltages are used, the penetration is very small so the diffraction occurs from the first layer or so; low-energy electrons are therefore a useful tool for studying atomic arrangements and electron densities at surfaces. This technique is called **low-energy electron diffraction**, or LEED. An example of a LEED pattern from a zinc crystal freshly cleaved in an ultrahigh vacuum (so the surface is unusually clean) is shown in Figure 11.5.1. If very high energies are used, the penetration is greatly increased, and at 10^5 V it is about 1 μ (10^{-6} m). Thus it is possible for electrons to transmit through thin foils. One can then not only get a diffraction pattern and hence the orientation of the foil, but by using a magnetic lense to magnify the image one can study certain imperfections within the crystal. This is called **transmission electron microscopy**.

Neutrons and other heavier particles also show the wave-particle duality of Equation (11.5.1). Thus one would expect that neutron beams are scattered from crystals (a neutron having kinetic energy of 0.08 eV would have a wavelength of 1.0 Å). We have already noted that photons and electrons are scattered by all the electrons associated with an atom or an ion; scattering by the nucleus is negligible. It is here that neutrons differ, and because of this, neutron diffraction can be considered as a tool which is complementary to X-ray diffraction since it provides information which X-ray diffraction could not. The reason is that neutrons are scattered by **nuclear** or **magnetic scattering**. The first mechanism is due to interaction of neutrons with the nucleus; the second is due to

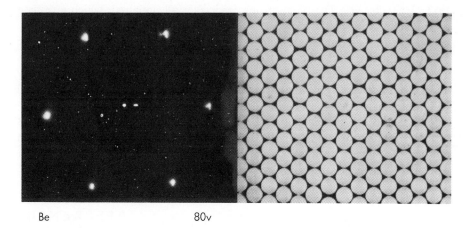

Be   80v

FIGURE 11.5.1. LEED pattern from zinc crystal compared with atoms at surface. (*Courtesy of J. Baker and J. M. Blakely, Cornell University, Ithaca, N.Y.*)

interaction of the permanent magnetic moment of the neutron with the permanent magnetic moment of atoms or ions. We consider now the structure of MnO. X-ray analysis shows that this is a NaCl structure (see Figure 9.2.2 for the unit cell). However, neutron diffraction shows that the Mn^{2+} ions have magnetic moments with the spins arranged alternately, as shown in Figure 11.5.2. We note from Appendix 8A that the Mn^{2+} ion has electronic structure

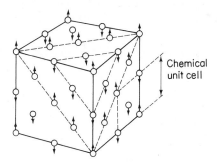

FIGURE 11.5.2. Magnetic structure of MnO as found by neutron diffraction. Only the Mn^{2+} ions are shown. Note that the "magnetic unit cell" has twice the length of the "chemical unit cell." [From C. G. Shull, E. O. Wollan, and W. A. Strauser, *The Physical Review*, **81**, 483 (1951).]

$1s^2 2s^2 2p^6 3s^2 3p^6 3d^5$. The $3d$ electrons are in fact all unpaired, which gives rise to a resultant magnetic moment. Three possibilities now arise. In ordinary **paramagnetic** materials the magnetic moments are random, in **ferromagnetic** materials the magnetic moments are regularly aligned with the spins all parallel, and in **antiferromagnetic** materials the magnetic moments are also regularly aligned but adjacent spins are opposite. It is clear from Figure 11.5.2 that MnO is antiferromagnetic.

The ability of neutrons to be scattered effectively by hydrogen atoms or ions while X-rays are hardly scattered at all is another advantage of neutron diffraction. Thus in compounds such as KHF_2, the hydrogen atoms are not ordinarily "seen" by X-ray diffraction but are readily seen by neutron diffraction!

REFERENCES

Bragg, W. H., *The Crystalline State*, Cornell University Press, Ithaca, N. Y. (1965). This book captures the excitement of the early years of X-ray analysis of crystals. The first correct structure determination is described by W. L. Bragg, *Proceedings of the Royal Society* (*London*), **A89,** 248 (1913).

Buerger, W. J., *Crystal Structure Analysis*, Wiley, New York (1960).

Bacon, G. E., *Neutron Diffraction*, Oxford, New York (1962).

Gevers, R., "Electron Diffraction", in *Interaction of Radiation in Solids* (R. Strumane et al., eds.), North-Holland, Amsterdam (1964).

Kittel, C., *Introduction to Solid State Physics*, Wiley, New York (1968). Chapter 2 treats additional topics of diffraction theory in quantitative detail—the Laue equations, the reciprocal lattice, the Ewald construction, and the geometric scattering factor.

Cohen, J. B., *Diffraction Methods in Materials Science*, Macmillan, New York (1966.)

Cullity, B. D., *Elements of X-ray Diffraction*, Addison-Wesley, Reading, Mass. (1956).

Barrett, C. S., and Massalski, T. B., *Structure of Metals—Crystallographic Methods, Principles and Data*, McGraw-Hill, New York (1966).

Bragg, Sir Lawrence, "X-ray Crystallography," *Scientific American* (July 1968) p. 58.

PROBLEMS

11.1. A photon is a finite wave packet, and when such a wave interacts with a crystal, it interacts with a large number of scattering centers. Show that the condition for reinforcement of the scattered wave from all of such scattering centers is given by Bragg's law.

11.2. Give the selection rules for diffraction lines for simple cubic, body-centered cubic, and face-centered cubic crystals. A selection rule states which *HKL* have finite intensity and which have zero intensity.

11.3. Assuming we have diffraction data for a cubic crystal (an element), how do we proceed to tell which crystal structure we have: sc, bcc, fcc?

11.4. Show that in a fcc crystal structure the distance between (100) planes is not a but $a/2$ and the distance between (110) planes is not $a/\sqrt{2}$ but $a/2\sqrt{2}$.

11.5. What set of planes is the farthest apart in
a. sc crystals?
b. bcc crystals?
c. fcc crystals?

11.6. What will be the first diffraction line (give *HKL* for the smallest experimental θ) for
a. sc crystals?
b. bcc crystals?
c. fcc crystals?

11.7. Write down the S factor for a fcc crystal of aluminum.

11.8. Suppose we have the following set of $\sin^2 \theta$ for a cubic crystal: 0.100, 0.200, 0.300, 0.400, 0.500, 0.600, 0.700, and 0.800. Is it sc or bcc?

11.9. A set of diffraction lines (θ's) was obtained with Cu K_α radiation: 13.70, 15.89, 22.75, 26.91, 28.25, 33.15, 36.62, 37.60, and 41.95 deg. Find the lattice parameter.

11.10. Calculate all of the diffraction angles for fcc silver if Cu K_α radiation ($\lambda = 1.54$ Å) is used. Silver has a lattice parameter of 4.08 Å at 20°C.

11.11. Electrons and neutrons sometimes behave as waves and sometimes as particles. Discuss.

11.12. What are the particular advantages of electron diffraction?

11.13. Give two examples where neutron diffraction yields structural determinations not possible with X-rays.

More Involved Problems

11.14. a. Derive the selection rules for the observed diffraction lines in germanium, which has a diamond cubic structure, assuming the electron concentration is spherical around the atoms.
b. Is this assumption true?

11.15. Write an essay on the uses of X-ray diffraction. Include five applications.

11.16. Write an essay on the uses of low-energy electron diffraction in studying surfaces.

11.17. a. Give the electron configuration of the K^+ ion and the Cl^- ion.

b. Although KCl has the NaCl-type structure (Figure 9.2.2), why does it appear, when studied by X-ray diffraction, to have a sc structure?

11.18. KBr has the NaCl structure. This is a fcc lattice plus a basis of a Cl^- ion at 000 and Na^+ ion at $\frac{1}{2}\frac{1}{2}\frac{1}{2}$.

a. What would be the basis if crystal structure = sc lattice + basis?

b. Write the S factor and determine the selection rule.

c. How would these be different for KCl if $f_{K^+} = f_{Cl^-}$? *Note:* Diffraction patterns for KBr and KCl are shown in C. Kittel, *Introduction to Solid State Physics*, Wiley, New York (1968) p. 64.

Sophisticated Problem

11.19. If the electron distribution is spherically symmetric about the origin, it can be shown that the atomic scattering factor is

$$f = 4\pi \int_0^\infty r^2 \rho(r) \frac{\sin \mu r}{\mu r} \, dr,$$

where $\mu = (4\pi/\lambda) \sin \theta$. For the hydrogen atom, $\rho(r) = \psi_{1s}^2(r)$. Set up a computer program to evaluate f vs. μ. What is f for $\theta = 0$?

Prologue

Atoms can be approximated as *rigid spheres* and crystals can often be considered as constructed of such rigid spheres. One of the most important geometric packings involves a *closest-packed* layer of spheres. Certain crystal structures such as the fcc crystal structure and the idealized hcp crystal structure can be constructed by stacking such closest-packed layers on top of each other in a specific *stacking sequence*. Another important packing concept is the *void*, i.e., the space between the rigid spheres. In metals such voids often can be occupied by the smaller atoms, H, B, C, N, and O. Thus we see that impurities can be dissolved *interstitially* in crystals. Other types of solid solutions, such as *substitutional solid solutions* and *superlattices*, are discussed. The void concept is particularly useful in describing ionic crystals and *ceramics*. Thus NaCl can be described as a fcc structure of chlorine ions with a smaller sodium ion in each of the octahedral voids.

We describe the *Pauling rules*, which enable us to predict the number of anions surrounding a cation based on the ratio of the cation radius to the anion radius. We note that many *intermetallic* compounds can be described in terms of these concepts. Finally, the structure of the *silicates*, which form the major part of the earth's crust, are discussed in terms of these packing concepts.

236

12

PACKING IN CRYSTALS

12.1 INTRODUCTION

We have already shown that atoms to a first approximation can be treated as relatively hard spheres. It is very useful to classify crystal structures in terms of the packing and coordination of spheres if we are always careful to consider the bonding characteristics. It is important, for example, in ionic solids to preserve electrical neutrality in each unit cell and to minimize ion-ion repulsion. The highly directional nature of the covalent bond is another important boundary condition. None of these factors are critical in metallic solids and it is in this class of solids that the packing concept is particularly fruitful. In this chapter we shall discuss the packing of spheres and voids in solids and the rules developed on this basis for the packing of atoms. We shall extend this concept to the discussion of real crystals in terms of whether the main bonding is of the metallic, ionic, covalent, or van der Waals type.

12.2 CLOSEST PACKING OF SPHERES

A **closest packing** is a way of arranging spheres in space so that the available space is filled most efficiently. Closest packing is achieved when each sphere is in contact with a maximum number of others. In a **closest-packed layer**, therefore, each sphere has *six* nearest neighbors and in a closest packed solid, *twelve* nearest neighbors. It can be easily shown (Figure 12.2.1), for example, that the circles occupy 90.7% of the area for a hexagonal layer and only 78.5% for a cubic layer. In three dimensions many closest packings are possible resulting from the different positions of the layers on top of each other (for every case, however, each sphere must touch 12 others for closest packing).

237

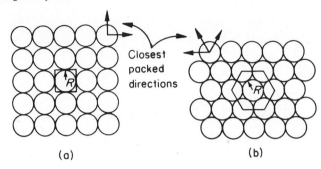

FIGURE 12.2.1. (a) Cubic layer. (b) Hexagonal layer.

Figure 12.2.2 illustrates how the layers are stacked. The order of the layers when repeated is referred to as the **stacking sequence**. The lowest symmetry that a three-dimensional structure built of closest-packed layers can have is that of a random stacking sequence.

The two-layer sequence ... ABAB ... is shown in Figure 12.2.3. This is an **ideal hcp crystal structure**. The sequence ... ABCABC ... is shown in Figure 12.2.4. This is a fcc crystal structure.

The closest-packed layer is the (0001) plane or the **basal plane** of the hcp structure. It is the (111) plane (or any of the {111} planes or **octahedral planes**) for the fcc structure.

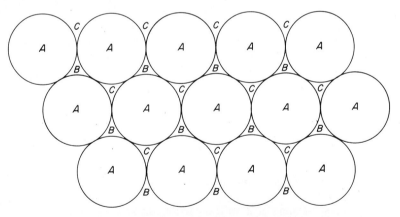

FIGURE 12.2.2. Consider the A layer of spheres in place. The second closest-packed layer of spheres sits on top of the A layer: It may be either the B layer or the C layer. Say that it is a B layer. (*Note:* An atom of the B layer then rests in the "valley" formed by three atoms of the first layer. This B layer atom will have the three atoms below it, six around it, and three above it when the third layer is added.) Then the third closest-packed layer of spheres can be an A layer or a C layer.

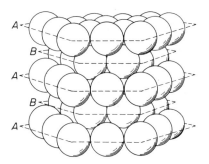

FIGURE 12.2.3. Closest-packed
solid (hcp).

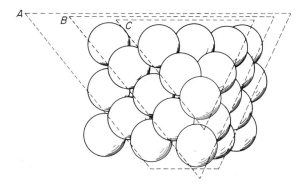

FIGURE 12.2.4. Closest-packed
solid (fcc).

EXAMPLE 12.2.1

Show that the (111) plane of the fcc crystal (see Figure 10.4.2) is a closest-packed plane.

Answer. We note that a characteristic of the closest-packed plane is that there are rows of atoms touching each other (see Figure 12.2.1). These rows are at 60 deg to each other. We note that in the fcc crystal atoms touch along the $\langle 110 \rangle$ directions, i.e., along the face diagonals of the cube. The three face diagonals $[\bar{1}01]$, $[01\bar{1}]$, and $[1\bar{1}0]$ lie in the (111) plane and these three directions are at 60 deg to each other.

Closest packings are also distinguished by the number of layers *n*, called the **identity period**, required to complete the stacking sequence. It is 2 for hexagonal closest packing, 3 for cubic closest packing, 4 for the sequence ABACA BAC . . ., etc.

12.3 VOIDS IN CLOSEST-PACKED STRUCTURES

Although a great variety of closest packings are possible, most actual closest-packed elemental structures in nature are either the hcp or fcc structure. In addition, most known inorganic compounds have structures that are either *closest-packed* or closely related to it. Many important classes of compounds, such as oxides, for example, have one atom closest-packed (oxygen) and the other present in voids. We shall discuss later the properties of alloys where the "foreign" atoms in the voids between the close-packed atoms influence important properties. The two kinds of voids that occur in closest packings are illustrated in Figure 12.3.1. If the void is surrounded by four spheres, it is called

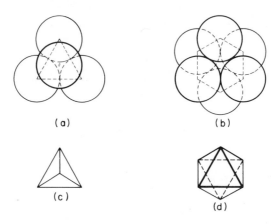

(a)   (b)

(c)   (d)

FIGURE 12.3.1. Voids in closest packing.

a **tetrahedral void**, and if surrounded by six spheres, an **octahedral void**. The number of spheres around a void is called the **coordination number** of the void and the space occupied by portions of the spheres coordinating the void is called the **coordination polyhedron**. A tetrahedron void and an octahedron void are illustrated in Figure 12.3.2.

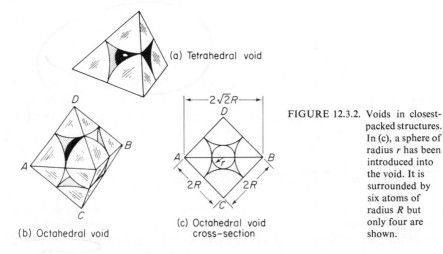

(a) Tetrahedral void

(b) Octahedral void

(c) Octahedral void cross-section

FIGURE 12.3.2. Voids in closest-packed structures. In (c), a sphere of radius r has been introduced into the void. It is surrounded by six atoms of radius R but only four are shown.

It is also important to know where these voids are with respect to the conventional cell. In the fcc cell of Figure 10.4.2, the center of the octahedral void is at the center of the cube edge, or if one considers a row of atoms in a $\langle 100 \rangle$ direction, then there is an alternating sequence of atoms and octahedral voids.

EXAMPLE 12.3.1

There are four atoms in the fcc cell of Figure 10.4.2. How many octahedral voids are there per unit cell?

Answer. As is clear from the previous discussion there is one void for each atom so there are four octahedral voids per unit cell. There is another way to show this: namely, the cube has 12 edges and each of these voids is one fourth within the cube. There is also a void at the center of the cube. Thus there are four octahedral voids or one void per atom.

The tetrahedral void is located at positions such as $\frac{1}{4}\frac{1}{4}\frac{1}{4}$, i.e., at the center of a cubic subcell, which has edges of length $a/2$ and a volume of $a^3/8$. There are eight such subcells per cell. Hence there are eight tetrahedral voids per fcc cell, or two tetrahedral voids per atom.

The order of the stacking sequence does not change the void per atom ratio. An important property of the voids is the ratio of radius r of the sphere which will just fit in the void to the radius R of the spheres making up the closest-packed structure. We call this ratio of radii the **radius ratio**. The results are shown in Table 12.3.1. These voids are extremely important in the technology of metals because the smaller atoms H, B, C, N, and O often can fit into these voids (although they may be squeezed some in the process). For example, iron has a fcc crystal structure at high temperatures and can dissolve several atomic percent of carbon.

Table 12.3.1. RADIUS RATIO FOR
CLOSEST PACKING

Void Type	Radius Ratio, r/R
Tetrahedral	0.225
Octahedral	0.414

EXAMPLE 12.3.2

Show that $r/R = 0.414$ for the octahedral void in the fcc crystal.

Answer. See Figures 10.4.2 and 12.3.2. The cube edge equals a. Spheres touch along the *face* diagonal whose length is $\sqrt{2}\,a$. Hence

$\sqrt{2}\,a = 4R$. Along the cube edge we have $2R + 2r = a$. Eliminating a from these two equations gives

$$R + r = \sqrt{2}\,R$$

and hence

$$r = 0.414R.$$

12.4   THE BODY-CENTERED CUBIC STRUCTURE

Iron has the bcc structures shown in Figure 12.4.1(a) at room temperature; so do tungsten, molybdenum, and sodium.

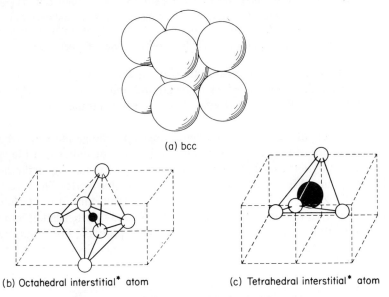

(a) bcc

(b) Octahedral interstitial* atom   (c) Tetrahedral interstitial* atom

FIGURE 12.4.1. The bcc structure and voids. *An interstitial is an atom in a void, which is not a normal site in the lattice.

The bcc structure is not closest-packed. It has eight nearest neighbors. The octahedral void and tetrahedral void, as shown in Figure 12.4.1(b) and

Table 12.4.1. RADIUS RATIOS FOR BCC
STRUCTURE

Voids	r/R
Tetrahedral	0.291
Octahedral	0.154

(c), respectively, are very important in determining the kind of "foreign" atom that can be accommodated in these voids when alloys are formed. The radius ratios for these voids are given in Table 12.4.1.

12.5 THE STRUCTURE OF ELEMENTAL METALS

Table 12.5.1 shows the crystal structure of metals. We note that the vast majority of these metals crystallize in the bcc, fcc, or hcp structures which we described previously. As we study this table we note that many metals crystal-

Table 12.5.1. CRYSTAL STRUCTURES OF METALS*

Li	Be											Zn	Ga	
bcc	hcp													
Na	Mg	Al												
bcc	hcp	fcc												
K	Ca	Sc	Ti	V	Cr	Mn	Fe	Co	Ni	Cu		Zn	Ga	
bcc	fcc	fcc	hcp	bcc	bcc	Cubic	bcc	hcp	fcc	fcc		hcp	Ortho.	
		bcc	hcp	bcc			fcc	fcc						
Rb	Sr	Y	Zr	Nb	Mo	Te	Ru	Rh	Pd	Ag		Cd	In	Sn
bcc	fcc	hcp	hcp	bcc	bcc	hcp	hcp	fcc	fcc	fcc		hcp	Tetr.	Tetr.
	hcp		bcc											
	bcc													
Cs	Ba	La	Hf	Ta	W	Re	Os	Ir	Pt	Au		Hg	Tl	Pb
bcc	bcc	hcp	hcp	bcc	bcc	hcp	hcp	fcc	fcc	fcc		Hex.	hcp	fcc
		fcc	bcc											bcc
Fr	Ra	Ac	Th	Pa	U	Np	Pu							
		fcc	fcc	Tetr.	Ortho.	Ortho.	Ortho.							
			bcc		Tetr.	Tetr.	fcc							
					bcc		Tetr.							

Rare earth metals:														
Ce	Pr	Nd	Pm	Sm	Eu	Gd	Tb	Dy	Ho	Er		Tm	Yb	Lu
hcp	hcp	Hex.		Hex.	bcc	hcp	hcp	hcp	hcp	hcp		hcp	fcc	hcp
fcc														

* L. V. Azaroff, *Introduction to Solids*, McGraw-Hill, New York (1960).

lize in more than one crystal structure; this is called **polymorphism**. As an example, iron has different crystal structures at atmospheric pressure depending on the temperature. These are

bcc	−273–910°C
fcc	910–1390°C
bcc	1390–1534°C (mp)

At very high pressures iron has an hcp structure. The existence of different

polymorphs is not only an interesting scientific phenomenon but has many different industrial applications, as we shall see later.

Another important example of polymorphism is the two well-known forms of carbon: graphite and diamond. At room temperature and pressure, graphite is stable relative to diamond. If the pressure were raised to above 14 kbars, diamonds would be the more stable. However, diamonds do persist at room temperature and pressure because the *rate* of their transformation into graphite is so very slow (the diamonds are said to be metastable). This would not be the case at 1700°C where the diamond would change to graphite. The study of the rates of such transformations is an important part of materials science which will be considered later.

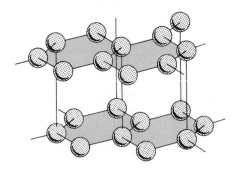

FIGURE 12.5.1. Graphite structure.

The graphite structure is shown in Figure 12.5.1. It is a layer structure with each carbon in the layer being strongly bonded to three other carbons in the layer, the bond angles being 120°. (The bonds can be considered sp^2 hybrids.) The fourth valence electron is (roughly) free to move in the layer. The layers, however, are bonded to each other by weak van der Waals' forces. This explains the enormous anisotropy of many properties of graphite. For example, it is a metal along the layers and a semiconductor normal to the layers. The other carbon polymorph, diamond, has tetrahedral bonds (see Figure 10.4.3). It is an insulator at room temperature but at high temperatures it is a semiconductor.

Polymorphism is not restricted to elements but is shown by many compounds such as SiO_2, which has several crystalline forms in addition to quartz.

We have noted that the metals can usually be considered as spheres packed together. Appendix 12A (at the end of this chapter) gives values of metallic radii. Covalent solids can also be described by packing of spheres. Covalent radii are given in Appendix 12B.

12.6   STRUCTURE OF SIMPLE ALLOYS

When different metals are mixed together as alloys, many additional, interesting structural features develop. These crystal structures are closely related to the behavior and usefulness of the alloys.

Metal atoms can replace one another in crystals in a variety of ways. When atoms of one kind replace atoms of another kind of similar size at the same lattice positions, the resultant structure is called a **substitutional solid solution**. There are many pairs of metals that form binary solid solutions in this way. The principal physical factors which define the degree of solid solubility have been given by *Hume-Rothery* in the following statements. They are derived from many empirical observations of the behavior of a large number of different alloys. To show extensive solubility,

1. *The atoms must have atomic radii that differ by less than 15%.*
2. *The two metals must have identical crystal structures except for the dimensions of the unit cell.*
3. *Two kinds of atoms with a difference in electropositive nature (valence can be substituted as a first approximation) will tend to form stable structures commonly referred to as intermediate phases rather than continuous solid solutions.*
4. *Continuous solid solutions can occur only between atoms having the same valency in the alloy.*

Iron and nickel form a solid solution called Hypernik on a fcc lattice, as indicated in Figure 12.6.1. Note that the Fe and Ni atoms are randomly distributed at lattice sites. Cu–Ni, Ag–Au, and Mo–W also form continuous series of solid solutions. Complete solid solubility is infrequent because all of the above criteria are seldom completely satisfied. When they are applicable, the change in unit cell dimensions is expected to vary in a linear manner with composition. This is called **Vegard's law**. It is seldom obeyed ideally by metals, although inorganic alloys such as NaCl–KCl are often well described by it.

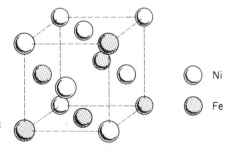

FIGURE 12.6.1. Fe–Ni solid
          solution.

The **interstitial type** of solution forms when the solute atoms are distributed among the voids generated by the close packing of the solvent atoms. An important example of this is the placing of carbon and other small atoms in the tetrahedral and octahedral voids previously discussed with reference to the polymorphs of iron. Since the presence of carbon in iron is very important in steel making, we shall digress briefly to describe some of the carbon-iron alloys.

As previously discussed with reference to polymorphism, iron can exist in three polymorphic modifications: body-centered cubic, α-Fe (below 910°C);

face-centered cubic, γ-Fe (910–1390°C); and body-centered cubic, δ-Fe (1390–1534°C). Because of the larger voids in the fcc structure, γ-Fe is the only polymorph that dissolves carbon interstitially to any large extent (up to 2.0 wt. %). It has also been shown that C atoms occupy the octahedral voids randomly in γ-Fe. This interstitial solid solution is commonly called **austenite**. The interstitial solid solution of C in α-Fe is called **ferrite**. Carbon solubility in ferrite is 0.025 wt. % at 723°C, which is called the eutectoid temperature, and it is much smaller than this at lower temperatures. Below this temperature, the compound **cementite**, Fe_3C, exists. The orthorhombic structure of Fe_3C is not related in any simple way to those of α-Fe or γ-Fe. Each carbon atom is coordinated by six iron atoms at the corners of the distorted triangular prism, as indicated in Figure 12.6.2. It is not known why cementite forms this asymmetric arrangement. Whereas the Fe–Fe bonds are metallic, it is suggested that the peculiar coordination of the Fe atoms around the C atom arises from the covalent character of the Fe–C bonds. The role of various phases and compounds in the alloys of iron is very important and will be discussed in detail later. It is significant for the present to observe the special features of their crystal structure.

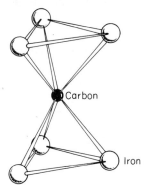

FIGURE 12.6.2. Cementite.

Let us return again to consider the general features of interstitial atoms in metals. The ratio of the radius of the largest void to the radius of the host atom in the fcc and hcp lattices is 0.41, whereas the same ratio for a bcc structure is 0.29. This limits the forming of interstitial solutions to very small atoms such as H, B, N, C, and O. The first two elements tend to occupy tetrahedral voids, whereas C and N tend to occupy the larger octahedral voids of the fcc structure. These elements also combine interstitially, usually by a "modified" covalent bonding with the transition metals, to form very stable **interstitial compounds** (hydrides, carbides, borides, nitrides, etc.).

The atomic array in most solid solutions is entirely **random** except for some short-range order (e.g., a slight preference for one nearest neighbor over another in a 50 at. % Cu–50 at. % Au alloy) usually present. However, at certain temperatures, a crystal structure may be more stable when the two types of atoms occupy different sets of lattice points. For example, a 50 at. % Au–50 at. % Cu alloy is random above 385°C [Figure 12.6.3(a)] and transforms to the ordered CuAu structure [Figure 12.6.3(b)] below this temperature. (*Note:* Atomic percent is related to weight percent through the expression:

$$\text{at.} \% \ A = \frac{\text{wt. } \% \text{ of } A/\text{molecular wt. of } A}{\text{wt. } \% \text{ of } A/\text{molecular wt. of } A + \text{wt. } \% \text{ of } B/\text{molecular wt. of } B} \times 100\%.$$

The 75–25 alloy transforms similarly to the ordered Cu_3Au structure below 390°C [Figure 12.6.3(c)]. The **ordered substitutional solutions** in (b) and (c) are examples of **superlattices**. It should be noted that the primitive cell is larger for the superlattice. (*Note:* Ordering in alloys can occur in other ways also, e.g., in the magnetic Mn–Cu alloy the electron spins between nearest neighbors are random above about 250°C and ordered in an antiparallel array below this temperature.) Note the vital role played by temperature. *High temperature favors disorder or randomness.*

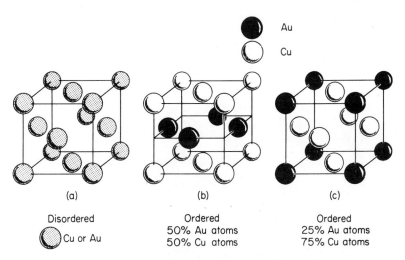

FIGURE 12.6.3. (a) Random substitutional solution of CuAu. (b) CuAu superlattice. (c) Cu_3Au superlattice.

There is another **Hume-Rothery solubility rule** (in addition to the four already given) which governs the extent of solubility of polyvalent elements in solid monovalent metals. Cu, Ag, and Au all have the fcc structures. It is found empirically (and there is also a quantum mechanical justification) that

5. *Polyvalent elements will dissolve (assuming sizes are compatible) in fcc monovalent metals until the ratio of the number of free electrons to the number of atoms reaches 1.4.*

EXAMPLE 12.6.1

What is the maximum zinc solubility in fcc copper?

Answer. Let x be the atom fraction of Zn which is divalent and hence contributes two electrons per atom to the electron gas. Then $1 - x$ is the atom fraction of copper which is monovalent. Hence $2x + (1 - x) = 1.4$ and $x = 0.4$. Thus up to 40% of the copper atoms can be replaced by zinc atoms and the structure will still be a fcc crystal. However, its lattice parameter will change as the percent Zn changes.

It should be emphasized that the Hume-Rothery electron-ratio rule is only an approximation. What we really expect in Example 12.6.1 is solubility in the neighborhood of 40% Zn.

12.7 STRUCTURE OF SIMPLE IONIC COMPOUNDS

SODIUM CHLORIDE. Consider Cl^- ions to be spheres packed together in the fcc structure. (You should think of this both as a fcc lattice with Cl^- ions at the lattice points and as a ... ABCABC ... stacking sequence of closest-packed layers.) Now place a smaller Na^+ ion in every octahedral void. Since the ratio of octahedral voids to Cl^- ions is 1:1, the chemical formula is NaCl. Moreover, since the $\langle 100 \rangle$ directions consist of alternating spheres and voids,

FIGURE 12.7.1. Model of sodium chloride. [From A. N. Holden and P. Singer, *Crystals and Crystal Growing*, Doubleday, New York (1960).]

this gives the crystal structure of NaCl shown in Figure 9.2.2. This crystal structure is perhaps best illustrated by the model shown in Figure 12.7.1. NiO, MgO, and other important ceramic materials have this structure. The radius ratio for the octahedral void is 0.414. This type of crystal structure tends to form when the actual radius ratio of cation to anion falls in the range 0.414–0.645. [From an empirical viewpoint we could say that the tendency is to push the large anions outward and to avoid having the smaller cation rattle around in the void.]

Values of ionic radii are given in Appendix 12C.

EXAMPLE 12.7.1

Calculate the radius ratio r_{Na^+}/R_{Cl^-}.

Answer. From Appendix 12C we have

$$r_{Na^+} = 0.95$$

and

$$R_{Cl^-} = 1.81.$$

Hence we have

$$r/R = 0.525.$$

CsCl, at room temperature and pressure, has the structure obtained by placing a basis of a Cs^+ at 000 and Cl^- at $\frac{1}{2}\frac{1}{2}\frac{1}{2}$ on a simple cubic lattice. This is referred to as the **CsCl crystal structure**. Many of the alkali halides have the NaCl-type structure at 1 atm but transform to the CsCl type at high pressures. NaCl itself requires a pressure of 141 kbars to transform (1 kbar = 10^9 dynes/cm² = 14,500 psi = 980 atm).

ZINC SULFIDE.   Consider the large ions S^{2-} to be spheres packed in the fcc arrangement. Place a small Zn^{2+} ion in every other tetrahedral void. Since there are two tetrahedral voids per large sphere, the chemical formula will be ZnS. Each Zn^{2+} appears to be tetrahedrally bonded to each S^{2-}; in fact it is true that there is considerable covalent bonding combined with ionic bonding in this case. The resultant crystal structure is called the **zinc blende** (or **sphalerite**) crystal structure. Such tetrahedral coordination tends to occur when r/R is in the range 0.225–0.414.

EXAMPLE 12.7.2

Calculate the radius ratio for Zn^{2+} ion to S^{2-} ion.

Answer. From Appendix 12C we obtain

$$\frac{r}{R} = 0.402.$$

In addition to ZnS there are a number of important semiconductor compounds with this crystal structure, e.g., β-SiC and InSb. Note that if both atoms were the same, e.g., if the sulfur was replaced by a carbon atom and the zinc by a carbon atom, the resultant structure would be the diamond cubic structure. This is also the crystal structure of germanium and silicon. Thus zinc blende can be formed from two interlocking fcc lattices (see Figure 10.3.4, #14) with Zn atoms on one lattice and S atoms on the other.

If we now start with the S^{2-} ions in the hexagonal closest-packed arrangement and place Zn^{2+} ions in alternate tetrahedral voids, we obtain the **wurtzite** structure.

RADIUS RATIO AND COORDINATION NUMBER. Relationships between coordination polyhedra, coordination numbers (CN), and minimum radius ratios are summarized in Table 12.7.1. What is meant by the **minimum radius ratio**? A void radius r to a sphere radius R is 0.225 for a tetrahedral void. It is

Table 12.7.1. MINIMUM RADIUS RATIOS

Coordination polyhedron	CN	Minimum Radius Ratio
Cubo-octahedron	12	1.000
Cube	8	0.732
Square antiprism	8	0.645
Octahedron	6	0.414
Tetrahedron	4	0.225
Triangle	3	0.155

0.414 for the next largest void, which is an octahedral void. The atom (or ion) will *tend* to go into the tetrahedral void if $0.225 < r/R < 0.414$; i.e., it will push the ions apart rather than fit loosely in the large octahedral void. Thus the void ratio in fact is a minimum radius ratio. The cubo-octahedron is the type of coordination polyhedron formed by the joining of the centers of 12 nearest neighbors surrounding an atom in the fcc crystal structure of Figure 12.2.4. The square antiprism is formed as follows: Consider two squares of edge L

Table 12.7.2. POSSIBLE POLYHEDRA OF COMPLEX IONS

Polyhedron	CN	Examples
Triangle	3	BO_3^{3-}, CO_3^{2-}, NO_3^{-}
Tetrahedron	4	AlO_4^{5-}, SiO_4^{4-}, PO_4^{3-}, SO_4^{2-}, ClO_4^{-}
Square	4	$Ni(CN)_4^{2-}$, PdO_4^{2-}, $PtCl_4^{2-}$
Octahedron	6	NaO_6^{5-}, MgO_6^{4-}, AlO_6^{3-}, TiO_6^{2-}

on parallel planes one directly above the other; rotate the second square by 45 deg; then join the corners of these two squares. These results can be used to predict the configuration of complex ions as in Table 12.7.2.

EXAMPLE 12.7.3

What coordination is expected for NaCl?

Answer. From Example 12.7.1, we have $r/R = 0.525$. Hence we expect octahedral coordination.

EXAMPLE 12.7.4

What coordination is expected for SiO_4^{4-}?

Answer. The radius of Si^{4+} is 0.65 (from Appendix 12C), while the radius of O^{2-} is 1.76. Hence the radius ratio is $r/R = 0.37$. Thus we expect tetrahedral coordination.

Complex ions are ionic groups of great stability and hence their polyhedra serve as building blocks in more complicated arrays. One of the most important of these is the tetrahedrally coordinated silicate ion, which we shall study later.

The coordination numbers and the radius ratio impose limitations on the possible relative arrangements of atoms of a given size and in this sense determine the observed structure, or equivalently, *for any given CN there is a range of permissible radius ratios*. The chemical formula of ionic and covalent solids is also related to the possible coordination numbers as follows. If there are a atoms of B around A and b atoms of A around B, then

$$\frac{\text{CN of A}}{\text{CN of B}} = \frac{a}{b} \qquad (12.7.1)$$

and the chemical composition is $A_b B_a$.

These coordination rules combined with the general requirement of electrical neutrality enable one to predict possible crystal structures. These rules are known as **Pauling's rules**. We used these rules in their simplest form in discussing NaCl and ZnS.

For a detailed discussion of these rules see L. V. Azaroff, *Introduction to Solids*, McGraw-Hill, New York (1960) p. 83.

An understanding of these coordination rules is particularly important in the study of ceramic materials. **Ceramic materials** are commonly understood to be compounds of metallic and nonmetallic elements. Thus NiO, MgO, Al_2O_3, $BaTiO_3$, and SiO_2 are simple ceramic compounds. Ceramic materials can also be very complicated and include clays, spinels, and common window glass. Many ceramic compounds are refractory materials (have very high melting points), as noted in Table 6.6.1. Ceramic materials have a wide range of application. They may be used as refractories (MgO), abrasives (alumina), ferroelectrics (BaTiO$_3$), piezoelectric transducers (SiO$_2$, quartz), magnets (spinels), building materials (concretes), surface finishes (vitreous enamel), etc.

12.8   THE STRUCTURE OF SIMPLE INTERSTITIAL COMPOUNDS

In many cases compounds are formed between transition metals and small atoms of nonmetals, H, B, C, N, and Si. Such compounds are called, respectively, hydrides, borides, carbides, nitrides, and silicides. They are called **interstitial compounds**. As a rule these compounds have many of the characteristics of metallic alloys such as opacity, high electrical and thermal conductivity, and metallic luster. [For example, they obey the Weidemann-Franz relation. See Equation (6.2.5).] However, these compounds are often *extremely* hard, have very high melting temperatures (hence are called refractories), and are often chemically unreactive except toward oxidizing agents. Some typical melting points are shown in Table 12.8.1.

Table 12.8.1. MELTING POINTS OF SOME
INTERSTITIAL COMPOUNDS

	mp, °K		mp, °K
TiC	3410	TiN	3220
HfC	4160	ZrN	3255
W$_2$C	3130	TaN	3630
NbC	3770		

Cutting tools for lathes, drills, etc., are often made of tungsten carbide because of its hardness. These tools are produced by powder metallurgy. In this case a fine powder of tungsten carbide is mixed with about 6% cobalt powder. The material is pressed together at high temperatures (hot pressed). It is then held at a still higher temperature where in the ideal case all the voids between the particles disappear (sintering). The resultant structure is known as a composite.

The structure of many of these compounds can be described by simple packing concepts. The metal atoms are arranged in a certain array and the small nonmetal atoms fill a certain fixed proportion of the voids. The structures of some of the interstitial compounds which may be derived on this scheme are indicated in Table 12.8.2. (A similar scheme may be drawn up for hexagonal closest-packed interstitial compounds.) It is likely that there is considerable covalent bonding.

Table 12.8.2. INTERSTITIAL STRUCTURES DERIVED
FROM CUBIC CLOSEST PACKING*

Nonmetal Atoms in	Proportion Occupied	Structure	Examples
Octahedral holes	All	NaCl	TiC, TiN, ZrC, ZrN, UC, HfC, VC, NbC, TaC
	$\frac{1}{2}$	—	W_2N, Mo_2N
	$\frac{1}{4}$	—	Mn_4N, Fe_4N
Tetrahedral holes	All	CaF$_2$†	TiH$_2$
	$\frac{1}{2}$	Zinc-blende	ZrH, TiH
	$\frac{1}{4}$	—	Pd$_2$H
	$\frac{1}{8}$	—	Zr$_4$H

* After A. F. Wells, *Structural Inorganic Chemistry*, Oxford, New York (1950).

† In this case the large Ca^{2+} ions are on a fcc lattice and the F$^-$ ions occupy all tetrahedral voids. The cubic unit cell has four Ca^{2+} and eight F$^-$ ions.

12.9   THE STRUCTURE OF BaTiO$_3$

A study of the structure of BaTiO$_3$ illustrates nicely Pauling's packing concepts and shows in a qualitative fashion how closely structure is related to properties, in this case ferroelectric behavior.

At high temperature, BaTiO$_3$ has a cubic crystal structure called the **perovskite structure**. As is clear from Table 12.7.1, the Ti^{4+} ion should have six O^{2-} ions surrounding it in the TiO$_6^{2-}$ complex. We next consider the coordination of Ba^{2+} with respect to oxygen. From Appendix 12C we have $r_{Ba^{2+}} = 1.53$ and $R_{O^{2-}} = 1.76$. Consequently $r/R = 0.87$. This is very close to 1, so that we would expect CN = 8 but in fact find CN = 12. The actual crystal structure is shown in Figure 12.9.1. It is only roughly consistent with Pauling's concepts of packing. It should be emphasized that Pauling's rules are only a rough approximation which provides helpful guidelines. When the temperature is lowered below 393°K the crystal structure changes to a tetragonal structure (the cell angles are still 90 deg but one length is different than the remaining two) with a c/a ratio of 1.04. We note that a cube edge consists of alternating Ti^{4+} and O^{2-} ions. When

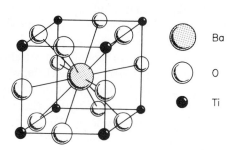

FIGURE 12.9.1. BaTiO$_3$ above
393°K.

the transformation occurs the Ti^{4+} ions on this edge move upward by 0.06 Å, while the 0^{2-} ions move downward by 0.09 Å (assuming that the center Ba^{2+} ion remains fixed). See G. Shirane, F. Jona, and R. Pepinsky, *Proceedings of the I.R.E,* **42**, 1738 (1955). Assume this happens along all the edges in the [001] direction only. The crystal now contains numerous electrical dipoles along the [001] direction and is therefore polarized along this direction, i.e., along the *c* axis.

The actual transformation from cubic to tetragonal crystals is not a uniform transformation as we have considered so far. Rather there are some regions of the original crystal which become polarized in the [100] direction, other regions which become polarized in the [010] direction, etc. These regions are called **ferroelectric domains** and the boundaries between the domains are known as **domain walls**. The total polarization may be zero. When an electrical field is applied, the domains whose polarization are most nearly parallel to the electrical field have the lowest energy. Thus these domains grow at the expense of the others (which results in a lowering of energy). This process does not take place reversibly; i.e., energy is dissipated when the domain walls move. Hence, a P-ξ hysteresis loop as shown in Figure 3.5.1 is obtained.

Below 278°K, $BaTiO_3$ has an orthorhombic structure and at 193°K it becomes rhombohedral. Both of these forms are also ferroelectric. Would you expect all ferroelectric materials to be piezoelectric?

12.10 STRUCTURE OF SILICATES

Silicates and silica compose a major fraction of our solid environment. They include metallic ores, mineral deposits, and inorganic building materials such as natural rock (granite), artificial materials (bricks, cement, mortar), ceramics, and glasses.

Silicates are based on the coordination tetrahedron SiO_4^{4-}. This has the shape shown in Figure 12.3.2(a) with the Si^{4+} in the void position. These tetrahedra can be connected to each other to make new **silicate structural units** as illustrated in Table 12.10.1. These units, when combined with positive ions in various ways, lead to the **silicates**. The simplest silicate structural units are the point units (the **islands**) SiO_4^{4-} and $Si_2O_7^{3-}$. Figure 12.10.1 shows a portion of a crystal structure formed from such an island structure.

Chain structures are formed if more than one corner of a tetrahedron is so joined. Joining of two corners of each tetrahedron to one corner of each of two other tetrahedra will form a **single chain**, whereas joining of three corners of every other tetrahedron to one corner of each of three other tetrahedra will form a **double chain.** See Table 12.10.1 where portions of such chains are shown. The chains can be very long. Parallel chains can be bonded to each other by placing positive ions such as Mg^{2+} between unshared corners of tetrahedra.

Table 12.10.1. SILICATE STRUCTURAL UNITS*

No. of Oxygens Shared	Structural Unit	Type of Structural Unit	Calculation of Charge of Structural Unit		Structural Formula
0		Island	$\mathrm{Si} +4$ O	$\dfrac{-8}{-4}$	$[\mathrm{SiO_4}]^{4-}$
1		Island	$\mathrm{Si} +8$ O	$\dfrac{-14}{-6}$	$[\mathrm{Si_2O_7}]^{6-} = [\mathrm{SiO_{7/2}}]^{3-}$
2		Single chain	$\mathrm{Si} +4$ O	$\dfrac{-6}{-2}$	$[\mathrm{SiO_3}]^{2-}$
$2\frac{1}{2}$		Double chain	$\mathrm{Si} +16$ O	$\dfrac{-22}{-6}$	$[\mathrm{Si_4O_{11}}]^{6-} = [\mathrm{SiO_{11/4}}]^{3/2-}$
3		Sheet	$\mathrm{Si} +8$ O	$\dfrac{-10}{-2}$	$[\mathrm{Si_2O_5}]^{2-} = [\mathrm{SiO_{5/2}}]^{1-}$
4	Three-dimensional network	Framework	$\mathrm{Si} +4$ O	$\dfrac{-4}{0}$	$[\mathrm{SiO_2}]^{0}$

* L. V. Azaroff, *Introduction to Solids*, McGraw-Hill, New York (1960).

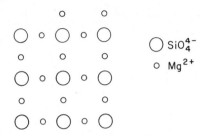

FIGURE 12.10.1. Island structures (two-dimensional schematic representation of Mg_2SiO_4). The magnesium atoms have supplied the necessary electrons to the $SiO_4{}^{4-}$ ions, and have become Mg^{2+} ions.

The mineral *asbestos* is an example of a long chain silicate polymer bonded in this manner. Its fibrous nature results from cleavage of the ionic bonds, which are weaker than the bonding in the tetrahedra which is partially covalent. The story of the physical and chemical behavior of many silicates rests essentially on the various *strengths* and *types of bonding* between the structural units such as *chains* and *sheets* in various arrangements with each other and with metal ions.

(A different kind of chain with a Si–O backbone is the silicone polymer:

$$
\begin{array}{ccccccc}
 & CH_3 & & CH_3 & & CH_3 & \\
 & | & & | & & | & \\
-O-&Si&-O-&Si&-O-&Si&-O- \\
 & | & & | & & | & \\
 & CH_3 & & CH_3 & & CH_3 &
\end{array}
$$

In this case there are deviations from the basic SiO_4^{4-} unit, and two methyl groups form two of the bonds to a silicon. There is only a weak van der Waals interaction between these chains. These silicone polymers have, of course, vastly different properties than the chains structures of this section. We point them out now to illustrate the close structural connection with the materials studied here and other polymer molecules which are discussed in Chapter 14.)

The three corners of *every* tetrahedron can be joined in turn to one corner of each of three other tetrahedra in a plane rather than along a line. This arrangement forms **sheet structures** typical of the ceramic minerals, *clays*, *micas*, and *talc*. The oxygen at the fourth corner of the tetrahedron is still capable of forming a covalent or ionic bond in the third dimension. Thus the oxygens on the bottom side of the sheet are saturated, while those on the top side are unsaturated. Two sheets may be brought together with their unsaturated sides facing each other. Positive ions placed between the sheets will then hold the pair together by ionic bonding. When such bonded pairs are placed on top of each other they in turn will be held together but only by weak secondary forces. It is a characteristic of the sheet structure that they either cleave readily parallel to the sheets (mica) or that the sheets readily slip over each other. As an example, the structure of talc is shown in Figure 12.10.2.

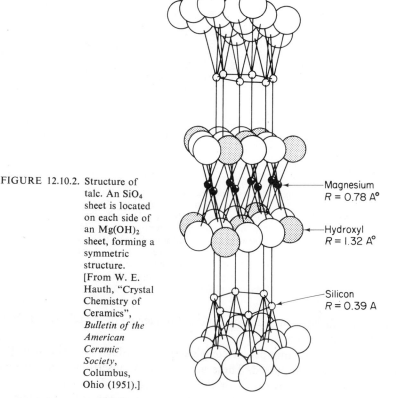

FIGURE 12.10.2. Structure of talc. An SiO_4 sheet is located on each side of an $Mg(OH)_2$ sheet, forming a symmetric structure. [From W. E. Hauth, "Crystal Chemistry of Ceramics", *Bulletin of the American Ceramic Society*, Columbus, Ohio (1951).]

Magnesium
$R = 0.78 A°$

Hydroxyl
$R = 1.32 A°$

Silicon
$R = 0.39 A$

Clay is a sheet silicate structure which is of interest to civil engineers, agronomists, and geologists. It is characterized by strong intralayer bonding but weak interlayer bonding. Moreover, polar molecules such as water are strongly attached to the surfaces of the layers causing swelling of the clay and easy slipping.

The *three-dimensional* structures are generated by coupling all four corners of each tetrahedron in various ways to each of the corners of four other tetrahedra. Infinite three-dimensional networks may be formed in this manner, leading to some complicated but at the same time most interesting structures. These are sometimes referred to as **framework** structures. A very simple proto-type of this structure, *tridymite*, is indicated in Figure 12.10.3. Note the high degree of symmetry in the arrangement. This fact and the preponderance of pure Si–O bonds makes this a very stable type of structure. There are three crystalline polymorphs of silica; *quartz, tridymite,* and *cristobalite.* All of these

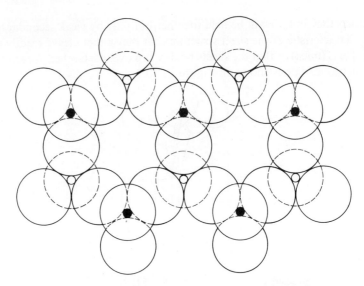

FIGURE 12.10.3. Silica structures. Note that the tetrahedra alternately point up and down. Similar layers are then directly bonded to each other. This is the SiO₂ polymorph-tridymite. The silicons in the tetrahedra pointing up (shown as solid circles) are above the plane of the paper while the silicons in the tetrahedra pointing down (shown as open circles) are below the plane of the paper.

form rings of six silicate ions but with different sequences of the relative orientation of the tetrahedral units.

There are a huge number of **complex silicates** which are related to the simpler structures just discussed. These are formed by insertion of metal ions (e.g., aluminum) into the network or in place of silicon. Examples based on three-dimensional networks are feldspars (the most important rock-forming minerals), zeolites (which can act as water softeners and as molecular sieves), and ultramarines (which are colored silicates used for paint pigments).

12.11   GLASSES

When the crystalline form of SiO_2 is melted a viscous liquid is formed. The bonding in this liquid does not have the highly regular bonding characteristic of the crystalline solids. If this liquid is not cooled extremely slowly, such a random network structure becomes frozen in [remember that the glass becomes more and more viscous as the temperature decreases; see Equation (2.8.5)]. The resulting structure is called **silica glass**. See Figure 12.11.1.

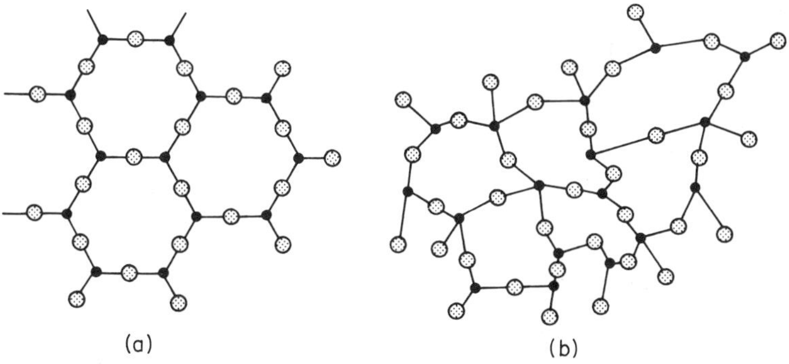

(a)   (b)

FIGURE 12.11.1. Schematic two-dimensional analogs, after Zachariasen, illustrating the difference between (a) the regularly repeating structure of a crystal and (b) the random network of a glass. The solid circles are oxygen atoms.

Glass is essentially a random network. Note in Figure 12.11.1 nonregular rings of five silica tetrahedra in the random network, while all the regular rings in the crystalline phase have only six tetrahedra.

Figure 12.11.2 shows the volume change versus temperature for formation of a crystalline solid from the melt and for the formation of a glass from the melt.

Upon very slow cooling a crystalline phase forms. Note the large volume *discontinuity* at the melting point. Upon very rapid cooling (there is no time for crystallization) the viscosity increases rapidly; as the temperature decreases the contraction is rapid (but continuous) because more efficient packing is occurring. Below a certain temperature, T_G, the **glass transition temperature**, the packing arrangement is frozen in; further contraction is a result of smaller thermal vibrations, just as in the crystalline phase in this temperature range. A **glass** is a material with the thermal expansion behavior of Figure 12.11.2; i.e., dV/dT is discontinuous but V is continuous, as the liquid cools. Recall that the volume

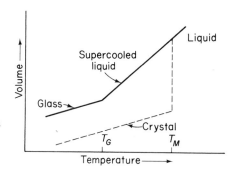

FIGURE 12.11.2. Volume changes upon cooling SiO_2 from the liquid.

thermal expansion coefficient is given by $\beta = d \ln V/dT$. Note that when the glassy state is reached, β for the glass and β for the crystal are nearly the same.

Inorganic compounds which form glasses are SiO_2, B_2O_3, GeO_2, BeF_2, and a number of trioxides and pentoxides of the fifth periodic group. Organic compounds such as glycerine as well as rubberlike materials and many polymers can exist as glasses at low temperatures.

Glasses have many engineering uses as containers, windows, and lenses and in combination with a variety of ceramics. The strongest bond in ordinary glass is based on the Si–O bond between silica tetrahedra. Many commercial glasses are alloys; e.g., soda glass contains Na_2O and SiO_2. The average bonding in this material is less strong than in pure silica glass so that soda glass has a much lower softening point. Pyrex contains boric oxide as an additive. In general in such glasses some of the silicon atoms have been replaced by other ions (such as boron). Metal ions which replace silicon atoms in this manner are called **network-forming ions**, in contrast to metal ions which fill the voids in the network and are called **network-modifying ions** (such as sodium). The color, viscosity, density, and electrical properties of glasses can be modified considerably by additions of metal ions.

The **degree of crystallinity** (the fraction which is crystalline) in glasses will increase slowly at temperatures sufficiently high to achieve mobility of atoms but below that at which thermal softening occurs. This process is called **devitrification**. Although glasses are important materials, their physics and chemistry are not well understood. There is extensive research currently under-way on the structure and properties of glasses.

APPENDIX 12A METALLIC VALENCE AND METALLIC RADII OF THE ELEMENTS IN ÅNGSTROMS FOR CN = 12*

(Each cell gives the element symbol, the metallic valence, and the metallic radius in Å.)

Li 1 1.549	Be 2 1.123	B 3 0.98											C 4 0.914	N 3 0.88 0.92	O 2 0.66	F 1 0.64
Na 1 1.896	Mg 2 1.598	Al 3 1.429											Si 4 1.316	P 3 1.28	S 2 1.27	Cl 1 0.99
K 1 2.349	Ca 2 1.970	Sc 3 1.620	Ti 4 1.467	V 5 1.338	Cr 2.90 5.78 1.357 1.267	Mn 4.16 5.78 1.306 1.261	Fe 5.78 1.260	Co 5.78 1.252	Ni 5.78 1.244	Cu 5.44 1.276	Zn 4.44 1.379	Ga 3.44 1.408	Ge 4 1.366	As 3 1.39	Se 2 1.40	Br 1 1.14
Rb 1 2.48	Sr 2 2.148	Y 3 1.797	Zr 4 1.597	Nb 5 1.456	Mo 6 1.386	Tc 5.78	Ru 5.78 1.336	Rh 5.78 1.342	Pd 5.78 1.373	Ag 5.44 1.442	Cd 4.44 1.543	In 3.44 1.660	Sn 2.44 4 1.620 1.542	Sb 3 1.59	Te 2 1.60	I 1 1.33
Cs 1 2.67	Ba 2 2.215	La 3 1.871	Hf 4 1.585	Ta 5 1.457	W 5.78 1.394	Re 5.78 1.373	Os 5.78 1.350	Ir 5.78 1.355	Pt 5.78 1.385	Au 5.44 1.439	Hg 4.44 1.570	Tl 3.44 1.712	Pb 2.44 1.746	Bi 3 1.70	Po 2 1.76	At 1
Fr 1	Ra 2	Ac 3	Th 4 1.795	Pa 5	U 5.78 1.516	Np	Pu	Am	Cm							

Ce 3.2 1.818	Pr 3.1 1.824	Nd 3.1 1.818	Pm	Sm 2.8 1.85	Eu 2 2.084	Gd 3 1.795	Tb 3 1.773	Dy 3 1.770	Ho 3 1.761	Er 3 1.748	Tm 3 1.743	Yb 2 1.933	Lu 3 1.738

*From Linus Pauling, Journal of the American Chemical Society, **69**, 542 (1947).

COVALENT RADII*

	Be	B	C	N	O	F
	1.07	0.89	0.77	0.70	0.66	0.64
	Mg	Al	Si	P	S	Cl
	1.40	1.26	1.17	1.10	1.04	0.99
Cu	Zn	Ga	Ge	As	Se	Br
1.35	1.31	1.26	1.22	1.18	1.14	1.11
				(1.21)	(1.17)	(1.14)
Ag	Cd	In	Sn	Sb	Te	I
1.53	1.48	1.44	1.40	1.36	1.32	1.28
				(1.41)	(1.37)	(1.33)
Au	Hg	Tl	Pb	Bi		
1.50	1.48	1.47	1.46	1.46		
				(1.51)		

* L. Pauling, *The Nature of the Chemical Bond*, Cornell University Press, Ithaca, N.Y. (1945); V. Schomaker and D. P. Stevenson, *Journal of the American Chemical Society*, **63**, 38 (1941).

PAULING IONIC RADII, Å*†

			H^-	He	Li^+	Be^{2+}	B^{3+}	C^{4+}	N^{5+}	O^{6+}	F^{7+}
			2.05	0.92	0.59	0.43	0.34	0.29	0.25	0.22	0.19
C^{4-}	N^{3-}	O^{2-}	F^-	Ne	Na^+	Mg^{2+}	Al^{3+}	Si^{4+}	P^{5+}	S^{6+}	Cl^{7+}
4.14	2.47	1.76	1.36	1.12	0.95	0.82	0.72	0.65	0.59	0.53	0.49
Si^{4-}	P^{3-}	S^{2-}	Cl^-	A	K^+	Ca^{2+}	Sc^{3+}	Ti^{4+}	V^{5+}	Cr^{6+}	Mn^{7+}
3.84	2.79	2.19	1.81	1.54	1.33	1.18	1.06	0.96	0.88	0.81	0.75
					Cu^+	Zn^{2+}	Ga^{3+}	Ge^{4+}	As^{5+}	Se^{6+}	Br^{7+}
					0.96	0.88	0.81	0.76	0.71	0.66	0.62
Ge^{4-}	As^{3-}	Se^{2-}	Br^-	Kr	Rb^+	Sr^{2+}	Y^{3+}	Zr^{4+}	Nb^{5+}	Mo^{6+}	
3.71	2.85	2.32	1.95	1.69	1.48	1.32	1.20	1.09	1.00	0.93	
					Ag^+	Cd^{2+}	In^{3+}	Sn^{4+}	Sb^{5+}	Te^{6+}	I^{7+}
					1.26	1.14	1.04	0.96	0.89	0.82	0.77
Sn^{4-}	Sb^{3-}	Te^{2-}	I^-	Xe	Cs^+	Ba^{2+}	La^{3+}	Ce^{4+}			
3.70	2.95	2.50	2.16	1.90	1.69	1.53	1.39	1.27			
					Au^+	Hg^{2+}	Tl^{3+}	Pb^{4+}	Bi^{5+}		
					1.37	1.25	1.15	1.06	0.98		

* L. Pauling, *The Nature of the Chemical Bond*, Cornell University Press, Ithaca, N.Y. (1945).
† A detailed description of how ionic radii are obtained is given in the article by Tosi [M. P. Tosi, in *Solid State Physics* (eds.), F. Seitz and D. Turnbull, Academic Press, New York, **16**, 1 (1964)]. Tosi arrives at values for the alkali metal ions and halide ions which are larger and smaller, respectively, than Pauling's values.

REFERENCES

Evans, R. C., *An Introduction to Crystal Chemistry*, Cambridge, New York (1964).

Azaroff, L., *Introduction to Solids*, McGraw-Hill, New York (1960). Chapters 3 and 4 give a good account of packing concepts.

Wells, A. F., *Structural Inorganic Chemistry*, Oxford, New York (1962). This gives an extensive list of compounds in the index for which the structure and coordination are given in the text.

Charles, R. J., "The Nature of Glass," *Scientific American* (Sept. 1967) p. 127.

Volf, M. B., *Technical Glasses*, Pitman, London (1961).

Van Vlack, L. H., *Physical Ceramics for Engineers*, Addison-Wesley, Reading, Mass. (1964). This book can be read with ease by students who have progressed through Chapter 16 of the present book.

Kingery, W., *Introduction to Ceramics*, Wiley, New York (1960).

Barrett, C., and Massalski, T. B., *Structure of Metals*, McGraw-Hill, New York (1966).

PROBLEMS

12.1. a. Define what is meant by a closest-packed layer.
        b. What is the stacking sequence for the fcc crystal structure and the hcp crystal structure?

12.2. a. Give the Miller indices for the closest-packed planes in the fcc crystal structure.
        b. Which set of planes are the farthest apart in the fcc structure?
        c. How far apart are they in terms of the lattice parameter a?

12.3. a. Give the Miller indices for the closest-packed planes in the hcp crystal structure.
        b. What word is often used to describe these planes?

12.4. a. Give the c/a ratio for four hcp metals.
        b. What would their c/a ratio be for the ideal case where the atoms were really spheres?

12.5. What is the identity period of the stacking sequence for hcp and fcc structures?

12.6. What is the ratio of tetrahedral voids to atoms in the
        a. fcc crystal structure?
        b. hcp crystal structure?

12.7. What is the ratio of octahedral voids to atoms in the
   a. fcc crystal structure?
   b. hcp crystal structure?

12.8. Derive the results given in Table 12.3.1.

12.9. Derive the results given in Table 12.4.1.

12.10. The carbon atom has a radius of 0.77 Å, while the iron atom has a radius of 1.26 Å. Fe has the bcc structure below 910°C, and the fcc structure from 910–1392°C. One of these structures dissolves several atomic percent carbon; the carbon goes into the voids even though it appears too large to do so. Which structure is most likely to dissolve the carbon, and in which void would it be found?

12.11. Along what direction (give the direction indices) do atoms touch in these crystals:
   a. fcc?   b. hcp?   c. bcc?

12.12. Suppose that iron atoms have a radius of 1.26 Å in both fcc and bcc iron. Which has the higher density?

12.13. a. Compare the bonding in graphite and diamond.
   b. Compare the electrical conductivity in these materials.
   c. Which polymorph is the stable one at room temperature and pressure?

12.14. Cu and Ni both have the fcc crystal structure. Discuss other reasons why they might be continuously soluble in the solid state.

12.15. Pb has a fcc crystal structure and Sn has a tetragonal crystal structure. Give reasons why you might expect some solubility but not continuous solubility of these elements in each other.

12.16. Give an example of
   a. A solid substitutional solution.
   b. A solid interstitial solution.
   c. An ordered solid solution.

12.17. Show that the expected crystal structure of MgO is the NaCl type.

12.18. Predict from packing concepts the crystal structure of InSb.

12.19. Why do you expect the CO_3^{2-} complex to be a planar triangle?

12.20. What are interstitial compounds, and what are their properties?

12.21. A large chunk of $BaTiO_3$ has the cubic crystal structure. It is cooled to room temperature. Describe what happens.

12.22. Discuss the formation of the following silicate structural units:
   a. Islands.
   b. Chains.
   c. Sheets.

12.23. Name a mineral based on the following silicate structures:
   a. Island.

b. Chain.
c. Sheet.

12.24. SiO_2 has several crystalline polymorphs.
   a. What is a polymorph?
   b. Describe the structure of tridymite.

More Involved Problems

12.25. Would you expect ZnS to be piezoelectric? Explain.

12.26. Would you expect tetragonal $BaTiO_3$ to be piezoelectric? Explain.

12.27. Is NaCl piezoelectric? Explain.

12.28. Dielectric behavior is described by a second-rank tensor (as is thermal conductivity).
   a. Show that the dielectric coefficient is the same in every direction for a cubic crystal.
   b. The dielectric constant at optical frequencies is related to the refractive index by $\kappa = n^2$. Is the velocity of light the same in all directions in a cubic crystal?

12.29. a. Using one or more of the Hume-Rothery rules, discuss the maximum extent of solubility of Ni, Zn, Al, Si, and Sb in fcc copper.
   b. Compare your results with the experimental results in *Metals Handbook*, American Society for Metals, Novelty, Ohio (1948).

Sophisticated Problem

12.30. The compound K_2SiF_6 has a cubic crystal structure. Using the Pauling rules, obtain the structure.

Prologue

Crystals are never perfect. There are always errors in the packing. These are described geometrically as *point, line, planar and spatial imperfections*. Various point imperfections can be introduced by bombardment with γ-rays, neutrons, electrons, or ions. They can also be produced by plastic deformation. Certain point imperfections can be present in sizable concentrations (10^{-4} mole fraction) in thermal equilibrium at high temperatures. The concentration of thermally produced defects decreases very rapidly as the temperature is decreased. It is important to study these point defects (and in particular the *vacancy*) because they play a vital role in determining many of the properties of solids. It is obvious that they are present in nuclear reactors because of the radiation. Vacancies also play a key role in creep and in the processing techniques used to make large-scale integrated electronic circuits. One of the important features of vacancies is their *energy of formation*.

The *dislocation* is the line imperfection which is involved in virtually all processes of plastic deformation of crystalline solids. Without dislocations, crystals would be very strong. However, without dislocations metals would not be malleable and the various production processes which are based on plastic deformation, such as wire drawing, would not work. *Crystal surfaces, grain boundaries*, and other planar defects are also discussed.

13

CRYSTALLINE IMPERFECTIONS

13.1 IMPERFECTIONS AND CRYSTAL PROPERTIES

Let us distinguish more clearly between the *idealized* crystals with which we have been concerned so far and the *real* crystals upon which we base our measurements. We have already defined the former in terms of *unit cells* arranged in a periodic *space lattice* generated by repetition of three *primitive lattice vectors* in space (and hence infinite in extent). All crystals can be classified in one of seven *crystal systems,* each system characterized by a typical minimum symmetry.

Many properties of actual crystals such as cohesion (binding) and elastic coefficients can be accounted for quite adequately on the basis of our definition of an ideal crystal. There are other properties, however, which deviate seriously from such predictions, indicating that real crystals differ in some critical way from our definition. This difference is attributed to **imperfections** which arise from deviations in arrangements of atoms from that typical of the ideal crystal. The properties associated with the corresponding real crystal are called **structure-sensitive** properties to distinguish them from properties that are not sensitively dependent on these deviations. Thus, even among samples with the same atomic composition we often find a bewildering variance in a particular structure-sensitive property. The pretreatment of each sample can be very critical in determining many of its important properties. Some typical structure-sensitive and structure-insensitive properties of crystals are listed in Table 13.1.1 to clarify this distinction.

Imperfections influence the structure-insensitive properties, too, but not to the degree to which, for example, the fracture strength or plasticity of solids is affected. Thus the tensile yield strength of copper can vary from 1 to

Table 13.1.1. STRUCTURE-SENSITIVE AND STRUCTURE-INSENSITIVE PROPERTIES

Type of property	Structure-Insensitive	Structure-Sensitive
Mechanical	Density Elastic moduli	Tensile yield strength Fracture strength Plasticity
Thermal	Thermal expansion (at high T) Melting point Specific heat Heat of fusion	Thermal conductivity, particularly at low temperatures
Electrical	Resistivity (metallic) at high temperatures, Electrochemical potential	Resistivity at low temperatures in semiconductors and metals
Magnetic	Paramagnetic and diamagnetic properties	Ferromagnetic properties (including magnetostriction)
Superconductivity	Transition temperature	Current carrying capacity

10^6 psi because of the presence of imperfections. Yet the presence of these same imperfections causes only a few percent change in density or in the elastic coefficients. Finally it should be realized that imperfections are not necessarily bad. There are many cases where a property may be optimized by adequate control of the defect structure during processing such as, for example, the optimization of the tensile yield strength of copper during drawing (cold working) which introduces billions of imperfections (called dislocations) per cubic centimeter.

We shall be primarily concerned here with the *geometric* characteristics of imperfections. Deviations from perfect crystalline order will be relatively small in terms of the total order in the structure so that the real crystals will still retain essentially the same overall structure and show the same structure-insensitive properties customarily associated with the ideal crystal.

Packing imperfections in crystalline structures are usually classified according to whether the disturbance is localized around a single lattice point, localized in a long narrow cylinder of related lattice points, localized around an internal or external surface, or whether the disturbance is spread throughout a sizable volume element. These are called, respectively, **point defects, line defects, planar defects**, and **spatial or volume defects**.

13.2 POINT IMPERFECTIONS

NONIONIC CRYSTALS. At a temperature near the melting point the stable structure for copper corresponds to one in which a significant number (10^{-4}) of lattice positions are vacant. These defects are called **vacancies**. The equilibrium vacancy concentration (fraction of vacant lattice sites) is given by

$$\frac{n}{N+n} \approx e^{-E_f/kT}. \qquad (13.2.1)$$

Here n is the number of vacancies, N is the number of atoms, E_f is the energy of formation of a vacancy, k is the Boltzmann constant, and T is the absolute temperature. The **energy of formation of a vacancy** is the increase in energy of the crystal when an atom is taken from one of its regular sites and placed on the surface. It is thus a characteristic of the crystal. For copper $E_f = 1.17$ eV.

EXAMPLE 13.2.1

Calculate the fraction of vacancies in copper just below the melting point of 1080°C.

Answer. The absolute temperature is 1353°K and $k = 8.62 \times 10^{-5}$ eV/°K. Hence

$$\frac{n}{N} = e^{-1.17/(8.62 \times 10^{-5} \times 1353)} = e^{-10.0} = 10^{-10.0/2.3} = 10^{-4.3}.$$

Equation (13.2.1) can be readily derived by the use of thermodynamics or statistical mechanics. See R. A. Swalin, *Thermodynamics of Solids*, Wiley, New York (1962) or see Chapter 17.

EXAMPLE 13.2.2

If a metal has a fractional concentration of 10^{-4} vacancies just below T_M, what is the vacancy concentration at $T_M/2$?

Answer. If

$$e^{-E_f/kT_M} = 10^{-4},$$

then

$$e^{-2E_f/kT_M} = 10^{-8}.$$

Note how *very* rapidly concentration varies with temperature.

Vacancy concentrations in excess of the equilibrium concentration can be created by radiation (with neutrons, electrons, γ-rays) and by mechanical deformation and quenching.

A **self-interstitial** is an atom not located at a lattice site in the lattice but squeezed between atoms. Its energy of formation is considerably larger than that of a vacancy. Self-interstitials can actually be directly observed in metals with high melting points using field ion microscopy, as shown by the sequence in Figure 13.2.1. A third important type of defect occurs when an atom is removed from a lattice site and is placed at an interstitial site. It is essentially a combination of the first two types and is called a **Frenkel defect**.

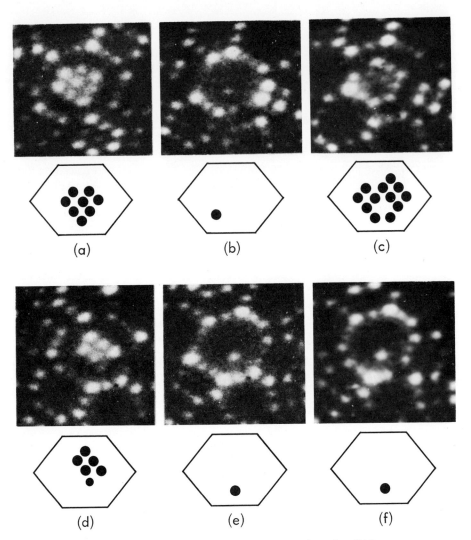

FIGURE 13.2.1. Field evaporation sequence through a (111) plane of a tungsten specimen which has been irradiated with 20 keV tungsten ions (W⁺). (a) A perfect (111) plane showing eight atoms on the plane. The line drawing below the micrograph schematically indicates the positions of all the atoms. (b) The same plane after 10 field evaporation pulses have removed the original eight atoms shown in (a). The atom which is protruding slightly is from the next (111) layer. (c) The application of 10 additional field evaporation pulses causes 12 atoms on this layer to become resolved. In addition the protruding atom in (b) has been preferentially field-evaporated leaving a vacant lattice site on the (111) plane. (d) After several more field evaporation pulses, five

Such defects are produced when crystals are bombarded with neutrons, electrons, and γ-rays. A fourth type results from an impurity atom and is called an **impurity defect**. They can occur in metallic, ionic, or covalent solids. Their concentration and movement are very important in many solid state processes such as self-diffusion, phase transformation, and electrical conductivity. It should be noted that the presence of a *single defect may alter the lattice over a significantly large volume*. Thus the lattice may be appreciably distorted several atom distances from a vacancy.

One of the most important properties possessed by interstitial ions and vacancies is that they usually can migrate about the crystal relatively easily when assisted by thermal fluctuations, i.e., vibrating neighboring atoms may exchange with the vacancy, and hence permit the *diffusion* of matter in solids. Different diffusion mechanisms are shown in Figure 13.2.2.

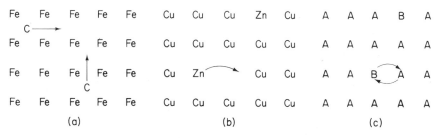

FIGURE 13.2.2. Mechanisms of diffusion. (a) Interstitial diffusion of carbon in iron. (b) Vacancy diffusion of zinc in copper. (c) Direct exchange, which occurs infrequently.

IONIC CRYSTALS. Examples of **Frenkel defects, Schottky defects**, and interstitial defects are shown in Figure 13.2.3 in the type of crystal in which they usually occur. An important constraint in the production of point defects in ionic crystals is the necessity of maintaining charge neutrality (the energy increase is enormous if electrical neutrality is not maintained). This can occur by balance of the (+) and (−) defects in a variety of ways.

lattice atoms remain on the plane, and the interstitial causing the lattice atom to protrude (b) can be seen as a dim spot just below these lattice atoms. (e) The five remaining lattice site atoms have been evaporated, leaving the self-interstitial atom, which remains as an extra bright spot. (f) Further field evaporation results in the removal of two atoms from the next (111) layer without field-evaporating the interstitial atom. (*Courtesy of D. N. Seidman and R. M. Scanlon, Cornell University.*)

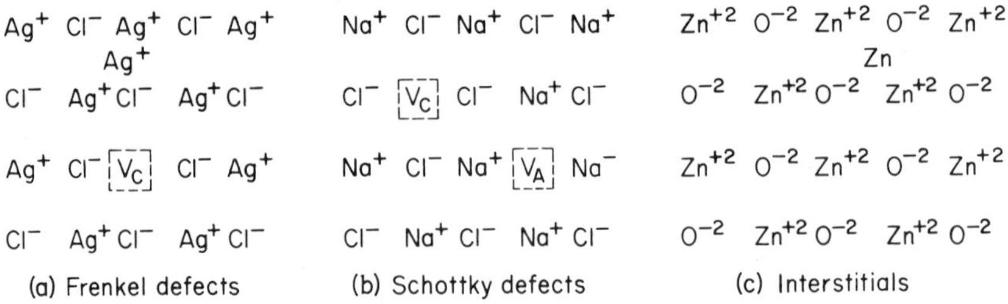

Ag^+ Cl^- Ag^+ Cl^- Ag^+ Na^+ Cl^- Na^+ Cl^- Na^+ Zn^{+2} O^{-2} Zn^{+2} O^{-2} Zn^{+2}
                    Ag^+ Zn
Cl^- Ag^+Cl^- Ag^+Cl^- Cl^- $\boxed{V_C}$ Cl^- Na^+ Cl^- O^{-2} Zn^{+2} O^{-2} Zn^{+2} O^{-2}

Ag^+ $Cl^-\boxed{V_C}$ Cl^- Ag^+ Na^+ Cl^- $Na^+\boxed{V_A}$ Na^- Zn^{+2} O^{-2} Zn^{+2} O^{-2} Zn^{+2}

Cl^- Ag^+Cl^- Ag^+Cl^- Cl^- Na^+ Cl^- Na^+Cl^- O^{-2} $Zn^{+2}O^{-2}$ Zn^{+2} O^{-2}

(a) Frenkel defects　　　　(b) Schottky defects　　　　(c) Interstitials

FIGURE 13.2.3. Balanced defects in ionic crystals. (a) Equal numbers of interstitial cations and cation vacancies. (b) Equal numbers of cation and anion vacancies. (c) Neutral interstitial atoms, which can be ionized at moderate or high temperature with release of electrons to the conduction band. [From Shockley, W.,ed. *Imperfection in Nearly Perfect Crystals*, John Wiley and Sons, Inc. (1952).]

There is evidence that groups of two or more point defects of the same kind form **complex point defects** in metals and in salts. Also a vacancy and a solute atom (either substitutional or interstitial) may form a *stable pair* in a metal. Note that vacancies formed at a positive ion site in an ionic salt have an effective *negative* charge and that vacancies formed at a negative ion site in an ionic salt have an effective *positive* charge. A coupled pair of vacancies of opposite sign in the alkali halides is shown in Figure 13.2.4. The coupling energy is relatively high (about 0.9 eV), indicating a fairly stable configuration.

For further information on the role of complex configurations see J. R. Reitz and J. L. Gammel, *Journal of Chemical Physics*, **19**, 894 (1951). It is also

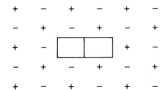

FIGURE 13.2.4. Coupled pair of vacancies of opposite sign in the alkali halides. This unit is believed to have the ability to diffuse very rapidly. The coupling energy is about 0.9 eV. [From J. R. Reitz and J. L. Gammel, *Journal of Chemical Physics*, 19, 894 (1951).]

apparent that there are many possible ways in which point defects may group to form clusters. How important the contribution of clusters is to the properties exhibited by the crystal is still not clear. It is established that point defects in nonionic crystals interact with each other at sufficiently high densities to form moderately stable complexes such as **divacancies**.

A pure crystal such as NaCl contains an equal number of cation and anion vacancies at a high temperature. The concentration is given by

$$\frac{n_a}{N}\frac{n_c}{N} \approx e^{-E_f^s/kT}, \tag{13.2.2}$$

where E_f^s is the energy of the Schottky pair. ($E_f^s = 2.0$ eV for NaCl.)

Equation (13.2.2) is simply the law of mass action, namely the concentration of anion vacancies C_{av} times the concentration of cation vacancies C_{cv} equals the equilibrium constant K:

$$C_{av} C_{cv} = K.$$

EXAMPLE 13.2.3

Calculate the fraction of cation vacancies at 900°K.

Answer. From (13.2.2) we have

$$\frac{n_c}{N} = e^{-E_f^s/2kT}$$

$$\frac{n_c}{N} = e^{-1.0/(8.62 \times 10^{-5} \times 900)} \simeq e^{-13} = 10^{-5.6} \simeq 2 \times 10^{-6}.$$

EXAMPLE 13.2.4

Suppose NaCl is doped with 0.2 at. % Ca^{2+} (which rests on Na^+ sites). What is the relationship between n_a/N and n_c/N now?

Answer. Whereas in the pure crystal $n_a/N = n_c/N$, in the present case because of the charge neutrality requirement

$$\frac{n_a}{N} + 0.002 = \frac{n_c}{N}, \tag{13.2.3}$$

i.e., there is a cation vacancy created for every thermally created anion vacancy and there is a cation vacancy created for every divalent calcium ion present.

To evaluate the number of cation vacancies in this case at 900°K solve (13.2.3) for n_a/N and substitute into (13.2.2). Ans. $n_c/N = 2 \times 10^{-3}$.

At high temperatures ionic crystals become good electrical conductors because these charged defects can move rapidly. In ultrapure NaCl (where $n_c = n_a$) the electrical conductivity is predominantly due to motion of the cation vacancy, which moves much faster than the anion vacancy.

EXAMPLE 13.2.5

What effect has the addition of $0.2\%\,Ca^{2+}$ on the electrical conductivity of NaCl at 900°K?

Answer. Since the addition of Ca^{2+} increases the charge carriers (the cation vacancy) by a factor of 10^3 (compare the answers to Examples 13.2.3 and 13.2.4), the conductivity also increases by this factor. Thus we see that electrical conductivity in NaCl owes its very existence to point defects and that it is strongly affected by small concentrations of impurities. Hence it is a structure-sensitive property. In more advanced studies, you will find that this calculation is complicated by the formation of divalent ion-positive ion vacancy pairs.

NONSTOICHIOMETRIC COMPOUNDS. **Wustite** has the same structure as NaCl. However, it has the chemical formula $Fe_{<1}O$; i.e., it deviates from the expected formula FeO. Such compounds are called **nonstoichiometric compounds**. Certain of the positive ion (Fe^{2+}) lattice sites must therefore be empty. However, this leaves an overall deficit of positive charge (which is not allowed). For every Fe^{2+} vacancy created, two Fe^{2+} can be converted into the Fe^{3+} state. Thus in the overall process three Fe^{2+} ions are destroyed and two Fe^{3+} ions and an ion vacancy are created. This process can be described by a chemical equation:

$$\frac{1}{2}O_2 + 3Fe^{2+} = 2Fe^{3+} + \boxed{v_c} + (Fe^{2+} + O^{2-}). \qquad (13.2.4)$$

As many as 14% of the normal cation sites can be empty in wustite.

13.3 LINE IMPERFECTIONS

Dislocations are line imperfections in crystals. There are many examples of their importance in crystals but one striking one is indicated in Table 13.3.1 in which the tensile strengths of a metal whisker (almost a perfect crystal) and an annealed single crystal are compared.

Table 13.3.1.

Type of Crystal	Effective Strength, lb/in.^{2}	Dislocation Density
Iron whisker	1,000,000	Nil
Ordinary crystal	100–1000	Approx. 10^8/in.^{2}

An example of an edge dislocation is shown in Figure 13.3.1 for the case of a simple cubic crystal. The symbol $\perp$ is placed at the dislocation line.

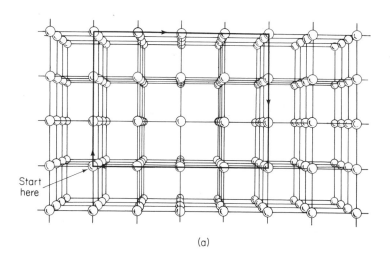

(a)

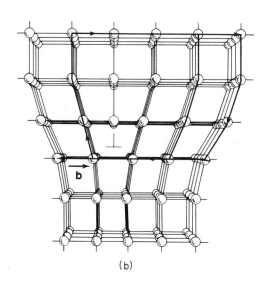

(b)

FIGURE 13.3.1. Description of a dislocation in terms of its Burgers vector. (a) A Burgers circuit in dislocation-free material. (b) The same Burgers circuit passing through dislocation-free material, but encircling a dislocation of unit Burgers vector *b*. [From A. G. Guy, *Elements of Physical Metallurgy* Addison-Wesley, Reading, Mass. (1959).]

Note in part (b) how the atomic packing is disturbed in the immediate vicinity of the dislocation line. Note, however, that far from the dislocation line the packing arrangement is correct, although the atoms there will be displaced slightly from their usual positions; i.e., there will be elastic strains. A dislocation line is described by its direction (a unit vector parallel to the line) and by its Burgers vector. Figure 13.3.1(a) shows a closed circuit of atom to atom steps. Suppose this circuit starts at the lower left corner of the loop. The circuit then consists of four steps upward, four to the right, four downward, and four to the left. If a similar circuit is made around a dislocated region, the circuit will not stop at the starting point. The additional vector that must be added to make the circuit close is called the **Burgers vectors b**. See Figure 13.3.1(b). Notice that the Burgers vector is perpendicular to the dislocation line: This *defines* an **edge dislocaton**. We note that the Burgers vector equals a translation vector of the lattice (in the direction of the Burgers vector). We note an extra plane of atoms appears to have been inserted in the upper half of the crystal of Figure 13.3.1(b). Electron microscopy can be used to actually observe the displacement field around an edge dislocation, as shown in Figure 13.3.2.

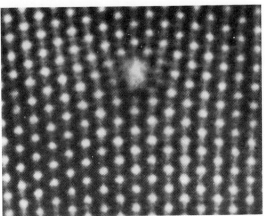

FIGURE 13.3.2. Electron micrograph moire pattern of edge dislocation in cadmium. The thin cadmium crystal was grown on molybdenum sulfide. The slight mismatch between the two crystals produced the moire effect, which makes possible the visualization of the pattern of an edge dislocation. ($\times 2,400,000$.) (*Courtesy of W. Menter and D. W. Pashley, Tube Investments Research Laboratory.*)

Figure 13.3.3 shows how such a dislocation can move under applied shear stresses.

Since only tiny motions of the atoms are necessary, dislocation motion can take place very easily, particularly in metals where the binding is long range. (In covalent materials such as diamond, covalent bonds will have to be broken

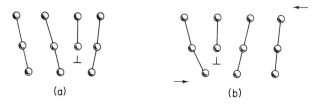

FIGURE 13.3.3. (a) Original dislocation position. (b) Dislocation moved one repeat distance to the left
under applied shear stresses.

and reformed and hence a much higher stress is needed to move the dislocations.)
In copper crystals of high purity, F. Young has found that shear stresses of
only 1 psi will make the dislocations move. If there are dissolved impurities
present or precipitate particles of a second phase in the copper or if other
dislocations are present which must be intersected, then a higher shear stress
will be needed to make the dislocation move. In a very pure single crystal of
copper the dislocation moves so easily because it is moving through a perfectly
periodic atomic array. Anything which disrupts the perfect periodicity will
make it more difficult for the dislocation to move.

Dislocation motion is so important because it is the mechanism whereby
plastic deformation occurs. This is illustrated in Figure 13.3.4.

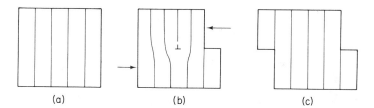

FIGURE 13.3.4. (a) Perfect crystals; (b) Crystal with dislocation introduced. The top half of the crystal has
an extra plane of atoms inserted above the
dislocation line; (c) The dislocation has moved
to the left out of the crystal. The crystal is now
permanently deformed.

The creation of dislocations by slip as in Figure 13.3.4(b) is difficult
compared to making them move when they are present. When materials crystallize from the melt at typical rates they usually contain a large number of
dislocations. Thus a typical copper single crystal may contain 10^{6-8} cm of
dislocation lines per cubic centimeter. When such materials are plastically
deformed the dislocations multiply and a highly deformed material may have
a dislocation density of 10^{11} cm/cm^3, or 10^{11} cm^{-2}. Figure 13.3.5 shows the
(111) face of a copper crystal whose original dislocation density was about
100 cm^{-2}. This crystal had been indented by a diamond after being carefully
polished. It was then etched with chemicals which dissolved away the highly

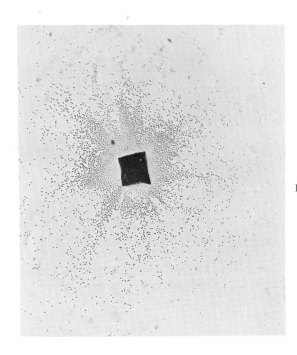

FIGURE 13.3.5. Etch pits associated with dislocations emerging from a copper crystal. The dislocations were formed by indentation on a (111) surface. (×100.) (*Courtesy of B. F. Addis, Cornell University.*)

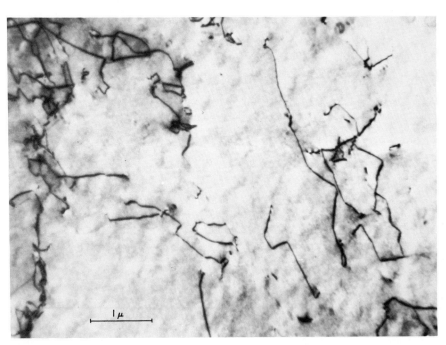

FIGURE 13.3.6. Dislocations in nickel by electron microscopy.

strained material around the dislocation more rapidly than the good material. Figure 13.3.6 shows a transmission electron micrograph of dislocations in nickel. Thin foils of the order of 1000–10,000 Å thick, depending on the material, are transparent to electrons. The transmission through the highly strained material along the dislocation line is different than that through the good material.

While many of the electron micrographs shown thus far were obtained using accelerating potentials of 100,000 V, more recent electron microscopes can use voltages in excess of 1,000,000 V. Hence they have larger penetration distances. An example of such a microscope is shown in Figure 13.3.7.

Dislocations can also be studied by other techniques. For example, impurities often segregate preferentially at dislocations so that dislocations *within* transparent materials can be directly observed. In crystals with low dislocations densities, the **Lang technique** can be used. This is based on the fact that very near to the Bragg diffraction angle, the transmission of X-rays through crystals is anomalously high (so-called anomalous transmission). Near the dislocation, the planes are bent so that these planes are not near the Bragg angle and the transmission is correspondingly low.

FIGURE 13.3.7. 1,200,000-V electron microscope in France. (*Courtesy of Professor G. Dupouy, Electron Optics Laboratory, Toulouse.*)

EXAMPLE 13.3.1

Which would have the higher yield strength, a copper bar with a dislocation density of 10^6 cm^{-2} or one of the same purity with 10^{11} cm^{-2}?

Answer. The latter. The reason for this is that when a dislocation moves through a crystal, it will be impeded by any disruption in the perfect periodic array (such as other dislocations). It is this interaction of dislocations with each other along with the increase in dislocation density during plastic straining which causes strain hardening. It should be noted that when large numbers of dislocations are present they exist in extremely tangled configurations.

A carefully grown copper crystal might have a dislocation density of only 10 cm^{-2}, while it is possible to grow tiny copper crystals (called **whiskers**) having a diameter of about 10^{-4} cm and a length of 2–3 × 10^{-1} cm which have no dislocations at all (or in some cases a single dislocation along the axis).

The bonding around a dislocation is not complete (the atoms do not have the appropriate number of neighbors), and the material farther away from the dislocation line is elastically strained. Energy is therefore needed to create a dislocation line. This energy is usually described in terms of the energy per atomic plane of length (about 2 Å or so) of the dislocation line. For an edge dislocation in copper this is about 8 eV/atomic plane.

SCREW DISLOCATIONS. There is another dislocation which is important in plastic deformation and in the growth of ordinary (nonperfect) crystals.

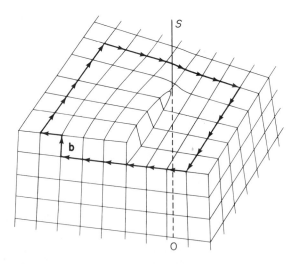

FIGURE 13.3.8. Burgers vector in simple cubic lattice for a screw dislocation.

This is the screw dislocation shown in Figure 13.3.8. The line $\overline{SO}$ is parallel to the dislocation line. Note that in this case the Burgers vector is parallel to the dislocation line: This is the defining characteristic of a **screw dislocation**. The slip motion of the screw dislocation (unlike the edge) is perpendicular to the Burgers vector. The crystal planes perpendicular to the dislocation line in Figure 13.3.8 actually form a spiral ramp (similar to the spiral ramp parking lot). Note that there is a ledge at the top of the crystal. When the crystal grows

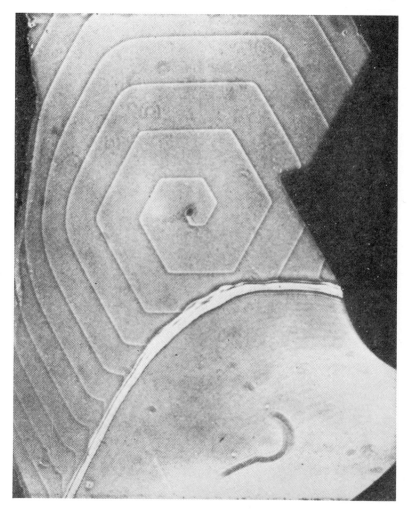

FIGURE 13.3.9. Regular hexagonal spiral on a SiC crystal originating from a group of screw dislocations at the center. ($\times$90.) [From A. R. Verma, *Crystal Growth and Dislocations*, Academic Press, New York (1953).]

by adding atoms from the gas or liquid, such atoms add at this ledge, which revolves around. This is one of the important mechanisms in the growth of nonperfect crystals. Figure 13.3.9 shows a growth spiral in SiC. The screw dislocation is in general just as important as the edge dislocation in plastic deformation processes. Moreover, the edge dislocation often plays a role in crystal growth.

13.4   PLANAR IMPERFECTIONS

Surfaces are the boundaries which form when the crystal originates during casting, condensation, or deposition. The structure and mobility of these boundaries exert a profound influence on many of the properties of the crystal.

The external surface is a major defect of the crystal in both composition and structure in a general sense. Similarly, there is a large group of planar defects associated with the interior of the crystal such as *grain boundaries* and *stacking faults* which range considerably in characteristics from the viewpoints of structure, energy, and relative dimensions.

Prior to the consideration of planar imperfections in more detail, let us get a perspective on their character as a group in terms of relative distinctions based on differences in structure and energy. Five of the most important types of

Table 13.4.1. PLANAR IMPERFECTIONS

Type of Planar Defect	Energy of Formation ergs/cm^2		Crystallography
	Copper*	Other Metals	
Coherent twin boundary	19	10–100	Difference in orientation with symmetry restriction
Stacking fault	40	20–400	Difference in stacking sequence of crystal layers
Low-angle grain boundary (1 deg)	80	10–250	Small orientation difference between two crystals of same phase
Grain boundary	550	100–1000	Relatively high disorientation between two crystals of same phase
External surface	1430	1000–9000	Surface exposed to equilibrium vapor pressure

* Specific values for copper are listed separately to provide a more precise basis for comparison.

planar defects are described in Table 13.4.1 in terms of their *crystallography* and *energy of formation*.

As indicated in Table 13.4.1 the coherent twin boundary typifies an extreme case for a planar defect in that the degree of disarrangement is minimal across the interface; i.e., the atoms are in sites of the same lattice on both sides. When crystals fit each other this way at a boundary they are said to be **coherent**. A **coherent twin** relationship indicates that the atom positions on one side of the interface are a mirror image of those of the other. Let us consider, for example, the closest-packed (111) plane of a fcc crystal, as indicated in Figure 13.4.1.

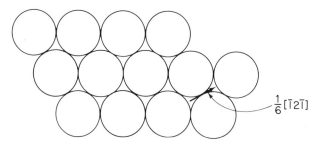

FIGURE 13.4.1. Projection on (111) plane of atom sites in fcc crystal. The C layer above the given B layer moves during twinning to an A position as shown by the vector.

Remember that one characteristic of the fcc lattice is that it can be generated by the ... ABCABC ... stacking sequence of such closest-packed planes.

A twinned fcc crystal has a stacking sequence of ... ABCABÅCBACBA .... The **twinning plane** (i.e., the mirror plane) is the (111) plane. Note that this twin could be *imagined* to occur not only by a growth process but by a mechanical process where successive atom layers above a given (111) plane are translated parallel to the (111) plane in the $[\bar{1}2\bar{1}]$ direction; the first such layer is translated by $\frac{1}{6}[\bar{1}2\bar{1}]$, the second layer by twice this distance in the same direction, etc. The direction $[\bar{1}2\bar{1}]$ is called the **twinning direction**. The twinning plane is typical for each crystal type, as indicated in Table 13.4.2. The actual formation of twins during mechanical deformation is called **mechanical twinning**.

Table 13.4.2. TWINNING PLANES

Structure	Twinning Plane
Face-centered cubic	{111}
Closest-packed hexagonal	{10$\bar{1}$2}
Body-centered cubic	{112}

This is an important process of deformation in bcc iron at low temperatures. A **stacking fault** in a fcc crystal corresponds to the sequence ... ABCĀB̄ABC .... A consequence of this is that the orientation across a stacking fault is the same on both sides, whereas it is not the same across a twinning plane. In the case of either stacking faults or twin boundaries each atom of the fcc crystal structure has the same 12 nearest-neighbor interactions as before. However, the twin contains only one layer (marked with *) which has a different next-nearest-neighbor layer, while the stacking fault contains two layers (marked with **) each of which has a different next-nearest-neighbor layer. This accounts for the factor of 2 (roughly) in energy.

A simple small-angle grain boundary is shown in Figure 13.4.2. This is a tilt boundary. It consists of a vertical array of edge dislocations and the angle of tilt θ is related to the distance between the dislocations D and the magnitude of the Burgers vector b by

$$\theta = \frac{b}{D}, \tag{13.4.1}$$

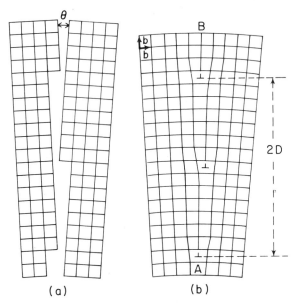

(a)   (b)

FIGURE 13.4.2. Simple grain boundary. The plane of the figure is parallel to a cube face and normal to the axis of relative rotation of the two grains. (a) Two grains have a common cube axis and an angular difference in orientation θ. (b) The two grains are joined to form a bicrystal. The joining requires only elastic strain except where a plane of atoms ends on the boundary in an edge dislocation, denoted by the symbol ⊥. [From W. T. Read, Jr., *Dislocations in Metals*, McGraw-Hill, New York (1953).]

if θ is small, or by

$$2 \tan \theta/2 = \frac{b}{D},$$ (13.4.2)

if θ is large. The fact that a **small-angle tilt grain boundary** is in fact an array of edge dislocations was first illustrated by using the etch pit technique to observe the dislocations (as shown in Figure 13.4.3) and using X-rays to measure the

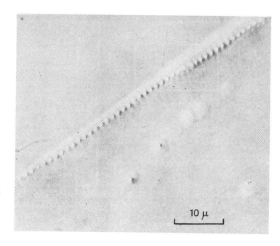

FIGURE 13.4.3. Etch pits that make up a grain boundary in germanium. [From F. L. Vogel, W. G. Pfann, H. E. Corey, and E. E. Thomas, *Physical Review*, **90**, 489 (1953).]

orientation of the crystal on each side of the boundary. In this way Equation (13.4.1) could be checked. With the more recent sophisticated developments in transmission electron microscopy, such arrays of dislocations can be observed with high resolution, as shown in Figure 13.4.4.

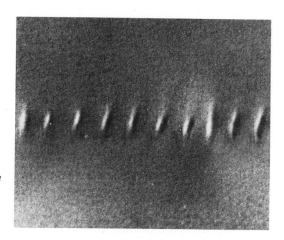

FIGURE 13.4.4. Dislocation array in aluminum. The light and very dark areas are dislocation stress fields. These dislocations are about 30 Å apart. (*Courtesy of Richard Parson and Larry Howe, Chalk River National Laboratory.*)

The energy of an array of such dislocations can be calculated so the energy of a simple tilt boundary can be computed. Such a calculated curve is shown in Figure 13.4.5.

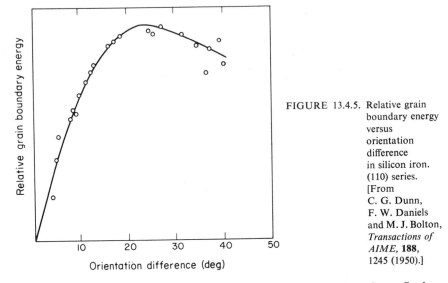

FIGURE 13.4.5. Relative grain boundary energy versus orientation difference in silicon iron. (110) series. [From C. G. Dunn, F. W. Daniels and M. J. Bolton, *Transactions of AIME*, **188**, 1245 (1950).]

Another fairly simple type of boundary is the **twist boundary.** Such a boundary is shown in Figure 13.4.6. It consists of a crossed grid of screw dislocations.

A large-angle grain boundary is one where the disorientation is too great for the simple dislocation model to apply. There is some evidence that the structure of the boundary depends on the disorientation of the crystals but in a much more complicated fashion. One very striking feature of large-angle grain boundaries is their tendency to move through the crystal if the driving force and the temperature are sufficiently high. This results in *grain growth* and corresponding modifications in the mechanical properties of the material. Another important characteristic is the marked tendency of foreign atoms to "segregate" at large-angle grain boundaries. This effect also may have marked consequences on mechanical behavior. Perhaps the most important feature is that *these boundaries represent a metastable condition of the material* and reflect both how the crystal was originally formed and to what forces and conditions of temperature (and pressure), etc., it was subsequently exposed. One of the first good models of a grain boundary was a raft of bubbles on the surface of a liquid. An example is shown in Figure 13.4.7.

More recently it has become possible to view grain boundaries directly. Figure 13.4.8 shows a field ion micrograph which illustrates a large-angle grain boundary in tungsten. Note that the boundary is very narrow, i.e., a few atoms wide.

The geometric properties of **external surfaces** of crystals also depend

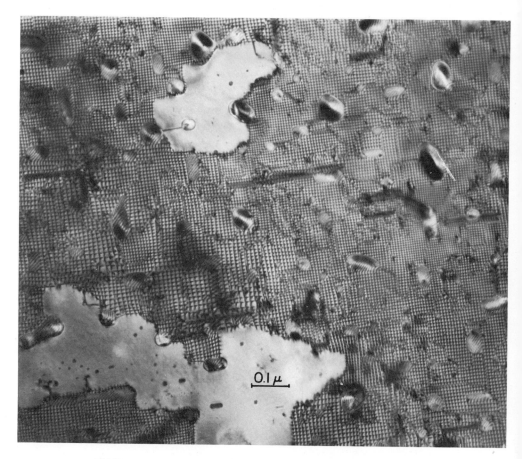

FIGURE 13.4.6. Twist boundary on the (100) plane of gold as
observed by thin film electron microscopy. In
some areas there is no boundary. (*Courtesy of
R. W. Balluffi and T. Schober, Cornell University.*)

strongly on their origin and treatment. It is characteristic of solid surfaces that
they usually do not represent equilibrium structures. Most treatments (such as
mechanical abrasion) of a crystal tend to intensify the defect structure of the
surface, whereas annealing a metal at high temperatures, especially in contact
with its own vapor, tends to reduce the defect structure of the surface. On the
other hand even the surface atoms of an otherwise perfect crystal at equilibrium
are displaced from the regular lattice sites by the unbalanced binding forces at
the surface of a solid.

The *atomistic theory of solid surfaces* is not nearly as advanced as that
for atomistic interactions inside crystal lattices. One important feature of all
solid surfaces arising from their metastable character is that they are *seldom*

FIGURE 13.4.7. Bubble raft illustrating grain boundaries. Note that the boundaries are only a few atoms thick. (*Courtesy of W. R. Day Jr., Mitron Research and Development Center.*)

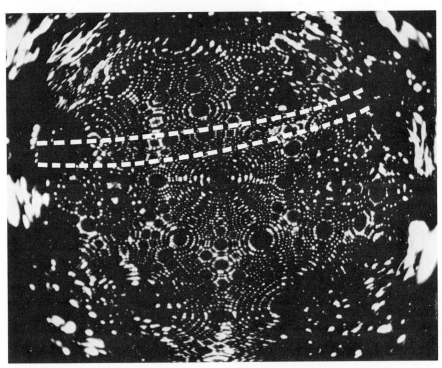

FIGURE 13.4.8. Field ion micrographs illustrating (a) grain boundary in tungsten and (b) no grain boundary. Note that the symmetry exhibited in (b) is destroyed in (a) by the grain boundary. (*Courtesy of D. N. Seidman and Y. C. Chen, Cornell University.*)

smooth. **Steps** of atomic dimensions or greater and of relatively great length occur on most crystal surfaces. On a larger scale, **crystal faceting textures** and **pitting structures** often occur resulting from the tendency of the solid surface to rearrange itself into a configuration of lower energy. Figure 13.4.9 shows a surface of silicon. An excellent tool for studying the grosser features of a surface

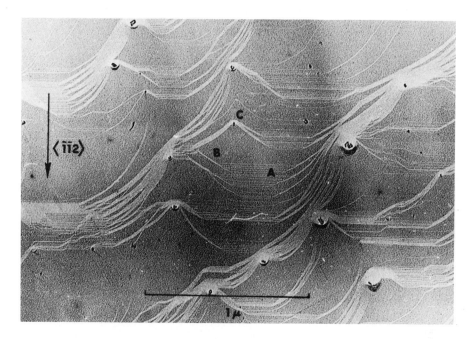

FIGURE 13.4.9. Electron micrograph of a replica of the step structure of a silicon layer. Steps at A are only one atom layer high. [H. C. Abbink, R. M. Broudy, and G. P. McCarthy, *Journal of Applied Physics*, **39**, 4673 (1968).]

because of its large depth of focus is the **scanning electron microscope**. An interesting scanning electron micrograph is shown in Figure 13.4.10. Finally, solid surfaces seldom represent the composition of the interior of the crystal. Like solid-solid interfaces, external surfaces often contain significant quantities of foreign atoms which are stabilized on or near the surface by the chemical and physical attractive forces which are unique to both interfaces and external surfaces.

EXAMPLE 13.4.1

Estimate the external surface energy of a (111) plane of a silver crystal.

Answer. We shall use the nearest-neighbor bond approximation. The coordination number of copper is 12; however, a surface atom has

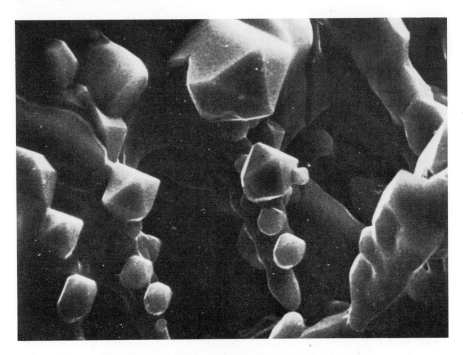

FIGURE 13.4.10. Melt interface of a silica brick from an open
hearth furnace. (*Courtesy of H. S. Sandhu
and R. Harmer.*)

only nine atoms around it. Hence it has three bonds missing. The binding
energy per atom is 3 eV. The bond energy is $\frac{1}{2}$ eV (since each neighbor
shares a bond). There are about 2×10^{15} atoms/cm² so the surface energy
is estimated to be

$$U_S = \left(3\,\frac{\text{bonds}}{\text{atom}} \times \frac{1}{2}\,\frac{\text{eV}}{\text{bond}}\right) \times 2 \times 10^{15} \text{ atoms/cm}^2.$$

Since $1 \text{ eV} = 1.6 \times 10^{-12}$ ergs, we have $U_S \sim 4800$ ergs/cm².

EXAMPLE 13.4.2

A powder of tiny particles of copper (10^{-4} cm in diameter) is pressed
together until the particles stick together. It is then held at a high temperature
below the melting point. After an appropriate time the material is removed
from the furnace. The porosity has disappeared and the density is that of solid
copper. Why did this occur?

Answer. Tiny particles have a large surface area to volume ratio.
Hence the system has a large surface energy. (Put another way, there are a
large number of surface atoms whose bonding could be better satisfied.)
During sintering almost all the surface atoms become interior atoms where

the bonding is more satisfactory. The material sinters to decrease the surface energy. The rate at which it sinters is determined by how fast the atoms move in the solid by interchanging with vacancies.

Figure 13.4.11 illustrates the sintering of copper wires. Note the formation of grain boundaries and, at longer times, grain coalescence and growth.

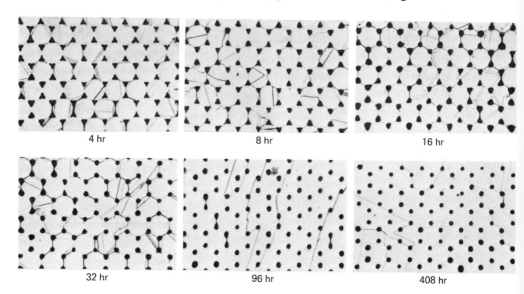

FIGURE 13.4.11. Sintering of copper wires of 0.0128 cm. dia. at 1075°C. (×50) [R. H. Alexander and R. W. Balluffi, *Acta Metallurgica* **5**, 666 (1957).]

The study of external surfaces is a large and important area of materials science. Surfaces of solids are important in catalysis, friction, growth of thin films on substrates, soldering, brazing, welding, adhesion, oxidation, etc. In the growth of thin films of one material on single crystal substrates of another material, it is often found that there is a definite crystallographic relationship between the substrate crystal and the film: this is **epitaxy**. Epitaxial growth usually, but not always, involves a close atomic match at the interface between the specific planes of the two crystals. In some cases epitaxial growth occurs on a certain face when impurities are adsorbed, but does not occur on a clean face. Sometimes, the epitaxial film is a polymorph which would not be stable in the bulk form. For a further discussion of epitaxy see D. W. Pashley, "The Growth and Structure of Thin Films" in *Thin Films*, American Society for Metals, Cleveland (1964).

Problems 13.24–13.30 represent a learning sequence on surface energy and surface tension.

13.5   SPATIAL IMPERFECTION

An alloy of Al with 4% Cu after age-hardening heat treatment consists of aluminum crystals (actually aluminum with a very small amount of copper dissolved) with tiny particles very rich in copper distributed in the crystals. Such macroscopic defects are very important in determining many properties and will be studied in more detail later. The copper-rich particles act as barriers for the motion of dislocations and hence increase the yield strength (see Table 2.1.2). (If the heat-treated aluminum alloy is used at a high temperature, the precipitate particles grow in size; their effectiveness as barriers to the motion of dislocations then decreases.)

It is also possible to have large voids in crystals.

REFERENCES

Newkirk, J. B., and Wernick, J. H., eds., *Direct Observations of Imperfections in Crystals*, Wiley-Interscience, New York (1962).

Amelinckx, S., *The Direct Observation of Dislocations*, Academic Press, New York (1964).

Cottrell, A. H., "The Nature of Metals," *Scientific American* (Sept. 1967) p. 90.

Tolansky, S., *Surface Microtopography*, Wiley-Interscience, New York (1960).

Kröger, F., *The Chemistry of Imperfect Crystals*, Wiley-Interscience, New York (1964).

Simmons, R. O., and Balluffi, R. W., "Measurements of Equilibrium Vacancy Concentrations in Aluminum," *Physical Review*, **117**, 52 (1960).

Germer, L. H., "The Structure of Crystal Surfaces," *Scientific American* (March 1965) p. 32.

Greenwood, N. N., *Ionic Crystals, Lattice Defects and Nonstoichiometry*, Butterworth, London (1968).

PROBLEMS

13.1.   a. What is the difference between a structure-sensitive and a structure-insensitive property?

b. In which class would elastic moduli and tensile yield strength be?

13.2.   a. About how many vacancies exist in a metal at thermal equilibrium just below the melting point?

b. At $T_M/4$?

13.3.   When a solid is bombarded with neutrons, large numbers of vacancies and self-interstitials are formed. Would you expect equal numbers of each?

13.4. The energy of formation of a vacancy in aluminum is 0.76 eV. Calculate the vacancy concentration at 900°K.

13.5. a. What is a Frenkel defect?
b. Give an example of one.
c. What is a Schottky pair?
d. Give an example of one.

13.6. What is the charge on a sodium ion vacancy in NaCl?

13.7. How does doping NaCl with Ca^{2+} ions affect the electrical conductivity of NaCl?

13.8. How large can the ratio of Fe^{3+} ions to Fe^{2+} ions be in wustite?

13.9. Give the dislocation density (cm/cm^3) in copper for
a. Perfect crystal.
b. Whiskers.
c. Very good high-purity crystal.
d. Ordinary crystal.
e. A cold-rolled sheet.

13.10. It is much easier to plastically deform a crystal by passing dislocations through it than it would be to slip the top half of the crystal (as a rigid block) over the bottom half. In an analogous situation, it is easier to move a rug by sending a wave down it than to attempt to drag the rug. Discuss this analogy further.

13.11. a. Define an edge dislocation and a screw dislocation.
b. What is the direction of motion of each during slip relative to the Burgers vector?

13.12. Suggest a simple general rule for changing the structure of a perfect crystal (with only one dislocation in it) so that a dislocation will be impeded in its motion through the crystal.

13.13. Give in electron volts for copper:
a. Cohesive energy (81 kcal/mole).
b. Energy of vacancy formation.
c. Energy of dislocation line per centimeter of length.

13.14. Why does a coherent twin boundary in copper have about half as much surface energy as a stacking fault? Copper has the fcc crystal structure.

13.15. Which plane in zinc (hcp) would you expect to be the twinning plane?

13.16. Derive Equation (13.4.2).

13.17. External surfaces seem to have the largest energy for surface formation of the various planar defects.
a. Why?
b. Why is the surface energy overestimated in Example 13.4.1?

13.18. In the absence of gravitational fields what will be the static equilibrium shape of a large volume (say a liter) of water resting in space in the presence of its equilibrium vapor pressure.

13.19. When uranium fissions, gases are often the products. Discuss the kind of spatial defects this might cause.

More Involved Problems

13.20. Suppose that 1 mole of water (18 cm³) is broken into droplets having a diameter 31.7 Å.
   a. What is the total surface energy if the surface energy per unit area is $\gamma_{LV} = 72.75$ ergs/cm² at 20°C?
   b. How does this compare with the heat of fusion of water (see Chapter 6)?
   c. The molecules in these droplets are clearly not bonded as well as the molecules in bulk water. Would you expect these droplets to have a higher, lower, or the same vapor pressure as bulk water?

13.21. An edge dislocation moves along on the (001) plane of a simple cubic crystal with $\mathbf{b} = [100]$ until it intersects a large-angle grain boundary. What happens then? (Suppose it is a tilt boundary.)

13.22. Assume that there are about 2×10^{15} copper atoms/cm² on the surface and that the volume per copper atom is 12×10^{-24} cm³.
   a. Calculate the fraction of atoms in the surface layer versus the radius r for particles having radii of 10^{-2}, 10^{-3}, 10^{-4}, 10^{-5}, 10^{-6}, and 10^{-7} cm.
   b. Plot the logarithm of the fraction versus the $\log_{10} r$.
   c. For what value of r does the surface energy become exceedingly important?
   d. A mole of copper is subdivided into spheres of radius 10^{-4} cm. Calculate the surface energy of the system and compare it to the cohesive energy of the solid, which is 3.5 eV/atom.

13.23. A copper crystal is deformed until its dislocation density is 10^{12} cm⁻².
   a. Calculate the energy owing to the presence of the dislocations in 1 cm³.
   b. How does this compare to the binding energy of 3.5 eV/atom? The volume per atom is 11.8 Å³.

Sophisticated Problems

13.24. Calculate the energy of formation of a vacancy in copper on the assumption (far from true in a metal) that a copper atom is bonded only to its nearest neighbor and does not interact with those farther away. The cohesive energy of copper is 3.5 eV.

13.25. A solid (or liquid) has assumed its equilibrium shape so that its total surface energy

$$\int \gamma \, dA$$

is a minimum. (Here γ, the surface energy per unit area, might vary with the surface area under consideration.) The condition that the integral be a minimum is equivalent to the statement that for a tiny variation in shape

$$\delta \int \gamma \, dA \longrightarrow 0.$$

This is the **Gibbs-Curie equation**.

a. Use this relation to find the equilibrium angle where a grain boundary intersects a surface:

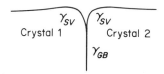

FIGURE P13.25

Hint: Consider an expanded view of the junction and imagine the junction to move downward by dx. (Consider the unit length of the boundary normal to the paper.) In the process, an area dx of the grain boundary is destroyed. This gives a contribution to the variation of the integral of $-\gamma_{GB} \, dx$. Proceed to calculate how much each of the surface-vapor interfaces increase in area. Since dx is very small, the angle between the dashed lines is very close to θ! Answer: $\gamma_{GB} = 2\gamma_{SV} \cos (\theta/2)$.

b. Suppose γ_{SV} is known. Outline an experiment which would enable us to obtain γ_{GB}.

13.26. Illustrate quantitatively how the creep experiment on very thin wires is used to measure γ_{SV}. If necessary, use a reference book on the theory of solid surfaces.

13.27. Write an article on whiskers and their growth.

13.28. Write a report on how vacancy concentrations in solids are measured.

13.29. Show that if a liquid completely wets a surface, i.e., a droplet placed on the surface spreads out uniformly, then

$$\gamma_{LV} + \gamma_{LS} < \gamma_{SV}.$$

Here γ_{SV} is the energy per unit area of the solid-vapor interface. Water on glass and Pb–Sn solder on clean copper completely wet the surface.

13.30. A liquid droplet placed on a surface retains its spherical shape (no wetting). Obtain an inequality such as in Problem 13.29 for this situation.

13.31. Suppose you are given a thin wire of gold and are told to measure its external surface energy. You are told that the wire actually shrinks in length when lying on an inert base at high temperatures, but that if placed in small tension the wire creeps at a constant strain rate. Devise a method to determine the surface energy.

13.32. In a cubic crystal with no defects present, measurement of fractional changes of lattice parameter $\Delta a/a$ and fractional changes of length $\Delta l/l$ due to temperature changes give the same result. This is not true, however, for a crystal containing an equilibrium concentration of vacancies. Assuming a and l are given at room temperature in copper (where there are virtually no vacancies present), describe how one could measure the equilibrium vacancy concentration at high temperatures.

Prologue

The most rapidly growing area of materials science is the study of polymers. Tiny molecules can be combined into giant molecules called *polymers*. Polymer molecules have many geometric shapes and forms and some of the possibilities, ranging from *linear chains* to *three-dimensional networks*, are described in this chapter. The so-called linear chain is not always a "straight" chain but often takes on a *random configuration*. The determination of the *molecular weight* of a polymer molecule is not as simple as for compounds such as CH_4 but it is important to know the molecular weight of the polymer. There are two polymerization processes called *addition polymerization* and *condensation polymerization* and examples of each of these are given. The conditions which determine whether a polymer will form a linear chain or a network are studied. It is interesting (and commercially important) to note that polymers are sometimes crystalline, sometimes amorphous, and sometimes part crystalline and part amorphous. One of the more interesting crystalline forms is the *folded-chain structure*. There is a brief introduction to the macromolecules in *living matter*, such as *cellulose* and the *proteins*.

14

POLYMERS

14.1 INTRODUCTION

Organic polymers are formed by the combination of many small mole-cules. One of the simplest of these molecules is formed from ethylene:

$$\begin{array}{c} H \\ \diagdown \\ H \diagup \end{array} C = C \begin{array}{c} H \\ \diagup \\ \diagdown H \end{array} \tag{14.1.1}$$

In ethylene the carbon-hydrogen bonds are single covalent bonds (one electron pair), while the carbon-carbon bond is a double covalent bond (two electron pairs). In the presence of heat, light, and an appropriate catalyst this molecule can be brought into an excited state:

$$\begin{array}{cc} H \bullet & \bullet H \\ \bullet C :: C \bullet \\ H \bullet & \bullet H \end{array} \longrightarrow \begin{array}{cc} H \bullet & \bullet H \\ \bullet C : C \bullet \\ H \bullet & \bullet H \end{array} \tag{14.1.2}$$

If such excited molecules come in contact it is possible to get addition of mole-cules (with single bonds between the carbons). When this process repeats itself, say 1000 times, we obtain a chain with a carbon backbone and a formula

$$H \left[\begin{array}{cc} H & H \\ | & | \\ -C-C- \\ | & | \\ H & H \end{array} \right]_n H \tag{14.1.3}$$

This is called **polyethylene**. The building unit or the **mer** is ethylene. We have assumed that after n ethylene units have added, the reaction is terminated by the addition of a hydrogen on each end. This process of polymerization is called **addition polymerization**.

The carbon-carbon bonds in polyethylene are single bonds at 109 deg to each other. Rotation can readily take place about single bonds *unless* the side groups are large and interfere with each other (in the present case the tiny hydrogens do not interfere). Figure 14.1.1 illustrates how side groups can prevent free rotation about the C—C bonds of the polymer chains backbone. This is called **hindered rotation**. Because polyethylene has relatively free rotation, the polyethylene chain takes on many configurations.

Large side groups

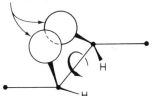

FIGURE 14.1.1. Illustration of hindered rotation.

Depending on the process used to make the polyethylene it may or may not have numerous **side chains**; these are short hydrocarbon chains attached to the longer chain. Polymers with side chains are called **branched polymers**. The possibility of packing nicely to form crystals is destroyed by the presence of

FIGURE 14.1.2. Polymer chain.

these side chains. When side chains prevent good packing the bonding is weak and the softening point is low. Polymers without side chains are called **linear polymers.** In this case the packing can be good so the bonding is stronger and the melting point is higher.

To develop a feeling for the nature of a polymer molecule a model of a polymer is shown in Figure 14.1.2. This shows how large a molecule would be if the mers were 1 cm in diameter, rather than several Ångstroms.

The industrial use of polymers has grown and is growing at a rapid pace, as shown in Figure 14.1.3. Polymers are used as plastics, adhesives, coatings, elastomers, fibers, etc. An understanding of them is important not only for many technological systems, but also in the life sciences.

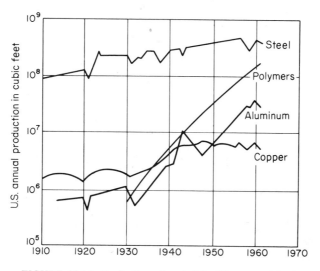

FIGURE 14.1.3. Production of materials. (*Courtesy of T. G. Fox, Mellon Institute.*)

The advances in polymer technology have led to such novel developments as the air-supported building shown in Figure 14.1.4 on page 302.

14.2   AN IDEALIZED RANDOM CHAIN

Let us consider a chain in which the bonds not only can rotate but can be bent to any angle (and not just restricted to 109 deg). A chain which has N such links, each of length l, would when completely stretched out have a length Nl. However, if each link were added in a random direction, then the more likely distance between the ends of a chain of N links is

$$L = \sqrt{N}\, l. \tag{14.2.1}$$

Equation (14.2.1) is derived from a random walk model in Section 17.3. Many

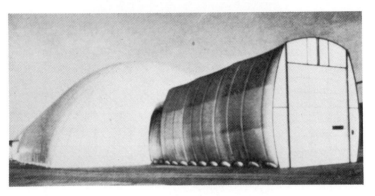

FIGURE 14.1.4. Air-supported warehouse. (*Courtesy of ILC Industries, Dover, Delaware.*)

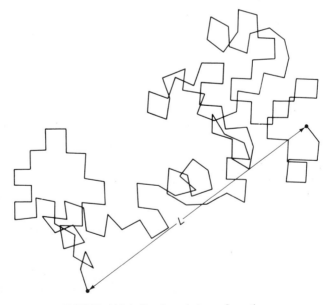

FIGURE 14.2.1. Random chain configuration.

302

polymer molecules and particularly rubberlike materials tend to take on such **random chain configurations** instead of packing nicely to form crystalline materials. See Figure 14.2.1.

14.3   DEGREE OF POLYMERIZATION

The **degree of polymerization** is the number of mers which have added together to form a polymer. It is directly related to the molecular weight. The molecular weight of polymers is very high. It is not possible to speak of the actual molecular weight (as for CH_4) but rather of the **average molecular weight** since a polymeric material contains polymer molecules of varying degree of polymerization (chains of varying length). The degree of polymerization can determine whether a polymer is a waxy substance or a plastic and it can strongly affect the softening and melting temperature and the degree of its solubility in specific solvents. For this reason we digress briefly to study one method of molecular weight determination.

The determination of the average molecular weight is not a simple matter, as it is with compounds such as CH_4. Average molecular weights are determined by physical measurements such as the osmotic pressure of the polymer in a liquid solution, the viscosity of the polymer in solution, or by light scattering of the polymers in solution.

One of the simplest techniques and the one first used to show that natural rubber is a polymer is osmotic pressure. Osmotic pressure was first observed by the Abbé Nollet in 1748; he placed "spirits of wine" in a cylinder which was then capped by an animal bladder and placed in water. He noted that the bladder swelled and sometimes burst. The bladder is a **semipermeable membrane**. Water passes through it but alcohol does not. The pressure increase in the cylinder was caused by the increased amount of water which passed through the membrane: It is called **osmotic pressure**. In 1885 J. H. van't Hoff showed from theoretical considerations that for dilute solutions the osmotic pressure is given by

$$\pi = CRT, \tag{14.3.1}$$

where C is the moles of solute per liter of solution. Hence if the osmotic pressure is measured for a solution containing a given weight of polymer, the molecular weight can be obtained.

EXAMPLE 14.3.1

Calculate the osmotic pressure of an 0.5-M solution of sucrose in water at 20°K.

Answer. We have

$$\pi = 0.5(0.082)293 \text{ atm}$$
$$= 12.0 \text{ atm.}$$

The measurements and theoretical basis of osmotic pressure are described in physical chemistry textbooks. See W. J. Moore, *Physical Chemistry*, Prentice-Hall, Englewood Cliffs, N. J. (1972).

EXAMPLE 14.3.2

If 10 g of a protein are dissolved in water to form a liter of solution at 7.0°C and the osmotic pressure is 8.7 mm Hg, what is the molecular weight?

Answer.

$$C = \frac{\pi}{RT} = \frac{8.7}{62.3 \times 280}.$$

Hence

$$MW = \frac{10 \text{ g/liter}}{C} \approx 20,000.$$

The process of reverse osmosis, where water is squeezed out of a salt solution by applying pressure of several hundred pounds per square inch, has great potential for desalting ocean water. See U. Merten, *Desalination by Reverse Osmosis*, M. I. T. Press, Cambridge, Mass. (1966).

14.4   VINYL POLYMERS

There are a number of polymers formed from compounds much like ethylene, e.g.,

$$\underset{H}{\overset{H}{\diagdown}} C = C \underset{R}{\overset{H}{\diagup}} \tag{14.4.1}$$

where R may be a halide, a benzene ring, etc. These are called **vinyl compounds**. If R is a chlorine the compound is called vinyl chloride and the resultant addition polymer is called poly(vinyl chloride) or (PVC). Table 14.4.1 illustrates some vinyl compounds. Polyethylene is a plastic widely used for packaging, soil mulching between plants, temporary walls exterior to buildings under construction, baby bottles, containers, etc. Poly(vinyl acetate) is used as the base for photographic materials, polyacrylonitrile when formed into fibers is called Orlon by du Pont, and polystyrene has many uses, one of which is for foamed plastics. We note that for vinyl compounds there are different topological arrangements possible depending on the manner of addition. Thus in vinyl chloride, chlorine may be found on alternate carbons only (head to tail addition or ... HTHT ...), chlorine may be found alternately on two adjacent carbons and be absent from the next pair of adjacent carbons (head to head, tail to tail addition or ... H HTT ...), or the chlorine may possibly be present in other ordered sequences or random sequences. This is a new topological feature which did not occur for straight chains of polyethylene. The possibility of packing closely would

Table 14.4.1. VINYL COMPOUNDS

Name	—R
Ethylene	—H
Vinyl chloride	—Cl
Propylene	—CH_3
Vinyl acetate	—O—C(=O)—CH_3
Acrylonitrile	—C≡N
Styrene (vinylbenzene)	(phenyl group)

be affected by such topology. The ... HTHT ... sequence is the most likely one to form, although vinyl polymers with other sequences have been made. The ... HTHT ... arrangement itself can take on different configura-

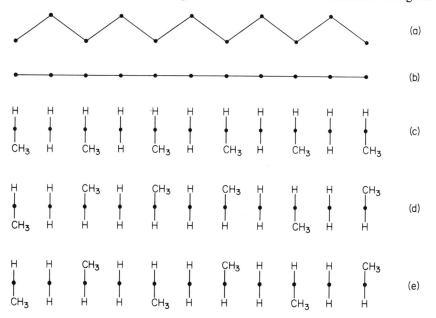

FIGURE 14.4.1. (a) Planar carbon backbone.
        (b) Top view of planar backbone.
        (c) Isotactic arrangement of CH_3.
        (d) Atactic (random) arrangement of CH_3.
        (e) Syndiotactic arrangement of CH_3.

tions. To illustrate these configurations for polypropylene consider the zigzag carbon backbone to lie in a plane as shown in Figure 14.4.1. A top view of this configuration would simply be a straight line as in (b). Here (c), (d), and (e) show the top view with the side atoms added. There is a methyl group on alternate carbons. Three possible spatial configurations are shown in Figure 14.4.1. These are called **stereoisomers**. In the **isotactic** isomer the CH_3 group is always on one side, in the **atactic** isomer the CH_3 group is randomly located, while on the **syndiotactic** isomer the CH_3 group alternates sides. [The reader should remember the actual zigzag nature of the chain; otherwise it appears that a simple rotation around the C–C bonds in (c) could give (d) or (e).] The atactic form is a waxy material at room temperature, while the isotactic form is a hard plastic because the chains can pack together more efficiently in the latter form and this leads to stronger binding.

The carbon backbone of the isotactic form does not have the planar zigzag structure of Figure 14.4.2. Rather, because of the mutual repulsion of CH_3 groups which are close together, the chain twists out of the planar zigzag by rotation about C–C bonds by 120°. This repeated twist in the same direction gives a **helical chain** with three propylene units per turn. These helical coils pack well and readily crystallize. The repulsion of large side groups, as in the case of the CH_3 groups here, is known as **steric hindrance**. Steric hindrance tends to prevent the free rotation of the chain. Such rotation is aided by thermal fluctuations and stops when the temperature gets too low. A change from rubbery behavior to rigid plastic behavior (glass transition temperature) is associated with this elimination of free rotation. The glass transition temperature will be higher in materials with large steric hindrance.

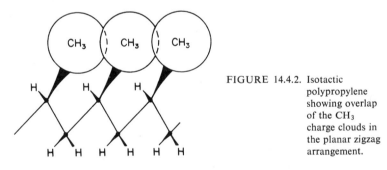

FIGURE 14.4.2. Isotactic polypropylene showing overlap of the CH_3 charge clouds in the planar zigzag arrangement.

14.5   OTHER ADDITION POLYMERS

POLYTETRAFLUOROETHYLENE. Another very important polymer is formed by addition from the compound

$$\begin{matrix} F \\ \diagdown \\ \diagup \\ F \end{matrix} C{=}C \begin{matrix} F \\ \diagup \\ \diagdown \\ F \end{matrix}$$

(14.5.1)

This molecule is called tetrafluoroethylene. The resultant polymer has a carbon backbone with fluorine atoms attached. It is called polytetrafluoroethylene (**PTFE**) or **Teflon**. It was discovered accidentally by R. J. Plunkett at du Pont while he was working with materials for refrigerants. Teflon is highly stable chemically. (Remember that fluorine itself is an extremely strong oxidizing agent.) Thus concentrated sulfuric acid can be boiled in Teflon. It has many other useful properties. Most materials do not wet it, i.e., do not stick to it. It therefore has a very low coefficient of friction and is often used as a sleeve bearing or in powder form as a lubricant.

1-4 ADDITION.   Consider a molecule such as isoprene:

$$
\begin{array}{c}
\quad\ \ \overset{CH_3}{\underset{|}{C}}\ \overset{H}{\underset{|}{C}} \\
\underset{H}{\overset{H}{>}}C{=}C{-}C{=}C\underset{H}{\overset{H}{<}} \\
\textcircled{1}\ \textcircled{2}\ \textcircled{3}\ \textcircled{4}
\end{array}
\qquad (14.5.2)
$$

This molecule will polymerize in the presence of heat and an appropriate catalyst by reaction at the $\textcircled{1}$ and $\textcircled{4}$ carbon positions to form **polyisoprene**,

$$
H\left[\ \overset{H}{\underset{H}{-C-}}\ \overset{CH_3}{\underset{}{C}}{=}\overset{H}{\underset{}{C}}{-}\overset{H}{\underset{H}{C-}}\ \right]_n H
\qquad (14.5.3)
$$

a long chain molecule in which unsaturated double bonds are still present. This is called **1-4 addition**. (Ordinary natural rubber is polyisoprene. It is interesting to note that Faraday, who is otherwise known for developments in electricity and magnetism, was the first to show that rubber is polyisoprene. Polyisoprene is a product of the rubber tree. However, the same material was produced synthetically by Giulio Natta in 1955.) The carbons associated with the double bonds ($\textcircled{2}$ and $\textcircled{3}$) are still reactive points and can be used for **crosslinking** from one polymer chain to another, e.g., by sulfur; this is **vulcanization**. All the polymer molecules therefore can be tied together at various points into a giant three-dimensional molecule. In soft rubber bands only a few weight percent of sulfur is used and the crosslinks may be a hundred units apart. Prior to and after the vulcanization the polyisoprene chains would be in a random chain configuration.

EXAMPLE 14.5.1

Calculate the extension ratio possible in rubber in which the number of chain links between crosslinks is 100.

Answer. Stretched out, the length is $100L$. In the random configuration the length is $\sqrt{100}L$. Hence the extension ratio $l/l_0 = 10$. This means that the rubber could be stretched 900 % before it would be necessary to stretch bonds.

Harder rubber materials such as tires and bowling balls contain more sulfur, and materials such as battery cases contain as much as 40 % sulfur. Such materials are much stiffer and have much smaller extension ratios than soft rubber.

The soft rubber materials such as in rubber bands are essentially liquids held together by the crosslinks. The chains between the crosslinks move quite readily over each other and are in constant thermal motion. These materials can be stretched several hundred percent. They are called **elastomers**. As the temperature is lowered the mobility of the chains decreases until a temperature, called the glass transition temperature, is reached (see Figure 12.11.2). At this temperature the mobility vanishes altogether and the configurations are frozen in place. The material is then rigid and brittle just as is ordinary window glass.

STEREOISOMERISM IN POLYISOPRENE.   In natural rubber the methyl group and hydrogen are on the same side of the C=C bond, as shown in Figure 14.5.1.

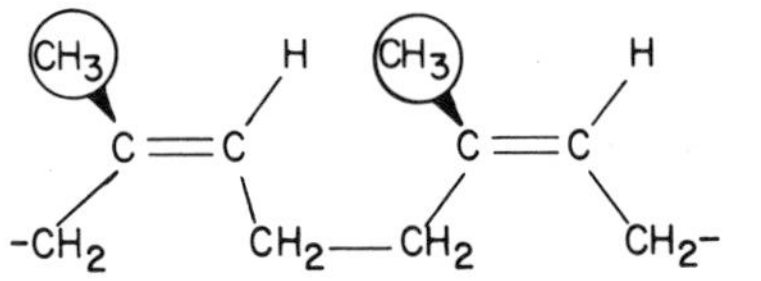

FIGURE 14.5.1. *Cis*-polyisoprene.

This is called the *cis*-**configuration** (same side). If the CH_3 and H are on opposite sides the configuration is called the ***trans*-configuration**. Molecules of this form pack well and readily crystallize; this is called **gutta percha**. The *cis* configuration of polyisoprene does not pack well and instead of crystallizing forms a rubber-like material with a (nearly) random chain configuration called **natural rubber**.

14.6   COPOLYMERS

It is often possible to combine two different mers into a polymer. Thus butadiene

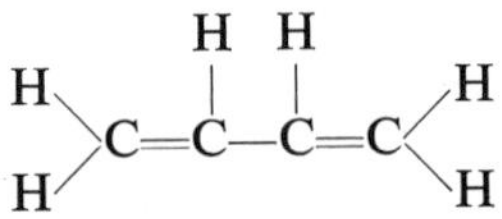

$$(14.6.1)$$

and styrene (see Table 14.4.1) can polymerize. The resultant polymer is known as a copolymer. Figure 14.6.1 shows schematic possible arrangements of such molecules. A **block polymer** consists of alternating chains of one polymer and a second polymer. A **graft polymer** consists of a backbone of one polymer with side chains of another.

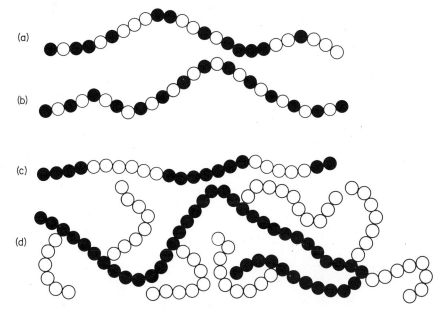

FIGURE 14.6.1. Copolymer arrangements. (a) A copolymer in which two different units are distributed randomly along the chain. (b) A copolymer in which the units alternate regularly. (c) A block copolymer. (d) A graft copolymer. [From J. Wulff et al., *The Structure and Properties of Materials*, Wiley, New York (1964).]

14.7 CONDENSATION POLYMERS

Organic acids have the structure

$$
\begin{array}{c}
\text{O} \\
\parallel \\
\text{R—C—OH}
\end{array}
\tag{14.7.1}
$$

where R is H, CH_3, C_2H_5, etc. Amines have the formula

$$
\text{R—N}\begin{array}{c}
\diagup\text{H} \\
\diagdown\text{H}
\end{array}
\tag{14.7.2}
$$

Adipic acid is dibasic; i.e., it has two of the –COOH groups, namely,

$$\text{HOOC—(CH}_2)_4\text{—COOH.}$$

Hexamethylene diamine has the formula

$$\text{H}_2\text{N—(CH}_2)_6\text{—NH}_2$$

These two compounds react under the appropriate conditions as

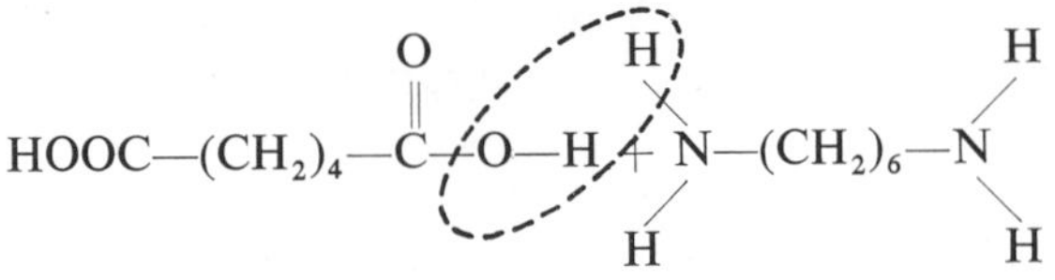

$$\text{HOOC—(CH}_2)_4\text{—}\overset{\overset{\text{O}}{\|}}{\text{C}}\text{—O—H   N—(CH}_2)_6\text{—N}$$

expelling H_2O and forming a C—N bond:

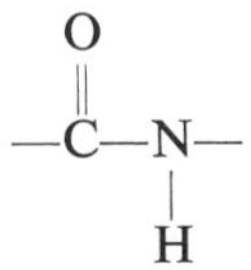

$$-\overset{\overset{\text{O}}{\|}}{\text{C}}\text{—}\underset{\underset{\text{H}}{|}}{\text{N}}-$$

(The remaining hydrogen on the nitrogen in this bond is not very reactive.) This reaction can carry on with alternate mers. It is called a **condensation polymerization** because of the by-product H_2O. The by-product of these reactions is usually a small molecule such as H_2O, NH_3, HCl, etc. The polymer, since it is formed of two mers, is called a **copolymer**. The specific molecule formed in the present case might be called polyhexamethylene adipamide but it is usually called by its generic name, **nylon 66** (the 66 refers to the six carbons in each mer). Although there were many earlier developments in the plastics industry, the synthesis of nylon by W. H. Carothers and its production by du Pont in 1939 signaled the beginning of the rapid growth in the plastics age.

Two of the important features of the nylon chain are the double-bonded oxygen and the hydrogen attached to the nitrogen. When two nylon chains are brought together a *hydrogen-bonded bridge* forms as shown in Figure 14.7.1. This is a particularly strong secondary bond.

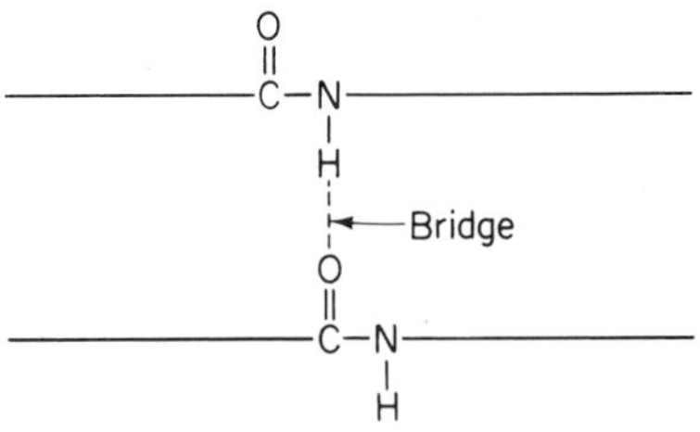

FIGURE 14.7.1. Hydrogen bridge in nylon.

When nylon fibers are made the molten nylon is forced out of pinholes and as it cools is stretched several hundred percent. This stretching tends to align the chains and to maximize the bonding between them. The resultant structure is highly crystalline and very strong (see Table 2.1.2). If the nylon molecules are not aligned in this fashion, the nylon has the more random structure of bulk nylon and is considerably weaker (see Table 2.1.2).

EXAMPLE 14.7.1

Explain why nylon fibers are so sensitive to water. For example, they are used in humidity measuring devices. Nylon drapes will alternately stretch and shrink several inches in climates in which the humidity varies greatly.

Answer. Water is also a strong hydrogen bond bridge former. Water breaks many of the bridges in nylon, and hydrogen bridges itself to a nylon chain. This causes swelling.

Nylon is only one example of a large class of linear condensation polymers.

SILICONES.   The compound

$$
\begin{array}{c}
R \\
| \\
HO\!-\!Si\!-\!OH \\
| \\
Z
\end{array}
\qquad (14.7.3)
$$

(where Z is either R′ or OH, and R and R′ are alkyl groups such as CH_3) can polymerize by a condensation process to yield

$$
\begin{array}{ccc}
R & R & R \\
| & | & | \\
-Si\!-\!O\!-\!Si\!-\!O\!-\!Si\!-\!O\!- \\
| & | & | \\
Z & Z & Z
\end{array}
\qquad (14.7.4)
$$

a linear chain silicone if Z = R′. If Z = OH, then primary Si–O bonds can form between the chains. **Silicone plastics** are characterized by linear chains, while crosslinked chains are found in **silicone rubber.** Note that the polymer backbone in this case is a silicate-type structure. The Si–O–Si type of bonds rotates more freely than the C–C–C bonds of natural rubber, so the glass transition temperature of silicone rubber can be much lower. Hence they have good low temperature properties. The backbone structure is quite resistant to heat (relative to the C–C backbone) and so the silicones are often used because of their heat resistance. They are also good electrical insulators which are much more resistant to dielectric breakdown by arcing than are the carbon backbone molecules (which form electrical conducting carbon when arcing occurs).

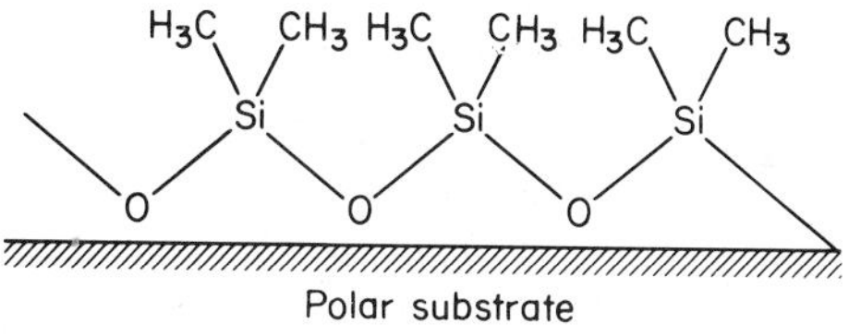

FIGURE 14.7.2. Silicone chain on a polar substrate.

Polar substrate

A silicone treatment for waterproofing is shown in Figure 14.7.2. Ordinarily the polar substrate such as glass, paper, or cloth would readily adsorb water. However, the treated surface is now hydrocarbon in character (and hence repels water). This will greatly increase the surface electrical resistivity of glass, which is usually low because of the adsorbed water.

14.8   NETWORK POLYMERS

Phenol has the structure

$$(14.8.1)$$

Asterisks are placed on hydrogens, which are reactive. Formaldehyde has the structure

$$(14.8.2)$$

Two asterisks are placed on the oxygen, which is reactive and has a double bond. Under appropriate conditions of heat and the like the molecules react as follows:

H_2O is a by-product.

This reaction can proceed to form a long chain, but because one of the mers has more than two reactive points, there is also the possibility of reactions at the third reactive point on the phenol. Hence chains can be linked to other chains by primary bonds. The resultant structure is a three-dimensional framework. It is a noncrystalline giant single molecule. It is called a **phenol-formal-**

dehyde plastic, a **phenolic**, or by its generic name **Bakelite**. (If the light switches in your room or lecture hall are brown, they are very likely to be Bakelite. Each switch is a molecule.)

Other important network polymers are the melamines and the epoxies. The epoxies are particularly useful as adhesives, which have high strengths at relatively high temperatures (for plastics).

14.9   THERMOPLASTICS AND THERMOSETTING RESINS

Some polymeric materials such as polyethylene can be reversibly softened by heating and conveniently formed by processes such as extrusion. These are called **thermoplastic** materials. The softening process here involves the breaking of *secondary* bonds.

Other polymeric materials, once formed, cannot be melted without actual decomposition. These are called **thermosetting** materials. An example is phenol-formaldehyde or Bakelite. Since this material is a three-dimensional network, *primary* covalent bonds would have to be broken. These softening characteristics are important properties both in fabricating the materials and in using them.

14.10   CRYSTALLINITY IN POLYMERS

Polymeric materials range from highly crystalline, ordered arrangements of chains to noncrystalline random chain configurations. Because of the distribution of chain lengths in a linear chain polymer, perfect crystals cannot be

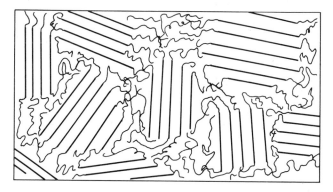

FIGURE 14.10.1. Diagram illustrating the proposed fringed micelle structure of semicrystalline polymers. [After P. J. Flory, *Principles of Polymer Chemistry*, Cornell University Press, Ithaca, N. Y. (1953) p. 49.]

found. One concept of a polymer showing crystallinity is the **fringed micelle structure** illustrated in Figure 14.10.1. In this case an individual polymer chain may extend through several crystallite regions (having a "diameter" of about 100 Å) and several amorphous regions. Single crystals of polyethylene can be grown from dilute solutions. Under these circumstances the individual chain is folded back and forth many times within the same crystals, as shown in Figure 14.10.2 (A.Keller, 1957). The fold length depends on the conditions present during crystallization. Numerous other polymers have since been grown from dilute solution in this **folded-chain structure**.

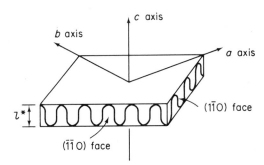

FIGURE 14.10.2. Folded-chain structure for polyethylene crystals.

At the present time it is thought that the structure of materials grown from the melt may be somewhere between the fringed micelle structure and the folded-chain structure. Figure 14.10.3 shows an example of a structure grown from the melt.

14.11   MACROMOLECULES IN LIVING MATTER

Cellulose in plants, and proteins in skin, hair, hoofs, toenails, muscle, and tendons in animals are all organic polymers. These naturally occurring polymers are important to us all.

CELLULOSE.   The building unit for cellulose can be considered to be glucose:

FIGURE 14.10.3. Spherulites in isotactic polystyrene (about halfway through the crystallization process) viewed under polarized light. ($\times$1500.) (*Courtesy of H. D. Keith, Bell Telephone Laboratories, Murray Hill, New Jersey.*)

The polymer **cellulose** is a linear chain

$$\text{(14.11.1)}$$

with a degree of polymerization of 1800–3500.

The linear polymer chains pack together surprisingly well, and there is strong secondary bonding between the chains owing to hydrogen bonding. Cellulose fibers are therefore very strong, having a tensile stress in excess of 100,000 psi.

EXAMPLE 14.11.1

Explain the process of ironing a cotton shirt.

Answer. The shirt is first moistened. The water tends to break the hydrogen bonds between the chains, forming its own hydrogen bonds to the chains. This weakens the fiber and makes it much more flexible and plastically deformable. The hot iron shapes the cloth and causes the water to evaporate so new hydrogen bonds are formed between the chains.

Starch is formed from a different isomer of glucose and in addition is highly branched. Hence the packing is poor so that starch swells readily in water.

We note that each glucose unit in the cellulose chain has three –OH groups. A number of cellulose derivatives can be made by reactions with one or more of these –OH groups. This includes cellulose acetate (rayon) and nitrocellulose (gun cotton). The degree of polymerization decreases when these compounds are formed. One of the first commercial plastics was celluloid, which is nitrocellulose plasticized with camphor; it was first produced and sold by John Wesley Hyatt in 1868.

PROTEINS. **α-amino acids** have the structure

$$\begin{array}{c} \text{H} \quad \text{O} \\ | \quad \parallel \\ \text{H}\!\!\diagdown \\ \diagup\!\!\text{N}\!-\!\text{C}\!-\!\text{C}\!-\!\text{OH} \\ \text{H} \quad\quad | \\ \text{R} \end{array} \qquad (14.11.2)$$

where R, called an alkyl group, may be H, CH_3, etc. Condensation reactions are possible resulting in a **peptide linkage**:

$$\begin{array}{c} \text{H} \quad \text{O} \quad\quad \text{H} \quad \text{O} \\ | \quad \parallel \quad\quad | \quad \parallel \\ \text{H}_2\text{N}\!-\!\text{C}\!-\!\text{C}\!-\!\text{N}\!-\!\text{C}\!-\!\text{C}\!-\!\text{OH} \\ | \quad\quad\quad | \quad | \\ \text{R} \quad\quad\quad \text{H} \quad \text{R} \end{array} \qquad (14.11.3)$$

Obviously the condensation process can continue with the same or other α-amino acids to yield a **polypeptide chain**. A **protein** is a polypeptide chain with a specific order of amino acids. If the alkyl groups are small, the protein chain will tend to form the extended zigzag configuration (silk takes on this configuration). There is then excellent hydrogen bonding between the chains (intermolecular), as illustrated in Figure 14.11.1. Bulky side groups cause steric hindrance and lessen the possibility of strong hydrogen bonds between chains in the extended zigzag configuration. Then polypeptide chains twist into helices,

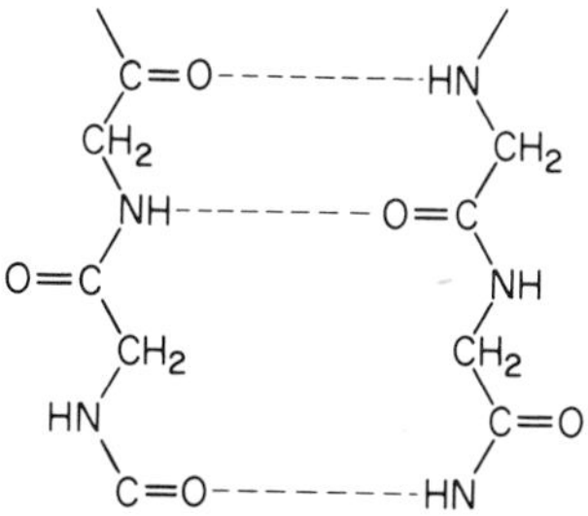

FIGURE 14.11.1. Hydrogen bonding between chains in polyglycine. About one half of the amino acids (on a mole basis) of silk is glycine.

$$C=O\text{-------}HN$$
$$CH_2 \qquad CH_2$$
$$NH\text{-------}O=C$$
$$O=C \qquad NH$$
$$CH_2 \qquad CH_2$$
$$HN \qquad C=O$$
$$C=O\text{-------}HN$$

in which case there is strong intramolecular hydrogen bonding within the chains with the bulky R groups extending outward. Wool (hair) has such a structure. Moreover, one of its important mers is the di-α-amino acid, cystine:

$$\text{HOOC}\underset{\underset{\text{H}}{|}}{\overset{\overset{\text{NH}_2}{|}}{\text{C}}}\text{—CH}_2\text{—S—S—CH}_2\text{—}\underset{\underset{\text{H}}{|}}{\overset{\overset{\text{NH}_2}{|}}{\text{C}}}\text{—COOH} \qquad (14.11.4)$$

This makes possible the formation of a primary bonded crosslink between chains (somewhat like the sulfur-sulfur crosslink in rubber). Dry wool is a hard plastic; wet wool, on the other hand, can be stretched elastically about 80%, and wool soaked in formic acid behaves in a rubberlike fashion.

EXAMPLE 14.11.2

Discuss the "home permanent."

Answer. If the S–S bonds are broken by a suitable chemical agent, the protein chains can readily slide over each other during the deformation process of curling (or straightening, as the case may be). The S–S bonds can then be reformed in new positions, which gives the hair fiber a new permanent configuration.

DNA. DNA, the heredity molecule, is a double helix of two sugar phosphate backbones twisted onto each other. The details of how this structure was obtained are given in a book by the Nobel Prize winner James D. Watson [*The Double Helix*, New American Library, New York (1969)], who was responsible in part for determining its structure.

REFERENCES

Mark, H. F., "The Nature of Polymeric Materials," *Scientific American* (Sept. 1967) p. 148. An hour of reading (plus perhaps more meditation) to establish contact with a lifetime of experience in the polymer area is suggested.

Alfrey, T., and Gurnee, E.F., *Organic Polymers*, Prentice-Hall, Englewood

Cliffs, N.J. (1967). This is a nice little monograph useful for structure studies now and for further study of viscoelasticity in Chapter 20.

Geil, P. H., *Polymer Single Crystals*, Wiley-Interscience, New York (1963).

Frazer, A. H., *High Temperature Resistant Polymers*, Wiley, New York (1968).

Wunderlich, B. "The Solid State of Polyethylene," *Scientific American* (Nov. 1964) p. 80.

Meyer, A. W., "Polymer Science Book List," *SPE Journal* (Jan. 1968). A compilation of books in the various areas of polymers.

Brydson, J. A., *Plastic Materials*, D. Van Nostrand, Reinhold, New York (1970). The structure of molecules is nicely related to chemical, mechanical, electrical, and optical behavior. There is also extensive discussion of specific polymers.

Winding, C. C., and Hiatt, G. D., *Polymeric Materials*, McGraw-Hill, New York (1961).

Rodriguez, R., *Principles of Polymer Systems*, McGraw-Hill, New York (1970).

PROBLEMS

14.1. Give an example of the formation of one-dimensional chains by
   a. Addition polymerization.
   b. Condensation polymerization.

14.2. a. Give three examples of vinyl polymers.
   b. Why are vinyl compounds with different side groups of interest?

14.3. Discuss the different kinds of topology along a chain that are possible with polypropylene.

14.4. a. What is meant by 1–4 addition?
   b. What feature of polyisoprene readily distinguishes it from poly-propylene?

14.5. A polymer chain with 10,000 links achieves a random configuration in solution. If you were betting on the distance between the ends, what would be your choice?

14.6. Is it possible, in a direct fashion, to measure the molecular weight of propane? How?

14.7. What methods are used to measure the average molecular weight of polymers?

14.8. List five different polymers from your everyday experience and suggest why each was used.

14.9. Ordinary vulcanized polyisoprene does not make a good gasoline hose. However, neoprene, a polymer whose mer is very much like isoprene except that the –CH$_3$ side group on the second carbon are replaced by –Cl, makes good gasoline hose. Explain why.

14.10. What must be true of one of the mers if a condensation polymer is to form a framework structure?

14.11. a. Give some reasons for the high tensile strength of nylon fibers.
b. Why is bulk nylon not so strong?

14.12. Discuss the use of silicone for increasing the surface resistivity of glass.

14.13. Give some reasons a thermoplastic might be preferable to a thermo-setting material.

14.14. a. Why is it topologically impossible (as a rule) to have perfectly crystalline polymers?
b. Discuss the two models of polymer crystallinity.

14.15. Give some examples of macromolecules in plants and animals.

14.16. Why is moist hair rubberlike?

More Involved Problems

14.17. Devise a method for measuring the osmotic pressure of a solution.

14.18. Devise a method for measuring the viscosity of a solution with a
a. Low viscosity.
b. High viscosity.

14.19. The root-mean-squared length of a random one-dimensional chain (the links either add or substract a length l along a straight line when added) is given by

$$\sqrt{x^2} = \sqrt{(l_1 + l_2 + l_3 + \cdots + l_N)^2}.$$

If each l_i is $\pm l$, show that

$$\sqrt{x^2} = \sqrt{N}\, l.$$

Sophisticated Problems

14.20. Write an essay on the measurement of viscosity, including the following types of instruments:
a. Ostwald.
b. Saybolt.
c. Couette.

14.21. Discuss quantitatively how the Newtonian viscosity coefficient (see Section 2.8) is obtained using the instruments described in Problem 14.20. See, e.g., G. W. Scott-Blair, *A Survey of General and Applied Rheology*, Pitman, New York (1944).

14.22. Discuss how measurements of viscosity coefficient can be used to obtain molecular weight. See, e.g., H. Tompa, *Polymer Solutions*, Academic Press, New York (1956).

Prologue

Real engineering materials are usually not single crystals. Rather they are *polycrystalline aggregates*. In the simpler cases only one kind of crystal is present. The size and shape of the crystals which make up the aggregate can be studied by *metallography*. More complex materials involve mixtures of crystals, such as crystals of Fe_3C mixed with crystals of bcc iron. These are known as *polyphase materials*. The size, shape, and orientation of these particles play a major role in determining the magnitude of many of the properties of a material. Such structures can be obtained by metallurgical processing and by mechanical means. Controlled structures obtained by mechanical means are called *composites*. The study of achieving controlled microstructures by either of these methods is an important part of materials science. From a practical viewpoint, this is where much of the payoff of materials research lies.

15

MICRO- AND MACROSTRUCTURE
OF MATERIALS

15.1 INTRODUCTION

It is convenient to classify materials according to states of aggregation: gases, liquids, and solids. We have already noted that solids can be separated into crystalline and amorphous solids. These solids are distinguished from each other by the degree of order of the atomic arrangements, i.e., on the **atomic scale**. There are numerous techniques for studying structure on this scale: X-ray, electron and neutron diffraction, field ion microscopy, etc. In this chapter, we are concerned more with arrangements on a somewhat larger scale, e.g., on a scale where the optical microscope (with a resolution of about 10^{-4} cm) and the electron microscope (with a resolution of 10^{-6}–10^{-7} cm) can be used for obtaining information. The structure of materials at this level is called **microstructure**. We shall also be concerned with different arrangements which are visible to the naked eye. This we call **macrostructure**.

15.2 THE REFLECTION MICROSCOPE

Figure 15.2.1 shows the polished and etched surface of aluminum as seen in the reflection microscope (also called the metallurgical microscope). The solid is a collection of single crystals of aluminum joined together at grain boundaries. This is called a **polycrystalline** material. Each crystal is essentially the same except for orientation, size, and shape. Because of their different orientations, different crystal planes coincide with the surface. [The surface is prepared by grinding, polishing with a wet slurry of very fine abrasive, and then etching. The different crystal planes react differently with chemicals which

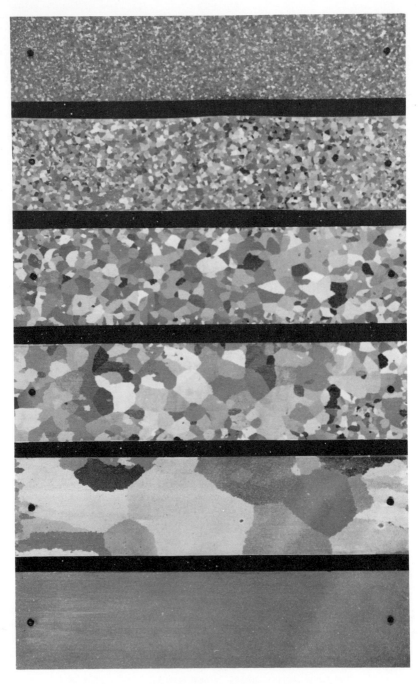

FIGURE 15.2.1. Micrograph of aluminum specimens. The crystals are much larger than would usually be found in commercial alloys. ($\times\frac{1}{2}$.) (*Courtesy of Stephen Sass, Cornell University.*)

attack the surface (**etchants**).] This results in each surface showing a different reflectivity. The resultant photograph of a surface is called a **micrograph**.

Figure 15.2.2 shows how a reflection microscope reveals the presence of a grain boundary after etching.

We have noted in Chapter 13 that the atoms in a grain boundary are not packed as perfectly as those within the crystal; hence their binding is not as good (this was the origin of the grain boundary energy). Grain boundaries tend to etch faster than the rest of the surface. Grain boundaries also can often be delineated by heating the prepared surface, in which case a groove forms at

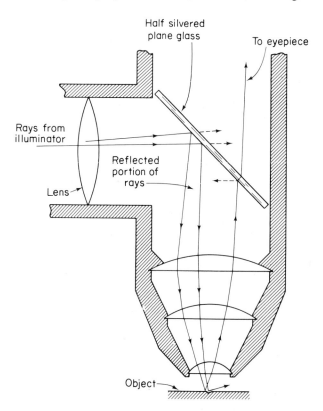

FIGURE 15.2.2. Light rays in a metallurgical microscope. The light from a bright point source enters through a lens at the left and is reflected downward onto the specimen. If the surface of the specimen is perpendicular to the axis of the microscope, it reflects the light to the eyepiece, with maximum intensity. If the surface is not perpendicular the rays are deflected and produce less effect at the eyepiece. [From B. A. Rogers, *The Nature of Metals*, M.I.T. Press, Cambridge, Mass. (1965).]

the grain boundaries **(thermal grooving)** in order to reduce the surface energy (see Problem 13.25). The atoms move by diffusion (see Chapter 18).

The study of the surface of materials by use of the reflection microscope is called **metallography**. [Metallography originally referred to the study of metals only, but now refers also to insulators (ceramics and plastics) and semiconductors—in other words, to all materials.] Metallography can be used for various aspects of structural characterization: grain size and shape, precipitate distribution, slip bands formed during plastic deformation, cracks formed during fatigue, voids formed during creep, voids or bubbles formed during radiation damage, ferroelectric domain size and shape, magnetic domain size and shape, etc.

Various micrographs will be shown in later chapters to illustrate the relation of microstructure to properties.

15.3   POLYCRYSTALLINE MATERIALS

While single crystals are often used in industry and research, the large bulk of commercial crystalline materials are polycrystalline. The reflection microscope can provide valuable data about such materials, e.g., the shapes of the **grains** (crystals in a polycrystalline aggregate) and their sizes. Grain boundaries are imperfections in crystalline materials which increase the energy of the system above that of the single crystal. Under true equilibrium conditions grain boundaries would not be present in a single-phase crystalline material. Grain boundaries are present usually as a result of the manner of solidification and other prior history; they often take up a **metastable** configuration in which it would be necessary for the material to pass through an intermediate state of higher energy before reaching true equilibrium (single crystal). An analog for a mechanical situation is shown in Figure 15.3.1. Such a metastable configuration for crystals in two dimensions, in which we have assumed that all the grain

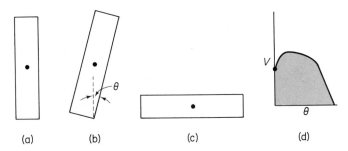

(a)   (b)   (c)   (d)

FIGURE 15.3.1. (a) Metastable equilibrium. (b) Intermediate state. (c) Stable equilibrium position. (d) Potential energy as a function of θ. Note the maximum.

FIGURE 15.3.2. Grain boundaries extending through a sheet. Special two-dimensional case.

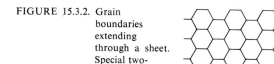

boundaries have equal surface energy per unit area, is shown in Figure 15.3.2; note that the angle between any two intersecting boundaries is 120 deg. Any small perturbation in the shape of some of these grains, as shown in Figure 15.3.3, would lead to an *increase* rather than a decrease in surface energy since it would lead to an increase in grain boundary area (prove this).

FIGURE 15.3.3. One grain attempts to grow at expense of neighboring grains. Dashed lines show the new grain boundaries.

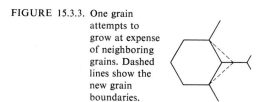

The corresponding situation is more involved in three dimensions because there is no regular polyhedron with plane faces which when packed together fills all space and which also meets the requirements of surface energy. The tetrakaidecahedrons (14 faces) of Figure 15.3.4 approximately satisfy these conditions (the sphere, of course, has minimum surface area for volume, but packed spheres do not fill space). The introduction of slight curvature in the faces allows the conditions for metastable equilibrium to be met.

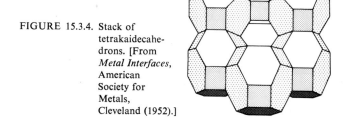

FIGURE 15.3.4. Stack of tetrakaidecahedrons. [From *Metal Interfaces*, American Society for Metals, Cleveland (1952).]

Figure 15.3.5 shows grains of a Ti alloy which have a striking similarity to the "grains" of Figure 15.3.4.

An important parameter in defining grains is the grain size. The **ASTM grain size** N is defined as follows: The number of grains in a square inch of micrograph whose magnification is 100 (linear) is 2^{N-1}. See Table 15.3.1. Grains which have nearly the same "diameter" D in all directions are called **equiaxed** grains.

FIGURE 15.3.5. These are individual grain groupings which parted from an arc-cast Ti Alloy billet under the blows of a hammer. These fragments have preserved the three bounding facets of the individual grains, which, belonging to a cast structure, are unusually large. Such perfect intercrystalline cleavage is rather rare and is usually associated with a thin intercrystalline film of a low-melting liquid phase. ($\times 1$.) [From W. Rostoker and J. R. Dvorak, *Interpretation of Metallographic Structure,* Academic Press, New York (1965).]

Table 15.3.1. ASTM GRAIN SIZE

ASTM Number	Grain/mm^2	Grains/mm^3
-3	1	0.7
-2	2	2
-1	4	5.6
0	8	16
1	16	45
2	32	128
3	64	360
4	128	1,020
5	256	2,900
6	512	8,200
7	1,024	23,000
8	2,048	65,000
9	4,096	185,000
10	8,200	520,000
11	16,400	1,500,000
12	32,800	4,200,000

Many of the properties of materials are closely related to the grain size, shape, and orientation. For example, the creep of equiaxed fine-grained MgO at high temperatures and low stresses is proportional to D^{-2}, where D is the grain diameter. Hence a large grain size would be desirable to reduce the creep rate. For other properties the opposite may be true. Thus the fracture stress of many polycrystalline materials which are brittle is proportional to $D^{-1/2}$. Orientation of the grains can also be very important since many properties of single crystals are highly anisotropic. It is often desirable to have **preferred orientation**, e.g., to have the $\langle 100 \rangle$ directions of crystals of cubic iron lie primarily along the magnetic transformer sheet and also lie normal to the sheet.

CASTINGS. We have already mentioned that polycrystallinity can be a result of solidification. Figure 15.3.6 shows a macroetched casting. Thermal effects are very important in solidification, because when a liquid freezes a large amount of heat (latent heat of solidification) is given off. This heat must

FIGURE 15.3.6. Typical ingot structure; a macroetched cross section through an iron-silicon alloy ingot. [From Cecil H. Desch, *Metallography*, McKay, New York (1937).]

be carried away either by conduction through the solid or by conduction and convection in the liquid. Many crystals begin to grow at the surface of the mold as shown in Figure 15.3.7, because the walls are initially cold and the liquid which comes in contact with these walls is rapidly supercooled. A large supercooling causes a rapid nucleation (initial formation) of crystals (we shall study later on why this is so). Their subsequent growth is controlled by two factors: motion of the atoms (or molecules) from the liquid to an appropriate site on the

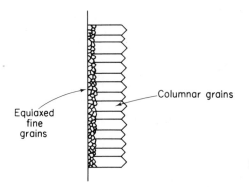

FIGURE 15.3.7. Crystals on surface of mold.

crystal surface (which may be on the revolving step of a screw dislocation) and the transfer of the heat released upon freezing. After the initial layers of equiaxed crystals are formed, the subsequent solidification is controlled by the transfer of heat through the equiaxed region to the walls. It is because of this heat transfer perpendicular to the walls that the growth of each grain is now inward and that **columnar grains** are formed. The description of the solidification process in a casting is very important industrially and is often studied in detail in advanced materials science and engineering courses. One of the important processes studied there is the formation of **dendrites**, skeleton-like or Christmas tree-like crystals whose formation depends upon the details of the rate of heat transfer from and material transfer to a growing branch.

EXAMPLE 15.3.1

On the basis of the above discussion, describe a technique to produce a single crystal by solidification.

Answer. As is clear from the preceding discussion only one tiny point in the mold should be lowered below the freezing point initially;

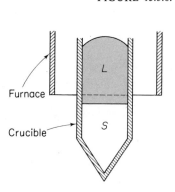

FIGURE 15.3.8. Bridgman method of growing single crystals. The crucible is lowered slowly from the furnace so that material at the tip solidifies first. The lowering continues slowly so that heat transfer takes place in a vertical direction.

moreover, the mold should be lowered slowly from the furnace so that the supercooling at the point at which freezing begins is small. This will enhance the probability that only one crystal forms there since the supercooling is very small. If the heat transfer is such that this crystal will grow essentially uniaxially (such as for the columnar grains of Figure 15.3.7), a single crystal will be obtained.

A method for growing crystals is shown in Figure 15.3.8. Often a single crystal is formed, although sometimes two or more columnar crystals will form. Single crystals are widely used in industry and in research. The **Bridgman technique** described here is only one of many techniques used to grow single crystals.

FIGURE 15.3.9. Crystal growth
apparatus.

In Figure 15.3.9 a student is shown with apparatus for growing a crystal from the melt. Here a seed crystal is carefully lowered partially into the melt and then slowly withdrawn. This is known as the **pulling technique** or the **Czochralski technique**. An example of a purposely grown bicrystal of nickel is shown in Figure 15.3.10. This crystal was grown to study the effect of grain boundaries on mechanical properties. Ionic crystals such as large NaCl crystals are often grown in this fashion. Crystals can also be readily grown from solution. The student is probably already familiar with growth of salt crystals (NaCl) from aqueous solution. Note that one obvious advantage is that such crystal growth is possible at much lower temperatures (at room temperature rather than at the melting point of NaCl, which is 801°C). The same technique is used when dia-

FIGURE 15.3.10. Nickel bicrystal grown by dipping two seed crystals into nickel melt and slowly withdrawing them. (*Courtesy of B. F. Addis, Cornell University*).

monds are *commercially* grown from carbon dissolved in liquid nickel at high pressures. Here conditions of about 1700°C and 70 kbars are needed; however, if attempts were made to grow directly from pure carbon melt, one would have to use temperatures in excess of 4000°C and pressures above 140 kbars. The solvent technique simplifies things considerably.

15.4 POLYPHASE MATERIALS

Polycrystalline materials such as those discussed in Section 15.3 may have been single-component materials such as pure aluminum or a multicomponent material such as 70–30 α-brass (70 wt. % Cu–30 wt. % Zn), which is a substitutional solution of Zn in fcc Cu. Such materials are single-phase materials. However, the bulk of engineering materials consists of more than one solid phase. A **phase** is the material in a region of space which in principle can be mechanically separated from other phases. Thus, if we have pure H_2O, we may under certain conditions have the three phases present: liquid, solid, and gas. (The student should not confuse phase with the state of aggregation!) Thus if a lead-tin alloy containing 40 wt. % Pb is studied at 400°C, it will consist of a single homogeneous liquid; this would be called the liquid phase. However, if this same alloy is cooled to room temperature, two solid phases exist. The

FIGURE 15.4.1. Pb–Sn eutectic showing layers of α phase and β phase. (×200.)

microstructure which results is shown in Figure 15.4.1. The α phase is a solid solution of Sn in fcc Pb; the β phase is a solid solution of Pb in tetragonal Sn. This alloy of Pb and Sn is common solder. We shall study the formation of this layer-like structure in Chapter 16.

A similar **lamellar structure** forms in eutectoid steel, iron containing approximately 0.8 wt. % C, upon slow cooling. This microstructure is shown in Figure 15.4.2 and its formation will be studied in Chapter 16. The yield strength of this steel increases as the spacing between the Fe_3C layers decreases, which is one reason for being concerned with the details of this microstructure. Methods to manipulate microstructure, in this case lamellae spacing, will be studied in later chapters; so will the reason that the yield strength is affected.

FIGURE 15.4.2. Lamellar structure in steel showing alternate layers of α-iron (nearly pure bcc iron with some carbon dissolved) and the compound Fe_3C. ($\times 1000$.)

Many other examples of micrographs of polyphase materials will be considered later. In the case just considered the second phase is in the form of planar sheets. In other examples it may be present as rods, cubes, or spheres or in much more complicated configurations. The size, shape, orientation, and distribution of these second-phase particles are often *very* important in determining the magnitudes of various properties such as yield strength, coercive force of hard magnetic materials, and the current carrying capacity of hard superconductors.

15.5   COMPOSITE MATERIALS

Composites are made from two or more different materials and the distribution of these materials is controlled by mechanical means. Thus the glass-reinforced plastic fishing rod is a composite. (The polyphase materials of Section 15.4 are essentially composites, but the material distribution is controlled by thermal and chemical means.) By combining materials in certain ways it is often possible to achieve a property which the individual materials did not possess (synergism). Thus fine individual glass fibers have a high tensile stiffness and a *very* high tensile strength. However, because of their small diam-

FIGURE 15.5.1. Cross section of composite.

eter, their bending stiffness is very small. Likewise if the rod was made of only epoxy plastic, it would have a low flexural stiffness. It could not be made of solid glass because bulk glass has a low tensile strength compared to fibers (see Table 2.1.2). However, when the fibers are placed in the epoxy plastic, as shown in Figure 15.5.1, the resultant structure has a high tensile stiffness, a high tensile strength, *and* a high bending stiffness. Such materials are called **glass-reinforced plastics (GRP)**.

EXAMPLE 15.5.1

Consider a bundle of glass filaments to be arranged in a composite to form a rod of circular cross section as shown in Figure 15.5.1. Let the filaments have radius r, the composite beam have radius R, and the volume fraction of filaments in the composite beam be $\frac{1}{2}$. Let $E_f = 10 \times 10^6$ psi for the filaments and $E_b = 5 \times 10^5$ psi for the plastic binder.

a. Calculate Young's modulus for the rod.

  Answer.

$$E_{\text{comp}} = \tfrac{1}{2}E_f + \tfrac{1}{2}E_b \doteq 5 \times 10^6 \text{ psi.}$$

The stiffness constant of a beam is proportional to its cross-sectional moment of inertia I times its longitudinal Young's modulus E. Suppose there are 500,000 continuous filaments in the beam. For a circular beam $I = \pi r^4/4$.

b. Calculate R.

  Answer.

$$\tfrac{1}{2}\pi R^2 = 500{,}000\,\pi r^2.$$

Hence $R = 1000r$.

c. What is the product EI for the bundle of fibers without binder?

  Answer.

$$500{,}000(10 \times 10^6) \times \frac{\pi r^4}{4}.$$

d. What is the product EI for the composite?

  Answer.

$$5 \times 10^6 \times \frac{\pi R^4}{4} = 5 \times 10^6 \times \frac{\pi r^4}{4} \times 10^{12}.$$

e. How much stiffer is the composite beam than the bundle?

  Answer. 10^6 times.

The assumption is made that the plastic carries the shear stresses which prevent slippage between the fibers.

f. If the tensile strength of the fibers is 500,000 psi, what is the tensile strength of the composite?

Answer. 250,000 psi.

The composite beam of Example 15.5.1 would deflect twice as much under a given load as a solid glass beam and it can carry a load 25 times as large. Hence the designer is faced with the problem of working with large deflections.

FIGURE 15.5.2. Boron filaments (diameter, 0.025 cm) in 6061-aluminum matrix. Matrix was applied by plasma spraying. (*Courtesy of Hamilton Standard Division of United Aircraft.*)

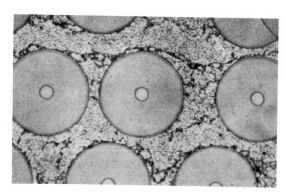

BORON FIBER COMPOSITES.   Figure 15.5.2 shows a cross section of boron filaments embedded in aluminum metal. Because of the high Young's modulus of the boron filaments, the resultant composite is extremely stiff. It also has a high

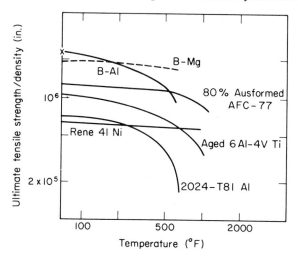

FIGURE 15.5.3. Comparison of strength to density ratio versus temperature for two composites and four alloys.

strength to density ratio. This is illustrated in Figure 15.5.3, where it is compared with alloys of aluminum, nickel, titanium, and steel. Such composite materials, because of their high stiffness to density and strength to density ratios, are being used at an increasing rate in aircraft (the economy of reducing structural weight has already been discussed in Chapter 2). The **boron filaments** are made by the decomposition of boron halides on a tiny hot tungsten wire. Under appropriate conditions this results in very strong and stiff boron filaments. Bulk boron is far inferior.

For a further discussion of these composites see *Metal Matrix Composites*, American Society for Testing and Materials, Philadelphia (1968). Composites can also be made of **whiskers** (short, fine, perfect, or nearly perfect single crystals) dispersed in an appropriate matrix.

EXAMPLE 15.5.2

Calculate the elastic strain energy density in a *future* filament composite if $E = 43.5 \times 10^6$ psi and $\sigma = 4{,}350{,}000$ psi.

Answer. The stored elastic energy per unit volume is, for a tensile loaded specimen,

$$w = \frac{\sigma^2}{2E}.$$

Since 14.5 psi $= 1$ bar $= 10^6$ dynes/cm^2, we have $E = 3 \times 10^{12}$ dynes/cm^2, $\sigma = 3 \times 10^{11}$ dynes/cm^2, and

$$w = 1.5 \times 10^{10} \text{ ergs/cm}^3 = 1500 \text{ J/cm}^3.$$

(The student may want to compare this with the energy of gunpowder.) In the future the designer must be concerned with such high stored energy situations.

CONCRETE. **Concrete** is a composite of rocks (coarse aggregate), sand (fine aggregate), hydrated Portland cement, and, in most cases, voids. Figure 15.5.4 is a photograph of a cross section of concrete. Portland cement is a mixture of calcium aluminum silicates such as

tricalcium silicate	$3CaO \cdot SiO_2$
dicalcium silicate	$2CaO \cdot SiO_2$
tricalcium aluminate	$3CaO \cdot Al_2O_3$.

When mixed with water the fine particles form a suspension; the water is absorbed into these particles, forming in time a rigid gel which is composed of a hydrate having approximately the formula $Ca_3Si_2O_7 \cdot 3H_2O$. A **gel** is a mixture of a solid and liquid (or gas) which behaves mechanically like a solid. One of the important properties of a wet concrete mix is the viscosity. Ordinarily much more water than is needed for the hydration is added to the concrete mix to

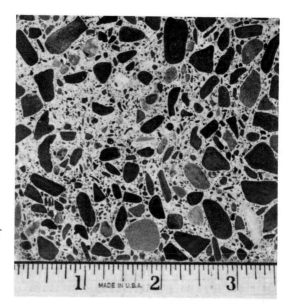

FIGURE 15.5.4. Cross section of
concrete.
(*Courtesy of
Floyd Slate,
Cornell
University.*)

decrease its viscosity so that the concrete will fill the mold. Hence when the
hydration is more or less complete there is excess water which either escapes,
leaving behind voids, or remains trapped in tiny capillaries.

It should be noted that the water in a hydrate is bonded by secondary
bonds only (as is water absorbed by hair).

The strength of concrete will be affected by the completeness of the
hydration, the aggregate size and distribution, the void volume, and the excess
water used. Concrete is usually weak in tension but fairly strong in compression
so that designers often consider the tensile strength to be zero.

OTHER COMPOSITES. Fine particles of tungsten carbide, which is extremely hard, are mixed with about 6% cobalt powder and sintered at high temperatures to obtain **sintered tungsten carbide** for use as cutters for machining,
as rollers, etc. Ordinary grinding wheels are composites of an abrasive with a
binder, possibly plastic or metallic. Walls used for portable housing are often
made of thin aluminum sheet epoxied on to polyurethane foam. The presence
of the latter provides excellent thermal insulation. The composite has high
structural rigidity which the thin aluminum sheets or the foam do not have
separately. The polyurethane foam itself is a composite of polyurethane and air.
Plywood is, of course, a composite also. Reinforced and prestressed concrete
is a composite of steel and concrete.

Composites are made for electrical as well as for mechanical applications.
Sodium metal, which is very reactive chemically, is enclosed in polyethylene
and used as an underground electrical cable. Nb_3Sn is deposited on copper
to make superconducting ribbon. Liquid lead is forced under pressure into

porous glass fibers to make synthetic hard superconductors. Microelectronic circuits are made from silicon, which can be oxidized to form a layer of an electrical insulator, SiO_2; this can be etched away in various places by hydrofluoric acid, and phosphorus can be diffused into the silicon (initially a *p*-type material, say) to make *p-n* or *p-n-p* junctions. Aluminum or another metal can be deposited at specified places to provide microconductors between points, etc. Hence a microelectronic circuit is a tailored composite.

15.6   QUANTITATIVE MICROSCOPY

The quantitative examination of microstructure is called **quantitative microscopy**. We have already noted in previous sections the need for quantitative data. The measurement of the grain size in a single phase material is an example of quantitative microscopy. The study of dislocation distributions in a pure copper foil is another example. The study of the size and shape of precipitate particles in an alloy is yet another example of quantitative microscopy. We shall give a brief illustration for the case in which spherical precipitates (called the α-phase here) are randomly distributed in a uniform matrix. The sample is sliced along an arbitrary plane, polished, etched, and observed in the microscope. For our purposes, we assume that a photograph of the observed image is made and that a square grid is placed upon this photograph as in Figure 15.6.1.

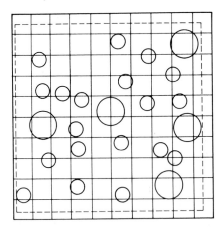

FIGURE 15.6.1. Square grid superimposed on a micrograph of a planar section of an alloy with spherical precipitate particles in a uniform matrix. Dashed line shows the edge of the photograph.

It can be shown [see R. T. Howard and M. Cohen, *Trans. AIME*, **172**, 413 (1947)] that

$$\frac{V_\alpha}{V} = \frac{A_\alpha}{A} = \frac{N_\alpha}{N} = \frac{L_\alpha}{L} \qquad (15.6.1)$$

if a sufficiently large sample is chosen. Here V_α/V is the volume fraction of the spheres; A_α/A is the area fraction of circles on the planar area; N_α/N is the fraction of grid intersections lying within circles; and L_α/L is the fraction of grid line length lying within circles. In the present case for that portion of the sample shown: $A_\alpha/A = 0.136$, $N_\alpha/N = 0.136$, and $L_\alpha/L = 0.130$. It is the purpose of quantitative microscopy to develop relations such as those of (15.6.1) which are applicable in specific cases and then to develop convenient experimental means of measuring the desired quantity. Thus, in the present case, counting of points (grid intersection points) is probably the most convenient.

For further discussions of quantitative microscopy the reader may study Rostoker and Dvorak, or Dehoff and Rhines (see References). For a discussion of the quantitative analysis of dislocation distributions see D. G. Brandon and Y. Komem, *Metallography*, **3**, 111 (1970).

REFERENCES

Kehl, G. L., *The Principles of Metallographic Laboratory Practice*, McGraw-Hill, New York (1949).

Rostoker, W., and Dvorak, J. R., *Interpretation of Metallographic Structures*, Academic Press, New York (1965).

Brandon, D. G., *Modern Techniques in Metallography*, Butterworth, London (1966).

Smallman, R. E., and Ashbee, K. H. G., *Modern Metallography*, Pergamon Press, New York (1966).

Cosslett, V. E., *Modern Microscopy*, Cornell University Press, Ithaca, N.Y. (1966). This is a somewhat more elementary book than the Smallman and Ashbee book above and can be easily read by students in this course.

Rabinowicz, E., "Polishing," *Scientific American* (June 1968) p. 91.

Smith, C. S., "Some Elementary Principles of Polycrystalline Microstructures," *Metallurgical Reviews*, **9**, 1 (1964).

Kelly, A., "The Nature of Composite Materials," *Scientific American* (Sept. 1967) p. 160.

Neville, A., *Properties of Concrete*, Wiley, New York (1967).

Brunauer, S., and Copeland, L. E., "The Chemistry of Concrete," *Scientific American* (April 1964) p. 80.

Dehoff, R. T. and Rhines, F. N., *Quantitative Microscopy*, McGraw-Hill, New York (1968).

PROBLEMS

15.1. Explain why etching takes place preferentially at grain boundaries.

15.2. Would grain boundaries be present under equilibrium conditions in a single-phase material?

15.3. Explain what is meant by metastable grain boundary configurations.

15.4. Describe briefly the various physical principles used in the design of the reflection microscope.

15.5. A metal has an ASTM grain size of 8. Show that there are 65,000 grains/mm^3.

15.6. What three features of grains in polycrystalline materials might determine properties?

15.7. a. Describe how grain size affects high-temperature creep of MgO.
b. Describe how grain size affects brittle fracture.

15.8. Describe two ways of growing NaCl crystals.

15.9. How is yield strength related to the lamellar spacing of α-Fe and Fe$_3$C in steel?

15.10. Define a composite and distinguish between the polyphase material of Figure 15.4.1 and the composite of Figure 15.5.2.

15.11. Give an example of synergism in composites.

15.12. Why are boron filament composites used in aircraft and space applications?

15.13. a. What is concrete?
b. Why does concrete ordinarily contain considerable nonhydrated water when "completely" cured?
c. What factors affect the strength of concrete?

15.14. List ten composites. Include at least two natural composites.

More Involved Problems

15.15. A second-phase precipitates as spherical particles all of the same size. The specimen is sliced, polished, and etched, and photomicrographs are made with a linear magnification of 500. Discuss quantitatively how you would measure the sphere diameter.

15.16. Consider Problem 15.15 for the case where there is a distribution of sphere diameter. Find an expression for the distribution function in terms of the measured diameters of the circles.

15.17. An important parameter in material is the grain size (because properties are often strongly affected by it). Suppose an industry produces metal with equiaxial grains. Design transparent overlays that could be used with an image projected from a microscope to estimate quickly the ASTM grain size number.

15.18. Three grain boundaries meet at a line as in Figure 15.3.2. Show that if their surface free energies are the same, then they must meet at 120 deg. (See Problem 13.25.)

Sophisticated Problems

15.19. An optical scanner is available for scanning reflected light intensity from a point. The scanner can be programmed (controlled by a digital computer) to scan continuously along a line $x =$ constant or $y =$ constant. Write a program which will solve Problem 15.15.

A book by R. E. Smallman and K. H. G. Ashbee, *Modern Metallography*, Pergamon Press, New York (1966) can be used as a starting reference for the following problems.

15.20. Describe the limit of resolution of the reflection microscope.

15.21. Describe the high-temperature microscope.

15.22. Describe the use of the interference microscope in studying surface topography.

15.23. Describe the electron microscope.

Prologue

Because the size, distribution, and composition of the different particles which make up an engineering material can have profound effects on the properties of materials, it is necessary to study *phase diagrams*. A two-dimensional map of pressure versus temperature can be used to show which phases are present under equilibrium conditions in a one-component system. A two-dimensional map of temperature versus composition can be used to show which phases are present under equilibrium conditions in a two-component system at constant pressure. The use of these phase diagrams and the application of the *lever rule* is described. Some of the more common types of phase diagrams involving *continuous solubility* and *limited solubility* are discussed. The meaning of *eutectic transformation* and *eutectoid transformation* is explained. Most transformations which occur in the real world take place under *nonequilibrium conditions*, and phase diagrams give us a hint of what will result. The *lamellar structure* which occurs during a nonequilibrium eutectic transformation is described; so is *coring*. A nonequilibrium transformation forms the basis of the *heat treatment* process known as *age precipitation hardening*.

We also describe some of the many microstructures possible in steel and cast iron because of the nonequilibrium processes which occur upon cooling in the Fe–C system. The eutectoid transformation and the *martensitic transformation* are introduced.

The process of *segregation*, which occurs during nonequilibrium freezing of a binary alloy, is described; this involves the use of the *distribution coefficient*. Similar concepts can be used to describe *zone refining*. How phase diagrams are determined is studied briefly. *Ternary diagrams* are introduced.

340

16

PHASE DIAGRAMS

16.1 THE PHASE RULE

In this chapter equilibrium in systems consisting of more than one phase is discussed. We begin with a discussion of the phase rule. This is followed by a study of equilibrium in a pure substance, first at constant pressure and then at variable pressure. Then equilibrium in systems in which composition may be a variable is described. We then proceed to discuss these systems when they are cooled under nonequilibrium conditions and we study the structures which are obtained. We are reminded that (1) structure determines properties and (2) the technological development of any age is limited by the properties of materials available.

We consider a number of components, c, distributed in a number of phases, p, in a closed system (whose composition is defined by the mole fractions x_i, where the subscript i refers to the ith component) in which the temperature everywhere is T and the pressure everywhere is P. (In the present discussion it is assumed that there are no fields other than the temperature and pressure fields which affect the materials.)

It is then possible to show from thermodynamics (which is itself based on phenomenology) or directly by experiment that

$$p + v = c + 2. \qquad (16.1.1)$$

This is the **Gibbs phase rule**; it forms the basis for our future discussions. Here v is the **variance** or the **degrees of freedom**, i.e., the number of variables (such as T, P, and x_i) which can be varied independently in the presence of p phases and c components (where p and c are fixed). Equation (16.1.1) applies to the situation in which temperature and pressure fields only are present and variable. The number of such external fields which are varied determines the constant in

Equation (16.1.1). (Other fields which are of importance in some cases are magnetic and electric fields.) If only temperature varies, i.e., pressure is kept constant, we have

$$p + v = c + 1. \tag{16.1.2}$$

The term *phase* was defined in Section 15.4. A component may be either an element or a compound. Thus if we are considering the system made of alcohol and water near room temperature, then alcohol is one component and water is the other. Earlier we discussed an interstitial solution of carbon in iron; there the components are the elements iron and carbon. The number of components is the number of distinguishable species of atoms or molecules that can move independently in the system under the experimental conditions.

16.2   PURE SUBSTANCES (ELEMENTS OR COMPOUNDS)

CONSTANT PRESSURE.   By the phase rule we now have $p + v = c + 1 = 2$ because pressure is fixed and $c = 1$. Thus only single-phase regions $(p = 1)$ are possible for which one variable can be varied independently $(v = 1$; this must be temperature). We can have two phases in equilibrium with each other $(p = 2)$ but then $v = 0$, which means that these two phases can coexist only at a *fixed* temperature. A typical example of this situation is the polymorphism of iron. Here the bcc structure exists alone, i.e., as a single phase on the range $0 - (1183 - \epsilon)°K$, where ϵ is a small number; a phase consisting of the bcc structure and a phase of the fcc structure coexist at $1183°K$; the fcc structure exists alone on the interval $(1183 + \epsilon) - (1672 - \epsilon)°K$; etc. The important point to note is that here two phases cannot exist in equilibrium over a range of temperature in a system of one component at constant pressure.

We might also have chosen the compound SiO_2 to illustrate this case. Then under equilibrium conditions we would have the *quartz* crystal structure at temperatures below $1148°K$; at $1148°K$, two phases, one of quartz, the other of the tridymite crystal structure; above $1148°K$ and below $1743°K$, the single phase having the tridymite structure; at $1743°K$, two phases, one of tridymite and one of the cristobalite crystal structure; above $1743°K$ and below the melting point of $1983°K$, only cristobalite; at $1983°K$, cristobalite and liquid SiO_2; and above $1983°K$, liquid SiO_2.

PRESSURE VARIABLE.   We previously noted that one of our principal goals in this chapter was to be able to describe the limits of stability of the various phases which a system can assume. These stability limits are conveniently recorded as curves on a set of Cartesian coordinates on which temperature and pressure are the variables. These curves are the loci of the points that represent those combinations of T and P at which two phases of a substance are in equilibrium. The *P-T* diagram for H_2O is shown in Figure 16.2.1. The

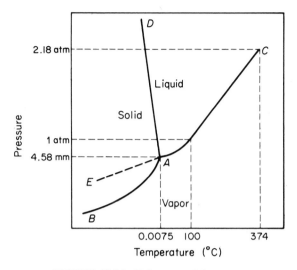

FIGURE 16.2.1. H_2O system (Schematic).

line AB separates the solid and vapor regions, the line AD separates the solid and liquid regions, and the line AC separates the liquid and vapor regions. It can be shown from thermodynamics that the slope dP/dT of the dividing line AC is given by

$$\frac{dP}{dT} = \frac{\Delta H_v}{T\,\Delta V_v},$$ (16.2.1)

where ΔH_v is the heat absorbed at constant pressure (enthalpy) and ΔV_v is the volume change when a mole of liquid is transformed to vapor at temperature T.

The quantities ΔH_v and $\Delta V_v = V_g - V_l$ can be measured independently of the determination of the P-T curve of Figure 16.2.1. The **heat of vaporization** ΔH_v can be measured calorimetrically. Thus two independent experiments can be used to obtain dP/dT. At sufficiently low pressures the gas obeys the ideal gas law and $V_l \ll V_g$, so we can write

$$\Delta V_v = \frac{RT}{P}.$$ (16.2.2)

Hence (16.2.1) can be written

$$\frac{dP}{P} \doteq \frac{\Delta H_v\,dT}{RT^2}.$$ (16.2.3)

Moreover, ΔH_v is nearly independent of P and T, and if we ignore any slight dependence, we can integrate (16.2.3) to obtain

$$P = A_0 e^{-\Delta H_v/RT}.$$ (16.2.4)

It can also be shown from thermodynamics that along line BA,

$$\frac{dP}{dT} = \frac{\Delta H_s}{T\,\Delta V_s},$$ (16.2.5)

where ΔH_s is the heat absorbed at constant pressure and temperature when a mole of solid is transformed to vapor (the heat of sublimation) and ΔV_s is the volume change. Similarly along line AD,

$$\frac{dP}{dT} = \frac{\Delta H_f}{T\,\Delta V_f}, \qquad (16.2.6)$$

where ΔH_f is the heat of fusion and ΔV_f is the volume change. This equation cannot be integrated as readily as was (16.2.1). Why? Thus we see that phase equilibrium curves provide important thermodynamic quantities such as the ratio of $\Delta H_f/\Delta V_f$ in addition to dividing *P-T* spaces into phase regions. An interesting and important example of the *P-T* diagram is provided by the one shown in Figure 16.2.2, which is for carbon. The equilibrium phase boundary between graphite and diamond has been the subject of search for several hundred years, but its determination had to await the development of modern high-pressure technology. The ability to make diamonds, the hardest known material, of industrial quality, has obvious commercial implications in this age when many very hard materials of great potential use can be machined only by use of a harder substance to wear away unwanted material.

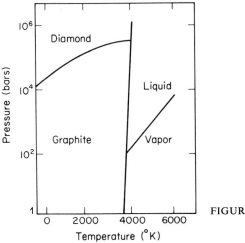

FIGURE 16.2.2. Carbon *P-T* equilibrium diagram.

Note that *within* the region labeled Graphite (and hence a single-phase region), we have by the phase rule a variance of 2; i.e., we can vary T and P independently and not change the number of phases. However, on the line separating the regions labeled Diamond and Graphite (along which we have two phases coexisting), we have a variance of 1; i.e., we can vary T (but variations of P cannot be independently made; rather we must vary P in a specified way; otherwise the number of phases will change from two to one).

The growth of diamonds from nickel solution depends on the presence of the narrow liquid plus diamond region from 1667-1728°K shown in Figure

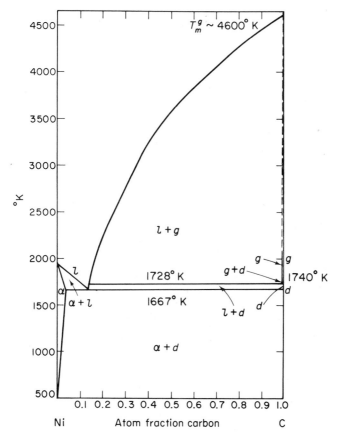

FIGURE 16.2.3. Nickel-carbon phase diagram at 54 kbars.
[From H. M. Strong and R. E. Hanneman,
Journal of Chemical Physics, **46**, 3668 (1967).]

16.2.3. Thus diamonds can form there in the liquid phase at modest temperatures just as salt crystals grow in water at room temperature. Figure 16.2.4 shows a diamond grown in this fashion in the author's laboratory. Note the well-developed {111} planes.

16.3   VARIABLE-COMPOSITION SYSTEMS

The addition of a second component to any thermodynamic system causes little difficulty, in a formal sense, in describing equilibrium. Our concern in such binary systems is primarily with the solid-liquid and solid-solid part of the system and not with the gaseous state. The number of phases and their volume proportions are usually observed as a function of temperature at *fixed*

FIGURE 16.2.4. Diamond grown from nickel solvent. (Photographed by C. C. Chao in author's laboratory.)

pressure for a number of fixed compositions in the system of interest and the results displayed on a temperature-composition (*T-x*) plot, commonly known as the equilibrium diagram or phase diagram. The systems which are discussed here will be at a pressure of 1 atm.

COMPLETE SOLUBILITY IN BOTH SOLID AND LIQUID. A typical system is illustrated in Figure 16.3.1. Here the components are the compounds NiO and MgO. Examples of other systems showing the general form of Figure 16.3.1

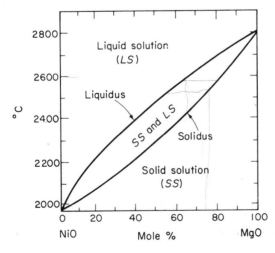

FIGURE 16.3.1. MgO–NiO system.

are given in Table 16.3.1. The major important difference between the metallic systems (under Elements in Table 16.3.1) and the ceramic systems (under Compounds) is the *rate* at which equilibrium is achieved. In the metals this is fairly rapid, while in the ceramics it is *very* slow. The fluidity coefficient of liquid metals at the melting point is about 50 P^{-1}, while for liquid silica at the melting point of cristobalite this is 10^{-7} P^{-1}. [Fluidity, defined in Equation (2.8.4), is a measure of the ability of atoms, molecules, or complexes (such as SiO_4^{4-}) to move.] The ratio of nearly 10^9 is an extreme case so that in this case silica tends to form glass when solidifying from the melt rather than forming an equilibrium crystal structure.

Table 16.3.1. SYSTEMS SHOWING
COMPLETE SOLUBILITY

Elements	Compounds
Au–Pt	CoO–CoS
Cu–Ni	CoO–MgO
Mo–W	FeO–MgO
Ag–Au	CoS–FeS
Ag–Pd	ThO_2–UO_2
Au–Pd	$PbCl_2$–$PbBr_2$
Pt–Rh	α-Al_2O_3–Cr_2O_3
Pt–Ir	$2MgO \cdot SiO_2$–$2FeO \cdot SiO_2$

There are three regions of interest in Figure 16.3.1. Above the **liquidus** line a single liquid solution is found (single phase). Below the **solidus** line a single solid solution is found (one phase). This solid phase has the NaCl structure. As MgO is added to NiO, the Mg^{2+} ions substitute for some of the Ni^{2+} ions to form a substitutional solid solution. Studies by X-ray diffraction yield the diffraction lines for one crystal structure only, an NaCl type. Continuous solid solutions (from pure NiO to pure MgO) are possible because both NiO and MgO have the NaCl crystal structure. Moreover, the ions have radii which are nearly the same. The third region is the two-phase region between the liquidus and the solidus. Here particles of solid solution will be suspended (assuming the gravitational field is zero) in liquid.

If an alloy of 80 mole % MgO is heated, it will exist as a single phase (crystalline with NaCl-type structure) up to 2600°C. Between 2600 and 2700°C it will exist as two phases, and above 2700°C there will only be liquid present.

Let us now examine the behavior at 2600°C as the composition varies from pure NiO to pure MgO. **Composition** means the mole fraction or percentage (or the weight fraction) of the component. The components in this case are NiO and MgO. With pure NiO there is only a single liquid phase; as MgO is added there is still only a single phase until a composition of 64 mole % MgO has been achieved. At this point the liquid (at 2600°C) has dissolved the maximum amount of MgO; it has reached its solubility limit. It cannot dissolve any more MgO. If more MgO is added to the system to bring its overall composition to 75% MgO, two phases will be present: liquid solution and solid solution.

EXAMPLE 16.3.1

What will be the composition of the liquid solution at 2600°C for an alloy of 25 mole % NiO–75 mole % MgO?

Answer. The liquid will be 64 mole % MgO. It cannot be more than this because that is the maximum solubility. It cannot be less because then only a single phase would be present and the overall composition could not be 75% MgO.

As additional MgO is added at 2600°C and the composition reaches 80% MgO the solidus will be intersected. Beyond 80% there is only one phase present, namely, the solid solution. The two-phase region extends from 64% MgO to 80% MgO at 2600°C. The liquid phase in this two-phase region has a composition of 64% MgO; the solid phase in this two-phase region has a composition of 80% MgO.

EXAMPLE 16.3.2

An alloy of 24% NiO–76% MgO is heated to 2600°C. What is the mole fraction of liquid present?

Answer. Let f_l be this unknown fraction. Then $1 - f_l$ is the mole fraction of solid solution. The MgO in the liquid plus the MgO in the solid solution must equal the total MgO. Let us suppose we start with 1 mole of alloy. Then we have 0.76 moles of MgO. Hence

$$(x_{\text{MgO in liquid}})f_l + (x_{\text{MgO in solid}})(1 - f_l) = \text{Overall mole fraction of MgO}$$

or

$$0.64f_l + 0.80(1 - f_l) = 0.76, \qquad (16.3.1)$$

which can be rearranged to give

$$f_l = \frac{0.80 - 0.76}{0.80 - 0.64} = 0.25. \qquad (16.3.2)$$

Hence there are 0.25 moles of liquid whose composition is 64% MgO and 0.75 moles of solid solution whose composition is 80% MgO.

Figure 16.3.2 is used to illustrate the calculation of Example 16.3.2 in another way. A line at constant temperature is drawn across the two-phase region. This is called the **tie line**. The composition at the liquidus side of the tie line gives the composition of the liquid in the *entire* two-phase region at 2600°C. Likewise the composition at the solidus side of the tie line gives the composi-

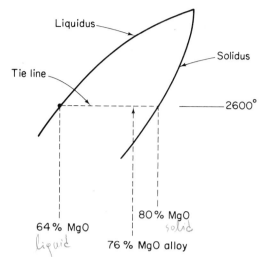

FIGURE 16.3.2. Expanded view
    of the two-phase
    region of Figure
    16.3.1.

tion of the solid solution in the *entire* two-phase region at 2600°C. The conservation of material described by Equation (16.3.2) leads to a simple rule called the lever rule. Since

$$f_l + f_s = 1,$$

Equation (16.3.1) can be rewritten as

$$0.64f_l + 0.80f_s = 0.76(f_l + f_s),$$

or as

$$(0.76 - 0.64)f_l = (0.80 - 0.76)f_s.$$

    This suggests the following mechanical balance which is an application of the principle of the lever: Consider a fulcrum placed at the given alloy composition and the tie line to be a lever, as in Figure 16.3.3, and the fraction of each

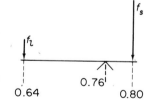

FIGURE 16.3.3. Illustration of
    the lever rule.

phase to be a force. Thus we have by the lever rule

$$f_l(0.76 - 0.64) = f_s(0.80 - 0.76)$$
$$= (1 - f_l)(0.80 - 0.76)$$

and hence

$$f_l = \frac{0.80 - 0.76}{0.80 - 0.64} = 0.25.$$

These results are the same as obtained in (16.3.2), which proves the **lever rule**, which states, "In a two-phase region the fraction of one phase equals the length of the lever opposite this phase divided by the length of the tie line."

Figure 16.3.4 shows a system with continuous solubility in both the liquid and solid phases but with a coincident minimum in the liquidus and solidus, called an **indifferent point**.

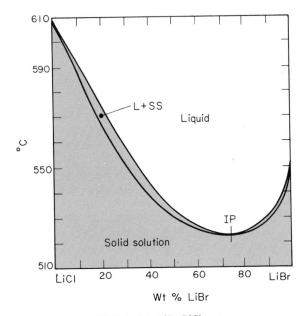

FIGURE 16.3.4. LiBr–LiCl system.

Other systems showing an indifferent point (labeled IP in Figure 16.3.4) are shown in Table 16.3.2. The conditions for forming a complete solid solution have been discussed in part in Section 12.6. Remember that it is absolutely necessary that the two components have the same crystal structure.

Table 16.3.2. SYSTEMS SHOWING COMPLETE SOLUBILITY WITH A COINCIDENT MINIMUM IN THE SOLIDUS AND LIQUIDUS CURVES

Elements	Compounds
Ni–Pd	KNO_2–$NaNO_2$
Cr–Mo	LiF–MgF_2
	UF_4–ZrF_4
	$CaO \cdot SiO_2$–$SrO \cdot SiO_2$

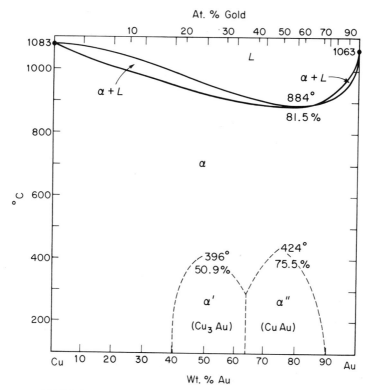

FIGURE 16.3.5. Cu–Au system. See Figure 12.6.3 for the crystal structure of ordered CuAu and Cu₃Au.

It is often a characteristic of binary systems with an indifferent point that they tend to form stable ordered structures at low temperatures even though their behavior at higher temperatures is characterized by the curve in Figure 16.3.4. This is illustrated in Figure 16.3.5.

Atomic percentage is related to weight percentage according to

$$\text{At. } \% \text{ Au} = \frac{(\text{Wt. } \% \text{ Au}/M_{\text{Au}})}{(\text{Wt. } \% \text{ Au}/M_{\text{Au}}) + (\text{Wt. } \% \text{ Cu}/M_{\text{Cu}})} \times 100 \%. \quad (16.3.3)$$

Here M_{Au} is the molecular weight of gold. When dealing with compounds, we relate the mole percent to weight percent in the same way. Other systems showing complete solubility and ordering phenomena are LiCl–NaCl, Cd–Mg, Cu–Pd, Cu–Pt, and Ni–Pt. Moreover, the Cr–Fe and V–Fe systems are similar to the above, although they show added effects as well.

COMPLETE SOLUBILITY IN THE LIQUID PHASE: SLIGHT SOLUBILITY IN THE SOLID PHASE. Because long-range disorder is a characteristic of liquids, it is often true that complete solubility is possible in the liquid phase even under conditions in which there is little solubility in the solid phase. This is illustrated in Figure 16.3.6. Solid Pb has a fcc crystal structure, while tin has a tetragonal crystal structure (although at lower temperatures Sn has a diamond cubic structure). Hence continuous solid solubility is not possible. (Figure 16.3.6 should be considered to be a modification of Figure 16.3.4.) The phase labeled α is a substitutional solid solution of Sn in Pb and it has a fcc crystal structure. The phase labeled β is a substitutional solid solution of Pb in Sn and it has a tetragonal crystal structure.

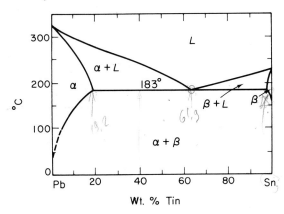

FIGURE 16.3.6. Pb–Sn diagram.

Note that at any constant temperature the number of phases present in a given region alternates from one to two to one, etc.

A point worthy of particular notice is the point at 61.9 wt. % Sn and 183°C. Here three phases coexist: solid α of 19.2% Sn in Pb, solid β of 2.5% Pb in Sn, and liquid with 61.9% Sn. Since the phase diagram under consideration is for constant pressure, we have $p + v = c + 1$, and with $c = 2$ and $p = 3$, we have $v = 0$. This point is hence an invariant, which in the present case is called a eutectic point. By definition a **eutectic point** is an invariant point involving a dynamic equilibrium:

$$\text{liquid} \underset{\text{heating}}{\overset{\text{cooling}}{\rightleftharpoons}} \text{solid } \alpha + \text{solid } \beta. \qquad (16.3.4)$$

The transformation from a liquid to two solids is called a **eutectic transformation,** and the temperature at which this occurs is called the **eutectic temperature** (183°C in the present case). The alloy of this composition is called a **eutectic alloy.**

EXAMPLE 16.3.3

Suppose you have an alloy with an overall composition of 40 wt. % Sn. Calculate the weight percent α and the weight percent liquid at $183 + \epsilon$°C, where ϵ is small.

Answer. Draw the tie line in the two-phase region $\alpha + L$ at $183 + \epsilon°C$. Then by the lever rule,

$$\text{wt. } \% \; \alpha = \frac{61.9 - 40}{61.9 - 19.2} \times 100\% = 51.3\%$$

$$\text{wt. } \% \; L = 100\% - 51.3\% = 48.7\%.$$

Note that these calculations are for *equilibrium* conditions; under actual continuous cooling conditions, equilibrium is not attained owing to the slowness of solid state diffusion. This will be discussed quantitatively in Chapter 18.

EXAMPLE 16.3.4

For the 40 wt. % Sn alloy, calculate the weight percent α at $183 - \epsilon°C$.

Answer. Now the tie line is in the two-phase region $\alpha + \beta$. Hence

$$\text{wt. } \% \; \alpha = \frac{97.5 - 40}{97.5 - 19.2} \times 100\% = 73\%.$$

EXAMPLE 16.3.5

Aside from such considerations as wetting, why would one use a eutectic Pb–Sn solder?

Answer. Because it melts at a lower temperature and is hence more convenient to use.

Table 16.3.3 gives some further examples of eutectic systems. We note that although LiF and LiBr have the same crystal structure there is a large variation in cell parameter; these are 4.0279 and 5.501 Å, respectively, and

Table 16.3.3. SIMPLE EUTECTIC SYSTEMS WITH LIMITED SOLUBILITY

Elements	Compounds
Bi–Cd	LiF–LiBr
	CaO–MgO
Ag–Cu	FeO–Na$_2$O·2SiO$_2$
Ag–Pb	Al$_2$O$_3$–Ca$_3$(PO$_4$)$_2$
Al–Si	KF–LiF
Au–Be	KF–NaF
Cd–Pb	KF–BaF$_2$
Cd–Zn	MgF$_2$–BeF$_2$
Pb–Sb	LiCl–LiCO$_3$
	CaMgSi$_2$O$_6$–2MgO·SiO$_2$
Sn–Zn	CaAl$_2$Si$_2$O$_8$–SiO$_2$
	HOCH$_2$CH$_2$OH–H$_2$O

hence differ by 37 or 30% depending on which would be considered solvent. Hence little solubility is expected.

COMPOUND FORMATION. The formation of compounds by two components which themselves are compounds in this case is illustrated in Figure 16.3.7. Here the new compound shows a **congruent melting point**; i.e., the solid compound $NaF \cdot MgF_2$ changes to a liquid of identical composition at 1030°C.

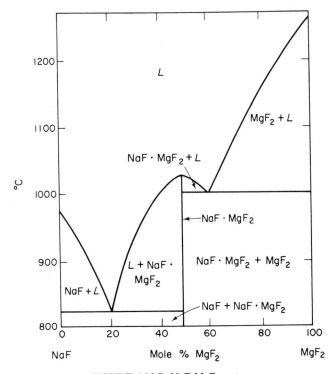

FIGURE 16.3.7. NaF-MgF₂ system.

Compound formation between the elements or between salts is extremely common, as is illustrated in Table 16.3.4. Thus in the Al–Ca system we find the stable compound Al_2Ca. Note that the phase diagram of Figure 16.3.7 could be considered to be two separate phase diagrams, one for the system NaF–$NaF \cdot MgF_2$ and another for the system $NaF \cdot MgF_2$–MgF_2.

OTHER CHARACTERISTICS. At this time, we shall list a number of invariant points: (1) inversion point, (2) congruent melting point, (3) eutectic point, (4) eutectoid point, (5) peritectic point, and (6) peritectoid point. The

Table 16.3.4. SYSTEMS SHOWING
COMPOUND FORMATION

Elements	Salts
Al–Ca	$BaO-TiO_2$
Ba–Pb	$CaO-TiO_2$
Fe–B	$CaO-ZrO_2$
Mg–Ca	$CaO\cdot SiO_2-CaO\cdot Al_2O_3$
Cu–P	$Li_2CO_3-K_2CO_3$
Mg–Si	$Li_2SO_4-K_2SO_4$
Mg–Sn	$PbO-PbSO_4$
	$NaF-AlF_3$

inversion point (between two polymorphs) has been described in Section 16.2, while (2) and (3) have already been described in this section. We now define the others in terms of transformations at invariant points:

Eutectoid: solid A $\underset{\text{heating}}{\overset{\text{cooling}}{\rightleftharpoons}}$ solid B + solid C   (16.3.5)

Peritectic: solid A + liquid $\underset{\text{heating}}{\overset{\text{cooling}}{\rightleftharpoons}}$ solid B   (16.3.6)

Peritectoid: solid A + solid B $\underset{\text{heating}}{\overset{\text{cooling}}{\rightleftharpoons}}$ solid C.   (16.3.7)

In all of the transformations given by (16.3.4)–(16.3.7) we shall be particularly interested in the kinetics of the process because, in general, at normal cooling rates the reactions are often so slow that we do *not* obtain an equilibrium structure. The effect of such nonequilibrium behavior will be discussed later.

16.4   NONEQUILIBRIUM TRANSFORMATIONS

EUTECTIC TRANSFORMATIONS.   From the viewpoint of equilibrium itself the eutectic point is not very interesting, i.e., it is merely an invariant point. However, from the kinetic viewpoint such a transformation is of great interest. Note that if a eutectic alloy is cooled sufficiently slowly through the eutectic transformation temperature or more simply the eutectic temperature, the melt would yield two crystals, one of α and one of β plus an interface between them. (If tiny crystals of α and β were formed, then considerable interfacial energy would be involved in the large interfacial area between the particles and the system would not be at equilibrium.) To obtain two large crystals would mean that nearly all the Pb atoms would have to transport themselves to one side of the crucible and nearly all the Sn atoms to the other. We cannot expect this to happen, however, if the melt is cooled rapidly; that is, we will *not* get an equilibrium structure. In the time in which the liquid solidifies, the atoms could move only a certain distance S and hence we get a **lamellar structure** (alternating layers of α and β with repeat distance S). This is called a **eutectic structure** and a micrograph of a typical eutectic structure is shown in Figure 15.4.1. The spacing S decreases as the rate of cooling increases. Mechanical properties such as yield strength are often strongly dependent on the lamellar spacing.

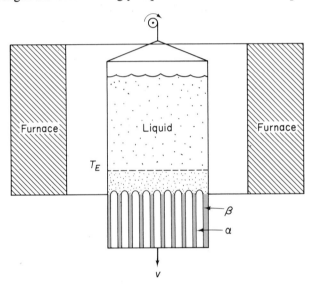

FIGURE 16.4.1. Crucible containing eutectic Pb–Sn liquid alloy is lowered from furnace at velocity v.

Figure 16.4.1 illustrates what happens if the liquid of eutectic composition is cooled unidirectionally by lowering a crucible at a velocity v from a furnace. We shall assume that the temperature is constant across the specimen (perpendicular to the motion of the crucible). There will be a slight supercooling; i.e., the eutectic transformation temperature T_E will be above the liquid-solid interface, which is at a temperature $T_E - \Delta T$. It is in this narrow liquid region that Pb atoms are moving toward the α phase and Sn atoms are moving toward the β phase. *Note:* The diffusion of atoms in liquids is rapid compared to the diffusion in solids (at the same temperature), and in fact the latter diffusion rate can often be considered negligible with respect to the former; moreover, the diffusion rate drops off very rapidly as temperature decreases; it is only 10^{-8} as fast at half the melting point as at the melting point in the solid metal.

It is found that the spacing S varies with velocity v according to

$$S^2 v = \text{constant.} \tag{16.4.1}$$

The derivation of this relation is left as a problem for the student. See Problem 18.45. Figure 16.4.2 illustrates this relationship.

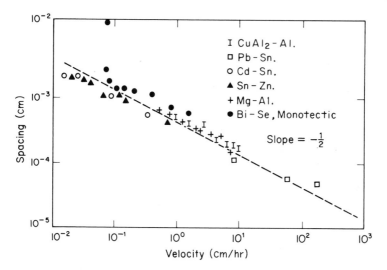

FIGURE 16.4.2. S^2v relationship for eutectic transformation. (C. Y. Li and H. Weart, *Materials Science Center Report #107,* Cornell University, Feb. 1963.)

EXAMPLE 16.4.1

A 40 wt. % Sn alloy of Pb–Sn is cooled rapidly. Discuss the microstructure which is obtained.

Answer. From Example 16.3.3 we note that the weight percent liquid = 48.7% at $183 + \epsilon°C$. This liquid has a composition arbitrarily close to 61.9% Sn, i.e., the eutectic composition. When it freezes (at $T_E - \Delta T$) it forms a eutectic structure, i.e., the lamellar structure or the "thumbprint" structure of α and β layers. We also note that at $183 + \epsilon°C$ the weight percent $\alpha = 51.3\%$. This remains as separated α phase when freezing occurs. The microstructure of the cold alloy consists of regions of α and regions of lamellar eutectic. There is 51.3 wt. % α and 48.7 wt. % of lamellar eutectic. *Note:* Actually these ratios would be slightly changed as would the composition of the α and eutectic because the solid forms under nonequilibrium conditions. Figure 16.4.3 illustrates the microstructure.

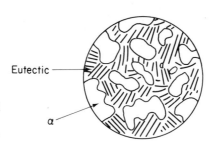

Eutectic

FIGURE 16.4.3. Microstructure of rapidly cooled 40 wt. % Sn–60 wt. % Pb solder.

α

The use of uniaxial solidification where the temperature gradient is unidirectional can result in a eutectic structure which is very similar to a filamentary composite. The Al–Al$_3$Ni phase diagram has a eutectic. When this eutectic alloy is cooled from the liquid, rod-like Al$_3$Ni forms in an Al matrix. When uniaxially solidified, the rods are all lined up in one direction. A cross section of such a solidified alloy is shown in Figure 16.4.4. The rods are many

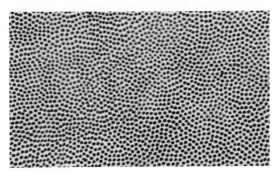

FIGURE 16.4.4. Fibrous structure: Al–Al$_3$Ni eutectic unidirectionally solidified at 2 cm/hr. Transverse section. ($\times$ 1500.) (*Courtesy of R. W. Kraft and L. D. Graham, Lehigh University.*)

times longer than their diameter. A goal of current research is to make the rods continuous for any desired length of a manufactured part. The unidirectional eutectic would then have the same structure as a filamentary composite such as the boron-reinforced aluminum discussed in Chapter 15. The possibility of growing eutectics uniaxially so that materials will have superior properties in

FIGURE 16.4.5. Lamellar structure of Al–CuAl$_2$ eutectic unidirectionally solidified at 1 cm/hr. Etchant 20% HNO$_3$, 80% H$_2$O. Dark phase is CuAl$_2$. ($\times$ 1000.) Transverse section. (*Courtesy of R. W. Kraft and L. D. Graham, Lehigh University.*)

certain directions is discussed in R. W. Kraft, "Controlled Eutectics," *Scientific American* (Feb. 1967) p. 86. A uniaxially solidified lamellar structure is shown in Figure 16.4.5.

SOLIDIFICATION IN A SYSTEM SHOWING CONTINUOUS SOLUBILITY. Figure 16.4.6 shows the Ni–Cu system. Consider an alloy with 22 wt. % Cu at 1500°C. This is a single-phase liquid until cooled to 1400°C; then the α phase begins to form. The α phase is a continuous solid solution of Cu in nickel with the fcc crystal structure. At 1400°C the tie line extends from 14 wt. % Cu to 22 wt. % Cu. The first tiny particles of α which form upon cooling have a composition

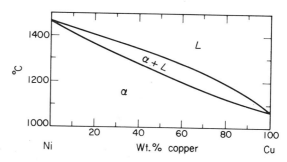

FIGURE 16.4.6. Ni–Cu system.

of 14 wt. % Cu. As the cooling proceeds these particles will grow. The material added to the outside of the particle will have a higher copper content. If there is no diffusion within the solid (or a negligible amount of diffusion), there will therefore be a variation of composition from the inside of the particle to the outside. This is known as **coring**. Recall that the average composition when solidification is complete must be 22 wt. % Cu; because the composition inside is only 14% Cu, the composition on the outside of this particle must be considerably above 22% Cu.

In coring, the center of the particle is richer in the higher melting component and has a higher melting point than the outside of the particle. Likewise the outside of the particle (where the copper content exceeds 22 wt. %) will have a melting point considerably *below* that of the homogeneous α-alloy of 22 wt. % Cu. Hence a bar of cored material would melt locally and hence be mechanically weak at a much lower temperature than the homogeneous solution.

To prevent or eliminate coring,

1. Cool very slowly.
2. Cool very rapidly to obtain coring on a fine scale and then anneal at a high temperature (so that solid state diffusion can speed up homogenization).

In Section 16.6 the one-dimensional solidification of a binary alloy with a single solid phase will be described quantitatively.

An example of a cored structure is shown in Figure 16.4.7.

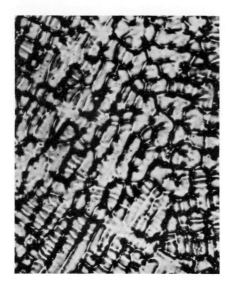

FIGURE 16.4.7. Micrograph of a
cored 30%
Ni–70% Cu
alloy. (×75.)

16.5 TWO IMPORTANT NONEQUILIBRIUM TRANSFORMATIONS

AGE HARDENING. Many commercial alloys depend upon a heat treatment known as age precipitation hardening to greatly enhance their mechanical
properties. We have already noted this in Chapter 2 (See Table 2.1.2) and have
pointed out the importance of this process to the aircraft industry. These
aluminum alloys basically contain about 4 wt. % copper plus small percentages
of other elements (which we will ignore for the sake of simplicity).

Relevant portions of the Al–Cu phase diagram are shown in Figure
16.5.1. We consider an alloy with 4 wt. % Cu at 550°C. If held at that temperature for a long time (so diffusion in the solid can occur—more about that in
later chapters), a single homogeneous phase called the κ phase will form. This

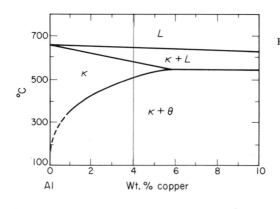

FIGURE 16.5.1. Phase diagram
for aluminum-
rich end of the
aluminum-
copper system.
The pure θ
phase (not
shown) contains
about 46 wt. %
Cu at room
temperature and
corresponds to
CuAl$_2$.

process is called **solutionizing**. The κ phase is a substitutional solid solution of Cu in Al. It has a fcc crystal structure. If it is cooled very slowly so that equilibrium always exists, then at room temperature there will be two phases: κ phase (nearly pure Al) and θ phase (essentially a compound $CuAl_2$). During such slow cooling large crystals of κ and large crystals of $CuAl_2$ form. Clearly, solid state diffusion of both copper and aluminum is necessary for this to occur. Solid state diffusion in metals depends on temperature according to

$$\text{rate} \propto e^{-Q_d/RT}, \tag{16.5.1}$$

where Q_d is the activation energy for diffusion which has a characteristic value for each material and R is the gas constant. The rate is only 10^{-8} as fast at $T_M/2$ as at T_M and only 10^{-16} as fast at $T_M/4$ as at $T_M/2$. Diffusion will be discussed in detail in Chapter 18.

EXAMPLE 16.5.1

If instead we cool from 550°C *very* rapidly (**quenching**) so that there is no time for such diffusion, what do we get at room temperature?

Answer. Cold κ phase.

We could also add that the κ phase is supersaturated with copper at room temperature. It would contain only a tiny fraction of 1 % of Cu at room temperature under equilibrium conditions. Instead it contains 4 %. The supersaturated κ phase is mechanically weak (has a low yield strength).

Thus we see from Example 16.5.1 that certain structures can be "frozen in." Such structures are nonequilibrium structures. If the supersaturated κ phase is heated up somewhat, say to 200°C, it is still greatly supersaturated. However, at this temperature atom motion is possible and tiny particles of Cu-rich precipitate begin to form. This treatment is called **aging**. (Because of the large supersaturation, many tiny particles form rather than one larger particle.)

EXAMPLE 16.5.2

a. A bottle of Seven-Up is placed in a freezer and frozen. What happens?

Answer. You might get one piece of ice forming initially and as cooling continues slowly this ice grows larger.

b. A bottle of Seven-Up is placed in a freezer and cooled nearly to the point where freezing begins. It is then removed from the freezer and opened. What happens?

Answer. First, the pressure is removed. The effect of the CO_2 pressure was to lower the freezing point. When the pressure is removed, the freezing point is raised. Hence the liquid is supercooled rapidly (compare with the quench which supersaturates the κ phase). *Millions* of tiny ice particles are instantaneously nucleated (so it appears). Large supersaturation or supercooling favors nucleation of large numbers of particles.

The age-hardening process was discovered in 1906 by Alfred Wilm. The first Cu-rich particle to form is coherent (see Section 13.4) with the κ lattice. Its size and shape can be studied when it is very small in diameter (≈ 10–20 Å) by X-ray scattering and transmission electron microscopy. As the particles grow larger they eventually become noncoherent with the κ lattice.

EXAMPLE 16.5.3

What good are the tiny Cu-rich precipitate particles?

Answer. They act as barriers for the motion of dislocations and hence greatly increase the yield strength and the hardness when present in sufficient numbers and optimum size. See Table 2.1.2.

Alloys heat-treated to obtain fine precipitates are called **age precipitation-hardened alloys**. The three-stage heat treatment process consists of:

1. Solutionizing.
2. Quenching.
3. Aging.

If the age-hardened alloy is heated for a longer time at high temperatures, the larger precipitate particles grow at the expense of the smaller, and eventually there are only a few large precipitate particles essentially in equilibrium with the κ matrix. This is called **overaging**. The resultant alloy is mechanically weak.

We note that the κ phase has a decreasing solubility of Cu in Al as temperature decreases. This is one of the important requirements of an age precipitation-hardenable alloy.

The Ni–Ti system has the same general form at nickel-rich composition

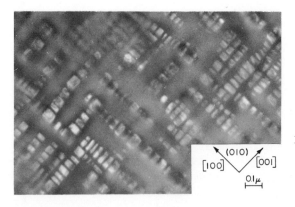

FIGURE 16.5.2. Precipitate in Ni–Ti. (*Courtesy of Stephen Sass, Cornell University.*)

as does the Al–Cu system at the aluminum-rich end. Figure 16.5.2 shows an electron transmission micrograph of precipitates in nickel. Note that at this stage of precipitation there is a definite orientation relationship between the precipitate and the matrix. Upon aging the precipitate will eventually grow into stable second-phase particles of Ni_3Ti.

Examples of age precipitation-hardenable alloys are given in Table 16.5.1. The age-hardened Mg-Al alloys are strong and very light in weight. The Cu–Be alloys which may contain only 1.7 wt. % Be can when aged-hardened have yield strengths near 200,000 psi (compare with annealed copper in Table 2.1.2). This alloy is nonsparking when it strikes a steel pipe and hence is used for wrenches in the petroleum industry. Because it is strong and a good electrical conductor, it is also used as a commutator spring in motors.

Table 16.5.1. AGE PRECIPITATION-
HARDENABLE ALLOYS

Solvent	Solute
Al	Cu
Al	Mg
Al	Si
Al	Zn
Ag	Cu
Au	Cu
Au	Ni
Cu	Be
Mg	Al
Ni	Ti
Pb	Te

THE Fe–C SYSTEM.   Figure 16.5.3 shows the Fe–C diagram. It also shows the Fe–Fe$_3$C diagram. Fe$_3$C is metastable and under equilibrium conditions decomposes virtually completely to iron and graphite. However, except in the presence of a catalyst such as silicon, Fe$_3$C forms when the iron-carbon alloy is cooled, at typical rates. By a typical rate we mean by cooling in air. Thus under such circumstances it is the Fe–Fe$_3$C diagram which concerns us even though one of the components is metastable (however, its lifetime in the absence of catalysts is of the order of millions of years). The phase designated by α has the bcc structure; the γ phase has the fcc structure. Carbon is dissolved interstitially in these structures as described in Section 12.3. The solubility of carbon in α-iron is about 0.025% C at the eutectoid temperature of 723°C. Let us suppose we cool down the γ phase for an 0.8% alloy starting at 800°C. At about 770°C we *would under equilibrium conditions* begin to find precipitated graphite; thus two phases would be present, γ and g, where g is graphite. This precipitation *would* continue until at 738°C the remaining γ (which now has a composition

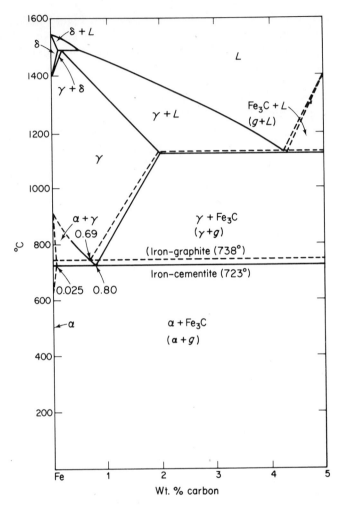

FIGURE 16.5.3. Fe–C and Fe–Fe₃C systems. (From *Metals Handbook*, American Society of Metals, Cleveland, 1918.)

of 0.69% C) would transform to $\alpha + g$. After this there would be only slight changes; on further cooling, additional slight amounts of graphite would precipitate out owing to the decreasing solubility of carbon in the bcc structure. What *actually* happens at typical cooling rates is that the 0.8% alloy remains as a single γ phase all the way down to 723°C, at which time it transforms to two phases $\alpha + Fe_3C$ by a eutectoid reaction. The Fe_3C (also called **cementite**) is metastable (i.e., it is thermodynamically unstable relative to iron and graphite but the reaction is so sluggish that no observable changes occur). Thus the fact that the metastable state is formed and persists is a result of the relative sluggishness of the reactions involving precipitation of graphite.

Let us consider the Fe–Fe$_3$C diagram further, in particular the eutectoid transformation. This is

$$\gamma \rightarrow \alpha + Fe_3C. \tag{16.5.2}$$

For simplicity, we can ignore the slight amount of carbon in the α phase and consider it simply as bcc iron. An additional nonequilibrium effect enters because of the finite cooling rate, which is an analog of the effect which occurred when the eutectic transformation occurred under nonequilibrium conditions. Clearly the structure should be one large crystal of α and one large crystal of Fe$_3$C in order to give a minimum total surface energy. Suppose, to begin with, we have a single γ crystal with 0.8 % C with a volume of 1 ft³. When this transforms to essentially pure iron and Fe$_3$C in the solid state, we see that in order to form but two crystals, the carbon atoms would have to migrate enormous distances. The time for this to occur below 723°C would be several thousand years (diffusion will be studied later). The net result is that we obtain instead many small crystals of α and Fe$_3$C; this increases the surface area and hence the free energy of the system; hence such a system is still further removed from equilibrium. This **eutectoid structure** in steel has a lamellar structure (alternate layers of Fe$_3$C and α) and is called **pearlite**. The lamellar spacing is determined by the temperature at which the eutectoid transformation occurs. As this temperature is lowered, the spacing decreases. This spacing profoundly affects the tensile yield strength, with smaller spacings corresponding to higher strengths. A micrograph of pearlite is shown in Figure 15.4.2.

EXAMPLE 16.5.4

a. A steel contains 0.4 wt. % C. Calculate the fraction of α and γ at $723 + \epsilon$°C, where ϵ is arbitrarily small.

Answer.

$$\% \alpha = \frac{0.80 - 0.40}{0.80 - 0.025} \times 100\% = 51.6\%$$

$$\% \gamma = 48.4\%.$$

Note. The γ has the eutectoid composition. See Figure 16.5.4(b) for the microstructure.

b. A steel contains 0.4 wt. % C. What do we see in the microscope if this is cooled to $723 - \epsilon$°C after the γ is transformed?

Answer. At $723 - \epsilon$°C, the α from the previous example is colder and otherwise unchanged, while the γ which has the eutectoid composition transforms to pearlite. Remember that pearlite contains both α and Fe$_3$C. See Figure 16.5.4(c) for the microstructure.

Figure 16.5.4 illustrates the microstructural changes upon cooling a hypoeutectoid steel (below 0.8 wt. % C). If steel transforms considerably below 723°C, **bainite**, a feathery arrangement of Fe$_3$C is formed. Bainitic steels have a high toughness.

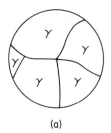

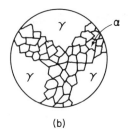

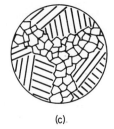

(a)   (b)   (c)

FIGURE 16.5.4. Microstructure of 0.4 wt. % steel at various
temperatures. (a) Austenized at 900°C. (b)
Slightly above the eutectoid temperature of
723°C. (c) Cooled to slightly below eutectoid
temperature; γ of (b) transforms to coarse
pearlite.

Let us next consider γ of eutectoid composition at 800°C; let us place
a sample of this in ice water. It is then cooled down so rapidly that the migration
of carbon atoms is negligible so that separate particles of $\alpha + Fe_3C$ cannot
form. At this temperature, however, iron prefers the bcc structure but with
negligible carbon; but here we have to dispose of 0.8% carbon.

. **EXAMPLE 16.5.5**

When the γ phase of any composition is quenched to room temperature
what is obtained?

Answer. We might answer cold γ by analogy with Example 16.5.1;
this is wrong! See the following.

When the γ phase is quenched below a certain temperature M_S (**martensite starts temperature**), **martensite** begins to form. This is a new crystalline form,

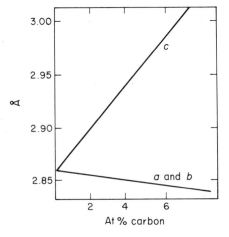

FIGURE 16.5.5. Lattice
parameters of
martensite
versus carbon
content.

a body-centered tetragonal crystal. The c/a ratio (ratio of lattice parameters) for the tetragonal form is proportional to the carbon content of the γ, as illustrated in Figure 16.5.5. Martensite can be considered to be a bcc crystal which is distorted by the carbon atoms (which are not soluble in the bcc crystal itself). The formation of martensite does not require diffusion. In fact the martensite platelets form by a shearing mechanism at velocities of the order of sound velocities in the material. This is called a **diffusionless transformation** (the eutectic, age precipitation, and eutectoid transformations considered previously here all took place by diffusion). Below the temperature M_F (**martensite finishes temperature**) no more austenite (γ phase) is transformed to martensite. There is usually some retained austenite. M_S and M_F vary with the carbon content, as shown in Figure 16.5.6. Additions of other alloying elements can displace these

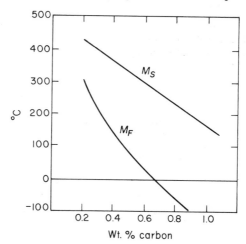

FIGURE 16.5.6. M_S and M_F temperatures for carbon steel.

curves. Martensite is *very* hard and strong. Dislocations move only with great difficulty through martensite because they interreact strongly with the carbon in martensite. The martensite crystals are in the form of convex lenses or plates. See Figure 16.5.7.

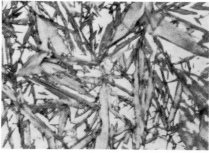

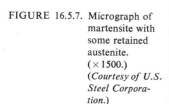

FIGURE 16.5.7. Micrograph of martensite with some retained austenite. ($\times 1500$.) (*Courtesy of U.S. Steel Corporation.*)

STEEL.   **Steel** is an iron-carbon alloy which involves carbon compositions up to 1.98 wt. % C. The usual sequence of heat treatment of steel (Section 6.7) has as its purpose the production of martensite in order to increase the strength of the steel. This is the three-step process:

1. Austenitizing.
2. Quenching.
3. Tempering.

Austenitizing is the process of converting the steel to austenite by annealing it at a sufficiently high temperature (in the stable γ-region). **Tempering** (heating to moderate temperatures after quenching) reduces the stresses induced by the martensite transformation and converts some of the martensite to carbide. While the Egyptians knew earlier that quenching caused hardness in a carburized iron, the Greeks by 400 B.C. had learned that tempering relieved the material of its brittleness.

EXAMPLE 16.5.6

A long, 12-in.-diameter solid cylinder of steel is quenched in oil at room temperature. Does martensite form?

Answer. Three possibilities exist: (1) The cooling rate is so slow that all of the γ phase decomposes by diffusion into pearlite (and either α or Fe_3C) before M_S is reached; (2) the cooling rate is so fast that none of the γ phase has decomposed by diffusion when M_S is reached and all of the γ phase transforms to martensite; (3) the cooling rate is intermediate so both pearlite and martensite are formed. Thus we need to be *quantitative*. Such quantitative discussions will be given in later chapters.

EXAMPLE 16.5.7

Alloying elements such as manganese, vanadium, and nickel in steel slow down the transformation of γ phase to pearlite by diffusion. How does this affect the possibility of obtaining martensite during quenching?

Answer. It *greatly* enhances it even when the cooling rate is low as along the axis of the cylinder of Example 16.5.6. This will be discussed quantitatively later.

When nickel (or certain other elements) is added in sufficient quantity to iron the austenite phase can be stabilized to room temperature and below.

To summarize, the types of steel which we have encountered are pearlitic, bainitic, martensitic, and austenitic.

CAST IRON.   The Fe–Fe$_3$C system contains a eutectic point at 4.3 wt. % C. Systems in the composition range shown by the eutectic line are known as **cast irons**. In white cast iron (lower part of the carbon range) the carbon is present in Fe$_3$C. In gray cast iron the carbon is present as graphite flakes (the addition of the catalyst silicon makes the decomposition to graphite take place).

16.6 SEGREGATION IN BINARY ALLOYS DURING SOLIDIFICATION

This section provides a quantitative discussion of the nonequilibrium freezing of binary alloys in special cases.

THE DISTRIBUTION COEFFICIENT.   Consider a portion of a binary phase diagram as shown in Figure 16.6.1.

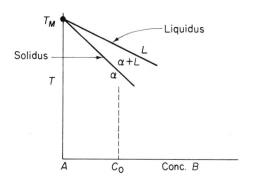

FIGURE 16.6.1. Portion of a binary phase diagram.

Define

$$m_s = \text{slope of solidus.}$$

$$m_l = \text{slope of liquidus.}$$

Both of these slopes are assumed to be constants in the discussion which follows. Then for the solidus we have

$$T_M - T = -m_s C_s, \qquad (16.6.1)$$

where C_s is the concentration (in units of mass per unit volume) of B in α at temperature T, and for the liquidus

$$T_M - T = -m_l C_l, \qquad (16.6.2)$$

where C_l is the concentration of B in L at temperature T.

Combining these equations gives

$$\frac{m_l}{m_s} = \frac{C_s}{C_l} \equiv K. \qquad (16.6.3)$$

K is the **distribution coefficient**.

EXAMPLE 16.6.1

An alloy of composition C_0 is cooled from the liquid phase of an alloy of the A-B system. The distribution coefficient is K. What is the composition of the first tiny particle of solid which forms?

Answer.

$$C_s = KC_0 \qquad (16.6.4)$$

since at the instant the first solid forms, the composition of the liquid is still C_0.

SEGREGATION.   Consider a liquid binary alloy of the A-B system shown in Figure 16.6.1 with composition C_0 which is dropped slowly into a cooled region as shown in Figure 16.6.2. The first drop to solidify at $x = 0$ has a con-

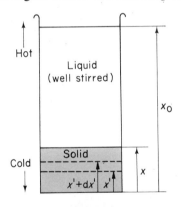

FIGURE 16.6.2. Solidification illustrating segregation.

centration $C_s(0)$; i.e., it is the drop formed at $x = 0$. Since the initial concentration of B in the liquid is C_0, we have

$$C_s(0) = KC_0. \qquad (16.6.5)$$

We show the liquid in Figure 16.6.2 as well stirred so that the concentration of the liquid is uniform; i.e., it does not vary with position. On the other hand very little change occurs within the solid because of the slowness of diffusion. We assume here that no further changes in composition take place in the solid. The freezing takes place at the interface at x. We use here the **concept of local equilibrium**; i.e., we assume that at the interface

$$C_s(x) = KC_l(x); \qquad (16.6.6)$$

this is an equilibrium assumption; it is assumed to be true at the interface even though the process is a dynamic one.

We now use a material balance on the solute B which enables us to complete our description of this problem (to simplify the mathematics we have ignored volume changes). We have

$$C_l(x)\left(1 - \frac{x}{x_0}\right)x_0 + \int_0^x C_s(x')\, dx' = C_0 x_0. \qquad (16.6.7)$$

EXAMPLE 16.6.2

State in words the material balance used to derive (16.6.7).

Answer. The amount of component B in the liquid (which has uniform composition at any instant) plus the amount of component B in the solid (in which the composition varies with position in the solid which is **segregated**) equals the total amount of component B present in the original alloy.

Using (16.6.6), (16.6.7) becomes

$$\frac{C_s(x)}{K}\left(1 - \frac{x}{x_0}\right)x_0 + \int_0^x C_s(x')\, dx' = C_0 x_0. \qquad (16.6.8)$$

This integral equation may be differentiated with respect to x to give

$$\frac{C_s'(x)}{K}\left(1 - \frac{x}{x_0}\right) - \frac{C_s(x)}{Kx_0} + \frac{C_s(x)}{x_0} = 0 \qquad (16.6.9)$$

and is as previously noted subject to (16.6.5). It is a linear first-order equation with separable variables. The solution to the problem is

$$C_s(x) = KC_0\left(1 - \frac{x}{x_0}\right)^{K-1}, \qquad x < x_0. \qquad (16.6.10)$$

We note that it is possible to have K less than or greater than 1 in different systems. In the case of Figure 16.6.1, $K < 1$. Suppose $K = \frac{1}{10}$ and $B_0 = 0.1$ wt. % and we wish to obtain A, which has less than 0.1 wt. %. Then by carrying out this process we can purify (by a factor up to 10) the first portion of this bar. After solidification we crop off the half near x_0 which contains most of the B impurity and melt the remaining material and repeat the process. After repeated steps of this sort we achieve nearly pure A (assuming no other impurities are present or become involved because of handling).

The segregation achieved in the previous discussion is a maximum. Had the liquid solution not been highly stirred, less segregation would have occurred. Had some diffusion taken place in the solid solution, less segregation would have occurred. Segregation such as this occurs at usual cooling rates in casting. An example is the variation of concentration across the individual crystals that grow from the melt, known as coring.

Note that in the problem considered above it was assumed that heat conduction was very rapid so that the problem of carrying away the heat of fusion could be completely ignored. This simplifies the picture considerably. In casting practice, the role that heat conduction plays is a vital one and must be properly accounted for.

16.7   ZONE REFINING AND LEVELING

A technique of great importance in the semiconductor industry achieves the purity for semiconductor crystals, about 10^{-8} at. % impurity, by moving a melted zone along a bar of the fairly pure material. It is clear that (1) if the melt is richer in one component and (2) if this zone is moved from one end of a bar to the other by moving a narrow furnace or other source of heat, some of this component will be moved to the far end. By repeated passes in the same direction the concentration of the component may be made arbitrarily low in principle. This technique is called **zone refining**. A few solutes concentrate in the solid and therefore move the other way. Some impurities (unwanted components) have $K = 1$ and are not cooperative. The phase diagrams for the systems give the essential information on distribution coefficients.

If the passes are run back and forth, homogenization of a nonuniform alloy may be achieved. This is called **zone leveling**. Often homogeneity of an alloy is very important.

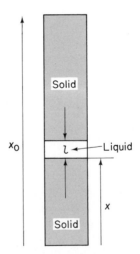

FIGURE 16.7.1. Zone refining.

The process of zone refining can be described quantitatively. Consider Figure 16.7.1. The analysis of this problem proceeds in the same fashion as the analysis of one-dimensional segregation in Section 16.6.

EXAMPLE 16.7.1

The two solid ends of the rod in Figure 16.7.1 can be held and rotated independently of each other in a high vacuum or in an inert atmosphere such as high-purity argon gas. The hot portion of the liquid (and nearby solid) does not touch a container in such a case and hence will not be contaminated. What holds the liquid in place?

Answer. Surface tension. Hence the length of the liquid zone is restricted unless the zone refining is carried out in a satellite where $g = 0$.

Figure 16.7.2 shows a tungsten rod undergoing zone refining. Induction heating is used to melt the refractory.

We continue here the discussion of Figure 16.7.1. A material balance

FIGURE 16.7.2. Zone refining of tungsten. $T_M = 3380°C$.
(*Courtesy of B. F. Addis, Cornell University.*)

of component B gives

$$\int_0^x C_s(x')\,dx' + lC_i(x) = (x + l)C_0, \qquad (16.7.1)$$

which can be written in the differential form [after (16.6.6) is used] as follows:

$$C_s(x) + \frac{l}{K}C_s'(x) = C_0. \qquad (16.7.2)$$

This is subject to the initial condition,

$$C_s(0) = KC_0. \qquad (16.7.3)$$

The solution of (16.7.2) and (16.7.3) is

$$C_s = C_0[1 - (1 - K)e^{-Kx/l}] \qquad \text{for } x \leq x_0 - l. \qquad (16.7.4)$$

In the case of Sb in Ge, $K = 4 \times 10^{-2}$. On the second and repeated passes it is necessary to account for the fact that the solid which melts and enters the liquid zone has variable composition.

EXAMPLE 16.7.2

Would the semiconductor industry as we know it today be possible without zone refining?

Answer. No. The development of this technique by W. G. Pfann at the Bell Telephone Laboratories was a great achievement since pure germanium was a necessary starting material for making good transistors.

The analysis of this section and Section 16.6 assumed a K which did not vary with concentration. This made it easy to get simple solutions. If K varies with C, then it is still fairly easy to carry out the calculations using a finite difference approach and a modern digital computer.

16.8   APPLICATION OF PHASE DIAGRAMS

Phase diagrams tell us more than just which phase(s) is present in a given P, T, x_i domain:

1. We saw in Section 16.2 that they yield certain thermodynamic data (this is also true of multicomponent systems—a subject for more advanced study).
2. We saw in Section 16.6 that binary phase diagrams could be used to describe segregation, and to predict coring.
3. We saw in Section 16.4 that we could by looking at a phase diagram (and with a qualitative understanding of kinetics) predict the existence of lamellar (or other periodic) structures in eutectic systems.
4. We saw in Section 16.5 that a quick look at a phase diagram would tell us if a system is a *possible* candidate for age precipitation hardening.

5. We saw in Section 16.7 that phase diagrams could be used to form a quantitative basis for the description of zone refining and zone leveling.

Is there even further information suggested by phase diagrams? Certain property relationships are suggested. For example, in the Pb–Sn system (Figure 16.3.6) at room temperature, the equilibrium structure of a 61.9% alloy is a mixture of two crystals of nearly pure Sn and nearly pure Pb. If a wire were

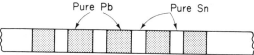

FIGURE 16.8.1. Wire cut from two-phase material.

cut from this material it might have the structures shown in Figure 16.8.1. The resistance of the wire is given by

$$R = \rho \frac{l}{A}, \tag{16.8.1}$$

where ρ is the effective resistivity of the mixture. The resistance of a series of resistances is given by

$$R = R_1 + R_2 + \cdots \tag{16.8.2}$$

and hence we can write

$$R = \frac{\rho_{Pb} l_{Pb}}{A} + \frac{\rho_{Sn} l_{Sn}}{A}, \tag{16.8.3}$$

and combining this with (16.8.1) we have

$$\rho = \frac{\rho_{Pb} l_{Pb}}{l} + \frac{\rho_{Sn} l_{Sn}}{l}. \tag{16.8.4}$$

Hence the resistivity of the mixture is determined by the individual resistivities weighted by the volume fraction. This is often referred to as the **law of mixtures**. Equation (16.8.4) can also be written as

$$\rho = \rho_{Pb} + (\rho_{Sn} - \rho_{Pb})x, \tag{16.8.5}$$

where x is the volume fraction of Sn; this is a linear relation.

Let us consider the Cu–Ni system (Figure 16.4.4). Here an alloy of 50% composition exists as a single-phase solid solution at room temperature (it is *not* a mixture of phases). The resistivity in this case behaves as shown schematically in Figure 16.8.2. Electrical resistivity is due to the scattering of electrons. If the electron moves through a nearly perfect periodic potential, little scattering occurs. As this periodicity is removed, e.g., by atomic vibrations, the scattering increases. The formation of a solid solution in which Ni atoms replace copper atoms in a random fashion also removes the periodicity and increases the scattering. The mean free path of the electron determines the electrical resistivity, which depends on the structure on an *atomic* scale.

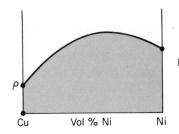

FIGURE 16.8.2. Schematic drawing of resistivity versus composition for a continuous solid solution.

We have noted earlier that the yield strength of pearlite increases as the lamellar spacing decreases (even though the composition is the same and the volume fraction of ferrite and cementite are the same). Dislocations move easily in ferrite but are stopped by the very hard cementite. The mean free path of the dislocations is determined by the lamellar spacing. Thus the yield strength depends on structure on the *microscale* (microscope level).

If we understand the origin of a physical property (electrical resistivity, yield strength, etc.), then we will know how it is affected by structure. Then to predict the behavior of a given material we must know the appropriate details of the structure. In later chapters we shall study the fundamental origin of properties.

The student should once again consider the triangle of knowledge in materials science in Figure 1.1, namely, phenomenology of properties, structure of materials, and the fundamental theory of properties.

16.9   DETERMINATION OF PHASE DIAGRAMS

Consider the Pb–Sn diagram of Figure 16.3.6. One of the most common ways of obtaining such a diagram is by the use of cooling curves. Thus an alloy of a certain composition is placed in a crucible and melted. A thermocouple is introduced into the melt and the alloy is allowed to cool. Temperature versus time curves (**cooling curves**) are recorded. Thus for pure lead we might obtain the curve shown in Figure 16.9.1(a). There is a **thermal arrest** at T_M because a finite quantity of heat is given off during freezing over an *infinitesimal* temperature range. In Figure 16.9.1(b), a cooling curve for a 40 wt. % Sn alloy is shown.

EXAMPLE 16.9.1

Why is there no thermal arrest when α starts appearing during cooling of the 40% alloy?

Answer. Because the freezing of α occurs over a *finite* temperature range (from the liquidus to the solidus) so that the heat of freezing is given off continuously over this range.

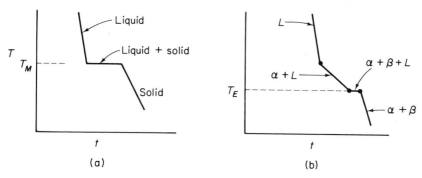

FIGURE 16.9.1. Cooling curves. (a) Pure Pb. (b) 60% Pb–40% Sn.

Note that the use of cooling curves involves a nonequilibrium test, i.e., a dynamic experiment, to obtain equilibrium behavior. If equilibrium is reached rapidly, fast cooling can be used. However, if equilibrium is reached very slowly (as is especially the case in ceramic systems) then *very* slow cooling rates must be employed.

Many tools are used to obtain phase diagrams in addition to the use of cooling curves. X-ray diffraction is widely used to find the boundary between solid phases. Electrical and magnetic measurements as a function of composition are also helpful in defining certain phase boundaries.

16.10 TERNARY DIAGRAMS

Important engineering materials are often multicomponent systems such as Al–Ni–Co in Alnico magnets, Fe–Cr–Ni in stainless steel, and $CaO–Al_2O_3–SiO_2$ in fire clay refractories and in Portland cement.

At constant pressure, a ternary system involves independent variations of temperature and two of three compositions. This system could be described by a three-dimensional figure. The usual way of making this representation is

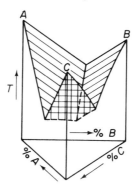

FIGURE 16.10.1. Representation of a ternary system at constant pressure.

shown in Figure 16.10.1. The surface of the liquidus is shown. Note the ternary eutectic. The base of this figure is an equilateral triangle. Composition is represented on the base of this figure; this representation makes use of the fact that the sum of the lengths of the three lines formed by dropping perpendiculars from a point within an equilateral triangle to the three base lines equals the altitude of the triangle. Hence composition is represented as in Figure 16.10.2.

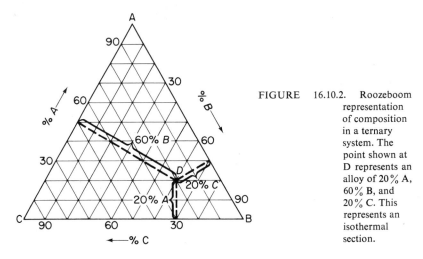

FIGURE 16.10.2. Roozeboom representation of composition in a ternary system. The point shown at D represents an alloy of 20% A, 60% B, and 20% C. This represents an isothermal section.

It is common to present a sequence of these isothermal sections rather than to use the representation of Figure 16.10.1. This is illustrated for the Pb–Sn–Bi system in Figure 16.10.3. The Sn–Bi eutectic is at 133°C, the Bi–Pb eutectic

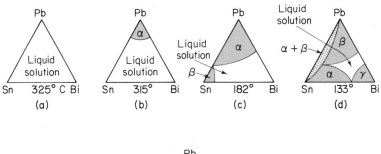

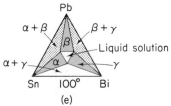

FIGURE 16.10.3. Isothermal sections of the Pb–Sn–Bi system.

is at 124°C, and the Pb–Sn eutectic is at 183°C. The ternary eutectic is at 96°C. This ternary eutectic is used in sprinkler systems. The ternary eutectic has a lower melting point than all the binary eutectics in the system.

THE Fe–Cr–Ni SYSTEM.   The Fe–Cr–Ni system is the basis of austenitic stainless steels. One of the more typical of these is the 18-8 stainless steel (18% Cr and 8% Ni). We note from Figure 16.10.4 that this alloy is the γ phase; this is fcc and nonmagnetic even at room temperature.

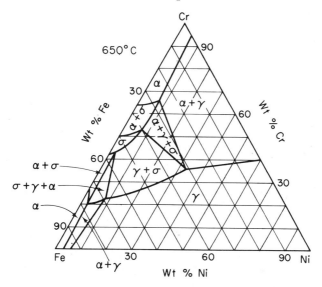

FIGURE 16.10.4. Isothermal section of the Fe–Cr–Ni system at 650°C.

THE CaO–Al$_2$O$_3$–SiO$_2$ SYSTEM.   The industrial significance of the CaO–Al$_2$O$_3$–SiO$_2$ system is shown in Figure 16.10.5. Ternary ceramic systems are

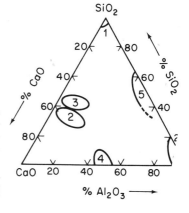

FIGURE 16.10.5. CaO-Al$_2$O$_3$-SiO$_2$ materials. (1) SiO$_2$ refractories. (2) Portland cement. (3) Blast furnace slag. (4) Low silica clay. (5) Fire-clay refractories. (6) High-temperature (>3500°F) Al$_2$O$_3$ refractories. [After L. Van Vlack, *Elements of Materials Science*, Addison-Wesley, Reading, Mass. (1959).]

also of great importance. A three-dimensional representation of the liquidus surface is shown in Figure 16.10.6.

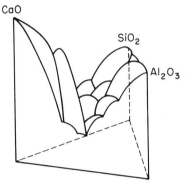

FIGURE 16.10.6. Liquidus of the CaO–Al₂O₃–SiO₂ system.

REFERENCES

Gordon, Paul, *Principles of Phase Diagrams in Materials Systems*, McGraw-Hill, New York (1968).

Rhines, F. N., *Phase Diagrams in Metallurgy*, McGraw-Hill, New York (1965).

Levin, E. M., Robbins, C. R., and McMurdie, H. F., *Phase Diagrams for Ceramists*, American Ceramic Society, Columbus, Ohio (1964). Contains numerous phase diagrams of ceramic systems (inorganic systems).

Hansen, M., *Constitution of Binary Alloys*, McGraw-Hill, New York (1958). This book and the two supplements by Elliott and Shunk contain much of what is known about binary systems involving the elements.

Elliott, R. P., *Constitution of Binary Alloys, First Supplement*, McGraw-Hill, New York (1965).

Shunk, F. A., *Constitution of Binary Alloys, Second Supplement*, McGraw-Hill, New York (1969).

Metals Handbook, American Society for Metals, Cleveland (1948). Contains a number of phase diagrams of particular importance in the metallurgical industry.

Pfann, W. G., *Zone Melting*, Wiley, New York (1966).

Pfann, W. G., "Zone Refining," *Scientific American* (Dec. 1967) p. 62.

Kraft, R. W., "Controlled Eutectics," *Scientific American* (Feb. 1967) p. 86.

PROBLEMS

16.1. a. How many variables (from P and T) can be varied independently along line AC in Figure 16.2.1?

b. How many variables can be varied independently in the region labeled LS in Figure 16.3.1?

c. How many variables (from T and weight percent Sn) can be varied independently along the liquidus line of Figure 16.3.6?

d. Along the eutectic line of Figure 16.3.6?

e. In the α region of Figure 16.3.6?

f. In the $\alpha + \beta$ region of Figure 16.3.6?

16.2. Give five types of congruent transformation temperatures.

16.3. For an alloy of 30 wt. % Sn and 70 wt. % Pb under equilibrium conditions,

a. What phases are present at 184°C?

b. What is the composition of each phase?

c. What is the weight fraction of each phase?

16.4. For an alloy of 30 wt. % Sn and 70 wt. % Pb under equilibrium conditions,

a. What phases are present at room temperature (300°K)?

b. What is the composition of each phase?

c. What is the weight fraction of each phase?

16.5. The ethylene glycol-water phase diagram has the general feature of the Pb–Sn diagram. Of what engineering importance is this?

16.6. In the NaF–MgF$_2$ system of Figure 16.3.7 under equilibrium conditions, calculate the weight fraction of each phase present at 810°C for a 40 mole % MgF$_2$ alloy.

16.7. Give a simple explanation of why a eutectic transformation might lead to a lamellar structure when cooled rapidly.

16.8. A 30 wt. % Sn, 70 wt. % Pb alloy is cooled rapidly (and, hence, under nonequilibrium conditions). Discuss as quantitatively as possible the microstructure which is obtained.

16.9. An alloy of 10 wt. % Sn–90 wt. % Pb is cooled *very* slowly (so slowly that diffusion in the solid state takes place over the entire sample).

a. What is the composition of the last drop of liquid which freezes?

b. At what temperature does this occur?

c. If the same alloy were cooled rapidly, what would be the composition of the last drop to freeze?

d. At what temperature would this occur?

16.10. An alloy contains 90 wt. % Ni–10 wt. % Cu.

a. If cooled very slowly, at what temperature does it become entirely solid?

  b. What is the composition of the last drop to freeze?

  c. If cooled very rapidly, what is the composition of the last drop to freeze?

  d. At what temperature does this occur?

  e. How might the rapidly cooled structure appear in a micrograph?

  f. What tensile load would this rapidly cooled alloy carry at 1200°C?

16.11. Name some of the necessary attributes of a system A-B if it is to show age-hardening characteristics.

16.12. Ordinary pure gold is very weak. Describe a mechanism for making it stronger for use in jewelry and the like.

16.13. The lead sheathing applied to the outside of electrical insulation on underground electrical cable has to have greater mechanical strength than pure lead. Suggest a way of achieving this.

16.14. a. What are the three steps in an age-hardening heat treatment?

  b. Describe what happens on the atomic scale in each step.

16.15. Steel contains a number of interesting phases which help make it useful, which helps explain why hundreds of millions of tons of steel are produced throughout the world annually.

  a. Is Fe_3C a stable phase?

  b. Is fine pearlite a stable two-phase configuration?

  c. Is bainite a stable two-phase configuration?

  d. Is martensite a stable phase?

16.16. a. Sketch a micrograph of a 1.5 wt. % C carbon steel at 725°C.

  b. At room temperature.

16.17. Why does pearlite form?

16.18. a. Why (not when, how, etc.) does martensite form?

  b. What characteristic of martensite makes it desirable?

16.19. Discuss the reason for each of the three steps in the heat treatment of steel.

16.20. Explain why an age-hardened Cu–Be alloy is a good electrical conductor.

16.21. Sketch the cooling curves for the Pb–Sn system at 10-wt. % intervals.

16.22. What is the principal difference between a binary ceramic system and a binary metal system?

16.23. Discuss several uses of phase diagrams.

16.24. The Na_2O–SiO_2 system has a eutectic. SiO_2 has a very high melting point, and pure SiO_2 glass has a high softening point. Explain why Na_2O is added.

16.25. Describe the nonequilibrium cooling of a peritectic alloy. Why is the structure called **surrounding**?

More Involved Problems

16.26. a. Use the data from the Ni–Cu phase diagram to get $C_s(B)$ and $C_l(B)$, where B represents the volume percentage of copper, at 5 vol. % Cu intervals. *Note.* To calculate the density of the alloys, assume that Vegard's law holds.
 b. Define

$$K = \frac{C_s}{C_l}.$$

 Write a computer program for describing one-dimensional segregation.
 c. Use your computer program to plot the composition versus distance for an alloy rod 10 cm long.

16.27. a. Show that Equation (16.6.10) is indeed a solution of (16.6.9).
 b. Discuss carefully all of the physical assumptions used in deriving (16.6.10).

16.28. Discuss the current status of zone refining techniques.

Sophisticated Problems

16.29. Assume that K is a constant and consider making repeated passes in zone refining. Let $C_s^n(x)$ be the concentration of the impurity in the solid at x after the nth pass.
 a. Show that the differential equation is

$$\frac{l}{K}\frac{dC_s^n}{dx} + C_s^n = C_s^{n-1}(x + l).$$

 The term $C_s^{n-1}(x + l)$ means C_s^{n-1} evaluated at $x+l$.
 b. Show that the appropriate initial condition is

$$C_s^n(0) = \frac{K}{l}\int_0^l C_s^{n-1}\,dx.$$

16.30. Can the solution to Problem 16.29 be obtained in closed form? Discuss.

16.31. Analyze carefully the advantages and disadvantages of carrying out zone refining and crystal growth in a satellite space station (where $g = 0$) and write a proposal to carry out such work.

16.32. Assume that both the liquidus and solidus in the Cu-Ni system can be fitted by parabolas.
 a. Do this. Assume a spherical container is full of 90 wt. % Ni and 10 wt. % Cu. Assume that only one nucleus is present at the center of the container and that this grows radially outward as the alloy is cooled.
 b. What is the composition of Cu as a function of radius?
 c. What is the composition of the last drop to solidify?
 d. Are the assumptions in (b) reasonable physically?

Prologue

Temperature causes atom motion, which in turn affects the behavior of materials. The study of the effect of temperature fields is known as *thermodynamics*. We begin this chapter with a study of the motion of atoms in a dilute gas. The *average kinetic energy* of such an assembly of gas molecules (at a constant temperature) can be used to define the *ideal gas temperature scale*. Because the *random walk process* is important in many areas of materials science, including the theory of ideal gases, it is introduced here. The random walk leads to the *random distribution function*. We study some of the important contributions made by *Maxwell* to the theory of gases and discuss the use of the *molecular beam* experiment to confirm his predictions.

We explain *internal energy* from both the *microscopic* and *macroscopic viewpoints*. We also study the concept of a *thermodynamic cycle* and the meaning of an *exact differential*. We introduce an important new concept, *entropy*, which is discussed from both the microscopic and macroscopic viewpoints. The *entropy of mixing* and also the *efficiency of a thermodynamic process* are described. We then discuss the concept of *thermodynamic equilibrium* at constant pressure and temperature. The preceding fundamental concepts are applied to calculation of equilibrium vacancy concentrations, calculation of the vapor pressure of a liquid, and a description of phase equilibria in two-component systems. The concepts of *ideal solution*, *regular solution*, and *order-disorder* are discussed.

There is an introduction to statistical thermodynamics showing how the internal energy and entropy of an assembly of atoms can be calculated. This is applied to the *specific heat theory of solids*. We study *Debye frequency*, *Debye temperature*, and *phonon distributions*. The origin of *thermal expansion* and *thermal conductivity* are examined. The important physical concept of the *mean free path* is introduced.

"A theory is the more impressive the greater the simplicity of its premises are, the more different kinds of things it relates, and the more extended is its area of applicability. Therefore the deep impression that classical thermodynamics made upon me. It is the only physical theory of universal content concerning which I am convinced that, within the framework of applicability of its basic concepts, it will never be overthrown."   (A. Einstein, 1949)

17

A DISCUSSION

OF STATISTICAL THERMODYNAMICS

17.1   INTRODUCTION

A **system** is a portion of matter which we choose to isolate for study from the rest of the **environment**. Thus we may be interested in studying the properties of a gas such as argon as a function of pressure and temperature. One mole of argon gas may be the system, and the cylinder and piston enclosing it plus a heater may be the environment.

The study of how temperature along with other fields affects the equilibrium behavior of such a system of matter is called **thermodynamics**. After this introduction to thermodynamics we shall show

1. Why vacancies exist in equilibrium concentration.
2. Why Sn dissolves in Pb at high temperatures (see Figure 16.3.6), although it does not at low temperatures.
3. Why ordered solid solutions become disordered at high temperatures (see Figure 12.6.3).
4. Why increasing the temperature favors melting and vaporization.
5. Why the vapor pressure above a liquid or solid depends exponentially on the negative reciprocal of temperature.
6. Why the air pressure depends exponentially on the negative of altitude.
7. Why the specific heat of solids is near $3R$ at high temperatures but varies as T^3 at low temperatures (below 30°K).

We will have formed the basis for explaining the exponential of reciprocal temperature dependence of conductivity in semiconductors noted in Chapter 3.

We will also have formed the basis for a later discussion of

8. Why high temperature destroys magnetism (which depends on the ordering of spins).

385

9. Why high temperature destroys ferroelectricity (which depends on the ordering of dipoles).

10. Why high temperature destroys superconductivity (which depends on the formation of paired electrons).

Finally, we will have formed the basis for the theory of reaction rates, which will be discussed in Chapter 18.

17.2   ATOM MOTION AND TEMPERATURE

Atoms are in constant motion. Energy is associated with this motion. In the case of an ideal gas the energy associated with the random motion can be conveniently measured.

An **ideal gas** is a gas in which there are no potential energy interactions between the atoms. Real gases at low pressures approximate ideal gases. The classical mechanics expression for the kinetic energy of an atom, such as an argon atom, is given by $\frac{1}{2}mv^2$. Thus the total kinetic energy of a system of N argon atoms is given by

$$U_K = \sum_{i=1}^{N} \tfrac{1}{2}mv_i^2, \qquad (17.2.1)$$

where v_i is the velocity of each atom. The velocity is a vector

$$\mathbf{v} = [v_{x_1}, v_{x_2}, v_{x_3}]$$

so that the square of the speed c is given by

$$c^2 = v^2 = v_{x_1}^2 + v_{x_2}^2 + v_{x_3}^2. \qquad (17.2.2)$$

Since the speeds of the atoms can be directly measured by the molecular beam experiment (this will be discussed later in this chapter), the quantity U_K is a directly measurable quantity. Let us assume that we have measured U_K for a mole of argon gas at the freezing point of water and at a pressure of 1 atm. We could then compare the *measured* kinetic energy with the calculated potential energy for a real gas owing to attractive and repulsive interactions between the atoms. The latter would be negligibly small compared to the former at low pressure as already discussed in Example 9.5.2.

It is also known that for experiments carried out on essentially ideal gases at constant temperature,

$$PV = \text{constant}, \qquad (17.2.3)$$

where P is pressure and V is volume; this is **Boyle's law**. Suppose we have now measured the product PV for a mole of argon gas and also U_K. Direct comparison shows that

$$PV = \tfrac{2}{3}U_K = \tfrac{2}{3}\sum_{i=1}^{N} \tfrac{1}{2}mv_i^2. \qquad (17.2.4)$$

Both of these quantities, i.e., PV and U_K, could also be measured at a higher temperature, say the boiling point of water. Equation (17.2.4) would still hold but U_K (and PV) would be larger. The random motion of the argon atoms, i.e., the random kinetic energy U_K, can be used as a quantitative measure of temperature. The **ideal gas temperature scale** is based on the definition

$$\tfrac{2}{3}U_K \equiv NkT = RT. \tag{17.2.5}$$

Here k is a constant, called the **Boltzmann constant**, whose value depends on the units used for U_K and T; thus

$$k = 1.38 \times 10^{-16} \text{ ergs/}^\circ\text{K}$$

if T is *defined* to be 273.16°K at the triple point of water (see Figure 16.2.1). (*Note:* T could have been defined to be equal to U_K of an ideal gas, where U_K is measured in ergs; then the gas law would be $PV = \tfrac{2}{3}T$ and the temperature unit would be the erg.)

Boyle's law (17.2.3) was first stated in 1660. It was correctly derived [meaning Equation (17.2.4) was theoretically proved] by Daniel Bernoulli in 1738 by considering the momentum change of atoms colliding with a container wall.

The actual measurement of the speeds was first made by A. A. Michelson in 1895 using Doppler broadening (studying the shift in frequency of radiation emitted from moving atoms) and by O. Stern in 1926 using molecular beam techniques.

We note that the ideal gas law for a mole of gas is an **equation of state**. Thus the volume is a uniquely determined function of the fields P and T; i.e., $V = V(P, T)$. The material may be taken through various P, V, and T cycles, but when it is returned to the original P and T (assuming it is still the same material and has not decomposed) it will have the same volume as before under equilibrium conditions. We say that V is a **state function** since it depends only on the field variables P and T (in this case) and not on the path (i.e., the sequence of pressures and temperatures) needed to reach V.

17.3   RANDOM DISTRIBUTION

Probability plays an important role in describing a system consisting of a large number of particles. It is also important in the insurance business, in quality control in manufacturing, in error analysis of experiments, in the motion of traffic, etc. In this section probability is studied with a view toward applying it to materials science, but the results are applicable to many fields (see Kac, References).

Consider a simple experiment called a **one-dimensional random walk**. A single particle undergoes successive displacements $l_1, l_2, l_3, \ldots$ in the x or $-x$ directions; all the displacements have the same magnitude, $|l_1| = |l_2| = \ldots l$. Each displacement is equally likely to be in the x or $-x$ direction. The following examples and discussion are concerned with describing the probable position of the particle after N steps.

EXAMPLE 17.3.1

If the number of steps $N = 4$, show that the total number of possible arrangements of plus and minus steps is $2^4 = 16$.

Answer.

$$
\begin{array}{llll}
+\,+\,+\,+ & -\,+\,+\,+ & +\,-\,-\,- & -\,-\,-\,- \\
+\,+\,+\,- & -\,+\,+\,- & +\,-\,-\,+ & -\,-\,-\,+ \\
+\,+\,-\,+ & -\,+\,-\,+ & +\,-\,+\,- & -\,-\,+\,- \\
+\,+\,-\,- & -\,+\,-\,- & +\,-\,+\,+ & -\,-\,+\,+.
\end{array}
$$

Thus one specific arrangement, for example, $+\,-\,+\,-$, is a fraction, $\frac{1}{16}$, of all arrangements.

The total number of possible arrangements of plus and minus sequences when there are N steps is

$$\omega = 2^N. \tag{17.3.1}$$

Thus one *specific* arrangement is a fraction $1/2^N$ of all the arrangements.

EXAMPLE 17.3.2

Suppose you have three boxes in a row and three different balls labeled A, B, and C. What are the total number of distinct arrangements if one ball is put in each box?

Answer. 3!.

They are:

$$
\begin{array}{ll}
\text{ABC} & \text{ACB} \\
\text{BAC} & \text{CAB} \\
\text{BCA} & \text{CBA.}
\end{array}
$$

For N boxes and N different balls the total number of arrangements would be $N!$.

If there are N boxes, $N - M$ red balls, and M black balls, then the number of distinct arrangements is

$$\frac{N!}{(N - M)!\,M!}. \tag{17.3.2}$$

EXAMPLE 17.3.3

Suppose $N = 3$ and $M = 2$. How many distinct arrangements are there?

Answer. 3, as predicted from $3!/(1!\,2!)$.

Note that in Example 17.3.2, if $B = C$ we would have

$$ABB \quad \text{ABB}$$
$$BAB \quad \text{BAB}$$
$$BBA \quad \text{BBA}.$$

We have crossed out arrangements already counted.

Let us return now to the random walk problem. Let N be the †
ber of steps and let us assume that N and p are even numbers; the
positive steps and $(N - p)/2$ negative steps would lead to a tot⸱
$x = pl$. There are then

$$\frac{N!}{[(N + p)/2]! \, [(N - p)/2]!} \tag{.3}$$

distinguishable arrangements. The probability of each specific arrangement is
$1/2^N$ so that the probability of finding the particle at $x = pl$ is

$$P(pl) = \frac{1}{2^N} \frac{N!}{[(N + p)/2]! \, [(N - p)/2]!}. \tag{17.3.4}$$

EXAMPLE 17.3.4

Suppose $N = 4$. What is the probability of $x = 2l$?. Use both Example
17.3.1 and Equation (17.3.4).

Answer. The possible arrangements from Example 17.3.1 for
$p = 2$ are

$$+++- \quad -+++$$
$$++-+ \quad +-++.$$

Thus there are four, and hence the fraction of all possible arrangements
which gives $x = 2l$ is $\frac{4}{16} = \frac{1}{4}$. Likewise,

$$P(2l) = \frac{1}{2^4} \frac{4!}{3! 1!} = \frac{1}{4}.$$

Suppose $N = 10$ and you were asked to plot $P(pl)$, where p is even, vs. p
[this would be an interesting exercise using (17.3.4)]. If $N = 100$, could you do
it if you had a digital computer at your disposal? If $N = 6.02 \times 10^{23}$, as it
would be if we were dealing with a mole of atoms, what would be your approach?
Fortunately, when N is large we can use the asymptotic expansion (students will
study this in advanced calculus or analysis courses) for $N!$, known as Stirling's
approximation:

$$N! \approx N^N e^{-N} \sqrt{2\pi N} \left(1 + \frac{1}{12N} + \frac{1}{288N^2} + \cdots \right). \tag{17.3.5}$$

For large N the term in square brackets is essentially equal to 1. Using this
approximation for each of the factorials of (17.3.4), we obtain

$$P(pl) = \sqrt{\frac{2}{\pi N}} e^{-p^2/2N}. \tag{17.3.6}$$

Since we are dealing with extremely large values of N, we might just as well replace the above function by a continuous function. $P(pl)$ could be regarded as the probability of the particle being located between $x = pl$ and $x = (p + 2)l$ (since p was assumed to be even). Hence the probability that the particle is located between x and $x + dx$ is

$$P(x)\,dx = P(pl)\frac{dx}{2l}$$

or

$$P(x)\,dx = \frac{1}{\sqrt{2\pi N l^2}}e^{-x^2/2Nl^2}\,dx. \qquad (17.3.7)$$

We define

$$\sigma^2 = Nl^2. \qquad (17.3.8)$$

Then

$$P(x)\,dx = \frac{1}{\sqrt{2\pi\sigma}}e^{-x^2/2\sigma^2}\,dx. \qquad (17.3.9)$$

The reader should sketch P vs. x to familiarize himself with the nature of this function.

EXAMPLE 17.3.5

Why is it true that

$$\int_{-\infty}^{\infty} P(x)\,dx = 1? \qquad (17.3.10)$$

Answer. This is simply a statement to the effect that the probability of finding the particle somewhere on the line is 1. A standard definite integral which the student may use to check (17.3.10) is

$$\int_{-\infty}^{\infty} e^{-y^2}\,dy = \sqrt{\pi}.$$

Note that (17.3.9) could be written as

$$P(x)\,dx = \frac{e^{-x^2/2\sigma^2}\,dx}{\displaystyle\int_{-\infty}^{\infty} e^{-x^2/2\sigma^2}\,dx}. \qquad (17.3.11)$$

The function $P(x)$ is called the **random distribution function**. Note that this is an even function of x.

EXAMPLE 17.3.6

Why does it follow that the average value of x is $\bar{x} = 0$?

Answer. By definition

$$\bar{x} = \int_{-\infty}^{\infty} xP(x)\,dx. \qquad (17.3.12)$$

[Note the comparison with the discrete case where n_i is the number of particles having displacement x_i. Let $\sum n_i = N$. Then

$$\bar{x} = \frac{\sum x_i n_i}{\sum n_i} = \sum x_i \frac{n_i}{N} = \sum x_i f_i, \qquad (17.3.13)$$

where f_i is the fraction of particles having displacement x_i.] The product $xP(x)$ is an odd function so this particular integral must be zero.

Although $\bar{x} = 0$, $\overline{x^2} \neq 0$ since

$$\overline{x^2} = \int_{-\infty}^{\infty} x^2 P(x)\, dx = \sigma^2. \qquad (17.3.14)$$

Use has been made of the fact that

$$\int_{-\infty}^{\infty} y^2 e^{-y^2}\, dy = \frac{\sqrt{\pi}}{2}. \qquad (17.3.15)$$

Then from (17.3.8) and (17.3.14),

$$\sqrt{\overline{x^2}} = \sqrt{N} l = \sigma. \qquad (17.3.16)$$

This is a result we discussed earlier in Chapter 14 for the root mean squared free length of a random polymer chain.

THE THREE-DIMENSIONAL RANDOM WALK. We consider now a random walk on a simple cubic lattice. Steps along cube edges only are allowed. We assume that for each plus or minus step along the x_1 direction, there is a plus or minus step along the x_2 direction and also a plus or minus step along the x_3 direction. Note that the probability of finding the particle between x_1 and $x_1 + dx_1$ is still given by (17.3.9), with x_1 replacing x. However, the same type of function applies to the probability of finding the particle between x_2 and $x_2 + dx_2$, etc. Thus the overall probability of finding the particle simultaneously between x_1 and $x_1 + dx_1$, between x_2 and $x_2 + dx_2$, and between x_3 and $x_3 + dx_3$, i.e., in the volume element $dx_1\, dx_2\, dx_3$, is given by the product of the independent probabilities. Hence

$$\begin{aligned} P(x_1, x_2, x_3)\, dx_1\, dx_2\, dx_3 &= P(x_1)\, dx_1\, P(x_2)\, dx_2\, P(x_3)\, dx_3 \\ &= P(x_1)P(x_2)P(x_3)\, dx_1\, dx_2\, dx_3 \\ &= \frac{1}{(2\pi\sigma^2)^{3/2}} e^{-(x_1^2 + x_2^2 + x_3^2)/2\sigma^2}\, dV, \end{aligned} \qquad (17.3.17)$$

where dV is the volume element.

We are now interested in the probability of finding the particle between two concentric shells of radius r and $r + dr$. Then $dV = 4\pi r^2\, dr$ and

$$P(r)\, dr = \frac{1}{(2\pi\sigma^2)^{3/2}} e^{-r^2/2\sigma^2}\, 4\pi r^2\, dr. \qquad (17.3.18)$$

$P(r)$ is the **radial distribution function**. The reader should now sketch $P(r)$ vs. r to familiarize himself with the nature of this function.

EXAMPLE 17.3.7

In the one-dimensional case the most probable value of x is at $x = 0$. What is the most probable value of r?

Answer. This corresponds to the value of r for which $P(r)$ is a maximum, i.e., for which $dP/dr = 0$. This gives $r = \sqrt{2}\,\sigma$.

The random distribution has application to the random distribution of velocity components in a gas (Section 17.4), to the random configuration of polymer chains and solid solutions, to diffusion of atoms in gases, liquids, or solids, etc.

17.4   VELOCITY DISTRIBUTION IN IDEAL GASES

Maxwell (Figure 17.4.1) postulated in 1859 that the distribution of velocity components in an ideal gas is a random distribution. In such a case the distribution of speeds would be given by

$$P(c)\,dc = \frac{1}{(2\pi\sigma^2)^{3/2}}e^{-c^2/2\sigma^2}\,4\pi c^2\,dc. \qquad (17.4.1)$$

[Note the analogy with (17.3.18), where $r^2 = x_1^2 + x_2^2 + x_3^2$; here $c^2 = v_{x_1}^2 + v_{x_2}^2 + v_{x_3}^2$.]

FIGURE 17.4.1. James Clerk Maxwell (1831–1879). Maxwell made fundamental contributions to the electromagnetic theory, macroscopic thermodynamics, and microscopic thermodynamics. Before becoming a professor at Cambridge University in 1871, he was at the University of Aberdeen in Scotland. [From G. Holton and D. Roller, *Foundations of Modern Physical Science*, Addison-Wesley, Reading, Mass. (1958).]

The kinetic energy in such a case would be given by

$$U_K = \frac{3N}{2}kT = N\frac{\overline{mc^2}}{2},$$ (17.4.2)

as noted in (17.2.4) and (17.2.5). Hence we need to evaluate $\overline{c^2}$. We have

$$\overline{c^2} = \int_0^\infty c^2 P(c)\, dc = 3\sigma^2.$$ (17.4.3)

Here we have used the fact that

$$\int_0^\infty c^4 e^{-c^2/2\sigma^2}\, dc = 3\left(\frac{\pi}{2}\right)^{1/2}\sigma^5.$$ (17.4.4)

Combining (17.4.2) and (17.4.3) gives

$$\sigma^2 = \frac{kT}{m},$$ (17.4.5)

and hence the root mean squared velocity,

$$c_{rms} = (\overline{c^2})^{1/2} = \left(\frac{3kT}{m}\right)^{1/2} = \left(\frac{3RT}{M}\right)^{1/2},$$ (17.4.6)

where M is the molecular weight.

The exponential term in $P(c)$ now takes the form

$$e^{(-mc^2/2kT)} = e^{-u/kT}.$$ (17.4.7)

We have noted in Chapter 6 the repeated occurrence of an exponential temperature dependence of this sort.

EXAMPLE 17.4.1

Calculate $\bar{c}$ for an ideal gas.

Answer.

$$\bar{c} = \int_0^\infty c P(c)\, dc.$$

Using the fact that

$$\int_0^\infty c^3 e^{-c^2/2\sigma^2}\, dc = \frac{\sigma^4}{2},$$

we have

$$\bar{c} = \left(\frac{8kT}{\pi m}\right)^{1/2} = \left(\frac{8RT}{\pi M}\right)^{1/2}.$$ (17.4.8)

Note that $\bar{c}$ is very nearly equal to c_{rms}. Some results are shown in Table 17.4.1.

THE MOLECULAR BEAM EXPERIMENT. In 1926 Stern carried out an experiment to determine $P(c)$ directly. His setup is shown in Figure 17.4.2. The Hg is heated to a constant temperature in the furnace at A. There is a small opening in

Table 17.4.1. AVERAGE MOLECULAR SPEEDS AT 0°C

Gas	$\bar{c}$, m/sec
Hg	170.0
Ar	380.8
O_2	425.1
N_2	454.2
He	1204.0
H_2	1692.0

this container which allows the vapor to escape. The vapor which passes through the slit forms a beam parallel to the axis which supports the rotating discs **B** and **C**. Each disc has a slit. For a given rotational speed only atoms whose speeds fall in a narrow range pass through and hit the cold mirror at **D**, where they

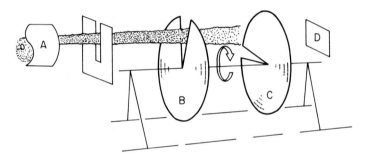

FIGURE 17.4.2. Stern's molecular beam experiment.

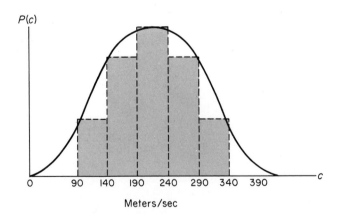

Meters/sec

FIGURE 17.4.3. Distribution of atom speeds for mercury vapor at 100°C. Dashed lines represent data. Solid curve is best fit, according to Equation (17.4.1).

condense and stick. After a given time, the experiment is stopped and the quantity of mercury on the mirror is measured. The experiment is then repeated at a different rotational speed where atoms with a different range of speeds get through. The results obtained by Lammert [B. Lammert, *Z. Physik* **56**, 244 (1929)] are shown in Figure 17.4.3. More recent high-precision measurements have been made by R. C. Miller and P. Kusch, *Physical Review* **99**, 1314 (1955).

17.5 INTERNAL ENERGY

MICROSCOPIC VIEWPOINT.   The **internal energy** U of a system is the sum of the kinetic energy and the potential energy of the atoms with the center of gravity of the entire system fixed and no rigid body rotation.

EXAMPLE 17.5.1

What is the internal energy of an ideal monatomic gas?

Answer. The potential energy is zero and the translational kinetic energy is given by U_K in Equation (17.2.4).

EXAMPLE 17.5.2

What is the internal energy of nitrogen gas at low pressure near room temperature?

Answer. This would approximate an ideal gas since the potential interactions between the molecules are very small. If the reference point for zero potential energy was the isolated nitrogen atoms, then the bond energy would be a potential energy term. Alternately the static energy of the molecule at absolute zero could be chosen as the zero potential energy.

The molecules are in constant motion so that there is translational kinetic energy, rotational kinetic energy, and vibrational energy (which would be both kinetic and potential since the total energy of a classic harmonic oscillator is $\frac{1}{2}mv^2 + \frac{1}{2}kx^2$). The sum of all of these energies is the internal energy.

The various molecular motions of a diatomic molecule are illustrated in Figure 17.5.1.

EXAMPLE 17.5.3

Discuss the internal energy of a copper crystal.

Answer. If the widely separated copper atoms are assigned zero potential energy, then the binding energy represents a potential energy term. We might evaluate this binding energy for the static case (atoms not

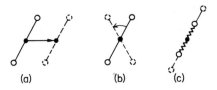

(a)   (b)   (c)

FIGURE 17.5.1. Motion of nitrogen molecule. Black dot shows the
center of gravity. (a) Translation of center of gravity.
(b) Rotation about center of
gravity. (c) Vibration about
center of gravity.

vibrating). However, the atoms in a crystal vibrate about their lattice
positions. If there are N atoms, there are $3N$ vibrations in all (each atom
can potentially vibrate in x, y, and z directions) and these can each be
approximated by a harmonic oscillator. A **harmonic oscillator** is a linear
spring plus a mass (see Section 2.4). It has an equation of motion

$$M\frac{d^2x}{dt^2} + kx = 0$$

and an energy

$$\frac{M}{2}\left(\frac{dx}{dt}\right)^2 + \frac{kx^2}{2}.$$

The problem of actually calculating the internal energy of these different
systems is left for Section 17.11. In the case of an ideal gas, the internal energy
depends only on the temperature. In the more general case it is a function of the
state of the system, which is determined by the fields, e.g., pressure and temperature (and other fields, if present).

MACROSCOPIC CONCEPT OF INTERNAL ENERGY.   The change in the internal
energy is simply the sum of all the different kinds of energy added to the system.
Thus the *increase in the internal energy* includes the heat ΔQ added to the system
and all the work ΔW done on the system. Thus

$$\Delta U = \Delta Q + \Delta W. \tag{17.5.1}$$

There are many ways of doing work on the system: compressing it, deforming
it in other ways, polarizing it, magnetizing it, etc. We are particularly interested
in those cases in which the changes are small and for which we can write

$$dU = đQ + đW. \tag{17.5.2}$$

[The symbol $đQ$ is a differential whose meaning is different than that of the
differential dU. This difference and the conditions for which (17.5.2) holds will
be illustrated next.]

We consider two experiments in which 1 mole of an ideal monatomic gas
is compressed. Assume that initially it is at P_0, V_0, T_0. We suppose that the gas

is in a cylinder whose wall is a perfect thermal insulator, for which the piston is a perfect insulator, and whose base can conveniently be made a perfect thermal insulator or a thermal conductor by flipping a switch. The entire assembly is placed in a constant temperature bath at temperature T_0. In the first experiment the base is a thermal conductor. The gas is then compressed very slowly (when an ideal gas is compressed rapidly its temperature rises) so that its temperature is always arbitrarily close to T_0. (Note the reason for the *slowness* of the compression. The gas has a finite thermal conductivity; the base has a finite thermal conductivity. Thus a finite time is required for heat to flow out of the system when the gas is compressed. If the compression is rapid, the gas must therefore heat up.) It is compressed to a final volume V_f. A process in which the temperature does not change is called an **isothermal process**. In the ideal gas case the pressure and volume per mole during the isothermal process are related by

$$PV = \text{constant} = RT_0. \tag{17.5.3}$$

In the second experiment the base is a thermal insulator. Thus when the gas is now compressed from P_0, V_0, T_0 to a final volume V_f there is no heat flow into or out of the system. Such a process involving no heat flow is called an **adiabatic process**. In an ideal gas the pressure and volume during an adiabatic process are related by

$$PV^\gamma = \text{constant} = P_0 V_0^\gamma. \tag{17.5.4}$$

For an ideal monatomic gas $\gamma = 1.6$. (The adiabatic compression must be *sufficiently slow* so that the pressure everywhere in the cylinder is at all times arbitrarily close to being uniform. This means that the piston velocity must be slow relative to the speed of sound. For example, if the piston moves at half the speed of sound, the gas at the base would remain at P_0, V_0, T_0 until the piston has moved to decrease the cylinder volume to $V_0/2$.)

We emphasize the fact that for the *P-V* behavior to be specified, as it is for these two processes by Equations (17.5.3) and (17.5.4), there is a *restriction on the rate at which these two processes can be carried out*. Since the specific paths are then followed in *either* direction [e.g., (17.5.3) applies along *OA* or

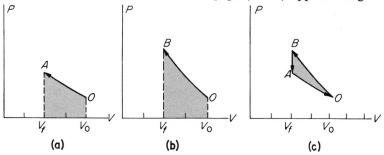

FIGURE 17.5.2. (a) Isothermal compression. (b) Adiabatic compression. (c) Cycle of adiabatic compression, constant volume decompression, and isothermal expansion.

along AO and (17.5.4) along OB or along BO], the processes are called **reversible processes**. The isothermal compression is illustrated in Figure 17.5.2(a) and the adiabatic compression in Figure 17.5.2(b). The shaded area in (a) is the work done during the isothermal process. The work done during the isothermal compression is

$$\Delta W_{\text{iso}} = -\int_{V_0}^{V_f} P \, dV = -RT_0 \int_{V_0}^{V_f} \frac{dV}{V}$$
$$= -RT_0 \ln \frac{V_f}{V_0}, \tag{17.5.5}$$

where we have used Equation (17.5.3). The work done during the adiabatic process is

$$\Delta W_{\text{ad}} = -\int_{V_0}^{V_f} P \, dV = -P_0 V_0^\gamma \int_{V_0}^{V_f} \frac{dV}{V^\gamma}$$
$$= \frac{P_0 V_0}{\gamma - 1}\left[\left(\frac{V_0}{V_f}\right)^{\gamma-1} - 1\right] \tag{17.5.6}$$
$$\Delta W_{\text{ad}} = \frac{RT_0}{\gamma - 1}\left[\left(\frac{V_0}{V_f}\right)^{\gamma-1} - 1\right].$$

This is the shaded area of Figure 17.5.2(b).

EXAMPLE 17.5.4

Compute ΔW_{ad} and ΔW_{iso} for $V_0/V_f = 10$ in units of RT_0 for $\gamma = 1.6$.

Answer.
$$\Delta W_{\text{iso}} = 2.3RT_0$$
$$\Delta W_{\text{ad}} = 4.97RT_0.$$

EXAMPLE 17.5.5

How does one carry out an adiabatic compression in the real world where perfect thermal insulation does not exist?

Answer. If the fluid is compressed fairly rapidly, the time available for heat to flow out is very small and hence the process is approximately adiabatic.

In the present case $\Delta W_{\text{ad}} > \Delta W_{\text{iso}}$. This is also true of nonideal gases, liquids, and solids, but the difference becomes progressively less and is quite small for solids.

EXAMPLE 17.5.6

Calculate the final temperature after the adiabatic compression of an ideal monatomic gas to $V_0/V_f = 10$.

Answer. We have

$$P_0 V_0^\gamma = P_f V_f^\gamma$$

or

$$(P_0 V_0)V_0^{\gamma-1} = (P_f V_f)V_f^{\gamma-1}.$$

But $P_0 V_0 = RT_0$ in the initial state and $P_f V_f = RT_f$ in the final state. Hence

$$T_f = T_0 \left(\frac{V_0}{V_f}\right)^{\gamma-1}. \tag{17.5.7}$$

For $V_0/V_f = 10$ we have $T_f = 3.98 T_0$.

Let us now consider the cycle shown in Figure 17.5.2(c). There is no work done on the system along the path BA since there is no volume change; $\Delta W_{BA} = 0$. The work done in the cycle is

$$\Delta W_{\text{cycle}} = \Delta W_{OB} + \Delta W_{BA} + \Delta W_{AO} = \Delta W_{OB} + \Delta W_{AO}.$$

Hence

$$\oint dW = \Delta W_{\text{cycle}} = \Delta W_{\text{ad}} - \Delta W_{\text{iso}} \neq 0. \tag{17.5.8}$$

The integral around the closed path is simply equal to the shaded region of Figure 17.5.2(c).

Since the internal energy of an ideal gas depends only on the temperature [see Equation (17.2.5)], it is clear that

$$\oint dU = \Delta U_{\text{cycle}} = \Delta U_{OB} + \Delta U_{BA} + \Delta U_{AO} = 0. \tag{17.5.9}$$

This is true not only for the ideal gas but for any system; i.e., the internal energy of a system depends only on the state of the system (P, V, T) and is independent of the path taken to achieve that state.

The differential dU is called an **exact differential** [since (17.5.9) applies], while the differential dW is not exact.

EXAMPLE 17.5.7

a. Suppose

$$dU = x\, dx + y\, dy.$$

Is dU an exact differential?

Answer. Note that we can write

$$dU = d\left(\frac{x^2 + y^2}{2}\right);$$

i.e., dU is the differential of a known function. When this is true dU is called an exact differential. It then follows that

$$\oint dU = 0.$$

b. Is

$$dQ = x^2 \, dx + yx \, dy$$

an exact differential?

Answer. No. But if we multiply this by the function $1/x$ we have $dQ/x = x \, dx + y \, dy$, which is an exact differential. We call $1/x$ an **integrating factor**.

c. Given a differential

$$dU = M(x, y) \, dx + N(x, y) \, dy,$$

it can be shown that dU is an exact differential if and only if

$$\left(\frac{\partial M}{\partial y}\right)_x = \left(\frac{\partial N}{\partial x}\right)_y. \tag{17.5.10}$$

Use this to show that the differential dU in part a is exact, while dQ in part b is not.

Answer. For part a we have

$$\left(\frac{\partial x}{\partial y}\right)_x = 0, \qquad \left(\frac{\partial y}{\partial x}\right)_y = 0$$

and hence (17.5.10) holds. For part b we have

$$\left(\frac{\partial x^2}{\partial y}\right)_x = 0, \qquad \left[\frac{\partial(yx)}{\partial x}\right]_y = y$$

and hence (17.5.10) does not hold.

It is easy to show that

$$\Delta Q_{\text{cycle}} = Q_{BA} + Q_{AO} \neq 0,$$

i.e.,

$$\oint dQ \neq 0, \tag{17.5.11}$$

inasmuch as $\oint dU = 0$ and $\oint dW \neq 0$ and since Equation (17.5.2) holds. Thus neither dW nor dQ are exact differentials, but dU is. Likewise, W and Q are not state functions, but U is. Thus the equation

$$dU = dQ + dW$$

applies to those processes which are carried out *reversibly*.

We note that dW is the product of a field (pressure in the present case) times the differential of a state function (volume). (The sign in $-P \, dV$ is negative only because of our convention in choosing to describe pressure as positive if it causes a volume decrease.) We are therefore led to ask, can dQ for a reversible process also be expressed as the product of a field (which implies temperature) times the differential of a state function? This is indeed the case, as will be discussed in Section 17.6.

17.6   RANDOMNESS AND ENTROPY

MICROSCOPIC VIEWPOINT.   Consider a system consisting of two tanks of gas at the same pressure and temperature, as in Figure 17.6.1. One tank contains helium, the other neon. Assume that $P = 1$ atm and that $T = 300°K$. Then each

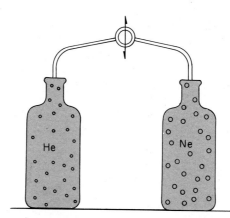

FIGURE 17.6.1. Two tanks of gas.

gas behaves as an ideal gas: the internal energy per mole of each of these monoatmic gases is simply $\frac{3}{2}RT$, which equals the kinetic energy (the potential energy is zero). The gas in each tank has the same value of $\bar{U}$ (internal energy per mole), the same P and T, and the same $\bar{V}$ (molar volume).

Suppose the valve between the two tanks is now opened. The two gases will eventually form a solution, as illustrated in Figure 17.6.2. (The solution is often called a mixture, and the process of interdiffusion, mixing.) There is no change in total pressure P or temperature T (pressure gauges and temperature sensors show no change). The internal energy U of the total system is also

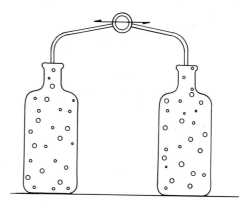

FIGURE 17.6.2. Solution of He and Ne gases.

unchanged; clearly no work was done (no displacements in the environment); moreover, no heat was added or subtracted; hence $\Delta U = 0$. Although U, P, and T remained constant, there was a profound change in the system! What did change? Why did the change occur?

EXAMPLE 17.6.1

Air contains about 4 parts of N_2 to 1 part of O_2. Some manufacturers go to a great deal of trouble (*work*) to separate these two components. Other manufacturers buy the oxygen gas for such applications as oxyacetylene welding. Others may buy the nitrogen and, after liquifying it, use it to maintain a constant temperature of 77°K. In either case the purchaser pays for the cost of the original separation.

What is lost if the separated gases both at P and T and in 4:1 volume ratio are allowed to diffuse together to form air?

Answer. The work done in the separation.

The fundamental change which occurred in mixing of the gases was the overall increase in randomness of the total system.

All students believe that the two gases of Figure 17.6.1 would diffuse together to form a solution as in Figure 17.6.2. (It seems to be accepted as a fact of life, like taxes and, perhaps, death.) There must therefore be a fundamental belief that *nature strives to attain maximum randomness* (at least in the case of the gas at room temperature).

To measure this randomness we define a quantity ω which is the total number of **distinct configurations** (or arrangements) of the system.

As an illustration consider the number of distinct arrangements of two pennies in a row which may be either heads or tails. The distinct arrangements are

HH TH

HT TT.

EXAMPLE 17.6.2

Four pennies are arranged in a row. Each may be either heads or tails. A student makes one arrangement which is recorded as HTTH. He then makes another arrangement HTTT. How many distinct arrangements are there in all?

Answer.

$$\omega = 2^4 = 16.$$

If he had used M pennies, there would be 2^M arrangements, which are distinct.

EXAMPLE 17.6.3

Molecules of NO (nitrous oxide) form a solid. Assume that there is no difference in the binding energy if a molecule occurs as NO or ON as marked by the asterisk. Suppose they form one-dimensional chains as shown below:

$$NO----NO----\overset{*}{NO}----NO$$

$$NO----NO----\overset{*}{ON}----NO.$$

The internal energy of these two configurations is the same. If there is a total of M molecules in a row, how many configurations are there in all?

Answer.

$$\omega = 2^M.$$

EXAMPLE 17.6.4

Suppose there are $N + n$ boxes and a different molecule is placed in each box. How many distinct arrangements are there in all?

Answer.

$$(N + n)!.$$

Illustration. Let $N + n = 2$. Let the molecules be He and Ne. Then there are 2! arrangements:

NeHe

HeNe

Let $N + n = 3$. Let the molecules be He, Ne, and Ar. Then there are 3! arrangements or

NeHeAr   HeNeAr   ArNeHe

NeArHe   HeArNe   ArHeNe

EXAMPLE 17.6.5

Suppose there are $N + n$ boxes, N Ne molecules, and n He molecules. One molecule is placed in each box. How many distinct arrangements are there?

Answer.

$$\omega = \frac{(N + n)!}{N! n!}.$$

Note: Interchanging Ne molecules with other Ne molecules and He molecules with other He molecules does not give new arrangements, which is why we divide by $N!$ and $n!$, respectively.

Consider the two tanks of Figure 17.6.1 to be of different initial volumes but the same initial P and T; then there would be N atoms of helium, say, and n atoms of neon. Suppose we divide the total volume V into elements of volume $V/(N + n)$; there are then $N + n$ such boxes. Each has a gas particle in it on the average. Hence the total number of configurations in Figure 17.6.2 is

$$\omega = \frac{(N + n)!}{N!n!},$$

(17.6.1)

just as in Example 17.6.5.

EXAMPLE 17.6.6

If $n = N$, what is the number of configurations?

Answer. This can be obtained from (17.6.1). We have

$$\omega = \frac{(2N)!}{N!N!}.$$

The **entropy** of a system can be shown to be

$$S = k \ln \omega,$$

(17.6.2)

where k is Boltzmann's constant. This quantity is a measure of randomness which is thermodynamically useful.

What is the entropy change upon mixing the two ideal gases each at pressure P and temperature T if there are N atoms of helium and n atoms of neon? Combining Equations (17.6.1) and (17.6.2) we have

$$\Delta S_{\text{mixing}} = k \ln \frac{(N + n)!}{N!n!}.$$

Using Stirling's approximation (17.3.5) it can be shown that

$$\Delta S_{\text{mixing}} = -k(N + n) \sum_{i=1}^{2} x_i \ln x_i,$$

(17.6.3)

where

$$x_1 = \frac{n}{N + n} \quad \text{and} \quad x_2 = \frac{N}{N + n}.$$

(17.6.4)

Hence x_1 and x_2 are mole fractions.

For 1 mole of gas (total),

$$\Delta S_{\text{mixing}} = -R \sum_{i=1}^{2} x_i \ln x_i.$$

(17.6.5)

The **gas constant** R is the product of Boltzmann's constant and Avogadro's number. A plot of ΔS_{mixing} vs. x_2 is shown in Figure 17.6.3.

EXAMPLE 17.6.7

Show that the initial slope of ΔS_{mixing} vs. x_2 is infinite.

Answer. From (17.6.5)

$$\frac{\partial \Delta S_{\text{mixing}}}{\partial x_2} = R \ln \frac{1 - x_2}{x_2}$$

and

$$\lim_{x_2 \to 0} R \ln \left(\frac{1 - x_2}{x_2}\right) = \infty.$$

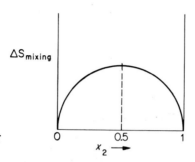

FIGURE 17.6.3. Entropy of mixing.

If two gases such as He and Ne are mixed together as in Figure 17.6.2, the work to separate them (to eliminate the randomness or the entropy) is at least $\Delta W = T \Delta S_{\text{mixing}}$, as will be shown later.

It is possible to get distinct configurations which depend not only on geometric configuration as with mixing but also on the distribution of particles among energy levels. This will be discussed later.

MACROSCOPIC VIEWPOINT OF ENTROPY.   In this subsection we shall show that when a process is carried out reversibly

$$đQ = T \, dS,$$

where S, the entropy, is a state function; i.e.,

$$\oint dS = 0.$$

In other words $1/T$ is an integrating factor for $đQ$, the product $đQ/T$ being an exact differential.

To show this we consider a special thermodynamic cycle called the **Carnot cycle**. This cycle is named after the French engineer Sadi Carnot (see Figure 17.6.4). A Carnot cycle is illustrated in Figure 17.6.5. It consists of an isothermal expansion at T_2 (A to B); an adiabatic expansion to T_1, so $T_1 < T_2$ (B to C); an isothermal compression at T_1 (C to D); and an adiabatic compression back to the original starting point at T_2 (D to A).

The work done on the system in this cycle is

$$\oint đW = \Delta W_{AB} + \Delta W_{BC} + \Delta W_{CD} + \Delta W_{DA} \neq 0$$

FIGURE 17.6.4. N. L. Sadi Carnot (1796–1832). In 1824, Carnot published a detailed analysis of heat engines in which he assumed that heat was just another form of energy. His work formed the basis of the development of the second law of thermodynamics by Clausius and Kelvin. [From Sadi Carnot, in *Reflection on the Motive Power of Fire* (E. Mendoza, ed.), New York (1960)]

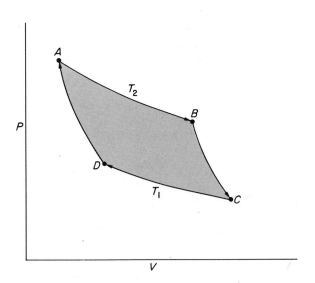

FIGURE 17.6.5. Carnot cycle.

since it equals the shaded area. The heat added to the system in the cycle is

$$\oint dQ = \Delta Q_{AB} + \Delta Q_{CD}$$

since ΔQ_{BC} and ΔQ_{DA} equal zero inasmuch as BC and DA are adiabatic processes. Let $\Delta Q_{AB} = Q_2$ and $\Delta Q_{CD} = Q_1$. Then since $\oint dU = 0$, we have

$$\oint dQ = -\oint dW = Q_2 + Q_1 \neq 0. \tag{17.6.6}$$

Here Q_2 is the heat added at T_2, while Q_1 is the heat added at T_1 (actually the system gives off heat at T_1 so Q_1 will have a negative value).

A heat engine is a machine to which we add heat at a high temperature and which then produces work. The heat added *to* the system is Q_2 and the work done *by* the system is $-\oint dW$. The efficiency of the engine is therefore

$$\text{efficiency} = \frac{-\oint dW}{Q_2} = \frac{\oint dQ}{Q_2} = \frac{Q_2 + Q_1}{Q_2}.$$

Since $\Delta U_{AB} = \Delta Q_{AB} + \Delta W_{AB} = 0$, we have, using (17.5.5),

$$Q_2 = -\Delta W_{AB} = -RT_2 \ln \frac{V_B}{V_A} \tag{17.6.7}$$

and, similarly,

$$Q_1 = -\Delta W_{CD} = -RT_1 \ln \frac{V_D}{V_C}. \tag{17.6.8}$$

We also have

$$\text{adiabatic processes:} \quad \begin{aligned} P_B V_B^{\gamma} &= P_C V_C^{\gamma} \\ P_A V_A^{\gamma} &= P_D V_D^{\gamma} \end{aligned}$$

$$\text{isothermal processes:} \quad \begin{aligned} P_A V_A &= P_B V_B \\ P_C V_C &= P_D V_D \end{aligned} \tag{17.6.9}$$

from which we can show that

$$\frac{V_B}{V_A} = \frac{V_C}{V_D}. \tag{17.6.10}$$

Hence

$$\text{efficiency} = \frac{Q_2 + Q_1}{Q_2} = \frac{T_2 - T_1}{T_2}. \tag{17.6.11}$$

This relation was quoted in Chapter 2 where we noted the reason engineers want to run heat engines at very high temperatures. It is emphasized that Equation (17.6.11) applies to the maximum possible efficiency of *every* process, e.g., the removal of salt from sea water by freezing, boiling, solvent extraction, or reverse osmosis (the result would be the same in all four cases). From (17.6.11) we have

$$\frac{Q_1}{Q_2} = \frac{-T_1}{T_2} \tag{17.6.12}$$

or

$$\frac{Q_2}{T_2} + \frac{Q_1}{T_1} = 0, \qquad (17.6.13)$$

which we might write as

$$\int_{AB} \frac{dQ}{T_2} + \int_{CD} \frac{dQ}{T_1} = 0$$

or as

$$\oint \frac{dQ}{T} = 0.$$

We define the new exact differential dQ/T as

$$dS = dQ/T.$$

The quantity S is called the **entropy**.

We can now write for the reversible process for which $dU = dQ + dW$ the expression

$$dU = T\,dS - P\,dV. \qquad (17.6.14)$$

EXAMPLE 17.6.8

Calculate the entropy change when 1 mole of an ideal gas is expanded isothermally from V_0 to V_f.

Answer. Since the internal energy of an ideal gas depends only on the temperature [Equation (17.2.5)], we have $dU = 0$, since the process is isothermal. Hence

$$dS = \frac{P\,dV}{T} = \frac{R\,dV}{V}$$

and

$$\Delta S = R \ln \frac{V_f}{V_0}. \qquad (17.6.15)$$

EXAMPLE 17.6.9

Consider two tanks of gas as in Figure 17.6.1 at the same initial pressure but with different volumes V_{He} and V_{Ne}, respectively. Calculate the entropy of mixing.

Answer. The final volume is $V_{He} + V_{Ne}$. Hence the entropy change of the neon is, from (17.6.15),

$$\Delta S_{Ne} = Rn_{Ne} \ln \frac{V_{He} + V_{Ne}}{V_{Ne}},$$

where n_{Ne} is the number of moles of neon. Similarly,

$$\Delta S_{He} = Rn_{He} \ln \frac{V_{He} + V_{Ne}}{V_{He}}.$$

Hence

$$\Delta S_{mixing} = \Delta S_{Ne} + \Delta S_{He} = -Rn \sum_{i=1}^{2} x_i \ln x_i, \qquad (17.6.16)$$

where $n = n_{He} + n_{Ne}$ and x_i is the mole fraction. Here, say,

$$x_1 = x_{He} = \frac{V_{He}}{V_{He} + V_{Ne}}$$

and

$$x_2 = x_{Ne} = \frac{V_{Ne}}{V_{He} + V_{Ne}}.$$

Compare (17.6.16) with (17.6.3), which in turn was based on (17.6.2). This suggests that (17.6.2) is the appropriate expression for entropy [recall that when (17.6.2) was given, we stated that it could be shown that this represents entropy].

17.7 EQUILIBRIUM AT CONSTANT TEMPERATURE AND PRESSURE

In a macroscopic mechanical system, such as the rectangular bar of Figure 15.3.1, the equilibrium position corresponds to a minimum of potential energy.

In a thermodynamic system which consists of a collection of atoms at constant pressure P and temperature T, internal equilibrium corresponds to a minimum of the Gibbs free energy (the proof of this statement follows). The **Gibbs free energy** G is *defined* as

$$G = U + PV - TS. \qquad (17.7.1)$$

This is also written as

$$G = H - TS, \qquad (17.7.2)$$

where

$$H = U + PV \qquad (17.7.3)$$

is the **enthalpy** of the system.

Thus equilibrium at a finite temperature corresponds to a situation in which H strives for a minimum, while S strives for a maximum (a maximum randomness).

From (17.7.1) we have

$$dG = dU + P\,dV + V\,dP - T\,dS - S\,dT,$$

which combined with (17.6.14) gives

$$dG = V\,dP - S\,dT. \qquad (17.7.4)$$

Thus for constant P and T (i.e., $dP = 0$, $dT = 0$) we have $dG = 0$, the condition for a minimum (or a maximum, but the former applies here). This is the reason the Gibbs free energy was defined according to (17.7.1).

EXAMPLE 17.7.1

What is the "chemical potential" which causes the two gases of Figure 17.3.1 to mix when the valve is opened?

Answer. In going from the unmixed state to the mixed state there is a decrease in the Gibbs free energy of $\Delta G_{mixing} = -T \Delta S_{mixing}$, where $\Delta S_{mixing} > 0$, since $\Delta U_{mixing} = 0$ and $P \Delta V_{mixing} = 0$.

The Gibbs free energy thus plays the same role in a chemical system at constant pressure and temperature that potential energy plays in macroscopic problems. In the macroscopic problem (see Figure 15.3.1) the potential energy decreases as the system approaches equilibrium. In the chemical system, the chemical potential (the Gibbs free energy per mole) decreases as the system approaches equilibrium.

EXAMPLE 17.7.2

What is the change in free energy when a mole of ideal gas is isothermally compressed from P_0 to P_f?

Answer. Since $dT = 0$ in (17.7.4),

$$\Delta G = \int_{P_0}^{P_f} V \, dP = RT \ln P_f/P_0. \qquad (17.7.5)$$

EXAMPLE 17.7.3

Water at 20°C has an equilibrium vapor pressure $P_0 = 0.023$ atm. Suppose water vapor in a container at 20°C with no liquid has a vapor pressure of $P = 2P_0$; i.e., it is supersaturated.

What is the "chemical potential" which tends to cause the vapor to condense?

Answer. Because of the low pressure the gas can be considered ideal. Hence, by (17.7.5), with $P_f = 2P_0$,

$$\Delta G = RT \ln \frac{P}{P_0} = RT \ln 2 = 0.69RT.$$

The **chemical potential** is the thermodynamic analog of a mechanical potential. It is the Gibbs free energy per molecule referred to some standard state. For an ideal gas the chemical potential is referred to 1 atm and hence

$$\mu = \mu_0 + kT \ln P/1 = \mu_0 + kT \ln P, \qquad (17.7.6)$$

where P is measured in atmospheres.

Note that this was obtained in the same way as was (17.7.5), which could be written as

$$G - G^0 = RT \ln P - RT \ln P_0$$

and if $P_0 = 1$, the last term goes to zero and G^0 is understood to mean the value of G at T and at $P = 1$. Then dividing by Avogadro's number N_0 gives

$$\mu - \mu_0 = \frac{G}{N_0} - \frac{G^0}{N_0} = kT \ln P.$$

The chemical potential of a component in one phase will be equal to the chemical potential of that component in a second phase if the whole system is in equilibrium.

This statement applies to multicomponent systems as well as to one-component systems. The Gibbs free energy of a phase which contains N_i molecules of the ith component is

$$G = \sum \mu_i N_i \qquad (17.7.7)$$

so that

$$\mu_i = \frac{\partial G}{\partial N_i}. \qquad (17.7.8)$$

We also write

$$G = \sum \bar{G}_i n_i, \qquad (17.7.9)$$

where n_i is the number of moles of the ith component. $\bar{G}_i$ is the molar chemical potential or the **partial molar Gibbs free energy** since

$$\bar{G}_i = \frac{\partial G}{\partial n_i}. \qquad (17.7.10)$$

EXAMPLE 17.7.4

Suggest a method for measuring the chemical potential of H_2O molecules in salt water at $300°K$ as a function of salt concentration C.

Answer. Measure the chemical potential of the water vapor in equilibrium with it. To do this measure the vapor pressure of the vapor which because it is so dilute will behave as an ideal gas. Then we have

$$\mu_v^{H_2O} = \mu_0 + kT \ln P.$$

Measure P vs. C and note that

$$\mu_v^{H_2O} = \mu_l^{H_2O}.$$

EXAMPLE 17.7.5

Make up a function F, called the **Helmholtz free energy**, which will have a minimum when V and T are constants. Start with (17.6.14).

Answer. We want to have

$$dF = 0 \qquad \text{when} \qquad dV = dT = 0.$$

Thus if we write

$$F = U - TS,$$

then
$$dF = dU - T \, dS - S \, dT,$$
which combined with (17.6.14) gives
$$dF = -P \, dV - S \, dT. \qquad (17.7.11)$$

STANDARD FREE ENERGY. The most stable form of an element in the standard state (1 atm, 25°C) is assigned a free energy of zero. The **standard free energy of formation** of a compound is the change in the Gibbs free energy when a mole of the compound in the standard state is formed from its elements in the standards state; thus
$$H_2(1 \text{ atm}) + \tfrac{1}{2}O_2(1 \text{ atm}) = H_2O(\text{gas}, 1 \text{ atm}).$$
Here $\Delta \bar{G}^0 = -54{,}638$ cal/mole. This means that H_2O vapor is more stable at 25°C by 54,638 cal/mole than the reactants. Extensive tables of standard free energies are available in reference books. See W. M. Latimer, *The Oxidation States of the Elements*, Prentice-Hall, Englewood Cliffs, N.J. (1952). The **free energy of a reaction** is defined as
$$\Delta G = \sum_i n_i \, \Delta \bar{G}_i(\text{products}) - \sum_j n_j \, \Delta \bar{G}_j(\text{reactants}), \qquad (17.7.12)$$
where n_i is the number of moles of the ith molecule.

EQUILIBRIUM CONSTANT. Consider a reaction involving only gases at sufficiently low pressures that they are ideal gases:
$$n_A A + n_B B \rightleftharpoons n_C C + n_D D. \qquad (17.7.13)$$
We have
$$\Delta \bar{G}_A = \Delta \bar{G}_A^0 + RT \ln P_A. \qquad (17.7.14)$$
[This is the equivalent of (17.7.6) but per mole instead of per molecule.] There are similar relations for $\Delta \bar{G}_B$, etc. We have by (17.7.12) for the reaction of (17.7.13),
$$\Delta G = n_C \, \Delta \bar{G}_C + n_D \, \Delta \bar{G}_D - n_A \, \Delta \bar{G}_A - n_B \, \Delta \bar{G}_B. \qquad (17.7.15)$$
Substitution of (17.7.14) and similar relations gives after some manipulation
$$\Delta G = \Delta G^0 + RT \ln K_P, \qquad (17.7.16)$$
where
$$\Delta G^0 = n_C \, \Delta \bar{G}_C^0 + n_D \, \Delta \bar{G}_D^0 - n_A \, \Delta \bar{G}_A^0 - n_B \, \Delta \bar{G}_B^0 \qquad (17.7.17)$$
and
$$K_P = \frac{P_C^{n_C} P_D^{n_D}}{P_A^{n_A} P_B^{n_B}}. \qquad (17.7.18)$$
If the reactants and products are in equilibrium, then $\Delta G = 0$. Therefore
$$K_P = e^{-\Delta G^0 / RT}. \qquad (17.7.19)$$

Since ΔG^0 is a constant (e.g., $-54{,}638$ cal/mole for the formation of water vapor from hydrogen and oxygen gas), K_P is a constant at a given temperature. K_P is called the **equilibrium constant**. We have just proved the **law of mass action**.

EXAMPLE 17.7.6

Find the equilibrium constant for the reaction

$$H_2 + CO_2 \rightleftharpoons H_2O + CO$$

at 25°C assuming only gases are involved.

Answer. We need to know ΔG^0 for the reaction:

$$\Delta G^0 = \Delta \bar{G}^0_{H_2O} + \Delta \bar{G}^0_{CO} - \Delta \bar{G}^0_{H_2} - \Delta \bar{G}^0_{CO_2}.$$

We would find the following values in a table:

Compound	$\Delta \bar{G}^0$, cal/mole
CO_2	$-94,240$
H_2O	$-54,638$
CO	$-32,790$

Hence $\Delta G^0 = 6812$ cal/mole (since $\Delta \bar{G}^0_{H_2} = 0$):

$$K_P = e^{-6812/1.987 \times 298}$$

Since $\Delta G^0 = \Delta H^0 - T \Delta S^0$ and since it can often be *assumed* that ΔH^0 and ΔS^0 do not vary with T, we can also calculate K_P at various temperatures if ΔH^0 and ΔS^0 are known.

Note that a table which contains $\Delta \bar{G}^0$ values for a few hundred compounds enables one to obtain the equilibrium constant for huge numbers of chemical reactions. This illustrates the function of classical thermodynamics (as stated by Pippard, see References), which is "to link together the many observable properties so that they can all be seen to be a consequence of the few."

17.8   APPLICATIONS OF THERMODYNAMICS

VACANCY CONCENTRATIONS

EXAMPLE 17.8.1

The creation of a vacancy at $P = 0$ in a metal crystal requires an energy E_f. If the crystal has N atoms, find the number of vacancies which will be present under equilibrium conditions at temperature T and $P = 0$.

Answer. Suppose we define $G = 0$ at the temperature T and at $P = 0$ for the perfect crystal. Then when n vacancies are added, since $G = U - TS$, we have

$$G = nE_f - kT \ln \omega. \qquad (17.8.1)$$

We now need to know what ω is. This is the total number of arrangements of N atoms and n vacancies on $N + n$ lattice sites. If these are random,

$\omega = (N + n)!/N!n!$ [see Equation (17.6.1)]. (Assuming the vacancy concentration is dilute and there is no interaction between vacancies, the distribution will be random.) Then

$$G = nE_f - kT \ln \frac{(N + n)!}{N!n!}. \tag{17.8.2}$$

Equilibrium corresponds to

$$\frac{\partial G}{\partial n} = 0 \tag{17.8.3}$$

or

$$\frac{E_f}{kT} = \frac{\partial}{\partial n} \ln \frac{(N + n)!}{N!n!}.$$

Use Stirling's approximation, Equation (17.3.5), and complete the problem to show that equilibrium corresponds to

$$\frac{n}{N + n} = e^{-E_f/kT}. \tag{17.8.4}$$

The derivation of (17.8.4) illustrates the function of statistical thermodynamics, which is to describe observable thermal behavior in terms of atomic models. Examples of E_f are shown in Table 17.8.1.

Table 17.8.1. ENERGIES OF VACANCY FORMATION

Metal	kcal/mole	Metal	kcal/mole
Al	15	Ag	25.1
Cu	21.6	Au	26.5

The *only* reason vacancies are present in equilibrium concentration different than zero is because of entropy. The entropy contribution is the largest at high temperatures and goes to zero at absolute zero. At absolute zero there would be no vacancies present in crystals under equilibrium conditions.

Actually, there is also an entropy change S_f owing to a change in the lattice vibrations when the vacancy is formed. There is also a volume change V_f when the vacancy is formed. Thus in place of (17.8.1) we have for $P \neq 0$,

$$G = n(E_f + PV_f - TS_f) - kT \ln \omega. \tag{17.8.5}$$

The equilibrium result then is

$$\frac{n}{N + n} = e^{S_f/k} e^{-(E_f + PV_f)/kT}. \tag{17.8.6}$$

This is often written in the form

$$\frac{n}{N + n} = e^{-G_f/kT}, \tag{17.8.7}$$

where $G_f = E_f + PV_f - TS_f$.

Since $S_f/k \sim 1$ in many cases and since PV_f is negligible at atmospheric pressure, Equation (17.8.6) can usually be replaced by Equation (17.8.4) to a good approximation.

PHASE EQUILIBRIUM IN A SINGLE-COMPONENT SYSTEM.   Consider the process in which a mole of liquid becomes a mole of vapor at constant P and T. The heat added to the liquid to cause this vaporization is ΔH_v. We have

$$\Delta G_v = \Delta H_v - T \Delta S_v. \tag{17.8.8}$$

However, there is no change in the Gibbs free energy per mole, i.e., $\Delta G_v = 0$; G_v and G_l are both at a minimum and must have identical values since the two phases are in equilibrium. Hence,

$$\Delta H_v = T \Delta S_v. \tag{17.8.9}$$

Thus phase changes such as solid to liquid, solid to vapor, or liquid to vapor all correspond to an increase in enthalpy (which means a decrease in binding) and an increase in entropy (which means an increase in disorder or randomness). Note that high temperature favors all of these phase changes. High temperature favors disorder.

Suppose that liquid and vapor are in equilibrium at P and T. The pressure is changed by dP and the temperature by dT. Then

$$dG_v = V_v\, dP - S_v\, dT$$

and

$$dG_l = V_l\, dP - S_l\, dT.$$

Also,

$$d(G_v - G_l) = (V_v - V_l)\, dP - (S_v - S_l)\, dT.$$

At equilibrium

$$d(G_v - G_l) = 0$$

and hence

$$\frac{dP}{dT} = \frac{S_v - S_l}{V_v - V_l} = \frac{\Delta S_v}{\Delta V_v}, \tag{17.8.10}$$

where ΔS_v is the entropy of vaporization, i.e., the change in entropy per mole of material when it goes from the liquid to the vapor phase. ΔV_v is defined similarly.

Using (17.8.9) and (17.8.10), we obtain

$$\frac{dP}{dT} = \frac{\Delta H_v}{T \Delta V_v} \doteq \frac{\Delta H_v}{RT^2/P}. \tag{17.8.11}$$

In the last step we have assumed that $V_v \gg V_l$ so that $\Delta V_v \doteq V_v$. We have also assumed that the vapor behaves as an ideal gas. Hence we have

$$\frac{dP}{P} = \frac{\Delta H_v}{RT^2}\, dT.$$

Consequently if ΔH_v does not vary with P or T,

$$P = P_0 \, e^{\Delta H_v / R T_0} \, e^{-\Delta H_v / RT} = A e^{-\Delta H_v / RT}, \qquad (17.8.12)$$

where A is a constant.

Note that Equations (17.8.11) and (17.8.12) were discussed in Chapter 16. The P-T curve for water is shown in Figure 16.2.1. We now have two ways of obtaining a P-T liquid-vapor curve: (1) Measure it directly as in Figure 16.2.1 or (2) measure ΔH_v and use (17.8.12). This exemplifies once again the *function of classical thermodynamics* (the macroscopic viewpoint) as stated earlier.

IMPURITY CONCENTRATIONS.   Let us try to understand the origin of the Pb–Sn phase diagram (Figure 16.3.6). It appears that near absolute zero the α phase would be pure Pb and the β phase pure Sn. We therefore conclude (since the entropy term makes no contribution to G as $T \to 0$) that solubility is not favored by the enthalpy term in G. From the binding viewpoint, Sn atoms prefer Sn neighbors and Pb atoms prefer Pb neighbors. This situation with regard to binding energy (and hence enthalpy) persists at higher temperatures. Yet as temperature increases, tin does dissolve in lead. Why? Because this means more configurations, more entropy, and hence a decrease in the Gibbs free energy.

17.9   PHASE EQUILIBRIA IN TWO-COMPONENT SYSTEMS

In this section we consider a simple model which helps us to understand the origin of binary phase diagrams.

NEAR ABSOLUTE ZERO.   As absolute zero is approached the Pb–Sn system shows no solubility; rather we have a mixture of Pb and Sn (see Figure 16.3.6). On the other hand, the Cu–Ni system shows complete solubility; thus we have a single phase at all compositions (see Figure 16.4.6).

We consider a simple model called the **quasichemical model** to describe this behavior. We assume all of the bonding is due to nearest neighbor bonds. We assume we have two atom types, A and B. Note in Figure 9.2.1 that there is a decrease in the energy when the bond forms. Since **bond energies** are quoted as positive when bonding occurs, we define H_{AA} to be the bond energy of an A—A bond under conditions of constant pressure. H_{BB} and H_{AB} are similarly defined. Note that inasmuch as $\Delta G = \Delta H - T \Delta S$, then equilibrium under conditions of constant pressure at $T = 0$ is determined by a minimum in ΔH. The enthalpy changes, when such bonds are formed, are given by $-H_{AA}$, $-H_{BB}$, and $-H_{AB}$. We assume the enthalpies are independent of environment. The total enthalpy in the solution will be compared with the total enthalpy of the mixture. Let us consider an alloy which contains N_A A atoms and N_B B atoms. Let us assume for simplicity here that pure A and pure B and the solution have the same crystal structure. Each atom has a total of Z nearest neighbors.

EXAMPLE 17.9.1

Calculate the total bonding energy of N_A atoms of pure A based on the above model.

Answer. Each A has Z neighbors. The number of A—A bonds in the crystal is thus $\frac{1}{2}ZN_A$. Hence

$$H_A^{\text{pure}} = \tfrac{1}{2}ZN_A H_{AA}. \tag{17.9.1}$$

The factor of $\frac{1}{2}$ is present so that each bond is not counted twice.

In a similar fashion,

$$H_B^{\text{pure}} = \tfrac{1}{2}ZN_B H_{BB}. \tag{17.9.2}$$

We now must calculate the binding energy for the solution. Clearly

$$H_{\text{soln}} = N_{AA}H_{AA} + N_{BB}H_{BB} + N_{AB}H_{AB}, \tag{17.9.3}$$

where N_{AA} is the number of A—A bonds in the solution, etc.

EXAMPLE 17.9.2

Evaluate N_{BB} in the solution if it is random.

Answer. The number of B atoms around a given B atom is Zx, where the mole fraction of B atoms is

$$x = \frac{N_B}{N_A + N_B} = \frac{N_B}{N} \tag{17.9.4}$$

and N is the total number of atoms. There are N_B atoms in all and hence

$$N_{BB} = \tfrac{1}{2}N_B Zx = \tfrac{1}{2}NZx^2. \tag{17.9.5}$$

In a similar fashion,

$$N_{AA} = \tfrac{1}{2}N_A Z(1 - x) = \tfrac{1}{2}NZ(1 - x)^2. \tag{17.9.6}$$

Likewise,

$$N_{AB} = N_A Zx = NZx(1 - x). \tag{17.9.7}$$

We can therefore write

$$H_{\text{soln}} = \tfrac{1}{2}NZ(1 - x)^2 H_{AA} + \tfrac{1}{2}NZx^2 H_{BB} + NZx(1 - x)H_{AB}. \tag{17.9.8}$$

The difference in binding energy between the solution and the mixture is

$$H_{\text{soln}} - H_{\text{mixture}} = H_{\text{soln}} - (H_A^{\text{pure}} + H_B^{\text{pure}}). \tag{17.9.9}$$

This can be shown to be

$$x(1 - x)ZN[H_{AB} - \tfrac{1}{2}(H_{AA} + H_{BB})].$$

This equals the negative of the enthalpy of mixing, ΔH_{mixing}, so

$$\Delta H_{\text{mixing}} = x(1 - x)ZN\mathscr{H}, \tag{17.9.10}$$

where

$$\mathcal{H} = [\tfrac{1}{2}(H_{AA} + H_{BB}) - H_{AB}]. \tag{17.9.11}$$

Note: Since $\Delta G_{mixing} \rightarrow \Delta H_{mixing}$ as $T \rightarrow 0$, the solution will be favored if its formation corresponds to a decrease in the Gibbs free energy or the enthalpy in this case; hence if $H_{AB} > (H_{AA} + H_{BB})/2$, a solution will be formed (the Cu–Ni case). However, if $H_{AB} < (H_{AA} + H_{BB})/2$, the mixture of two phases will be stable (the Pb–Sn case).

This is summarized in Table 17.9.1.

Table 17.9.1. CRITERIA FOR SOLUBILITY AT $T = 0$

$\mathcal{H} = \tfrac{1}{2}(H_{AA} + H_{BB}) - H_{AB}$	Phase Present
Approx. zero, Negative	Solid solution (1 phase)
Positive	Mixture (2 phases)

EXAMPLE 17.9.3

For the Cu–Zn system near 50 at. %, $H_{AB} > (H_{AA} + H_{BB})/2$. The crystal structure of the 50 at. % alloy is based on a bcc lattice with one atom at each lattice point. Will the solution be ordered or not?

Answer. The A—B bond energy is larger than the average of A—A and B—B bonds; hence A atoms prefer to be bonded to B atoms.

Similarly, Cu would prefer to be bonded to Zn only and vice versa. This can be arranged if the basis is Cu at 000, Zn at $\tfrac{1}{2}\tfrac{1}{2}\tfrac{1}{2}$ on a simple cubic lattice. This is an ordered solid solution.

TEMPERATURE EFFECTS.   A solution is said to be an **ideal solution** if $\Delta H_{mixing} = 0$. If such a solution is considered at $T > 0$, there is an entropy contribution to the Gibbs free energy. Since there is no preference for neighbors of one type or another based on binding energy, the solution will be a random solution and the entropy can be calculated from (17.6.5). Hence

$$\Delta G_{mixing} = -T \Delta S_{mixing} \quad \text{(ideal solution).} \tag{17.9.12}$$

An ideal solution also has $\Delta V_{mixing} = 0$. In addition the partial vapor pressure P_A of component A is $x_A P_A^0$, where x_A is the mole fraction of A in the solution and P_A^0 is the vapor pressure of pure A. This is known as **Raoult's law**. The chemical potential of the component A is, by (17.7.6),

$$\mu_A = \mu_A^0 + kT \ln x_A P_A^0.$$

This can also be written in the form

$$\begin{aligned} \mu_A &= \mu_A^0 + kT \ln P_A^0 + kT \ln x_A \\ &= \mu_A^{0\prime} + kT \ln x_A. \end{aligned} \tag{17.9.13}$$

EXAMPLE 17.9.4

Starting with $G_{soln} = G_{mixture} + \Delta G_{mixing}$, derive (17.9.13) for an *ideal* solution.

Answer. Here $G_{mixture}$ means the total free energy of the two separate phases; the free energy of each phase is the product of the number of moles of that phase and the molar free energy of that phase. We have $G_{soln} = \sum_i n_i G_i^0 + RT \sum_i n_i \ln x_i$. This can be written in the form

$$G_{soln} = \sum_i N_i \mu_i^0 + kT \sum_i N_i \ln x_i$$

and hence

$$\mu_i = \frac{\partial G_{soln}}{\partial N_i} = \mu_i^0 + kT \ln x_i.$$

If $\Delta H_{mixing} \neq 0$ but ΔS_{mixing} is still given by (17.6.5), the solution is called a **regular solution**. For a regular solution in which $\Delta H_{mixing} > 0$, the solubility will increase as temperature increases because of the entropy of mixing term. The solution will be a random solution.

For the regular solution we have for $\Delta H_{mixing} > 0$ by (17.9.10) and (17.6.16):

$$\Delta G_{mixing} = x(1 - x)ZN\mathscr{H}$$
$$+ NkT[x \ln x + (1 - x) \ln(1 - x)]. \qquad (17.9.14)$$

(*Note:* There can be an added entropy change owing to a change in vibrational entropy. We ignore this here.) The first term in (17.9.14) is positive for all x, while the second is negative for all x. A sketch of ΔG is shown in Figure 17.9.1.

EXAMPLE 17.9.5

If the solubility is small ($x \ll 1$), find the solubility versus temperature curve.

Answer.

$$\Delta G_{mixing} \doteq xZN\mathscr{H} + NkTx \ln x$$

since

$$1 - x \doteq 1.$$

Since $\Delta G_{mixing} = 0$ at the solubility limit

$$Z\mathscr{H} = -kT \ln x$$

or the solubility limit is given by

$$x = e^{-Z\mathscr{H}/kT}; \qquad (17.9.15)$$

i.e., it depends exponentially on the negative reciprocal of absolute temperature.

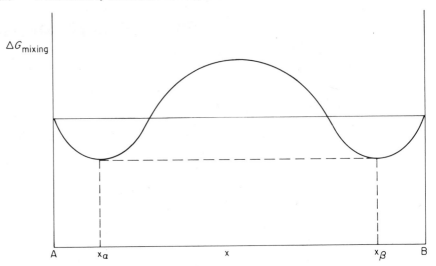

FIGURE 17.9.1. Sketch of ΔG_{mixing} vs. x for a given temperature for $\Delta H_{\text{mixing}} > 0$ shown by solid line. Dashed line shows ΔG_{mixing} for mixture of two phases with composition x_α and x_β. The predicted results are: for $0 \lesssim x \lesssim x_\alpha$, solution (call it α); for $x_\beta \lesssim x \lesssim 1$, solution (call it β); for $x_\alpha < x < x_\beta$, mixture of two phases of composition x_α and x_β.

EXAMPLE 17.9.6

Consider vacancies in copper as a dilute binary alloy. Let vacancy $= B$, and copper atom $= A$. Show that the fraction of vacant sites is

$$x = e^{-\Delta H_f / kT}.$$

Answer. In copper $Z = 12$. When a vacancy is formed, 12 A—A bonds are broken, but when the atom is placed on the surface a new interior atom is created and 6 A—A bonds are formed (see Example 17.9.1). Hence

$$\Delta H_f = 6H_{\text{AA}}.$$

In the present case $Z\mathcal{H} = \frac{12}{2}(H_{\text{AA}} + 0) - 0 = 6H_{\text{AA}}$ so that $Z\mathcal{H} = 6H_{\text{AA}} = \Delta H_f$ and from (17.9.15) we have

$$x = e^{-\Delta H_f / kT}.$$

ORDER-DISORDER. For a solution in which $\Delta H_{\text{mixing}} < 0$ (or $\mathcal{H} < 0$) ordering will occur at low temperatures (see Example 17.9.3), but the entropy effect will favor randomness at high temperatures, so there will be a change from an ordered to a disordered or random solid solution as temperature increases. Above a certain temperature T_c, the solution will be completely random.

Consider a solution of Cu–Zn alloy with 50 at. % Cu. In the perfectly ordered 50–50 alloy, Cu atoms are at 000 sites and Zn atoms are at $\frac{1}{2}\frac{1}{2}\frac{1}{2}$ sites of the bcc lattice. Let the ordering be imperfect. Then let r_{Zn} be the fraction of Zn atoms at $\frac{1}{2}\frac{1}{2}\frac{1}{2}$. Define

$$\mathscr{L} = \frac{r_{Zn} - \frac{1}{2}}{1 - \frac{1}{2}}. \tag{17.9.16}$$

We call $\mathscr{L}$ the **long-range order parameter**.

EXAMPLE 17.9.7

    a. What is $\mathscr{L}$ for perfect order?

    Answer.

$$r_{Zn} = 1 \quad \text{and} \quad \mathscr{L} = 1$$

    b. What is $\mathscr{L}$ for complete randomness?

    Answer.

$$r_{Zn} = \frac{1}{2} \quad \text{and} \quad \mathscr{L} = 0.$$

It can be shown that the enthalpy of mixing is

$$\Delta H_{mixing} = \Delta H_{mixing}(\mathscr{L}, \mathscr{H}),$$

while

$$\Delta S_{mixing} = \Delta S_{mixing}(\mathscr{L}),$$

so that

$$\Delta G_{mixing} = \Delta G_{mixing}(\mathscr{L}, \mathscr{H}, T). \tag{17.9.17}$$

See R. A. Swalin, *Thermodynamics of Solids*, Wiley, New York (1962) p. 117. Since $\mathscr{H}$ is a fixed parameter for a given binary system, the Gibbs free energy varies only with $\mathscr{L}$ and T.

At a given temperature, equilibrium corresponds to a minimum of ΔG_{mixing}; i.e.,

$$\frac{\partial \Delta G_{mixing}}{\partial \mathscr{L}} = 0.$$

This enables us to calculate $\mathscr{L}(T)$. There is a critical T, called T_c or the **disordering temperature**, above which $\mathscr{L} = 0$. It can be shown that

$$T_c = \frac{-4}{k} \mathscr{H}. \tag{17.9.18}$$

(Recall that $\mathscr{H} < 0$ in the case under consideration.)

It can also be shown that

$$\ln \frac{(1 + \mathscr{L})}{(1 - \mathscr{L})} = 2\mathscr{L}\frac{T_c}{T}. \tag{17.9.19}$$

$\mathscr{L}$ vs. T/T_c is plotted in Figure 17.9.2.

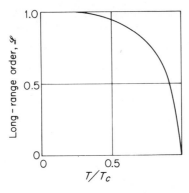

FIGURE 17.9.2. Long-range order parameter versus T/T_c for an AB alloy.

Figure 17.9.2 should be *qualitatively* compared with Figure 4.2.5 (ferromagnetism) and Figure 5.2.1 (superconductivity). Ordering of solutions, ferromagnetism, ferroelectricity and superconductivity are all examples of long-range ordering phenomena (**cooperative phenomena**) which are destroyed above some critical temperature T_c.

There is a contribution to the specific heat associated with the increase in disorder. Since

$$C_p = \left(\frac{\partial H}{\partial T}\right)_p,$$

this contribution can be calculated for the model used here. One finds that C_p increases as T increases, and near to T_c, C_p increases very rapidly and approaches infinity as $T \to T_c$. The value of C_p then drops rapidly above T_c and becomes nearly a constant for higher T. The theory given here is only a crude approximation to cooperative phenomena. See Stanley, References.

17.10   CALCULATION OF INTERNAL ENERGY

Suppose an atom has energy levels $u_1, u_2, u_3, \ldots$, which are known from quantum mechanics. Suppose there are N similar atoms. How does one calculate the total internal energy U of such a system under equilibrium conditions? This would be easy if the number of atoms n_i occupying energy levels u_i were known since

$$U = \sum_i u_i n_i, \tag{17.10.1}$$

where

$$N = \sum_i n_i. \tag{17.10.2}$$

It can be shown that

$$\frac{n_i}{N} = \frac{n_i}{\sum_i n_i} = \frac{e^{-u_i/kT}}{\sum_i e^{-u_i/kT}}. \tag{17.10.3}$$

Note that if all the u_i are known, then the n_i can be evaluated. Equation (17.10.3) is the **distribution formula for Boltzmann statistics.**

EXAMPLE 17.10.1

Suppose a molecule has two energy levels, namely, a ground state $u_0 = 0$ and one excited state $u_1 = 1$ eV.

a. What fraction of such molecules will be in the excited state under *equilibrium conditions* at room temperature?

Answer.

$$f_1 = \frac{n_1}{n_0 + n_1} = \frac{e^{-u_1/kT}}{1 + e^{-u_1/kT}}.$$

At room temperature $kT \approx 1/40$ eV. Hence we can write

$$\frac{n_1}{n_0 + n_1} \doteq e^{-u_1/kT} = e^{-40} = 10^{-40/2.3}.$$

(Note that if there were other excited states, $u_2 = 5$ eV, say, etc., this result would be virtually unchanged.) The average energy would be very close to the ground state and could be calculated from (17.10.1).

b. What fraction is in the ground state?

Answer.

$$f_0 = \frac{n_0}{n_0 + n_1} = \frac{e^{-0/kT}}{1 + e^{-u_1/kT}} \doteq 1 - e^{-u_1/kT}.$$

c. What is the average energy?

Answer.

$$\bar{u} = f_0 u_0 + f_1 u_1 = 0 + e^{-40} \text{ eV}.$$

AN HEURISTIC DEVELOPMENT OF EQUATION (17.10.3).   Let us assume that air has molecules of constant mass m, that it is at a constant temperature T, and that it is attracted to the earth by a constant g. Then it can be shown that the pressure varies with height h above sea level as

$$P = P_0 e^{-mgh/kT}, \tag{17.10.4}$$

where P_0 is the pressure at sea level and k is Boltzmann's constant. This is known as the **barometric formula**.

EXAMPLE 17.10.2

Derive (17.10.4).

Answer. Consider a small volume element of cross section A located between h and $h + dh$. It is acted on by forces as shown in Figure

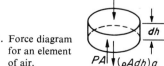

FIGURE 17.10.1. Force diagram for an element of air.

17.10.1, where ρ is the density. A balance of force gives

$$(P + dP)A + \rho A \, dh \, g = PA$$

or

$$dP = -\rho g \, dh,$$

and since for a mole of gas, $PV = N_0 kT$ (which we assume is independent of gravitational forces), we have

$$\rho = \frac{M}{V} = \frac{mP}{kT}.$$

Hence

$$\frac{dP}{P} = -\frac{mg}{kT} \, dh.$$

Integration of this equation subject to $P \longrightarrow P_0$ as $h \longrightarrow 0$ gives (17.10.4).

If we define $\bar{n}$ as the number of molecules per unit volume, we have from (17.10.4)

$$\bar{n} = \bar{n}_0 e^{-mgh/kT}. \tag{17.10.5}$$

Let dn be the number of molecules between h and $h + dh$ in a unit cross-sectional area. Then

$$dn = \bar{n} \, dh,$$

and the total number of molecules in a column of unit cross-sectional area is

$$N = \int dn = \int_0^\infty \bar{n} \, dh = \bar{n}_0 \frac{kT}{mg}.$$

Hence we have

$$\frac{dn}{\int dn} = \frac{e^{-mgh/kT} \, dh}{\int_0^\infty e^{-mgh/kT} \, dh}.$$

We might write this in terms of the potential energy:

$$\frac{dn}{\int dn} = \frac{e^{-u/kT} \, du}{\int_0^\infty e^{-u/kT} \, du}. \tag{17.10.6}$$

If the energies were discrete, i.e., n_i molecule exactly at height h_i or with energy u_i,

$$\frac{n_i}{\sum_i n_i} = \frac{e^{-u_i/kT}}{\sum_i e^{-u_i/kT}}. \tag{17.10.7}$$

We have not proved that (17.10.3) is a general formula. Rather we have given one physical example where it does apply.

A MORE RIGOROUS DEVELOPMENT OF EQUATION (17.10.3).   If it is assumed that it is equally likely that any energy state be occupied, then the number of distinct configurations ω is

$$\omega = \frac{N!}{\prod_i n_i!},$$
(17.10.8)

where the denominator means the product of $n_i!$ terms, i.e., $n_1!n_2!n_3! \cdots n_k!$.

Here we have assumed that there are N identical particles, n_0 in the energy level u_0, n_1 in the level u_1, etc. Permuting the particles within the level u_0, say, does not produce a new distribution, which is why we divide by $n_0!$, etc.

It is usually stated without proof in elementary texts and most intermediate texts that equilibrium corresponds to the most probable configuration. This means we must find the maximum of ω subject to the constraints imposed by (17.10.1) and (17.10.2). This is a fairly common type of problem called a constrained maxima problem. The result [see, e.g., W. J. Moore, *Physical Chemistry*, Prentice-Hall, Englewood Cliffs, N.J. (1955) p. 349] is (17.10.3). The rigorous proof that equilibrium corresponds to the most probable configuration is given in advanced statistical mechanics texts.

17.11   VIBRATIONAL ENERGY

The potential energy curve for binding between a pair of atoms is shown for a typical case in Figure 9.2.1. An expanded version of the region of the minimum of the curve is given in Figure 17.11.1. Near the minimum the curve is

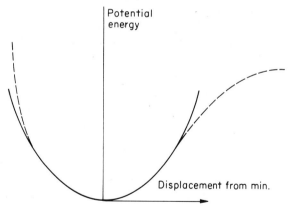

FIGURE 17.11.1. Dashed line represents interatomic potential. Solid line represents parabolic approximation.

fitted by a parabola in which the potential energy is proportional to the square of the displacement from the equilibrium position or the minimum; i.e.,

$$V \propto x^2.$$

We recall that a harmonic oscillator (linear spring plus mass; see Section 2.4) has the property that the stored energy or the potential energy is given by

$$V = \int F \, dx = \int kx \, dx = \frac{kx^2}{2}. \tag{17.11.1}$$

Thus to a good first approximation the vibration of the atoms in a molecule such as nitrogen is described by the harmonic oscillator. However, we cannot use the classical expression $u = \frac{1}{2}mv^2 + \frac{1}{2}kx^2$ for the total energy of such as oscillator.

Rather we must use the quantum mechanical expression

$$u_i = (i + \tfrac{1}{2})hv, \tag{17.11.2}$$

where $i = 0, 1, 2, 3, \ldots$, and where

$$v = \frac{1}{2\pi} \sqrt{k/m}. \tag{17.11.3}$$

EXAMPLE 17.11.1

What is the ground state of the harmonic oscillator and what is the significance of this?

Answer. $u_0 = \frac{1}{2}hv$. A molecule would be constantly vibrating in the ground state at absolute zero.

THE EINSTEIN MODEL OF A VIBRATING CRYSTAL.   Consider a crystal of N_0 atoms such as a copper crystal. This will have $3N_0$ harmonic oscillators with each in the most general case having a *different fundamental* frequency v. Einstein (1906) assumed that to a rough approximation a solid consists of $3N_0$ harmonic oscillators of the *same* fundamental frequency v. The total vibrational energy would then be, by Equations (17.10.1), (17.10.3), and (17.11.2),

$$U = 3N_0 \sum_{i=0}^{\infty} \frac{(i + \tfrac{1}{2})hve^{-(i+1/2)hv/kT}}{\sum_{i=0}^{\infty} e^{-(i+1/2)hv/kT}}. \tag{17.11.4}$$

This can be shown to be

$$U = \frac{3N_0hv}{2} + 3N_0hv(e^{hv/kT} - 1)^{-1}. \tag{17.11.5}$$

EXAMPLE 17.11.2

Let $e^{-x} = y$. Then show that

$$\sum_{i=0}^{\infty} y^i = \frac{1}{1 - y}$$

and hence that

$$\sum_{i=0}^{\infty} e^{-ix} = \frac{1}{1 - e^{-x}}.$$

Answer.

$$\sum_{i=0}^{\infty} y^i = 1 + y + y^2 + y^3 + \cdots.$$

Long division of $1/(1 - y)$ leads to this same series. This can be used to form the sum in the denominator of (17.11.4).

The specific heat of a solid is due to the atom vibrations. From (6.3.1) we have

$$C_V = \left(\frac{\partial U}{\partial T}\right)_V, \qquad (17.11.6)$$

and from (17.10.5) we then have

$$C_V = 3N_0 k \left(\frac{hv}{2kT} \operatorname{csch} \frac{hv}{2kT}\right)^2. \qquad (17.11.7)$$

Figure 6.3.1 illustrates how closely the Einstein function approximates the actual specific heat (the agreement is poor at low temperatures, however). Note that at high temperatures

$$hv/2kT \ll 1$$

and

$$\operatorname{csch} hv/2kT \doteq 2kT/hv.$$

Under these conditions we note that

$$C_V \doteq 3N_0 k \doteq 6 \text{ cal/mole-}°K \qquad (17.11.8)$$

and that all the solids shown in Figure 6.3.1 approach this value. (There are deviations from this value at high temperatures in a few solids owing to the fact that the harmonic oscillator approximation does not suffice when the vibrational amplitudes become large.) It had already been observed in 1819 by Dulong and Petit that the molar heat capacity of most metals was around 6 cal/mole-$°$K.

EXAMPLE 17.11.3

Show that (17.11.8) follows from (17.11.7) if the temperature is high.

Answer. At high temperature $hv/2kT \ll 1$. The term in parentheses in (17.11.7) can then be approximated by 1 since

$$x \operatorname{csch} x = \frac{x}{\sinh x} = \frac{x}{x + 0(x^3)} \doteq 1.$$

Hence $C_V \doteq 3N_0 k \doteq 6$ cal/mole-$°$K.

The frequency which gives the best fit of Equation (17.11.7) to the experimental data is called the Einstein frequency. It has values in the neighborhood of $v \approx 10^{13}$ sec^{-1}.

PHONON DISTRIBUTIONS. The quantum of energy in a lattice vibration is called a **phonon**. We have already noted that there are many fundamental modes. These may be represented by a distribution as shown in Figure 17.11.2 for sodium.

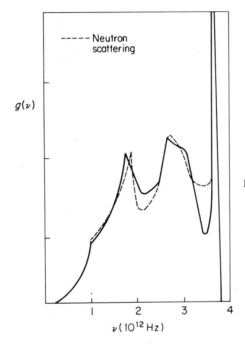

FIGURE 17.11.2. Comparison of the calculated and measured phonon frequency distribution for Na. [P.S. Ho and A.L. Ruoff, *Journal of Physics and Chemistry of Solids*, **29**, 2101 (1968).]

Here $g(v) \, dv$ means the number of phonons with frequencies between v and $v + dv$.

Such a curve can be obtained experimentally from a neutron scattering experiment. It can also be calculated, at least in principle, from binding theories. Each solid has its specific phonon spectrum.

EXAMPLE 17.11.4

At very low temperatures many of the lattice vibrations are in the ground state. Which ones?

Answer. Those for which $hv \gg kT$ would be almost entirely in the ground state since the fraction of these in the excited state is $e^{-hv/kT} \doteq 0$.

DEBYE MODEL OF SPECIFIC HEAT. Debye's model of the specific heat of solids is based on the assumption that the actual phonon spectrum is roughly represented by

$$g(v) = Av^2, \qquad 0 < v < v_D$$
$$ = 0, \qquad\quad v > v_D. \tag{17.11.9}$$

This is the correct function for small v and hence (see Example 17.11.4) should give the proper results at low temperatures. The cutoff frequency v_D is called the **Debye frequency**. The temperature Θ_D, called the **Debye temperature**, is defined by

$$k\Theta_D = hv_D. \tag{17.11.10}$$

Since a crystal of N atoms has $3N$ vibrational modes,

$$\int_0^{v_D} g(v)\, dv = 3N;$$

it can therefore be shown that

$$A = \frac{9N}{v_D^3}.$$

From (17.11.5) the energy for a single oscillator is

$$u = \frac{hv}{2} + \frac{hv}{e^{hv/kT} - 1}. \tag{17.11.11}$$

Since $U = \sum_i n_i u_i$ and since we are dealing with a continuous distribution, we can write $U = \int u\, dn = \int ug(v)\, dv$, so

$$U = \int_0^{v_D} \left[\frac{hv}{2} + \frac{hv}{e^{hv/kT} - 1} \right] g(v)\, dv. \tag{17.11.12}$$

This can be used to show after considerable manipulation that at low temperatures ($kT \ll hv_D$)

$$C_V \propto T^3, \tag{17.11.13}$$

which is in agreement with the experimental results. See Figure 6.3.1. Values of the Debye temperature are shown in Table 17.11.1.

Table 17.11.1. DEBYE TEMPERATURES
AND FREQUENCIES

Substance	Θ_D, °K	v_D, sec^{-1}
Be	1160	2.4×10^{13}
Si	658	1.4×10^{13}
Al	418	9.0×10^{12}
Cu	339	7.3×10^{12}
NaCl	281	6.0×10^{12}
Ag	215	4.6×10^{12}
Na	159	3.4×10^{12}
K	100	2.1×10^{12}

17.12   VIBRATIONAL ENTROPY

If all the *possible* energy levels u_i of a system are known, then not only the total energy of a system can be calculated by (17.10.1) and (17.10.3) but also all the thermodynamic quantities $(U, H, S, G,$ and $F)$.

EXAMPLE 17.12.1

If we define

$$z = \sum_i e^{-u_i/kT}, \tag{17.12.1}$$

show that

$$\bar{u} = kT^2 \frac{\partial \ln z}{\partial T}. \tag{17.12.2}$$

Here $\bar{u}$ is the average internal energy per "particle."

Answer. We have

$$\frac{\partial \ln z}{\partial T} = \frac{1}{z}\frac{\partial z}{\partial T} = \frac{\sum_i (u_i/kT^2)e^{-u_i/kT}}{\sum_i e^{-u_i/kT}}.$$

If this is multiplied by kT^2, we have

$$\frac{\sum u_i e^{-u_i/kT}}{\sum e^{-u_i/kT}},$$

which by (17.10.1) to (17.10.3) is U/N and hence $\bar{u}$.

It can be shown that the Helmholtz free energy F $(F \equiv U - TS)$ is given by

$$\bar{f} = -kT \ln z. \tag{17.12.3}$$

Here $\bar{f} \equiv F/N$. This proof makes use of a thermodynamics formula

$$\left[\frac{\partial}{\partial T}\left(\frac{F}{T}\right)\right]_V = \frac{-U}{T^2}. \tag{17.12.4}$$

Thus the entropy can be calculated from

$$\bar{s} = \frac{\bar{u} - \bar{f}}{T}. \tag{17.12.5}$$

EXAMPLE 17.12.2

Calculate $\bar{s}$ at 300°K for a harmonic oscillator whose Einstein temperature is 50°K.

Answer. Here $h\nu_E = k\Theta_E$, but $h\nu_E \ll kT$. Hence

$$z = \frac{e^{-h\nu_E/2kT}}{1 - e^{-h\nu_E/kT}} \doteq \frac{1}{h\nu_E/kT}.$$

Then using (17.12.5), (17.12.2), and (17.12.3) gives

$$\bar{s} = k \ln \frac{kT}{h\nu_E} = k \ln \frac{T}{\Theta_E} = k \ln 6.$$

The entropy at $300°K$ of a solid with $3N$ such oscillators and an Einstein temperature of $50°K$ is $S = 3R \ln 6 = 10.8$ cal/mole-°K. This is a good approximation for sodium ($S = 12.3$ cal/mole-°K).

At this point the reader is inclined to ask how this is related to the expression $S = k \ln \omega$ which we previously wrote for entropy. Recall that at that time we discussed ω in terms of the *geometric* arrangements of balls in boxes. In the present case we have energy levels corresponding to boxes and individual harmonic oscillators corresponding to the balls. The entropy then is the result of the number of ways the oscillators can be placed in the different energy levels. At absolute zero all the oscillators would be in the ground state; hence there would be only one distinct configuration and the vibrational entropy would be zero.

It could be shown theoretically that a mole of ideal monatomic gas (sodium vapor) at standard conditions (1 atm, 25°C) would have an entropy of 29.8 cal/mole-°K.

MEASUREMENT OF ENTROPY. Entropy can be evaluated experimentally from

$$S = \int_0^T \frac{C_p \, dT}{T}$$

over regions for which there is no phase change. At a phase change there is a discontinuous jump in entropy

$$\Delta S = \frac{\Delta H}{T},$$

where ΔH is the heat added at constant temperature. Thus if a solid phase is bcc from $T = 0$ to $T = T_M$ and liquid thereafter, we have for $T > T_M$,

$$S = \int_0^{T_M} \frac{C_p^s \, dT}{T} + \frac{\Delta H_f}{T} + \int_{T_M}^T \frac{C_p^l \, dT}{T}. \tag{17.12.6}$$

Table 17.12.1 gives values of the entropies calculated from measured specific heats and measured heats of transformation. For comparison purposes of the entropy in different states of aggregation, systems with the same number of atoms per molecule should be used. We would expect the greatest randomness (entropy) for gases and the least for solids, and this is confirmed. The entropies for the gases can also be easily calculated with high accuracy. See W. J. Moore, *Physical Chemistry*, Prentice-Hall, Englewood Cliffs, N.J. (1955).

Table 17.12.1. ENTROPIES (SUBSTANCES IN THE STANDARD STATE AT 25°C)*

Substance	S^0_{298}, cal/deg-mole	Substance	S^0_{298}, cal/deg-mole
		Gases	
H_2	31.2	CO_2	51.1
D_2	34.4	H_2O	45.2
He	29.8	NH_3	46.4
N_2	45.8	SO_2	59.2
O_2	49.0	CH_4	44.5
Cl_2	53.2	C_2H_2	48.0
HCl	44.7	C_2H_4	52.5
CO	47.3	C_2H_6	55.0
		Liquids	
Mercury	17.8	Benzene	41.9
Bromine	18.4	Toluene	52.4
Water	16.8	Diethylether	60.5
Methanol	30.3	n-Hexane	70.6
Ethanol	38.4	Cyclohexane	49.2
		Solids	
C (diamond)	0.6	K	16.5
C (graphite)	1.4	I_2	14.0
S (rhombic)	7.6	NaCl	17.2
S (monoclinic)	7.8	KCl	19.9
Ag	10.2	KBr	22.5
Cu	8.0	KI	23.4
Fe	6.7	AgCl	23.4
Na	12.3	Hg_2Cl_2	46.4

* From W. J. Moore, *Physical Chemistry*, Prentice-Hall, Englewood Cliffs, N.J. (1955).

17.13   THERMAL EXPANSION

The potential energy curve of Figure 17.11.1 can be represented by

$$U(x) = K\frac{x^2}{2} - bx^3. \tag{17.13.1}$$

(Note that the cubic term correctly describes the direction of deviation of the true potential curve from the curve of the harmonic oscillator. The cubic term is called an anharmonic term.)

The average atom position is

$$\bar{x} = \int_{-\infty}^{\infty} xP(x)\, dx, \tag{17.13.2}$$

where

$$P(x) = \frac{e^{-U(x)/kT}}{\int_{-\infty}^{\infty} e^{-U(x)/kT}}.$$

It can then be shown, if the cubic term is small relative to the square term, that

$$\bar{x} = \frac{3b}{K^2} kT. \tag{17.13.3}$$

For a harmonic oscillator $\bar{x} = 0$; i.e., there is no thermal expansion. For an **anharmonic solid** of the type used here the thermal expansion is a constant. Thus thermal expansion is an anharmonic property of the crystal.

EXAMPLE 17.13.1

Evaluate

$$\int_{-\infty}^{\infty} xe^{-U(x)/kT}$$

for the case where the anharmonic term $bx^3 \ll kT$.

Answer. We have

$$e^{-Kx^2/2kT} e^{bx^3/kT} \doteq e^{-Kx^2/2kT}\left(1 + \frac{bx^3}{kT}\right).$$

Hence the integral is

$$\int_{-\infty}^{\infty} xe^{-Kx^2/2kT}\, dx + \int_{-\infty}^{\infty} \frac{bx^4}{kT} e^{-Kx^2/2kT}\, dx.$$

The first term is zero. Why? To evaluate the second term we note that

$$\int_{-\infty}^{\infty} z^4 e^{-z^2}\, dz = \frac{3\sqrt{\pi}}{4}.$$

17.14 THERMAL CONDUCTIVITY

GASES. We consider thermal conductivity in gases first. Consider two layers of gas a distance λ apart, where λ is the average distance between collisions or the **mean free path** (see Figure 17.14.1).

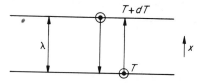

FIGURE 17.14.1. Layers of gas showing heat flow.

Let u be the heat exchanged when one molecule at T goes to the layer $T + dT$ and one at $T + dT$ goes to T. If c_v is the heat capacity per gram, then mc_v is the heat capacity per molecule and

$$\Delta u = -mc_v\, dT \doteq -mc_v \frac{\partial T}{\partial x} \lambda. \tag{17.14.1}$$

The number of molecules per unit volume is n and their average speed is $\bar{c}$, so the number crossing unit area (normal to the x direction) in unit time is $\frac{1}{3}n\bar{c}$. Hence the flux of heat in the x direction is

$$J_x = \tfrac{1}{3}n\bar{c}\,\Delta u = -\tfrac{1}{3}n\bar{c}mc_v\,\lambda\frac{\partial T}{\partial x}. \qquad (17.14.2)$$

But

$$J_x = -k_T\frac{\partial T}{\partial x}$$

is the Fourier heat conduction law so that

$$k_T = \tfrac{1}{3}n\bar{c}mc_v\lambda. \qquad (17.14.3)$$

The heat conductivity is thus found to be proportional to the concentration of heat carriers n and the mean free path. (It is shown in Section 18.4 that $\lambda \propto 1/n$ for an ideal gas so that k_T is independent of pressure for an ideal gas.)

SOLIDS. In the case of solids which are electrical insulators, heat is transmitted by phonons. An equation similar to (17.14.3) can be used to describe the conductivity. The velocity will be the sound velocity in the solid. The important quantity in the theory of thermal conductivity of solids is the mean free path λ. In a perfectly crystalline harmonic solid, λ would be determined only by scattering at surfaces. In an anharmonic solid, different phonons are coupled and this limits λ. This value of λ increases as the temperature decreases. In very good crystals at low temperature, λ is determined by surface scattering and hence by the size of the specimen.

Recall that a vibrational frequency is of the form $\omega = \sqrt{k/m}$, so that if locally either k or m changes, then we can expect additional scattering of phonons. Thus the addition of isotopes, impurities, vacancies, interstitials, dislocations, grain boundaries, etc., causes scattering of phonons and a decrease in λ and hence in the thermal conductivity of solids. The structure-sensitive nature of the thermal conductivity of solids is clearly illustrated in Figures 6.2.1 and 6.2.2. The quantitative study of thermal conductivity in solids is a complex subject which is currently studied extensively by solid state physicists.

EXAMPLE 17.14.1

In Figure 6.2.1, it appears that k_T decreases rapidly toward zero as T approaches zero. What factor in Equation (17.14.3) varies in such a way that it might be responsible for this behavior?

Answer. $c_v \propto T^3$ at low temperatures. See (17.11.13)

In metals heat can be carried by an electron gas. Properties of the electron gas then enter (17.14.3).

EXAMPLE 17.14.2

Why in Figure 6.2.1 do the thermal conductivities of all the copper specimens appear to have the same value at high temperatures?

Answer. Thermal conductivity in copper is due to the motion of electrons. The scattering of electrons at high temperatures is dominated by phonon scattering. However, at lower temperatures this scattering is small compared to scattering by imperfections.

Electrical conductivity obeys Ohm's law $\mathbf{J} = \sigma \boldsymbol{\xi}$. The conductivity coefficient will later be shown to be

$$\sigma = \frac{nq^2\lambda}{mv_F}.$$

Thus both conductivity coefficients are proportional to n and to λ. This explains the origin of the **Wiedemann-Franz ratio** [Equation (6.2.5)].

REFERENCES

Kac, M., "Probability", *Scientific American* (September 1964). An excellent elementary discussion of the random walk.

Mark, M., *Thermodynamics, An Auto-Instructional Text*, Prentice-Hall, Englewood Cliffs, N.J. (1967). A programmed learning approach.

Sears, F. W., *Thermodynamics*, Addison-Wesley, Reading, Mass. (1952). A good discussion of the fundamentals of thermodynamics, kinetic gas theory, and the elements of statistical mechanics.

Pippard, A. B., *The Elements of Classical Thermodynamics*, Cambridge, New York (1957).

Darken, L. S., and Gurry, R. W., *Physical Chemistry of Metals*, McGraw-Hill, New York (1953).

Swalin, R. A., *Thermodynamics of Solids*, Wiley, New York (1962).

Kittel C., *Thermal Physics*, Wiley, New York (1969).

Fast, J. D., *Entropy*, McGraw-Hill, New York (1962).

Stanley, H. E., *Introduction to Phase Transitions and Critical Phenomena*, Oxford University Press, New York (1971). For the instructor.

PROBLEMS

17.1.   a. Can it be shown directly by experiment that $PV = \frac{2}{3}U_K$ for a dilute argon gas?
      b. If so, explain how.

17.2.   Discuss the ideal gas temperature scale.

17.3.   List three applications of the random walk to
      a. Problems in materials science.

b. Problems not involving materials science.

17.4. How many different bridge hands are there?

17.5. What is the most probable distance between the ends of a random polymer chain of N links?

17.6. What is the most probable speed of an ideal gas molecule?

17.7. a. What are the necessary conditions for carrying out a reversible process?
b. Give examples of two.

17.8. A rod of copper, as shown in Figure 15.3.1, is falling toward the table.
a. What is its external energy when the angle is θ?
b. What is the internal energy of the copper?

17.9. a. Define internal energy from the microscopic viewpoint.
b. Define entropy from the microscopic viewpoint.

17.10. Does the statement $\oint dU = 0$ depend on the system, e.g., is it true only for an ideal gas?

17.11. Why is it true that the internal energy of an ideal monatomic gas is given by the translational kinetic energy?

17.12. What is the expression for dU for a reversible process in which the only work is PV work?

17.13. Show that dQ for a reversible process is $T\,dS$. (*Note:* You may use an ideal gas for your system, although the result is perfectly general and applies to any system.)

17.14. a. Define change in internal energy from a macroscopic viewpoint.
b. Define change in entropy from the macroscopic viewpoint.

17.15. Derive the expression for the entropy of mixing of two ideal gases from
a. The microscopic viewpoint.
b. The macroscopic viewpoint.

17.16. Two ideal gases are mixed together, and their entropy of mixing is ΔS_{mix}. What is the minimum work that has to be done (assuming 100% efficiency) to separate these two gases at temperature T?

17.17. Explain the reason for defining the Gibbs free energy according to (17.7.1).

17.18. A system is in equilibrium at
a. Constant P and T if _____ is a minimum.
b. Constant V and T if _____ is a minimum.
c. Constant V and S if _____ is a minimum.
d. Constant U and V if _____ is a maximum.

17.19. Explain the analogy between chemical potential, mechanical potential, and electrical potential.

17.20. a. Plot the function G vs. $n/(N+n)$ in Equation (17.8.2).
b. Explain why vacancies exist in equilibrium concentration at high temperatures.

17.21. Explain why Sn dissolves in Pb at high temperatures.

17.22. Why do ordered solutions become disordered at high temperatures?

17.23. Why do materials vaporize at high temperatures?

17.24. A piece of Pb and another piece of radioactive Pb* are welded together as shown in the accompanying figure. After a sufficiently long time at high temperature (but below T_M) the Pb* is uniformly distributed through the couple.
   a. Are Pb and Pb* similar chemically?
   b. Do Pb and Pb* crystals have the same vibrational internal energy?
   c. What is the potential which causes the diffusion to occur?

FIGURE P17.24

17.25. Describe the possibility of an altimeter based on the barometric formula.

17.26. Write the Schrödinger equation for a one-dimensional harmonic oscillator.

17.27. What is $g(v)$ for a Debye solid?

17.28. Both the Einstein and Debye models of the specific heat of solids involve one parameter.
   a. Which gives the better agreement with experiment?
   b. Why?

More Involved Problems

17.29. A gas is placed in a centrifuge and rotated at an angular velocity ω for a long time. Calculate the gas pressure distribution versus radius r, assuming the chamber has an inner radius a and an outer radius b. Suggest values for a, b, and ω which would make this a suitable device for bringing about enrichment of $^{235}UF_6$ in a gas solution also containing $^{238}UF_6$.

17.30. What is the average potential energy of a vertical column of gas, assuming the barometric formula applies?

17.31. When a sound wave travels through a media the compression and expansion are adiabatic.
   a. Show that the adiabatic bulk modulus for an ideal gas is $K^s = \gamma P$.
   b. Hence, show that the sound velocity for an ideal gas is
   $$v_s = \sqrt{\frac{\gamma RT}{M}}.$$
   c. How does v_s compare with $\bar{c}$ for argon?
   d. Does the fact that $v_s \sim \bar{c}$ bother you?

17.32. Using the expression for $dU = T\,dS - P\,dV$, we have
$$dS = \frac{dU}{T} + \frac{P}{T}\,dV.$$

Since $C_v = (\partial Q/\partial T)_V$ and since $dQ = dU + P\,dV$, we have

$$\left(\frac{\partial Q}{\partial T}\right)_V = \left(\frac{\partial U}{\partial T}\right)_V.$$

Hence the equation for dS can be written

$$dS = \frac{C_V\,dT}{T} + \frac{P}{T}\,dV.$$

If we are dealing with an ideal gas,

$$dS = \frac{C_V\,dT}{T} + \frac{R\,dV}{V}.$$

Starting with the expression for $dH = T\,dS + V\,dP$, obtain

$$dS = C_p\frac{dT}{T} - R\frac{dP}{P}$$

for an ideal gas. Use these two equations for dS to show that for an adiabatic process ($T\,dS = 0$), for an ideal gas,

$$PV^\gamma = \text{constant}$$

where

$$\gamma = C_p/C_v.$$

17.33. From Problem 17.32, we have $T\,dS = C_v\,dT + P\,dV$ and $T\,dS = C_p\,dT - V\,dP$.
   a. Find a general expression for $C_p - C_v$.
   b. Show that $C_p - C_v = R$ for an ideal gas.
   c. State what the *function* of classical thermodynamics is and how the derivation of $C_p/C_v = \gamma$ (see Problem 17.32) and $C_p - C_v = R$ illustrate this function.

17.34. Develop a generalized thermodynamics in which the work is represented by

$$dW = \sum_i F_i\,dx_i$$

instead of just by the $-P\,dV$ term.
   a. Define $G(F_i, T)$.
   b. Define $F(x_i, T)$.

17.35. A differential $dg(x, y)$ is exact if, given $dg = M(x, y)\,dx + N(x, y)\,dy$, it is true that

$$\left(\frac{\partial M}{\partial y}\right)_x = \left(\frac{\partial N}{\partial x}\right)_y.$$

Prove this.

17.36. An often used modification of the ideal gas law is the **van der Waals** equation

$$\left(P + \frac{a}{V^2}\right)(V - b) = RT.$$

   a. Explain the origin of the **van der Waals** b and show that it is equal to $4NV_m$, where V_m is the actual volume of the not quite rigid gas molecule. Experimental measurement of b gives the molecular volume and hence the **molecular diameter** d. Examples are shown:

Molecule	d
Ar	2.86
N_2	3.14
O_2	2.90

b. Explain the origin of the a/V^2 term in the van der Waals equation. See Example 9.5.1.

c. The value of b leads to a molecular diameter of 2.86 Å for argon, while the value of the diameter of an argon atom in crystalline argon is 3.83 Å. What is the basis of this discrepancy?

17.37. A chemical analyst finds the following percentages of Cl in a sample:

i	y_i
1	35.18
2	35.30
3	35.06
4	35.44
5	35.02

a. What is $\bar{y}$?

b. Define $x_i = y_i - \bar{y}$.

c. Then $\bar{x} = 0$ and

i	x_i
1	−0.02
2	+0.10
3	−0.14
4	+0.24
5	−0.18

Compute the value of $\overline{x^2}$. If the errors are random, then $\overline{x^2} = \sigma^2$ by (17.3.14). The quantity σ is called the **standard deviation**.

Sophisticated Problem

17.38. a. What is the Doppler shift?

b. A mercury atom at 100°C in an excited state emits radiation with a wavelength of 5461 Å. If the atom is moving with a speed of $\bar{c}$ away from the observer, what is the wavelength?

c. Discuss Doppler broadening in general.

Prologue

Interstitial atoms, such as carbon in iron, jump from one site to another owing to thermal energy. Atoms located at regular lattice sites move through the lattice by interchanging with vacancies. In both cases the atoms move over a *potential energy barrier*. The height of this barrier is characterized by a quantity called the *activation energy*. The number of jumps per second is described by the *Arrhenius equation*. The process of motion of atoms through a crystal from regions of high concentration to low concentration is known as *diffusion*. Diffusion is described by *Fick's law*, which involves a *diffusion coefficient*. We briefly study the mathematics of diffusion and the *root mean square diffusion distance*. We derive an expression for the diffusion coefficient based on the random walk model.

The nature of the diffusion coefficient of *gases*, *liquids*, and *solids* is examined. We study diffusion of carbon in steel, which involves the *error function* and the *carburization depth*. This is followed by a description of the diffusion of vacancies in solids and *self-diffusion* in metals, ionic crystals, and polymers. There is a brief discussion of *chemical interdiffusion* and of the Kirkendall effect. Some examples of diffusion which have commercial importance (see Table 18.12.1) are given, after which the processes used in semiconductor doping in the making of microelectronic circuits are studied in detail. The kinetics of the *nucleation process*, which is often the first step in the occurrence of a phase transformation, is described. Finally, it is noted that some transformations can occur without nucleation by *spinodal decomposition*.

"There is nothing permanent except change."   Heraclitus, 513 B.C.

440

18

KINETICS

18.1   INTRODUCTION TO KINETICS

A large and important part of materials science is the study of **kinetics** or **rate processes**, i.e., the study of how fast processes occur. Such processes might involve single chemical reactions between molecules, diffusion of vacancies in solids, diffusion of interstitial atoms in solids, viscous flow of liquids, creep of metals or ceramics or polymers, oxidation of metals, diffusion bonding of metals, age precipitation hardening, eutectic transformations, eutectoid transformations, grain boundary migration, and recrystallization.

THE REACTION PATH.   We consider the process of vacancy motion in a fcc crystal such as copper. Suppose the vacancy is initially at the center of the cube face, as shown in Figure 18.1.1; suppose that the reaction process, which in this case gives rise to vacancy motion, consists of an exchange of the atom at 000 with the vacancy at $0 \frac{1}{2} \frac{1}{2}$. The atom might move along the face diagonal [011] as shown by the path p_1 in (a) or it might conceivably move along path p_2 as shown in (b). These are called **possible reaction paths**. Suppose that it is possible to calculate the potential energy of the crystal for any position of the atom along p_1. Such a potential energy curve along p_1 is shown in Figure 18.1.2. The minima correspond to the *equilibrium vacancy* positions. The corresponding curve for reaction path p_2 is shown in Figure 18.1.3. Note that for path p_2 the maximum is much higher than for path p_1. The reaction would proceed along the easier path, i.e., along path p_1. It is analogous to climbing across a mountain range; we tend to take the lower path lying between peaks if we are interested in getting across most rapidly. Let us assume that the path p_1 involves the smallest

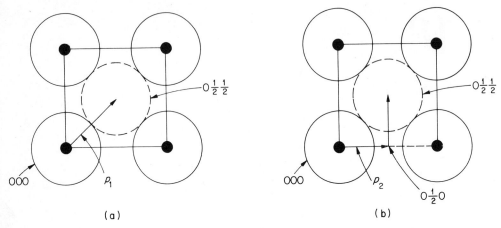

FIGURE 18.1.1. (a) Reaction path p_1. (b) Reaction path p_2 in (100) plane of a fcc crystal. Vacancy shown by broken circle.

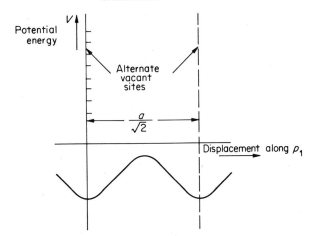

FIGURE 18.1.2. Potential energy versus possible reaction path for vacancy motion. The path p_1 considered here is the straight line from $0\,0\,0$ to $0\frac{1}{2}\frac{1}{2}$.

potential energy peak. The barrier over which the atom must pass is called the **potential energy barrier**.

THE ACTIVATED STATE. Let the increase in energy from the minimum along the actual reaction path to the maximum be ΔU^*, as in Figure 18.1.4. This is called the **activation energy** (for vacancy motion in the present case). Although the maximum corresponds to a metastable equilibrium position rather than a stable equilibrium position (refer to Figure 15.3.1), it is usually assumed

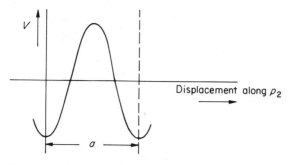

FIGURE 18.1.3. Potential energy versus possible reaction path
for vacancy motion. The path p_2 considered
here is the pair of straight lines $0\,0\,0$ to $0\frac{1}{2}0$
and $0\frac{1}{2}0$ to $0\frac{1}{2}\frac{1}{2}$.

FIGURE 18.1.4. Activation
energy.

that it can be treated as an excited energy level called the **activated state**, which
is an equilibrium-like state. When we consider a crystal with a large number of
vacancies, the fraction of vacancies which are in the activated state is given by

$$f = e^{-\Delta U^*/kT}. \qquad (18.1.1)$$

This follows from either (17.10.7) or by the same argument used to obtain
(17.8.7). (Actually we should use

$$f = e^{-\Delta G^*/kT},$$

where ΔG^* is the Gibbs free energy, but we leave this improvement to more
advanced discussion.) The fraction f can also be given the following interpretation: It is the fraction of time a single vacancy is in the activated state. Since an
atom is in the process of exchanging with the vacancy, it is also the fraction of
time that atom is in the activated state.

THE ATTEMPT FREQUENCY.   Atoms in crystals are vibrating. The high-
energy vibrations have frequencies of about 10^{13} sec^{-1}. The atom at 000 would
usually tend to have displacements which would keep it near the minimum of
Figure 18.1.4. However, for a fraction f of these vibrations the atom would move
to the activated state where it could then pass over the barrier to the new position.
The jumps per second for such a process are given by

$$\Gamma = \nu e^{-\Delta U^*/kT}, \qquad (18.1.2)$$

where ν is the vibrational frequency along the reaction path. (It can be assumed to be roughly equal to the Debye frequency; see Section 17.11.) The theory of reaction rates is primarily due to H. Eyring (see Glasstone, et al, References).

EXAMPLE 18.1.1

a. Assuming that aluminum has $\nu = 0.8 \times 10^{13}$ sec^{-1} and $\Delta U^* = 0.6$ eV, calculate the number of vacancy jumps per second at 300°K.

Answer.

$$\Gamma = 8 \times 10^{12} 10^{-0.6/2.3 \times 8.62 \times 10^{-5} \times 300}$$
$$= 570 \text{ jumps/sec.}$$

b. Calculate the number of vacancy jumps in the crystals when $T = T_M = 943°$K.

Answer.

$$\text{jumps/sec} = 4.7 \times 10^7.$$

GENERALIZATION.   The above discussion applies not only to vacancy motion but to chemical reactions in general. In most cases the barrier is not symmetric, and the products have a lower energy than the reactants, as shown in Figure 18.1.5. In this case the activation energy for the reaction $A \to B$ is much less than the activation energy for the reaction $B \to A$.

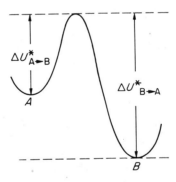

FIGURE 18.1.5. Typical potential barrier.

EXAMPLE 18.1.2

Graphite is more stable than diamond at atmospheric pressure and temperature. Yet diamond persists. Why?

Answer. Diamond corresponds to A and graphite to B in Figure 18.1.5 (however, the barrier would be many times higher than the small energy difference between the two minima). Because of the large barrier, the *rate* at which diamond transforms into graphite is extremely small at room temperature. This would not be the case at 2000°K.

From an historical viewpoint the chemist Svante Arrhenius had noted empirically that chemical reactions proceed at a rate which depends exponentially on the negative reciprocal of absolute temperature :

$$\text{rate} = Ae^{-\Delta U^*/RT}. \tag{18.1.3}$$

In this case the activation energy ΔU^* has units of energy per mole and hence we use R instead of k. Equation (18.1.3) is called the **Arrhenius rate equation**. A is a pre-exponential parameter.

18.2 INTRODUCTION TO DIFFUSION

If a tank of argon gas is connected to a tank of helium gas at the same pressure and temperature, the two gases will diffuse into each other until a uniform composition is reached throughout. Similarly, two miscible liquids such as water and alcohol, if placed in contact with each other, will diffuse into each other until a uniform composition is reached throughout. Likewise, if a block of copper and a block of nickel are clamped together and held at 1000°C for a long time, they will diffuse into each other and eventually a uniform composition will be reached throughout (copper and nickel form a continuous solid substitutional solution; see Figure 16.4.6). From the macroscopic viewpoint the only difference between these processes is the rate: Diffusion is fastest in the gas and slowest in the solid.

MACROSCOPIC VIEWPOINT. In 1855 Adolph Fick proposed empirically that diffusion is described (in one dimension) by

$$J_x = -D\frac{\partial C}{\partial x}. \tag{18.2.1}$$

Here J_x is the flux in the x direction (quantity of material crossing a unit area, whose normal is in the x direction, in unit time). C is the concentration of the diffusing species and x measures distance, so that $\partial C/\partial x$ is a gradient. D is the diffusion coefficient. This equation is now known as **Fick's law**, or, more properly, as **Fick's first equation**. Note that it is a direct analog of Fourier's heat conduction equation and Ohm's electrical conduction equation. If C is given in mole fraction, x in centimeters, and t in seconds and flux is measured in terms of cubic centimeters per square centimeter-second, then the unit of D is square centimeters per second. This is the most common unit. According to (18.2.1), atoms will diffuse from regions of high concentration to regions of low concentration (in multicomponent systems there can be exceptions to this rule but that is a more involved topic which need not concern us here). There are often cases in which D does not vary with x or with C, in which case (18.2.1) becomes a *linear* relation.

STEADY-STATE DIFFUSION. In some problems involving diffusion the concentration of the diffusing atom at a given position does not vary with time. These are called **steady-state diffusion** problems. As an example, hydrogen gas can be purified by passing it through a palladium foil. Hydrogen dissolves to a considerable extent in palladium (the molecule dissociates and the hydrogen in the palladium is present as an atom). The solubility increases as the pressure increases, so if there is a high gas pressure on one side of the foil, the concentration there will be high, while if there is a low gas pressure on the other side, the concentration there will be low. It can be safely assumed that the rate at which the hydrogen enters or leaves the foil is very fast compared to the rate of diffusion through it. (*Note:* Other gases which might be present in the hydrogen as impurities dissolve only slightly if at all in the palladium and diffuse much more slowly than the tiny hydrogen atoms, which can readily move from interstitial position to interstitial position.) The hydrogen flow rate can then be written:

$$Q = J_x A = \frac{D(C_2 - C_1)A}{L}. \qquad (18.2.2)$$

The process is illustrated in Figure 18.2.1. Here A is the foil area, C_2 is the dissolved hydrogen concentration on the high pressure side and C_1 on the low pressure side, while L is the foil thickness.

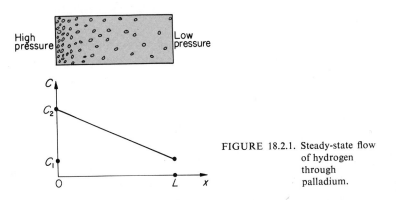

FIGURE 18.2.1. Steady-state flow of hydrogen through palladium.

TIME-DEPENDENT DIFFUSION IN ONE-DIMENSION. Consider a volume element of material lying between x and $x + dx$ and having a unit cross-sectional area as in Figure 18.2.2. The flux of a given component into the element minus the flux out of the element equals the rate of accumulation in the element:

$$J_x - J_{x+dx} = \frac{\partial \bar{C}}{\partial t} dx. \qquad (18.2.3)$$

Here $\bar{C}$ is the average concentration in the element so that $\bar{C} \, dx$ is the total amount of the component in the element at time t.

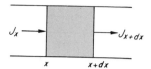

FIGURE 18.2.2. Element involving diffusion.

We expand J_{x+dx} about x in a Taylor series:

$$J_{x+dx} = J_x + \frac{\partial J_x}{\partial x}dx + \frac{\partial^2 J_x}{\partial x^2}\frac{dx^2}{2} + \cdots. \tag{18.2.4}$$

It follows that in the limit as $dx \rightarrow 0$, Equation (18.2.3) has the form

$$-\frac{\partial J_x}{\partial x} = \frac{\partial C}{\partial t}. \tag{18.2.5}$$

This is the **equation of continuity**. Note that it is simply a statement of the conservation of matter.

    If the equation of continuity and Fick's first equation are combined, we
obtain

$$\frac{\partial}{\partial x}\left(D\frac{\partial C}{\partial x}\right) = \frac{\partial C}{\partial t}. \tag{18.2.6}$$

If D does not vary with x, we have

$$D\frac{\partial^2 C}{\partial x^2} = \frac{\partial C}{\partial t}. \tag{18.2.7}$$

If, furthermore, D does not vary with t, this is a *linear* partial differential equation. It is often known as **Fick's second equation**, or as the **parabolic diffusion
equation**.

    Since few students at this stage are familiar with the methods of solution
of partial differential equations, we shall give the solutions for a few specific
cases in later sections.

18.3   THIN LAYER OF TRACER ATOMS ON A THICK SLAB

    Figure 18.3.1 shows a thin layer of radioactive tracer atoms on a thick
slab of metal, e.g., Ag* on Ag. We let the quantity of tracer per unit area be α.
The solution to the differential Equation (18.2.7) subject to the condition that
all of the tracer plate is initially at $x = 0$ (which is an approximation to the
actual case [Figure 18.1.1(b)] where the plate has thickness x_0) is

$$C^*(x, t) = \frac{\alpha}{\sqrt{\pi Dt}}e^{-x^2/4Dt}, \qquad x > 0, t > 0. \tag{18.3.1}$$

You can verify that this is a solution by direct substitution.

    Experimentally, to obtain the tracer coefficient D, we plate a very thin
layer on a clean surface of the specimen. Then the specimen is raised to a high

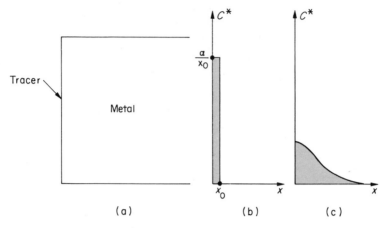

FIGURE 18.3.1. Geometry for diffusion of tracer plate into a slab.

temperature and held at a fixed diffusion temperature for a period of time t. (This anneal may last for ten days for diffusion in a solid near its melting point.) The solid is then cooled to room temperature. The most common and most accurate method to follow next is to slice the specimen very carefully, i.e., to remove layers perhaps 0.001 in. thick. Then the radioactivity in each of these layers is measured. Finally, we usually plot $C^*(x, t)$ vs. x^2 on semilog paper. As seen from (18.3.1) we have

$$\ln C^*(x, t) = \ln\left(\frac{\alpha}{\sqrt{\pi Dt}}\right) - \frac{x^2}{4Dt}. \tag{18.3.2}$$

Thus the slope of the straight line is $-(1/4Dt)$. Since t is known and the slope of the straight line is known, D can be obtained. This is known as the **slicing method**. By repeating the experiment at several different diffusion temperatures we can obtain the temperature dependence of diffusion.

We have assumed here that, in the solid silver, diffusion takes place at a reasonable rate near the melting point and *very* slowly (relatively) at room temperature (the temperature at which slicing occurs). Because the temperature dependence of tracer diffusion in solids or liquids (but not gases) is given by

$$D = D_0 e^{-Q_D/kT} \tag{18.3.3}$$

and because $D \approx 10^{-8}$ cm²/sec near T_M, this is justified. Q_D is called the **activation energy for diffusion**.

Because a tracer atom of Ag* has the same electron configuration as Ag (the system differs only through the nuclear mass and hence the vibrational frequencies will be slightly different), the measurement of the rate of diffusion of tracer silver atoms essentially gives the rate of diffusion of silver atoms, i.e., the **self-diffusion** rate.

THIN LAYER OF TRACER MATERIAL BETWEEN TWO THICK SLABS. If the layer can be considered as completely concentrated at $x = 0$, we have as a solution to the diffusion problem

$$C^*(x, t) = \frac{\alpha}{2\sqrt{\pi Dt}} e^{-x^2/4Dt}. \tag{18.3.4}$$

The initial position of all the tracer atoms is at $x = 0$. The average position at time t is $\bar{x} = 0$, while the average squared position is

$$\overline{x^2} = \frac{\int_{-\infty}^{\infty} x^2 C^*(x, t)\, dx}{\int_{-\infty}^{\infty} C^*(x, t)\, dx}. \tag{18.3.5}$$

This is so since $C^*(x, t)\, dx$ is simply the quantity of tracer material between x and $x + dx$, and

$$\frac{C^*(x, t)\, dx}{\int_{-\infty}^{\infty} C^*(x, t)\, dx}$$

is the *fraction* of *all* the tracer material between x and $x + dx$. Evaluation of the definite integral gives $\overline{x^2} = 2Dt$ so that the root mean squared position is

$$L = \sqrt{\overline{x^2}} = \sqrt{2Dt}. \tag{18.3.6}$$

This equation is one of the most important equations in diffusion. Because of its simplicity, it is very useful to use in making estimates of how far an atom (or molecule) diffuses in a time t. Engineers and scientists must develop the ability to make rapid estimates.

EXAMPLE 18.3.1

Estimate the diffusion distance in solid silver at the melting point assuming that $D = 10^{-8}$ cm²/sec and $t = 10$ days.

Answer. Since 10 days $\approx 10^6$ sec, we have by (18.3.6),

$$L \approx 0.14 \text{ cm.}$$

The actual concentration profiles for several diffusion times are shown in Figure 18.3.2.

EXAMPLE 18.3.2

Given that $D_0 = 1$ cm²/sec for Ag, calculate D at $T_M/3$ and use this to estimate L for $t = 10$ days at $T_M/3$.

Answer. Since by (18.3.3) we have at $T = T_M$

$$D = D_0 e^{-Q_D/RT_M} \approx 10^{-8} \text{ cm}^2/\text{sec}$$

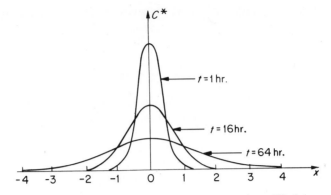

FIGURE 18.3.2. Concentration profiles for various diffusion times.

and we have at $T_M/3$

$$D = e^{-3Q_D/RT_M} = [e^{-Q_D/RT_M}]^3 = 10^{-24} \text{ cm}^2/\text{sec},$$

then

$$L \approx 1.4 \times 10^{-9} \text{ cm} \approx 0.$$

EXAMPLE 18.3.3

If the tracer experiment described by (18.3.2) involved Ag* in solid Ag near T_M, how thin is a *thin* layer and how thick is *thick* slab?

Answer. The layer is thin if it is very small relative to the diffusion distance $L = 0.14$ cm (Example 18.3.1). Thus a layer of 0.001-cm thickness would be thin. The slab is thick if its thickness is very large relative to $L = 0.14$ cm. Here a thickness of only 1.4 cm would suffice.

In the future we shall have many occasions to refer to Equation (18.3.6). We shall illustrate the usefulness now in an example.

EXAMPLE 18.3.4

During a given eutectic transformation the liquid spends 1 sec between T_E and $T_E - \Delta T$. It freezes at $T_E - \Delta T$. If the effective diffusion coefficient in the liquid at T_E is 10^{-5} cm²/sec, estimate the resultant lamellar spacing.

Answer. The diffusion takes place in the *thin* liquid layer between T_E and $T_E - \Delta T$, in a direction normal to the lamellae. Hence

$$L \approx \sqrt{2 \times 10^{-5} \times 1} \approx 0.0045 \text{ cm}.$$

18.4   THE MICROSCOPIC VIEWPOINT OF DIFFUSION

THE RANDOM WALK.   Diffusion in a crystal can be described in terms of a random walk. The simplest case is the one-dimensional case. Here the atom jumps from one position to another a distance $\pm l$ along the x axis. The probability of a step in the positive x direction equals the probability in the negative x direction. Hence

$$x = l_1 + l_2 + l_3 + \cdots + l_N \tag{18.4.1}$$
$$= (\pm l) + (\pm l) + (\pm l) + \cdots + (\pm l). \tag{18.4.2}$$

Thus for a large number of steps N, $\bar{x} = 0$, but

$$\overline{x^2} = \overline{[l_1^2 + l_2^2 + l_3^2 + \cdots + l_N^2]} + \overline{(2l_1 l_2 + 2l_1 l_3 + \cdots + 2l_1 l_N)} + \cdots. \tag{18.4.3}$$

We then have

$$\overline{x^2} = Nl^2,$$

since only the term in square brackets contributes, so that

$$L = \sqrt{\overline{x^2}} = \sqrt{N}\, l. \tag{18.4.4}$$

[The quantity in parentheses in (18.4.3), and similar quantities like it, are zero, since each term in this quantity may have the value $-2l^2$ or $2l^2$ with equal probability.]

From (18.3.6) we have $L = \sqrt{2Dt}$ which combined with (18.4.4) gives

$$D = \frac{N}{t} \frac{l^2}{2} = \frac{1}{2}\Gamma l^2 \tag{18.4.5}$$

since N/t is the number of jumps per second Γ.

In a liquid or solid but not in a gas, Γ is given by (18.1.2), i.e.,

$$\Gamma = v e^{-\Delta U^*/RT}.$$

This explains why the diffusion coefficient has the form

$$D = D_0 e^{-Q_D/RT} \tag{18.4.6}$$

in liquids and solids.

The actual distribution of atom positions in a one-dimensional random walk is given by Equation (17.3.7), and if this is combined with (18.4.5), it leads to Equation (18.3.4) with $\alpha = 1$. Thus the connection between the phenomenological description of diffusion and the random walk fundamental theory is clearly established. (The student should note that Section 17.3 can be read independently of the rest of Chapter 17 if he has not yet covered that.)

LIQUIDS.   Self-diffusion in liquids is an activated state process with a temperature dependence as given by (18.4.6). The diffusion coefficient for simple organic liquids at the melting point is about 10^{-4}–10^{-5} cm²/sec. Numerous values are given in S. Glasstone, K. J. Laidler, and H. Eyring, *The Theory of Rate Processes*, McGraw-Hill, New York (1941).

GASES.   In a dilute gas an atom (or molecule) moves freely through space until it collides with another atom (or molecule) and then it moves off in some other direction.

Let λ be the mean free path (average distance between collisions) and $\bar{c}$ the average speed of the molecules. It can be shown that the mean free path is given by

$$\lambda = \frac{1}{\sqrt{2}\,\pi n d^2},\tag{18.4.7}$$

where n is the number of molecules per unit volume and d is the molecular collision diameter. When a test molecule of diameter d moves along, the centers of other atoms must be a distance d from the center of the test molecule. Hence, it sweeps a cylinder of diameter $2d$ and area πd^2 and moves through a volume

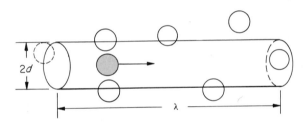

FIGURE 18.4.1. Molecule (shown darkened) moves through cylinder without colliding.

$\pi d^2 \lambda$ before it undergoes a collision as illustrated in Figure 18.4.1. Hence the number of molecules per unit volume is

$$n = \frac{1}{\pi d^2 \lambda}.$$

An exact derivation accounts for the fact that all the molecules are moving and gives the extra factor $1/\sqrt{2}$ in (18.4.7). For a derivation of (18.4.7), see E. H. Kennard, *Kinetic Theory of Gases*, McGraw-Hill, New York (1938). From (18.4.5) we would write

$$D = \frac{1}{2}\left(\frac{\bar{c}}{\lambda}\right)\lambda^2 = \frac{1}{2}\bar{c}\lambda.$$

Kennard also shows that this is not quite correct and that the effective jump distance is only $2\lambda/3$. Then

$$D = \tfrac{1}{3}\lambda\bar{c}. \qquad (18.4.8)$$

Molecular collision diameters can be obtained from the van der Waals b (see Problem 17.36). Thus for argon $d = 2.86$ Å. From this a mean free path of $\lambda = 635$ Å is obtained for 0°C and 1 atm. Using $\bar{c}$ from Table 17.4.1 (381 m/sec) we have $D = 0.08$ cm²/sec. This is a typical value for self-diffusion of a gas.

EXAMPLE 18.4.1

Show that diffusion in a gas is similar to that of a solid in which $\Gamma = \nu$ (i.e., the activation energy is zero) with very large jump distances.

Answer. We note that $\Gamma = \bar{c}/\lambda$ or 0.6×10^{10} but $\Gamma = \nu$ if the activation energy is zero. This value of Γ is much less than the upper vibrational frequency of solids.

AN ALTERNATE DERIVATION OF EQUATION (18.4.5). We consider atoms located at x and at $x + l$ and calculate the net flux across $x + (l/2)$. The concentration at x is C.

EXAMPLE 18.4.2

What is the concentration at $x + l$?

Answer. Assuming that C is a well-behaved function of x, we can write

$$C(x + l) = C(x) + \left(\frac{\partial C}{\partial x}\right)l + \left(\frac{\partial^2 C}{\partial x^2}\right)\frac{l^2}{2} + \cdots.$$

Since l is so very small and since C does not vary extremely rapidly with x (so all the derivatives can be assumed small), we can write

$$C(x + l) \doteq C(x) + \frac{\partial C}{\partial x}l.$$

We assume jumps can occur only in the x direction. An atom at x makes Γ jumps/sec. These may be in either the positive or the negative direction. The number in the positive x direction is $\Gamma/2$. The flux in the positive x direction across $x + (l/2)$ is

$$J_+ = \frac{\Gamma}{2}Cl.$$

Note that $(\Gamma/2)l$ is the velocity of atoms in the positive x direction.

Similarly, the flux across $x + (l/2)$ in the negative direction is

$$J_- = \frac{\Gamma}{2}C(x + l)l = \frac{\Gamma}{2}\left(C + \frac{\partial C}{\partial x}l\right)l.$$

The net flux across $x + (l/2)$ in the positive direction is

$$J_x = J_+ - J_- = -\frac{\Gamma}{2}l^2\frac{\partial C}{\partial x}.$$

This is Fick's law with the diffusion coefficient given by (18.4.5). This could be generalized to include jumps in various directions.

18.5   INTERSTITIAL IMPURITY DIFFUSION

The diffusion of carbon in steel is an example of interstitial diffusion. The rate of a eutectoid transformation is often governed by the rate of carbon diffusion. So is the rate of carburization of steel (here carbon is diffused in from the surface to produce a steel with higher carbon content near the surface; upon quenching, such a steel attains a high hardness near the surface). The carbon atoms are located in the octahedral voids (see Figure 12.4.1), although the tetrahedral void is larger. This must be due to the fact that carbon prefers the octahedral coordination in solution just as it does in forming compounds (see Figure 12.6.2). The octahedral void position in bcc iron is at the center of faces and at the center of edges. A specific interstitial atom may move into any one of four positions a distance $a/2$ away, where a is the lattice parameter.

It is left for Problem 18.37 to show that

$$D = \frac{\Gamma a^2}{6} \tag{18.5.1}$$

and

$$D = D_0 e^{-Q_D/RT}. \tag{18.5.2}$$

The accurate calculation of D_0 for this process by C. M. Zener was one of the early successes of diffusion theory. Measured values for D_0 and Q_D are given in Table 18.5.1.

Table 18.5.1 INTERSTITIAL IMPURITY DIFFUSION

Solvent	Solute	D_0, cm^2/sec	Q_D, kcal/mole	Ref.
Fe bcc	C	2.0×10^{-2}	20.1	Wert
Fe fcc	C	0.21	33.8	Guy
Fe fcc	N	1.07×10^{-1}	34.0	Smithells
Fe fcc	S	4.8×10^{-6}	23.4	Barrer
Fe fcc	P	4.5×10^{-2}	43.3	Barrer
Al fcc	H	3.4×10^{2}	24.6	Smithells
Fe bcc	H	1.6×10^{-2}	9.2	Barrer
Ni fcc	H	2.04×10^{-3}	8.7	Barrer
Pd fcc	H	1.5×10^{-2}	6.8	Smithells

EXAMPLE 18.5.1

How many jumps per second does a carbon atom make in iron at room temperature?

Answer.

$$\text{jumps/sec} = \Gamma = \nu e^{-Q_D/kT}.$$

We can solve for ν since we have $D_0 = 2.0 \times 10^{-2}$ cm²/sec from Table 18.5.1 and $a = 2.86 \times 10^{-8}$ cm from Table 10.6.1. We have

$$\nu = \frac{6D_0}{a^2} \approx 1.5 \times 10^{14} \text{ sec}^{-1}.$$

Hence

$$\text{jumps/sec} = 1.5 \times 10^{14} 10^{-20,100/2.3 \times 1.987 \times 300}$$
$$\approx \tfrac{1}{2} \text{ jump/sec.}$$

EXAMPLE 18.5.2

What is the diffusion coefficient for carbon in bcc iron at room temperature?

Answer. From Table 18.5.1 and Equation (18.5.2)

$$D \approx 10^{-16} \text{ cm}^2/\text{sec.}$$

Note how small this is compared to the value for argon gas at room temperature and atmospheric pressure but how large it is compared to silver [room temperature is about $T_M/4$ so $D = 10^{-32}$ cm²/sec for silver at room temperature (see Example 18.3.2)].

Note that the rate would be 10^{10} times as fast just below the eutectoid temperature.

DIFFUSION INTO A SEMI-INFINITE SLAB WITH CONCENTRATION FIXED AT THE SURFACE. We are often concerned with one-dimensional diffusion of an impurity into a solid in which the initial impurity concentration is

$$C(x, t) = C_0 \qquad \text{for } x > 0 \text{ at } t = 0 \tag{18.5.3}$$

and in which the concentration at the surface is very rapidly (assume instantaneously) brought to a fixed concentration C_s so that

$$C(x, t) = C_s \qquad \text{for } x = 0 \text{ at } t > 0. \tag{18.5.4}$$

We are therefore interested in solving

$$D\frac{\partial^2 C}{\partial x^2} = \frac{\partial C}{\partial t} \tag{18.5.5}$$

subject to (18.5.3) and (18.5.4). The solution is

$$\frac{C(x, t) - C_0}{C_s - C_0} = 1 - \text{erf}\left(\frac{x}{2\sqrt{Dt}}\right). \tag{18.5.6}$$

Here we have the function erf (y), which is called the **error function of y**, the Gaussian error function, or the normalized probability integral:

$$\text{erf } y = \frac{2}{\sqrt{\pi}} \int_0^y e^{-z^2} \, dz. \tag{18.5.7}$$

This function, like many other well-known functions such as $\sin \theta$, $\cos \theta$, etc., is tabulated. Table 18.5.2 gives a few results.

Table 18.5.2. THE ERROR FUNCTION

y	erf y	y	erf y
0.0	0.	0.8	0.742
0.1	0.113	0.9	0.797
0.2	0.223	1.0	0.843
0.3	0.329	1.2	0.910
0.4	0.428	1.4	0.952
0.5	0.521	1.6	0.976
0.6	0.604	1.8	0.989
0.7	0.679	2.0	0.995

EXAMPLE 18.5.3

If the table of error functions were not available, how would you evaluate erf y?

Answer. Expand the integrand in a power series and integrate term by term. For small y only a few terms would be needed. For large y, however, this method would become tedious. There is an asymptotic formula which could be used for large y.

EXAMPLE 18.5.4

Suppose an 0.2 wt. % carbon steel is carburized at 950°C. Assume the surface carbon content under these conditions reaches a carbon content of 1.4 wt. %. What is the depth at which the increased carbon content equals one half of the increased carbon content at the surface?

Answer. The question is really, what is $x/2\sqrt{Dt} = y$ when

$$\frac{C(x, t) - C_0}{C_s - C_0} = \frac{1}{2}?$$

From Equation (18.5.6), we have erf $(x/2\sqrt{Dt}) = \frac{1}{2}$. From Table 18.5.2,

$y \approx \frac{1}{2}$. Hence we have

$$\frac{x}{2\sqrt{Dt}} \approx \frac{1}{2}$$

or

$$x \approx \sqrt{Dt}.$$

This could be used as a measure of the **carburization depth**. Since D could be calculated from the data of Table 18.5.1 and from (18.4.6), the carburization depth can be readily calculated once t is known. The high carbon steel at the surface when quenched produces very hard martensite. Thus, the resultant component has a hard surface but a tough (not brittle) interior. Processes for producing such surfaces also include nitriding and siliciding of steels, i.e., diffusion of nitrogen and silicon into steels.

18.6   THE DIFFUSION OF VACANCIES

VACANCY FLOW.   If there is a flux of vacancies in one direction, there must be a flux of atoms in the opposite direction, i.e., a net transport of material. A net flux of vacancies is very important in many problems such as creep and void formation during deformation. The random motion of vacancies (no net flux of vacancies) is also important since it can cause interdiffusion of atoms. Thus if pieces of Pb* and Pb are placed together, they will eventually end up as a uniform solution of Pb* in Pb. The interdiffusion is caused by the random motion of vacancies *only*. If Cu and Ni pieces are clamped together at high temperatures, they will also interdiffuse, but in this case there will be both a random motion of vacancies and a net flux of vacancies due to the fact that copper atoms exchange positions with vacancies more rapidly than nickel atoms exchange positions with vacancies.

Figure 18.6.1 shows a fcc crystal with a vacancy surrounded by 12 nearest-neighbor atoms. Atoms 1, 2, 3, and 4 and 9, 10, 11, and 12 are at face centers.

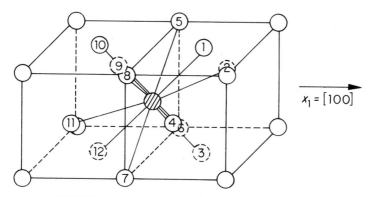

FIGURE 18.6.1. Vacancy in fcc crystal (shown hatched).

By calculating the net number of jumps of this vacancy across the plane at $x_1 = a/4$ (where a is the lattice parameter) we can show that the diffusion coefficient for vacancies is given by

$$D_v = va^2 e^{-E_m/kT}, \qquad (18.6.1)$$

where E_m is the activation energy for vacancy motion. The actual derivation of (18.6.1) is left as Problem 18.38. A similar expression but with a different geometric factor applies for bcc or diamond cubic crystals. In general for cubic crystals

$$D_v = D_0 e^{-E_m/kT}. \qquad (18.6.2)$$

Some values of the activation energy for vacancy motion are given in Table 18.6.1. These are primarily from the work of R. W. Balluffi.

Table 18.6.1. ENERGIES FOR VACANCY MOTION

Metals	kcal/mole
Al	14
Au	20
Ag	19
Cu	23

The description of net vacancy flow therefore involves the flux of vacancies from one region to another. This is described by Fick's law and by the equation of continuity.

VACANCY SOURCES AND SINKS.   We note that there must be a source and a sink for vacancies since there is a *net* flux of vacancies in the above case. Vacancies can be readily created or destroyed at **kinks** on surface **steps** (surfaces as a rule are not smooth). Figure 18.6.2 shows how a vacancy within the crystal could exchange with the shaded atom and the kink would move one unit toward negative x_1. The vacancy would be destroyed.

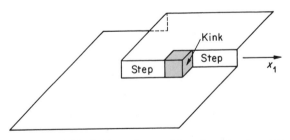

FIGURE 18.6.2. Portion of a crystal surface showing steps and kinks.

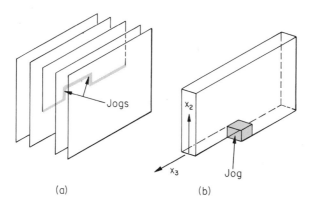

FIGURE 18.6.3. (a) Jogs on an edge dislocation. (b) Extra half-plane of atoms with dislocation line at bottom and jog on dislocation line.

Vacancies can also be readily created at jogs on dislocations. Figure 18.6.3 shows the extra plane of atoms inserted to create an edge dislocation (see Figure 13.3.1). If a vacancy interchanges with the shaded atom, the vacancy is destroyed and the jog moves in the negative x_3 direction. This is called **disloca-tion climb**. Note that by repeated operations of this sort the edge dislocation line climbs upward (in the positive x_2 direction). This type of dislocation motion is different from the motion of slip or glide discussed in Chapter 13. Dislocation climb is an important process in high-temperature creep. Near the melting point the equilibrium vacancy concentration in a metal is 10^{-4} (see Example 13.2.1). However, this drops off rapidly with temperature so that at $T_M/4$ the vacancy fraction is only 10^{-16} (none for most practical purposes). Consider now a rod of silver with essentially no vacancies at room temperature. If this is heated quickly almost to T_M, how long will it take to nearly reach the equilibrium con-centration (which might mean only half the equilibrium concentration)? If there are no dislocations in the crystal, then the vacancies would have to come from the surface. If x is the distance from the surface to the center of the rod, then we write

$$x \approx 2\sqrt{D_v t},$$

and if D_v, the vacancy diffusion coefficient, is known, t can be calculated.

Usually, there are numerous dislocations in the crystal and the time would then be determined by half the distance between the dislocations rather than half the rod diameter. *Note:* It can usually be assumed that at the vacancy source (surface or dislocation) the equilibrium concentration is very rapidly reached and maintained.

The motion of self-interstitials is also important in processes such as the annealing following radiation damage of a metal (radiation produces both vacancies and interstitials).

18.7   SELF-DIFFUSION AND TRACER DIFFUSION

SELF-DIFFUSION.   If all n vacancies in a copper crystal make one jump, each of the N atoms on the average make only $n/(N + n)$ jumps since the chance of there being a vacancy next to any copper atom at any instant is $n/(N + n)$. Hence the diffusion coefficient for atoms is

$$D_a = \frac{n}{N + n} D_v. \tag{18.7.1}$$

This is called the **self-diffusion coefficient**.

Here $n/(N + n)$ is given by (17.8.4) and D_v is given by (18.6.2). Thus

$$D_a = D_0 e^{-(E_m + E_f)/kT}, \tag{18.7.2}$$

where E_m is the energy of vacancy motion and E_f is the energy of vacancy formation. Thus

$$Q_D = E_m + E_f. \tag{18.7.3}$$

These three quantities can be measured independently so that (18.7.3) can be directly verified by experiment. Much of the work in this area is due to R. W. Balluffi and is discussed in P. G. Shewmon, *Diffusion in Solids*, McGraw-Hill, New York (1963) p. 71. It is currently felt that tracer diffusion in metals takes place by the vacancy mechanism.

TRACER DIFFUSION.   It can be shown that the tracer diffusion coefficient in systems in which the tracer atom moves by exchange with a vacancy is

$$D_T = f D_a. \tag{18.7.4}$$

The **tracer diffusion coefficient** is the diffusion coefficient measured by using a tracer atom, as discussed in Section 18.3. Here f, called the **correlation factor**, is approximately given by

$$f \approx 1 - \frac{2}{z}, \tag{18.7.5}$$

where z is the number of nearest neighbors. The need for the correlation factor is illustrated in Figure 18.7.1. In this case $D_T = 0$ since the tracer atom at position p moves to $p + 1$ and back to p. Hence $D_T = 0$ and $f = 0$. The correlation factor arises because of the fact that after the jump the chance that a vacancy is next to the tracer in a position to cause a reverse jump is not $n/(N + n)$ but is initially 1.

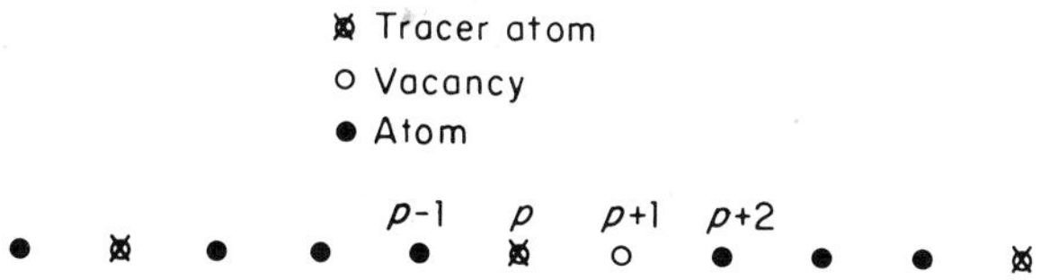

       ⊠ Tracer atom
       ○ Vacancy
       ● Atom

$p-1$ p $p+1$ $p+2$

●   ⊠   ●   ●   ●   ⊠   ○   ●   ●   ●   ⊠

FIGURE 18.7.1. Illustration of the lack of tracer diffusion by a vacancy mechanism on a one-dimensional lattice. The tracer atom at p is doomed to stay at p or $p + 1$ regardless of how many jumps the vacancy makes.

The evaluation of the correlation factor in other cases is more involved but has been done for several cases by J. Bardeen and C. Herring and later by K. Compaan and V. Haven, *Transactions of the Faraday Society*, **52**, 786 (1956). The results are shown in Table 18.7.1. The student may want to derive an approximate answer for the square lattice.

Table 18.7.1 CORRELATION FACTORS

Crystal Structure	Correlation Factor
Linear chain	0
Square lattice	0.467
Hexagonal planar lattice	0.560
Diamond cubic	0.500 . . .
Simple cubic	0.65
Body-centered cubic	0.72
Face-centered cubic	0.78

Tracer diffusion data for some metals are shown in Table 18.7.2. There are some rough correlations which are helpful—first, in the solid at the melting point $D_T \approx 10^{-8}$ cm²/sec. Second, there is a rough correlation giving

$$Q_D \approx 36 T_M \text{ cal/mole} \qquad (18.7.6)$$

where T_M is the melting point in degrees Kelvin.

18.8   IONIC CRYSTALS

Pure ionic crystals such as NaCl will have sodium ion vacancies and chloride ion vacancies in equal concentrations. Tracers can be used to obtain the tracer diffusion coefficients for the sodium and the chlorine. Usually in ionic crystals one of these atoms diffuses much faster than the other. In the case of NaCl, the smaller sodium ion diffuses much faster. The electrical conductivity

Table 18.7.2. SELF-DIFFUSION DATA*

Metal	Melting tempera- ture, °C	Crystal Structure	Temp. Range of diffusion data, °C	Values of diffusion constants	
				D_0, cm²/sec	Q_d, kcal/mole
Sodium	98	bcc	0–95	0.242	10.45
Lithium	186	bcc	0–86	0.23	13.2
β-Thallium	302	bcc	241–275	0.7	20.0
γ-Uranium	1150	bcc	807–1069	1.2×10^{-3}	26.5
α-Iron	1535	bcc	800–900	2300	73.2
				5.8	59.7
β-Zirconium	1857	bcc	Not stated	10^{-4}	27.0
Chromium	1890	bcc	950–1250	10^{-4}	53.0
Niobium	2467	bcc	1585–2120	12.4	105.0
Tantalum	2996	bcc	1827–2527	2.0	110.0
Indium	156.5	fc tetragonal	50–150	1.02	17.9
α-Thallium	302	hcp	149–227	0.4 (‖ to c)	22.9 (‖ to c)
				0.4 (⊥ to c)	22.6 (⊥ to c)
Cadmium	321	hcp	127–309	0.075	18.6
Lead	337	fcc	174–322	0.28	24.2
Zinc	419	hcp	240–410	0.19	22.7
Magnesium	651	hcp	468–627	1.2	32.0
Silver	960.5	fcc	600–948	0.724	45.5
Gold	1063	fcc	704–1048	0.09	41.7
Copper	1083	fcc	658–1062	0.2	47.1
β-Uranium	1150	tetragonal	700–755	1.35×10^{-2}	42.0
Nickel	1455	fcc	870–1248	1.27	66.8
β-Cobalt	1495	fcc	1000–1300	0.2	62.0
γ-Iron	1535	fcc	1000–1300	0.44	67.0
Platinum	1773	fcc	1325–1600	0.33	68.0
α-Zirconium	1857	hcp	Not stated	5×10^{-8}	22.0
Graphite	3737	Complex hex.	2150–2347	0.4 to 14.4	163.0
β-Tin	232	tetragonal		8.2 (‖ to c)	25.6 (‖ to c)
				1.4 (⊥ to c)	23.3 (⊥ to c)
Germanium	958	Diamond	766–928	7.8	68.5

* After D. Lazarus in *Solid State Physics*, Vol 10, ed. by F. Seitz and D. Turnbull, Academic Press, New York (1959).

(ac) of ionic crystals is due to this ion movement and can be shown to be equal to

$$\sigma = \frac{D_v C_v q^2}{kT} = \frac{D_T q^2}{f k T} \qquad (18.8.1)$$

for the case where one ion of charge q diffuses much *faster* than the other (otherwise it is a sum of the contribution of each). Here C_v is the sodium ion vacancy concentration. (See Problem 18.39 for the derivation of this expression.) Thus, assuming that we know the mechanism so that f can be calculated, there are two independent ways of measuring the self-diffusion coefficient for sodium ions: the use of tracer atoms and the measurement of ionic conductivity.

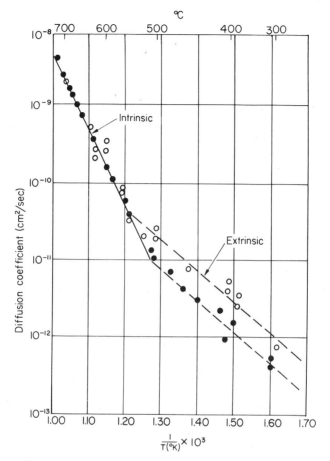

FIGURE 18.8.1. Logarithm of the tracer diffusion coefficient versus reciprocal of the temperature as measured by radioactive sodium (open circles) and as calculated from conductivity (full circles) for sodium chloride. [From D. Mapother, H. N. Crooks, and R. Maurer, *Journal of Chemical Physics*, **18**, 1231 (1950).]

A comparison between the results of these two methods is shown in Figure 18.8.1. Note from the figure that D_v depends exponentially on the negative reciprocal of temperature and that C_v does likewise in pure enough crystals or at high enough temperatures, but in impure crystals (divalent impurities such as Ca^{2+}) C_v does not vary with temperature (see Examples 13.2.3 and 13.2.4). Hence at high temperatures the activation energy is half the energy of formation of the vacancy pair $E_f/2$ plus the energy of motion E_m. This is called the **intrinsic region** of conductivity since the conductivity depends on carriers which are present in a perfectly pure material; i.e., they occur intrinsically. However in an impure sample at low temperatures the temperature dependence of D_T

Table 18.8.1. DIFFUSION DATA FOR Na* IN NaCl

Temperature	Q_d	Q_d, kcal/mole	D_0, cm²/sec
High	$E_m + E_f/2$	41.5	3.1
Low	E_m	17.7	†

† Depends on sample purity

involves E_m only, as noted in Table 18.8.1. This is called the **extrinsic region** of conductivity since the charge carriers are present as a result of the presence of impurity atoms. The conductivity of ionic crystals, like the conductivity of semiconductors (Figure 3.3.1), is a structure-sensitive property at low temperatures.

Diffusion data for some ionic crystals are shown in Table 18.8.2.

Table 18.8.2. DIFFUSION DATA FOR IONIC CRYSTALS

Compound	Cation		Anion	
	D_0, cm²/sec	Q_D,* kcal/mole	D_0, cm²/sec	Q_D, kcal/mole
KCl	16.1	43		
		18		
NaCl	3.1	41		
		18	56.0	49
CaF₂		46		
		18		11–18
BeO	10^{11}	150		
	10^{-4}	115		
UO₂†	10^{-4}	29	3×10^{-4}	14
			2.6×10^{-5}	30 (interstitial O²⁻)
CoO†	8×10^{-5}	11		
Fe₃O₄†	0.4	22		
MgO†	5×10^{-4}	27		
CaO†	1.5×10^{-3}	29		
Al₂O₃	10^{-8}	180		
TiO₂			1.6	65
Zr₀.₈₅Ca₀.₁₅O₁.₈₅			10^{-2}	28 (O²⁻)

* High and low temperature values, respectively, are given in several cases.
† Approximate values from W. Kingery, *Introduction to Ceramics*, Wiley, New York (1960) p. 232.

Those processes which involve a net flow of NaCl from one position in the crystal to another (creep is such an example, as we shall see later) will be controlled by the diffusion of the *slowest*-moving ion since both ions must move.

18.9   DIFFUSION AT SURFACES, IN GRAIN BOUNDARIES, AND ALONG DISLOCATIONS

In general, diffusion is much faster along plane or line imperfections than through the good crystal. The study of diffusion along imperfections is an active area of research. There are no extensive data available. A rough rule is that the activation energy for diffusion along the imperfection is about half of the activation energy for diffusion in the good crystal.

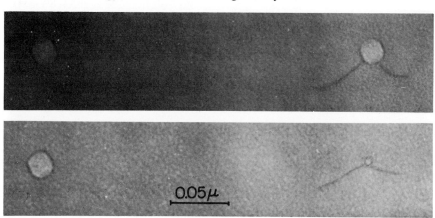

FIGURE 18.9.1. Annealing of voids in a thin film electron microscope specimen of aluminum. Void on left is an isolated void, while void on right is hooked up to the surfaces of the specimen by a dislocation. Above: specimen partially annealed at 180°C. Below: same as above except 29 min. later. The shrinkage of the void connected to the dislocation is greater than that of the isolated void. The dislocation ends at the surface of the foil. (T. E. Volin and R. W. Balluffi, Cornell University, Ithaca, New York.)

Figure 18.9.1 provides a direct experimental verification of the fact that diffusion of vacancies along a dislocation is faster than in the good crystal.

Consider a single crystal with ρ cm of dislocation per cubic centimeter and consider the dislocation line to be a pipe of cross-sectional area a along which diffusion occurs. Let D_P be the diffusion coefficient in the pipe and D_F be the diffusion coefficient in the dislocation free cyrstal. What is the total diffusion coefficient?

The fraction of volume which is in dislocations is $g = \rho a$, which is usually much less than 1 (the area a might be 100 Å²). Hence the total D is

$$D = (1 - g)D_F + gD_P \doteq D_F + gD_P. \tag{18.9.1}$$

The process is completely analogous to a situation in which all the people on the east coast headed for the west side of the Mississippi River (to escape pollution?) with a tiny fraction traveling rapidly along highways and all of the rest walking across all the remaining area.

EXAMPLE 18.9.1

If $a = 100$ Å^2 and $p = 10^8$ cm^{-2}, at what temperature do the two effects become equal, i.e., $(1 - g)D_F = gD_P$, if the activation energy in the good crystal is 46,000 cal/mole and the activation energy in the pipe is 23,000 cal/mole and if the D_0's are the same in each case?

Answer.

$$(1 - pa)D_0 e^{-46,000/RT} = D_0 pae^{-23,000/RT}.$$

Since $pa = 10^{-6}$, $1 - pa \doteq 1$ and hence

$$\log_{10} pa = -\frac{23,000}{2.3 \times RT}.$$

Hence

$$T = 833°\text{K} = 560°\text{C}.$$

Below this temperature pipe diffusion would be the more important process.

18.10   CHEMICAL DIFFUSION

Blocks of nickel and copper are clamped together at high temperatures to form a diffusion couple as shown in Figure 18.10.1. Small inert wires are introduced at the interface and also near the Cu and Ni ends of the couple.

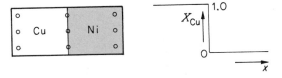

FIGURE 18.10.1. Chemical diffusion couple.

Examination of the phase diagram for this system shows that copper and nickel form a system having continuous solubility in the solid state (see Figure 16.4.6). When the couple is held at a constant high temperature, diffusion occurs, after which the concentration profile (atom fraction X_{Cu} versus position x) appears as in Figure 18.10.2. The interface $x = 0$ is defined for the case where the two integrals A_1 and A_2 are equal for each time t. This defines the **Matano interface**. What is particularly interesting about this system is that the original

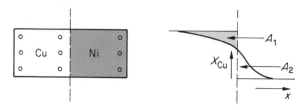

FIGURE 18.10.2. Couple after diffusion has occurred.

interface markers move according to a law,

$$x_M^2 \propto t. \tag{18.10.1}$$

This implies that copper atoms move faster than nickel atoms, a situation which implies a net flux of vacancies from the nickel side to the copper side of the moving marker interface and a net flux of atoms from the copper side to the nickel side of the moving marker interface. This is known as the **Kirkendall effect** (1947). It was the first good evidence that diffusion in substitutional alloys occurs via the vacancy mechanism rather than by direct interchange of a pair of atoms.

EXAMPLE 18.10.1

What is x_M for a direct interchange mechanism?

Answer. Zero [if correction for deviation from Vegard's law (Section 12.6) is made].

It can be shown that the diffusion coefficient is given by

$$D = X_{Ni}D_{Cu} + X_{Cu}D_{Ni}. \tag{18.10.2}$$

A similar equation applies to the interdiffusion of gases, a result which was given by Meyer in 1865. The corresponding equation for diffusion in solids was given by Darken in 1948 following the Kirkendall effect discovery.

18.11   DIFFUSION IN POLYMERIC MATERIALS

Diffusion occurs when small molecules (soluble in the polymer under the conditions of the experiment) pass through the "voids" between the polymer molecules. Here, the term "void" has, to a limited extent, the same meaning as does the term vacancy in a crystal. (Thus when a crystal melts to a liquid, there is usually a volume increase of about 5%, which can be crudely attributed to the presence of voids.) The rate of diffusion will depend to a large extent on the size of the small molecules and the size and mobility of the voids. The size and mobility of the voids will depend upon the state of the polymer, i.e., whether it is crystalline, glassy, or rubberlike. The diffusion rate will be high in the rubber-like state because the packing is poor in the amorphous state, and because the highly flexible chain segments can move out of the way of a diffusing molecule. The rate will be lower in the same material in the glassy state (below the glass transition temperature for the polymer-small molecule solution) because there is a reduction in volume (see Figure 12.11.2) and flexibility of the chain. Finally, the molecular packing in the crystalline material is still better and the chain flexibility is also low, so diffusion will be the slowest of all in the crystalline state.

There are many examples where a high diffusion rate is desirable and many others where a low diffusion rate is required. For example, in packing fruit, the film should allow outward passage of carbon dioxide but should not permit the inward flow of oxygen. When making cellulose acetate film by casting from solution, the removal of solvent in the film takes place by diffusion. Low permeability, however, is desired for tubes or tubeless tires.

The permeation of molecules through a polymer, i.e., steady-state diffusion, depends upon two factors, namely, the solubility and the diffusion coefficient. The solubility depends upon chemical factors. In the simplest form the rule of solubility is that like tends to dissolve like. There are semiempirical formulas for solubility based in part on the theories of regular solutions discussed in Section 17.9.

EXAMPLE 18.11.1

Rubber gloves of thickness L are attached to a chamber containing pure argon (crystals of sodium are growing in the chamber). Show that the flow of oxygen through the gloves is determined by the coefficient

$$p = DC_s \tag{18.11.1}$$

where D is the diffusion coefficient and C_s is the solubility limit of the oxygen in the rubber in the presence of air at 1 atm.

Answer. For steady-state diffusion

$$J = -D\frac{dC}{dx};$$

J is a constant and $dC/dx = -(C_s - 0)/L$. We have assumed that the oxygen pressure inside the chamber is virtually zero, so the dissolved oxygen at the inner face is zero. Hence

$$J = \frac{DC_s}{L} = \frac{p}{L}.$$

If the solubility C_s is directly proportional to the gas pressure (Henry's law), then the flux rate will be directly proportional to the pressure difference across the membrane; i.e.,

$$J = -k_p\frac{dP}{dx}. \tag{18.11.2}$$

This is Darcy's equation. Both p and k_p are called permeability coefficients in the literature. Some values of k_p are given in Table 18.11.1. (Recall that 1 bar = 10^6 dynes/cm^2 = 0.9869 atm.)

Table 18.11.1. PERMEABILITY DATA AT 30°C, (10^{-8} cm^2/sec-bar)

Polymer	Gas			Nature of polymer
	N_2	O_2	CO_2	
Polyvinylidene chloride (Saran)	0.007	0.041	0.21	Crystalline
Polytetrafluoroethylene	0.02	0.075	0.54	Crystalline
Nylon	0.07	0.28	1.2	Crystalline
Polyvinylchloride	0.30	.90	7.5	Partially crystalline
Cellulose acetate	2.1	5.9	51	Glassy
Polyethylene (high density)	2.1	7.9	27	Crystalline
Polyethylene (low density)	14	41	264	Partially crystalline
Butyl rubber	2.3	9.7	39	Rubberlike
Polybutadiene	48	143	55	Rubberlike
Natural rubber	60	175	982	Rubberlike

EXAMPLE 18.11.2

The butyl rubber gloves of a dry box have an area of 4000 cm^2 and a thickness of 0.05 cm; the chamber has a volume of 10^5 cm^3. Assuming the dry box initially contains pure argon at 1 atm and the air pressure outside is 1 atm, what is the partial oxygen pressure in the box at the end of one day?

Answer.

$$Q = JAt = k_p \frac{\Delta P}{L} A \, \Delta t$$

$$= 10^{-7} \times \frac{0.20}{0.05} \times 4000 \times 10^5$$

(The oxygen pressure in air is about 0.2 bar.)

$$Q = 160 \text{ cm}^3$$

The partial pressure is $160/10^5 = 1.6 \times 10^{-3}$ atm. assuming no argon flows.

In desalinization by **reverse osmosis**, salt water is placed under pressure in the presence of a semipermeable membrane which passes water but not solute. The applied pressure has to exceed the osmotic pressure of the solution in order for water to be squeezed out. The additional pressure above the osmotic pressure determines the flow rate. Since the flux rate is proportional to the pressure, high pressures are used and the membranes are made thin. To obtain economical rates the membrane must pass 10 gal/ft^2-day for P no greater than 1500 psi. The membrane can be supported by porous materials. Cellulose acetate is presently one of the best membrane materials. (There is a need for better membrane materials with better long time stability and cheaper porous backup materials.)

In **electrodialysis** two membranes are used for desalinization, one of which passes cations, the other of which passes anions as illustrated in Figure 18.11.1. Once again the properties of the membrane material are critical. They must be permeable, have high electrical conductivity, have reasonable mechanical strength, and resist chemical attack. As a rule, *ion exchange resins* have poor mechanical strength, so they must be reinforced, usually with glass fabric.

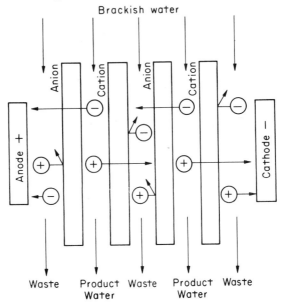

FIGURE 18.11.1. Desalinization by electrodialysis.

A resin such as styrene divinylbenzene can be made permeable to cations by treating it to contain sulfonic acid groups. Alternatively, it can be made permeable to anions by treating it so that it contains quaternary amino groups.

Diffusion through polymeric membranes is an important process in biology. There are many biomedical materials problems (such as the synthetic kidney) which depend upon diffusion through membranes.

18.12   ILLUSTRATION OF APPLICATIONS

DISCUSSION.   There are many examples of the diffusion process. Two of these are the problems of crystal growth and those problems associated with solidification. A partial list of processes in which diffusion plays a vital role is given in Table 18.12.1.

Table 18.12.1. SOME DIFFUSION-
CONTROLLED PROCESSES

Semiconductor doping by diffusion
Oxidation
Homogenization
Graphitization
Age precipitation hardening
Eutectic and eutectoid lamellar formation
Vacuum melting and casting
Carburization and the like
Annealing of radiation damage
Sintering
Creep
Recovery
Recrystallization

EXAMPLE—MAKING TRANSISTORS. A method called **oxide masking** can be applied in the fabrication of transistors in which the geometry of the emitter region is controlled by diffusing phosphorus through a hole in the oxide film which has been grown on the silicon. This is possible because the oxide film is adherent and because the diffusion coefficient of the phosphorus in the oxide film is orders of magnitude less than in the silicon itself.

We consider here two aspects of the problem:

1. The oxidation process which yields the oxide film.
2. The diffusion of phosphorus into silicon.

The oxidation process can be described on the basis of Figure 18.12.1. The assumption is made (1) that once the oxide film has become several atomic layers thick a steady-state concentration is reached at the gas interface. This

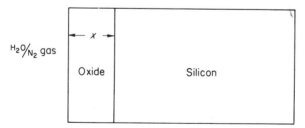

FIGURE 18.12.1. Oxidation of silicon.

means that a fixed concentration difference ΔC exists across the oxide layer of thickness x. We assume (2) that a steady-state concentration exists within the oxide layer as well. The rate of the flux of silicon atoms through the oxide is proportional to the overall gradient $\Delta C/x$ and hence equals $D\,\Delta C/x$; the flux is also proportional to dx/dt. Hence we have

$$\frac{dx}{dt} \propto \frac{D\,\Delta C}{x}, \qquad (18.12.1)$$

which by integration gives

$$x^2 \propto Dt. \qquad (18.12.2)$$

Hence

$$x^2 = Kt, \qquad (18.12.3)$$

where K is a constant determined by temperature and the gas pressure. This is the **parabolic growth** law which is characteristic of the rate of adherent oxide film formation and is illustrated in Figure 18.12.2. [The oxidation of metals is similar and is an important area of materials study. See, e.g., Carl Wagner,

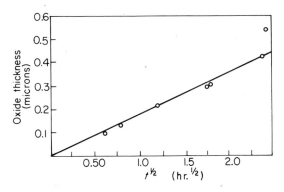

FIGURE 18.12.2. Oxide thickness versus the square root of oxidation time. The films were grown in wet N_2 at 1250°C with a gas flow of 1.5 liters/min. [From R.B. Allen, H. Bernstein, and A.D. Kurtz, *Journal of Applied Physics*, **31**, 334 (1960).]

"Diffusion and High Temperature Oxidation of Metals," in *Atom Movements*, American Society for Metals, Cleveland, Ohio (1951) p. 153.] The study of the formation of protective films on materials by chemical reaction and diffusion is also an important area of materials research. As an example, molybdenum, which at high temperatures forms an oxide which readily vaporizes, can be protected by formation of a nitride at the surface.] Let us now etch a hole in the oxide layer as shown in Figure 18.12.3. Hydrofluoric acid readily etches SiO_2 but not Si.

Let us next consider the diffusion of the phosphorus into the silicon at the hole. Here a gas with controlled concentration of P_2O_5 is passed over the surface, which rapidly becomes saturated with phosphorus; i.e., the surface

FIGURE 18.12.3. Hole in oxide
layer on silicon.

concentration of phosphorus reaches a value C_s. (The P_2O_5 decomposes at the surface.) We assume here that this saturation happens essentially immediately. We assume one-dimensional diffusion. Hence our problem is to find a solution to

$$D_p \frac{\partial^2 C_p}{\partial x^2} = \frac{\partial C_p}{\partial t}, \tag{18.12.4}$$

where we have assumed that D_p does not vary with C_p, subject to

$$C_p(x, t) = 0 \qquad \text{for } x > 0, \, t = 0 \tag{18.12.5}$$

$$C_p(x, t) = C_s \qquad \text{for } x = 0, \, t > 0. \tag{18.12.6}$$

The solution is from (18.5.6):

$$C_p(x, t) = C_s \left[1 - \operatorname{erf} \frac{x}{2(D_p t)^{1/2}} \right]. \tag{18.12.7}$$

The values of D_p were measured by C. S. Fuller and J. A. Ditzenberger, *Journal of Applied Physics*, **27**, 544 (1956). The result is

$$D_p = 10.5 e^{-(85,000 \text{ cal/g-mole})/RT} \text{ cm}^2/\text{sec}. \tag{18.12.8}$$

At 1200°C this gives a diffusion coefficient of about 3×10^{-12} cm²/sec.

The diffusion of phosphorus in the silicon oxide layer has been studied by R. B. Allen et al, *Journal of Applied Physics*, **31**, 334 (1960). Their value is

$$D = 3.9 \times 10^{-12} e^{-(32,000 \text{ cal/g-mole})/RT} \text{ cm}^2/\text{sec}. \tag{18.12.9}$$

At 1200°C this gives $D \approx 7 \times 10^{-17}$ cm²/sec.

Now let us return to our original problem. We wish to diffuse the phosphorus into the silicon and obtain a phosphorus concentration near to C_s at a depth of 10^{-3} cm but a concentration very much less than C_s at twice this depth. A good estimate of this depth is $2\sqrt{D_p t}$. Hence we can solve for the approximate diffusion time ($t = 83{,}000$ sec, nearly a day). Let us ask how thick an oxide layer would be needed to essentially completely mask the silicon from the phosphorus in the region surrounding the hole in the oxide layer. We want the concentration of the phosphorus to be only a tiny fraction of C_s at the oxide-silicon interface. To insure that the required ratio is 0.005, we let the thickness of the oxide film be

$$x = 4\sqrt{Dt} = 1 \times 10^{-5} \text{ cm}.$$

As is clear from Figure 18.12.2, it would take about 0.25 hr of oxidation time to prepare a "mask" of this thickness.

Suppose that the actual starting material was silicon doped with boron [this is a *p*-type semiconductor, so called because the charge carriers are positive (Chapter 28)]. When sufficient phosphorus is diffused into this material, it becomes an *n*-type semiconductor (negative charge carriers) down to a depth of about 10^{-3} cm. The resultant *p-n* junction acts as a rectifier. With slight modifications to the procedure mentioned here *n-p-n* junctions and transistors (which amplify) can be produced. Materials science research has provided the basis for large scale integration (LSI) in which 100,000 or more transistors, resistors, etc., can be placed on 1 in.² of silicon wafer area [see F. C. Heath, "Large Scale Integration in Electronics," *Scientific American* (Feb. 1970) p. 22].

It should be noted that doping with trivalent or pentavalent elements can also be achieved by ion implantation, i.e., by bombardment of the crystal with high energy ions.

OTHER EXAMPLES.   Certain diffusion-controlled processes such as eutectic lamellar formation and carburization have been given as examples throughout the chapter. The last three topics in Table 18.12.1, creep, recovery, and recrystallization, are intimately connected with deformation and so will be discussed later.

The process of thermal grooving (the formation of a groove at a grain boundary) takes place by either surface or volume diffusion, with the driving force being surface energy [see, e.g., W. W. Mullins and P. G. Shewmon, *Acta Metallurgica* **7**, 163 (1959)]. The process of sintering in powder metallurgy is similar [see C. Herring in *The Physics of Powder Metallurgy*, ed. by W. E. Kingston, McGraw-Hill, New York (1951)]. There are many opportunities for further developments in the powder metallurgy area.

18.13   GENERALIZED DIFFUSION THEORY

MOBILITY.   An object moving under the influence of a fixed force **F** through a *linear* viscous medium first accelerates but eventually reaches a terminal velocity **v** such that

$$\mathbf{v} = B\mathbf{F}. \qquad (18.13.1)$$

(If the viscous medium is nonlinear, then the **mobility coefficient** B would vary with the magnitude of **F**.)

The student has probably already encountered terminal velocity or drift velocity in problems involving the descent of a man on a parachute, a sailboat moving in a constant wind, etc.

If the moving object is a diffusing molecule and there are large numbers of such molecules such that a concentration C can be defined, then the flux of such molecules is given by

$$\mathbf{J} = \mathbf{v}C = BC\mathbf{F}. \qquad (18.13.2)$$

THE ELECTRICAL FORCE.   The motion of charges in the presence of an applied field in a conductor may be described by Equation (18.13.2). Then the **electrical force** is

$$\mathbf{F}_{\text{elect}} = q\xi = -q \text{ grad } V, \tag{18.13.3}$$

where q is the charge, ξ is the electric field, and V is the electrical potential. The current flux is given by

$$\mathbf{J}_{\text{charge}} = q\mathbf{J}_{\text{ions}} = BCq^2\xi. \tag{18.13.4}$$

This is Ohm's law. If there are several charged species diffusing,

$$\mathbf{J}_{\text{charge}} = \sum_i B_i C_i q_i^2 \xi. \tag{18.13.5}$$

The conductivity is then

$$\sigma = \sum_i B_i C_i q_i^2. \tag{18.13.6}$$

THE CHEMICAL FORCE.   A flux of charged or noncharged particles is also caused by a chemical force. An example is the interdiffusion of two ideal gases, e.g., He and Ne at constant pressure and temperature. The **chemical force** is given by

$$\mathbf{F}_{\text{chem}} = -\text{grad } \mu, \tag{18.13.7}$$

where μ is the chemical potential [note the analogy with the electrical force of (18.13.3)]. For an ideal solution (one whose heat of mixing is zero) the chemical potential is, by (17.9.13),

$$\mu = \mu_0 + kT \ln C. \tag{18.13.8}$$

Hence

$$\text{grad } \mu = \frac{kT}{C} \text{ grad } C, \tag{18.13.9}$$

and combining (18.13.2), (18.13.7), and (18.13.9) gives

$$\mathbf{J} = -BkT \text{ grad } C. \tag{18.13.10}$$

Hence the diffusion coefficient is given by the **Nernst equation** (1885),

$$D = BkT. \tag{18.13.11}$$

EXAMPLE 18.13.1

Derive the general form of Darcy's equation for a reverse osmosis process. (*Hint*: Start with the expression $dG = VdP - SdT$.)

Answer. For an isothermal process, $dG = VdP$. Let us write this in the form $d\mathscr{G} = \mathscr{V}dp$, where script $\mathscr{G}$ and $\mathscr{V}$ refer to a given quantity of water, e.g., 1 cm³ of water at 1 atm and at the temperature T. We then have for the driving force

$$\mathbf{F}_{\text{chem}} = -\text{grad } \mathscr{G} = -\mathscr{V} \text{ grad } P.$$

Hence from Equation (18.13.2),

$$\mathbf{J} = -BC\mathscr{V} \ \text{grad} \ P$$

which from (18.13.11) can be written as

$$\mathbf{J} = -\frac{D}{kT}C\mathscr{V} \ \text{grad} \ P.$$

The permeability coefficient is

$$k_p = \frac{DC\mathscr{V}}{kT}. \tag{18.13.12}$$

In general D, the concentration of solvent C and $\mathscr{V}$ may also depend on pressure; if they vary negligibly with pressure, then k_p is a constant, and the flux relation is linear.

ANISOTROPY.   In liquids and gases the diffusion coefficient does not vary with direction: the behavior is said to be isotropic. However, in single crystals the diffusion coefficient varies with direction so that diffusion is an anisotropic property (the exception is the cubic crystal which because of its symmetry behaves isotropically). In the general anisotropic case, the velocity $\mathbf{v} = [v_1, v_2, v_3]$ is related to the generalized force $\mathbf{F} = [F_1, F_2, F_3]$ by

$$
\begin{aligned}
v_1 &= B_{11}F_1 + B_{12}F_2 + B_{13}F_3 \\
v_2 &= B_{21}F_1 + B_{22}F_2 + B_{23}F_3 \\
v_3 &= B_{31}F_1 + B_{32}F_2 + B_{33}F_3.
\end{aligned}
\tag{18.13.13}
$$

It can be shown from the principle of microscopic reversibility (which is studied in more advanced courses) that

$$B_{ij} = B_{ji},$$

i.e., $B_{12} = B_{21}$, etc. Hence there are six independent mobilities in the general case (the same could be said of diffusion coefficients) of a triclinic crystal. The array of nine B_{ij}'s is known as a second-rank tensor and (18.13.13) in tensor notation is

$$v_i = B_{ij}F_j. \tag{18.13.14}$$

18.14   NUCLEATION

Examples of phase transformations which occur were discussed in Chapter 16. We are concerned here with the appearance of the new phase. A process called nucleation is usually, but not always, the first step in the formation of a new phase.

EXAMPLE 18.14.1

Suppose we have pure water vapor but no liquid or solid H_2O and no other phases around in a large region at 300°K. What are the conditions under which liquid water droplets will form?

Answer. Our first conclusion would be that if the water vapor pressure is increased to the equilibrium vapor pressure of water (which can be read off a phase diagram such as Figure 16.2.1 or found in a handbook) liquid water droplets should form. We would be wrong. They do not, and they should not. Why?

FREE ENERGY OF A DROPLET. When we measure the so-called equilibrium vapor pressure p_0 of water we use a large body of water as in a liter flask. Here we have a surface with essentially zero curvature. Moreover, a negligible fraction of the atoms are at the surface. The molar Gibbs free energy of the liquid at temperature T under equilibrium pressure p_0 is $\bar{G}_l^0$ and of the vapor is $\bar{G}_v^0$, where

$$\bar{G}_v^0 = \bar{G}_l^0. \tag{18.14.1}$$

Let us next suppose that we have only vapor at a pressure p at the same temperature T. Its molar Gibbs free energy is

$$\bar{G}_v = \bar{G}_v^0 + RT \ln \frac{p}{p_0}. \tag{18.14.2}$$

The Gibbs free energy of the bulk liquid (if it were present at p and T) is

$$\bar{G}_l = \bar{G}_l^0 + \Delta\bar{G}_l, \tag{18.14.3}$$

but $\Delta\bar{G}_l$ is negligible compared to $RT \ln p/p_0$.

EXAMPLE 18.14.2

Show that $\Delta\bar{G}_l$ is negligible compared to $\Delta\bar{G}_v = RT \ln p/p_0$.

Answer. Recall that $d\bar{G} = \bar{V} dP - \bar{S} dT$. This follows from (17.7.4) if we consider molar Gibbs free energy $\bar{G}$, molar volume $\bar{V}$ and molar entropy $\bar{S}$. Inasmuch as we are considering changes in pressure at constant temperature, $d\bar{G} = \bar{V} dP$. For water at 300°K, $\bar{V}_l = 18 \text{cm}^3/\text{mole}$ and $p_0 = 31.8$ mm Hg. Suppose $p = 2p_0$. (It would be easy to show using the bulk modulus for H_2O in Table 2.3.2 that $\bar{V}_l$ decreases little when $dP = p - p_0 = p_0$.) Thus an *upper* bound on $\Delta\bar{G}_l = \bar{G}_l - \bar{G}_l^0$ is

$$\Delta\bar{G}_l < \bar{V}_l(p - p_0).$$

A *lower* bound on $\Delta\bar{G}_v$ is obtained by integrating $d\bar{G}_v = \bar{V}_v dP$ with $\bar{V}_v$, taking the smallest value on the range of integration. Thus

$$\Delta\bar{G}_v < \bar{V}_v(p)(p - p_0),$$

with $p = 2p_0$, $\bar{V}_v = 22{,}400 \times \frac{300}{273} \times \frac{760}{63.6}$ cm³/mole. Clearly $\Delta\bar{G}_l < 10^{-4}\,\Delta\bar{G}_v$.

Let us now calculate the free energy of a tiny spherical droplet of water of radius r which is formed from water vapor at a pressure p at temperature T. The droplet has volume $V = 4\pi r^3/3$, and the number of moles of water in the droplet is $V/\bar{V}_l$. When a mole of vapor is transferred to the liquid there is a change in the Gibbs free energy of

$$\Delta\bar{G} = \bar{G}_l - \bar{G}_v. \tag{18.14.4}$$

The overall Gibbs free energy change in forming the droplet is

$$\Delta G = \frac{V}{\bar{V}_l}\Delta\bar{G} + \gamma A, \tag{18.14.5}$$

since surface area A is also created when the droplet is formed. We have for the Gibbs free energy of formation of a droplet

$$\Delta G = \mathscr{G}V + \gamma A, \tag{18.14.6}$$

where $\mathscr{G}$ is the increase in the Gibbs free energy per unit volume when the droplet is created ($\mathscr{G} = \Delta\bar{G}/\bar{V}_l$) and γ is the increase in the Gibbs free energy per unit area when the droplet is created.

EXAMPLE 18.14.3

What is the form of $\mathscr{G}$ in the present case?

Answer.

$$\mathscr{G} = \frac{\Delta\bar{G}}{\bar{V}_l} = -\frac{RT}{\bar{V}_l}\ln\frac{p}{p_0}, \tag{18.14.7}$$

where we have used $\bar{G}_l \doteq \bar{G}_l^0 = G_v^0$.

We assume that the vapor is supersaturated, i.e., that $p > p_0$. Then $\mathscr{G}$ is negative.

EXAMPLE 18.14.4

If $\mathscr{G}$ is negative and since γ is positive, *discuss* or sketch the shape of the ΔG vs. r curve.

Answer. You do not have to know the magnitude of $\mathscr{G}$ or γ to sketch ΔG. Since

$$\Delta G = \mathscr{G}\frac{4\pi r^3}{3} + \gamma 4\pi r^2, \tag{18.14.8}$$

at sufficiently small r the second term dominates and ΔG is positive, while at sufficiently large r the first term dominates and ΔG is negative.

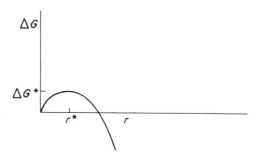

FIGURE 18.14.1. Free energy of formation of a droplet.

The ΔG vs. r curve has the form shown in Figure 18.14.1.

What are the conditions for which the droplet would neither tend to grow nor evaporate? This would be the case in which ΔG did not change for a slight change in r; i.e.,

$$\partial \, \Delta G / \partial r = 0 \qquad \text{at } r = r^*. \qquad (18.14.9)$$

This leads to

$$r^* = -\frac{2\gamma}{\mathscr{G}}, \qquad (18.14.10)$$

and to

$$\Delta G^* = \frac{16\pi\gamma^3}{3\mathscr{G}^2} = \frac{4}{3}\pi\gamma r^{*2}. \qquad (18.14.11)$$

EXAMPLE 18.14.5

What is r^* for a water droplet in terms of the vapor pressure p? Evaluate r^* for $p/p_0 = 2.7$ at $300°K$.

Answer.

$$r^* = \frac{2\gamma \bar{V}_l}{RT \ln p/p_0}. \qquad (18.14.12)$$

This is called the **Kelvin equation**.

The surface tension of water is about 72 ergs/cm² and $\bar{V}_l$ is about 18 cm³/mole. Moreover, $R = 8.3 \times 10^7$ ergs/mole-°K so $r^* = 10.4 \times 10^{-8}$ cm or 10.4 Å. This corresponds to a droplet with about 157 molecules of water.

NUCLEATION OF A WATER DROPLET.   We return now to the question asked in Example 18.14.1. What do we have to do to get water droplets? Clearly we must have supersaturated vapor. Let the pressure be p, where $p > p_0$; then suppose we have a droplet of radius r, where $r < r^*$. What happens? Clearly this droplet has a Gibbs free energy of formation greater than zero which will *decrease* if the droplet evaporates. Since such a decrease corresponds to an approach to equilibrium, the droplet will evaporate. Thus we have a dilemma:

since a droplet has to be small before it is big, it will evaporate before its radius equals r^*. On the other hand, should a droplet by some wild chance ever attain a radius in excess of r^*, then it would tend to grow because this would result in a decrease in the Gibbs free energy of formation. Such a droplet is called a **critical nucleus** and r^* is called a **critical radius**. If $p/p_0 = 2.7$, this wild chance corresponds to the probability that 157 molecules simultaneously are found within the droplet volume (in the gas phase they would occupy a volume 20,000 times as large). The probability of such an event is

$$e^{-\Delta G^*/kT}. \tag{18.14.13}$$

(Students often question why we do not divide by RT instead of kT. The reason is that we are considering the probability of formation of one droplet, not a mole of droplets, and ΔG^* is the free energy of formation of *one* droplet.) The concentration of such critical size droplets is

$$n^* = ne^{-\Delta G^*/kT} \tag{18.14.14}$$

where n is the concentration of single water molecules (the number of single molecules per unit volume). This droplet becomes super-critical if one more molecule is added. It can be shown from the kinetic theory of gases that the number of gas particles striking unit area per unit time is $\frac{1}{4}n\bar{c}$ where $\bar{c}$ is the average speed. The area in the present case is the surface area of the spherical droplet, $4\pi r^{*2}$. Therefore the time required for one molecule to hit the droplet is

$$\frac{1}{n\bar{c}\pi r^{*2}}. \tag{18.14.15}$$

Hence, the number of nuclei formed per unit time is

$$I = \pi n^2 \bar{c} r^{*2} e^{-\Delta G^*/kT}$$

$$I = \pi n^2 \bar{c} r^{*2} e^{-(4/3)\pi\gamma r^{*2}/kT}. \tag{18.14.16}$$

Since r^* is a known function of the pressure [see (18.14.12)] and since the ideal gas law is $p = nkT$, the nucleation rate is given in terms of the pressure.

EXAMPLE 18.14.6

Show that the pre-exponential factor in (18.14.16) is about 10^{28} for $p/p_0 = 2.7$ at $300°$K.

Answer. From Example 18.14.2, $p_0 = 31.8$ mm Hg $= (31.8/760)$ atm. From this it is possible to calculate using the ideal gas law that for $p/p_0 = 2.7$ at $300°$K, $n = 2.77 \times 10^{18}$ cm^{-3}. Hence using $\bar{c} = 5.94 \times 10^4$ cm/sec and $r^* = 1.04 \times 10^{-7}$ cm (from Example 18.14.5) we have

$$\pi n^2 \bar{c} r^{*2} = 10^{28.2}$$

Note that n increases as p increases while r^* decreases as p increases. The pre-exponential factor is nearly independent of p.

Table 18.14.1 gives results for I depending on the **supersaturation ratio** $p/p_0 = \alpha$. Note that increasing the supersaturation ratio by 100% (from $\alpha = 2$ to 4) changes I by 10^{54}. This is a *strong* dependence!

Table 18.14.1. RATE OF NUCLEATION AT 300°K

p/p_0	I, nuclei/cm^3-sec	t*
2	10^{-43}	$\sim 10^{35}$ yr
3	1	1 sec
4	10^{11}	10^{-11} sec

* t = time for the appearance of one nucleus in a volume of 1 cm^3.

OTHER EXAMPLES OF NUCLEATION.   The nucleation of the water droplet as just described is called **homogeneous nucleation**. It may happen that dust (or some other particle) is present. Adsorbed layers of water on this dust particle would effectively make a water droplet available for further condensation. This is called **heterogeneous nucleation**. This process can take place at much lower supersaturations.

The formation of solid crystals from the melt (e.g., ice in water) is another example of nucleation. The motion of the molecules in the water to the crystal involves diffusion in the water or diffusion on the surface to kinks in ledges on the surface. Thus the term in the nucleation rate corresponding to (18.14.14) would be different, but the probability term of (18.14.13) would still be present. However, the latter would in general have to be modified to account for the fact that the surface tension γ varies with direction. If this correction is ignored then ΔG^* can be shown to be equal to

$$\Delta G^* = \frac{16\pi}{3} \frac{\gamma^3 T_m^2}{\mathscr{H}^2(T_m - T)^2}, \qquad (18.14.17)$$

where $\mathscr{H}$ is the enthalpy change in the bulk per unit volume during freezing. $T_m - T$ represents the supercooling.

The formation of a precipitate such as $CuAl_2$ in aluminum is also a nucleation process. This differs from the water droplet nucleation in at least three ways. First, the mechanism of aggregation of $CuAl_2$ or its equivalent concentration fluctuation involves the diffusion of Cu into and Al out of a critical radius region. Second, the surface energy may vary with direction and with the concentration and the concentration gradient. Third, because solids support static shear stress, elastic strains may be introduced. The details of analysis in this case are therefore more involved than with the single-component liquid-vapor system. However, Equation (18.14.17) may be used as a first approximation to describe such processes if T_m is replaced by T_0, the equilibrium temperature.

If a spherical precipitate forms from the solid phase then the rate of nucleation is given by

$$\frac{4\pi r^{*2}}{a}\Gamma n^{*}$$

where a is the cross sectional area per molecule, Γ is the jump frequency and n^{*} is the concentration of critical size nuclei. Therefore

$$I = \frac{4\pi r^{*2}}{a} v n e^{-(\Delta G_B + \Delta G^{*})/kT}. \qquad (18.14.18)$$

Here v is the vibrational frequency, ΔG_B is the activation energy for motion of the atom in the interface between the precipitate particle and the matrix, and n is the number of atoms per unit volume. The maximum value of I occurs at $T \approx T_0/2$.

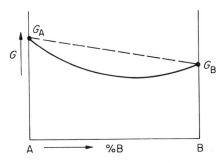

FIGURE 18.14.2. (a) Gibbs free energy plot of a solution of B in A at high temperature. (b) Dashed line is the Gibbs free energy of a mixture of pure A and pure B.

In principle there are two types of precipitation processes. These can be illustrated by the Gibbs free energy diagram of Figure 18.14.2 for a hypothetical solution of an A-B alloy (we assume that both A and B have the same crystal structure such as is the case for Cu and Au). Recall that equilibrium at constant T and P corresponds to a minimum of the Gibbs free energy G. Clearly, the solution is favored at all compositions. At low temperatures the curves appear as in

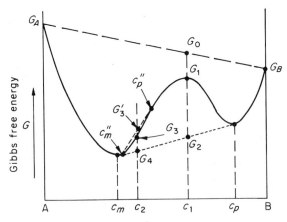

FIGURE 18.14.3. Free energy versus composition at low temperature illustrating nucleation and growth transformation.

Figure 18.14.3. We have drawn the tangent common to the two valleys in the curve. The points of tangency (*m* and *p*) are *not* the minima.

EXAMPLE 18.14.7

a. What is (are) the stable phase(s) between $c = 0$ and $c = c_m$ (% B)?
b. Between $c = c_p$ and $c = 100\%$ B?
c. Between $c = c_m$ and $c = c_p$?

Answers. a. A *solution* of B in A, i.e., a single phase; call it the α phase. Depending on composition this has the free energy given by the curve G_A to *m*. Note that a mixture of pure A and pure B would have a free energy given by the dotted line; so we definitely get a solution rather than a mixture.

b. A *solution* of A in B, i.e., a single phase. Call it the β phase.

c. This is a *mixture* of the α phase and β phase. The α phase has the composition c_m and the β phase has the composition c_p. Suppose you have an alloy of composition c_1. Then the mixture of α and β will have the Gibbs free energy G_2, which is less than that of the solution G_1, which is less than that of the mixture of pure A and pure B, G_0.

Let us next start with the alloy of composition c_2 at high temperature where only one phase is present and cool to the low temperature (see Figure 18.14.3). The single untransformed phase would have free energy G_3. However, this should decompose into α and β phases of energy G_4. But to do this, it must first pass through states such as c''_m and c''_p which have energy $G'_3 > G_3$. This is a process which requires nucleation. The second type of process is described in Section 18.15.

The continued growth of *perfect* crystals from the melt or vapor necessarily also involves repeated nucleation. Thus, suppose you have a perfect planar surface. Further growth of a new plane involves the nucleation of a patch (disc) on the planar surface which can then grow outward until this new plane is complete. See Figure 18.14.4. Then nucleation is required again. Such nucleation requires sizable supersaturation. Ordinary crystal growth involves growth at imperfections such as at screw dislocations so that nucleation is not required.

The formation of lamellar, folded-chain polymer crystals (see Figure

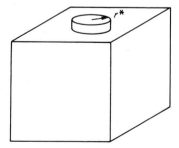

FIGURE 18.14.4. Patch of critical radius r^* nucleated on the planar surface of the crystal. Atoms will add readily at the step provided.

14.10.2) is an analogous example of a nucleation controlled transformation. In this case repeated nucleation is needed when each polymer chain is laid down.

When a crystalline material is highly deformed (so that its dislocation content is very high), there is a large stored energy which could be released if the dislocations were eliminated. This happens when the crystal is heated above an appropriate temperature and results in **recrystallization**. New crystals are formed. In this case $\mathscr{G}$ is the stored energy and γ is the grain boundary energy in (18.14.11). There is a large decrease in the dislocation density and hence a mechanical softening of the material when recrystallization occurs.

18.15   SPINODAL DECOMPOSITION

Figure 18.15.1 shows the same free energy behavior as Figure 18.14.3.

Suppose that we now cool the alloy of composition c_1 from the high temperature, where only one phase is present, to the low temperature, where two phases are present. The two components tend to diffuse in opposite directions,

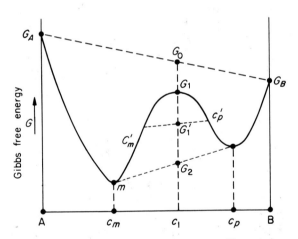

FIGURE 18.15.1. Free energy versus composition at low temperature illustrating spinodal decomposition.

forming a mixture c'_m and c'_p which has a free energy $G'_1 < G_1$. No nucleation is needed; this is called a **spinodal decomposition**. When the two phases form (by a process of reverse diffusion), strain fields are set up because of the difference in volume and elastic constants of the two phases. This determines the three-dimensional periodic distribution of the phases [see, e.g., J. D. Cahn, *Acta Metallurgica* **9**, 795 (1962)], which is a characteristic of the spinodal decomposition.

REFERENCES

Shewmon, P. G., *Diffusion in Solids*, McGraw-Hill, New York (1963). A good discussion of the mechanics of diffusion.

Girifalco, L. A., *Atomic Migration in Crystals*, Blaisdell, Waltham, Mass. (1964).

Glasstone, S., Laidler, K. J., and Eyring, H., *The Theory of Rate Processes*, McGraw-Hill, New York (1941).

Jost, W., *Diffusion in Solids, Liquids, Gases*, Academic Press, New York (1952).

Barrer, R. M., *Diffusion in and through Solids*, Macmillan, New York (1941).

Crank J., *The Mathematics of Diffusion*, Oxford, New York (1956).

Carslaw, H. S., and Jaeger, J. C., *Conduction of Heat in Solids*, Oxford, New York (1959). The differential equations for diffusion are analogous to those for heat flow.

Lazarus, D. "Diffusion in Metals" in *Solid State Physics*, Vol. 10, ed. by F. Seitz and D. Turnbull, Academic Press, New York (1959).

Peterson, N. L. "Diffusion in Metals" in *Solid State Physics*, Vol. 22, Ed. by H. Ehrenreich, F. Seitz and D. Turnbull, Academic Press, New York (1968).

Turnbull, D., "The Undercooling of Liquids," *Scientific American* (Jan. 1965) p. 38.

Fine, M. E., *Phase Transformations in Condensed Systems*, Macmillan, New York (1964).

Christian, J. W., *The Theory of Transformation in Metals and Alloys*, Pergamon Press, New York (1965). A comprehensive treatise.

Askill, J., *Tracer Diffusion for Metals, Alloys and Simple Oxides*, Plenum, New York (1970).

PROBLEMS

18.1.   a. Give five examples of diffusion in the processing of materials.

   b. Give five examples in which diffusion is important in the use of materials; e.g., the lifetime is determined by $\sqrt{Dt}$ and hence by the temperature and time.

18.2.   A good engineer or scientist is often capable of taking limited data and using these to calculate (almost instantaneously) good approximate values (factor of 2, order of magnitude, etc.). What two diffusion equations would the engineer have memorized?

18.3.   Which two equations enable one to formulate the macroscopic description of diffusion?

18.4. Use two different methods to show that the root-mean-squared diffusion distance for a one-dimensional random walk is $\sqrt{N}l$.

18.5. Use two different methods to derive $D = \frac{1}{2}\Gamma l^2$. Explain why (18.3.4) and (18.3.1) differ by a factor of 2.

18.6. a. Based on an oversimplified one-dimensional random walk model in a gas, $D = \frac{1}{2}\Gamma\lambda^2$. If λ is the mean free path in a gas and $\bar{c}$ is the velocity, then what is the jump frequency?

b. What is the pressure and temperature dependence of diffusion in an ideal gas?

c. Is the variation of D with temperature for an ideal gas more or less rapid than the variation of D with temperature for a solid?

18.7. Three processes which can be used for desalinization are distillation, freezing, and reverse osmosis. If the salt content is reduced from a fixed initial concentration to a fixed final concentration by any of the three processes, the energy required if the processes are carried out reversibly (at maximum efficiency) is the same for all. However, when carried out at a finite rate, additional energy is required. In the case of reverse osmosis, for what is the additional energy required?

18.8. Discuss the origin of the equation $\Gamma = \nu e^{-\Delta G^*/RT}$ for a specific reaction rate between a hydrogen molecule and a deuterium atom, such as $H_2 + D \longrightarrow H\text{-}D + H$.

18.9. When a steel is rapidly cooled to, say, 500°C, the austenite decomposes to ferrite plus cementite instead of ferrite and graphite, although the latter is more stable. Explain why with illustrations.

18.10. a. Is a *p-n* junction a thermodynamically stable solid or is it metastable?

b. Are diamonds stable or metastable?

18.11. To achieve certain behavior, the materials specialist often generates structures which are thermodynamically unstable. Give four examples and the reasons these are made.

18.12. a. Compare the diffusion coefficients of gases, liquids, and solids.

b. Compare the diffusion coefficients for diffusion along dislocations, along interfaces, and in good crystal.

c. Compare the diffusion coefficients for diffusion of interstitial atoms and substitutional atoms.

18.13. Give a process in ionic crystals which is controlled by the diffusion rate of

a. The fastest-moving ion. Why?

b. The slowest-moving ion. Why?

18.14. Explain how surfaces, grain boundaries, and dislocations (edge and screw) can act as sources or sinks for vacancies.

18.15. Suppose you are given the cored structure of Figure 16.4.7. You find that you can anneal this at a temperature such that $D = 10^{-10}$ cm^2/sec. Estimate the time needed to homogenize the material.

18.16. A vacancy is created on a dislocation line by dislocation climb. It then makes 10,000 jumps through the good crystal. What is the most probable distance from the origin?

18.17. Calculate the time it would take at 1750°F to diffuse carbon into iron to a penetration depth of 0.005 in.

18.18. Discuss the role which an adherent oxide film plays in oxidation resistance.

18.19. The kinetics of the formation of precipitates from a supersaturated solution of Al–4% Cu is 10^8 times as fast as expected from lattice diffusion under equilibrium conditions. Give two possible explanations.

18.20. Suppose a large number of vacancies (10^6, say) are placed side by side within a circle of radius r within the crystal on the (111) plane of copper.
  a. What is the result?
  b. Under what sort of physical conditions might this be achieved?

18.21. A new steel being marketed for automotive applications consists of a low alloy steel core and a high Cr surface layer diffused into the steel for increased corrosive resistance. Calculate how long a heat treatment would be required at 1000°C to diffuse Cr far enough into the steel so that the composition is 18% Cr 0.020 cm below the surface and 100% Cr at the surface. Assume $D_0 = 0.47 \text{ cm}^2/\text{sec}$ and $Q_D = 79.3 \text{ kcal}/$ mole for diffusion of chromium in iron.

18.22. Silicon is somewhat transparent in the infrared. In what application is it desirable to have a *p-n* junction near to a surface?

18.23. What characteristic of the oxide is necessary in order that a parabolic diffusion law holds?

18.24. The oxidation of a certain iron alloy boiler tube follows the law $X^2 = Kt$, where X is the thickness of the oxide and $K = e^{-17,500/T(°K)} \text{ cm}^2/\text{sec}$. If $T = 600°C$, calculate how much time has to elapse before an oxide layer of 0.1 cm forms on the surface.

18.25. You have an assignment to measure the self-diffusion coefficient of copper. Describe in detail (with equations) how you would do this. Include the size of the sample and the like.

More Involved Problems

18.26. a. What is the flux of atoms of H (hydrogen) across a foil of Pd 1.5×10^{-2} cm thick at 750°K if the concentration of H on the side of the foil on which the pressure is high is $1.1 \times 10^{22} \text{ atoms/cm}^3$ and on the low-pressure side the concentration is only $0.1 \times 10^{22} \text{ atoms/cm}^3$.
  b. How long would you have to run this system, assuming the foil has a cross section of 1 cm² to produce 1 mole of pure hydrogen?
  Note: Assume that the dissolved hydrogen concentration is fixed in time at each side of the foil. It will, in fact, be determined by the pressure

according to $C(H) = K(P_{H_2})^{1/2}$. This is another example of the *assumption of local equilibrium*. In this case, the assumption is made that the transport of the gas to the surface and the decomposition and dissolution of the gas are very rapid compared to the rate of diffusion through the solid, so that the concentration at the surface is the same as if no diffusion were occurring.

18.27. A salt water solution has a freezing point of $-4.0°C$. Assume that the specific heat of the water is 1 cal/g-°K and of ice, 0.5 cal/g-°K, and that the fusion energy is 80 cal/g.

    a. Calculate the energy needed to form pure water reversibly (calories per gram).

    b. Convert this to kilowatt hours per liter.

    c. Look in the literature for diffusion data of H_2O in polymers or other potential membranes for reverse osmosis.

    d. Using these data, design a reverse osmosis system in which the total energy needed is just twice the reversible energy. Assume all pumps are 100% efficient.

18.28. Given erf x, discuss how you would evaluate this for

    a. $x \ll 1$.

    b. $x \gg 1$.

Although the value is small, we are interested in an accurate value. Assume that tables and digital computers are not available.

18.29. In Section 16.6, we noted that, in the case of one-dimensional segregation, we could ignore the diffusion in the solid state. Analyze this as quantitatively as possible and comment on the justification.

18.30. When a gold wire is cooled very rapidly, there is a supersaturation of vacancies. However, it is usually assumed that in the good crystal material near to grain boundaries, free surfaces, and dislocations, the concentration very quickly reaches the equilibrium concentration. Criticize this assumption of *local equilibrium vacancy concentration*.

18.31. The comment was made in Section 16.4 that to eliminate coring easily, a sample is cooled down very rapidly and then annealed. Explain as quantitatively as possible the role of rapid cooling here and the role of annealing.

18.32. Consider diffusion on a simple cubic lattice. Show that $\overline{r^2} = 6Dt$.

18.33. For diffusion on a simple cubic lattice, show that the most probable distance from the origin is at $r_{mp} = \sqrt{4Dt}$.

18.34. Calculate the time necessary at 1100°C to get a penetration depth of 10^{-3} cm of phosphorus in silicon.

18.35. Read the article by Hittinger and Sparks, *Scientific American* (Nov. 1965) p. 56, on the preparation of microelectronic circuits. Discuss the various materials problems encountered.

18.36. Read the monograph *Gas Carburizing*, American Society for Metals, Cleveland (1964) and write a critical paper on the carburization of steels.

18.37. Derive the expression $D = \frac{1}{6}\Gamma a^2$ for diffusion of carbon in α-iron. Here a is the lattice parameter.

18.38. Derive the expression $D = \Gamma a^2$ for diffusion of vacancies in fcc metal. Here a is the lattice parameter.

18.39. Derive the expression for electrical conductivity of an ionic crystal, such as NaCl:

$$\sigma = \frac{C_v D_v q^2}{kT}.$$

Hint: The flux of sodium ion vacancies is given by $\mathbf{J}_{\text{vacancies}} = B_v C_v \mathbf{F}$, where $\mathbf{F}$ is the force which causes the flux of vacancies. In the present case, the vacancy has a charge q (which is negative) and the force acting is an electrical one.

Hint: How is the flux of charge $\mathbf{J}_{\text{charge}}$ related to the flux of vacancies?

Hint: Use the Nernst relationship $B_v = D_v/kT$.

18.40. Derive the Nernst relation $B = D/kT$. Start with $\mathbf{v} = B\mathbf{F}$ and express this as a flux equation. Then let the force be the general chemical force $\mathbf{F}_{\text{chem}} = -\text{grad } \mu$.

Hint: For an ideal solution, $\mu = \mu_0 + kT \ln C/C_0$.

Sophisticated Problems

18.41. a. Given that the entropy of vacancy formation is $1.0k$, calculate the vacancy concentration in gold at T_M and at $T_M/4$.

b. Comment on the measurability of the vacancy concentration in an 0.005 in.-diameter gold wire 30 in. long if vacancies contribute a resistivity change of $1.5\mu\,\Omega$-cm/at. % at T_M.

c. Do the same if the wire is quenched from T_M to $T_M/4$ and no vacancies are lost.

18.42. Assume that one-dimensional coring has occurred in a sample such that $C = C_1 + C_2 \sin(\pi x/L)$. Such a cored specimen is held at a high temperature at which we can assume that the chemical diffusion coefficient is D and that D does not vary with C. Obtain an expression for the homogenization of this sample, i.e., $C(x, t)$.

Hint: Use separation of variables.

18.43. An 0.8 wt. % steel is rapidly cooled to 500°C and held at that temperature. Explain what happens in terms of the free energy diagram; i.e., what determines the lamellar spacing?

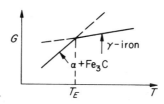

FIGURE P18.43.

18.44. Hydrogen can readily be purified by passing through Pd. Invent similar separation or purification procedures for He and O_2.

18.45. Assume a uniaxial eutectic transformation is taking place as shown in Figure 16.4.1 and the accompanying figure. The diffusion which causes the separation into α and β takes place between T_E and $T_E - \Delta T$ in a thin layer (thin with respect to the average diffusion distance). On this basis, assuming one-dimensional diffusion, derive Equation (16.4.1). Assume that the height h of the supercooled liquid is independent of the velocity.

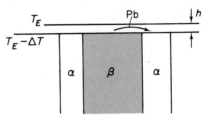

FIGURE P18.45.

Prologue

We begin the study of the origin of mechanical behavior of materials by studying the origin of elastic behavior and the origin of viscous behavior, the two extremes of mechanical behavior. The elastic constants of single crystals are described in terms of the *generalized Hooke's law*. We study the form of the *elastic compliance matrix* for an elastically isotropic solid. Crystals are in general *elastically anisotropic*, and examples of such anisotropy are given for cubic crystals. We observe that the elastic constants of crystalline materials are determined primarily by the binding energy. We find that rubberlike materials behave differently elastically than crystalline (or even glassy) materials, exhibiting a behavior known as *rubberlike elasticity*. We derive expressions for the elastic behavior of rubberlike materials based on the *entropy of random chain configurations*. Composites are also often elastically anisotropic; the calculation of the state of stress in loaded composite structures is a challenging problem. Until recently most of the effort of elasticians had been directed toward the study of elastically isotropic materials.

In studying the viscosity of gases, we see how the viscosity coefficient of gases depends on a *momentum transfer process*. We note that the viscosity of liquids depends on *activation energy processes*. We find that at very high-temperature crystals exhibit *Newtonian viscosity* and that this is due to diffusion of vacancies.

492

19

EXTREMES OF MECHANICAL BEHAVIOR

19.1 INTRODUCTION

The two extremes of mechanical behavior are purely elastic behavior and purely viscous behavior. The first of these is entirely reversible or conservative behavior, while the second is entirely dissipative. Both of these phenomena may be linear or nonlinear as shown in Table 19.1.1. Depending on the rate of

Table 19.1.1. EXTREMES OF MECHANICAL BEHAVIOR

	Elastic	Viscous
Linear		
Behavior	Stress $\propto$ strain, Eqs. (2.3.4) and (2.3.8)	Drag stress $\propto$ velocity gradient, Eq. (2.8.3)
Name of behavior	Hookean	Newtonian
Macroscopic example	Deflection $\propto$ load of beam, Eq. (2.1.5)	Rate of flow, $\propto$ pressure drop in pipe, Eq. (2.8.1)
Nonlinear		
Behavior	Complex relations	Complex relations
Name of behavior	Non-Hookean	Non-Newtonian

loading, time, stress, and temperature, solids may exhibit elastic or viscous behavior, liquids may exhibit purely elastic behavior (if supercooled to glasses) or purely viscous behavior, while gases exhibit elastic behavior under hydrostatic pressure only, and in the presence of shear stresses gases exhibit viscous behavior.

Real materials often behave as combinations of linear elastic and viscous elements, as will be discussed in Chapter 20.

19.2   ELASTIC BEHAVIOR OF CRYSTALS UNDER HYDROSTATIC PRESSURE

The discussion of elastic behavior may be divided into a discussion of the dependence on hydrostatic pressure (which is discussed in this section) and a discussion of the dependence on more general stress states (discussed in Section 19.3). We shall be concerned with the atomic origin of elastic behavior.

In the case of an ideal gas the origin of the P, V, T behavior is the kinetic energy of the gas particles. Since the equation of state is completely known, this form of elastic behavior is entirely understood. For denser gases (and liquids), potential energy interactions between the atoms become important and the calculation of these interactions and of an equation of state (P, V, T relation) becomes extremely involved. This is an active area of current research.

In the case of solids the pressure-volume arrangement for the static crystal is nearly the same as for the crystal with atomic vibrations. Hence the pressure-volume relationship is determined by the binding forces. The bulk modulus K is defined by

$$dP = -K\frac{dV}{V}. \tag{19.2.1}$$

For solids

$$K = K_0 + K_0'P. \tag{19.2.2}$$

At small pressures the second term can be neglected so that

$$\frac{V - V_0}{V_0} = \frac{\Delta V}{V} = \frac{-P}{K_0}. \tag{19.2.3}$$

Hence the P, V behavior is known if K_0 is known. In the general case the calculation of K_0 is a quantum mechanical problem. However, for an ionic crystal such as sodium chloride the Born model (Chapter 9) can be used. We note from Equation (9.2.4) that K_0 is determined by two parameters, r_0, the distance of closest approach of the ions under equilibrium conditions, and n, the exponent in the repulsive inverse power term of Equation (9.2.2). The quantity r_0 can be obtained from the lattice parameter. It can be shown from theory after tedious effort (which is an extension of Problem 9.15) that the pressure derivative of the bulk modulus evaluated at zero pressure is for a Born model

$$\lim_{P \to 0} \frac{dK}{dP} = K_0' = \frac{n + 7}{3}. \tag{19.2.4}$$

Hence experimental values of K_0' can be used to evaluate the parameter n of the Born theory. Then K_0 can be calculated from Equation (9.2.4). The theoretical result for KBr at room temperature is $K_0 = 1.33 \times 10^{11}$ dynes/cm² (133 kbars), while the experimental value is 142 kbars. The agreement is within 7%.

Some crystals such as argon (which have a fcc structure) are bonded together by secondary bonds, so-called van der Waals forces (actually London dispersion forces; see Section 9.5) in which the attractive term varies as $1/r^6$ and the repulsive term as $1/r^{12}$. The potential energy of such crystals can be represented by the **6–12 potential**:

$$U = -\frac{a}{r^6} + \frac{b}{r^{12}}, \qquad\qquad (19.2.5)$$

which is often called the **Lennard-Jones potential**.

The parameter a can be calculated from quantum mechanical considera-tion. The parameter b can be evaluated from the equilibrium condition $\partial U/\partial r = 0$ at $r = r_0$. The bulk modulus can be calculated as in Problem 9.15. Results in fair agreement with experiment are obtained.

For some simple metals and covalent crystals the bulk modulus can also be calculated to accuracies of this sort using quantum mechanics. Such results are shown in Table 19.2.1. However, for most solids, the mathematical com-plexities are such that even with the help of giant digital computers, successful calculations have not been made. This is an active area of current research.

Table 19.2.1. BULK MODULI (KBARS)

Material	Theory	Experiment
Li	104	102
Na	62	58
K	29	31
Rb	15	17
Cs	13	16

Except for cubic crystals, single crystals, when subjected to hydrostatic pressure, show different percentage length changes in different directions. This effect could be used, for example, to make a bimetallic pressure measuring device. It is for this reason that the theory discussed above was related to cubic crystals in each case (NaCl and argon); it was easier.

EXAMPLE 19.2.1

Polycrystalline bismuth is pressurized in a liquid medium at 10 kbars. (Bismuth has a rhombohedral crystal structure.) Is the stress state within a given bismuth crystal a hydrostatic pressure?

Answer. No. Each bismuth crystal by itself would contract differently in different directions under the pressure. Since each crystal is connected to other crystals with different orientations, additional stresses appear.

19.3   ELASTIC CONSTANTS OF CRYSTALS

The calculation of the state of stress and the deflections of a beam, a plate, or some other structural component is determined by the size and shape of the component, the equation of equilibrium, the forces applied to the component, and the *elastic behavior* of the material. In this section we shall study the elastic behavior of crystalline materials.

The state of stress is defined by nine stress components. See Equation (7.2.1) and Figure 7.2.1. In the absence of body couples this array becomes a symmetric array,

$$\begin{matrix} \sigma_{11} & \sigma_{12} & \sigma_{13} \\ \sigma_{12} & \sigma_{22} & \sigma_{23} \\ \sigma_{13} & \sigma_{23} & \sigma_{33}; \end{matrix}$$

i.e., there are only six independent stress components. In the notation used here σ_{ij} means a stress component owing to a force component F_i (i.e., in the x_i direction) acting on a plane whose normal is in the x_j direction. (Stress is discussed in Sections 7.2 and 7.3.) We now use a contracted notation:

$$\text{Normal stresses:} \quad \begin{cases} \sigma_{11} \longrightarrow \sigma_1 \\ \sigma_{22} \longrightarrow \sigma_2 \\ \sigma_{33} \longrightarrow \sigma_3 \end{cases}$$

$$\text{Shear stresses:} \quad \begin{cases} \sigma_{23} \longrightarrow \sigma_4 \\ \sigma_{13} \longrightarrow \sigma_5 \\ \sigma_{12} \longrightarrow \sigma_6. \end{cases} \tag{19.3.1}$$

We use a similarly contracted notation for strain components:

$$\text{Extension strains:} \quad \begin{cases} \epsilon_1 = \epsilon_{11} \\ \epsilon_2 = \epsilon_{22} \\ \epsilon_3 = \epsilon_{33} \end{cases}$$

$$\text{Shear strains:} \quad \begin{cases} \epsilon_4 = 2\epsilon_{23} \\ \epsilon_5 = 2\epsilon_{13} \\ \epsilon_6 = 2\epsilon_{12}. \end{cases} \tag{19.3.2}$$

Such extension strains are defined by Equation (2.3.1) and the shear strains by Equation (2.3.5). Thus ϵ_4 is the shear strain as illustrated in Figure 19.3.1. For a further study of tiny or infinitesimal strains, see Problem 19.32.

The **generalized form of Hooke's law** assumes that each stress component is a linear combination of the strain components. Hence

$$\sigma_q = \sum_{r=1}^{6} C_{qr} \epsilon_r, \qquad q = 1, 2, \dots, 6, \tag{19.3.3}$$

where the C_{qr} are the **elastic stiffnesses** of the crystal.

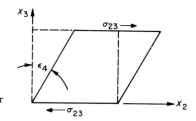

FIGURE 19.3.1. Small shear strain.

Thus for σ_1 we have

$$\sigma_1 = C_{11}\epsilon_1 + C_{12}\epsilon_2 + C_{13}\epsilon_3 + C_{14}\epsilon_4 + C_{15}\epsilon_5 + C_{16}\epsilon_6.$$

There are similar relations for the other five stress components. Thus if the only strain present is ϵ_1, then the stress $\sigma_1 = C_{11}\epsilon_1$.

The differential strain energy per unit volume is

$$dw = \sum_{q=1}^{6} \sigma_q d\epsilon_q. \tag{19.3.4}$$

From Hooke's law this is

$$w = \sum_{q,r} \tfrac{1}{2} C_{qr}\epsilon_q \epsilon_r, \qquad \text{where } q, r = 1, 2, \ldots, 6. \tag{19.3.5}$$

Note that if we expand the strain energy in a Taylor series about the unstrained state, the quadratic terms are the first of the nonvanishing terms. The Taylor series expansion of the strain energy about the unstrained state in terms of the stress is

$$w = w(0) + \sum_r \left(\frac{\partial w}{\partial \epsilon_r}\right)_{\epsilon_r=0} \epsilon_r + \sum_{r,q} \left(\frac{\partial^2 w}{\partial \epsilon_r \partial \epsilon_q}\right)_{\epsilon_r, \epsilon_q=0} \frac{\epsilon_r \epsilon_q}{2} + \cdots.$$

The term $w(0)$ is, of course, zero in the unstrained state. Likewise, $\partial w/\partial \epsilon_r = 0$ at equilibrium. Thus we see that the ordinary elastic stiffness coefficients equal the second derivation of the strain energy with respect to the strain. Cubic and higher terms correspond to nonlinear elasticity.

There are 36 elastic constants in all, but it can be shown (see Problem 19.24) that $C_{qr} = C_{rq}$, so that there are in general only 21 independent constants (which is the situation for a triclinic crystal). For a cubic crystal there are only 3 independent elastic constants (because of symmetry), while an isotropic medium has only 2 independent elastic constants.

EXAMPLE 19.3.1

Give an example where the elastic constants of a single crystal are necessary in an engineering application.

Answer. The resonance frequencies of a quartz crystal (which might be one of 10^8 crystals used annually as transducers in sonar, as frequency controls, etc.) depend on the density of the quartz, on the *elastic constants* of the quartz, and on the dimensions of the crystal.

The array of elastic stiffness constants for a cubic crystal, referred to $\langle 100 \rangle$ axes, is

$$
\begin{pmatrix}
C_{11} & C_{12} & C_{12} & 0 & 0 & 0 \\
C_{12} & C_{11} & C_{12} & 0 & 0 & 0 \\
C_{12} & C_{12} & C_{11} & 0 & 0 & 0 \\
0 & 0 & 0 & C_{44} & 0 & 0 \\
0 & 0 & 0 & 0 & C_{44} & 0 \\
0 & 0 & 0 & 0 & 0 & C_{44}
\end{pmatrix}
\tag{19.3.6}
$$

and for the special case of isotropic materials

$$
\begin{pmatrix}
\lambda + 2G & \lambda & \lambda & 0 & 0 & 0 \\
\lambda & \lambda + 2G & \lambda & 0 & 0 & 0 \\
\lambda & \lambda & \lambda + 2G & 0 & 0 & 0 \\
0 & 0 & 0 & G & 0 & 0 \\
0 & 0 & 0 & 0 & G & 0 \\
0 & 0 & 0 & 0 & 0 & G
\end{pmatrix},
\tag{19.3.7}
$$

where λ is the **Lamé parameter** and G is the shear modulus.

Here $\lambda = Ev/(1 + v)(1 - 2v)$, where E is Young's modulus and v is Poisson's ratio. Hooke's law for the strains is given by

$$
\epsilon_i = \sum_{j=1}^{6} s_{ij}\sigma_j,
\tag{19.3.8}
$$

where the s_{ij} are called **elastic compliance coefficients**. It is also true that $s_{ij} = s_{ji}$. The array for a cubic crystal is

$$
\begin{pmatrix}
S_{11} & S_{12} & S_{12} & 0 & 0 & 0 \\
S_{12} & S_{11} & S_{12} & 0 & 0 & 0 \\
S_{12} & S_{12} & S_{11} & 0 & 0 & 0 \\
0 & 0 & 0 & S_{44} & 0 & 0 \\
0 & 0 & 0 & 0 & S_{44} & 0 \\
0 & 0 & 0 & 0 & 0 & S_{44}
\end{pmatrix}.
\tag{19.3.9}
$$

The array for the C_{ij} has the same form as the array for the s_{ij}.

The elastic constants of crystals can be calculated from the theory of binding. The strain energy of a deformed crystal is calculated, and the elastic stiffnesses are then obtained from

$$
C_{ij} = \frac{\partial^2 w}{\partial \epsilon_i \, \partial \epsilon_j}.
\tag{19.3.10}
$$

Results for NaCl are shown in Table 19.3.1.

The elastic constants of crystals as a rule depend only slightly on tempera-

Table 19.3.1. ELASTIC STIFFNESSES OF NaCl (kbars)

Constant	Theory	Experiment
C_{11}	450	585
C_{12}	98	110
C_{44}	153	134

ture (they decrease slightly as temperature increases). There are a few exceptions to this. One is shown in Figure 19.3.2. Studying *unusual* behavior of this sort is an active area of current research. V_3Si has a phase transformation at about 21°K. It appears that $C' = (C_{11} - C_{12})/2$ is rapidly decreasing toward zero and even negative values. It can be shown that a cubic crystal will spontaneously deform if $C' < 0$, if $C_{44} < 0$, if $C_{11} < 0$, or if the bulk modulus $B < 0$. [A more general statement is that a crystal will be **elastically unstable** if the elastic constant matrix, such as given by (19.3.6) for the cubic crystal, is not positive definite.

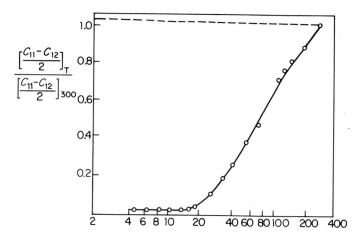

FIGURE 19.3.2. Temperature dependence of $(C_{11} - C_{12})/2$ of V_3Si relative to room-temperature value. Dashed line shows typical behavior of solids. [From L. R. Testardi and T. B. Batemen, *Physical Review*, **154**, 402 (1967).]

(See Problem 19.30.) This leads, for the cubic crystal, to the four conditions just stated.] It should be noted that the origin of the elastic constants of crystalline materials arises nearly entirely from the binding forces between the atoms in a static lattice.

For an isotropic material the elastic compliance array has nearly the same form as for the cubic crystal with $s_{11} = 1/E$, where E is Young's modulus; $s_{12} = -v/E$, where v is Poisson's ratio; and $s_{44} = 1/G$, where G is the shear modulus. The array of compliances for an isotropic material is

$$
\begin{pmatrix}
\dfrac{1}{E} & \dfrac{-v}{E} & \dfrac{-v}{E} & 0 & 0 & 0 \\[2mm]
\dfrac{-v}{E} & \dfrac{1}{E} & \dfrac{-v}{E} & 0 & 0 & 0 \\[2mm]
\dfrac{-v}{E} & \dfrac{-v}{E} & \dfrac{1}{E} & 0 & 0 & 0 \\[2mm]
0 & 0 & 0 & \dfrac{1}{G} & 0 & 0 \\[2mm]
0 & 0 & 0 & 0 & \dfrac{1}{G} & 0 \\[2mm]
0 & 0 & 0 & 0 & 0 & \dfrac{1}{G}
\end{pmatrix}
\tag{19.3.11}
$$

We note that for an isotropic material

$$
v = \frac{E}{2G} - 1. \tag{19.3.12}
$$

The elastic constants of a crystal could be described in terms of an E (Young's modulus) and a G which vary with direction. Thus for a cubic crystal, G takes on values between C_{44} and $(C_{11} - C_{12})/2$ depending on the orientation. For copper $C_{44} = 818$ kbars and $(C_{11} - C_{12})/2 = 256$ kbars, so G varies by more than a factor of 3 with orientation. Figure 19.3.3 illustrates the orientations of shear stress relative to the cubic crystal which leads to C_{44} and $(C_{11} - C_{12})/2$.

Polycrystalline aggregates with random orientations of crystals show isotropic elastic behavior on a macroscopic scale, i.e., a scale many times larger than the grain size.

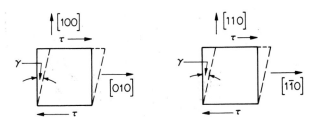

FIGURE 19.3.3. Shear in cubic crystals. $\tau = G\gamma$. In (a), $G = C_{44}$. In (b), $G = (C_{11} - C_{12})/2$.

EXAMPLE 19.3.2

Discuss the variation of G with direction in NaCl.

Answer. From Table 19.3.1 we see that G varies from 153 kbars to $(450 - 98)/2 = 176$ kbars. Thus NaCl single crystals are much more nearly isotropic than copper crystals.

The variation of E and G with direction can be represented by elastic bodies. The elastic body for E is a surface such that the length of a vector from the origin to a point on the surface represents the magnitude of E in the direction of that vector. Examples are shown in Figure 19.3.4. Composite materials are often highly anisotropic elastically. The application of elasticity theory to anisotropic plates, shells, beams, etc., is currently an active area of research.

FIGURE 19.3.4. Elastic bodies for iron. [From E. Schmidt and W. Boas *Plasticity of Crystals,* Chapman and Hall, London (1968).]

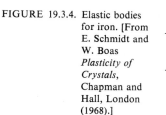
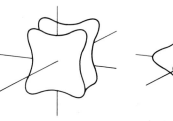
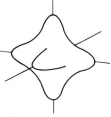

Young's modulus   Shear modulus

19.4   RUBBERLIKE ELASTICITY

Rubberlike materials or **elastomers** are characterized by

1. High (reversible) extension ratios ($l/l_0 \gg 1$).
2. A shear modulus which is directly proportional to the absolute temperature.
3. A Poisson's ratio $\nu = \frac{1}{2}$.
4. Very small shear modulus and Young's modulus.

[The fact that $\nu = \frac{1}{2}$ means that the material is incompressible ($K \rightarrow \infty$). This, of course, is not so, and rubber has a bulk modulus about the same as any hydrocarbon liquid. However, when loaded under tension or torsion, it does indeed exhibit $\nu \doteq \frac{1}{2}$ and E (or $G) \ll K$.]

These unique elastic properties of rubber make it an ideal material for many applications ranging from blimps to tires. A tire must be able to undergo large deflections under small stress, and rubber is ideally suited for this application. Moist human hair is a rubberlike material, as are muscles, so the theory of rubberlike elasticity has biological applications. Examples of various elastomers are shown in Table 19.4.1.

At a sufficiently low temperature there is a transition from the rubberlike state to a glass hard state; the above materials then behave elastically similar to ordinary solids. Some examples of this transition temperature are given in Table 19.4.2. This transition temperature can be strongly affected by many variables, including the presence of a plasticizer, the presence of stress, and vulcanization. Figure 19.4.1 shows the effect of vulcanization of natural Hevea rubber.

A substance such as Hevea rubber can therefore occur either as a rubber or a plastic. We usually think of a rubber as a material which has rubberlike properties at room temperature. Thus poly(methyl methacrylate) (transition

Table 19.4.1. SOME ELASTOMERS

$$-CH_2-\underset{\underset{\displaystyle}{|}}{\overset{\overset{\displaystyle CH_3}{|}}{C}}=CH-CH_2-$$ Polyisoprene (natural rubber)

$$-CH_2-\underset{\underset{\displaystyle}{|}}{\overset{\overset{\displaystyle Cl}{|}}{C}}=CH-CH_2-$$ Polychloroprene (neoprene)

$$-CH_2-\underset{\underset{\displaystyle CH_3}{|}}{\overset{\overset{\displaystyle CH_3}{|}}{C}}-$$ Polyisobutylene

$$-CH_2-\underset{\underset{\displaystyle}{|}}{\overset{\overset{\displaystyle Cl}{|}}{C}H}-$$ Poly(vinyl chloride) (at high temperatures)

$$-CH_2-CH=CH-CH_2-\underset{\underset{\displaystyle \bigcirc}{|}}{C}H-CH_2-$$ Butadiene-styrene (GRS)

$$-CH_2-\underset{\underset{\displaystyle COOCH_3}{|}}{\overset{\overset{\displaystyle CH^3}{|}}{C}}-$$ Poly(methyl methacrylate) (at high temperature)

$$-NH-\underset{\underset{\displaystyle}{|}}{\overset{\overset{\displaystyle R}{|}}{C}H}-CO-$$ Protein (wool, silk, muscle) (in the presence of water)

$$-\underset{\underset{\displaystyle CH_3}{|}}{\overset{\overset{\displaystyle CH_3}{|}}{S}i}-O-$$ Poly(methyl siloxane) (silicone rubber)

Table 19.4.2. GLASS TRANSITION
TEMPERATURES

	°C
Polyisobutylene	−74
Natural rubber	−73
Poly(vinyl chloride)	80
Poly(methyl methacrylate)	120

temperature, 120°C) is usually known as a plastic (called Lucite) even though it does have a rubberlike state above 120°C. Wool or human hair, for example, when completely dry extends only a few percent, but when immersed in water it extends about 80% but with considerable hysteresis, whereas when immersed in formic acid it behaves like a perfect elastomer. The plasticizing effect of the formic acid is due to breaking of the interchain hydrogen bonds leaving only the sulfur crosslinks to hold the wool together.

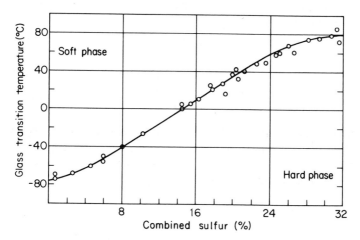

FIGURE 19.4.1. Glass transition of natural rubber versus sulfur content. Data of R. F. Boyer and R. S.
Spencer, *Advances in Colloid Science*, **2**, 1
(1946).

We now proceed to show that the elasticity of rubber is purely due to an
entropy change [which we shall later see is due to the change in the randomness
of a polymer chain (see Figure 14.2.1) when it is stretched]. Rubber is essentially
a liquid which is held together in solid form by crosslinks. In the case of vulcanized rubber these crosslinks are sulfur crosslinks. The chains between the
crosslinks have considerable mobility since there is little attractive force between
the molecules. Hence the chains are free to take on many configurations.

THERMODYNAMICS OF RUBBER.   The differential internal energy of a system which behaves reversibly is given by

$$dU = T\,dS + dW. \tag{19.4.1}$$

In the present case we consider a rubber specimen of length l being stretched
isothermally by a force f. Then $dW = f\,dl$ and

$$dU = T\,dS + f\,dl. \tag{19.4.2}$$

(Note that there is no $-P\,dV$ work because $dV \rightarrow 0$.) Solving for f we have an
expression for the force

$$f = \left(\frac{\partial U}{\partial l}\right)_T - T\left(\frac{\partial S}{\partial l}\right)_T \tag{19.4.3}$$

when the tensile specimen has a length l at the temperature T. It can be shown by
purely thermodynamical means (as we shall do shortly) that

$$f = \left(\frac{\partial U}{\partial l}\right)_T + T\left(\frac{\partial f}{\partial T}\right)_l \tag{19.4.4}$$

Now f, T, and $(\partial f/\partial T)_l$ are readily measurable quantities. So next we do an experiment and we find that

$$f \doteq T\left(\frac{\partial f}{\partial T}\right)_l,\qquad(19.4.5)$$

or in other words that

$$\left(\frac{\partial U}{\partial l}\right)_T = 0,\qquad(19.4.6)$$

at least for extensions up to a few hundred percent. Returning to Equation (19.4.3) we see that f is determined essentially only by the entropy. (For crystalline solids we would find instead that the elastic behavior is determined almost entirely by the internal energy only.)

EXAMPLE 19.4.1

Prove that

$$\left(\frac{\partial f}{\partial T}\right)_l = -\left(\frac{\partial S}{\partial l}\right)_T.$$

Hint: Start with a quantity F defined by $F \equiv U - TS$ and find an expression for dF. See also Example 17.5.7(c).

Answer.

$$dF = dU - T\,dS - S\,dT,$$

but

$$dU - T\,dS = f\,dl$$

so

$$dF = f\,dl - S\,dT.$$

Since dF is an exact differential it follows that

$$\left(\frac{\partial f}{\partial T}\right)_l = -\left(\frac{\partial S}{\partial l}\right)_T.\qquad(19.4.7)$$

MODEL OF RUBBERLIKE ELASTICITY.   The entropy of a polymer chain is determined by

$$s = k \ln \omega,\qquad(19.4.8)$$

where ω is the number of distinct configurations. A completely stretched out chain has $\omega = 1$. A random chain of considerable length has numerous configurations. One configuration is shown in Figure 14.2.1. In the present case a chain is the portion of the polymer between crosslinks. A model which can be used to describe these configurations is the random walk model of Section 17.3. Let one end of the chain be considered as a reference point at the origin of a Cartesian coordinate system and let the other end be located within a volume element (spherical shell) $4\pi r^2\,dr$. The probability of this configuration according to

(17.3.18) is

$$P(r)\, dr = \frac{1}{(2\pi\sigma^2)^{3/2}} e^{-r^2/2\sigma^2} 4\pi r^2 \, dr, \qquad (19.4.9)$$

where

$$\sigma^2 = \frac{N}{3} l^2 \qquad (19.4.10)$$

and N is the total number of links in the chain. [Considered as a random walk on a simple cubic lattice, there would be $N/3$ steps in each of the x_1, x_2, and x_3 directions. The most probable distance between the ends is $r = \sqrt{2}\,\sigma = \sqrt{\frac{2}{3}N}\,l$ (see Example 17.3.7).]

The number of configurations of a chain, whose mobile end falls between r and $r + dr$, is proportional to the probability of that configuration:

$$\omega = AP(r)\, dr.$$

Hence by (19.4.9)

$$s = c + k \ln \frac{r^2}{2\sigma^2} - k \frac{r^2}{2\sigma^2}. \qquad (19.4.11)$$

EXAMPLE 19.4.2

Find the radial force on such a chain.

Answer. From (19.4.2), with $dU = 0$, we have

$$f_r = -T\left(\frac{\partial s}{\partial r}\right),$$

and using (19.4.11) we get

$$f_r = kT\left(\frac{r}{\sigma^2} - \frac{2}{r}\right).$$

Note that $f_r = 0$ when $r^2 = 2\sigma^2$, which corresponds to the most probable distance, i.e., the most probable configuration.

Note that this is *not* the expression one would get if a one-dimensional tension test were performed since then the mobile end would be restricted to move along a given axis.

We next proceed to calculate the average entropy per chain (assuming we have a large number of chains). This is given by

$$\bar{s} = \int_0^\infty s(r)P(r)\, dr. \qquad (19.4.12)$$

[Recall that $P(r)\, dr$ is the fraction of chains with the mobile end between r and $r + dr$.] We can write

$$\bar{s} = \int_0^\infty cP(r)\, dr + \int_0^\infty k \ln \frac{r^2}{2\sigma^2} P(r)\, dr - \int_0^\infty k \frac{r^2}{2\sigma^2} P(r)\, dr$$
$$= I_1 + I_2 - I_3.$$

It can be shown that $I_2 \ll I_3$ (plot the integrands in each case; it is convenient to let $z^2 = r^2/2\sigma^2$). Hence for the evaluation of $\bar{s}$ we can neglect the second term of (19.4.11) and use

$$s = c - \frac{kr^2}{2\sigma^2}. \tag{19.4.13}$$

We shall assume that a unit cube of rubber is deformed in such a way that it becomes a rectangular parallelepiped with dimensions λ_1, λ_2, and λ_3 along x_1, x_2, and x_3, respectively. See Figure 19.4.2. We assume that the sample deformation is homogeneous (the same in all parts of the sample). Hence a point

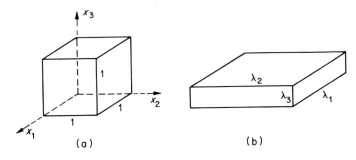

FIGURE 19.4.2. Unit cube of rubber (a) before deformation and (b) after deformation.

at x_1, x_2, x_3 in the undeformed state moves to $\lambda_1 x_1$, $\lambda_2 x_2$, $\lambda_3 x_3$ in the deformed state. Consequently a chain which has one end at $r = 0$ and the other at r, where

$$r^2 = x_1^2 + x_2^2 + x_3^2, \tag{19.4.14}$$

now has that other end at r' relative to the first end, where

$$r'^2 = \lambda_1^2 x_1^2 + \lambda_2^2 x_2^2 + \lambda_3^2 x_3^2. \tag{19.4.15}$$

Hence

$$s' = c - \frac{kr'^2}{2\sigma^2}. \tag{19.4.16}$$

We are interested in the entropy change of n chains (where n is the number of chains per unit volume) owing to deformation. This is given by

$$\Delta S = n(\bar{s}' - \bar{s}). \tag{19.4.17}$$

We now express both s' and s in terms of x_1, x_2, and x_3. Then we evaluate the averages using $P(x_1, x_2, x_3)\, dx_1\, dx_2\, dx_3$ from Equation (17.3.17). We have

$$\Delta S = n \int_{-\infty}^{\infty} \int_{-\infty}^{\infty} \int_{-\infty}^{\infty} (s' - s)P(x_1, x_2, x_3)\, dx_1\, dx_2\, dx_3. \tag{19.4.18}$$

This equals

$$\Delta S = -\frac{nk}{2}(\lambda_1^2 + \lambda_2^2 + \lambda_3^2 - 3). \tag{19.4.19}$$

The stored elastic energy per unit volume, under conditions of constant volume and temperature, is the Helmholtz free energy, which in the present case is

$$w = -T \Delta S$$
$$= \frac{nkT}{2}(\lambda_1^2 + \lambda_2^2 + \lambda_3^2 - 3). \qquad (19.4.20)$$

This is subject to the constant volume condition

$$\lambda_1 \lambda_2 \lambda_3 = 1. \qquad (19.4.21)$$

Equation (19.4.20) may be used to show that a *linear* shear stress-strain relation is obtained; i.e.,

$$\tau = nkT\gamma = G\gamma. \qquad (19.4.22)$$

The derivation is easy if one understands pure shear, simple shear, and the correspondence between them. To explain this now is outside the scope of this text. It is discussed in elasticity texts. See J. C. Jaeger, *Elasticity, Fracture and Flow*, Wiley, New York (1956) p. 32. Note that $G \propto T$, which has already been mentioned as an experimental fact. The experimental curve is nearly linear even for large γ ($\gamma = 5$) as predicted.

Equation (19.4.20) may also be used to show that the nominal tensile stress versus extension ratio for a simple tension test is *nonlinear*:

$$\sigma = G\left(\lambda - \frac{1}{\lambda^2}\right). \qquad (19.4.23)$$

EXAMPLE 19.4.3

Derive (19.4.23).

Answer. Let the tensile extension ratio be $\lambda_1 = \lambda$. Then $\lambda_2 = \lambda_3$, and from Equation (19.4.21), $\lambda_2^2 = 1/\lambda$. Hence

$$w = \frac{nkT}{2}\left(\lambda^2 + \frac{2}{\lambda} - 3\right). \qquad (19.4.24)$$

Since $dw = \sigma \, d\lambda$, we obtain the desired expression from $\sigma = dw/d\lambda$.

EXAMPLE 19.4.4

Equation (19.4.23) holds only to a certain large strain, e.g., to $\lambda \approx 3$ for chains with 67 links. Why?

Answer. In the unstretched state these chains would have the most probable length $\sqrt{\frac{2}{3}Nl}$, while in the stretched state the length would be Nl. This suggests that the rubber should get rapidly stiffer as $\lambda \longrightarrow Nl/\sqrt{2N/3l} = \sqrt{3N/2}$. For $N = 67$ this gives $\lambda = 10$. Beyond this ratio the elasticity of stretched chains is due to stretching and bending covalent bonds. However, it should be noted that in the equilibrium state not all chains have the most probable length. Some chains are more fully extended and hence have less extensibility than estimated here. The result $\lambda = \sqrt{3N/2}$ should hence be taken as an upperbound for easy extension. Actually, the random chain model fails long before this extension ratio is reached because highly stretched chains cannot be random.

EXAMPLE 19.4.5

Sketch $\sigma(\lambda)$ for a tensile test on rubber with $N = 33$.

Answer.

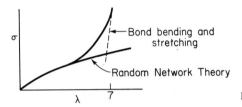

FIGURE E19.4.5.

The actual stress will dip below the random theory line for large λ (say, $\lambda > 3$) because chains are becoming aligned, which increases the probability of crystallization.

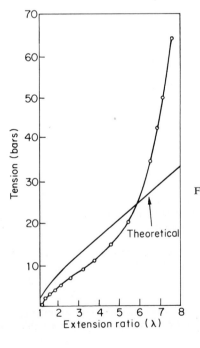

FIGURE 19.4.3. Stress-strain curve for a simple tension test in soft material rubber. $G = 4$ bars. [L. R. G. Treloar, *The Physics of Rubberlike Elasticity*, Oxford, New York (1958).]

Figure 19.4.3 shows the stress-strain curve for a simple tension test. The initial slope of the experimental curve is used to obtain G and the theoretical curve then follows (19.4.23). Note that $G = 4$ bars (1 bar $= 10^6$ dynes/cm² $= 14.5$ psi). Compare with the G values in crystalline materials (see Table 2.2.1).

19.5   VISCOSITY OF GASES

Gases, liquids under certain conditions, and even solids under special circumstances exhibit linear viscous behavior.

It is our intention in this section to derive for gases **Newton's equation of viscosity**,

$$\tau = -\eta \frac{\partial v}{\partial x};\qquad(19.5.1)$$

i.e., the shear stress (drag), τ, equals the negative of the velocity gradient normal to the flow directions, $-\partial v/\partial x$, times the viscosity coefficient, η. In so doing we shall also obtain an expression for the magnitude of η for gases. The viscosity coefficient of gases is an important parameter in the design of airplanes, air bearings, etc. Consider Figure 19.5.1, which shows two planes whose normal is x_3,

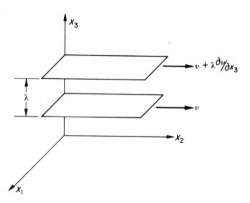

FIGURE 19.5.1. Viscosity of gases.

and which are a distance λ apart, where λ is the **mean free path** of the gas molecules (distance between collisions). The molecules in the planes are moving with different *net* velocities as shown in the figure. Molecules jump back and forth between these planes. (A parallel can be drawn with two trains traveling side by side at different velocities in which people jump back and forth in equal numbers. Those that jump from the slower-moving train to the faster tend to slow it down and vice versa.) When an atom of mass m moves from the upper layer and interchanges with an atom from the lower layer, there is net momentum transfer $-m\lambda(\partial v/\partial x_3)$ to the upper layer. The number moving up or down across a square centimeter area per second is $\frac{1}{3}n\bar{c}$, where n is the number of molecules per unit volume and $\bar{c}$ is their average speed. Hence the momentum change of unit area of the upper layer per second is

$$\frac{dp_2}{dt} = -\frac{1}{3}n\bar{c}m\lambda\frac{\partial v}{\partial x_3}. \tag{19.5.2}$$

Newton's equation is $\mathbf{F} = d\mathbf{p}/dt$ so $F_2 = dp_2/dt$, and since we are dealing with unit area, $dp_2/dt = \sigma_{23}$, the shear stress.

Hence we have

$$\eta = \tfrac{1}{3}nm\bar{c}\lambda.$$

It was noted in Equation (18.4.7) that

$$\lambda = \frac{1}{\sqrt{2}\,\pi nd^2},$$

where d is the molecular diameter, so that

$$\eta = \frac{m\bar{c}}{3\sqrt{2}\,\pi d^2}. \tag{19.5.3}$$

[From Equation (17.4.8) we see that $\bar{c} = \sqrt{8kT/\pi m}$.] This important result, first given by Maxwell, has several interesting features:

1. η does not vary with the density of the gas. This was later shown to be experimentally true.
2. η varies as $T^{1/2}$. Actually it varies somewhat faster, perhaps as $T^{3/4}$.
3. η provides a convenient measure of molecular diameters which may be compared with values obtained by other techniques. These results are illustrated in Table 19.5.1. Values of η are given in Table 2.8.1.

Table 19.5.1. MOLECULAR DIAMETERS (Å)

Molecule	From Viscosity*	From the van der Waals b†	From Closest Packing‡
Ar	2.86	2.86	3.83
O_2	2.96	2.90	3.75
N_2	3.16	3.14	4.00

* Calculated from measured viscosity coefficients and Equation (19.5.3).
† See Problem 17.36.
‡ From packing of spheres in crystals as in Chapter 12.

We note that

$$\eta D = \frac{8}{9\sqrt{2}\,\pi^2}\left(\frac{\lambda}{d^2}\right)kT, \tag{19.5.4}$$

where D is the self-diffusion coefficient given by Equation (18.4.8).

19.6   VISCOSITY OF LIQUIDS

It is our intention here to describe an atomic model which leads to Newton's equation of viscosity for liquids.

The viscosity of liquids is a flow process involving the passage of a molecule (or a segment of a molecule if the molecule is large) over a potential energy barrier. This is illustrated in Figures 19.6.1 and 19.6.2. A liquid may be

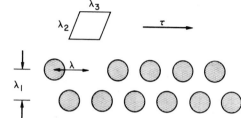

FIGURE 19.6.1. Flow in a liquid under applied shear stress.

considered to a *rough* approximation to have numerous vacancies since the volume increase upon melting is usually about 5%. The moving molecule has a stress τ acting on it on an area $\lambda_2 \lambda_3$ helping to move it to the top of the barrier, i.e., a distance $\lambda/2$. This applied force does work $\tau(\lambda_2 \lambda_3 \lambda/2)$. Thus the potential barrier becomes asymmetric in the presence of an applied shear force or stress. Hence the rate of jumps in the direction of the force will be larger than in the

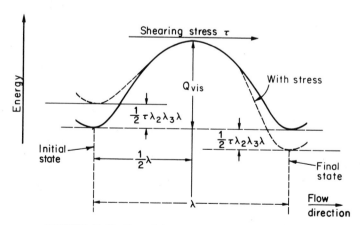

FIGURE 19.6.2. Potential energy barrier for viscous flow, with and without shearing force. [From S. Glasstone, K. J. Laidler, and H. Eyring, *The Theory of Rate Processes*, McGraw-Hill, New York (1941).]

reverse direction. The rate of an individual jump in the absence of a stress is given by

$$\Gamma = v e^{-Q_{vis}/kT}.\tag{19.6.1}$$

Here Q_{vis} is the height of the activation barrier and v is the vibrational frequency. The rate of forward jumps over the barrier at $\lambda/2$ is

$$\Gamma_f = v e^{-[Q_{vis}-(\tau/2)\lambda_2\lambda_3\lambda]/kT},$$

and the rate of backward jumps is

$$\Gamma_b = v e^{-[Q_{vis}+(\tau/2)\lambda_2\lambda_3\lambda]/kT}.$$

The net rate of flow is

$$\Delta v = (\Gamma_f - \Gamma_b)\lambda = 2\lambda\Gamma \, \sinh\left(\frac{\tau\lambda_2\lambda_3\lambda}{2kT}\right).$$

Since the viscosity coefficient is defined by

$$\eta \equiv -\frac{\tau}{\partial v/\partial r} = \frac{\tau\lambda_1}{\Delta v},\tag{19.6.2}$$

we have the **Eyring equation**:

$$\eta = \frac{\lambda_1\tau}{2\lambda\Gamma \, \sinh(\tau\lambda_2\lambda_3\lambda/2kT)}.\tag{19.6.3}$$

EXAMPLE 19.6.1

If a molecule is small, then $\lambda_2\lambda_3\lambda \approx 10^{-23}$ cm³. It is usually found for ordinary liquids (e.g., in pumping gasoline) that τ is of the order of 1 dyne/cm². What is the value of the argument of the hyperbolic sine under these conditions?

Answer.

$$\frac{\tau\lambda_2\lambda_3\lambda}{2kT} = \frac{10^{-23}}{2 \times 1.38 \times 10^{-16} \times 300} \sim 10^{-10}.$$

This is very small with respect to 1.

We can write $\sinh x \doteq x$ if $x \ll 1$. For these conditions (19.6.3) becomes

$$\eta = \frac{\lambda_1 kT}{\Gamma\lambda_2\lambda_3\lambda^2}.\tag{19.6.4}$$

This can be written in the form

$$\eta = \eta_0 e^{Q_{vis}/kT},\tag{19.6.5}$$

where we set

$$\eta_0 = \frac{\lambda_1 kT}{v\lambda_2\lambda_3\lambda^2},$$

so the dependence is the same as is empirically found [see Equation (2.8.5)].

Inasmuch as the self-diffusion coefficient of the liquid is proportional to Γ, we can write from (18.4.5) with $l = \lambda$ and (19.6.4) for a liquid in which $\lambda = \lambda_1 = \lambda_2 = \lambda_3$ that

$$\eta D \approx \frac{2kT}{\lambda}. \tag{19.6.6}$$

For small organic molecules it is found that $Q_{vis} \approx (1/3.5)\Delta H_v$, where ΔH_v is the heat of vaporization. However, for much larger molecules Q_{vis} is a much smaller fraction of ΔH_v, and in fact it appears that as the chain length increases in a hydrocarbon, Q_{vis} reaches a constant value, which suggests a moving unit containing perhaps 20–25 carbon atoms. Thus we envisage the motion of a polymer molecule as a coordinated movement of units of about that size.

For liquid metals $Q_{vis} \sim \frac{1}{20}\Delta H_v$, which suggests that the ion (rather than the atom) is the moving unit since the ion is much smaller than the atom. It is for this reason that liquid metals have a low viscosity. The viscosity coefficient of all liquid metals at the melting point is about 0.02 P. This low viscosity helps to explain why fairly rapid cooling can be used to obtain cooling curves and equilibrium diagrams in metal systems, and why metals almost always crystallize rather than form glasses even at rapid rates of cooling.

The viscosity of liquid NaCl at the melting point is about 0.015 P so that we might also expect this material to crystallize rather than to supercool.

The viscosity of many of the glass-forming oxides at the melting point is very high (liquid silica at the melting point of cristobalite has a viscosity coefficient of 10^7 P), which is one reason that they tend to form glasses rather than crystals. At the ordinary cooling rate the flow units just do not move rapidly enough to the crystallization sites and a supercooled liquid results. At lower temperatures—at the glass transition temperature—the viscosity becomes so high (10^{13} P) that even local rearrangements become unlikely so that below that temperature the material behaves like a solid (see Figure 12.11.2). The viscosity above the glass transition temperature is nearly Newtonian, although $\log_{10} \eta$ vs. $1/T$ is not exactly linear (see Figure 19.6.3 for the viscosity coefficient of a soda glass). There is an inflection point at T_G, and the behavior below T_G is no longer that of a liquid.

EXAMPLE 19.6.2

The pressure P in a bubble of radius R is related to the circumferential stress σ in the bubble wall which has thickness t by

$$P = \frac{2\sigma t}{R}.$$

What must be the viscosity coefficient of a glass if $t = 0.2$ cm, $R = 10$ cm, and a glass blower wants to increase the size of the spherical bottle at reasonable rates?

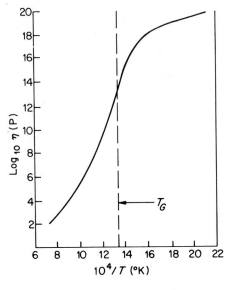

FIGURE 19.6.3. Measured viscosity coefficient of a soda glass (Na_2O and SiO_2) versus reciprocal temperature.

Answer. Newton's law of viscous flow is $\tau = \eta\dot{\gamma}$ (here τ is the shear stress and $\dot{\gamma}$ is the shear strain rate). In terms of longitudinal strain, $\sigma \equiv 3\eta\dot{\epsilon}$, as we shall see later. Hence

$$3\eta\dot{\epsilon} = \frac{PR}{2t} \quad \text{or} \quad \eta = \frac{PR}{6\dot{\epsilon}t}.$$

Estimating $\dot{\epsilon} \approx 1$ sec^{-1} and the pressure as about 0.1 atm ≈ 0.1 bar $\approx 10^5$ dynes/cm^2, we have $\eta \approx 10^6$ P.

Note that the Eyring equation does not always lead to a linear viscosity relationship. If the argument of the hyperbolic sine is large enough, the viscosity coefficient decreases as the stress increases. This is an example of non-Newtonian behavior. High polymer melts which are tangles of nearly random chain polymers often show this behavior. These melts are characterized by high viscosity, which is one reason why large single polymer crystals are not obtained when a melt cools. The slowness of nucleation is another possible reason.

It should also be noted that the high flexibility of a polymer chain decreases as the temperature is lowered (because rotation about the chain bonds is a thermally activated process). The moving unit in viscous flow appears to involve a segment having about 25 bonds when the chain is highly flexible. As the chain becomes less flexible the activation energy for flow increases. The viscosity therefore increases much more rapidly than for normal behavior.

The viscosity coefficient of liquids is important in the design of ships, in the transport of water and other liquids through pipes, soil, etc., and in lubrication. Figure 19.6.4 shows a wheel of a railroad car rotating on the axle.

FIGURE 19.6.4. Wheel and axle.
(a) Large angular
velocity ω.
(b) Small ω.

Hydrodynamicists can show that for $\omega \rightarrow \infty$, the shaft will be centered. If $\omega \neq 0$, the shaft will be closer to the sleeve, but if the angular velocity is large enough, it will not touch (a pressure P builds up in the lubricant in the narrow opening which lifts the shaft). The magnitude of the pressure P depends on the product $\eta\omega$ [see D. D. Fuller, *Theory and Practice of Lubrication for Engineers*, Wiley, New York (1956) Chapter 6] so that the viscosity coefficient must be large enough to build up a sufficient pressure to lift the axle. Pressures in bearings can be quite large (5 kbars, 70,000 psi) and so the pressure dependence of the viscosity often becomes important. Viscosity often depends on pressure according to

$$\eta(P) = \eta(0)e^{PV_{vis}/kT}, \tag{19.6.7}$$

where V_{vis} is a property of each liquid. See Problem 2.30.

The large pressure present on the sleeve causes radial tension stresses there. Since the sleeve rotates, it is subjected to cyclic stressing and hence fatigue. This was responsible for fatigue failures in early high speed trains.

19.7   VISCOSITY OF CRYSTALS

Linear viscous flow of polycrystalline materials can take place at high temperatures and low stresses.

We consider a single crystal to which shear stresses have been applied as in Figure 19.7.1(a). The two stress states of Figure 19.7.1 are equivalent. We prefer to consider (b). In the absence of stress the material at high temperature would have an equilibrium vacancy concentration of C_v^0. We recall from Section 18.6 that the surface of a crystal is an excellent source and sink for vacancies. We shall assume in the present case that there are no sources or sinks within

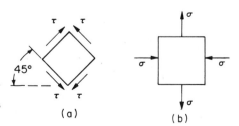

FIGURE 19.7.1. Equivalent
states of stress
with $\tau = \sigma$.

the crystal (although we noted in Section 18.6 that dislocations could also act as sources and sinks, it is probably true that they do not do so at the tiny stresses considered here). We also assume that under the tiny stresses present here the dislocations do not glide; i.e., they have formed a network which is kinetically metastable (a similar situation was mentioned for grain boundaries in Chapter 15).

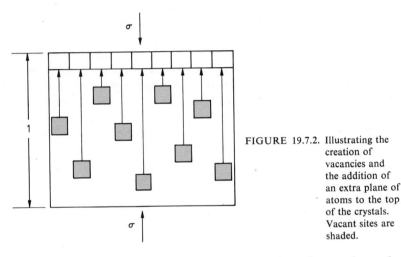

FIGURE 19.7.2. Illustrating the creation of vacancies and the addition of an extra plane of atoms to the top of the crystals. Vacant sites are shaded.

Let Ω be the atomic volume. Consider the formation of vacancies at the top surface of the crystal shown in Figure 19.7.2. These vacancies then diffuse into the crystal and atoms are deposited on the surface. We assume that one layer of atoms is deposited in the presence of the compressive stress σ. We assume for simplicity of discussion that the atoms are cubes of edge b. Then if the crystal is a unit cube the extra energy needed to create the layer of atoms, and hence the equivalent number of vacancies, in the presence of the stress σ is

$$\Delta w = \sigma \times \text{unit area} \times b = \sigma b.$$

The number of atoms added to the surface and hence vacancies created is $N = \text{unit area}/b^2 = 1/b^2$ so the extra work to create one vacancy is

$$\frac{\Delta w}{N} = \sigma b^3 = \sigma \Omega.$$

Hence the energy of creating a vacancy becomes

$$E_f + \sigma \Omega,$$

where E_f is the formation energy of a vacancy in the absence of stress. Since the concentration of vacancies in the absence of stress is

$$C_v^0 = e^{-E_f/kT},$$

the concentration near to the surface where the compressive stress is acting is

$$C_v = e^{-(E_f+\sigma\Omega)/kT} = C_v^0 e^{-\sigma\Omega/kT}. \tag{19.7.1}$$

Similarly, for a tensile stress, the concentration near the surface is

$$C_v = C_v^0 e^{\sigma\Omega/kT}. \tag{19.7.2}$$

We assume that such *local equilibrium* concentrations exist at the boundaries of the crystal. Under the conditions of small stress,

$$C_{source} = C_v^0 e^{\sigma\Omega/kT} \doteq C_v^0\left(1 + \frac{\sigma\Omega}{kT}\right) \tag{19.7.3}$$

and

$$C_{sink} = C_v^0 e^{-\sigma\Omega/kT} \doteq C_v^0\left(1 - \frac{\sigma\Omega}{kT}\right). \tag{19.7.4}$$

Hence there is a difference of vacancy concentration which sets up a vacancy flux, as shown in Figure 19.7.3. The atom flux is necessarily opposite to this and this results in an elongation of the specimen. (See Problem 19.23.)

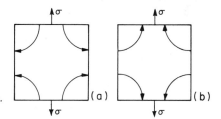

FIGURE 19.7.3. (a) Vacancy flux. (b) Atom flux.

If we assume that we have a polycrystalline specimen made up of spheres of radius R (this is not quite possible, of course; see Section 15.3), then the concentration at the surface of the sphere can be written down and the steady-state flow of vacancies can readily be calculated. From this the strain rate can be calculated. A spherical grain changes to an ellipsoidal grain. The longitudinal strain rate is

$$\dot{\epsilon} = \frac{8}{3} D_v C_v^0 \frac{\Omega}{kT} \frac{\sigma}{R^2}, \tag{19.7.5}$$

where D_v is the diffusion coefficient for vacancies. Note also that $D_v C_v^0 = D$, the self-diffusion coefficient. (If only a simple tensile stress were present, the rate would be half of this.) This could also be written in terms of the shear strain rate:

$$\tau = \frac{3kTR^2}{16D\Omega} \dot{\gamma}. \tag{19.7.6}$$

Since $\dot{\gamma}$ is the velocity gradient, we can define a viscosity coefficient:

$$\eta = \frac{3kTR^2}{16D\Omega}. \tag{19.7.7}$$

Once again, as with liquids [see (19.6.6)],

$$\eta D = \frac{3R^2 kT}{16\Omega}. \tag{19.7.8}$$

Creep of the kind described here which involves only vacancy flow (and does not involve dislocation motion) is called **diffusional creep**. The theory was worked out by C. Herring, *Journal of Applied Physics*, **21**, 437 (1950). Certain geophysicists believe that diffusional creep is the process whereby continental drift occurs.

Measurement of the viscosity coefficient in a material of a given grain radius enables one to calculate the diffusion coefficient D, which can also be measured by tracer diffusion. In the case of ionic solids, creep will be determined by the diffusion rate of the slowest-moving ion. Such results are shown in Figure 19.7.4.

It should be emphasized that linear viscous behavior in crystalline materials holds only for very small stresses (10 psi in metals, 1000 psi in ceramics). At higher stresses the behavior is nonlinear and involves the motion of dislocations.

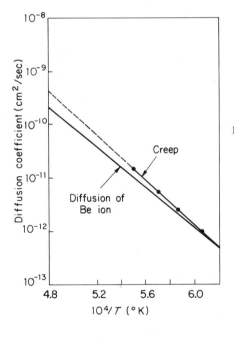

FIGURE 19.7.4. Comparison of diffusion coefficients in BeO. (a) Be ion diffusion. (b) Calculated from compressive creep data and the Herring equation for unrelaxed boundary. [From A. L. Ruoff, *Journal of Applied Physics*, **36**, 2903 (1965).]

REFERENCES

Jaeger, J. C., *Elasticity, Fracture and Flow*, Wiley, New York (1956). Stress and strain are described in simple but thorough fashion.

Nye, J. F., *Physical Properties of Crystals*, Oxford, New York (1960). Has an excellent discussion of stress and infinitesimal strain in terms of Cartesian tensors. The linear theory of elasticity is discussed in detail.

Kittel, C., *Introduction to Solid State Physics*, Wiley, New York (1968) Chapter 4. The background provided by the Nye reference (above) would be helpful in reading this chapter, which is primarily concerned with elastic waves in cubic crystals.

Treloar, L. R. G., *The Physics of Rubberlike Elasticity*, Oxford, New York (1958).

Sears, F. W., *Thermodynamics*, Addison-Wesley, Reading, Mass. (1952). A good elementary discussion of kinetic gas theory.

Glasstone, S., Laidler, K. J. and Eyring, H., *The Theory of Rate Processes*, McGraw-Hill, New York (1941). A good discussion of viscosity in liquids.

Cottrell, A. H., *The Mechanical Properties of Matter*, Wiley, New York (1964).

PROBLEMS

19.1. What is the origin of the elastic constants of
 a. Crystalline materials?
 b. Rubberlike materials?

19.2. A cube of solid homogeneous material is subjected to three normal stresses only: σ_1, σ_2, and σ_3. It can be shown that the maximum shear stress present in the sample is one of the following:

$$\frac{\sigma_1 - \sigma_2}{2}, \quad \frac{\sigma_2 - \sigma_3}{2}, \quad \frac{\sigma_1 - \sigma_3}{2}.$$

The solid is subject to hydrostatic pressure. What is the shear stress?

19.3. Show that the stress state

$$\begin{array}{ccc} \sigma & 0 & 0 \\ 0 & -\sigma & 0 \\ 0 & 0 & 0 \end{array} \quad \text{referred to } x_1 x_2 \text{ axes}$$

is equivalent to

$$\begin{array}{ccc} 0 & \tau & 0 \\ \tau & 0 & 0 \\ 0 & 0 & 0 \end{array} \quad \text{referred to } x_1' x_2' \text{ axes.}$$

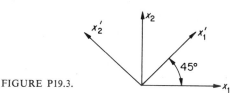

FIGURE P19.3.

19.4. An isotropic solid is subjected to hydrostatic pressure.
   a. Write the expressions for ϵ_1, ϵ_2, and ϵ_3.
   b. Use this to obtain the initial value of the bulk modulus K_0, which is defined by $P = -K_0(\Delta V/V_0)$.

19.5. Show that for an isotropic material a Poisson ratio of $v = \frac{1}{2}$ corresponds to no volume change.

19.6. Compare the shear modulus of rubber with that of steel. Why is there such a large difference?

19.7. In stating that a polymer chain is random, we make some awesome assumptions. List at least two important reasons a polymer chain (even in a dilute solution) cannot be a random chain.

19.8. a. From (19.4.23) show that $\sigma = 3G\epsilon$, when ϵ is small. Here ϵ is the normal tensile strain.
   b. How should E and G be related for an isotropic elastic solid with Poisson's ratio $v = \frac{1}{2}$?

19.9. How can thermodynamics be used to design an entropy "meter" for rubberlike materials?

19.10. Discuss the various physical aspects of the random chain model for rubberlike elasticity.

19.11. By what percentage can rubber with 24 links per chain be stretched?

19.12. Outline the procedures used to obtain the stored elastic energy function for rubber.

19.13. Show how (19.4.18) leads to (19.4.19).

19.14. Pure shear is defined by $\lambda_1 = \lambda$, $\lambda_2 = 1/\lambda$, and $\lambda_3 = 1$.
   a. What is the strain energy density for pure shear?
   b. It can be shown that the simple shear strain γ is related to pure shear λ by $\gamma = \lambda - (1/\lambda)$. Write the strain energy for simple shear.
   c. Use this to derive $\tau = nkT\gamma$.

19.15. Suppose rubber has a density of 0.9 g/cm³.
   a. How many isoprene units are there per cubic centimeter?
   b. If each isoprene unit is regarded as one link and there are 33 links per chain, what is G?
   c. What would be the estimated maximum extension ratio of this material?
   d. Compare with the material of Figure 19.4.3.

19.16. A randon chain molecule is tested in tension in a uniaxial tension test. Show that $f = (kT/\sigma^2)r$.

19.17. a. Give the essential physical features of the Maxwell theory of viscosity of ideal gases.
   b. Give the essential physical features of the Eyring theory of viscosity of liquids.

c. Give the essential physical features of the Herring theory of linear viscous flow of solids.

19.18. a. Explain in words why the viscosity coefficient of ideal gases does not vary with the density of the ideal gas.

b. Using the van der Waals d of Table 19.5.1, and $\bar{c}$ from Table 17.4.1, calculate the viscosity of nitrogen gas at 1 atm and 20°C and compare with the value in Table 2.8.1.

19.19. Explain why in Table 19.5.1 the molecular diameters obtained from closest-packing studies are consistently higher than the other values.

19.20. Calculate the viscosity coefficient for solid silver (having a stable grain size of 10^{-3} cm) at the melting point. Compare with the viscosity coefficient of liquid silver at the melting point. Compare with the viscosity coefficient of silver vapor at the boiling point.

19.21. It can be shown from the theory of linear viscosity that the rate at which a ball falls through a viscous media is

$$v = BF$$
$$= \frac{1}{6\pi\eta r}F,$$

where $F = \frac{4}{3}\pi(\rho - \rho_0)r^3g$. If the sphere is small and the density difference is small, the sphere will be a colloidal particle.

a. Evaluate the diffusion coefficient for this colloidal particle.

b. Use this result to derive an expression similar to (19.6.6).

19.22. In Equations (19.5.4), (19.6.6), and (19.7.8) we find $\eta D \approx kT/L_c$, where L_c is an effective length parameter, being d^2/λ, λ, and d^3/D_g^2, respectively. Here d is the molecular diameter, λ is the mean free path, and D_g is the grain size. Actually, except for fine numerical detail, all of the theory of linear viscosity is embodied in this relation. Discuss.

More Involved Problems

19.23. The concentration difference between the side and top of Figure 19.7.3(b) is $\Delta C = C_v^0(2\sigma\Omega/kT)$. If the particle edge is L, then the diffusion path is about L also. Here the gradient is $\Delta C/L$.

a. Estimate the atom flux J_a to the top surface. The velocity of motion of the top edge is related to J_a by $J_a = vC_a \doteq v = \frac{1}{2}(dL/dt)$ [since $C_a = N/(N + n) \doteq 1$].

b. Use this to show that the longitudinal strain rate is $\dot{\epsilon} \propto (D/L^2)\sigma$.

19.24. Starting with

$$dw = \sum_{q=1}^{6} \sigma_q d\epsilon_q,$$

it can be shown for a Hookean solid that

$$\frac{\partial^2 w}{\partial\epsilon_r\partial\epsilon_q} = C_{qr}.$$

a. Show that this is so.
b. Start with

$$dw = \sum_{r=1}^{6} \sigma_r d\epsilon_r$$

and proceed to express C_{rq} as a second derivative of strain energy with respect to strains.
c. Show that $C_{qr} = C_{rq}$.

19.25. Beginning with

$$U = -\frac{NAe^2}{r} + \frac{B}{r^n}$$

for a NaCl crystal ($A = 1.74755$), where N is the number of atoms, e is the charge, and B and n are unknown parameters, derive the expression for K_0'.

Hint: Recall the definition of K and recall how P is related to U for a static lattice; i.e., use the thermodynamic expression for dU and apply the static lattice condition.

19.26. Derive the bulk modulus K_0 in a cubic crystal
a. In terms of the C_{ij}.
b. In terms of the s_{ij}.

19.27. Show rigorously that for the integral of Section 19.4, $I_2 \ll I_3$.

19.28. Instead of (19.6.1), we should use

$$\Gamma = \nu e^{-\Delta G_{vis}/kT}$$

a. What pressure dependence does this lead to for the viscosity coefficient?
b. In what practical application is the pressure dependence of viscosity likely to be important?

Sophisticated Problems

19.29. An infinitely long straight rod of radius r_0 is being dragged along the center of a cylinder of radius R through a Newtonian fluid at a constant velocity v. Let F be the force which must be applied to a length L to overcome the frictional drag. Calculate F. Use the results of the above problem to estimate how long a polymer molecule can exist in a very dilute solution for which $\eta = 10^{-2}$ P if the solution in a beaker is stirred with a glass rod at a velocity of 1 cm/sec. Assume a bond strength in the polymer chain and design your own criteria for the breaking of this bond.

19.30. What must be the conditions on the elastic constant matrix in order that the strain energy be positive? What are these conditions for a cubic crystal?

19.31. Assume a polycrystalline material consists of spherical grains of radius R. Let there be a grain boundary diffusion coefficient D_B and let the grain boundary have a thickness d. Calculate the rate of diffusional creep by this grain boundary mechanism assuming that there is no diffusion in the bulk. Calculate the viscosity coefficient.

19.32. So-called infinitesimal strains are often defined in terms of the displacement vector $\mathbf{u} = [u_1, u_2, u_3]$. A point initially located at $\mathbf{r}$ in the unstrained state is located at $\mathbf{r} + \mathbf{u}$ in the strained state. The state of strain is defined by the array of components

$$
\begin{matrix}
\epsilon_{11} & \epsilon_{12} & \epsilon_{13} \\
\epsilon_{21} & \epsilon_{22} & \epsilon_{23} \\
\epsilon_{31} & \epsilon_{32} & \epsilon_{33},
\end{matrix}
$$

where

$$
\epsilon_{ij} = \frac{1}{2}\left(\frac{\partial u_i}{\partial x_j} + \frac{\partial u_j}{\partial x_i}\right).
$$

Thus $\epsilon_{11} = \partial u_1/\partial x_1$ and $\epsilon_{12} = \frac{1}{2}(\partial u_1/\partial x_2 + \partial u_2/\partial x_1)$.

a. Show that ϵ_{11} is simply the engineering tensile strain ϵ_1, while $\epsilon_{12} = \frac{1}{2}\epsilon_6$ (see an elasticity textbook for help).

b. Show that ϵ_{ij} is a second-rank Cartesian tensor.

Prologue

The combination of a viscous element with an elastic element leads to a new type of material behavior, *viscoelastic behavior*. A spring being stretched within a highly viscous liquid would exhibit such behavior. A single material such as a block of rubber exhibits viscoelastic behavior, which can be visualized as being caused by springs and dashpots (on a molecular level) in the material. In this chapter we shall study the origin of *internal friction* in materials and shall compute the *mechanical loss angle* in different materials.

20

ANELASTICITY AND VISCOELASTICITY

20.1 INTRODUCTION

We noted in Chapter 2 that springs, although nearly elastic, exhibit a small damping. We call such behavior **anelastic behavior** or viscoelastic behavior. Such behavior is important in the design of material used in dynamic applications, such as tires. Such damping characteristics are also important features of materials used for turbine blades and shafts. In such applications we wish to make the internal friction large, while in others, such as in tuning forks, bells, etc., we wish to make the internal friction small. The deviation from elastic behavior is often very large in rubber, plastics, asphalt, dried lacquer and paints, baking dough, etc. Viscoelasticity is particularly important in the manufacture of plastics by extrusion and the like as well as in the behavior of plastics under load. It is an important area of current research. Many aspects of viscoelastic behavior can be described by combining the linear elastic and viscous elements of Chapter 19.

20.2 THE TWO-ELEMENT MODEL

One of the simplest models combines a linear (Hookean) spring with a linear (Newtonian) dashpot as shown in Figure 20.2.1. This is called **a Voigt model**. The spring exerts a force on the mass equal to $-kx$ and the **dashpot** exerts a force on the mass equal to $-\mu\dot{x}$. Here μ is the **frictional coefficient** of the viscous element (the **dashpot**). It is assumed here that the drag force is proportional to the velocity, which will be the case if the viscous flow is New-

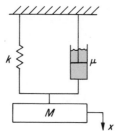

FIGURE 20.2.1. Parallel spring-dashpot model with mass attached.

tonian. The equation of motion is therefore

$$M\ddot{x} = -kx - \mu\dot{x}$$

or

$$M\ddot{x} + \mu\dot{x} + kx = 0. \tag{20.2.1}$$

If we extend the system to $x = x_0$ and release, we have for the case where $\mu = 0$ a solution showing simple sinusoidal motion with a frequency ω_0, called the natural frequency, given by

$$\omega_0 = \sqrt{k/M}. \tag{20.2.2}$$

(The student can show that $x = x_0 \cos \omega_0 t$ is a solution.)

If μ is small (meaning $\mu/2M\omega \ll 1$), we obtain a **damped natural frequency**,

$$\omega_d = \sqrt{\frac{k}{M} - \frac{\mu^2}{4M^2}}. \tag{20.2.3}$$

[See Problems 2.39–2.42 to show that this and the following results can be obtained from (20.2.1).] In this case the **amplitude** of the oscillations would decrease with time according to

$$x_0 e^{-t/\tau}, \tag{20.2.4}$$

where

$$\tau = \frac{2M}{\mu} \tag{20.2.5}$$

is the **relaxation time** (see Problem 2.40).

The force $F(t)$ which is exerted by the moving mass on the material (spring and dashpot) and the displacement $x(t)$ can be shown (see Problems 2.41 and 2.42) to be out of phase with each other by an angle δ, called the **loss angle**, where for small damping

$$\delta \doteq \frac{\mu}{M\omega_d} \doteq \frac{\mu}{M\omega_0}. \tag{20.2.6}$$

The fractional decrease in the amplitude per cycle (the **logarithmic decrement**) is

$$\frac{\Delta x_0}{x_0} = \pi\delta \tag{20.2.7}$$

(see Example 2.4.2) and the fractional decrease in the energy per cycle is

$$\frac{\Delta u}{u} = 2\pi\delta. \tag{20.2.8}$$

If there is an *external* force $F_0 \cos \omega t$ acting on the mass so that the differential equation becomes

$$M\ddot{x} + \mu\dot{x} + kx = F_0 \cos \omega t, \tag{20.2.9}$$

it can be shown (see Problems 2.43 and 2.44) that

$$x(t) = \frac{F_0}{\sqrt{(k - M\omega^2)^2 + (\mu\omega)^2}} \cos(\omega t - \beta), \tag{20.2.10}$$

where

$$\beta = \tan^{-1}\frac{\mu\omega}{k - M\omega^2}. \tag{20.2.11}$$

For very small damping, the amplitude of x reaches a maximum when $\omega \doteq \omega_d$ (which in turn can be approximated by ω_0). Hence

$$x(t) \approx \frac{F_0}{\mu\omega_0} \cos \omega t. \tag{20.2.12}$$

The curves of the amplitude versus frequency are shown in Figure 20.2.2.

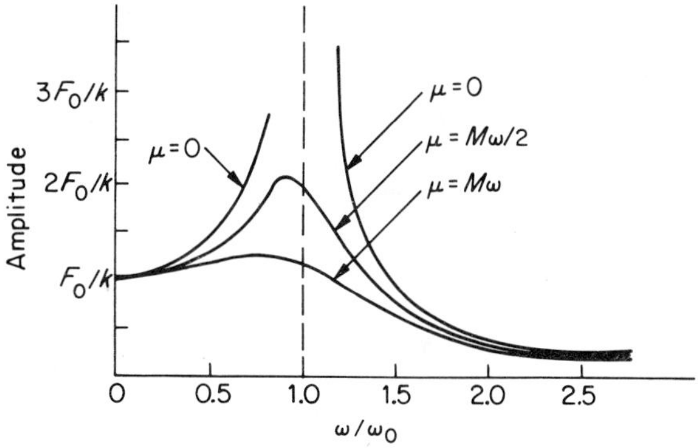

FIGURE 20.2.2. Amplitude of a forced oscillator system. Note how the resonance frequency approaches ω_0 as μ approaches zero.

The energy absorbed per cycle is

$$W_{\text{cycle}} = \oint F\,dx = \oint F\dot{x}\,dt,$$

and hence the energy absorbed is

$$W_{\text{cycle}} = \frac{\mu\omega F_0^2 \pi}{(k - M\omega^2)^2 + (\mu\omega)^2},$$

which under resonance conditions for small damping becomes

$$W_{\text{cycle}}^{\max} = \frac{F_0^2 \pi}{\mu\omega_0}. \tag{20.2.13}$$

This can also be written in the form

$$W_{\text{cycle}}^{\max} = \frac{F_0^2 \pi}{k} \frac{1}{\delta} \tag{20.2.14}$$

or

$$W_{\text{cycle}}^{\max} = \frac{F_0^2 \pi}{2k}(\omega_0 \tau). \tag{20.2.15}$$

Thus the linear Voigt model exhibits *all* the properties of damping mentioned in Section 2.4. (From a pedagogical viewpoint it will often be true that the student did not have the mathematical background to readily solve Problems 2.39–2.43 when he studied Chapter 2 but has since achieved such capability.)

20.3   LOSS FACTOR IN RUBBER

An example of the loss factor in rubber is shown in Figure 20.3.1. In this case the driving frequency is constant, but the resonance frequency is changed by varying the mass. The agreement found for one given driving frequency is excellent.

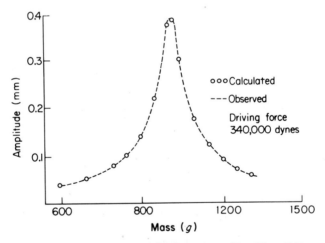

FIGURE 20.3.1. Resonance curve for gum rubber. [From S. D. Gehman, D. E. Woodford, and R. B. Stambaugh, *Industrial and Engineering Chemistry*, **33**, 1032 (1941).]

EXAMPLE 20.3.1

Calculate the loss angle from the data of Figure 20.3.1.

Answer. Our problem is to find the half-width of the resonance peak at half the maximum amplitude. Starting with Equation (20.2.10) note that for resonance (in the case where mass is varied) $k = M\omega^2$. At resonance the argument of the square root of Equation (20.2.10) is $(\mu\omega)^2$. The amplitude would be reduced to one half of the resonance value if the argument $\{[k - (M + \Delta M)\omega^2]^2 + (\mu\omega)^2\} = 4(\mu\omega)^2$. Solving for $\Delta M/M$ we obtain

$$\frac{\Delta M}{M} = \sqrt{3}\frac{\mu\omega}{M\omega^2} = \sqrt{3}\,\delta, \tag{20.3.1}$$

where δ is obtained from (20.2.6).

From the graph $\Delta M/M = 70/1020$ so we obtain $\delta \approx 0.04$.

If the behavior is studied at several driving frequencies the damping coefficient varies with frequency according to $\mu\omega = $ constant. This suggests that our simple model based on a single relaxation time τ (or a single damping coefficient μ) is insufficient.

The elastic behavior of rubber is associated with the stretching of random polymer chains. The viscous behavior is associated with the flow of segments of these chains (20–25 carbons per segment) over each other as in a liquid hydrocarbon. (Recall that rubber is really a liquid with the crosslinks providing rigidity.)

It can be shown (if the applied force causes compression or tension) that

$$\mu = \frac{3\eta A}{x_0} \tag{20.3.2}$$

and

$$k = \frac{EA}{x_0}, \tag{20.3.3}$$

where η is the viscosity coefficient, E is Young's modulus, A is the area on which the force acts, and x_0 is the initial thickness of the specimen. [*Note:* To derive (20.3.2) note that $\tau = \eta\dot\gamma$ and that $\tau = \sigma/2$ and $\gamma = \frac{3}{2}\epsilon$. Here τ and γ are shear stress and strain, respectively, and σ and ϵ are tensile stresses and strains, respectively.]

In this notation the loss angle [see Equation (20.2.6)] can be written

$$\delta \doteq \frac{3\eta\omega_0}{E} = \frac{\eta\omega_0}{G}. \tag{20.3.4}$$

Thus the loss angle δ can be expressed in terms of the material properties η and G (rather than component properties μ and k).

EXAMPLE 20.3.2

Gum rubber is a long chain hydrocarbon and as such might be expected to have the viscosity of heavy machine oil. Calculate the loss angle for gum rubber for $\omega = 60 \text{ sec}^{-1}$.

Answer. From Table 2.8.1, $\eta = 5000$ P, and from Section 19.4, $G = 4 \times 10^6$ dynes/cm². Hence

$$\delta \doteq 0.075.$$

When rubber is used as an energy absorber it is used at frequencies far below ω_0. The addition of a filler such as carbon black increases both η and G of the rubber.

EXAMPLE 20.3.3

Show that the Voigt model (ignoring inertial effects) obeys a relation

$$\sigma = a\epsilon + b\dot{\epsilon}.$$

Answer. We note that the sum of the forces is given by

$$F = kx + \mu\dot{x} \quad \text{and} \quad \sigma = \frac{F}{A} = \frac{kx_0}{A}\epsilon + \frac{\mu x_0}{A}\dot{\epsilon} = a\epsilon + b\dot{\epsilon}, \quad (20.3.5)$$

where A is the area on which the force acts and x_0 is the initial specimen length.

20.4   GENERALIZED VISCOELASTICITY

Although the two-element model is very successful in explaining certain phenomena, it is not consistent with others. For example, the Voigt model shows no purely elastic behavior upon rapid loading. It likewise shows no

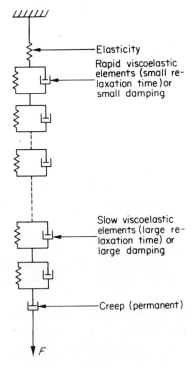

FIGURE 20.4.1. Generalized viscoelastic model.

permanent creep. Moreover, the use of a single relaxation time is not a realistic description of solids (recall the experimental results for rubber which give $\eta\omega$ = constant). Consequently, a more general description as in Figure 20.4.1 is needed.

For many applications (excluding creep) the three-element model of a spring in series with a Voigt model describes material behavior (see Figure 20.4.2).

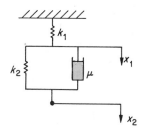

FIGURE 20.4.2. Three-element model.

EXAMPLE 20.4.1

A material describable by the three-element model of Figure 20.4.2 is pulled very rapidly in tension.

a. What is its stiffness constant?

Answer. $k_\infty = k_1$, since the dashpot does not respond to rapid loading.

b. What is its stiffness constant if the material is pulled *very* slowly?

Answer. $k_0 = k_1 k_2/(k_1 + k_2)$, since the dashpot exerts no force, i.e., flows freely under these conditions. Hence we have effectively two springs in series.

EXAMPLE 20.4.2

Write the equations of motion for the three-element model of Figure 20.4.2 assuming there is an applied force $F(t)$ and assuming the system has no mass.

Answer.

$$k_1 x_1 = F(t)$$

$$k_2(x_2 - x_1) + \mu(\dot{x}_2 - \dot{x}_1) = F(t).$$

The equations given in Example 20.4.2 can be written in terms of the operator $\mathscr{D} = d/dt$ as follows:

$$k_1 x_1 = F(t) \qquad (20.4.1)$$

$$(\mu\mathscr{D} + k_2)x_2 - (\mu\mathscr{D} + k_2)x_1 = F(t). \qquad (20.4.2)$$

Multiply the first of these equations by $\mu\mathscr{D} + k_2$ and the second by k_1. Add the results to get

$$k_1(\mu\mathscr{D} + k_2)x_2 = (\mu\mathscr{D} + k_2)F(t) + k_1 F(t).$$

This can be written in the form

$$k_1 \mu \dot{x}_2 + k_1 k_2 x_2 = \mu \dot{F} + (k_2 + k_1)F.$$ (20.4.3)

In terms of strains and stresses

$$a_1 \dot{\epsilon} + a_2 \epsilon = b_1 \dot{\sigma} + b_2 \sigma.$$ (20.4.4)

This material is called a **standard linear solid**.

EXAMPLE 20.4.3

Suppose the standard linear solid described by (20.4.3) is rapidly stretched to x_2^0 at $t = 0$. The force needed to retain this extension will decrease with time; this is called **stress relaxation**. Find the expression for it.

Answer. We have $F = k_1 x_2^0$ at $t = 0$. We also have

$$\mu \dot{F} + (k_1 + k_2)F = k_1 k_2 x_2^0.$$

This differential equation has a solution

$$F = Be^{-(k_1 + k_2)t/\mu} + \frac{k_1 k_2}{k_1 + k_2} x_2^0,$$

where B is an integration constant. Since F is known at $t = 0$, we can evaluate B. Hence,

$$F = (k_1 - k_0)x_2^0 e^{[-(k_1 + k_2)/\mu]t} + k_0 x_2^0,$$

where k_0 is defined in Example 20.4.1. The exponential decay then could be written in terms of the decay time τ_σ as

$$e^{-t/\tau_\sigma},$$

where

$$\tau_\sigma = \frac{\mu}{k_1 + k_2}$$ (20.4.5)

is the **stress relaxation time**.

If a fixed force F_0 is applied at $t = 0$, then there will be an instantaneous displacement followed by continual stretching called **strain relaxation**. It can be shown that

$$x_2(t) = \left(\frac{1}{k_1} - \frac{1}{k_0}\right) F_0 e^{(-k_2/\mu)t} + \frac{F_0}{k_0}$$

The strain relaxation time is defined by

$$\tau_\epsilon = \frac{\mu}{k_2}.$$ (20.4.6)

The equation of motion could also be written out for the case where a mass is attached to the three-element model. One could then solve for $x_2(t)$ for the case where the applied force is zero but the specimen is given an initial displacement x_2^0 (for a long time) and then released. This is an example of a free damped vibration (similar to that shown in Figure 2.4.3). For the case of small damping it can be shown that there is a loss angle having the form

$$\delta(\omega) \doteq A_0 \frac{\omega\tau}{1 + (\omega\tau)^2}, \qquad (20.4.7)$$

where

$$A_0 = \frac{k_\infty - k_0}{(k_\infty k_0)^{1/2}}, \qquad (20.4.8)$$

where k_0 and k_∞ are as given in Example 20.4.1, and τ is a characteristic relaxation time

$$\tau = \sqrt{\frac{\mu^2}{(k_1 + k_2)k_2}} = \sqrt{\tau_\sigma \tau_\epsilon}. \qquad (20.4.9)$$

Note that δ has a maximum when

$$\omega\tau = 1. \qquad (20.4.10)$$

It can also be shown that the effective stiffness constant is

$$k(\omega) = k_\infty - \frac{(k_\infty - k_0)}{1 + (\omega\tau)^2}. \qquad (20.4.11)$$

It would reinforce the student's understanding of the behavior of δ and k if he now took time to sketch $\delta(\omega)$ and $k(\omega)$. If instead of spring stiffness we used elastic stiffness, a similar set of relations would apply. Then $(E_\infty - E_0)/E_0$ would be known as the **modulus defect**.

20.5 INTERNAL FRICTION IN SINGLE CRYSTALS OWING TO MOTION OF INTERSTITIAL ATOMS

Figure 20.5.1 shows the (100) face of an α-iron crystal. Carbon atoms may be located in positions such as A, B, or C, i.e., at edge centers or face centers.

We recall that the carbon atom is highly compressed in these interstitial positions. If the crystal is stretched as in Figure 20.5.1(b), the carbon atom would

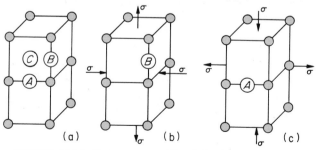

(a)   (b)   (c)

FIGURE 20.5.1. Carbon atom positions in bcc iron. (a) Carbon has equal probability of being at A, B, or C sites. (b) Carbon prefers B site. (c) Carbon prefers A site.

tend to be further compressed in A but less compressed in B. Recall that motion from A to B involves an activation process. Hence, in the present case the activation barrier has the form shown in Figure 19.6.2. With the stress state as in (b) more carbon atoms will jump from A to B during tension than from B to A. Just the opposite will happen during the stress state of Figure 20.5.1(c). This is **stress-assisted diffusion**. The applied stress does work in helping the atom jump over the barrier. Note that the state of stress in Figure 20.5.1(b) and (c) corresponds to a state of shear at 45 deg.

Note that when the stress is applied there will be an immediate elastic deflection followed by an additional deflection as the carbon atoms jump into the new preferred positions. This is exactly the response of the three-element model of Section 20.4.

EXAMPLE 20.5.1

What evidence do we have that B sites would be preferred in Figure 20.5.1(b) over A sites?

Answer. First the volume of α-iron expands when carbon is added to α-iron by about 6 cm³/mole added. Thus carbon expands the lattice and hence the carbon atom must be in compression. Second, when martensite forms from supersaturated α-iron the carbon atoms tend to go preferentially in one set of sites, leading to the c/a ratio of Figure 16.5.5.

If C_0 is the concentration of carbon atoms in the absence of stress, then the concentration of B atoms during long-term application of stress as in Figure 20.5.1(b) is $C_0 \exp(\sigma \Omega/RT)$ *under equilibrium*, while the concentration of A atoms is $C_0 \exp(-\sigma \Omega/RT)$. Here Ω is the partial molar volume of carbon atoms in iron.

(The problem is analogous in some sense to the diffusional creep problems of Section 19.7. It is different in the following way: There are no sources or sinks for carbon as there were for vacancies.) If the frequency of the stress cycle is *much* higher than the atom jump frequency, the concentration at both A and B will be very nearly C_0. Hence, there will be virtually no damping. If the frequency is *very* low relative to the atom jump frequency, then the concentration at both A and B sites is always very near to the equilibrium concentration for the instantaneous value of the applied stress. This corresponds to reversible isothermal loading. Thus during a full cycle no net energy is absorbed. The question then arises: At what frequency ω_m should there be a maximum absorption? The stress will be most effective when the frequency of the oscillation is the same as the atom jump frequency, i.e., for

$$\omega_m = 6\Gamma.$$

(*Note:* The frequency of an individual jump is Γ and a given carbon atom has six possible jumps. We assume all neighboring positions are unoccupied so the concentration of carbon must be low.)

EXAMPLE 20.5.2

If ω_m is measured as a function of temperature, T, what fundamental information is obtained?

Answer. Since $\Gamma = \nu e^{-Q_D\, KT}$, the activation energy for diffusion Q_D can be obtained from the above experiment. Note that D_0 of the diffusion coefficient can also be obtained.

It is usually easiest to work with a fixed frequency torsional pendulum and to vary the temperature and obtain an internal friction peak as shown in Figure 20.5.2. This peak has the general form of Equation (20.4.7). The $\delta(T)$ relationship follows from (20.4.7) if τ is replaced by $1/6\Gamma$ where

$$\Gamma = \nu e^{-Q_D/RT}.$$

The maximum amplitude of the resonance peak is proportional to the carbon concentration in the α-iron. Similar peaks are found for interstitial elements in other bcc crystals, e.g., oxygen in tantalum.

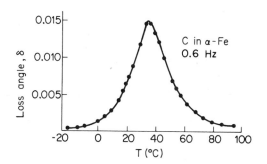

FIGURE 20.5.2. Damping peak in iron owing to carbon diffusion.

20.6   OTHER DAMPING MECHANISMS IN CRYSTALS

PAIR INTERCHANGE.   In substitutional alloys such as α-brass (fcc phase of Zn in Cu) in which there is a sizable difference in atomic size, the presence of stress tends to cause a redistribution of neighbors around a given atom; since the atoms move by diffusion (a dissipative process), damping is found (**Zener relaxation**). The peak in α-brass is at 300°C at 1 Hz.

GRAIN BOUNDARY PEAKS.   Grain boundaries are imperfections in crystalline materials. They are regions perhaps several Angstroms thick in which the packing is in a sense more like a liquid than a crystal. In this region the atoms have a high mobility. When the materials are stressed, there is elastic deformation within the crystals and viscous deformations at the grain boundaries, leading to internal friction in the polycrystalline material which is absent in the single crystal. The internal friction peak due to grain boundaries in α-brass is near 425°C at 1 Hz.

DISLOCATION DAMPING.   When a stress is applied to a crystal the dislocations in the crystal move. The dislocation may be pinned at various points by impurity atoms and other imperfections. The dislocation line may then bow out under the applied stress and even break away from weak pinning points as shown in Figure 20.6.1.

FIGURE 20.6.1. Dislocation bowed out under applied stress. Solid circles are strong pinning points. Open circles are weak pinning points.

The motion of the dislocation under an alternating applied stress is very much like the vibration of a string in a viscous medium. [In this case the thermal phonons (lattice vibrations) are scattered by the moving dislocations.]

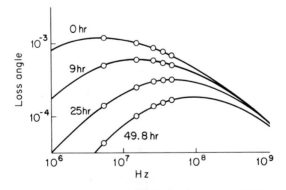

FIGURE 20.6.2. Internal friction losses owing to dislocation damping in irradiated copper. The times shown are the times the sample was subjected to γ-ray radiation from a 6000-Curie cobalt source. [From R. M. Stern and A. V. Granato, in *Internal Friction Due to Crystal Lattice Imperfections* (P. J. Leurgans, ed.), Pergamon Press, New York (1962).]

It has been shown by J. Koehler, by J. Weertman, and by A. Granato and K. Lucke that an internal friction peak of the form of Equation (20.4.7) should be obtained. The constant A should be proportional to the square of the distance between the pinning points. Figure 20.6.2 shows a plot of the loss angle for copper irradiated by a 6000-Curie cobalt source. The γ-rays produce point defects in the solid which move to the dislocations and somehow pin them. It has also been shown that the relaxation time should be proportional to the square of the length. Since the irradiation changes the length between pinning points, it affects both the frequency at which the loss angle is a maximum and the maximum value of the loss angle.

MAGNETO-MECHANICAL DAMPING. Ferromagnetic materials are characterized by domains [regions in which the magnetization is in a single direction (Chapter 32)]. When a field is applied, the domains tend to align with the field. It is the motion of these domain walls which leads to the *B-H* hysteresis loop. An applied stress also induces domain motion (Chapter 32). The resulting internal friction associated with a periodic applied stress is called **magneto-mechanical damping**. The loss angle for this process can be very high ($\delta \sim 0.1$) at low frequencies. It is therefore an important industrial property with applications in the heavy machinery industry. A damping curve is shown in Figure 20.6.3.

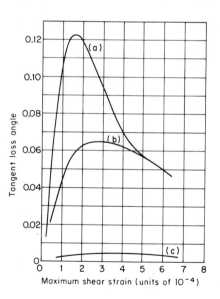

FIGURE 20.6.3. Tangent of the loss angle versus amplitude of shear strain at 20 cycles/sec as a function of the field. (a) $H = 0$, (b) $H = 153$ amp/m, (c) $H = 1210$ amp/m. [From the data of G. Sumner and K. M. Entwisele, *Journal of the Iron and Steel Inst.* **192**, 238 (1959)].

20.7 VISCOELASTICITY OF POLYMERIC MATERIALS

We note that polymeric materials may in general be described by the model of Figure 20.4.1. The molecular architect can in fact fashion polymer systems with varied properties. The *polymer state* may be: a highly crystalline fiber, a rigid plastic (below the glass transition temperature, see Figure 12.11.2), or a rubberlike material. The factors which primarily determine which of these three states a polymer system takes are:

1. chain flexibility
2. interchain interactions
3. regularity of the polymer
4. temperature.

If the nature of the polymer chains (because of irregularity of molecular structure) is such that the material does not crystallize, then the amorphous polymer

Table 20.7.1. GLASS TRANSITION TEMPERATURES OF POLYMERIC
MATERIALS*

Polymer	Structure	T_G (°C)	Special Molecular Feature
Poly(dimethyl siloxane)	$-O \cdot Si(CH_3)_2-$	-123	siloxane bonds very flexible
Polybutadiene	$-CH_2 \cdot CH=CH \cdot CH_2-$	-85	C—C bonds flexible
*Cis*polyisoprene	$-CH_2 \cdot C(CH_3)=CH \cdot CH_2-$	-70	stiffening effect of methyl group
Poly(*n*-heptyl butadiene)	$-CH_2 \cdot C \cdot (C_7H_{15})=CH \cdot CH_2-$	-83	chain separation balanced by entanglement
Poly(*n*-decyl butadiene)	$-CH_2 \cdot C \cdot (C_{10}H_{21})=CH \cdot CH_2-$	-53	side chain entanglement and crystallization
Butadiene–styrene copolymer	75% Bu–25% St.	-55	copolymerisation
Polypropylene	$-CH_2 \cdot CH(CH_3)-$	-27	stiffening effect of methyl group
Polyisobutylene	$-CH_2 \cdot C(CH_3)_2-$	-65	increased flexibility through reduction of dipole moment
Poly(3-methyl butene-1)	$-CH_2 \cdot CH \cdot [CH(CH_3)_2]-$	$+50$	stiffening effect of isopropyl group
Polyoxymethylene	$-CH_2-O-$	-73	flexibility of C—O bonds
Poly(ethylene adipate)	$-(CH_2)_2OOC \cdot (CH_2)_4COO-$	-70	flexibility of C—O and C—C bonds
Poly(ethylene terephthalate)	$-(CH_2)_2OOC \cdot C_6H_4COO-$	$+67$	stiffening effect of phenylene group in backbone
Polycarbonate of (bisphenol A)	$-C_6H_4 \cdot C(CH_3)_2 \cdot C_6H_4 \cdot O \cdot CO \cdot O$	$+149$	stiffening effect of phenylene groups
Poly(vinyl chloride)	$-CH_2 \cdot CH(Cl)-$	$+80$	dipole attraction of chlorine atoms
Poly(vinylidene chloride)	$-CH_2 \cdot C \cdot (Cl)_2-$	-17	reduction of dipole moment
Polystyrene	$-CH_2 \cdot C(C_6H_5)-$	$+100$	stiffening effect of attached benzene ring

* [J. A. Brydson, *Plastic Materials*, D. Van Nostrand Co., Inc. Princeton, N.J. (1966)].

may be present in either the glassy state or the rubberlike state. The glass transition temperature is governed by the mobility of the chains which in turn is determined by the freedom of rotation. The latter flexibility disappears when the temperature is sufficiently low that thermal fluctuations do not readily allow the energy barriers to rotation about bonds in the chain backbone to be surmounted. Table 20.7.1 shows the glass transition temperature for some polymeric materials as well as the structural reason for having a particular transition temperature. Below the glass transition temperature, the loss angle is usually less than 0.01, often being as low as 0.001. Above the glass transition temperature the loss angle is of the order of 0.1.

EXAMPLE 20.7.1

The number of crosslinks in natural rubber (polyisoprene) appears to be directly related to the sulfur content of the vulcanized rubber. How does the glass transition temperature of rubber vary with the sulfur content?

Answer. Here T_G increases as %S increases because the crosslinks decrease the mobility of a chain. Actual results are shown in Table 20.7.2.

Table 20.7.2. DEPENDENCE OF THE GLASS TRANSITION OF RUBBER ON THE SULFUR CONTENT

Sulfur (%)	T_G (°C)
0	−65
0.25	−64
10	−48
20	−24

The addition of certain liquids, and in a few cases solids, which are high boiling point solvents for a polymer, will produce a polymeric material which is softer and more flexible than the polymer alone. Such liquids are called **plasticizers**. The general rule for solvents is that like tends to dissolve like. Thus natural rubber is readily swollen by simple hydrocarbons such as gasoline, while a highly polar polyamide (such as nylon 66) requires a highly polar solvent such as dimethyl formamide. Since the solvent increases the mobility of the chains, it lowers the glassy transition temperature.

REFERENCES

Zener, C. M., *Elasticity and Anelasticity of Metals*, University of Chicago Press, Chicago (1948).

Wert, C., "The Metallurgical Use of Anelasticity," in *Modern Research Techniques in Physical Metallurgy*, American Society for Metals, Cleveland (1953).

Alfrey, T., and Gurnee, E. F., *Organic Polymers*, Prentice-Hall, Englewood Cliffs, N.J. (1967) Chapters 4–6. Contains a very good discussion of viscoelasticity applied to polymers.

Houwink, R., *Elasticity, Plasticity and Structure of Matter*, Dover, New York (1958).

Leurgans, P. J., ed., *Internal Friction Due to Crystal Lattice Imperfections*, Pergamon, New York (1962).

Ferry, J. D., *Viscoelastic Properties of Polymers*, Wiley, New York (1970). This is a detailed, fairly sophisticated treatment of the subject.

Nowick, A. S., and Berry, B. S. *Anelastic Relaxation in Crystalline Solids*, Academic Press, New York (1972). An excellent sophisticated study for the advanced student and the instructor.

PROBLEMS

20.1. Would resonance be more of a problem in a welded airplane wing than in a riveted one? Explain your answer.

20.2. Discuss various damping mechanisms in metals and what procedure you might use in designing metals for high internal friction at 1000 Hz at room temperature.

20.3. a. Explain why in dislocation damping you would expect the loss angle to increase as the length between pinning points increased.
   b. Explain why you would expect the frequency at which the loss angle attained a maximum to decrease as the length between pinning points increased.

20.4. Many cooks have noticed that certain dough climbs up the mechanical stirrer rod, while ordinary liquids do the opposite. Suggest some reasons for this.

20.5. Suggest two problems each in which viscoelasticity is important to
   a. The highway engineer.
   b. The automobile engineer.
   c. The plastics production engineer.
   d. The paint and coatings engineer.

20.6. Describe five types of internal friction measurements which the materials scientist can use to study the properties of materials.

20.7. a. Show that the maximum shear stress in a tensile specimen is related to the tensile stress by $\tau = \sigma/2$.
   b. Show that the maximum shear strain rate is related to the tensile strain rate in rubber by $\dot{\gamma} = \frac{3}{2}\dot{\epsilon}$.

20.8. a. Obtain the general expression for the energy produced per cycle for the linear system of Equation (20.2.9) for all frequencies.
   b. For what value of ω is this a maximum?

20.9. When rubber is used in applications which require damping, the frequency is considerably less than the resonance frequency.
   a. Why?
   b. What is the expression for the energy absorption at such frequencies?

20.10. In addition to changing η and G, the addition of carbon black to tires might change some other properties. Suggest one or more properties which might be changed by carbon black additions and justify your answer for one of these properties.

20.11. A cube of gum rubber with edge = 1 cm is cycled at a frequency ω.
   a. Calculate the thermal diffusivity of rubber.
   b. Estimate the thermal diffusion time.
   c. Estimate the frequency above which rapid heating of the specimen will occur when the rubber is cyclically loaded even though two opposite surfaces are kept at low temperatures.

20.12. Design a demonstration experiment to measure the loss angle of a rubber pad assuming all the equipment you have is a yardstick and a steel ball.

20.13. A material is represented by a Voigt model with $G = 10^6$ dynes/cm² and $\eta = 10^7$ P.
   a. It is tested at a strain rate of $\dot{\epsilon}_{11} = 0.001$ sec⁻¹. Calculate σ_{11} at the end of 20 sec.
   b. If $\dot{\epsilon}_{11} = 10^{-5}$ sec⁻¹, calculate σ_{11} at the end of 2000 sec.

20.14. A material is represented by the three-element model of Figure 20.4.2. It is pulled in a tension machine at a constant cross-head speed, until the slope of the force-time curve approaches its limiting value. The cross-head motion is then reversed. Sketch the stress-strain curve.

20.15. Starting with Equation (20.4.3) derive the expression $x_2(t)$ for strain relaxation and obtain the strain relaxation time.

20.16. Describe how internal friction measurements in α-iron can be used to measure the limit of solubility of carbon in α-iron. [If you need help, see C. Wert, *Journal of Applied Physics*, **21**, 1196 (1950).]

20.17. Using the c/a data of Figure 16.5.5 estimate the volume change per mole of carbon added to α-iron.

20.18. At what temperature would the carbon diffusion resonance peak be for 1000-Hz oscillation of α-iron?

20.19. a. Estimate from internal friction data the bulk diffusion coefficient for α-brass at 300°C.
   b. Estimate the activation energy for diffusion in α-brass.

More Involved Problems

20.20. If the activation energy for diffusion in grain boundaries is much less than the activation energy for bulk diffusion, why is the grain boundary internal friction peak for α-brass at 425°C, while the Zener relaxation peak is at 300°C?

20.21. Suppose α-iron contains 10^{-3} wt. % carbon. Estimate the shear strain owing to the carbon motion when the crystal is stressed as in Figure 20.5.1(c). Compare this with the instantaneous elastic shear strain using the elastic constants of polycrystalline steel.

20.22. Show that the three-element model below, by appropriate adjustment of the constants, gives a response to any loading identical to that of the model of Figure 20.4.2.

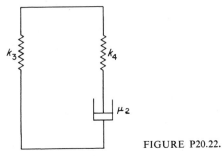

FIGURE P20.22.

Sophisticated Problems

20.23. a. Estimate the strain in a Ford automobile tire when the car moves down a smooth road. Justify your estimates.
   b. Use this information to estimate the energy per unit volume of heat generated per cycle.
   c. The car is driven at 50 mph. What is the rate of heat production?
   d. Estimate the temperature of the tire.

20.24. Assume a mass is attached to a three-element material.
   a. Write the equation of motion.
   b. Solve for the case in which the system is released from a fixed displacement which was made just before release.
   c. Find the expression for the loss angle assuming it is small.

20.25. A cantilever beam is made of a viscoelastic material (Voigt) with $G = 10^6$ dynes/cm^2 and $\eta = 10^7$ P. It supports a concentrated load P located near the free end. Calculate the deflection of the free end versus time.

Prologue

Most engineering materials are complex multiphase, polycrystalline aggregates. However, it is also true that to develop an understanding of such complex systems, it is necessary to understand the behavior of the single crystals which make up such aggregates. Plastic flow in crystals takes place by *slip* on distinct types of planes called *slip planes* in specific directions called *slip directions*. The slip planes are usually those which are the farthest apart, and the slip directions are those along which the atomic repeat distance is the least. Slip begins when the shear stress on a slip plane in the slip direction reaches a critical value called the *critical resolved shear stress*. This shear stress is far below that predicted for perfect crystals by *Frenkel*, namely, $\tau = G/2\pi$, where G is the shear modulus. The actual critical stress is strongly structure sensitive and is affected by impurities, thermal history, previous plastic straining, etc. Fracture of crystals occurs by slip or by *cleavage*. Cleavage fracture occurs at a characteristic tensile stress normal to the fracture plane. Deformation of crystals also involves a process known as *twinning*.

21

PLASTIC FLOW OF CRYSTALS

21.1 INTRODUCTION

Crystalline materials are an important class of materials. There are two areas where an understanding of the plastic deformation processes in these materials has practical importance. In one case, flow is desirable since we are interested in shaping an object by forging, extruding, rolling, drawing, etc. In the second case, flow is (usually) undesirable since we are interested in using the material as a component in a device and we would like to have the component exhibit reversible behavior. In this case we do not want plastic flow, creep, fatigue, etc. It is still necessary to understand the plastic flow process in the second set of circumstances since understanding can lead us to suggest ways of strengthening the material so that it can carry higher stresses without exhibiting plastic flow. When crystals deform plastically as a result of shear stresses the slip takes place usually on a definite set of planes and in general in a definite crystallographic direction. The slip in a specific crystal starts on a certain plane in a certain direction when the shear stress on that plane and in that direction reaches a critical value—called the **critical resolved shear stress**. This quantity is an extremely structure-sensitive property; it depends on what impurities are present and how the crystal was grown and how it was handled before testing and in what environment it is tested; it may vary by a factor of 10^5 and probably more in a specific crystal, say, copper. In this chapter we shall study the nature of slip as well as other aspects of the deformation of crystals.

21.2   SLIP

Slip (or **glide**) in single crystals is illustrated in Figures 21.2.1 and 21.2.2. It can be compared to the shear distortion of a pack of cards. The individual segments shown in these figures are called **glide packets**.

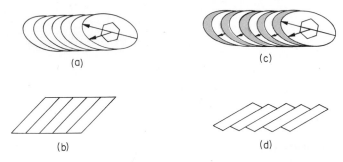

(a)   (c)

(b)   (d)

FIGURE 21.2.1. The mechanism of slip. (a) and (b), initial state. (c) and (d), after slip. [From E. Schmid and W. Boas, *Plasticity of Crystals*. F. A. Hugher, London (1953).]

The displacement of the glide packets takes place on crystallographic planes called **slip planes** or **glide planes** and in a crystallographic direction called the **slip direction**. Table 21.2.1 gives examples.

Table 21.2.1 SLIP SYSTEMS

Structure	Type of Material	Material	Plane	Direction
Diamond cubic	Homopolar	Si	(111)	[1$\bar{1}$0]
Wurzite	Homopolar	InSb	(111)	[1$\bar{1}$0]
NaCl	Ionic	NaCl	(110)	[1$\bar{1}$0]
NaCl	Ionic	KCl	(110)	[1$\bar{1}$0]
fcc	Metallic	Al	(111)	[10$\bar{1}$]
fcc	Metallic	Cu	(111)	[10$\bar{1}$]
fcc	Metallic	Ni	(111)	[10$\bar{1}$]
bcc	Metallic	α-Fe	(101)	[11$\bar{1}$]
bcc	Metallic	α-Fe	(112)	[11$\bar{1}$]
bcc	Metallic	α-Fe	(123)	[11$\bar{1}$]
bcc	Metallic	Mo	(112)	[11$\bar{1}$]
bcc	Metallic	Na	(112)	[11$\bar{1}$]
hcp	Metallic	Zn	(0001)	[2$\bar{1}$$\bar{1}$0]
hcp	Metallic	Mg	(0001)	[2$\bar{1}$$\bar{1}$0]
Rhombohedral	Metallic	Bi	(111)	[10$\bar{1}$]

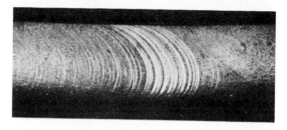

FIGURE 21.2.2. Appearance of glide lines on metal crystals. [From Frhrn. von Goler and G. Sachs, *Zeitschrift für Physik*, **55**, 581 (1929).]

The **glide packets** might have thicknesses of 10^{-3}–10^{-4} cm. Often the glide packets themselves are made up of many smaller lamellae. We have restricted the examples in Table 21.2.1 to the simpler structures. The most prominent characteristic of slip is that the slip direction is universally the direction along which the atoms are closest packed, while the slip plane is a plane in which the atoms are closely packed (but not necessarily the most closely packed plane in that crystal). *This means that the slip distance is as short as possible while the slip planes tend to be as far apart as possible.*

Consider the work needed to cause a displacement of one repeat distance. Work is given by $W = \int F\,dx$. It seems logical that F would be the smallest for the planes which are the farthest apart and of course the displacement is the least for the closest-packed direction. We shall return to this point later. We note that slip does sometimes occur on planes other than those listed in Table 21.2.1 for a specific crystal; this is particularly true at high temperatures.

EXAMPLE 21.2.1

On which planes would you expect slip to occur in the sc, bcc, and fcc crystal structures?

Answer.

sc	(100)
bcc	(110)
fcc	(111).

Do you remember that these are the planes which give the first X-ray diffraction line, *HKL*, in these crystals?

The combination of a slip plane and a slip direction is known as a **slip system**. Face-centered cubic metals have 12 slip systems; there are four slip planes {111} with three slip directions ⟨110⟩ on each plane. Hexagonal closest-packed zinc will have only three slip systems. The possibility of gliding confers plasticity to a crystal. Crystals with low symmetry and hence few glide systems will tend to have low plasticity and tend to be more brittle.

EXAMPLE 21.2.2

Prove that there are 12 slip systems in the fcc crystal structure.

Answer. There are four slip planes: (111), ($\bar{1}$11), (1$\bar{1}$1), and (11$\bar{1}$). The additional four planes, ($\bar{1}\bar{1}\bar{1}$), (1$\bar{1}\bar{1}$), ($\bar{1}$1$\bar{1}$), and ($\bar{1}\bar{1}$1), obtained by permutation are not new slip planes inasmuch as they are, respectively, parallel to the previous set. Note that these eight planes together form an octahedron and the {111} planes are called **octahedral** planes. There are three slip directions in each of these planes, the general slip direction being ⟨110⟩ or a face diagonal. Since in a cubic system an (*hkl*) plane has as a normal [*hkl*], it is easy to choose the specific three face diagonals for a given plane. Thus the ($\bar{1}$11) plane has [110], [101], and [01$\bar{1}$]. Note that the normal is [$\bar{1}$11], while the three slip directions lie in the plane and hence the dot product of the normal vector and a slip vector must be zero. There are thus $4 \times 3 = 12$ slip systems. The octahedral plane is the *closest*-packed plane (see Section 12.2 and note that these planes are in the ABC sequence).

The slip systems in the hcp crystal are shown in Figure 21.2.3. Slip predominates on the basal plane or the closest-packed plane and such slip is exhibited by Zn, Mg, Cd, Be, and Ti. Recall from Section 12.2 that an ideal hcp crystal is formed by an AB stacking sequence of closest-packed layers.

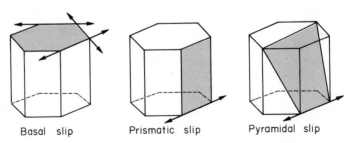

Basal slip   Prismatic slip   Pyramidal slip

FIGURE 21.2.3. Slip systems in hcp crystals.

Titanium also ordinarily exhibits prismatic slip on the prismatic planes $\{10\bar{1}0\}$ and pyramidal slip on the $\{10\bar{1}1\}$ planes.

At a certain critical stress a crystal begins to deform rapidly. When differently oriented crystals of the same material with the same history are loaded, slip will appear under the conditions given by **Schmid's law**: *Slip begins when the stress resolved on the slip plane in the slip direction reaches a certain value called the critical resolved shear stress.* In Figure 21.2.4 we show a crystal of uniform

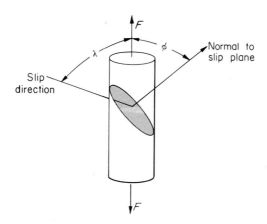

FIGURE 21.2.4. Single crystal in tension.

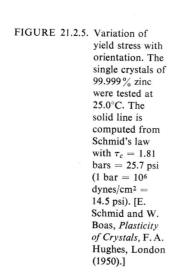

FIGURE 21.2.5. Variation of yield stress with orientation. The single crystals of 99.999 % zinc were tested at 25.0°C. The solid line is computed from Schmid's law with $\tau_c = 1.81$ bars = 25.7 psi (1 bar = 10^6 dynes/cm² = 14.5 psi). [E. Schmid and W. Boas, *Plasticity of Crystals*, F. A. Hughes, London (1950).]

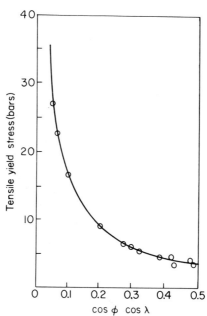

cross section A loaded in tension. The shear stress resolved in the slip direction is

$$\tau = \frac{F}{A} \cos \phi \cos \lambda. \qquad (21.2.1)$$

To get a large range of λ and ϕ when making such studies, crystals having few slip systems are chosen. Here zinc is a good example since slip occurs on the single basal plane in three directions. Figure 21.2.5 illustrates how well Schmid's law applies.

The critical resolved shear stress is extremely sensitive to impurities, surface conditions, etc. It is a structure-sensitive property. We shall return to these points later. In any case we note that τ_c, the critical resolved shear stress, is small.

21.3   THE STRESS REQUIRED FOR SLIP

Table 21.3.1 shows some values for the shear stress at which gross macroscopic flow begins. Recall that 1 bar $= 0.98$ atm and that the pressure at the deepest place in the ocean is about 1 kbar. A brief historical account is in order. The first values for monocrystals were obtained in 1922 [H. Mark, M. Polanyi, and E. Schmid, *Zeitschrift für Physik*, **12**, 58 (1922)]. The values for polycrystalline specimens were well known by then. The first whiskers (which are perfect or nearly perfect tiny crystal fibers) and measurements on them were reported in 1952 [C. Herring and J. K. Galt, *Physical Review*, **85**, 1060 (1952)], while the theoretical prediction of the shear strengths of perfect crystals, $G/2\pi$, had already been made by J. Frenkel in 1926 [*Zeitschrift für Physik*, **37**, 572 (1926)]. We shall now discuss the theoretical shear strength of perfect crystals following the classic method put forth by Frenkel.

Table 21.3.1. SHEAR STRESS IN KILOBARS* AND, IN PARENTHESES, POUNDS PER SQUARE INCH AT WHICH FLOW BEGINS

Crystal Structure	Material	"Pure" Monocrystals	Commercial-Purity Polycrystalline Aggregates	Strongest Whiskers	$\dfrac{G}{2\pi}$
fcc	Ag	Less than 0.006 (0.8×10^2)	0.4 (0.6×10^4)	14 (0.2×10^6)	41 (0.6×10^6)
bcc	Fe	Less than 0.3 (0.4×10^4)	0.6 (0.95×10^4)	62 (0.9×10^6)†	130 (1.9×10^6)
hcp	Zn	Less than 0.002 (0.3×10^2)	1.1 (1.5×10^4)	14 (0.2×10^6)†	55 (0.8×10^6)

* 1 kbar $= 10^9$ dynes/cm^2 $= 14{,}500$ psi.
† This is the shear stress when the specimen fractures; slip may not yet have begun.

We assume that one half of the crystal slides as a rigid body over the other half. Now we need a force law. We know that for small displacements Hooke's law is obeyed across the glide plane. We also know that the force law is periodic in the sense that if the atoms move one slip distance, the atomic arrangement is just as before. Now consider Figure 21.3.1 where the atoms in a simple cubic array are noted.

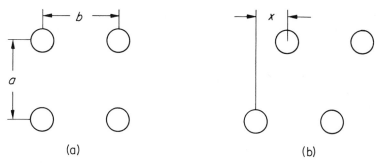

(a)   (b)

FIGURE 21.3.1. (a) Undeformed crystal. (b) Deformed crystal.

EXAMPLE 21.3.1

Sketch the potential energy and the force as the top half of the crystal of Figure 21.3.1 slips over the lower half.

Answer. Clearly, the potential is a minimum at the equilibrium positions, namely, $x = 0$, $\pm b$, $\pm 2b$. Also, the midway positions $[x = \pm(b/2), \pm(3b/2) + \cdots]$ must correspond to a metastable equilibrium or a maximum potential. Thus the potential curve must have some general periodic form as in Figure 21.3.2(a). The stress which causes this displacement is the derivative of the potential and hence has the form shown in Figure 21.3.2(b).

Every student who has progressed to this point knows that the simplest periodic function that satisfies the conditions listed in Example 21.3.1 is

$$\tau = k \sin\left(\frac{2\pi x}{b}\right). \qquad (21.3.1)$$

(You do not have to be a great scientist such as Frenkel to see that!) Most students will probably be concerned with the justification for using the sinusoidal function. Since periodic functions can be expressed in a Fourier series (sum of sine and cosine terms), the present approximation corresponds to dropping all but the first term of the series. When the displacements are very small we must have

$$\tau = G\gamma = \frac{Gx}{a}, \qquad (21.3.2)$$

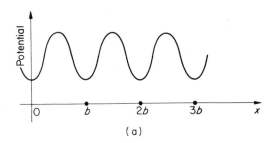

(a)

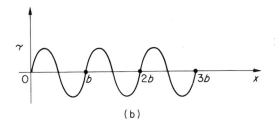

(b)

FIGURE 21.3.2. (a) Potential energy variation as top half of crystal slides over bottom half. (b) Corresponding stress variation.

where G is the shear modulus and γ is the shear strain. But by (21.3.1) we have for small x

$$\tau = k\frac{2\pi x}{b} \tag{21.3.3}$$

so that we conclude that $k = Gb/2\pi a$ and that

$$\tau = \frac{Gb}{2\pi a}\sin\left(\frac{2\pi x}{b}\right) = \tau_{max}\sin\left(\frac{2\pi x}{b}\right), \tag{21.3.4}$$

which when $a = b$ gives simply

$$\tau_{max} = \frac{G}{2\pi}. \tag{21.3.5}$$

Thus according to **Frenkel's model** we will have flow if the applied stress equals $G/2\pi$. This is an estimate of the critical resolved shear strength of *perfect* crystals. By the results for whiskers which followed just 26 years later we see that this is within a factor of 2 or so.

As single crystals become purer, as the methods of growing them become better, and as the methods of detecting small plastic strains improve, the value of τ_c for metal crystals continues to plummet downward. The present value for copper crystals is about $10^{-6}G$. Real crystalline materials have such relatively low values of the critical resolved shear stress because of the presence of dislocations. We shall see in Chapter 22 why and how dislocations lead to flow at such low stresses.

21.4   THE STRUCTURE-SENSITIVE NATURE OF SLIP IN CRYSTALS

IMPURITY EFFECTS.   The critical resolved shear stress can depend strikingly on the impurity content, on previous deformation, and on the thermal cycles to which a material has been subjected. Figure 21.4.1 shows a strong impurity effect on τ_c (at $-60°C$) in mercury monocrystals containing silver. An

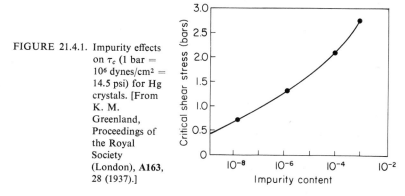

FIGURE 21.4.1. Impurity effects on τ_c (1 bar $=$ 10^6 dynes/cm^2 $=$ 14.5 psi) for Hg crystals. [From K. M. Greenland, *Proceedings of the Royal Society (London)*, **A163**, 28 (1937).]

impurity content of 1 part in 10^4 raises τ_c by a *factor* of 5. (Compare this effect with the change in elastic modulus which would be only a fraction of a percent.) Results are also shown for Cu–Ni monocrystals in Figure 21.4.2.

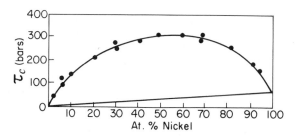

FIGURE 21.4.2. Compositional variation of critical resolved shear stress in the Cu–Ni System. Lower line would be expected for a mixture. [From E. Osswald, *Zeitschrift für Physik* **83**, 55 (1933).]

AGE PRECIPITATION (THERMAL HISTORY).   We have already discussed the Al–Cu system (Figure 16.5.1) in which precipitates resembling $CuAl_2$ were formed. By forming many fine precipitates it is possible to raise τ_c by a factor of 5 or more compared to a crystal containing large precipitates but of the same overall composition.

STRAIN HARDENING (STRESS HISTORY).   The value of τ_c for imperfect crystals is always increased by prior cold work. In certain crystals this strengthening effect is sensational. Factors as large as 10^3 are observable and even larger values can be expected. Examples of strain hardening are shown in Figure 21.4.3.

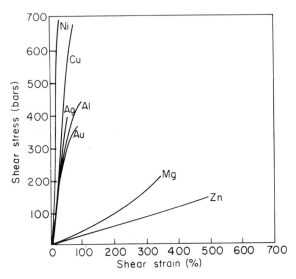

FIGURE 21.4.3.   Effects of cold working on the critical resolved shear stress of single crystals.

Here the stress is plotted versus the strain (elastic strains form only a negligible part of the strains listed). Consider now the curve for magnesium. A new single crystal will be sheared 200% by a shear stress of 90 bars (this includes an elastic strain of about 0.02%). If the load causing the stress of 90 bars is removed, the elastic shear strain will be reduced to zero, while the plastic shear strain will remain. If now the load is reapplied in increments, the specimen will deform elastically until the stress once again reaches 90 bars. A further stress increment will cause new plastic deformation. Thus the critical shear stress for this cold-worked crystal will be 90 bars, whereas for the original crystal, $\tau_c = 8$ bars.

Note the large differences in the rate of strain hardening between the two classes of materials, namely, the fcc and hcp crystals. The significant structural difference is that slip occurs on only one plane (the *basal plane*) in the hcp crystals (this is called simple slip), whereas it can occur on up to four intersecting planes in the fcc crystals (multiple slip). Strain hardening is very small when slip is confined to one system, e.g., cadmium, and is greater when two systems operate, and still greater when three systems operate. At the same time that τ_c increases, other changes are taking place in the crystal. The area under the real stress-strain curve (to a given strain) represents the work per unit volume needed to deform the crystal. This includes three terms:

1. Stored elastic energy. This is reversible and, for large γ, is relatively small. It disappears if the external stress is released.
2. Energy dissipated as heat.
3. Stored energy of plastic deformation. It remains if the external stress is released.

The energy of plastic deformation (which is stored in the absence of any external load on the specimen) is about 7% of the energy of cold working, the remainder of the work done on the specimen being dissipated as heat [A. Seeger and H. Kronmuller, *Philosophical Magazine*, **7**, 897 (1962)]. Now we might ask the next question, What is the nature of the stored energy of plastic deformation? If we examine the glide packets carefully, we will find that not only have they slid over each other but that they are individually twisted and bent. Thus if we obtain a Laue back reflection X-ray photograph, the spots (instead of being sharp as with a perfect crystal) are elongated. It can be shown by the theory of diffraction that Laue spots from bent planes should be elongated.

The bending and twisting of the glide packets involves internal *elastic* strains with each strain component showing variation across the crystal, but of such a nature that it averages to zero across the entire crystal. The internal elastic energy, however, depends on the *square* of the local value of the strain components and hence cannot be zero if any of the strain components are nonzero. Thus the stored energy of plastic deformation is internal elastic energy. As we shall see in Chapter 22 these strains may vary slowly over the glide packets, or there may exist in the glide packet a number of lines (straight or curved) about which the variation in strain is very large, such that the strain reaches values as high as $1/2\pi$ near the line but drops off considerably with distance from the line. These lines about which intense strain (and hence stress) gradients exist are called dislocations. (We have already studied dislocations in crystals from a geometric viewpoint in Section 13.3 and we noted there that they are imperfections in stacking which are centered about a line.)

RECOVERY (THERMAL HISTORY). We have now illustrated further the structure-sensitive nature of τ_c in that it depends on cold working. Let us now see what happens when we anneal such a cold-worked pure crystal by holding it at high temperatures for a period of time. First, if given a proper anneal, it is found that the stored energy of plastic deformation disappears, more or less. Let us suppose that this anneal is carried out in such a way that we still have a single crystal with the same orientation and that after the anneal the crystal is then stressed. We find that the resolved shear stress has dropped, perhaps to a value as low as that before cold working. This is illustrated for the case of zinc

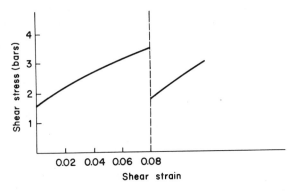

FIGURE 21.4.4. Deformation-anneal-deformation cycle in single crystals of 99.999% zinc. [From R. Drouard, J. Washburn, and E. R. Parker, *Transactions of the AIME*, **197**, 1227 (1953).]

in Figure 21.4.4, where the specimen has been sheared to a strain $\gamma = 0.08$ and then unloaded and annealed at 10°C for 700 min and then reloaded. For small annealing times the recovery would be less. It is also found that the rate of recovery is strongly temperature-dependent.

21.5   FRACTURE OF CRYSTALS

Fracture of crystals occurs by two distinct mechanisms: shear fracture and cleavage or brittle fracture.

In **shear fracture** so much slip occurs that the glide packets finally slide apart. How this might occur is obvious from Figure 21.2.1. It is found for the hcp crystals of zinc, cadmium, and magnesium that shear fracture occurs at nearly a constant value of the resolved shear stress on the basal plane. Glide separation also occurs in the bcc and fcc metals, although here multiple slip occurs.

Single crystals may part *across* crystallographic planes (with or without prior slip) leaving two crystal halves with plane faces. This is called **cleavage**. It is most easily studied in hexagonal crystals where fracture occurs across the basal plane. Thus a tension test with the tensile axis along the c axis of the crystal would give a maximum tensile stress across the basal plane and a zero shear stress along it.

It is found experimentally that there is a critical resolved normal stress σ_c for fracture across a given plane. This is **Sohncke's law**. Thus in a tensile test (see Figure 21.2.4) we have

$$\sigma = \frac{F}{A} \cos^2 \phi = \frac{F}{A} \sin^2 (90 - \phi) \qquad (21.5.1)$$

and hence the tensile stress for fracture of a specimen whose tensile axis is at an angle ϕ relative to the normal to the fracture plane is

$$\frac{F}{A} = \frac{\sigma_c}{\sin^2 (90 - \phi)} = \frac{\sigma_c}{\cos^2 \phi}. \qquad (21.5.2)$$

Figure 21.5.1 shows results for F/A vs. $(90 - \phi)$ for bismuth crystals. The solid curve is for $\sigma_c = 32$ bars or about 470 psi. This value of σ_c is low by a factor of several hundred from what is expected for a perfect crystal. Similar results occur for other materials. A simple estimate of the cleavage strength follows. Suppose we cleave a crystal across a plane lying normal to the tensile axis. We create new surface area $2A$, where A is the cross-sectional area. Let the surface energy be γ_s.

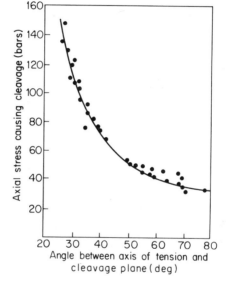

FIGURE 21.5.1. Tensile stress for cleavage of bismuth at 20°C. [From M. Georgieff and E. Schmid, *Zeitschrift für Physik.*, **36**, 759 (1926).]

Then the work necessary is $W = 2\gamma_s A$. Now suppose that in causing fracture an average force F_C acts over a distance δ. The work done is $W = F_c \delta$. Hence we have, since $\sigma_c = F_c/A$,

$$\sigma_c = \frac{2\gamma_s}{\delta}. \qquad (21.5.3)$$

EXAMPLE 21.5.1

Estimate σ_c for a perfect crystal.

Answer. A reasonable estimate for δ is about 4 Å and γ_s is about 2000 ergs/cm^2. Hence we have as a rough estimate, $\sigma_c = 10^{11}$ dynes/cm^2 (10^5 bars), which is over 1,000,000 psi. We already have noted (see Table

21.3.1) that crystals (whiskers) can be very strong; e.g., iron whiskers have withstood a tensile stress of up to 1.4×10^5 bars (1.9×10^6 psi) before fracture. We note from Figure 9.2.1 that the equilibrium distance between atoms is about 2 Å. Hence if they are separated by 4–8 Å, the interactions are essentially negligible.

The cleavage planes in crystals are usually planes with large interplanar spacing (planes of low Miller indices). Thus the cleavage plane in NaCl is {100}, as is also the case in α-Fe. A further rule is that crystals with layered structures cleave across the layers because the forces between the layers are much smaller than those within a layer. Examples are graphite, MoS_2, and mica. In a sense the same rule applies to the hexagonal closest-packed crystals Zn, Mg, and Cd and to the rhombohedral crystals (which can also be considered as belonging to a hexagonal system) Hg, Sb, and Bi; however, these crystals will also cleave on other planes but at a higher value of σ_c. Thus bismuth cleaves across the (111) plane with $\sigma_c = 32$ bars and across (11$\bar{1}$) with $\sigma_c = 69$ bars. In the case of these crystals, either slip on the layer plane, the (111) plane for Bi, or fracture across it occurs depending on the angle ϕ.

The cubic valence-bonded crystals such as diamond fracture by cleavage along either the cube or octahedral planes with little plastic deformation. The ionic crystals are also readily cleaved, but depending on the orientation, may first deform by a few percent. Metallic crystals are in general very ductile and will deform several hundred percent; in general there is a smaller tendency to cleave, although many but not all metallic crystals have been cleaved.

A tension test on a crystal which is also acted on by a sufficiently high hydrostatic pressure will eventually result in slip and finally shear fracture rather than cleavage, even for crystals which are normally considered to be extremely brittle (cleavage fracture cannot occur unless there is a sufficiently large net tensile stress somewhere in the crystal).

EXAMPLE 21.5.2

Suppose a crystal fractures normal to the tensile axis when the tensile stress reaches 32 bars and slips on an inclined slip plane when the tensile stress reaches 218 bars. Suppose the tension specimen is submerged in the ocean at a depth which gives a pressure of 1000 bars. Describe the behavior of the specimen in a tension test.

Answer. When the tensile stress (owing to the tension machine) reaches 218 bars, slip will occur, assuming that τ_c does not depend on pressure P. It will be necessary to increase this tensile stress to above 1000 bars before there is any net tensile stress and to 1032 bars before cleavage occurs, assuming that σ_c is not affected by prior slip or by P. Does this suggest a process for extrusion of brittle metals?

21.6   DEFORMATION TWINNING

Another important aspect of plastic flow is **deformation twinning**, which consists of shearing movements of individual atomic planes over each other. It consists of an essentially homogeneous shear (with exceptions to be noted later). This is illustrated in Figure 21.6.1. The plane ABC is called the **twinning plane**. The origin of the terminology "twinning" is now apparent. The twinned

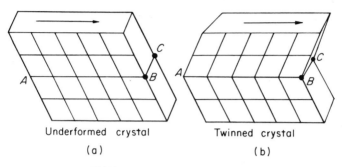

Underformed crystal	Twinned crystal
(a)	(b)

FIGURE 21.6.1. Mechanical twinning.

crystal is actually two crystals: These two crystals have a definite mutual orientation (in the present case they are mirror images with the plane ABC as the mirror plane). Deformation twinning occurs in crystals of all types of bonding: metallic, ionic, and homopolar. It occurs in a specific crystallographic direction on a specific plane for each crystal structure, as illustrated in Table 21.6.1.

Table 21.6.1. TWIN PLANES AND DIRECTIONS

Crystal Structure	Material	Plane	Direction
hcp	Cd, Zn, Mg	(10$\bar{1}$2)	[$\bar{1}$011]
bcc	α-Fe, Ta	(112)	[11$\bar{1}$]
fcc	Cu	(111)	[11$\bar{2}$]
Diamond cubic	Ge, Si	(111)	[11$\bar{2}$]
Diamond cubic	Ge, Si	(123)	[$\bar{4}\bar{1}$2]

Twins can form extremely rapidly (microseconds) even though the overall strain rate of the crystal is small. Because of the rapidity at which they are formed, a loud click or "cry" is often heard when a twin forms. Another result of the rapidity of twinning is that stress-strain curves will be jagged. In the case of cadmium, at least, a critical resolved shear stress for twinning is seen to exist [R. King, *Nature (London)*, **169**, 543 (1952)]. In general this is difficult to establish since if slip is also taking place the critical resolved shear stress may change.

The twinning plane in a bcc crystal is illustrated in Figure 21.6.2. The

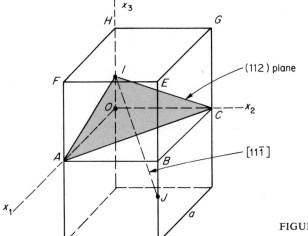

FIGURE 21.6.2. Twinning plane in a bcc crystal.

actual motion of the atoms during twinning is illustrated in Figure 21.6.3. Here the plane of the paper is the (1$\bar{1}$0) plane. The plane through *HB* and normal to the paper is the (112) twin plane. Shearing motion of the atoms in one part of the crystal takes place as shown by the arrows.

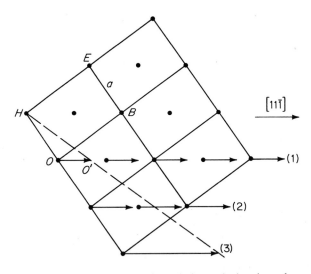

FIGURE 21.6.3. Atom motions during twinning in a bcc crystal. The points H, E, 0 and B refer to Figure 21.6.2. The plane shown here is the (1$\bar{1}$0)plane which is normal to the (112) plane. The mirror plane of the twinned crystal passes through HB. Thus E and 0′ reflect into each other.

Note that the strain is a simple homogeneous shear strain. The atoms in row (1) each move a distance $s = \overline{00'}$ as shown; the atoms in the second row move a distance $2s$; etc. The atoms do not all shift homogeneously during twinning in some crystals, e.g., in hcp crystals such as cadmium.

REFERENCES

Schmid, E., and Boas, W., *Plasticity of Crystals*, F. A. Hughes, London (1950); Translation of *Kristallplastizitat*, Springer, Berlin (1936).

Elam, C. F., *The Distortion of Metal Crystals*, Oxford, New York (1936).

Barrett, C., and Massalski, T. B., *Structure of Metals*, McGraw-Hill, New York (1966).

PROBLEMS

21.1. Assuming that the following crystal structures are made of spheres, in which direction will the spheres be touching?
   a. bcc.
   b. fcc.
   c. Ideal hcp.

21.2. Discuss the following: The production engineer is interested in the plastic flow of materials from a different viewpoint than the designer who uses the material as a component (although he may also be interested in the ease of making the component).

21.3. Derive the mathematical expression of Schmid's law.

21.4. What is a good estimate of the shear strength of a dislocation-free crystal?

21.5. Derive the expression $\tau_c = G/2\pi$ for yield of a perfect crystal.

21.6. A copper crystal is pulled in tension along the [100] direction. What is the tensile stress (in terms of τ_c) when slip begins?

21.7. A copper crystal is tested in tension along the [110] direction. What is the tensile stress (in terms of τ_c) when slip begins?

21.8. How are the slip planes and slip directions of fcc and hcp crystals related?

21.9. Name five or more examples of the structure-sensitive nature of τ_c.

21.10. What are the components of the integral $\int_0^{3.0} \tau \, d\gamma$ of a crystal and what is the order of magnitude of these quantities?

21.11. Discuss ways in which the following affect critical resolved shear stress:
   a. Impurities.
   b. Cold working.
   c. Thermal history.

21.12. The bulk modulus of crystalline materials can be represented by $K = K_0 + K_0'P$, where P is the hydrostatic pressure. $K_0' \approx 5$ for all crystals. Suppose a perfect crystal is tested in hydrostatic tension.
   a. Estimate its maximum fracture strength.
   b. Write this in terms of the Young's modulus E when $v = \frac{1}{3}$.

21.13. Calculate the shear strain which occurs when a bcc crystal twins.

21.14. Calculate the shear strain which occurs when a fcc crystal twins.

21.15. Calculate the shear strain which occurs when a hcp crystal with the ideal c/a ratio twins.

More Involved Problems

21.16. Suppose you had been given Schmid's law as a hypothesis. Describe in detail the experimental steps which you would take to prove or disprove this hypothesis.

21.17. Write a computer program for a fcc cubic crystal structure which will generate a tensile yield strength surface which is defined as follows: The vector from the origin to the surface in a given direction represents the magnitude of the tensile yield strength of a tensile specimen tested in the direction of the vector.

21.18. Find a recent review paper in the literature on whiskers and write a synopsis of the paper.

21.19. Suppose a perfect crystal fractures when a given tensile strain reaches a critical value. If such a crystal is found to fracture under hydrostatic tension, $\sigma_c = K_0/5$, what would be the stress for fracture under a single tension stress? Here K_0 is the bulk modulus at zero pressure.

Prologue

Plastic deformation of crystals takes place by the motion of dislocations (except for the case of linear viscous creep discussed in Chapter 19). Some of the geometrical aspects of dislocations were studied in Chapter 13 and should be reviewed at this time. It is now necessary to consider dislocations in a more quantitative fashion: to study how they interact with each other and other defects, such as vacancies, and how they move through a crystal and how this motion leads to plastic deformation.

We begin this chapter with a brief history of dislocations and then we show how macroscopic plastic deformations (plastic shear strain) are related to the density of dislocations, their Burgers vector, and the distance which they move. The *stress fields* associated with screw and edge dislocations in elastically isotropic crystals are studied in detail. From here we proceed to discuss the *self-energy* of a dislocation. This provides the necessary background to then discuss *dislocation reactions*.

We then derive the expression for the force on a dislocation line and discuss *climb force* and *glide force*. We then describe the climb process. The glide process is discussed in some detail. Included are discussions of the *Peierls-Nabarro stress* (the stress needed to slowly move an isolated dislocation through a crystal), *dislocation intersections, dislocation multiplication, and dislocation velocities*.

We next discuss the forces between dislocations. This provides the necessary background to discuss dislocation arrays such as the *pileup* and the *simple tilt boundary*. Finally, the interaction of impurity atoms with dislocations is discussed, and it is noted that a stress is needed to tear the dislocation away from the solute atom.

564

22

PROPERTIES OF DISLOCATIONS

22.1 INTRODUCTION AND HISTORY

It was realized from the work of Frenkel (1926) (see Section 21.3) that slip in real crystals does *not* take place by the *simultaneous* slipping of all the atoms on a slip plane. An alternative mechanism must be that it takes place by *consecutive* slipping in which only a relatively small number of atoms are directly involved at any one time. G. I. Taylor (1934) suggested that edge dislocations were responsible and illustrated in detail how they could accomplish this process [see G. I. Taylor, *Proceedings Royal Society*, **A145**, 362 (1934)]. The geometry of an edge dislocation as a stacking imperfection has been introduced in Chapter 13. J. M. Burgers (1939) introduced the screw dislocation, the general concept of the slip vector (now called the Burgers vector), and the general dislocation (which is part edge and part screw). He also showed that a small angle tilt boundary could be considered as an array of edge dislocations and he treated this problem in quantitative detail for an elastically isotropic crystal. While it was apparent that the stress required to move a dislocation was relatively small (in a crystal in which long-range many-body forces were involved) it was not until 1947 that F. R. N. Nabarro was able to treat the problem quantitatively. The assumption made in Nabarro's paper [*Proceedings Physical Society* **59**, 256 (1947)] was improved upon in the paper by A. J. E. Foreman, M. A. Jaswon and J. K. Wood [*Proceedings Physical Society* **64**, 156 (1951)]. The stress predicted to move an isolated dislocation in an fcc crystal was very small ($\ll 10^{-6}G$). Another significant theoretical development owing to F. C. Frank (1950) was the growth theory of crystals involving screw dislocations.

The experimental verification of many of the theoretical predictions of dislocation theory lagged behind. Striking confirmation of Frank's theory came

FIGURE 22.1.1   Electron micrograph of a crystal of *n*-paraffin
($C_{36}H_{74}$) after shadow casting with palladium,
showing a right-handed spiral originating
from a single dislocation. [A. R. Verma,
Crystal Growth and Dislocations, Butterworth
Scientific Publ., London (1953).]

quickly. Figure 22.1.1 illustrates a growth spiral on crystalline $C_{36}H_{74}$ obtained
by I. M. Dawson and V. Vand (1951) using electron microscopy. The step height
was shown to be equal to the size of the unit cell. Then in 1952 C. Herring and
J. K. Galt discovered whiskers. The strength of these tiny crystals approached
the value predicted by Frenkel's theory. Whiskers may contain no dislocations
or they may contain a single screw dislocation along the axis. However, when
tested in tension or bending, such a dislocation would not tend to move and the
material would appear to be perfect. Some whiskers contain several dislocations
and are weaker. The stress to cause a single dislocation in a large copper crystal
to move has been shown by F. W. Young, Jr. (1962) to be at least as small as
$10^{-6}G$. The method used by Young involves the etch pit technique in which the
highly stressed region around the dislocation is preferentially etched away so
that a pit on the surface marks the position of the dislocation. When the dislocation moves, the pit remains behind but upon further etching changes its
shape while a new pit forms at the new position of the dislocation. This technique
was also used by F. L. Vogel, W. G. Pfann, H. E. Corey, and E. E. Thomas
(1953) who showed directly that a small angle grain boundary consists of an
array of dislocations as shown in Figure 13.4.3. Direct observations of dislocations have been made by a number of methods. Dislocation networks were

first observed by transmission electron microscopy by P. B. Hirsch; a network of dislocations is illustrated in Figure 13.3.6.

Finally, it is possible to observe directly the lattice distortion at a dislocation. When two thin crystals, one with a dislocation normal to the surface, and one without, are rotated slightly, and electrons are transmitted through the pair, a Moire pattern results. The resultant magnification makes it possible to observe the pattern from an edge dislocation as shown in Figure 13.3.2.

CONSERVATION PRINCIPLE. We state without proof two **conservation principles**:

I. A dislocation cannot end within a crystal.

II. The Burgers vector of a dislocation is everywhere the same.

Thus a dislocation line might be a closed loop. There may be pure edge, mixed, and pure screw dislocation character at various points along the dislocation loop. However, the Burgers vector is everywhere the same.

RELATION OF MACROSCOPIC STRAIN TO SLIP. Let us consider a unit cube. Consider a single dislocation which moves entirely across this cube. Then one part of the crystal is displaced by b relative to the other part and the shear strain is $\gamma = b/1$ as shown in Figure 22.1.2. If p parallel dislocations in the unit cube

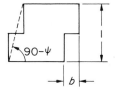

FIGURE 22.1.2. Shear strain in a unit cube owing to passage of a dislocation; $\gamma = \tan \psi = b$.

have moved similarly, then the shear strain is pb and if instead they have moved a distance l, then the shear strain is

$$\gamma = pbl. \tag{22.1.1}$$

Let us suppose that a group of straight edge dislocations having Burgers vectors b move with velocity v (in the direction of the Burgers vector) on a set of parallel slip planes. Let the density of dislocations be p. In this case the density is simply the number of dislocations which cross a unit area of a plane whose normal is parallel to the set of straight dislocation lines. (More generally we define the dislocation density as the total length of dislocation lines per unit volume.) The shear strain rate is given by

$$\dot{\gamma} = bpv. \tag{22.1.2}$$

We note that γ and $\dot{\gamma}$ are macroscopic quantities which are assumed to be continuous functions. The slip does not take place in a continuous fashion. We have, however, made the assumption that there is a high enough density of dislocations distributed sufficiently randomly so that a macroscopic density of dislocations can be defined.

22.2   THE STRESS FIELDS OF DISLOCATIONS

SCREW DISLOCATION.   The screw dislocation in a crystalline solid is shown in Figure 13.3.8. It can be imagined to be formed by a shear motion as shown in Figure 22.2.1. The resultant dislocation is shown in Figure 22.2.2(a).

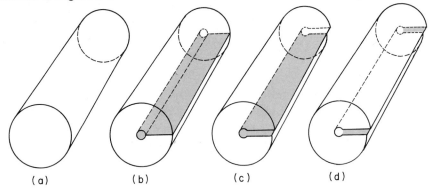

(a)   (b)   (c)   (d)

FIGURE 22.2.1. Formation of a screw dislocation (thought experiment). (a) Perfect materials. (b) Slice made parallel to the axis and removal of a small cylinder of material around the tip of the cut of the axis. (c) Application of forces of opposite sign parallel to axis to cause a rigid displacement b across the cut in the direction of the axis. (d) The material across the cut [shown shaded in (b) and (c)] is "welded" together, leaving the dislocation.

We note that if we unfold a cylindrical element lying between r and $r + dr$ it appears as shown in Figure 22.2.2(c). The deformation is a simple shear with an engineering shear strain given by $b/2\pi r$. Thus, in polar coordinates ($x_1 = r \cos \theta$, $x_2 = r \sin \theta$, $x_3 = z$), the shear strain is given by

$$\gamma_{z\theta} = \frac{b}{2\pi r}. \qquad (22.2.1)$$

EXAMPLE 22.2.1

In the screw dislocation, how is the slip vector **b** related to the unit tangent vector **v** of the dislocation line?

Answer. They are parallel.

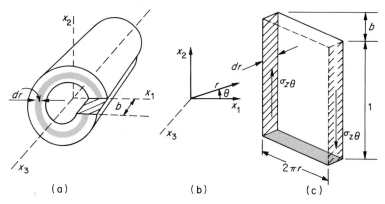

FIGURE 22.2.2. (a) Screw dislocation in an elastic media. (b) Cylindrical coordinates. (c) Unfolded cylindrical element of thickness dr. Shaded face corresponds to shaded area in (a). Cross-hatched faces correspond to crosshatched face in (a).

EXAMPLE 22.2.2

When the screw dislocation moves, what is the direction of motion of the dislocation line relative to the slip vector **b**?

Answer. Perpendicular.

Actual crystals are elastically anisotropic. However, for simplicity we shall assume that they behave as if they were elastically isotropic. The state of stress is given in cylindrical coordinates by the array of nine stress components:

$$\begin{pmatrix} \sigma_{rr} & \sigma_{r\theta} & \sigma_{rz} \\ \sigma_{\theta r} & \sigma_{\theta\theta} & \sigma_{\theta z} \\ \sigma_{zr} & \sigma_{z\theta} & \sigma_{zz} \end{pmatrix}.$$

Thus $\sigma_{z\theta}$ means a shear stress component owing to a force acting in the z direction on a plane whose normal is in the θ direction.

We then have by Hooke's law

$$\sigma_{z\theta} = G\gamma_{z\theta} = G\frac{b}{2\pi r}. \tag{22.2.2}$$

Except for the complementary shear stress components $\sigma_{z\theta} = \sigma_{\theta z}$, all other stress components have a zero value. We note that the shear stress field has radial symmetry. Note that the result (22.2.2) would be obtained with any slip plane parallel to the x_3 axis. Therefore, the screw dislocation in an isotropic medium cannot be identified with a particular slip plane.

Note that according to the present analysis the stress goes to infinity as $r \rightarrow 0$. We say that there is a singularity in $\sigma_{\theta z}$ at $r = 0$. To remove this mathematical difficulty we assume that a core of material is removed from the center of the dislocated region.

THE CORE.   The cylinder of material which was removed from the elastic media because of the presence of a singularity is known as the **dislocation core**. In the case of a real crystal the core is represented by atoms in an environment varying greatly from that in the ideal crystal. The detailed arrangement of the atoms in the core and the manner in which this affects the overall stress field and properties of a dislocation are not known. The approximate position of the core atoms has been calculated using the Born model for central force interactions. There is a net increase of volume of the crystal of about two atomic volumes per atom length of dislocation line. The core region can therefore be roughly visualized as a region with very high hydrostatic pressure. The linear elastic theory used in this section leads to no volume change.

ANISOTROPY.   The stress fields for dislocations lying along certain directions with particular Burgers vectors have been worked out for a number of cases involving anisotropic elasticity, particularly in cubic crystals, following a classic paper by J. D. Eshelby, W. T. Read, and W. Shockley, *Acta Metallurgica*, **1**, 251 (1953). In many applications it is very important to take anisotropy into account. In a first study only isotropic elastic behavior is considered.

EXAMPLE 22.2.3

Assuming that linear elasticity theory holds, explain why the stress is proportional to the deflection.

Answer. The force causing the element to deflect by an amount b in Figure 22.2.2 is the stress $\sigma_{z\theta}$ times the area 1 dr, i.e.,

$$\sigma_{z\theta} 1 \, dr.$$

For small deflection of a linear system, as noted in Sections 2.2 and 2.3, the deflection is proportional to the force.

EXAMPLE 22.2.4

Is it legitimate to assume that linear theory holds for r as small as b?

Answer. This means $\gamma \approx 1/2\pi \approx \frac{1}{6}$. The linear elastic theory of solids is applicable only in those cases where $\gamma^2 \ll \gamma$ (e.g., for $\gamma = 0.01$). Thus, it is not applicable to such large strains as $\gamma = \frac{1}{6}$.

EXAMPLE 22.2.5

The change in volume of a crystal is, according to linear elasticity theory, given in terms of the bulk modulus (Section 2.3) by

$$\frac{\Delta V}{V} = \frac{\sigma_{11} + \sigma_{22} + \sigma_{33}}{3K}.$$

What are σ_{11}, σ_{22}, and σ_{33} for the screw dislocation of Figure 22.2.1, assuming linear isotropic elastic theory?

Answer. They are zero ($\sigma_{rr} = \sigma_{zz} = \sigma_{\theta\theta} = 0$ since only $\sigma_{z\theta} = \sigma_{\theta z} \neq 0$). Thus $\Delta V/V = 0$ for a screw dislocation according to this model. However, if the nonlinear elastic behavior near and in the core is taken into account, it is found that there is a volume increase in the crystal of about two atomic volumes per atom length of dislocation line. A rough approximation to the stress state can be made by assuming linear elasticity with the cylindrical core filled with a liquid at pressure P. A pressure of $P \approx K/6$, where K is the bulk modulus, would be required to account for the dilatation of two atomic volumes per atomic length of dislocation. (This is also true for edge dislocations.)

EDGE DISLOCATION. Consider the edge dislocation shown in Figure 22.2.3. This edge dislocation could be imagined to be formed as shown in Figure 22.2.4. The reason for removing the small cylinder of material from around the

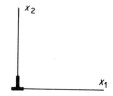

FIGURE 22.2.3. Edge dislocation lying along the x_3 axis. The positive x_3 axis is toward the reader in the right-handed coordinate system which we shall use. The symbol illustrates that the extra half-plane of atoms lies on the x_2x_3 plane on the positive x_2 side. The x_1x_3 plane is the glide plane.

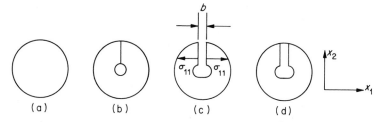

(a) (b) (c) (d)

FIGURE 22.2.4. Formation of an edge dislocation (thought experiment). (a) The perfect crystal. (b) Vertical cut plus removal of a small cylinder of material around the tip of the cut. (c) Rigid displacement of the faces of the cut in the x_1 direction. (d) Insertion of a slab of similar material into the cut after which the faces of the cut and the slab are welded together.

tip of the cut is this: We wish to use linear elasticity; a singularity in the stresses exists along the x_3 axis as in the case of the screw dislocation.

The resultant edge dislocation is identical to that given in Figure 13.3.1. We can also imagine that we form an edge dislocation by slip as shown in Figures 22.2.5 and 22.2.6.

FIGURE 22.2.5. Formation of an edge dislocation in a simple cubic crystal by slip.

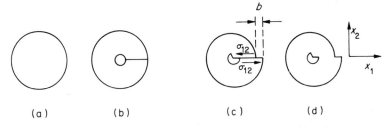

(a)   (b)   (c)   (d)

FIGURE 22.2.6. Formation of an edge dislocation (thought experiment) (a) Perfect solid. (b) Horizontal cut plus removal of a small cylinder of material around the tip of the cut. (c) Rigid shear displacement along the cut. (d) Two faces of the cut are welded together.

The two edge dislocations of Figures 22.2.4 and 22.2.6 are identical; however, it is important to note that their mode of formation was distinctly different.

EXAMPLE 22.2.6

When the edge dislocation of Figures 22.2.3–22.2.6 moves by slip, in what direction does it move?

Answer. In the direction of the Burgers vector $\mathbf{b} = [b, 0, 0]$ in the present case, i.e., in the x_1 direction as in Figure 22.2.6.

EXAMPLE 22.2.7

Is it possible for the edge dislocation to slip in the x_2-direction?

Answer. No. But there is another kind of motion which involves the motion of vacancies to (or from) the dislocation line called **climb**, which causes an edge dislocation to move in a direction perpendicular to the plane defined by the **unit tangent vector to the dislocation line, v**, and **b**; this is the x_2 direction here. Note that the dislocation of Figure 22.2.4 was created by the insertion of the extra half plane of atoms. Diffusion of vacancies to the line and diffusion of these atoms away from the line would tend to eliminate the lower part of this extra inserted plane. The dislocation line would *climb* upward.

The elastic description of the edge dislocation proceeds as follows. We consider the dislocation to extend indefinitely along the x_3 axis. Hence there is no variation of the displacements with x_3. The problem is therefore one of *plane deformation*. (Many students have studied the deformation of a very long thick-walled cylinder with internal pressure in strength of materials; that problem was also a plane deformation problem.) At this point we consider the solutions only. Moreover, they are put here for later use, not for memorization! We have

$$\sigma_{11} = -D\frac{x_2(3x_1^2 + x_2^2)}{(x_1^2 + x_2^2)^2}, \tag{22.2.3}$$

$$\sigma_{22} = D\frac{x_2(x_1^2 - x_2^2)}{(x_1^2 + x_2^2)^2}, \tag{22.2.4}$$

$$\sigma_{12} = D\frac{x_1(x_1^2 - x_2^2)}{(x_1^2 + x_2^2)^2}, \tag{22.2.5}$$

where

$$D = \frac{Gb}{2\pi(1 - v)}. \tag{22.2.6}$$

Here G is the shear modulus, v is Poisson's ratio, and b is the Burgers vector. Note that σ_{11} and σ_{22} are even functions of x_1 and odd functions of x_2; note that this is precisely what we would expect, e.g., from Figure 22.2.4. Note, likewise, that σ_{12} is an odd function of x_1 and an even function of x_2 and that this is what we would have expected from Figure 22.2.6.

It is convenient to use polar coordinates. Then $x_1 = r \cos\theta$, $x_2 = r \sin\theta$, and $x_3 = z$ and the stresses are given by

$$\sigma_{rr} = \sigma_{\theta\theta} = -D\frac{\sin\theta}{r} \tag{22.2.7}$$

$$\sigma_{r\theta} = D\frac{\cos\theta}{r}. \tag{22.2.8}$$

22.3   FREE ENERGY OF FORMATION OF A DISLOCATION

SCREW DISLOCATION. The stress acting along the shaded face in Figure 22.2.1 in the x_3 direction is $\sigma_{z\theta}$. Note that if the two faces were not welded together, then this stress would have to be applied along the two faces to keep them fixed in this position. If we consider a unit length of dislocation then the total force along the face will have to be

$$F_l = \int_{r_0}^{r_1} \sigma_{z\theta} \, dr, \tag{22.3.1}$$

where r_1 is the outside radius of the crystal and r_0 is the core radius. The work done on a *linear* elastic solid when the force F_l causes a displacement b is

$$w_l = \tfrac{1}{2} F_l b \tag{22.3.2}$$

inasmuch as F_l builds up linearly from zero to the final value F_l, as the displacement increases from zero to b. This can be seen from Figure 22.2.1(b) and (c); as the force which causes the displacement is applied, the displacement increases linearly with the force (since we are using linear elastic theory). To the approximation that $\sigma_{z\theta}$ is given by (22.2.2) we have for the energy of formation of unit length of screw dislocation

$$U_l = \frac{Gb^2}{4\pi} \ln \frac{r_1}{r_0}. \tag{22.3.3}$$

EXAMPLE 22.3.1

Estimate the energy of a screw dislocation in copper.

Answer. For a copper crystal with $r_1 \approx 1$ cm and $r_0 \approx 10^{-7}$ cm we have $U_l = 3.3 \times 10^{-4}$ erg/cm (since $G = 4 \times 10^{11}$ dynes/cm², $b = 2.5 \times 10^{-8}$ cm). This corresponds to about 5.3 eV/atomic length. (Here atomic length means atomic diameter.)

We shall now show that the entropy introduced is negligible. Consider a dislocation line of N atomic lengths. Suppose that it threads its way through the crystal along a random path (which it does not, but this will give us the greatest entropy). Let us start to build up such a dislocation line. On each step of the random walk it could take up 12 new positions; i.e., it could go to any of the 12 nearest neighbor positions in the fcc structure of copper. Thus each atomic length has a total of 12 configurations or in all there are 12^N configurations after N steps. The entropy is given by $S = k \ln 12^N$ and the Helmholtz free energy by $F = U - TS$ or

$$F = 5.3(\text{eV})N - kTN \ln 12. \tag{22.3.4}$$

Now

$$5.3 \text{ eV} \gg kT_m \ln 12 \approx 0.1 \text{ eV}$$

so that F increases as N increases. Since F is always positive, the dislocation is *not* a thermodynamically stable defect as a vacancy might be. Because the entropy plays such a minor role, it is common to ignore this contribution; it is sufficient to consider only the self energy of the dislocation.

EDGE DISLOCATION. The analysis for an edge dislocation is similar. It should be noted that the only energy term discussed here is the elastic energy.

Other terms may be very important, particularly in covalent crystals; there the changing of bond angles and distances in the core can be very important. Likewise, charge effects can be very important in ionic crystals. In metals, the elastic term is likely to be of overwhelming importance. In the discussion which follows we shall focus our attention on the elastic terms.

It should be noted that the self energy of the dislocation involves the square of the Burgers vector: You would predict this to be so for a linear elastic solid without getting involved in a great mass of calculations. Why? You will similarly predict the proportionality to G, the shear modulus. Why?

Using the same techniques as used for the screw dislocation one can readily show that

$$U_l = \frac{Gb^2}{4\pi(1 - v)} \ln \frac{r_1}{r_0}.$$ (22.3.5)

In either case for a normal size crystal it is approximately true that

$$U_l \approx Gb^2.$$ (22.3.6)

22.4   DISLOCATION REACTIONS

MULTIPLE DISLOCATIONS.   A dislocation which has a Burgers vector exactly equal to one repeat distance is called a **total dislocation** (usually it is called a dislocation and deviations from it are given modifiers). Thus, if two edge dislocations of Burgers vector b are pushed together to form a new dislocation of Burgers vector $2b$, the new dislocation is called a **multiple dislocation**. We can use the energy equation ($U \propto b^2$) to decide whether such a reaction is possible. In general

$$\mathbf{b}_1 + \mathbf{b}_2 \xrightarrow{?} (\mathbf{b}_1 + \mathbf{b}_2)$$ (22.4.1)

and

$$U_{\text{product}} \propto (\mathbf{b}_1 + \mathbf{b}_2)^2$$ (22.4.2)

while

$$U_{\text{reactant}} \propto b_1^2 + b_2^2.$$ (22.4.3)

The reaction proceeds if $U_p < U_r$. (22.4.4)

SESSILE DISLOCATIONS.   The closest packed direction in a bcc crystal structure is the [111] direction. The distance of closest approach of the atoms is $a/2$ [111] where a is the lattice parameter. Thus, an edge dislocation with Burgers vector $a/2$ [111] could be responsible for the observed slip. Consider now the possibility of the reaction:

$$\frac{a}{2}[111] + \frac{a}{2}[1\bar{1}\bar{1}] \longrightarrow a[100].$$ (22.4.5)

First, it is vectorially true. Second, consider the energetics:

$$U_p \propto a^2$$

$$U_r \propto \frac{a^2}{4}(1^2 + 1^2 + 1^2) + \frac{a^2}{4}(1^2 + \bar{1}^2 + \bar{1}^2) = \frac{3}{2}a^2.$$

Thus, the product is favored: two dislocations can react to form a third dislocation which is more stable. The resultant dislocation is immobile with respect to glide because its Burgers vector is not in a $\langle 111 \rangle$ direction. A dislocation which cannot glide is called a **sessile dislocation**. (There is some controversy over the question of whether this dislocation is sessile.) Sessile dislocations, because they cannot glide, can act as barriers or blocking dislocations in a pileup as will be seen later.

PARTIAL DISLOCATIONS.   The slip directions in an fcc crystal such as copper are the $\langle 110 \rangle$ directions, i.e., the face diagonals. The repeat distance (distance between the closest packed directions) is, e.g., $a/2\,[110]$.

EXAMPLE 22.4.1

Give reasons why the following reaction might occur for a screw dislocation in an fcc crystal:

$$\frac{a}{2}[110] \longrightarrow \frac{a}{6}[21\bar{1}] + \frac{a}{6}[121]. \tag{22.4.6}$$

Answer. First, it is vectorially true. Second, it is energetically feasible since $U_p \propto a^2/3$ and $U_r \propto a^2/2$.

The resultant dislocations in (22.4.6) are called **partial dislocations**. The packing of atoms along the plane connecting these partials must be disrupted, i.e., there must be a **stacking fault** (this will be discussed further following Figure 22.4.1). Its energy has to be considered in order to calculate the separation of the partials. Let us call γ the energy per unit area of stacking fault. Suppose the two parallel partials are separated by a distance r. Then the energy change per unit length of dislocation in the reaction is

$$\Delta U = U_p + \gamma r - U_r. \tag{22.4.7}$$

Here, U_p is the energy per unit length of the two partials (including their interaction energy which would go to zero if they were widely separated), and U_r is the energy per unit length of the total dislocation. The equilibrium separation is given by

$$\frac{\partial \Delta U}{\partial r} = 0.$$

Because the resultant dislocation is part edge and part screw (though primarily screw), the calculation is somewhat involved. If we assume the partials are pure screw dislocations, we can show that the equilibrium distance of separation is

$$r^* = \frac{G\mathbf{b}_2 \cdot \mathbf{b}_3}{2\pi\gamma}, \tag{22.4.8}$$

where $\mathbf{b}_2$ and $\mathbf{b}_3$ are the Burgers vector of the products in (22.4.6).

EXAMPLE 22.4.2

Using the data for stacking fault energy from Table 13.4.1, shear modulus in Table 2.2.1, and lattice parameter from Table 10.6.1, calculate r^* for copper.

Answer. Since $\gamma = 40$ ergs/cm^2, $a = 3.61$ Å, $G = 4 \times 10^{11}$ dynes/cm^2 and

$$\mathbf{b}_2 \cdot \mathbf{b}_3 = \frac{a^2}{36}(2 + 2 - 1) = \frac{a^2}{12}$$

we have

$$r^* = \frac{4 \times 10^{11} \times 3.6 \times 10^{-8}}{12\sqrt{2}\,\pi\gamma}\left(\frac{a}{\sqrt{2}}\right)$$

$$\approx 7\left(\frac{a}{\sqrt{2}}\right).$$

The quantity $a/\sqrt{2}$ is the atomic diameter or the length of the total Burgers vector.

Thus, in materials such as copper the dislocation breaks up into two partials separated by r^* with a stacking fault in between. We noted earlier that a screw dislocation in an isotropic elastic media (unlike an edge dislocation) has no preference for glide planes. However, if the screw dislocation is effectively now a

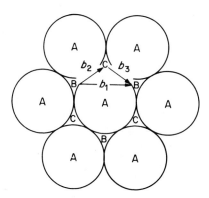

FIGURE 22.4.1. Stacking faults in thin foils of stainless steel as seen by electron microscopy. There are several striations on each stacking fault due to special interference effects. Stainless steel has a very low stacking fault energy and hence the stacking faults separating the partial dislocations are wide. ($\times 20,000$). [M. J. Whelan, P. B. Hirsch, R. W. Horne and W. Bollmann, *Proc. Roy. Soc.* **A240**, 524 (1957).]

planar defect, changing glide planes (**cross slip**) becomes more involved. We show the total and partial dislocations given in Equation (22.4.6) in Figure 22.4.2. Note that the motion of the B-layer of atoms via the partials would be from B to C and C to B, a process which would be easier than the straight line motion from B to another B.

Consider a perfect crystal with the ABC stacking sequence. Now, imagine that the b_2 *partial* dislocation has moved through the crystal just above the A-plane shown in Figure 22.4.2. Then the B-plane (above the paper) goes to a C-position, the C-plane above it goes to an A-position, and the A-plane above it goes to a B-position, etc. The resulting stacking sequence is as shown:

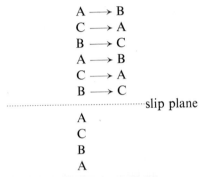

$$A \longrightarrow B$$
$$C \longrightarrow A$$
$$B \longrightarrow C$$
$$A \longrightarrow B$$
$$C \longrightarrow A$$
$$B \longrightarrow C$$

·······························slip plane

A
C
B
A

FIGURE 22.4.2. Slip in a fcc crystal.

There is a stacking fault at the slip plane. If now the b_1 partial dislocation were to move through the crystal on the same slip plane, the original perfect stacking sequence would be restored.

EXAMPLE 22.4.3

Give a reason why partial dislocations are important.

Answer. We noted earlier that total screw dislocations were not restricted to glide motion on a single glide plane (as an edge dislocation). However, if the screw dislocation splits into two partial dislocations with a stacking fault between, the combination is a plane defect which will have difficulty changing glide planes. Such a dislocation can more easily be stopped by barriers, and hence, a material might be expected to be stronger if such dislocations are present.

If a number of vacancies condense into a penny-shaped layer one atom thick on a (111) plane of an fcc crystal, the disc will collapse. The boundary of this collapsed disc will be an edge dislocation, as shown in Figure 22.4.3, whose Burgers vector is in the [111] direction which is not a slip direction in an fcc crystal. The resulting sessile dislocation is called a **Frank sessile dislocation.**

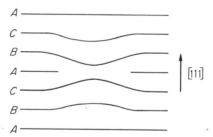

FIGURE 22.4.3. A sessile dislocation in a fcc crystal.

22.5   THE FORCE ON A DISLOCATION LINE

Let us consider a screw dislocation whose Burgers vector is parallel to the z axis. Suppose a macroscopic stress $\sigma_{z\theta}$ (owing to external stresses) is also present. This will tend to make the dislocation move (note that a stress $\sigma_{z\theta}$ is necessary in causing the dislocation of Figure 22.2.1 to form; here x_3 corresponds to z).

When the dislocation line of unit length moves by a distance dr, the area swept out is dr [see Figure 22.5.1(a)]. The displacement of material across the area swept out is b. This material displacement is in the z direction. The force in the z direction is the product of the stress $\sigma_{z\theta}$ times the area swept out dA. Hence the work done is given by

$$dW = (\sigma_{z\theta}\, dA)b = (\sigma_{z\theta}\, dr)b$$
$$= \sigma_{z\theta}b\, dr. \tag{22.5.1}$$

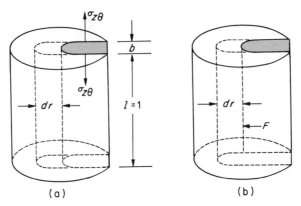

FIGURE 22.5.1. Force per unit length of dislocation line.

We would also like to describe the work done in terms of the displacement of the dislocation line itself (rather than the displacement of one part of the crystal with respect to the other).

To do this we imagine a force F acting normal to a unit length of the dislocation line as shown in Figure 22.5.1(b) such that when the dislocation line undergoes a displacement dr the work done is

$$dW = F\,dr.$$

This force F is called the **Mott-Nabarro force**. By equating the two expressions for dW we find

$$F = \sigma_{z\theta}b. \tag{22.5.2}$$

The presence of the external stress $\sigma_{z\theta}$ tends to make the screw dislocation glide in the radial direction.

EXAMPLE 22.5.1

Would the application of a stress $\sigma_{11} = -|\sigma|$ cause a force on the edge dislocation of Figure 22.2.3 ?

Answer. This can best be explained by reference to Figure 22.2.4. Such a stress is the opposite of the one used to go from (b) to (c). This compressive stress would therefore tend to squeeze out the extra plane of atoms added in (d). There is therefore a **climb force** on the edge dislocation tending to make it move in the positive x_2 direction:

$$F_{\text{climb}} = |\sigma|b.$$

EXAMPLE 22.5.2

Is there a simple stress state which will make the edge dislocation of Figure 22.2.3 tend to slip or glide ?

Answer. This can best be visualized by examining Figure 22.2.6. Note that in going from (b) to (c) we must apply a shear stress $\sigma_{12} = \tau$. The continued presence of such an externally applied stress as shown below would cause the dislocation to move in the negative x_1 direction.

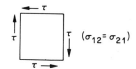

FIGURE E22.5.2

The dislocation motion would be by slip or glide. The **glide force** would be

$$F_{\text{glide}} = -\tau b.$$

SIGN CONVENTION. Let **v** be a unit tangent vector to the dislocation. Take the Burgers circuit (see Figure 13.3.1) in the direction of rotation of a right-handed screw advancing along the dislocation. The Burgers' vector b goes from the end of the Burgers circuit to the start.

Thus, the dislocation shown in Figure 22.2.2 has $v = [0, 0, 1]$ and $b = [0, 0, -b]$. The dislocation shown in Figure 22.2.4 has $v = [0, 0, 1]$ and $b = [-b, 0, 0]$.

GENERAL EQUATION OF FORCE ON A DISLOCATION. Let (σ) be the stress matrix at a given point along a dislocation where the unit tangent vector is **v**. Then the force on the length ds of the dislocation is

$$d\mathbf{F} = \mathbf{v} \times [(\sigma) \cdot \mathbf{b}] \, ds. \tag{22.5.3}$$

This is known as the **Peach-Koehler equation**. It is a generalization of the Mott-Nabarro equation. Here (σ) might be due to another dislocation, an externally applied stress, etc.

22.6   DISLOCATION CLIMB

We have thus far concerned ourselves primarily with dislocation glide. We have, however, noted the possibility of an edge dislocation moving in a nonconservative manner which involved the creation or annihilation of vacancies or interstitials (see Figure 18.6.3). Consider an edge dislocation line as shown in Figure 22.6.1. The dislocation line lies along x_3 except for a jog PP'. A **jog** is the segment of a dislocation line which connects a dislocation line on one slip plane to a dislocation line on another slip plane. The Burgers vector is parallel to x_1. The extra plane of inserted atoms is shown by the shading. If an additional atom is added at PP' (resulting in the creation of a vacancy which diffuses away)

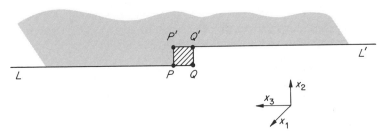

FIGURE 22.6.1. Edge dislocation line with a jog. Extra half-plane of atoms is in the $x_2 x_3$ plane.

the jog moves from PP' to QQ'. By many such additions the dislocation line would move from $P'L'$ downward. The extra plane of atoms would grow. We call this manner of dislocation motion **climb**. If a tensile stress acts across the extra plane of atoms, as in Figure 22.6.2, it will tend to cause the dislocation to climb downward (toward negative x_2). If an atom is added at the jog of the dislocation resulting in the formation of a vacancy, the applied stress does work $\sigma_{11}\Omega$, where Ω is the atomic volume. Consequently, the energy of formation of

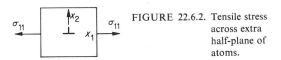

FIGURE 22.6.2. Tensile stress across extra half-plane of atoms.

a vacancy in the presence of the stress component σ_{11} is given by

$$E_f(\sigma_{11}) = E_f - \sigma_{11}\Omega. \tag{22.6.1}$$

Consequently, if the vacancies are in equilibrium with the dislocation line they will have a concentration

$$C_v(\sigma_{11}) = e^{-E_f(\sigma_{11})/kT} = C_v^0 e^{\sigma_{11}\Omega/kT}. \tag{22.6.2}$$

Here C_v^0 is the equilibrium concentration in the absence of stress. The detailed arguments used to derive (22.6.1) and (22.6.2) are similar to those used in deriving (19.7.2) and will not be repeated here. When this concentration is reached, there is no net force on the dislocation, i.e., it will neither climb upward nor downward. If σ_{11} is tensile, we see that the equilibrium vacancy concentration is greater than the equilibrium concentration C_v^0 in the stress free case. Therefore, with a positive σ_{11} applied, the dislocation acts as a vacancy source until the concentration of (22.6.2) is reached. With a negative σ_{11}, the dislocation acts as a vacancy sink until the concentration given by (22.6.2) is reached. There is considerable evidence which suggests that if a stress σ_{11} is applied at a temperature near to the melting point (so that point defects can be readily created and can readily move), the vacancy concentration in the bulk near to the dislocation reaches very readily the concentration given by (22.6.2), although elsewhere in the crystal the concentration may be very different; we say that a **local equilibrium** exists. It must be emphasized that climb, because it involves the motion of point defects, is a high temperature process. This is of importance in the theory of the Kirkendall effect (binary diffusion) and in high-temperature deformation theory involving creep, recovery, and relaxation.

22.7   THE PEIERLS-NABARRO STRESS

The shear stress which must be applied to an isolated dislocation to make it move from one equilibrium position (Figure 13.3.3) to another is called the **Peierls-Nabarro stress**. Nabarro [F. R. N. Nabarro, *Proceedings Physical Society,* **59**, 256 (1947)] has shown (using the concepts of a distributed dislocation introduced by Peierls) that this stress is given by

$$\sigma_{PN} = \frac{2G}{1-v}e^{-(4\pi\zeta/b)} \tag{22.7.1}$$

for an edge dislocation where ζ, called the **dislocation width**, is

$$\zeta = \frac{a}{2(1-v)}. \tag{22.7.2}$$

Here b is the Burgers vector, a is the distance between glide planes, G is the shear modulus, and v is Poisson's ratio. The dislocation width is a measure of the distance over which the atom displacements are more than a fixed quantity.

We note that σ_{PN} decreases as the distance a between the planes increases, and decreases as the slip distance, b, decreases. Both of these predictions are consistent with Schmid's law.

A. J. E. Foreman, M. A. Jaswon, and J. K. Wood, *Proceedings Physical Society,* **64**, 156 (1951), noting that the actual width is closely related to the interaction potentials between the atoms, have shown for the case where $a = b$ that

$$\sigma_{PN} = \frac{2G}{1-v}e^{-2\pi/(1-v)}e^{-k}\left(1 + k + \frac{k^2}{6}\right), \tag{22.7.3}$$

where

$$k = 2\pi(A - 1)/(1 - v) \tag{22.7.4}$$

and A is a parameter which depends on the interaction potential. For copper, $A \approx 6$.

For the case where $a = b$, Equation (22.7.1) gives for $v = \frac{1}{3}$

$$\sigma_{PN} = 3.6 \times 10^{-4}G,$$

while for copper Equation (22.7.3) gives for $v = \frac{1}{3}$

$$\sigma_{PN} = 6.1 \times 10^{-22}G,$$

where G is the shear modulus. This large discrepancy simply illustrates that this is an exceedingly difficult quantity to estimate theoretically. Because **core effects** undoubtedly play a vital role in determining σ_{PN}, the theoretical estimates are even more difficult. However, we can conclude, for cases where the primary contribution is expected to be representable by elastic effects, that σ_{PN} is indeed only a tiny fraction of G. The smallest experimental values found to date are those of Young who worked with copper crystals having only 10^2 dislocations/ cm². He found that dislocations moved over macroscopic distances for shear stresses of about 1 psi ($\approx 10^5$ dynes/cm²) [F. W. Young, *Journal Applied Physics* **33**, 963 (1962)].

In covalent or homopolar crystals such as silicon, where bonds are broken and reformed as the dislocation moves, fairly high stresses are required to move the dislocation, although this is not predicted by the above model. The use of the elastic continuum analysis, which leads to (22.7.1) or (22.7.3), may give good results in metals and ionic crystals where the attractive interactions are long range. It would not give a reasonable result for homopolar crystals in which the bonds are essentially nearest neighbor pair bonds.

22.8   DISLOCATION INTERSECTIONS

Figure 22.8.1 shows a number of possible intersections which result in various kinks and jogs. A kink can be formed (or eliminated) by glide alone in the absence of other dislocations. A **kink** is the portion of a dislocation line which connects two parallel segments of a dislocation line lying on the same slip plane but not coincident. A **jog** is the portion of a dislocation line which connects two segments of a dislocation line lying on *different* slip planes. Assuming that the jog is a single jog (one atomic plane separation between the two slip planes, which in a simple cubic crystal is a distance b), we can estimate the **jog energy**

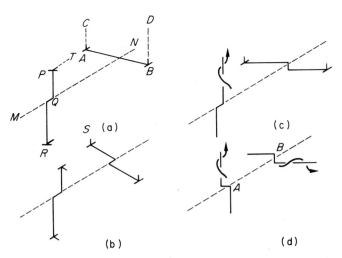

FIGURE 22.8.1. Crossings of dislocation: (a) edge-edge; (b) edge-edge; (c) edge-screw; (d) screw-screw. [From A. H. Cottrell, *Dislocations and Plastic Flow in Crystals*, Oxford, New York, (1953).]

as Gb^3 [from (22.3.6)] inasmuch as we have created a new length b of dislocation line. This energy is strongly affected by the separation of extended dislocations. A jog is formed by intersections at Q in (a) while two kinks are formed in (b). A jog is formed, by intersection, on the edge dislocation in (c). The step formed in (c) on the screw dislocation may be either a kink or a jog, depending upon what the orientation of the glide plane is. If the glide plane of the screw is parallel to the dashed line, then a kink is formed. If the glide plane of the screw is perpendicular to the dashed line, then a jog is formed. The jog on a screw dislocation has a Burgers vector (by necessity—conservation of Burgers vector) parallel to the main dislocation and is hence a short segment of an edge dislocation (since this line segment is perpendicular to the Burgers vector). Hence, the jog can glide up and down the dislocation, but if the screw dislocation itself glides, then nonconservative motion of the jog is necessary (unless the dislocation changes its glide plane by 90 deg, which converts the jog into a kink). There is still some question about how easily screw dislocations can change their glide planes. Certainly if they tend to be extended dislocations, then they will not do this readily; hence, if they contain jogs, they will move nonconservatively. This is shown in Figure 22.8.2. The jog is a pure edge component. The extra plane of atoms is shown shaded. As the *entire* dislocation line moves forward (in the x direction) one repeat distance it will be necessary to add atoms at the bottom of the extra plane and hence vacancies will have to be created.

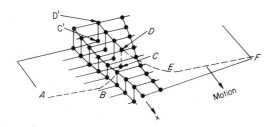

FIGURE 22.8.2. Motion of a jogged screw dislocation. Dislocation shown by dashed line ABCDEF.

An alternative is that the jog is pinned and point defect formation does not take place. Then as the rest of the dislocation line moves in the x direction, two new parallel nearby dislocation lines of opposite sign (lying in the x direction) are created. This is called a **dislocation dipole**. In either case the motion is nonconservative. We can conclude that the jogs on screw dislocations are impediments to glide motion.

If a jog is formed on a dislocation line which is part edge (edge component $b \sin \delta$) and part screw (screw component $b \cos \delta$) then the jog can move conservatively at a velocity $v \csc \delta$ if the dislocation line itself glides at velocity v. Clearly, when δ is small (when the dislocation is primarily screw) the jog glide velocity must be many times higher than v. If v approaches the speed of sound, the jog must then by necessity move nonconservatively. Figure 22.8.3 illustrates the debris left behind moving jogged screw dislocations.

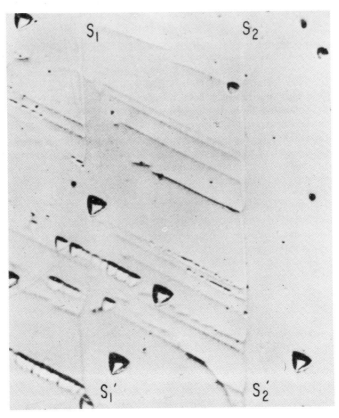

FIGURE 22.8.3. Debris left by moving screw dislocations SS' in silicon. (*Courtesy of W. C. Dash.*)

22.9   MULTIPLICATION

We have already noted that single crystals could be grown with only 10 cm of length of dislocation line per cubic centimeter, but that this dislocation density increases rapidly during deformation, reaching 10^{11} cm/cm^3 in a highly cold worked crystal. Processes whereby such multiplication occurs are discussed in this section.

FRANK-READ SOURCE.   Figure 22.9.1 illustrates the **Frank-Read multiplication mechanism**. The details of the mechanism are described in the legend. The

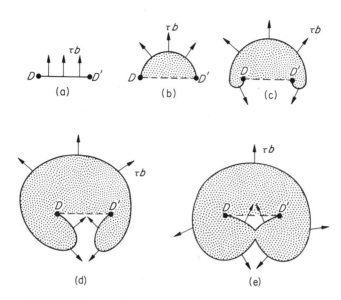

FIGURE 22.9.1. Frank-Read source. The plane of the figure is the slip plane of dislocation DD'; the dislocation leaves the plane of the figure at the fixed points D and D'. An applied stress produces a glide force τb on the dislocation and makes it bulge. The initially straight dislocation (a) acquires a curvature proportional to τ. If τ is increased beyond a critical value corresponding to position (b), where the curvature is a maximum, the dislocation becomes unstable and expands indefinitely. The expanding loop doubles back on itself, (c) and (d). Unit slip occurs in the (shaded) area swept out by the bulging loop. In (e) the two parts of the slipped area have joined; now there is a closed loop of dislocation and the section DD' is ready to bulge again and give off another loop. [From W. T. Read, *Dislocation in Crystals* McGraw-Hill, New York (1953).]

closed loop may expand until it passes out of the crystal in which case we can get unlimited slip from a single dislocation, or it may eventually become blocked by obstacles on the glide plane such as precipitates, grain boundaries, etc., so that we obtain a pileup of n loops where n depends on the shear stress τ. Theory suggests that the number of such loops is

$$n = e^{2r\tau/Gb}, \tag{22.9.1}$$

where r is the radius of the outer loop. Figure 22.9.2 illustrates a Frank-Read source in silicon.

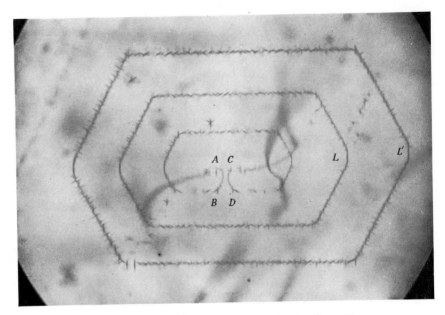

FIGURE 22.9.2. Frank-Read source in action in silicon. The
dislocations are decorated with copper precipitates. [W. D. Dash, *J. Appl. Phys.* **27**, 1193
(1956).]

CROSS SLIP.   Inasmuch as the formation of the screw dislocation does not
involve the addition or subtraction of atoms, it is clear that such a dislocation
is not geometrically restricted to a single glide plane as is the edge dislocation. (However, it may be energetically difficult to change glide planes, particularly if the dislocation tends to be spread out on the slip plane and form
extended dislocations. This occurs particularly in fcc crystals with low stacking
fault energies.) If a portion of the screw dislocation does change glide planes we
call the process **cross-slip**. If soon after it makes a second change to a third
glide plane which is parallel to the first we call the process **double-cross-slip**.
In metals such as austenitic stainless steel which has a low stacking fault energy
(≈ 13 ergs/cm^2) there is little cross slip. The reason for this is that the dislocations are dissociated into two partials with a stacking fault in between; to have
cross slip it is necessary first to push the partials together over some region;
then the total dislocation formed in this region can change slip planes. In
aluminum, which has a very high stacking fault energy so that very little dissociation occurs, cross-slip occurs very readily.

Dipole formation (discussed in Section 22.8), cross-slip, and Frank-Read
source operations are some of the possible mechanisms whereby the dislocation
density in a crystal is increased by mechanical deformation.

22.10 DISLOCATION VELOCITY

We have already noted that a nonzero stress is needed to move a single dislocation (arbitrarily slowly) through an otherwise perfect crystal (the Peierls-Nabarro stress). If a straight dislocation moves at a finite velocity, it will experience a viscous drag. There are many mechanisms possible for such drag. Only one of these will be discussed here. We have already noted that there is effectively a large hydrostatic stress component associated with the core. When the core moves rapidly to a new position, the core material will be rapidly compressed and its temperature will rise. Heat will tend to flow from this hot region to the cold. Heat flow is a dissipative process; hence, the motion is damped. This is called **thermoelastic damping**.

(Of course, if the dislocation moves in an otherwise imperfect crystal, it may experience additional drag due to interaction with impurity atoms, interaction with precipitates, interactions with other dislocations, etc. If the dislocation is a jogged screw dislocation, the drag due to its nonconservative motion will also be appreciable.) Recall from Section 22.1 that the strain rate is propor-

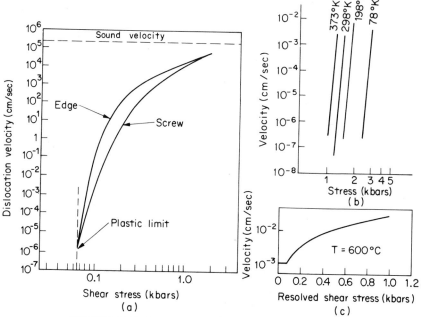

FIGURE 22.10.1. Dislocation velocity vs applied stress for: (a) LiF [W. G. Johnston and J. J. Gilman, *J. Appl. Phys.* **30**, 129 (1959)], (b) 3% Si-Fe [D. F. Stein and J. R. Low, Jr., *J. Appl. Phys.* **31**, 362 (1960)], (c) Ge [A. R. Chaudhuri, J. R. Patel and L. G. Rubin, *J. Appl. Phys.* **33**, 2736 (1962)].

tional to the dislocation velocity. Without going into experimental details, we present some data in Figure 22.10.1 for single crystals of an alkali halide, an alloy, and a semiconductor.

The effect of changing the "hardness" of a crystal by various means, including temperature changes, addition of impurities, and irradiation, is generally to shift the dislocation velocity-applied stress curve along the applied stress axis. This is evident in Figure 22.10.1(b) for 3% Si-Fe examined at various temperatures; similar results are shown for LiF in Figure 22.10.2.

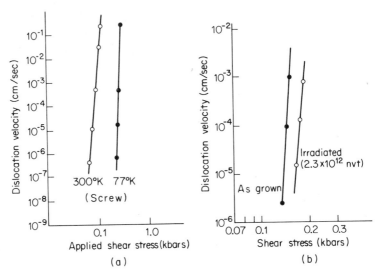

FIGURE 22.10.2. Dislocation velocity versus applied stress for LiF crystals. (a) Effect of temperature; (b) Effect of radiation damage. [W. G. Johnston and J. J. Gilman, *J. Appl. Phys.* **30**, 129 (1959).]

22.11   THE FORCES BETWEEN DISLOCATIONS

In Section 22.5 we noted that there was a force on a dislocation line due to the presence of an applied stress field. If this stress field is due to the presence of a second dislocation, then we say that the second dislocation exerts a force on the first. Two examples involving parallel dislocations will be discussed.

THE FORCE BETWEEN PARALLEL SCREW DISLOCATIONS. Consider two parallel screw dislocations of the same sign as in Figure 22.11.1. The stress

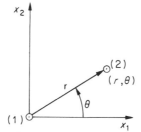

FIGURE 22.11.1. Parallel screw
dislocations of
the same sign.

field of the first screw dislocation (at the origin) is

$$\sigma_{z\theta} = \frac{Gb_1}{2\pi r}.$$

This stress field is the external stress field acting on dislocation (2). Thus the force
acting on dislocation (2) is

$$F_r = \sigma_{z\theta}b_2 = \frac{Gb_1 b_2}{2\pi r}. \tag{22.11.1}$$

Since the stress field $\sigma_{z\theta}$ is due to the presence of dislocation (1), F_r is the force
on dislocation (2) owing to dislocation (1) (and vice versa, although we have not
proved that).

Thus two parallel screw dislocations of the same sign repel each other
in proportion to the inverse distance between them.

EXAMPLE 22.11.1

Suppose the two screw dislocations of Figure 22.11.1 were pushed
together to form a dislocation of the Burgers vector $2b$. Discuss from an energy
viewpoint why they would tend to dissociate.

Answer. From (22.3.6) the energy is seen to be proportional to the
square of the Burgers vector:

One of $2b$, $U_I \propto (2b)^2$
Two of b, $U_{II} \propto 2(b)^2$.

Since $U_{II} < U_I$ the dissociated pair would be preferred.

EXAMPLE 22.11.2

Suppose the two screw dislocations of Figure 22.11.1 are of opposite
sign. What is the force then?

Answer. Attractive and of the same magnitude as in (22.11.1). If
these two dislocations come together the resultant Burgers vector is zero.
The material is perfect and the dislocations have annihilated each other.
This is called **dislocation annihilation**.

PARALLEL EDGE DISLOCATIONS.   Consider two edge dislocations as shown in Figure 22.11.2. The stress field owing to dislocation (1) located at the origin is given by Equations 22.2.3–22.2.5.

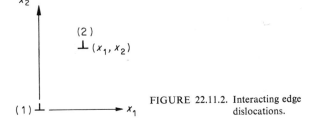

FIGURE 22.11.2. Interacting edge dislocations.

Dislocation (1) exerts a glide force per unit length on dislocation (2) in the x_1 direction given by

$$F_1 = \sigma_{12}b_2 = \frac{Gb_1b_2}{2\pi(1-v)} \frac{x_1(x_1^2 - x_2^2)}{(x_1^2 + x_2^2)^2}. \qquad (22.11.2)$$

This is a *glide force acting on the dislocation.*

Dislocation (1) also exerts a climb force per unit length on dislocation (2) in the x_2 direction given by

$$F_2 = -\sigma_{11}b_2 = \frac{Gb_1b_2}{2\pi(1-v)} \frac{x_2(3x_1^2 + x_2^2)}{(x_1^2 + x_2^2)^2}. \qquad (22.11.3)$$

This is a *climb force.* Note that *glide* takes place without adding or subtracting atoms from the extra half-plane while *climb* can occur only upon the addition or subtraction of atoms from the extra half-plane. We say that glide involves **conservative motion** while climb involves **nonconservative motion**. We note that two dislocations of like sign as considered above exert no glide force on each other when the position of dislocation (2) is $(0, x_2)$ (i.e., when they are directly above each other). Moreover, this is a stable (not a metastable) position.

EXAMPLE 22.11.3

Show that the force between two parallel edge dislocations with the same glide plane is a central force.

Answer. Equation (22.11.2) with $x_2 = 0$ gives

$$F_1 = \frac{Gb_1b_2}{2\pi(1-v)} \frac{1}{x_1}; \qquad (22.11.4)$$

i.e., the force depends only upon the distance between the pairs. $F_2 = 0$.

DISLOCATION-DISLOCATION INTERACTIONS.   It can be shown that the force that one dislocation exerts on a second is the same as that which the second exerts on the first. This, however, need not be true of the torques exerted by two dislocations upon each other.

22.12 ARRAYS OF DISLOCATIONS

There are certain arrays of dislocations for which it is possible to describe the energy of the system and the forces on the individual dislocations with comparative ease.

PILEUPS. Suppose that parallel edge dislocations lie on the xz plane with $b = [-b, 0, 0]$. Let one of these be locked at $x = 0$. See the **pileup** in Figure 22.12.1. Let there be an externally applied load which results in a uniform applied shear stress $\sigma_{xy} = -\sigma$.

FIGURE 22.12.1. Pileup.

By Equations 22.2.5 and 22.2.6 the shear stress components σ_{xy} at x owing to the ith dislocation which is located at x_i is

$$\sigma_i = \frac{D_i}{x - x_i}, \tag{22.12.1}$$

where

$$D_i = \frac{Gb_i}{2\pi(1 - v)}.$$

The glide force per unit length on the jth dislocation located at x_j and having Burgers vector b_j owing to the ith dislocation *and* the applied stress is

$$F_j = (\sigma_i - \sigma)b_j, \quad i \neq j. \tag{22.12.2}$$

Now the $n - 1$ free dislocations will move until the net force on each of them vanishes, i.e., until

$$\sum_{\substack{i=0 \\ i \neq j}}^{n-1} \sigma_i b_j - \sigma b_j = 0, \quad j = 1, \ldots, n - 1. \tag{22.12.3}$$

In the problem at hand all the b_j are equal to b. Hence the above condition becomes

$$\sum_{\substack{i=0 \\ i \neq j}}^{n-1} \frac{D_i}{x_j - x_i} = \sigma, \quad j = 1, \ldots, n - 1. \tag{22.12.4}$$

We thus have a set of $n - 1$ equations with $n - 1$ unknowns. If n is 2, the solution is simply

$$\frac{D}{x_1 - 0} = \sigma,$$

so that one dislocation is at $x = 0$ and the other at $x = D/\sigma$. It is important to note that the shear stress at the locked dislocation is a combination of the applied stress σ and the stress owing to the dislocation at x_1, i.e.,

$$\sigma_c = -\sigma + \frac{D}{0 - x} = -2\sigma \quad \text{for } n = 2.$$

Thus a stress concentration factor of 2 exists.

EXAMPLE 22.12.1

Show that for a pileup of n dislocations the shear stress on the obstacle exhibits a stress concentration factor of n relative to the applied shear stress.

Answer. Suppose we have the $n - 1$ free dislocations in their equilibrium positions. Suppose we now move the locked dislocation by δx. Then each of the dislocations must move by δx in going to their new equilibrium position. The work done by the applied stress $(-\sigma)$ is $n(-\sigma)b\delta x$. Hence the stress on the barrier is $-n\sigma$.

Not only do dislocation pileups cause shear stress multiplication, but they also intensify the tensile stress present, as Koehler first noted. To obtain the maximum tensile stress, it is necessary to know the equilibrium position of each of the dislocations.

For larger n the direct approach becomes cumbersome although a rapidly converging relaxation method can be used conveniently with an electronic digital computer. The cases of interest are when n is quite large. Here an asymptotic solution can be obtained in a simple form. This gives the number (assumed to be a continuous function) of dislocations along the slip plane as a function of x. The length of such a pileup can be shown to be

$$L = \frac{2nD}{\sigma}. \tag{22.12.5}$$

The tensile stress at the point P in Figure 22.12.1 is given by

$$\sigma_{\theta\theta} = \sigma(L/r)^{1/2} f(\theta), \quad r \ll L. \tag{22.12.6}$$

[See A. N. Stroh, *Proceedings Royal Society* **A223**, 404 (1954).] In an elastically isotropic material $f(\theta)$ will have a maximum of $2/\sqrt{3}$ when $\theta = 70.5$ deg.

Pileups play a vital role in fracture mechanics which will be discussed in more detail later. Examples of pileups are shown in Figure 22.12.2.

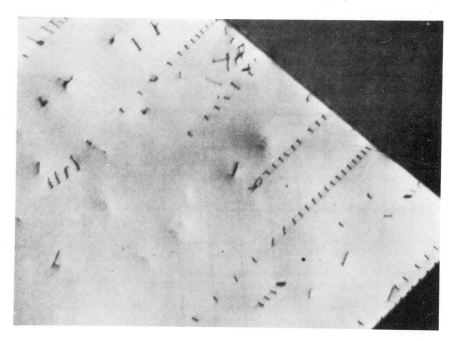

FIGURE 22.12.2. Thin foil electron micrograph of pileups in a slightly deformed 18-8 stainless steel. ($\times$ 20,000). [M. J. Whelan, P. B. Hirsch, R. W. Horne and W. Bollmann, *Proc. Roy. Soc.* **A240**, 524 (1957).]

SMALL ANGLE GRAIN BOUNDARIES. We have already noted (see Figure 13.4.2) that a small angle simple tilt boundary can be considered as a vertical array of dislocations. One of the major early successes of dislocation theory was that the energies of such boundaries could be quantitatively calculated directly from dislocation theory and were found to be in agreement with measured values (see Figure 13.4.5). A rigorous calculation following Burgers is given in A. H. Cottrell, *Dislocations and Plastic Flow in Crystals*, Oxford, New York (1956) pp. 89–98 for isotropic materials. The problem is considered in much detail in the book by W. T. Read, *Dislocations in Crystals*, McGraw-Hill, New York (1953) Chapters 11–13, where elastic anistropy is taken into account. The expression for the grain boundary energy of a simple tilt boundary of angle θ is, where A is a constant,

$$\gamma_B = \frac{Gb\theta}{4\pi(1-v)}(A - \ln \theta). \qquad (22.12.7)$$

This should hold only to about $\theta \approx 6$ deg since overlap of the cores will certainly occur around $\theta \approx 12$ deg and the equation ignores such overlap.

22.13 INTERACTION OF IMPURITY ATOMS WITH DISLOCATIONS

It is possible by considering size effects alone to calculate approximately the interaction between an impurity atom and an edge dislocation. Consider a solute atom located at the point P in Figure 22.13.1. If we consider elastic interactions only, then it is necessary to take into account the fact that the solute atom has a larger (or smaller) radius than the atom which is replaced; moreover,

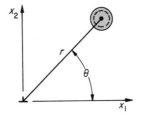

FIGURE 22.13.1. Solute atom-dislocation interaction.

it has a different elastic modulus. (It is quite common to consider the atoms as elastic spheres and to ascribe to this sphere the same elastic behavior possessed by a solid aggregate of such atoms. This is a type of approximation one is driven to when there is no better theory available.) If, moreover, the impurity atom is near to the dislocation, it would be necessary to take into account the variation of the stress field (due to the dislocation) across the solute atom. Great simplifications result if we assume that the impurity atom is spherical and remains spherical. Thus the radial strain is given by

$$\epsilon = \frac{\rho' - \rho}{\rho}, \tag{22.13.1}$$

where ρ' is the radius of the solute atom and ρ the radius of the solvent atom.

The change in energy for the case where the elastic modulus of the solute atom is the same as that of the solvent and where the strain energy in the solute atom is taken into account is

$$U_{\mathrm{I}} = 4Gb\epsilon\rho^3 \frac{\sin\theta}{r}. \tag{22.13.2}$$

EXAMPLE 22.13.1

According to (22.13.2) where would a substitutional atom which is smaller than the solvent atom go?

Answer. In this case ϵ is negative so to make U_{I} a minimum we choose $\theta = 90$ deg. The solute atom would go to a point near to the bottom of the extra plane of inserted atoms in Figure 22.2.5. This appears obvious without equations since the smaller atom would go to a region of high compressive stress.

In the case of interstitial atoms such as carbon in iron considerably stronger binding results. Here the maximum binding energy (interaction energy) is about 0.75 eV/carbon atom. [See, e.g., A. W. Cochardt, G. Schoek, and H. Wiedersich, *Acta Metallurgica*, **3**, 533 (1955).] The details of the tetragonal distortions caused by carbon atoms in the bcc lattice will not be given here. The student has noted earlier how this causes martensite formation. Thus impurity atoms which will be randomly distributed in good lattice will tend to segregate at dislocations to decrease the overall energy of the system. It will be necessary to overcome this binding energy when the dislocation attempts to move.

REFERENCES

Weertman, J., and Weertman, Julia R., *Elementary Dislocation Theory*, Macmillan, New York (1964).

Hull, D., *Introduction to Dislocations*, Pergamon Press, New York (1965).

Read, W. T., *Dislocations in Crystals*, McGraw-Hill, New York (1953).

Cottrell, A. H., *Dislocations and Plastic Flow in Crystals*, Oxford, New York (1953).

Hirth, J. P., and Lothe, J., *Theory of Dislocations*, McGraw-Hill, New York (1968). A somewhat more sophisticated discussion than in the other books listed here.

Nabarro, F. R. N., *Theory of Dislocations*, Oxford, New York (1967).

PROBLEMS

22.1. Discuss four of the ways by which edge dislocations can be observed.

22.2. How are the unit tangent vector and the Burgers vector related for the screw and edge dislocations?

22.3. Derive the stress field of a screw dislocation in an elastically isotropic media.

22.4. a. What is the volume change of a crystal for a screw dislocation in an isotropic media according to linear elasticity theory?
  b. An edge dislocation?
  c. Either an edge or screw in a real crystal?

22.5. a. How is the direction of glide motion of an edge dislocation related to the unit tangent vector to the dislocation line?
  b. A screw dislocation?

22.6. Using a plus sign to designate a positive stress component and a minus sign for a negative stress component, for an edge dislocation in an isotropic crystal:

a. Make a sketch showing the signs of σ_{11} in the four quadrants of the $x_1 x_2$ plane.

b. Same for σ_{22}.

c. Same for σ_{12}.

22.7. a. Discuss two methods by which the edge dislocation could be imagined to be made.

b. What does this suggest about the possible motion of an edge dislocation?

22.8. a. An edge dislocation in aluminum is 1 cm long. Estimate its self-energy and compare this with the energy of a vacancy in aluminum.

b. Explain why a dislocation cannot be an equilibrium defect while a vacancy can be.

22.9. Explain why the Peierls-Nabarro stress is low for copper and high for germanium.

22.10. What is meant by the Mott-Nabarro force on a dislocation line?

22.11. a. Which external stress component(s) must be nonzero in order that there be a glide force on the edge dislocation of Figure 22.2.3?

b. Climb force?

22.12. Two parallel edge dislocations lie on the same plane with the same Burgers vector. Show that there is a central force acting between them.

22.13. a. Plot F_1 vs. x_1 for the dislocation (2) of Figure 22.11.2 assuming $x_2 > 0$ and is fixed.

b. Tell where the stable position is and also the metastable position.

22.14. a. What is meant by a pileup of edge dislocations?

b. Why are pileups important?

22.15. Show that a pileup causes a shear stress concentration factor of n, where n is the number of dislocations in the pileup.

22.16. One of the important equations of dislocation theory relates the shear strain to the number of dislocations and the distance they move. Derive this relationship.

22.17. What simple relationship can be used to decide whether or not a dislocation will dissociate?

22.18. a. What is a sessile dislocation?

b. Give an example.

c. Can sessile dislocations be formed by vacancy condensation?

22.19. a. Discuss the formation of partial dislocations in fcc crystals.

b. Why is the formation of such dislocations important?

22.20. a. Describe the process of dislocation climb.

b. Why is climb important?

22.21. Discuss the concept of the local equilibrium concentration of vacancies around a stressed edge dislocation line.

22.22. If a screw dislocation line becomes jogged by intersection, why must further motion of the jog be nonconservative, assuming that the direction of motion of the screw remains unchanged?

22.23. a. Discuss the binding of impurity atoms to dislocation lines.
b. Why does this make the motion of dislocations more difficult?

22.24. a. Describe the operation of the Frank-Read source.
b. What other mechanisms of dislocation multiplication exist?

22.25. a. In general one might say that the dislocation velocity in a stressed crystal depends strongly (weakly) on stress?
b. What is the upper velocity limit of a dislocation line?

More Involved Problems

22.26. Discuss the use of X-ray topography in studying crystals having a low density of dislocations [R. E. Smallman and K. H. G. Ashbee, *Modern Metallography*, Pergamon Press, New York (1966)].

22.27. Derive the self-energy of a screw dislocation using a different technique than used in the text.

22.28. Write a computer program for evaluating the positions of $n + 1$ edge dislocations in a pileup in which one of the dislocations is locked at $x = 0$.

22.29. Describe two different techniques for evaluating the positions in Problem 22.28 for $n = 2, 3, \ldots, 10$. Do not actually carry out the evaluation.

22.30. Show that the energy of a simple tilt boundary in an elastically isotropic media has the form

$$\gamma_{GB} = \frac{Gb}{4\pi(1 - \nu)}\theta(A - \ln\theta),$$

where θ is the angle of tilt. For a heuristic derivation, see W. T. Read Jr., *Dislocations in Crystals*, McGraw-Hill, New York (1953) p. 160, and for an exact derivation see A. H. Cottrell, *Dislocations and Plastic Flow in Crystals*, Oxford, New York (1956) p. 89.

22.31. Discuss recent work from the literature on grain boundary energy versus angle dependence in simple boundaries. (See *Acta Metallurgica* or *Transactions of the AIME*.)

Sophisticated Problems

22.32. Derive the stress field for an edge dislocation in a linear isotropic solid. Before working this problem, the student should understand the definitions of strain in terms of the displacement and the compatibility equations for plane strain which he can get from a text on elasticity. He can then proceed to p. 32 of *Dislocation and Plastic Flow of Crystals* by A. H. Cottrell, Oxford, New York (1956).

22.33. Derive the Peach-Koehler equation.

22.34. Although the force that one dislocation exerts on a second must be equal to the force that the second exerts on the first, this is not always true of the torques. Prove this.

Prologue

The fundamental concepts of dislocation theory developed in Chapter 22 can be used to explain, at least in a qualitative fashion, many of the mechanical properties of materials. *Yield point phenomena* in steel are discussed in this chapter. *Strain aging* is explained by the interactions of carbon with dislocations. Processes of *recovery* from cold work are explained in terms of dislocation annihilation and rearrangement. The process of *recrystallization* of a highly cold worked material is described in terms of a nucleation process in which the driving force is provided by the stored energy of the dislocations. Dislocation concepts are also used to explain the effect of temperature on yield strength, and to explain creep and fracture of brittle crystals. The concepts of Weertman and Dorn are developed in the study of high temperature creep. *Superplasticity* is discussed. The theoretical fracture stress of perfect crystals is described. Methods of crack nucleation are discussed. The concept of crack growth is described in terms of the *Griffith model* which is applicable not only to brittle crystals but also to brittle amorphous materials such as window glass. *Ductile-to-brittle transition temperatures* and *fatigue* are also discussed.

23

ON UNDERSTANDING DEFORMATION
AND ANNEALING IN CRYSTALLINE
MATERIALS

23.1 INTRODUCTION

Figure 23.1.1 shows dislocation tangles in polycrystalline copper. The tangles in this case tend to concentrate in **cell walls**. Often at low temperatures there is some fluctuation in dislocation density without any obvious **cell formation**. The tendency for cell formation to occur increases as the temperature increases, as we shall see later. The cell walls also become narrower and better defined at higher temperatures.

An annealed copper crystal might contain 10^6–10^7 dislocations/cm^2. A cold-worked copper crystal may contain 10^9–10^{11} (and possibly more) dislocations/cm^2, depending on the degree of cold work. The complete description of the deformation of such a solid in terms of dislocation mechanics would require the calculation of the forces and torques acting on each segment of each dislocation at every instant, plus the calculation of the dynamics of the motion of each segment of these dislocations. This problem is impossible to solve. Therefore simplified models are used which illustrate the general features of the macroscopic model. It is the purpose of this chapter to examine several properties in terms of simplified models so that we have at least a qualitative understanding of the deformation processes in crystals. We note that dislocations, grain boundaries, etc., are nonequilibrium defects and so we will be concerned with various processes by which these defects are removed from the crystal including recrystallization of cold-worked specimens, grain growth, etc., all of which lower the stored energy of cold work.

FIGURE 23.1.1. Tangles of dislocations in a copper alloy containing 2.6 vol. % SiO_2 examined by thin foil electron microscopy after 34% deformation at 77°K. The SiO_2 particles which can be seen in the cell walls are probably responsible for cell formation in this case. [A. Kelly, *Strong Solids*, Oxford University Press, New York (1966).]

23.2 YIELD POINTS

Figure 23.2.1 shows the **yield point phenomena** in an iron single crystal. (The drop in yield strength after yielding is a characteristic of many impure bcc crystals. It is also a characteristic of LiF, Si, and Ge as well as many other compounds and elements.) The deformation that occurs during the yield point extension is heterogeneous; i.e., the strain varies from position to position. At the **upper yield point** (see Figure 23.2.1), a discrete band of deformed metal (the **Lüders band**) appears (see Figure 23.2.2). (Sometimes several appear simultaneously; we consider here the case where only one occurs.) The band then propagates along the length of the specimen. The appearance and the propagation of the Lüders band can often be directly observed visually. The use of stresscoat or Photostress with polarized light will make observation simple. When these bands have propagated over the entire test section at the stress corresponding to the **lower yield point**, strain hardening will begin. In the specimen of Figure 23.2.1 the yield point elongation is about 5%. The yield point phenomena are

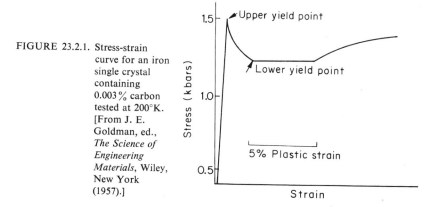

FIGURE 23.2.1. Stress-strain curve for an iron single crystal containing 0.003 % carbon tested at 200°K. [From J. E. Goldman, ed., *The Science of Engineering Materials*, Wiley, New York (1957).]

of considerable practical importance in deep drawing operations (automobile fenders) where stretcher strains are formed; here it is a very undesirable effect which is to be avoided if at all possible.

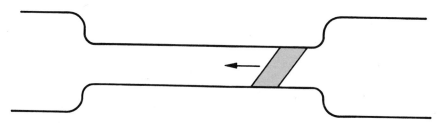

FIGURE 23.2.2. Lüders band propagation.

One of the early successes of dislocation theory was the explanation of this phenomenon. As shown experimentally [J. R. Low and M. Gensamer, *Transactions of AIME*, **158**, 207 (1944)], the effect disappears if essentially all the carbon or nitrogen is removed from the steel. We have noted earlier that impurity atoms interact with dislocations and that carbon or nitrogen interstitial atoms in bcc crystals produce tetragonal distortions and hence interact strongly with both edge and screw dislocations. There is direct experimental evidence for this interaction. It has been shown [A. H. Cottrell and A. T. Churchman, *Journal of the Iron and Steel Institute*, **162**, 271 (1949)] that 10^{-3}–10^{-2} wt. % carbon or nitrogen is removed from random solution as a result of moderate cold working of iron. More recent quantitative evidence in steel containing nitrogen has been given by H. A. Wriedt and L. S. Darken, *Transactions of AIME* **223**, 122 (1965). In general we can say that a high concentration of impurity atoms segregates along the dislocations, say three carbon atoms per atomic length of dislocation line; these regions of high impurity content which surround the dislocation are known as **atmospheres**.

It was first suggested by Nabarro that the yield point phenomenon was due to the dislocations tearing away from their atmospheres. The problem was treated in more detail by A. H. Cottrell and B. A. Bilby, *Proceedings of the Physical Society (London)*, **A62**, 49 (1949) and later by others. A more likely mechanism suggested by Hahn is that initially there are only a few dislocations free to move and these multiply rapidly. Since the strain rate involves the product of the dislocation density and their velocity, a lower stress will be needed (see Section 22.10) to maintain a given strain rate if the dislocation density is higher. (Tension tests are usually carried out at nearly constant strain rate.) Hence, the stress drops off as the dislocations multiply. The yield point phenomena are closely related to the phenomena of **strain aging**. If in the specimen of Figure 23.2.1 we had stopped the test at, say, 3% strain and waited for a considerable time and then reloaded, we would find that the new yield stress would be the upper yield stress; this is called strain aging. During the waiting period the carbon was able to diffuse to the dislocations at their new sites and create a new atmosphere. The rate at which this process occurs is therefore diffusion-controlled.

EXAMPLE 23.2.1

Iron containing an atom fraction of 5×10^{-4} carbon is cold-worked until its dislocation density is 2.5×10^9 dislocations/cm²; this corresponds to a 3% strain. Estimate the time required for the return of the upper yield strength at room temperature.

Answer. A cylinder of one atom length containing $1/(5 \times 10^{-4}) = 2000$ iron atoms would contain one carbon atom, while one with an area of 6000 would contain three carbon atoms. This is an area of about 24,000 Å², with a radius of $\sqrt{24,000/\pi}$ Å (this is the distance the carbon atoms have to move). The carbon jump distance is about 2 Å and the jump time is about 1 sec, so (using $x \approx \sqrt{N}l$ for the total displacement) we have for the number of jumps $N \approx 2000$ and a time $t \approx 2000$ sec. See Example 18.5.1.

The Cottrell-Bilby expression for the fraction f of original solute which has segregated to dislocations in the time t is

$$f = 3\left(\frac{\pi}{2}\right)^{1/2} \rho\left(\frac{U_m D t}{kT}\right)^{2/3}. \qquad (23.2.1)$$

This expression takes into account the fact that the carbon flux is affected by the stress field gradient around the dislocation. Here p is the dislocation density, U_m is the maximum binding energy, and D is the diffusion coefficient for carbon diffusion in iron. This expression is true only for the earlier stages of aging. The fraction of carbon atoms remaining in solution can be measured by internal friction techniques. By studying the temperature dependence of the rate of strain aging, the activation energy for diffusion of carbon in iron can be obtained. The value obtained is 20,000 cal/mole [S. Harper, *Physical Review*, **709**, 83 (1951)]. This agrees well with the value given by Wert for self-diffusion (see Chapter 18). The time dependence also agrees with that predicted by Equation (23.2.1). Thus in this case both the correct time dependence and the correct temperature dependence are obtained. There are certain factors which can be estimated only and so appear in the equation as parameters; these are U_m and p. The values obtained for them are, however, consistent with various other theoretical and experimental estimates.

23.3   FLOW STRESS AND DISLOCATION DENSITIES

There is a fairly general empirical relationship between the total dislocation density p and flow stress σ of the form

$$\sigma = \sigma_0 + AGb\sqrt{p}, \tag{23.3.1}$$

where σ_0 and A are constants, G is the shear modulus, and b is the magnitude of the Burgers vector of the total dislocation. This relationship seems to be applicable, more or less, not only for room temperature situations [as has been accurately shown by J. D. Livingston, *Acta Metallurgica*, **10**, 229 (1962)] but for high temperature deformation as well, as shown in Figure 23.3.1, where the equation of the dashed line is

$$p(\text{cm/cm}^3) = 1.8 \times 10^{12} \, [\sigma(\text{bars})]^2 \tag{23.3.2}$$

(1 bar $= 10^6$ dynes/cm$^2 = 14.5$ psi). Because of the general nature of this result we present here one simple model which leads to it although there are several models which are consistent with this behavior. We consider an edge dislocation intersecting a forest of edge dislocations as shown in Figure 23.3.2. The forest, for simplicity, is considered as a square net.

If we consider a displacement by b, during which a jog is created as shown, then the work done per mesh length l is the product of $b\sigma$ (which is the force per unit length acting on the line) times l times the displacement b, or

$$(b\sigma)bl. \tag{23.3.3}$$

Now the energy of creating a jog has been shown to be equal to about Gb^3 (see Section 22.3), so we have

$$\sigma = \frac{Gb}{l}. \tag{23.3.4}$$

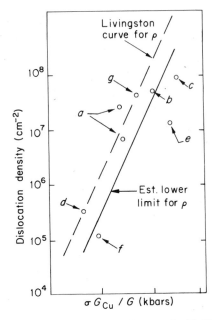

FIGURE 23.3.1. Dislocation densities vs. $\sigma G_{Cu}/G$. Here G_{Cu} is the shear modulus of copper and G is the shear modulus of the other material given in the table. The Livingston curve gives total dislocation densities ρ in copper after deformation at room temperature. The points shown indicate approximate lower limits to the dislocation densities within the subgrains, ρ, of various metals after deformation at elevated temperatures. (a) Silver deformed at 500 and 600°C; upper point, 600°C; lower point, 500°C. (b) α-iron deformed at 600°C, obtained by transmission electron microscopy. (c) Copper deformed at 500°C. (d) Aluminum deformed at 647°C. (e) Nickel deformed at 700°C. (f) Aluminum deformed at 621°C. (g) α-iron deformed at 806°C. [From R. W. Balluffi and A. L. Ruoff, *Journal of Applied Physics*, **34**, 1634 (1963).]

If we put l in terms of dislocation densities we have

$$l = \frac{1}{\sqrt{\rho}} \qquad (23.3.5)$$

so that

$$\sigma = Gb\sqrt{\rho}. \qquad (23.3.6)$$

In the present case there is no backstress acting so that $\sigma_0 = 0$ in Equation (23.3.1). In other theories involving pileups, etc., there is a backstress which

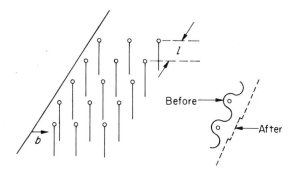

FIGURE 23.3.2. Edge dislocation cutting through a forest of
edge dislocations and forming jogs.

tends to cancel the effect of the applied stress. Other theories are referred to in
L. S. Darken's review paper [*Transactions of the American Society for Metals,* **54,**
599 (1961)]. It is also left as an exercise for the student to suggest other ways in
which work hardening occurs, i.e., ways in which the presence of dislocations
makes the motion of other dislocations more difficult.

A highly cold-worked crystalline material can contain up to about 10^{12}
dislocations/cm². It will have a stored energy (owing to cold work) equal to
about 10^9 ergs/cm³ since the energy per atomic plane of dislocation line is several
electron volts. The stored energy per unit volume is roughly equal to the total
length of dislocation lines per unit volume times the self-energy per unit length
of the dislocation line.

23.4   ANNEALING OF COLD-WORKED MATERIALS

In general, a cold-worked material can be softened by annealing. This
softening is extremely important in design considerations since such parameters
as yield strength can be altered in a major way. It is also important in processing.
For example, in wire drawing the wire is annealed after each drawing operation.
It is the purpose of this section and Section 23.5 to consider such annealing.
We might roughly classify the *recovery* from the cold-worked state by two processes: first, a stress relief or partial recovery and, second, a complete recrystallization. It is with the first of these processes that we shall be concerned here.
Recovery, in general, involves annihilation as well as rearrangement of disloca-
tions into arrays of lower energy; it is strongly temperature-dependent.

EXAMPLE 23.4.1

Describe a mechanism whereby parallel edge dislocations of opposite sign can annihilate each other.

Answer. If each climbs toward the other by the addition of atoms as shown, the result will be an extra plane of atoms extending all the way across the crystal, i.e., no dislocation at all.

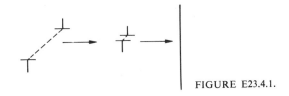

FIGURE E23.4.1.

Figure 23.4.1 illustrates the release of stored energy owing to rearrangement of the interacting dislocations and owing to their annihilation.

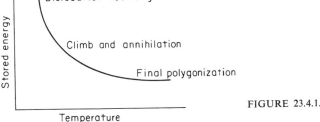

FIGURE 23.4.1. Recovery of cold work.

POLYGONIZATION.   The best understood of the recovery processes is the final arrangement of dislocations into walls. This process is called **polygonization** and is illustrated by Figure 23.4.2, which depicts the case of a bent crystal of zinc with a constant radius of curvature. We show only edge dislocations. Note that initially edge dislocations of both signs are present but that there is a net excess of positive edge dislocations. This net number stays the same during the annealing process. The curvature K is given by

$$K = \Delta\rho b, \tag{23.4.1}$$

where $\Delta\rho$ is the excess density of positive edge dislocations. (The gross curvature tensor in the case of an arbitrary distribution of dislocations of any kind can also be written down in terms of the net excess.) The explanation of polygoniza-

(a)   (b)

FIGURE 23.4.2. (a) More or less random distribution of dislocations of different sign (but with net excess)
in a cold-bent crystal. (b) After polygonization
the dislocations have formed a sub-boundary
structure by the process of rearrangement and
annihilation. Each pair of vertical arrays of
dislocations encloses a subcrystal.

tion or subgrain (polygon) formation within a grain in the above terms was first
given by R. W. Cahn, *Journal of the Institute of Metals*, **79**, 129 (1949), and was
motivated by an attempt to explain the fact that the streaks in X-ray photographs
of cold-worked specimens (which streaks are due to the planes being bent) broke
up into a series of dots on annealing (the dots arise from individual subgrains).
Following Cahn's work a number of investigators used etch pit techniques to
illustrate polygonization. Figure 23.4.3 illustrates the process. The dislocation
content of the cold-worked crystal was 3.3×10^7 cm^{-2} prior to annealing.

The actual alignment process, i.e., partial polygonization, starts below
700°C and involves glide motion of dislocations freed from barriers. It is
sometimes true that fairly extensive polygonization can occur by this process.

EXAMPLE 23.4.2

Assume that the initial dislocation density of the bent crystal of Figure
23.4.3 is 2.5×10^7 cm^{-2}. Estimate the temperature at which noticeable climb
occurs in 1 hr assuming that a vacancy has to travel the average distance
between dislocations.

Answer. We use $x^2 \approx 4Dt \approx 1/\rho$. Here we estimate that the diffusion distance is about equal to the average distance between dislocations.
We use the value of D_0 and Q_d for α-iron given in Chapter 18 and solve for
the absolute temperature T. We obtain $T \approx 1050°$K $\approx 775°$C. Experimentally, climb is first observed at 700°C.

The structure shown in Figure 23.4.3 is due to slip on a single plane only.
When multiple slip has occurred, as it necessarily must in polycrystalline aggregates, subwalls form in various directions leading to polygon-shaped subgrains. Extensive polygonization also occurs during high-temperature creep.

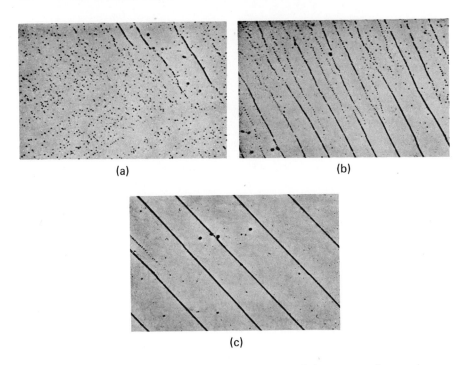

(a)   (b)

(c)

FIGURE 23.4.3. Dislocation rearrangement during recovery of cold-worked silicon iron. Slip was on the $(01\bar{1})$ plane only. Photographs taken after annealing 1 hr at (a) 700°C, (b) 875°C, and (c) 1060°C. [From W. R. Hibbard, Jr., and C. G. Dunn, *Creep and Recovery*, American Society for Metals, Cleveland (1957) p. 52.]

23.5   RECRYSTALLIZATION

Heavy cold working produces high dislocation densities $(10^{10}–10^{12}$ cm$^{-2})$. Since these are nonequilibrium defects, the free energy of the cold-worked material is considerably above the equilibrium state.

EXAMPLE 23.5.1

Calculate the energy stored in a cubic centimeter of copper in which there are 10^{11} dislocations.

Answer. The energy per atomic length of dislocation (about 2 Å) is about 8 eV (see Section 22.3). Thus the stored energy is about $8 \times 5 \times 10^7 \times 10^{11} = 4 \times 10^{19}$ eV, or about 6.4×10^7 ergs/cm^3.

EXAMPLE 23.5.2

Suppose after recrystallization that the dislocation density within the grains is only 5×10^7 cm^{-2} but the grain size is 5×10^{-3} cm. Estimate the total grain boundary energy per cubic centimeter in copper. Assume that the grains are cubes for simplicity of calculation.

Answer. The grain boundary area is about 600 cm^2/cm^3 and the grain boundary energy is about 550 ergs/cm^2 (see Chapter 13) so the energy is about 3.3×10^5 ergs/cm^3.

We note from Examples 23.5.1 and 23.5.2 that the stored energy would decrease by a factor of about 200 if the cold-worked crystal were to crystallize into a polycrystalline material. Nucleation theory (see Section 18.14) is often used (with some misgiving) to estimate the rate of appearance of new crystals, which then grow by grain boundary migration, i.e., atom motion in the grain boundary. The critical nucleus radius is given by (18.14.10)

$$r^* = \frac{-2\gamma}{\mathscr{G}},$$

where γ is the grain boundary energy and $\mathscr{G}$ is the difference in energy per unit volume between the annealed crystal and the cold-worked crystal. The probability that a nucleus of critical size exists is

$$e^{-\Delta G^*/kT},$$

where ΔG^*, the activation energy for critical nucleus formation, is given by (18.14.11). Such a particle will become supercritical if one atom jumps across the grain boundary.

The number of atoms in the grain boundary is $4\pi r^{*2}/a$, where a is the cross-sectional area per atom. The number of atoms per cubic centimeter is n. Hence the rate of nucleation is

$$I(\text{nuclei/cm}^3\text{-sec}) = \frac{4\pi r^{*2} n}{a} \Gamma e^{-\Delta G^*/kT}, \qquad (23.5.1)$$

where Γ is the jump frequency. In the present case

$$\Gamma = \nu e^{-Q_b/kT},$$

where Q_b is the activation energy for grain boundary diffusion.

The process of recrystallization is a nucleation process followed by a growth process both of which are strongly temperature-dependent as well as strongly dependent on the amount of cold work.

EXAMPLE 23.5.3

Does the nucleation rate at a given temperature increase or decrease with the severity of cold work?

Answer. The stored energy increases and hence ΔG^* decreases [see (18.14.11)] and I clearly increases.

To obtain nucleation rates in agreement with experiment, it is necessary that r^* be very small (of the order of several Ångstroms); this requires a high stored energy density $\mathscr{G}$. We note that the dislocation distribution in cold-worked materials is highly nonuniform. Hence there are tiny regions where the local stored energy density must be very much higher than the average.

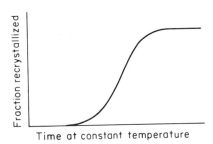

FIGURE 23.5.1. Recrystallization kinetics.

The nature of recrystallization is illustrated in Figure 23.5.1. Because of the very strong temperature dependence of the recrystallization rate, it is possible to speak of a recrystallization temperature of a given cold-worked material. Such a temperature is often near $T_M/3$, although it depends on the amount of cold work, on the purity, etc.

23.6   GRAIN GROWTH

When a strain-free polycrystalline material is heated to a temperature sufficient for atomic diffusion to occur at a "reasonable rate," grain growth is observed to occur through the shrinking and eventual disappearance of some grains and the simultaneous growth, at comparable rates, of the remaining grains. The thermodynamic driving force for grain growth is the decrease in total grain boundary surface energy that results from grain growth; i.e., grain growth is accompanied by decreasing grain boundary area and hence decreasing total free energy. Because grain growth (like recrystallization and recovery) creates a thermodynamically more stable condition, it is not a reversible process; i.e., "grain shrinkage" does not appear upon cooling.

For a given grain, the curvature of the grain boundary determines the direction of movement of the boundary. Consider the "triple point" intersection

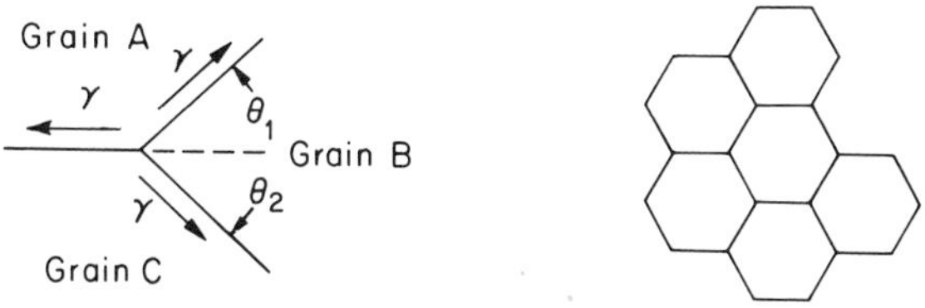

Grain A

Grain B

Grain C

(a) Triple point

(b) Kinetically
stable structure

FIGURE 23.6.1. Grain boundaries.

of three grain boundaries in two dimensions given in Figure 23.6.1. The surface energies are represented as surface tensions, as was done in Chapter 13. For the junction to be stable, i.e., the sum of the surface tensions to be zero, simple algebra demands that $\theta_1 = \theta_2 = 60$ deg. Because of this, and again using the two-dimensional case as an example, the idealized hexagonal structure in (b) is kinetically stable, as was discussed in Section 15.3. In this case if grain growth were to occur, there would be an increase in surface area and energy before there would be a decrease. Hence major fluctuation would be needed to get grain growth.

Real structures are not geometrically ideal, however, and will consist of grains of varying sizes and number of sides. Consider an eight-sided grain with curved boundaries meeting in 120-deg triple junctions as in Figure 23.6.2.

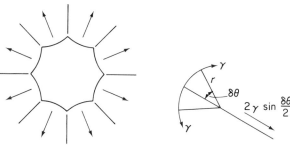

FIGURE 23.6.2. Nonequilibrium
grain shape.

(a) Eight-sided grain with
120° junctions

(b) Resulting unbalanced
forces

From geometry it is evident that a net unbalanced force of $2\gamma \sin (\delta\theta/2)$ acts upon an increment of grain boundary length and in the direction of the center of curvature. Here $\delta\theta$ is an increment of angle θ. This grain will then tend to grow, with the triple point junctions maintained at 120 deg. By a similar argument it can be shown that, in two dimensions, a grain with less than six sides will tend to shrink. A schematic view of grain disappearance is given in Figure 23.6.3.

A crude kinetic model for grain growth can be constructed in the following

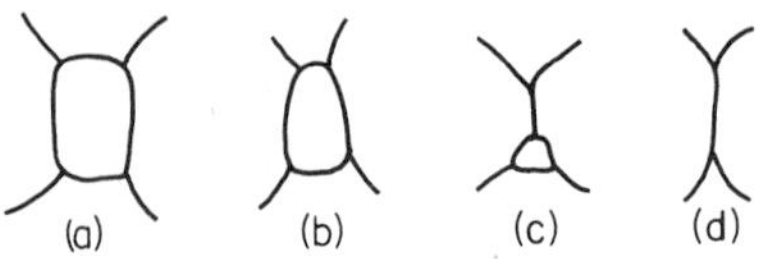

FIGURE 23.6.3. Elimination of four- and three-sided grains during growth of neighboring grains.

(a)   (b)   (c)   (d)

manner. If it is assumed that the grain growth rate is proportional to the total grain boundary energy per unit grain volume, then

$$\frac{dD}{dt} \propto \frac{\gamma D^2}{D^3}, \qquad \begin{aligned} D &= \text{grain diameter,} \\ t &= \text{time,} \end{aligned} \qquad (23.6.1)$$

and upon integrating,

$$D^2 - D_0^2 = kt. \qquad (23.6.2)$$

The constant k contains a host of parameters, including temperature, geometry, grain boundary energy, and grain boundary diffusion coefficient. Because of this the equation is obeyed only occasionally, and it is customary to use the following empirical generalization to represent data:

$$D^{1/n} - D_0^{1/n} = ct. \qquad (23.6.3)$$

Observed values of n range from 0.1 to 0.5.

EXAMPLE 23.6.1

Why is grain size important?

Answer. Certain physical quantities such as yield strength and fracture strength are strongly related to grain size. Grain boundaries are often the nucleating sites (heterogeneous transformation) for phase transformations. For example, the γ-grain size in steel affects the pearlite transformation and hence plays a role in the determination of whether steel forms martensite or pearlite when it is cooled.

CONTROLLING GRAIN GROWTH. Zener in 1948 showed theoretically that grain growth in a fine grained structure could be inhibited by the presence of a finely dispersed second phase. A number of such systems have now been made. There is currently a considerable amount of research underway on developing methods for obtaining such structures. These structures apparently are necessary in metals if they are to exhibit superplasticity (see Section 23.8).

23.7   YIELD STRENGTH VERSUS TEMPERATURE

Ordinarily the yield strength of crystalline materials decreases slowly with temperature from near absolute zero to about $T_M/4$ or $T_M/3$, at which tem-

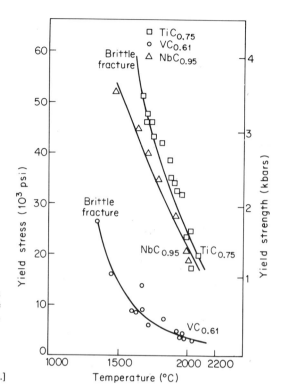

FIGURE 23.7.1. Temperature dependence of the yield strength of polycrystalline metallic carbides. [From D. J. Rowcliffe, Ph.D. thesis, Cambridge University (1965).]

perature it begins to decrease rapidly. Figure 23.7.1 illustrates the behavior of yield strength with temperatures.

In the example shown here brittle fracture rather than flow takes place below a certain temperature. (The melting point of NbC is 4043°K.) Figure 23.7.2 shows similar results for metals. We note that the yield strength of the bcc metals drops at about $0.1 T/T_M$. There is another large drop off in these materials at $0.5 T/T_M$. The fcc metals (Cu and Ni) show a slowly decreasing yield strength; however, they exhibit a rapid drop near $0.5 T/T_M$ (not shown).

There are several mechanisms which impede the motion of dislocations in addition to cold working. These various strengthening mechanisms will be discussed in Chapter 25. In all cases the motion of dislocations becomes easier at higher temperatures, because dislocation pinning is less effective, because the impediments are broken down as with recovery, annihilation, etc., because various nonconservative processes, such as dislocation climb, can readily occur, or because additional slip planes become active.

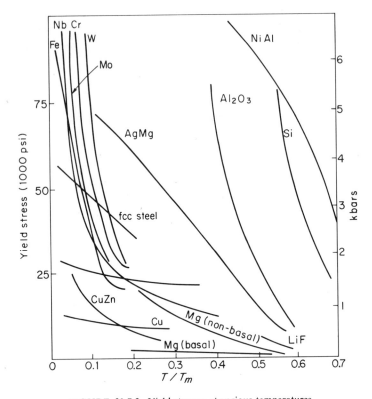

FIGURE 23.7.2. Yield stresses at various temperatures.

23.8   HIGH-TEMPERATURE CREEP

Suppose we apply a load which gives a stress considerably below the yield strength of a specimen. At low temperatures the specimen is deformed only elastically. However, at high temperatures we find that the specimen deforms as a function of time; i.e., it creeps. Moreover, the effect depends strongly on temperature.

The dislocations must be locked against barriers; otherwise they would glide under the applied stress and cause deformation. The barriers might be particles of a second phase, sessile dislocations, dislocations on parallel glide planes, etc. At high temperatures there must be some mechanism which enables the dislocations to free themselves from the barriers; clearly they will eventually come up against another barrier from which they will again have to break free. Figure 23.8.1 illustrates how vacancy motion to a dislocation with resultant climb makes this possible. In this case it is necessary for the edge dislocation to move upward by five glide planes at which time it is free to pass the barrier by

(a)   (b)

FIGURE 23.8.1. Climb of dislocations over a barrier. (a) Pileup
without climb. (b) Climb of dislocations. The
first dislocation has climbed upward and glided
past the barrier.

glide. Since the barrier is often a stress field owing to other dislocations, the
effective height of the barrier must depend on the shear stress. In the case
shown the edge dislocation changes its glide plane by the diffusion of vacancies
to the dislocation and hence by diffusion of atoms away from the dislocation.
The atoms in question are from the lower edge of the extra inserted plane of
atoms that must be present with an edge dislocation. At high temperatures it
can be assumed that vacancies are in *local equilibrium* with the dislocation
line. Strictly speaking the climb process takes place only at the jog and so the
local equilibrium would seem to hold there only. However, vacancies may
diffuse from the lattice to the dislocation pipe. (The dislocation pipe is the
small region around the dislocation where the atoms are badly misplaced.)
It is known that they can then diffuse very rapidly along the dislocation relative
to their diffusion rate in the lattice (see Section 18.9). Consequently, it is possible
to maintain the *local equilibrium* value everywhere along the dislocation. Let us
consider a simple process where diffusion occurs between two dislocations,
one of which is acting as a vacancy source and the other as a vacancy sink as in
Figure 23.8.2.

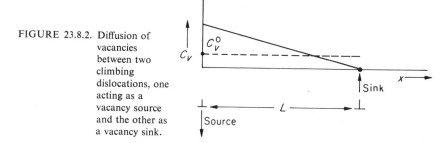

FIGURE 23.8.2. Diffusion of
vacancies
between two
climbing
dislocations, one
acting as a
vacancy source
and the other as
a vacancy sink.

The concentration of vacancies in the lattice near the dislocation is (see
Section 22.6)

$$C_v^{\pm} = C_v^0 e^{\pm n(\sigma)\sigma\Omega/kT}. \tag{23.8.1}$$

Here C_v^0 is the equilibrium concentration, σ is the stress, Ω is the atomic volume, and $n(\sigma)$ is a stress concentration factor (which in the general case might be different for the sources and the sinks). The plus sign applies to the sources. We assume that during steady-state creep (where the strain rate $\dot{\epsilon}$ does not vary with t), a steady-state diffusion process occurs. If we consider this, for simplicity, as a one-dimensional diffusion problem, the vacancy flux is given by

$$J_v = \frac{D_v(C_v^+ - C_v^-)}{L} = \frac{D_v C_v^0}{L}[e^{n(\sigma)\sigma\Omega/kT} - e^{-n(\sigma)\sigma\Omega/kT}]. \tag{23.8.2}$$

We note that

$$D_v C_v^0 = D, \tag{23.8.3}$$

where D is the self-diffusion coefficient. When $n(\sigma)\sigma\Omega \ll kT$, we can write for the vacancy flow Q_v into a unit length of dislocation of effective cross section S:

$$Q_v = J_v S = \frac{a}{L} Dn(\sigma)\sigma, \tag{23.8.4}$$

where

$$a = \frac{2\Omega S}{kT}. \tag{23.8.5}$$

To surmount the barrier the dislocation must climb a height h. The net input number of vacancies needed per unit length of dislocation to accomplish this is hb/Ω, where b is the Burgers vector. The time interval to accomplish this is

$$\Delta t = \frac{hb}{Q_v \Omega}.$$

If when the dislocation breaks away from the barrier it moves a distance d, then the increment of strain caused by each dislocation breaking away once is given by

$$\Delta \gamma = b\rho d,$$

where ρ is the dislocation density. Thus the strain rate is given by

$$\dot{\gamma} = \frac{\Delta \gamma}{\Delta t} = \frac{\rho d Q_v \Omega}{h} \tag{23.8.6}$$

or by

$$\dot{\gamma} = a\Omega \frac{\rho d}{hL} n(\sigma)\sigma D. \tag{23.8.7}$$

With the exception of D we could assume that the other factors of (23.8.7) depend only weakly on temperature. Neglecting this dependence we can write

$$\dot{\gamma} = B(\sigma)e^{-Q_d/kT}. \tag{23.8.8}$$

Thus the activation energy for high-temperature creep is the same as for self-diffusion.

Experimental activation energies for creep are compared with activation energies for self-diffusion in Figure 23.8.3. This excellent agreement shows that the rate-determining step in high-temperature creep is the diffusion of vacancies.

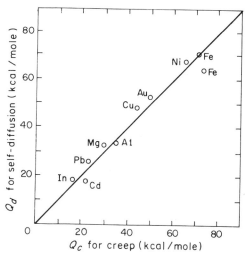

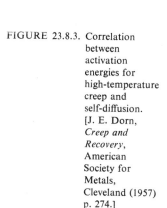

FIGURE 23.8.3. Correlation between activation energies for high-temperature creep and self-diffusion. [J. E. Dorn, *Creep and Recovery*, American Society for Metals, Cleveland (1957) p. 274.]

From a detailed theoretical consideration of the magnitude of p, d, h and L in (23.8.7), one can calculate B in (23.8.8), obtaining a value roughly in agreement with experiment. It therefore appears that the diffusion of vacancies occurs between one dislocation where the equilibrium concentration is slightly greater than the equilibrium value to another dislocation where the concentration is slightly less than the equilibrium value. We define **high-temperature creep** in pure crystals to be creep above $0.55 T_m$ at stresses such that the strain rate is less than 10^{-5} sec^{-1}. By considering the nature of p, d, h, and L and using a more sophisticated model, one could show that B is, under the above conditions, given by

$$B \propto \sigma^n, \tag{23.8.9}$$

where n is 4 to 5 and this is found to be roughly in agreement with experiment [J. Weertman, *Transactions of the Metallurgical Society of AIME*, **218**, 207 (1960)]. On the other hand it has not yet been possible to calculate the creep rate directly. (The exception to this is the pure diffusional creep at very low stresses which does not involve dislocation motion; this was discussed in Chapter 19.)

At lower temperatures the creep process is much more complex and several mechanisms are possible, all of which appear to be thermally activated processes.

In polycrystalline materials it would be expected that the parameter d would be no larger than the grain size so that there would be a grain size dependence of creep. In general the **Dorn relationship** (23.8.8) can be written as

$$(\dot{\gamma} e^{Q_d/kT}) = f(\sigma, H). \tag{23.8.10}$$

The term in parentheses is called the **temperature-reduced strain rate**. We note that if Q_d is the activation energy per atom, we divide by kT, while if Q_d is the activation energy per mole, we divide by RT. The function f is a function of the stress and the history H of the specimen. If all specimens have the same history, then f varies with σ only. A knowledge of $f(\sigma)$ then enables one to predict the strain rate-stress-temperature behavior. This is extremely important as the engineer often has to extrapolate from short-time experimental data (perhaps one year) to expected lifetimes of 30 years. We can also write

$$(te^{-Q_d/kT}) = F(\sigma, H), \tag{23.8.11}$$

where the quantity in parenthesis is called the **temperature-compensated time** and t is the time to reach a given strain (which is a criterion for failure).

A closely related function involves the **Larson-Miller parameter**, $T(C + \log_{10} t)$, so that

$$T(C + \log_{10} t) = g(\sigma), \tag{23.8.12}$$

where T is the temperature in degrees Rankine, t is the time in hours, and $C = 20$. See A. J. Kennedy, *Processes of Creep and Fatigue in Metals*, Wiley, New York (1963). Larson-Miller plots for some superalloys are shown in Figure 23.8.4. Such materials are used in special high temperature applications.

SUPERPLASTICITY.   **Superplasticity** refers to a state of a material in which it can undergo very large tensile extensions (1000%) at relatively low stresses,

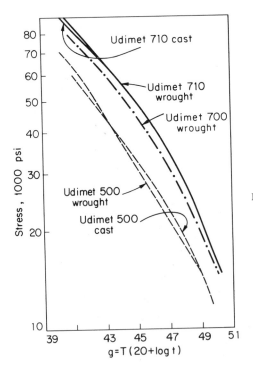

FIGURE 23.8.4. Larson-Miller plot for some Ni-Cr-Co superalloys. Udimet 710 has a composition: 55Ni-18Cr-15Co-5Ti-3Mo-2.5Al-1.5W-0.2B-.07C. Here T is the temperature in degrees Rankine and t is the time in hours.

but with moderate (not slow) strain rates. Glass when heated is superplastic and a glass rod can be readily drawn into a fine fiber. The glass in such a state is simply a highly viscous liquid which obeys the Newtonian relationship

$$\tau = \eta \dot{\gamma}. \tag{23.8.13}$$

Thermoplastics when softened by heating behave similarly. A wide variety of processing techniques can be used to shape materials in the superplastic state. Superplasticity in metals was discovered by C. E. Pearson of the University of Durham in 1934 [C. E. Pearson, *J. Inst. Metals* **54**, 111 (1934)]. In the past several years extensive research has uncovered many metal systems which are superplastic, and further, has found the prerequisite conditions for super-plasticity. It is necessary to have a **microduplex structure,** i.e., a very fine grained structure of two phases. The second phase inhibits the process of grain growth which would ordinarily take place. The grain size of the microduplex structures is a few microns. An example is illustrated in Figure 23.8.5. The Pb-Sn eutectic alloy when cast as a 1″ diameter rod gives a lamellar structure. This is then heavily cold worked by cold rolling to 1/4″ and swaging to 1/8″ diameter. The cold working transforms the lamellar structure to the microduplex structure. (In the case of the aluminum-zinc system, the microduplex structure is obtained by appropriate heat treatment.) We note that the grain size in the microduplex structure is of the order of the subcell size in deformed metals.

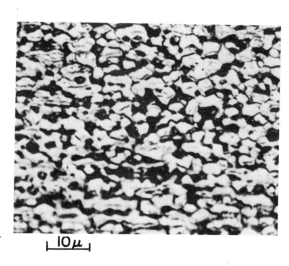

FIGURE 23.8.5. Microduplex structure of lead-tin eutectic alloy. (*Courtesy of W. B. Morrison, Edgar C. Bain Laboratory, U. S. Steel Corp.*).

⌐ 10μ ⌐

Superplasticity is a process of high temperature creep although the exact mechanism has not been proven. The grain boundaries probably serve as good sources and sinks for vacancies and dislocations and it appears that pileups, tangles, and subcells do not form in the tiny microduplex grains and hence do not act as barriers to dislocation motion. See H. W. Hayden, R. C.

Gibson and J. H. Brophy, "Superplastic Metals", *Scientific American* (March, 1969) p. 28 and R. Johnson, "Superplastic Metals", *Design Engineering* (March, 1969).

It is usually found that

$$\dot{\gamma} \propto \tau^{m}, \tag{23.8.14}$$

where $m = 1$ to 3.

23.9   FRACTURE OF MATERIALS

FRACTURE STRENGTH OF PERFECT CRYSTALS.   The fracture stress of real crystals usually is far below the theoretical limit for perfect crystals which is of the order of $E/5$. Estimates of this can be made in various ways. One has already been illustrated by Equation (21.5.3). Another method which leads to a similar result is given in Problem 21.12. Also, we noted when considering the hydrogen molecule (Chapter 9), that the distance between atom centers was 14% less than twice the Bohr radius. Hence if we were to stretch this bond by 14% it could be considered broken. Assuming $\sigma = E\epsilon$ we have $\sigma_c = 0.14E$. Finally, more exact calculations could be made for an ionic crystal based on the Born model (Chapter 9). The student could show that if such a crystal is stretched equally in all directions, the maximum stress which the crystal can withstand will occur at the inflection point of the potential energy curve, i.e., when $\partial^2 U/\partial r^2 = 0$. This occurs when

$$\frac{r}{r_0} = \left(\frac{n+1}{2}\right)^{1/(n-1)}$$

where n is the exponent in the repulsive potential term B/r^n. Usually, $n \approx 10$, in which case $r/r_0 \approx 1.2$, hence $\epsilon \approx 0.2$ which leads to a predicted fracture stress of $E/5$.

BRITTLE CRYSTALS.   In the general case the problem of fracture is a rather complicated one, particularly when complex engineering materials are involved. However, in the case of single crystals of ionic nature (such as MgO) which are nearly brittle materials, it has been possible to describe in more or less general detail the nature of the dislocation pileups and the fracture initiation, i.e., crack formation.

Three types of pileup groups of dislocations have been experimentally observed in MgO. These are illustrated in Figure 23.9.1. The *single pileup* group has already been discussed in Chapter 22. It was shown there that the tip stress would be raised to 1000σ, where σ is the applied stress, if the pileup consisted of 1000 dislocations. The mathematical details of the **90-deg double pileup** and also the **120-deg double pileup** taking anisotropy into account are treated elsewhere [Y.T. Chou and R. W. Whitmore, *Journal of Applied Physics*, **32** 1920

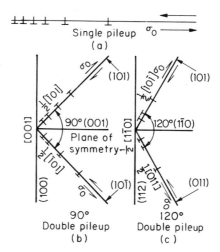

FIGURE 23.9.1. Three pileup groups in MgO leading to crack formation. [Y. T. Chou and R. W. Whitmore, *Journal of Applied Physics,* **32**, 1920 (1961).]

(1961)]. There are in a sense some unresolved difficulties with the 90-deg double pileup as regards the formation of a stable arrangement of this type; if core effects are ignored, the dislocation formed at the junction by the interaction of two dislocations on intersecting glide planes by the reaction

$$\tfrac{1}{2}[\bar{1}01] + \tfrac{1}{2}[101] \longrightarrow [001] \tag{23.9.1}$$

does not involve a lowering of the elastic energy of the crystal. However, core effects may be such that the [001] dislocation is more stable than the pair. In the case of the 120-deg double pileup the $\tfrac{1}{2}[110]$ dislocation is elastically more stable and hence this type of pileup is *expected* to form when double slip occurs as shown in Figure 23.9.1. It should be noted that the stress-concentration factor at this double pileup is as follows: For the tensile tip stress to reach 1000σ, the 120-deg double pileup requires only 62 dislocations. The experimental fracture stress σ is about 10,000 psi. A pileup for which the tip stress reaches 1000σ in such a case is such that the nearest dislocation to the locked dislocation is less than one lattice spacing away. This is taken as the criterion for the onset of a microcrack, as shown in Figure 23.9.2.

CRACK GROWTH IN BRITTLE MATERIALS. We are now concerned with the propagation of the above crack through the material. This is essentially the same as crack propagation in glass. In glass no dislocation structure is present inasmuch as glass is amorphous. Thus at low temperatures there is no mechanism for plastic deformation. At high temperature, diffusion takes place at a more rapid rate and the glass behaves in a viscous manner. Here we are concerned with its mechanical behavior at low temperatures. In particular we are concerned with its low fracture strength (10,000 psi). We have known (Griffith, 1921) that tiny glass fibers that are freshly drawn have high strengths (of the order of 10^6

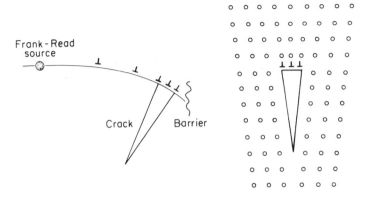

FIGURE 23.9.2. Incipient crack at a pileup.

psi). However, upon aging, these, too, become weak. An additional factor is the size dependence of the strength; the strength decreases as the size increases.

Griffith explained these phenomena on the basis that cracks are created in the material (mechanically, chemically, or as a result of devitrification). [The density of the glass is less than that of the crystalline phase. Hence if at the surface there is a small quantity of material which transforms from the glassy state to the crystalline state (**devitrification**), a crack might appear to compensate for the volume change.] Let us now briefly discuss the **Griffith theory** [A. A. Griffith, *Philosophical Transactions of the Royal Society of London*, **A221**, 163 (1921)]. Suppose we have within a plane body of *unit thickness* a very sharp crack of length $2c$ (or alternatively a crack of depth c at the surface). The surface energy of the crack is

$$U_s = 4c\gamma. \tag{23.9.2}$$

The difference in elastic energy between this body with a crack and the same body without the crack when a tensile stress σ is applied normal to the crack is

$$U_e = -\frac{\pi c^2 \sigma^2}{E}, \tag{23.9.3}$$

a result first given by C. E. Inglis [*Transactions of the Institute of Naval Architects*, **55**, Part 1, 219 (1913)]. The *increase* in energy when a crack is formed is

$$U = 4c\gamma - \frac{\pi c^2 \sigma^2}{E}. \tag{23.9.4}$$

The crack will grow spontaneously if $\partial U/\partial c < 0$, so that critical stress for a given crack can be calculated by setting $\partial U/\partial c = 0$. We have

$$\sigma_{\text{crit}} = \left(\frac{2E\gamma}{\pi c}\right)^{1/2}. \tag{23.9.5}$$

The density of cracks throughout the volume will be of such a nature that the probability of having a crack of a critical length at a given stress in a tiny specimen will be very low. Thus the size effect can be estimated.

Once the stress reaches the critical value for the weakest crack present, that crack will begin to propagate. It can be shown by the theory of elasticity that its maximum velocity is the velocity of a sound wave. For comparison with experiment, see D. K. Roberts and A. A. Wells, *Engineering*, **178**, 820 (1954).

It has not been possible to test the Griffith equation precisely because of the smallness of the cracks on the surface. It is thought that for glasses and other materials for which slip does not occur, Equation (23.9.5) is essentially correct.

EXAMPLE 23.9.1

If a glass has $E = 7 \times 10^{11}$ dynes/cm² (10^7 psi) and $\gamma = 300$ ergs/cm² and $\sigma_{crit} = 7 \times 10^8$ dynes/cm², calculate the depth of the crack causing fracture.

Answer. From (23.9.5) we find $c \approx 2 \times 10^{-4}$ cm.

BIAXIAL STRESS. The previous discussion was limited to a macroscopic stress state in which only one normal *tensile* stress component was present.

If the cracks are randomly distributed in direction and if the crack propagates when the tensile stress normal to the crack at the tip of the crack reaches the values which would cause fracture in a tension test normal to the crack, then the Griffith theory can be applied to biaxial stress states. [See J. C. Jaeger, *Elasticity, Fracture and Flow*, Wiley, New York (1962) p. 167.] Let σ_1, σ_2, and σ_3 be principal stresses. Then a **biaxial stress** state is one for which $\sigma_1 \neq 0$, $\sigma_2 \neq 0$, and $\sigma_3 = 0$. Principal stress and axes are discussed in Section 7.3.

The fracture stress criteria depends on both σ_1 and σ_2 in the biaxial stress case and it can be shown that:

if $3\sigma_1 + \sigma_2 > 0$, then the fracture stress is given by

$$\sigma_1 = \left(\frac{2E\gamma}{\pi c}\right)^{1/2}, \tag{23.9.6}$$

and if $3\sigma_1 + \sigma_2 < 0$, the fracture stress is given by

$$(\sigma_1 - \sigma_2)^2 + 8\left(\frac{2E\gamma}{\pi c}\right)^{1/2}(\sigma_1 + \sigma_2) = 0. \tag{23.9.7}$$

EXAMPLE 23.9.2

What is the fracture stress for pure compression?

Answer. Here $\sigma_1 = -\sigma$ and $\sigma_2 = 0$. Hence $3\sigma_1 + \sigma_2 = -3\sigma < 0$ and (23.9.7) applies and

$$\sigma_1 = -8\left(\frac{2E\gamma}{\pi c}\right)^{1/2}.$$

Thus the *compressive fracture stress* is eight times larger than the *tensile fracture stress*. This is found to be the case for glass and is also approximately true for sintered carbides. Such materials can be considered truly brittle materials. Gray cast iron which is partially ductile has a compressive fracture stress which is about three times as large as the tensile fracture strength.

EXAMPLE 23.9.3

A piece of chalk is twisted in torsion only. Using the biaxial criteria for brittle fracture, how should this break?

Answer. In this case one has shear stresses as shown but this is equivalent to the biaxial stress state with equal compressive and tensile stresses as shown. In this case $3\sigma_1 + \sigma_2 = 3\tau - \tau = 2\tau > 0$, so (23.9.6) applies. The fracture surface will be at 45 deg to the rod axis. The student should experimentally check this.

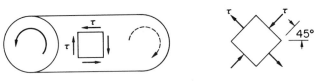

FIGURE E23.9.3.

CRACKS IN DUCTILE MATERIALS.   There are shear stresses as well as tensile stresses present at the tip of a crack. These shear stresses can cause plastic flow in the material surrounding the crack tip. The energy expended in plastic deformation may be several thousand times as large as the surface energy. Under these conditions we write

$$\sigma_{\text{crit}} = \left(\frac{2E(\gamma + p)}{\pi c}\right)^{1/2},\tag{23.9.8}$$

where p is the **plastic work term**, i.e., the plastic energy per unit area of new surface created. Here p may be of the order of 10^6 ergs/cm² for ductile materials.

The ability to flow plastically clearly enhances the fracture strength. Suppose that the crack is moving into an area free of dislocations. Can plastic flow still occur? The answer in some cases is "yes," because it is possible to nucleate new dislocations in the presence of the very high stresses at the crack tip. This nucleation process could occur at the surface or inside the material, in which case a pair of dislocations of opposite sign would be created. This is the reverse of annihilation by glide mentioned earlier [see J. P. Hirth and J. Lothe, *Theory of Dislocations*, McGraw-Hill, New York (1968)]. If dislocations can be nucleated rapidly at the tip of a crack or if those already there can multiply sufficiently rapidly, the crack will be impeded by the plastic energy;

otherwise, the crack will propagate elastically. Because the plastic deformation processes (nucleation of dislocation, dragging of jogged screw dislocations, etc.) become more difficult and, hence, less likely at lower temperatures, we might expect that at a sufficiently low temperature the crack would propagate elastically.

EXAMPLE 23.9.4

A steel is tested in tension. In one case the crack propagates elastically, in the other, plastically. If $\gamma = 10^3$ ergs/cm² and $p = 10^6$ ergs/cm², what is the ratio of the crack size which can be tolerated for a given applied stress in the two cases?

Answer.

$$\frac{c_{\text{plastic}}}{c_{\text{elastic}}} = 1001.$$

NUCLEATION OF CRACKS. Where do cracks begin? In some materials, weak second phase particles such as graphite in iron are sites at which cracks nucleate. Voids are also often present within materials as a result of the manufacturing process and these are also sites at which fracture begins. Vacancies can be produced during deformation by dragging of jogs on screw dislocations, and these vacancies can diffuse together to form voids. The voids in turn can coalesce to form cracks. Cracks can also begin at the tips of pileups against barriers such as grain boundaries and at pileups at sessile dislocations. Twin deformation followed by twin deformation on intersecting planes can also lead to crack formation. This is one of the important crack-generating mechanisms in bcc metals. Such metals show an increase in twin deformation at low temperatures

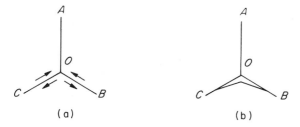

FIGURE 23.9.3. Crack nucleation at grain boundaries. If the state of stress is shear as shown in (a), then the tensile axis is OA.

and an increase in brittleness at low temperatures. Figure 23.9.3 shows how sliding grain boundaries at high temperatures can lead to crack nucleation. An example of such a crack is shown in Figure 23.9.4.

FIGURE 23.9.4. Intergranular crack in an aluminum alloy creep specimen. [From H. C. Chang and N. J. Grant, *Transactions of AIME*, **206**, 544 (1956).]

ENVIRONMENTAL EFFECTS ON FRACTURE.   A freshly drawn glass fiber may have a strength of 1.5×10^6 psi (10^{11} dynes/cm^2) and if kept in an ultravacuum would maintain that strength for a long time. However, if water vapor is present, the strength will quickly decrease. Since silica glass shows a small solubility in water, it would appear that a chemical dissolution process takes place and that this forms cracks. Ordinary glass has a strength of 10^4 psi (7×10^8 dynes/cm^2), but when the surface is etched in agitated hydrofluoric acid the strength can increase by a factor of 5.

EXAMPLE 23.9.5

   If the fracture of glass is due to cracks on the surface, suggest ways to increase the strength.

   Answer. There are several methods. One method depends on the introduction of compressive stresses in the surface. There are several ways of doing this. One involves a heat treatment. Another involves ion bombardment in which potassium ions are driven into the glass and result in an expanded silica network. The resultant expansion of the outer layer means that the glass there will be in compression. Although seemingly simple and obvious, techniques such as this are important engineering innovations and illustrate the creative aspect of engineering.

   High-strength steels in which cracks are present and which are under continued stress are highly susceptible to liquid water. It is found that at a given stress level cracks will grow slowly in the presence of water (and eventually reach a critical crack length), although they will not grow in air. Oxygen has a beneficial effect in that it tends to retard crack growth. Hydrogen is harmful. Since hydrogen is used as a fuel in rockets, NASA often adds about $\frac{1}{2}\%$ oxygen to the hydrogen, and this inhibits crack growth.

Liquid metals are often desirable materials for heat transfer agents. However, liquid metals often penetrate grain boundaries fairly readily, although showing little solubility in the bulk phase. Mercury, for example, readily penetrates the grain boundaries of steel at a pressure of several kilobars (several thousand atmospheres) and causes intergranular fracture. Aluminum alloy grain boundaries are readily penetrated by mercury at room temperature and pressure, and the individual crystals can be completely separated, as was illustrated for titanium in a similar situation in Figure 15.3.5.

DUCTILE-TO-BRITTLE-TRANSITION. Body-centered cubic metals undergo a transition from ductile to brittle fracture when

1. The temperature is decreased sufficiently.
2. The strain rate is increased sufficiently.
3. The specimen is notched.

All of these factors cause a decrease in ductility of the material of which the most common is steel (excepting fcc stainless steels).

EXAMPLE 23.9.6

Explain how a notch decreases ductility.

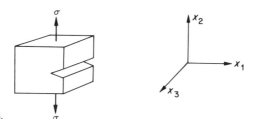

FIGURE E23.9.6.

Answer. The application of the macroscopic stress σ will cause a tensile stress component at the tip of the crack, σ_{22}, which is very large with respect to σ.

This tensile component would tend to cause a lateral contraction along x_3. However, except for the material very near to the tip of the crack, the stress normal to x_3 is relatively quite small and so the material does not contract laterally along x_3. The material at the tip is therefore hindered in its attempt to contract laterally and must therefore be in tension. Detailed analysis shows that we have a state of triaxial tension at the tip of the crack. In the extreme case of hydrostatic tension there are no shear stresses present and hence there is no ductility. Recall that plastic deformation is caused by *shear* stresses only.

The decrease in ductility at low temperature (or high strain rate) is due to the decreased slip activity and increased twinning.

The **ductile-to-brittle transition temperature** may be measured in several ways:

1. Change in energy absorbed.
2. Change in ductility.
3. Change in the fracture appearance.
4. Change in lateral contraction at the tip of the notch.

Each of these on a given specimen type gives nearly the same transition temperature. Impact tests are commonly made on V-notched or keyhole-notched Charpy specimens. The latter, since it has a rounded tip, gives a lower transition temperature.

Examples of experimental data are shown in Figure 23.9.5. The fracture

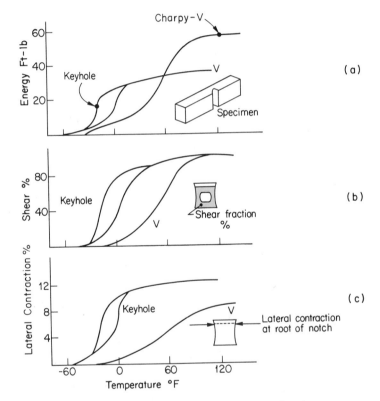

FIGURE 23.9.5. Transition temperature curves based on energy absorbed, fracture appearance, and notch ductility for a semikilled steel: 0.18% C, 0.54% Mn, 0.07% Si. (a) Energy transition. (b) Fracture transition. (c) Lateral contraction transition. [From W. S. Pellini, *ASTM Special Technical Publication*, **158**, 222 (1954).]

appearance is based on the fibrous (shear) or cleavage nature of the fracture. The lateral contraction at the root is another measure of ductility.

FRACTURE APPEARANCE. The appearance of a fracture surface changes rather abruptly in the brittle to the ductile transition. The brittle fracture surface appears relatively smooth (though not on a fine scale) and bright, while the ductile fracture surface appears ragged and dull. In the case of noncrystalline materials such as glasses, which are often major components of ceramic materials, the fracture surface appears **conchoidal** (this refers to the approximately circular-shaped steps on the surface, with the center of the circle at the crack). The same behavior is shown by brittle polymers far below the glass transition temperature. In the case of brittle crystalline materials, fracture generally occurs by cleavage across particular cleavage planes (rather than in the direction of maximum normal stress) although at high temperatures intergranular fracture may occur. If the fracture is associated with sizeable plastic flow, voids can be formed and shear deformation can cause these voids to join, leading eventually to ductile fracture. Thus, if a tough pitch copper specimen which contains oxide inclusions is tested in tension, necking will occur. The extensive plastic deformation will cause voids to form in the necked area at the inclusion, most profusely at the center of the neck where the triaxial stresses are highest (see Section 24.2). This leads to the **cup** and **cone fracture**. At the bottom of the cup the surface is jagged and the mode is called fibrous fracture. After the large center crack forms by void coalescence, the cone side of the cup approximately follows the surface of maximum shear stress.

IMPACT STRENGTH OF POLYMERS. Considerably below the glass transition temperature, polymers break with a brittle fracture. As the glass transition temperature is approached, they become tougher (above T_G they are rubberlike). A high degree of crystallinity and the presence of very large spherulites will lead to low impact strength. Spherulite size can be controlled by varying the nucleation to growth rate. The impact strength of brittle polymers such as polystyrene can be increased by incorporating rubbery materials into the glassy resins.

ENGINEERING DESIGN. Engineering designers often specify that a steel must show a V-notch Charpy impact energy of at least 15 ft-lb at all temperatures of application.

How can the engineer avoid brittle fractures? He can avoid notches which result from poor design (use fillets with a large radius) or cracks and notches which result from poor manufacturing processes. For example, poor welding practice often results in inclusions and in residual stresses. The engineer can also use a different material (e.g., fcc stainless steels are used for cryogenic vessels), but he must constantly remember economic factors. As a rule you cannot decide to replace a steel costing 25 cents/lb with one costing $3.00/lb if you are building large ships (except in very special cases).

23.10   FATIGUE

We are concerned here with the cyclic stressing of a sample to a stress such as $\sigma_0/3$; i.e., we apply a stress such as

$$\sigma = S \cos \omega t, \qquad (23.10.1)$$

where, say, $S = \sigma_0/3$.

We then find that for a sufficient number of cycles N the specimen fractures. If experiments are carried out at different values of S, and N is measured for each, and S is plotted versus $\log_{10} N$, we obtain an *S-N* curve. For $N = 1$, S is the tensile strength. Usually the sample can undergo a few hundred cycles before S decreases appreciably. One then obtains a continued decrease in S as N increases, as was already shown in Figure 2.6.1.

In some materials S decreases indefinitely as N increases. Age precipitation-hardened aluminum alloys exhibit this behavior. In other materials S decreases to a limiting value S_∞ but decreases no further as N increases. The quantity S_∞ is called the **endurance limit**. Steel exhibits such behavior.

Very small stresses can cause dislocations to bow out between pinning points. At somewhat larger stresses the dislocation breaks away from some of these pinning points. However, considerably larger stresses are needed to produce macroscopic flow, e.g., a flow stress defined by a 1% yield strain. If extremely sensitive methods of measuring strain are used ($\epsilon \approx 10^{-6}$, **microstrain**), then plastic deformation can be observed and hence dislocation motion must occur at much lower stresses.

We are concerned here with how alternating stressing can cause additive plastic strain and cumulative damage which leads to fatigue failure. One such mechanism involves the motion of jogged screw dislocations (see Figure 22.8.2). In such a case there is nonconservative dislocation motion which either results in excess point defects (vacancies, say) which can nucleate voids or in dipole formation. As these dislocations undergo small motions they come to new barriers but avoid these by cross-slip. When the stress is alternated they return on a different glide plane, in which case the opposite defect is produced (interstitials, say). The presence of these defects can cause dislocation climb and a softening of the material and a simultaneous decrease in the ultimate tensile strength of the material as shown in Figure 23.10.1. It should also be noted that fatigue deformation is not homogeneous and that a specimen surface which was originally *smooth* has extrusions and intrusions, as shown in Figure 23.10.2. It is clear that cracks are originating at the free surface.

Fatigue behavior ordinarily begins at the surface, and thus the nature of the surface is very important. A crystal at a surface must show different mechanical behavior than an interior crystal. The latter is usually surrounded by 14 or

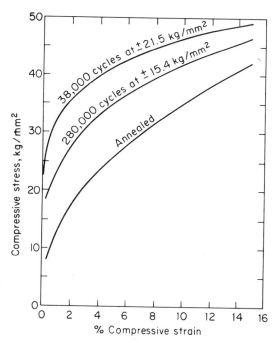

FIGURE 23.10.1. Cyclic stressing effects on the stress-strain curve of nickel [N. Polakowski and A. Polchoudhuri, *Trans. ASTM* **54**, 701 (1954)].

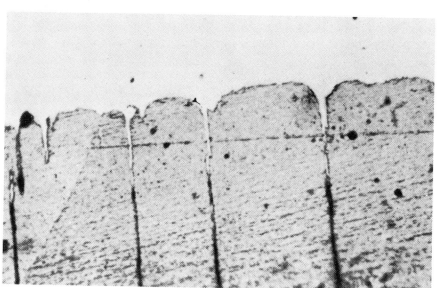

FIGURE 23.10.2. A taper section through a surface of copper fatigued to $\frac{1}{10}$ of its life showing slip bands developing into intrusions. Original photographs were at $\times 1000$ but the taper magnifies the surface behavior by another factor of 10. [W. A. Wood, *Phil. Mag.* **46**, 1028 (1955).]

so other crystals, and the deformation of an interior crystal is in part determined by the neighbors. A crystal at a surface has a face whose deformation is not restricted by the presence of other crystals. One would expect that the surface would therefore be effectively weaker than the interior, an expectation which is proven true by Figure 23.10.2. Fatigue failure, which follows when the fatigue cracks grow to a critical size, can be eliminated by periodically etching away the surface. While this is not a practical way of preventing fatigue, it proves unequivocally that fatigue cracks originate at surfaces.

EXAMPLE 23.10.1

Suggest a way of increasing fatigue strength.

Answer. From the preceding discussion, it is clear that we should harden the surface of the material. This can be done by local cold working (shot peening) or by carburizing or nitriding. In this way compressive stresses are introduced into the surface layer.

REFERENCES

Cottrell, A. H., *The Mechanical Properties of Matter*, Wiley, New York (1964).

Fisher, J. G., Johnston, W. G., Thompson, R., and Vreeland, T., Jr., eds., *Dislocations and Mechanical Properties of Crystals*, Wiley, New York (1957).

Byrne, J. G., *Recovery, Recrystallization and Grain Growth*, Macmillan, New York (1965).

Himmel, L., ed., *Recovery and Recrystallization*, Wiley-Interscience, New York (1963).

Kennedy, A. J., *Processes of Creep and Fatigue in Metals*, Wiley, New York (1963).

Garafalo, F., *Fundamentals of Creep and Creep-Rupture in Metals*, Macmillan, New York (1965).

Tetelman, A. S., and McEvily, A. J., Jr., *Fracture of Structural Materials*, Wiley, New York (1967).

Burke, J. J., Reed, N. L., and Weiss, V., *Fatigue, An Interdisciplinary Approach*, Syracuse University Press, Syracuse, N.Y. (1965).

PROBLEMS

23.1. a. An ordinary annealed crystal might have what dislocation density?
b. A heavily cold-worked crystal?

23.2. a. What is meant by a yield point?
b. Which materials exhibit them?
c. Why are they of commercial importance?

23.3. a. What is a dislocation atmosphere?
b. What is strain aging?
c. Explain the mechanism of strain aging.

23.4. Derive at least one work-hardening mechanism which leads to (23.3.1).

23.5. What is meant by recovery?

23.6. a. What is dislocation annihilation?
b. Describe how it can occur by glide.
c. By climb.

23.7. Describe the process of polygonization.

23.8. What is the driving force for recrystallization of a cold-drawn wire?

23.9. To obtain typical nucleation rates (new crystals formed from cold-worked material) r^* must be of the order of several Angstroms.
a. Illustrate this quantitatively.
b. Is this possible?

23.10. a. What is the driving force for grain growth?
b. Describe the overall process.

23.11. Give a quantitative example of why yield strength in fcc crystals falls off rapidly at about $0.5T_M$.

23.12. Discuss the quantitative aspects of high-temperature creep.

23.13. Why does high-temperature creep have the same activation energy as self-diffusion?

23.14. Discuss two of the ways used by the engineer to extrapolate creep data to longer times.

23.15. a. Discuss the formation of a crack by slip in MgO.
b. Discuss the rapid growth of a crack in MgO.

23.16. Explain the low fracture stress of ordinary window glass.

23.17. Describe in detail Griffith's theory of the critical crack size for crack propagation.

23.18. A window glass has a tensile fracture stress of 10,000 psi. According to Griffith's theory the compressive strength should be what?

23.19. A sintered tungsten carbide fractures at a tensile stress of 125,000 psi.
a. Estimate the compressive fracture stress.
b. Suggest methods for increasing the strength.

23.20. How is the Griffith theory modified if plastic flow precedes the crack front?

23.21. A steel tested in tension has a reduction in area of 50% at fracture.
a. What is the reduction in area if tested under the stress $\sigma_1 = \sigma_2 > 0$, $\sigma_3 = 0$?
b. For $\sigma_1 = \sigma_2 = \sigma_3 > 0$?

23.22. Give five different possible origins of crack nucleation.

23.23. A steel pressure vessel contains liquid mercury as a pressure media. It fractures at a stress much below that predicted for the steel on the basis of tension tests performed in air. Explain why.

23.24. What factors decrease ductility in bcc metals?

23.25. a. Why is the state of stress at the tip of a crack a state of triaxial tension?
b. Why is this significant?

23.26. Name some methods of measuring the ductile-to-brittle transition temperature.

23.27. Design an experiment which shows that fatigue failure is governed by plastic deformation of the surface.

23.28. Metals fatigue at stresses considerably below the 0.2% offset yield strength. How can this be?

23.29. Suggest several ways of increasing yield strength.

23.30. Suppose the fatigue strength of aluminum could be tripled. What would the annual impact of this be on the aircraft industry?

More Involved Problems

23.31. Give three mechanisms of work hardening which lead to (23.3.1).

23.32. Discuss a quantitative model of high-temperature creep.

23.33. Give an heuristic derivation of (23.9.3). (You may not get the exact proportionality constant.)

23.34. Give several examples of disastrous mechanical failures owing to ductile-to-brittle transition in steel [see E. Parker, *Brittle Fracture of Engineering Structures*, Wiley, New York (1957)].

23.35. How can internal friction studies be used to measure the carbon present in solution in α-iron?

23.36. If high-temperature creep occurs because shear stresses are present in the presence of a large hydrostatic pressure P, how does the creep rate depend on pressure?

Sophisticated Problems

23.37. Derive Equation (23.2.1).

23.38. Derive Equation (23.9.7).

Prologue

The basis for yielding criteria for polycrystalline aggregates is discussed. Both the *von Mises criteria* and the *Tresca criteria* are discussed. Examples of how these are used by engineers are given. The *idealized plastic material* is discussed and an example is given of how the engineer sometimes allows a prescribed amount of plastic yielding to occur in a component. The concepts of *generalized stress* and *generalized strain* are introduced and are used to correlate plastic deformation in the *tension test* and *torsion test*. It is shown that cold-working processes can be correlated in terms of the generalized stress-strain curve. The relationship between *hardness* and the tensile yield strength is discussed.

24

PLASTICITY OF POLYCRYSTALLINE AGGREGATES

24.1   YIELDING

We have previously noted that crystals are highly anisotropic with respect to plastic deformation and, relatively speaking, somewhat anisotropic with respect to elastic properties. An actual bar of an engineering material (assuming it is made up of a crystalline substance) contains millions of small and, usually, randomly oriented crystals. Such a material, at least if it consists of cubic crystals, often behaves in an isotropic manner not only in regard to elastic properties but to plastic properties as well. The latter fact, which has both theoretical and experimental bases, leads to the von Mises criteria of yielding; thus a single value for a yield strength obtained in a simple tension test is sufficient to describe the onset of yielding in a complex multiaxial stress situation. It is likewise possible to define a generalized stress-strain curve so that all types of deformation lead to the same curve, or so that, if it is possible to overcome the mathematical problems, the deformation for any loading can be calculated from this generalized curve.

When a polycrystalline specimen which contains a large number of randomly oriented grains is plastically deformed it usually deforms homogeneously on the macroscopic scale (at least for the first several percent of strain). We are tempted to conclude that the strain is homogeneous throughout the specimen; this would require that five distinct slip systems be operative in each crystal, as was proved by R. von Mises [see C. F. Elam, *Distortion of Metal Crystals*, Oxford Press, New York (1935)]. A fcc crystal has three slip systems on one (111) plane but only two of these are distinct since slip in the [1$\bar{1}$0] direction could effectively be achieved by a combination of [10$\bar{1}$] and [0$\bar{1}$1] slip. It has been proved experimentally by F. Koch that between four and five slip systems are operative.

641

In certain crystals all of the slip systems could be of the same type (such as {111} ⟨110⟩ in aluminum), whereas in other crystals as a result of low symmetry all of the slip systems could be of different types. The discussion which follows applies essentially to cubic systems only.

The fact that it is necessary to have several operative slip systems means in general that the applied stresses will have to be quite high before the critical resolved shear stress will be reached on some of the slip systems which must operate. The earliest attempt to calculate the properties of a polycrystalline aggregate from the properties of a single crystal were due to G. I. Taylor (who first suggested the edge dislocation). The details of this problem are still not fully resolved. However, it is generally accepted that yielding occurs in cubic metals when the elastic energy of distortion reaches a critical value. If the principal stresses are σ_1, σ_2 and σ_3 (we shall assume that the state of stress is always referred to the principal axes, so only normal stress components, called the principal stresses, are involved), yielding therefore occurs when

$$\sigma_0 = \sqrt{\frac{(\sigma_1 - \sigma_2)^2 + (\sigma_2 - \sigma_3)^2 + (\sigma_3 - \sigma_1)^2}{2}}; \qquad (24.1.1)$$

any combination of these principal stresses which makes the right side of (24.1.1) equal to σ_0 causes yielding. Equation (24.1.1) is known as the **von Mises criteria for yielding** although M. T. Huber used it earlier [see A. Föppl and L. Föppl, *Drang and Zwang*, Munich, 2nd ed., Vol. 1, p. 50 (1924)].

If a material which has flowed a considerable amount [so that the right side of (24.1.1) equals some value of σ_0] is unloaded (and if it can still be assumed to be plasticly isotropic), then, when stresses are reapplied (perhaps in a different combination than previously), yielding will occur when the right-hand side of (24.1.1) again reaches the value σ_0. We call the right-hand side of (24.1.1) the **generalized flow stress**.

G. Sachs has shown that the form of (24.1.1) is very near to that obtained by averaging the shear stress necessary for yielding over all possible orientations of a crystal.

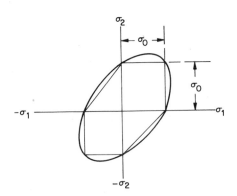

FIGURE 24.1.1. The von Mises yield ellipse and the Tresca yield hexagon for plane stress.

Engineers also often use the **Tresca** or **maximum shear stress criteria** of yielding. The maximum shear stress is one of the following three quantities which has the largest absolute value:

$$\frac{\sigma_1 - \sigma_2}{2}, \qquad \frac{\sigma_2 - \sigma_3}{2}, \qquad \frac{\sigma_1 - \sigma_3}{2}.$$

According to this criteria, yielding occurs when this shear stress exceeds $\sigma_0/2$, the maximum shear stress in a tension test. The two yield criteria are compared in Figure 24.1.1 for the case of plane stress ($\sigma_3 = 0$). For the Tresca criteria, yielding will not occur for those points σ_1, σ_2 within the hexagon but will occur otherwise. It can be seen that the Tresca condition is more conservative.

EXAMPLE 24.1.1

A chemical reaction is carried out in a cylindrical pressure vessel at high pressures. The pressure vessel behaves elastically. The inner radius is a, the inside pressure is P_a, the outer radius is b, and the outside pressure is P_b. The three principal stresses can be obtained from the theory of elasticity:

$$\sigma_{\theta\theta} = A - \frac{B}{r^2},$$

$$\sigma_{rr} = A + \frac{B}{r^2}, \qquad a \leq r \leq b,$$

$$\sigma_{zz} = 2\nu A,$$

where ν is Poisson's ratio ($\nu \sim 0.3$) and

$$A = -P_a + \frac{P_a - P_b}{1 - (a^2/b^2)}$$

and

$$B = \frac{P_b - P_a}{(1/a^2) - (1/b^2)}.$$

Find the maximum shear stress for $b/a = 5$. When will yielding occur according to Tresca if $P_b = 0$?

Answer. It can be shown that the maximum shear stress is

$$\frac{\sigma_{\theta\theta} - \sigma_{rr}}{2} = \frac{P_a - P_b}{1 - (a^2/b^2)};$$

i.e., it occurs at the inside of the pressure vessel.

Yielding will occur according to the Tresca criteria when

$$\frac{P_a - P_b}{1 - (a^2/b^2)} = \frac{\sigma_0}{2}$$

or if $P_b = 0$ at a pressure of

$$P_a = \frac{\sigma_0}{2}\left(1 - \frac{1}{25}\right) = \frac{12}{25}\sigma_0.$$

Thus the pressure at which yielding just begins is about half the tensile yield strength. Note that making the pressure vessel larger would help us very little in achieving higher pressures. The von Mises criteria would give an allowable pressure several percent higher. Note that if P_b is large and $P_a = 14.7$ psi, we have the case of a cylindrical deep sea submergence vessel.

IDEALIZED PLASTIC MATERIAL.   In an **idealized plastic material** there is no strain handening. See Figure 24.1.2. Engineers sometimes design components to operate beyond the elastic limit and into the plastic zone. This is illustrated in Example 24.1.2.

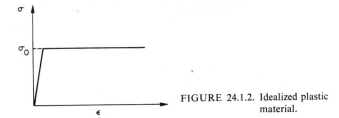

FIGURE 24.1.2. Idealized plastic material.

EXAMPLE 24.1.2

Consider the pressure vessel of Example 24.1.1 when it is operating in the plastic region. Find the maximum attainable pressures.

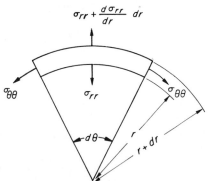

FIGURE E24.1.2.

Answer. We consider a small element of unit length along the *z* direction. Summing all the *forces* acting in the radial direction gives

$$-\sigma_{rr} r\, d\theta + \left(\sigma_{rr} + \frac{d\sigma_{rr}}{dr} dr\right)(r + dr)\, d\theta - 2\sigma_{\theta\theta}\, dr \sin\frac{d\theta}{2} = 0.$$

For small $d\theta/2$, $\sin d\theta/2 \rightarrow d\theta/2$ and hence we have

$$r\frac{d\sigma_{rr}}{dr} = \sigma_{\theta\theta} - \sigma_{rr}.$$

According to Tresca's criteria, yielding begins when

$$\frac{\sigma_{\theta\theta} - \sigma_{rr}}{2} = \frac{\sigma_0}{2},$$

as we saw in Example 24.1.1, so we have

$$r\frac{d\sigma_{rr}}{dr} = \sigma_0,$$

which is true as yielding continues because of the nature of the stress-strain diagram shown in Figure 24.1.2. Hence

$$\int_{-P_a}^{-P_b} d\sigma_{rr} = \sigma_0 \int_a^b \frac{dr}{r}.$$

Hence

$$P_a - P_b = \sigma_0 \ln b/a.$$

For the conditions of Example 24.1.1,

$$P_a = \sigma_0 \ln 5.$$

The pressure vessel can clearly withstand higher pressures when yielding is allowed. However, the vessel stretches permanently on each loading so this is not as practical as it may seem. It can be shown, however, that if the internal pressure is increased only to twice the pressure at which plastic deformation begins, then it will behave elastically during unloading and also will behave elastically during reloading up to this pressure. This pressure is known as the **shakedown pressure**. In this case the inner part of the pressure vessel would be plastically deformed during the first loading, while the outer part always behaves elastically. This process is known as **autofrettaging**. It is often used to strengthen pressure vessels and gun barrels.

PLASTIC FLOW. When materials deform plastically it is usually assumed that the volume is a constant. Barring the formation of cracks and cavities in the solid, this is essentially true. The condition for constant volume is

$$\epsilon_1 + \epsilon_2 + \epsilon_3 = 0, \tag{24.1.2}$$

where the ϵ's are real principal strains. A real strain under conditions of homogeneous deformation is defined by

$$d\epsilon = \frac{dl}{l} \tag{24.1.3}$$

or

$$\epsilon = \ln \frac{l}{l_0}. \tag{24.1.4}$$

(In this chapter ϵ is the symbol for *real* strain instead of ϵ_r, as used in Chapter 2. This should cause no confusion because only real strains are used in this chapter.)

This is to be compared with the nominal principal strain

$$\epsilon_n = \frac{l - l_0}{l_0} = \frac{l}{l_0} - 1 = e^\epsilon - 1. \qquad (24.1.5)$$

It is found *experimentally* that the ratios of the differences between principal stresses are directly proportional to the differences between the increments of principal strain:

$$\frac{\sigma_2 - \sigma_1}{\sigma_1 - \sigma_3} = \frac{d\epsilon_2 - d\epsilon_1}{d\epsilon_1 - d\epsilon_3} \quad \text{and} \quad \frac{\sigma_2 - \sigma_3}{\sigma_1 - \sigma_2} = \frac{d\epsilon_2 - d\epsilon_3}{d\epsilon_1 - d\epsilon_2}. \qquad (24.1.6)$$

This is **St. Venant's hypothesis.**

If we consider those deformation conditions in which the principal axes of stress and strain coincide *during* deformation, then (24.1.6) becomes

$$\frac{\sigma_2 - \sigma_1}{\sigma_1 - \sigma_3} = \frac{\epsilon_2 - \epsilon_1}{\epsilon_1 - \epsilon_3} \quad \text{and} \quad \frac{\sigma_2 - \sigma_3}{\sigma_1 - \sigma_3} = \frac{\epsilon_2 - \epsilon_3}{\epsilon_1 - \epsilon_3}. \qquad (24.1.7)$$

We could write $\sigma_2 - \sigma_1 = \alpha(\epsilon_2 - \epsilon_1)$ and $\sigma_1 - \sigma_3 = \alpha(\epsilon_1 - \epsilon_3)$, where α is a proportionality constant which depends only on the state of deformation. Equation (24.1.7) may be written in the form

$$\frac{2\sigma_2 - \sigma_1 - \sigma_3}{\sigma_1 - \sigma_3} = \frac{2\epsilon_2 - \epsilon_1 - \epsilon_3}{\epsilon_1 - \epsilon_3} \qquad (24.1.8)$$

or

$$\mathscr{L} = \mathscr{R}, \qquad (24.1.9)$$

where $\mathscr{L}$ equals the left-hand side of (24.1.8) and $\mathscr{R}$ the right-hand side. An experimental "proof" is given in Figure 24.1.3. On this basis we accept (24.1.7).

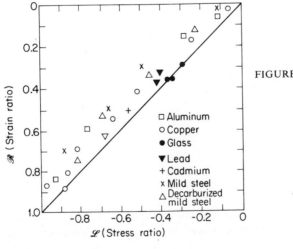

FIGURE 24.1.3. Lode's diagram for the relation between stress and strain during plastic flow. [From G. I. Taylor and H. Quinney, *Philosophical Transactions of the Royal Society of London*, A230, 323 (1931).]

We shall next define a generalized strain and we shall then show that the generalized stress is a function of generalized strain. This means that one would get the same generalized stress-generalized strain relationship from a tension test as one would get from a shear test or various other tests. We note that by (24.1.7) and (24.1.1) we have

$$\sigma_0 = \sqrt{\alpha^2[(\epsilon_1 - \epsilon_2)^2 + (\epsilon_2 - \epsilon_3)^2 + (\epsilon_1 - \epsilon_3)^2]}, \qquad (24.1.10)$$

where α is a proportionality constant. In simple cases (of no rotation)

$$\epsilon_2 = k\epsilon_1, \qquad (24.1.11)$$

where k is some constant, and by (24.1.2)

$$\epsilon_3 = -(1 + k)\epsilon_1 \qquad (24.1.12)$$

so that (24.1.10) becomes

$$\sigma_0 = \sqrt{3\alpha^2[1 + k^2 + (1 + k)^2]\epsilon_1^2} \qquad (24.1.13)$$

or

$$\sigma_0 = \sqrt{3\alpha^2[\epsilon_1^2 + \epsilon_2^2 + \epsilon_3^2]}. \qquad (24.1.14)$$

Here α depends only on the state of deformation and hence on σ_0, as noted earlier. We note that for a simple tension test, since $\sigma_2 = 0$ and $\sigma_3 = 0$, we have by (24.1.1) $\sigma_0 = \sigma_1$. We now *define* a **generalized strain**

$$\epsilon_0 = \sqrt{\tfrac{2}{3}[\epsilon_1^2 + \epsilon_2^2 + \epsilon_3^2]} \qquad (24.1.15)$$

such that for a simple tension test where the three principal strains during plastic flow are

$$\epsilon_1, \qquad -\tfrac{1}{2}\epsilon_1, \qquad -\tfrac{1}{2}\epsilon_1$$

we obtain

$$\epsilon_0 = \epsilon_1. \qquad (24.1.16)$$

Consequently (24.1.14) states that

$$\sigma_0 = \sigma_0(\epsilon_0). \qquad (24.1.17)$$

Thus the generalized stress *during* plastic deformation (in which the principal axes of stress and strain coincide and for which St. Venant's conditions apply) is related to the generalized strain in the same way that the real tensile stress (in an ordinary tensile test) is related to the real tensile strain (at least up to the necking point).

It should be pointed out that the conditions of (24.1.7) do not hold in situations in which the direction of plastic deformation is reversed. (When a specimen is plastically deformed to a real stress σ_0' and this stress is removed and a compressive stress is applied, yielding in compression occurs at a stress whose magnitude is less than σ_0'. This is known as the **Bauschinger effect**.)

Examples of generalized flow curves obtained under various conditions of biaxial stress are given in Figure 24.1.4.

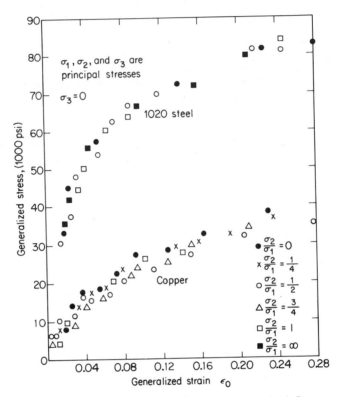

FIGURE 24.1.4. Effect of biaxial stress systems on plastic flow.
[From E. A. Davis, *Journal of Applied Mechanics*, **12**, A13 (1945).]

24.2   THE TENSION TEST

A typical ductile material will deform homogeneously in a tension test for strains up to about 20%; for larger deformation a region of the specimen will become necked, as shown in Figure 24.2.1. When this condition applies, the generalized flow curve is determined by the plastic flow which occurs in the narrowest part of the neck. The generalized flow stress is no longer P/A, where P is the load and A is the area of the narrowest cross section. Likewise the strain is no longer given by $\ln l/l_0$; the strain, however, can be readily determined using the condition of constancy of volume during the deformation; i.e., $V = Al =$ constant. Hence

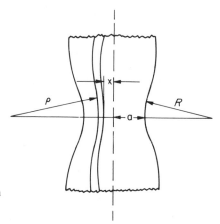

FIGURE 24.2.1. Necked portion
of a tension
specimen.

$$\frac{dl}{l} = -\frac{dA}{A},$$ (24.2.1)

and now we have for the real longitudinal strain at the neck

$$\epsilon = \ln \frac{A_0}{A},$$ (24.2.2)

where A is the cross-sectional area at the neck. Consequently, we can measure the real longitudinal strain at the neck; hence we can obtain the generalized real strain.

The state of stress in the neck is not simple tension. The material which has not necked down restricts the deformation of the necked region and so causes a triaxial state of stress.

The problem of describing the state of triaxial stress at the neck is more

FIGURE 24.2.2. Comparative
curves for
tension and
torsion. 1,
Torsion curve.
2, Tension curve
without
correction. 3,
Tension curve
with Davidenkov
correction. [After
N. N.
Davidenkov and
N. I.
Spiridonova,
*Proceedings of
ASTM,* **46,**
1147 (1946).]

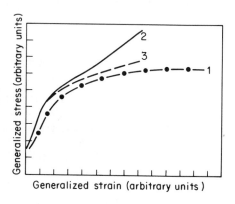

difficult. Two approaches have been used. These are described in detail in P. Bridgman, *Transactions of the American Society for Metals,* **32,** 553 (1944) and N. N. Davidenkov and N. I. Spiridonova, *Proceedings of ASTM,* **46,** 1147 (1946). An example of a flow curve is shown in Figure 24.2.2. The details of the manner in which these curves are obtained are too lengthy to be discussed here; the original papers should be consulted. However, the essential agreement indicates that we can to a fair approximation represent various types of deformation by a single curve. Figure 24.2.3 shows the distribution of stresses across the neck of a steel specimen.

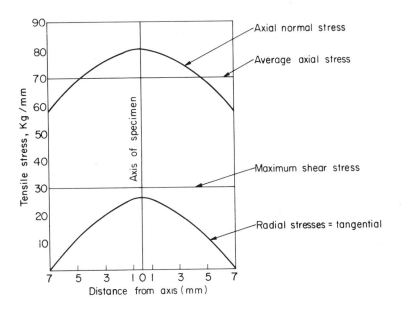

FIGURE 24.2.3. Distribution of stresses in the neck of a ductile steel specimen which has been reduced in area by 68% at the neck.

EXAMPLE 24.2.1

Show that the mean normal stress $\bar{\sigma}_n$ at the axis of the necked specimen (Figure 24.2.3) is about doubled because of triaxiality.

Answer. If we assumed that a state of uniaxial stress persists in the neck, then from Figure 24.2.3, $\sigma_{zz} = 70$, $\sigma_{rr} = \sigma_{\theta\theta} = 0$, and $\bar{\sigma}_n = (\sigma_{rr} + \sigma_{\theta\theta} + \sigma_{zz})/3 = 23$. In the actual triaxial case, $\sigma_{zz} = 80$, $\sigma_{\theta\theta} = \sigma_{rr} = 27$, and $\bar{\sigma}_n = (82 + 26 + 26)/3 = 45$. This triaxial tension state is undoubtedly the reason less ductility is shown in tension than in torsion (see Figure 24.2.2).

P. W. Bridgman has shown that when hydrostatic pressure is superimposed on a tension test there can be enormous increases in ductility [*Studies in Large Plastic Flow and Fracture*, McGraw-Hill, New York (1952)]. This behavior is illustrated in Figure 24.2.4 (note the analogy with the Charpy impact test of Figure 2.6.4). This has great significance to the production engineer. Forming of components under hydrostatic pressure is an active area of current research.

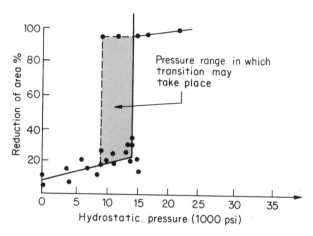

FIGURE 24.2.4. Brittle-ductile transition in zinc. [From H.L.D. Pugh in *Irreversible Effects of High Pressure and Temperature on Materials* (ASTM Special Technical Service Publication No. 374) American Society for Testing and Materials, Philadelphia (1965) p. 68.]

24.3   COLD-WORKING PROCESSES

It is often true that the generalized flow curve can be represented by an expression involving two constants

$$\sigma = K\epsilon^n, \tag{24.3.1}$$

where n, the strain-hardening exponent, is often about 0.25. It was shown in Chapter 2 that in a tension test the real strain which corresponds to the maximum of a load versus strain (real or nominal) curve is $\epsilon^* = n$.

Tensile strength (TS) is the maximum load divided by the original area and is hence given by

$$\text{TS} = \frac{\sigma A}{A_0} = \sigma e^{-\epsilon} = K n^n e^{-n}. \tag{24.3.2}$$

The equations covered in this section are extremely useful in calculating the results of such cold-working processes as the cold rolling of bars and sheets and the drawing of wire from rods.

EXAMPLE 24.3.1

Suppose we have a supply of 1-in. diameter normalized steel rods for which $n = 0.20$ and $\sigma(\epsilon^*) = 88{,}000$ psi. We wish to supply $\frac{1}{4}$-in. diameter rods of 76,000 psi yield strength. Show that this might be done.

Answer. Using the given values of n and $\sigma(\epsilon^*)$, we find by (24.3.1) that $K = 121{,}000$ psi. We now proceed to find what initial diameter of normalized steel when cold-rolled to $\frac{1}{4}$ in. gives a yield strength of 76,000 psi. From (24.3.1) we have that

$$76{,}000 = 121{,}000\epsilon^{0.20}$$

and hence $\epsilon = 0.10$. We note that

$$\epsilon = \sqrt{\tfrac{2}{3}(\epsilon_t^2 + \tfrac{1}{4}\epsilon_t^2 + \tfrac{1}{4}\epsilon_t^2)} = \epsilon_t,$$

where $\epsilon_t = \ln(A_0/A)$. Hence $A_0/A = e^{0.10} = 1.105$ and

$$\frac{d^2}{(1/4)^2} = 1.105$$

and

$$d = 0.264 \text{ in.}$$

The procedure to follow would be to hot-roll the 1-in. diameter bar to 0.264 in. (the properties of the bar would then be the same as for the original annealed bar) and then to cold-roll the bar to a 0.250-in. diameter.

The concepts of plasticity are particularly important in production processes of materials which depend on material movement, e.g., rolling and extrusion, and on material removal by mechanical means.

24.4   THE HARDNESS TEST

Hardness is defined as resistance to penetration. Usually in a hardness test we force a small sphere, pyramid, or cone into a body by means of a certain applied load. In principle, the penetration can be calculated if the generalized flow stress is known. However, the mathematical analysis is difficult. The hardness, which is a definite number, is related to the depth or diameter or area of the indentation. A hardness test is to a large extent a simple substitute for a tension test. There are a number of standard hardness tests such as the Rockwell, Brinnell, and Vickers tests. In addition, microhardness testers are available which can be used for studying hardness variations across single crystals in a polycrystalline aggregate.

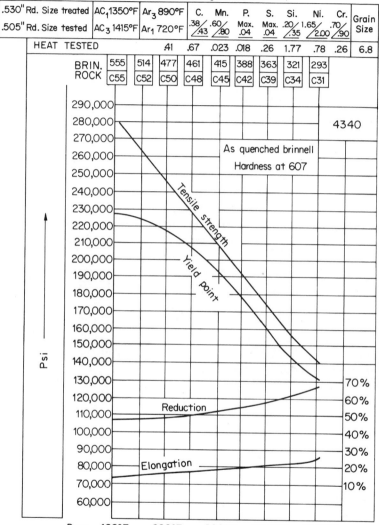

AISI – 4340 (oil quenched)
PROPERTIES CHART (Single Heat Results)

.530" Rd. Size treated	AC₁ 1350°F	Ar₃ 890°F	C.	Mn.	P.	S.	Si.	Ni.	Cr.	Grain Size
.505" Rd. Size tested	AC₃ 1415°F	Ar₁ 720°F	.38/.43	.60/.80	Max. .04	Max. .04	.20/.35	1.65/2.00	.70/.90	
HEAT TESTED			.41	.67	.023	.018	.26	1.77	.78	.26	6.8

BRIN. | 555 | 514 | 477 | 461 | 415 | 388 | 363 | 321 | 293
ROCK | C55 | C52 | C50 | C48 | C45 | C42 | C39 | C34 | C31

4340

As quenched brinnell
Hardness at 607

Tensile strength

Yield point

Reduction

Elongation

Psi

Draw   400°F   600°F   800°F   1000°F   1200°F
Normalized   at   1600° F, reheated to 1475° F, and
quenched in agitated oil.

FIGURE 24.4.1. Properties of heat-treated AISI-4340 steel.
[From *Modern Steels and Their Properties*,
Bethlehem Steel Corporation, Bethlehem,
Pa. (1967).]

In the case of steels there is a general relationship in which the Brinell hardness number (BHN) is related roughly to the tensile strength (psi) as follows:

$$TS = 500 \, BHN. \qquad (24.4.1)$$

The properties of an alloy steel (which had been originally normalized at 1600°F before machining, was then heated to 1475°F, quenched in agitated oil, and then annealed at the temperatures shown on the abscissa) are shown in Figure 24.4.1. (The supplier of specialty steels provides a similar data sheet for each of his materials.) Note the interrelationship between Rockwell C hardness, yield point, and elongation as the function of the annealing temperature after quench (draw).

24.5   ANISOTROPIC PLASTICITY

Many materials currently being produced are both elastically and plastically anisotropic and the theories of Tresca and von Mises do not hold. Likewise the fracture strength varies greatly with direction. Examples of such materials are glass-reinforced plastics (GRP) and boron fiber epoxy composites. Because such anisotropic materials have such vast potential and a rapidly growing rate of use, considerable research work on their anisotropic elastic, plastic, and fracture properties is underway presently. Many cold-worked metals such as rolled zirconium sheet are also highly anisotropic. In such materials the crystals are not oriented randomly but have a **preferred orientation** because of the cold-working process. Preferred orientation may lead to enhanced properties or undesirable effects, and it is therefore important that the materials scientist learn to produce specific orientations.

REFERENCES

Dieter, G., *Mechanical Metallurgy*, McGraw-Hill, New York (1961). Part Three contains a good discussion of applications of theory to materials testing and Part Four is a good introduction to the applications of theory to plastic forming of materials.

Lubahn, J. D., and Felgar, R. P., *Plasticity and Creep of Metals*, Wiley, New York (1961).

Hollomon, J. H., *The Problem of Fracture*, American Welding Society, New York (1946). Contains an excellent discussion of yielding and flow criteria.

Nadai, A., *Theory of Flow and Fracture of Solids*, McGraw-Hill, New York (1950).

McClintock, F. A., and Argon, A. S., eds., *Mechanical Behavior of Materials*, Addison-Wesley, Reading, Mass. (1966).

PROBLEMS

24.1. Prove that the maximum shear stress in a tension test is half the tensile stress.

24.2. In Example 24.1.1, it is stated that making the pressure vessel larger (increasing *b*) would help little.
a. Discuss physically in terms of $\sigma_{\theta\theta}$ why this is so.
b. Suggest two ways to make the pressure vessel stronger.

24.3. A deep sea submergence vessel is made of steel with a tensile yield strength of 100,000 psi. It has an i.d. of 7 ft and is a long cylinder. Ignore end effects and assume the stress state is the same as in Example 24.1.1. Can an o.d. be obtained for which there will be no yielding at a depth of 34,000 ft?

24.4. a. Solve Problem 24.3 for the case where $\sigma_0 = 300,000$ psi.
b. Will the resultant vessel need ballast or flotation devices?

24.5. Explain the origin of the Bauschinger effect in terms of dislocation theory.

24.6. Justify, if possible, the following: Thus the ductility with biaxial tensile stresses is less than with uniaxial tension, while the ductility with triaxial tensile stresses is less than with biaxial stresses. Clearly, a specimen tested in pure triaxial tension (if this were practical) would fracture without any plastic deformation; its ductility would be zero. Its toughness would be only its elastic resilience, which would be low.

24.7. Attempt to justify or discredit the following: Crystalline materials are inherently ductile and can, if intelligence is used, be plastically formed into useful components.

24.8. We mentioned earlier in the text that glass airplanes were not to our liking. Why might we be less hesitant about a glass sphere for a deep sea submergence vessel? (Ignore hatch opening stresses for the time being.)

24.9. Why is a hardness test related to yield strength for a class of material with a given elastic modulus, such as steels?

24.10. List various quantities which are measures of ductility (see Chapter 2 if necessary).

24.11. Discuss experimental evidence for St. Venant's plasticity hypothesis.

24.12. Which is in general the more conservative criteria of yielding, Tresca or von Mises?

24.13. Discuss the state of stress in a necked tension specimen.

24.14. Figure 24.2.4 has been likened to Figure 2.6.4 or 23.9.5. Fundamentally, however, they are different. Discuss the origins of these behaviors.

24.15. A sheet of metal is being rolled into a thinner sheet. Assuming that the sheet does not widen, obtain the real strain in terms of the ratio of thickness t/t_0.

More Involved Problems

24.16. Assume that a spherical thick-walled shell is to be used as a deep sea submergence vessel. Assume the pressure inside is negligible and the pressure vessel behaves as an ideal plastic material with a flow stress σ_0. What maximum external pressure can the vessel withstand when it just becomes fully plastic? Assume the outside radius is b, the outer pressure is P_b, and the inner radius is a.

24.17. A glass rod is heated to a temperature at which it flows when pulled in a tension test. It never necks even when pulled several hundred percent. Explain why.

24.18. A spool of $\frac{1}{16}$-in. diameter annealed copper wire is placed on an axle within a pressure vessel so that it is free to rotate. The end of the wire has been drawn down to 0.002-in. diameter and this is led through a die which forms the bottom piston. The die and wire are tapered as shown in the enlarged view and initially sealed to each other with heavy grease. Describe what happens (as a function of pressure) when the pressure is increased.

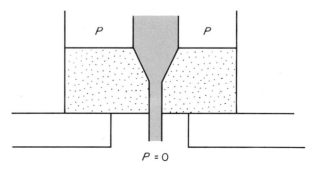

FIGURE P24.18.

Sophisticated Problem

24.19. A thick-walled spherical vessel has an internal pressure P_a and an external pressure P_b. It has internal and external radii of a and b, respectively. *Describe* a systematic approach which you would use to eventually find the radial and circumferential stresses assuming the vessel behaves elastically. Do not actually write any equations. After you complete this part, go over it with your instructor. Then proceed to use mathematical expressions and to obtain solutions.

Prologue

The yield strength, fracture strength, fatigue strength, and creep strength of materials are far below the ultimate strength possible. In this chapter, some of the techniques used by the materials scientist to make solids stronger are discussed. The *grain size* can be decreased. *Precipitates* can be introduced by age-hardening mechanisms. Second phase particles which are chemically more stable can also be introduced by chemical or other means (*dispersion hardening*). Such second phase particles act as barriers to dislocations. The production of *lamellar eutectics* or *eutectoids* by nonequilibrium cooling can also lead to the formation of barriers to dislocation motion, as in the case of *pearlite*. Steels can be greatly strengthened by the production of martensite. Polymers can be strengthened by increasing the crystallinity and by adding appropriate fillers to form a composite. Numerous types of *composites* can be made to achieve strengthening. We are reminded that enormous developments are possible in this field.

25

STRENGTHENING MECHANISMS

25.1 INTRODUCTION

It is the purpose of this chapter to discuss in a systematic fashion how solids are made mechanically stronger. In addition to making crystals free of defects or glasses free of flaws there is one other fundamental mechanism which is used repeatedly for strengthening solids.

FUNDAMENTAL THEOREM OF STRENGTHENING. To strengthen an imperfect solid, disrupt the continuum.

EXAMPLE 25.1.1

Give some examples of application of this theorem to crystalline materials.

Answer.
1. Add solute atoms.
2. Add dislocation forests.
3. Introduce grain boundaries.
4. Introduce second phase particles.

EXAMPLE 25.1.2

Give examples of application of this theorem to noncrystalline sol'ds.

Answer.
1. Place glass fibers in a plastic matrix.
2. Place asbestos fibers in phenolics.

659

COROLLARY TO THE FUNDAMENTAL THEOREM OF STRENGTHENING. The strengthening increases as the mean free path between the disruptions decreases.

We have already seen examples of solute strengthening and dislocation strengthening (strain hardening) in Chapter 23. Various other mechanisms will be discussed in this chapter.

It should be noted that the same fundamental theorem which applies to making materials mechanically hard (high yield strength) applies in a general way to making materials magnetically hard (high coercive force) or to making harder superconductors (high current-carrying capacity).

25.2 GRAIN SIZE

Grain size strongly influences yield stress, fracture stress, and ductility, as shown in Figure 25.2.1. These data illustrate a rather general point, that decreasing grain size is accompanied by increasing strength and by increasing

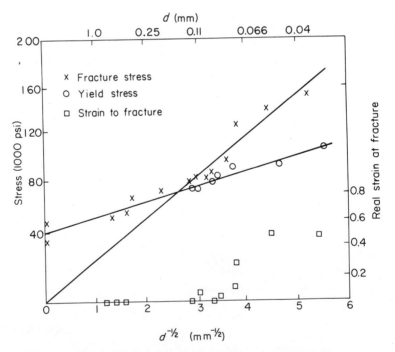

FIGURE 25.2.1. Yield and fracture stresses at $-195°C$ as a function of grain size for a low carbon steel. Single crystal cleavage stresses plotted at $d^{-1/2} = 0$. [J. R. Low in *Relation of Properties to Microstructure*, American Society for Metals, Cleveland (1954) p. 167.]

ductility. This may be accounted for in a general way on the basis that grain boundaries resist the passage of dislocations; in a fine-grained material there are more boundaries and hence more barriers, and therefore the material is stronger. A similar strength-grain size relationship exists for nearly every polycrystalline material that has been investigated.

With respect to fracture, the grain size diameter can be considered as a potential crack length, so that according to the Griffith theory the fracture stress could be expected to depend on the inverse square root of the grain size.

EXAMPLE 25.2.1

Assuming that in the high ductility region of Figure 25.2.1, the plastic crack energy is $p = 5 \times 10^5$ ergs/cm², calculate the fracture stress of the material with $d = 0.04$ mm.

Answer. We use Equation (23.9.8) with $d = c$:

$$\sigma = \left(\frac{2Ep}{\pi c}\right)^{1/2} = \left(\frac{2 \times 2 \times 10^{12} \times 5 \times 10^5}{\pi \times 4 \times 10^{-3}}\right)^{1/2}$$
$$= 13 \times 10^9 \text{ dynes/cm}^2$$
$$= 13 \text{ kbars}$$
$$= 180{,}000 \text{ psi.}$$

25.3   PRECIPITATION HARDENING

We consider a dislocation line of Burgers vector b which is pushed against precipitate particles by an applied stress τ as shown in Figure 25.3.1. As the dislocation line becomes curved, its length increases. By analogy with a surface

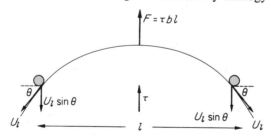

FIGURE 25.3.1. Dislocation, under an applied shear stress τ, stopped by precipitate particles.

(energy per unit area or force per unit length) there is an increase in energy per unit length of line created or a dislocation line tension (force) U_l tending to keep the line from stretching. A simple force balance gives

$$\tau b l = 2U_l \sin \theta. \qquad (25.3.1)$$

We have noted earlier [see Equation (22.3.6)] that $U_l = Gb^2$ (or about

5.3 eV/atom plane length for copper) so we have

$$\tau = \frac{2Gb}{l} \sin \theta. \tag{25.3.2}$$

The motion of a dislocation past such particles is shown in Figure 25.3.2. The maximum stress occurs when $\sin \theta = 1$ so the shear yield stress is given by

$$\tau_0 = \frac{2Gb}{l}. \tag{25.3.3}$$

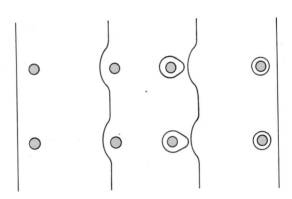

FIGURE 25.3.2. Progress of a dislocation along a slip plane containing second phase particles. Successive stages shown in four views, from left to right.

The tensile yield strength would be twice as large. This analysis assumes that the particles cannot be cut by the dislocation. Some particles, for example, coherent precipitate particles, can be cut if the stress on the dislocation is sufficiently high.

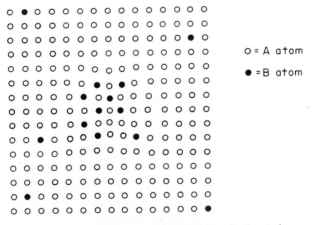

FIGURE 25.3.3. Coherent precipitate with elastic strains. Clustering takes place prior to formation of coherent precipitate; the eventual precipitate may have a different crystal structure.

EXAMPLE 25.3.1

Estimate in units of b the value of l needed to obtain $\tau_0 = 40,000$ psi. for aluminum.

Answer. Since $G = 4 \times 10^6$ psi (see Table 2.2.1) we have $l = 200b$.

An aluminum alloy contains 1.7 at. % copper. See Figure 16.5.1. This forms upon quenching from the κ range a penny-shaped precipitate called a GPII precipitate (a Guinier-Preston zone of the second type; this is a copper-rich region and is apparently an ordered structure which is coherent to the lattice) with a thickness of 50 Å and a diameter of 500 Å. A second phase forms a **coherent boundary** with the matrix if, and only if, only elastic strains are present with no dislocations present. Figure 25.3.3 is illustrative of the formation of a tiny coherent precipitate particle on a hypothetical square lattice. The strain field around precipitate particles can be studied by transmission electron microscopy as shown in Figure 25.3.4.

EXAMPLE 25.3.2

Estimate the distance l between precipitate particles when GP zones as described above are formed in an aluminum alloy containing 1.7 at. % copper. The composition of the precipitate can be taken to be roughly $CuAl_2$ (although it is not the compound) and it can be assumed that all the copper is present in these precipitates.

Answer. As an approximation we shall assume that atom fractions and volume fractions are equal. Then the volume fraction of precipitate is 0.051. Let N be the number of precipitate particles per cubic centimeter and v the volume of one particle. Then $Nv = 0.051$, and since we can calculate v from the given dimensions, we have $N = 5 \times 10^{15}/cm^3$. Then $l \approx 1/N^{1/3} \approx 6 \times 10^{-6}$ cm $= 600$ Å, and since the Burgers vector of aluminum is 2.8 Å, we have $l \approx 200b$. We would expect from (25.3.3) that such an alloy would have a shear yield stress of about 40,000 psi. The actual value for a single crystal is about 20,000 psi, compared to a value of a few thousand in a pure annealed single crystal.

To obtain sizable increases in strength the precipitate particles must have a spacing of less than 1000 Å. Figure 25.3.5 shows precipitate particles in MgO acting as barriers to dislocations.

EXAMPLE 25.3.3

Are the fine precipitates of Example 25.3.2 in equilibrium and if not, why not?

Answer. The 5×10^{15} particles/cm^3 have a total interfacial area of about 5×10^5 cm^2, while if all the "$CuAl_2$" were present as one particle, its surface area would be about 1 cm^2. The interfacial energy is thus about

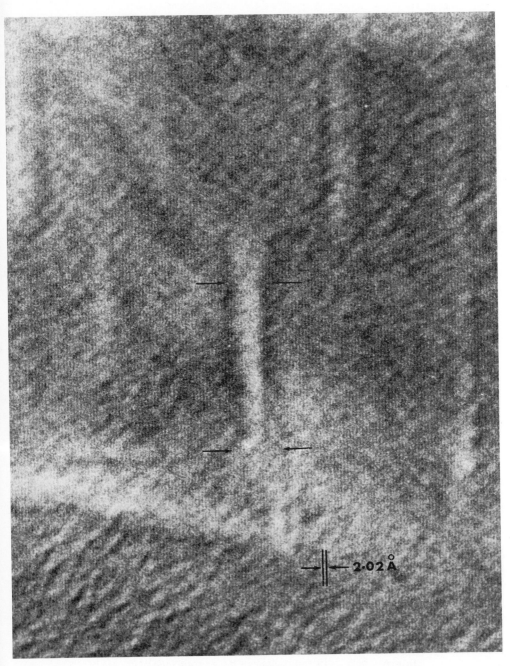

FIGURE 25.3.4. Transmission electron micrograph of GP-I
zones in Al + 4% Cu. (*Courtesy of Richard
Parson and Larry Howe, Chalk River*).

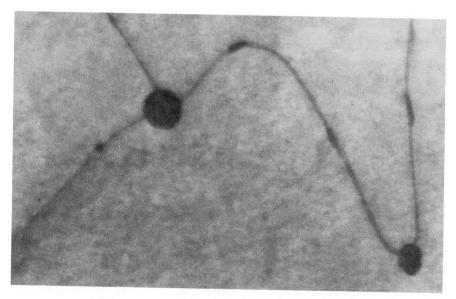

FIGURE 25.3.5.   Electron micrograph illustrating age precipita-
tion hardening of MgO. (*Courtesy of G.
Thomas and J. Washburn, University of
California.*)

10^9 ergs/cm^3 in the one case and negligible in the other. The fine precipitates have a higher free energy. The free energy of the systems can
decrease by particle growth (larger ones grow at the expense of smaller
ones). This involves diffusion. An estimate of the diffusion distance
squared is $x^2 = 4Dt$ (see Section 18.3). Hence additional growth might
occur if

$$l^2 \approx 4Dt,$$

where l is the interparticle distance, D is the diffusion coefficient, and t is
the diffusion time. If $l = 500$ Å and $t = 3$ years ($\approx 10^8$ sec), this means
that D must be less than 6×10^{-20} cm^2/sec. At room temperature $D \approx$
10^{-22} cm^2/sec for diffusion of aluminum in aluminum. In the present case
both Cu and Al must diffuse in opposite directions and the process would
be controlled by the diffusion of the slowest moving atom.

Clearly, age-hardened alloys are not of much use at elevated temperatures because **overaging** occurs. In overaging some precipitate particles grow
larger, while many others become smaller and disappear. As the distance between
the fewer larger particles becomes larger, their ability to stop dislocation movement decreases.

Cyclic straining also causes continued weakening of these alloys (there
is no endurance limit). This is probably due to point defect production during

cyclic straining which can cause overaging and easy nonconservative dislocation motion.

Note that useful precipitate particles are smaller than the wavelength of light and so cannot be seen in the optical microscope. When the particles are large enough for such observations the material is overaged. At such time they have transformed to $CuAl_2$ compound and are incoherent with the aluminum matrix.

EXAMPLE 25.3.4

If the alloy of Example 25.3.2 is aged until the $CuAl_2$ particles are spheres with a diameter of 2 μ, estimate the distance between particles.

Answer. Using $Nv = 0.05$ and $v = \frac{4}{3}\pi(10^{-4})^3$ we have

$$N \approx 10^{10} \quad \text{and} \quad l \approx 3 \times 10^{-4} \text{ cm} \approx 10^4 b.$$

This will give no effective hardening; i.e., dislocation hardening, etc., would be much larger.

EXAMPLE 25.3.5

Suppose you invent a new cheap method for obtaining yield strengths of 200,000 psi in aluminum. Discuss the economic impact of this discovery on the production of 2000 airplanes weighing 500 tons each when loaded. See Chapter 1.

Answer. We noted in Chapter 1 that each pound of material not needed for structure means 1 lb of cargo and $10. Let us assume that 30% of the loaded plane is structure and that two thirds of this is aluminum, i.e., 20% of 500 tons. The total aluminum involved in this application is $2000 \times 500 \times 2000 \times \frac{1}{5} = 4 \times 10^8$ lb. If only one third as much aluminum is needed as at present (note the yield strength of present-day heat-treated aluminum alloys in Table 2.1.2), we would have an economic contribution of

$$\tfrac{1}{3} \times 4 \times 10^8 \text{ lb} \times \$10/\text{lb}$$

or

$$\$1.3 \times 10^9$$

(even if 1 lb of your alloy costs as much as 3 lb of the present alloys). Such an improvement is in principle possible.

25.4 DISPERSION-HARDENED ALLOYS

Aluminum powder very rapidly forms an oxide coating. Such powder when sintered forms **sintered aluminum powder**, or **SAP**. The strengthening mechanism is the same as with the age precipitation-hardening alloys. There is one very important difference. The Al_2O_3 particles are extremely stable thermally (compared to the GP zones in the Cu–Al system) and the rate at which one Al_2O_3

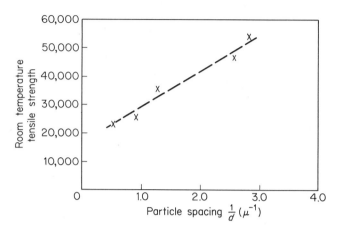

FIGURE 25.4.1. Variation of tensile strength of Al–Al$_2$O$_3$ alloys with particle spacing. [E. Gregory and N. J. Grant, *Trans. AIME* **200**, 247 (1954).]

particle grows at the expense of another is very slow. Al$_2$O$_3$ has a melting point of 2050°C, and oxygen has a very low solubility in aluminum. An illustration of the strengthening of Al–Al$_2$O$_3$ alloys is shown in Figure 25.4.1 where a progressive increase in strength with decreasing interparticle spacing is established. In this system the dispersed phase (Al$_2$O$_3$) is an incoherent dispersion; the same qualitative effect is obtained with coherent precipitates, but the theoretical considerations are somewhat different.

Nickel hardened by a dispersion of ThO$_2$ also has good strength at high temperatures. It can be produced by starting with a mixture of nickel oxides and thorium oxides; the nickel is reduced to metal but the more stable ThO$_2$ remains. It is called **TD nickel**.

25.5 PERIODIC STRUCTURES

LAMELLAR STRUCTURES. When a eutectoid steel (see Figure 16.5.3) is cooled below the eutectoid temperature at 723°C, it forms pearlite, a structure consisting of alternate layers of hard Fe$_3$C and relatively soft α-iron or ferrite. We would expect the dislocation movement to take place primarily in the ferrite. Hence decreasing the lamellar spacing decreases the mean free path of the dislocations; i.e., they can effectively move no farther than the ferrite thickness. Thus we expect the strength to increase as the lamellar spacing decreases. This is shown in Figure 25.5.1. The pearlite spacing decreases as the isothermal transformation temperature decreases.

Patented steel wire is an alloy of about 0.9% C, 0.4% Mn, and 0.2% Si which is transformed to pearlite at 500°C (from austenite at 1000°C). This

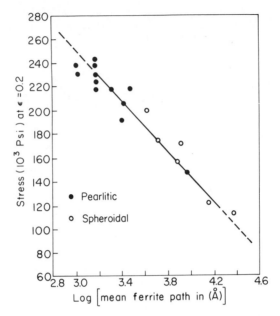

FIGURE 25.5.1. Relation between pearlite lamellar spacing and strength. [J. D. Holloman and L. D. Jaffee, *Ferrous Metallurgical Design*, Wiley, New York (1945).]

results in a very fine pearlite, which is then drawn to a real strain $\epsilon = 4$ (note the enormous ductility of fine pearlitic steels). The resultant material has a strength of about 600,000 psi.

SPINODAL DECOMPOSITION HARDENING.   Figure 25.5.2 shows the phase diagram for a 67.2 Cu–30Ni–2.8Cr alloy.

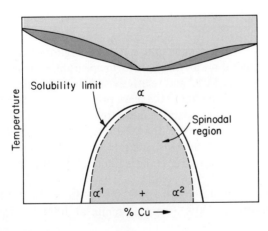

FIGURE 25.5.2. Phase diagram for 67.2Cu-30Ni-2.8Cr showing spinodal regions.

We note that over a certain composition range the high temperature α solution (a single fcc phase of nickel and chromium dissolved in copper—a substitutional solid solution) decomposes at lower temperatures into a copper poor region designated α^1 and a copper rich region designated α^2. Each of these new phases has fcc crystal structures; however, they differ in composition (the variation of composition is similar in form to the cored structure shown in Figure 16.4.7 but on a much finer scale. Recall that the cored structure resulted from freezing of a liquid). The decomposition process involves a spinodal decomposition (see Section 18.15).

There are no definite interfaces between α_1 and α_2 but instead a gradual compositional variation; hence, the two phases are coherent. This coherency gives rise to high lattice strains. The periodic variation of these high strains causes the strengthening which we call here **spinodal decomposition hardening**. Heat treating simply involves cooling from the α-region after hot working or annealing. The heat-treated alloy has a yield strength of 155,000 psi compared to an ordinary hot-rolled 70–30 Cu–Ni alloy for which the yield strength is 50,000 psi. The alloy is extremely tough, having a Charpy V-notch impact strength of over 150 ft-lbs at cryogenic temperatures (materials with fcc crystal structures as a rule do not show the drastic drop in impact strength exhibited by bcc metals and hcp metals as the temperature is lowered).

25.6   MARTENSITE FORMATION

Figure 25.6.1 illustrates the transformations which take place when a eutectoid steel is cooled very rapidly from the austenitic (γ-iron) range to a fixed temperature and is then held at the temperature. The resultant curve is called a **time-temperature-transformation curve** (**TTT-curve**). Suppose the steel is cooled rapidly from 1400 to 1200°F. After 3 sec at 1200°F some pearlite has formed. Some arbitrary definition, such as the presence of 5% pearlite, is used to define the "transformation starts" line. To carry out this analysis, the experimentalist would rapidly quench the specimen in water when $t = 3$ sec and would subsequently polish and etch the tiny specimen and examine it microscopically for the presence of pearlite (no additional pearlite will form during the rapid quench or subsequently at room temperature). If a specimen is held for 30 sec at 1200°F, it nearly completely transforms to pearlite. The presence of 95% pearlite can be arbitrarily set to define the "transformation ends" line.

This diagram indicates two basic and radically different mechanisms of austenite transformation; at high temperatures the reaction products, pearlite and/or bainite, form with increasing time at constant temperature. The low-temperature product martensite usually increases in amount with decreasing temperature and except in a few cases does not increase in amount at constant temperature. When the temperature is lowered to a given temperature below the M_s temperature, a fraction of the material is transformed to the martensitic

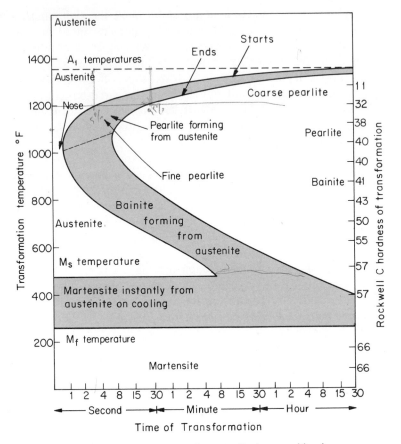

FIGURE 25.6.1. TTT curve for austenite decomposition in a eutectoid steel.

phase extremely rapidly, with the interface velocity between the phases being nearly the speed of sound.

Martensite is a very hard structure. Its hardness increases as the carbon content increases. It is a highly dislocated tetragonal crystal structure. The martensite is present as tiny lenticular platelets, as shown in Figure 16.5.7.

EXAMPLE 25.6.1

Why is the engineer interested in producing martensite?

Answer. Martensitic steels can have yield strengths approaching 300,000 psi (20 kbars). However, they are nearly nonmachinable at that strength level (except by grinding). The component is therefore machined when the steel is in a coarse pearlitic form, i.e., when its strength and hardness are relatively low. It is then austenized and then quenched to produce martensite. It is then given an anneal (temper) to remove residual stresses and to increase ductility and toughness, as shown in Figure 24.4.1. Because there is some distortion as a result of the heat treatment, the component is finally machined by grinding.

In actual heat-treating practice, when a steel is quenched the transformation takes place continuously at various temperatures rather than at a fixed temperature. Such behavior is shown in Figure 25.6.2. As is clear from this diagram, 99% martensite could be produced throughout the specimen only if the specimen is very thin because the inside of the specimen would have to cool to 450°F in 10 sec or less if no pearlite is to be formed. From a practical viewpoint we are interested in producing martensite in very large sections, often several inches in diameter.

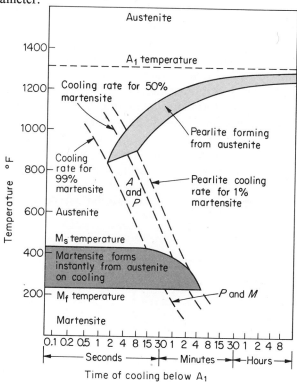

FIGURE 25.6.2. An approximate CCT (continuous cooling transformation) diagram for eutectoid carbon steel.

EXAMPLE 25.6.2

Estimate the order of magnitude of time required for the temperature at the center of a 6-in. diameter rod to drop from 1400°F to below M_s.

Answer. We can use

$$t \approx \frac{L^2}{4D_T}$$

(see Section 6.2), where D_T is the thermal diffusivity given by (6.2.2). We can use the data for iron for k_T (see Table 6.2.1) and for ρ (Table 2.7.1) and the fact that $C_p \approx 6$ cal/mole-°K for metals at high temperature. We have $D_T \approx 10^{-3}$ cm²/sec and hence

$$t \approx 10{,}000 \text{ sec.}$$

As is clear from Example 25.6.2 the steel of Figure 25.6.2 would have no martensite at the center of a 6-in. diameter rod. There is, however, a practical solution to this problem, although it adds to the cost of the steel. The addition of alloying elements invariably moves the cooling curves to longer times. Vanadium and manganese are particularly effective in this regard. The presence of these elements slows down the nucleation rate of pearlite at the austenite boundaries and may also slow the overall diffusion rate. Thus it is indeed possible to make alloy steels which would upon quenching have 50% martensite at the center of a 6-in. round. The ability of a steel to form martensite during a quench is called the **hardenability**. It is often measured in terms of the diameter in inches of a round bar which will form 50% martensite at the center when quenched. Hardenability should not be confused with hardness since hardenability is the ability of a steel to obtain hardness.

A convenient hardenability test is the **Jominy test**. Here a specimen (see Figure 25.6.3) is taken from the furnace at the austenitizing temperature, placed in a holder, and quenched at one end by a jet of water. Heat losses to the air are negligible. (The specimen initially has two flat parallel ground surfaces on

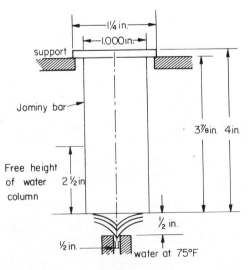

FIGURE 25.6.3. Jominy end quench hardenability test.

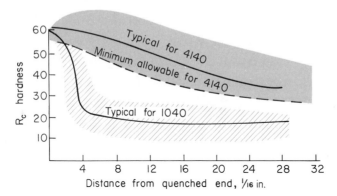

FIGURE 25.6.4. Jominy curves for two 0.40% C steels. The last two digits in the alloy number give the carbon content.

opposite sides of the cylinder.) When the specimen is cold, Rockwell-C hardness measurements are made at $\frac{1}{16}$-in. intervals from the quenched end. The results are shown for two steels, each of which contain 0.4% C, in Figure 25.6.4. We note that the maximum hardness in *each* case is about R_c-58. The maximum hardness is obtained when 100% martensite is produced and is determined by the carbon content, as shown in Figure 25.6.5. The 1040 steel is a plain carbon steel with 0.40% C. It has a low hardenability. The 4140 steel is a low alloy steel (0.80–1.10% Cr, 0.18–0.25% Mo, and 0.40% C). The American Iron and Steel Institute (AISI) and the Society of Automotive Engineers (SAE) have established standard four-numeral designations for carbon and alloy steels, as shown in Table 25.6.1. The hardenability of 4140 steel is considerably higher than that of 1040 steel. Figure 25.6.6 shows the hardness versus cross section of quenched rods of these materials. Steel, depending on the heat treatment and processing, can have varied strengths, as shown in Figure 25.6.7. It should be noted that all of the strengthened steels are strengthened by the presence of nonequilibrium structures and hence will lose their strength at high temperatures and will be susceptible to cyclic stress softening. For example, carbides are precipitated from martensite upon tempering.

In general, as the overall strength of a steel increases, the cost increases. The actual choice of material is based on the total economic factor.

MARTENSITE TRANSFORMATIONS. The term **martensitic transformation** is usually reserved for transformations which occur by a displacement mechanism (such as shear) in which individual atoms execute well-defined and correlated movements. This shape change causes an initially smooth surface to become serrated as a result of the transformation. The nature of the surface relief suggests

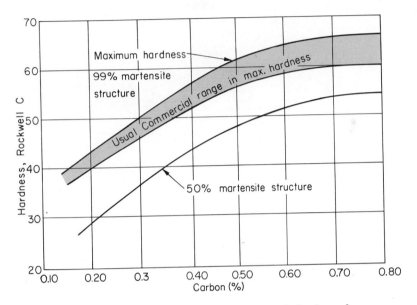

FIGURE 25.6.5. Relation of carbon content to the hardness of plain carbon steel. Data apply approximately to low and medium alloy steels. [*Metals Handbook*, American Society for Metals, Novelty, Ohio (1948) p. 497.]

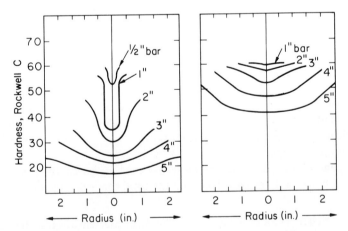

FIGURE 25.6.6. Hardness versus cross section for hardened bars of various diameters. (a) SAE 1040 quenched in water. (b) SAE 4140 quenched in water.

Table 25.6.1. COMBINED AISI AND SAE STANDARD DESIGNATIONS FOR CARBON AND ALLOY STEELS

Carbon Steels

$10 \times \times$	Nonresulfurized basic open-hearth and acid Bessemer carbon steel
$11 \times \times$	Resulfurized basic open-hearth and acid Bessemer carbon steel

Low Alloy Steels

$13 \times \times$	Manganese (1.60–1.90%) steels
$23 \times \times$	Nickel (3.25–3.75%) steels
$25 \times \times$	Nickel (4.75–5.25%) steels
$31 \times \times$	Nickel (1.10–1.40%)–chromium (0.55–0.75% or 0.70–0.90%) steels
$33 \times \times$	Nickel (3.25–3.75%)–chromium (1.40–1.75%) steels
$40 \times \times$	Molybdenum (0.20–0.30%) steels
$41 \times \times$	Chromium (0.80–1.10%)–molybdenum (0.15% or 0.18–0.25%) steels
$43 \times \times$	Nickel (1.65–2.0%)–chromium (0.40–0.60% or 0.70–0.90%) steels–molybdenum (0.20–0.30%) steels
$46 \times \times$	Nickel (1.65–2.0%)–molybdenum (0.15% or 0.20–0.27% or 0.30%) steels
$48 \times \times$	Nickel (3.25–3.75%)–molybdenum (0.20–0.30%) steels
$50 \times \times$	Chromium (0.20–0.35% or 0.55–0.75%) steels
$51 \times \times$	Chromium (mean percent 0.80, 0.90, or 1.05) steels
$5 \times \times \times \times$	Chromium steel (mean percent 0.50, 1.00, or 1.45) with carbon 0.95–1.10%
$61 \times \times$	Chromium (mean percent 0.80 or 0.95)–vanadium (0.10 or 0.15% min.) steel
$86 \times \times$	Nickel (0.40–0.70%)–chromium (0.40–0.60%)–molybdenum (0.15–0.25%) steels
$87 \times \times$	Nickel (0.40–0.70%)–chromium (0.40–0.60%)–molybdenum (0.20–0.30%) steels
$92 \times \times$	Manganese (0.85% mean)–silicon (1.8–2.2%) steels
$93 \times \times$	Nickel (3.0–3.50%)–chromium (1.0–1.4%)–molybdenum (0.08–0.15%) steels
$94 \times \times$	Manganese (1.0% mean)–nickel (0.30–0.60%)–chromium (0.3–0.5%)–molybdenum (0.08–0.15%) steels (formerly N.E. type)
$97 \times \times$	Nickel (0.40–0.70%)–chromium (0.1–0.25%)–molybdenum (0.15–0.25%) steels (formerly N.E. type)
$98 \times \times$	Nickel (0.85–1.15%)–chromium (0.70–0.90%)–molybdenum (0.20–0.30%) steels (formerly N.E. type)
$99 \times \times$	Nickel (1.00–1.30%)–chromium (0.40–0.60%)–molybdenum (0.20–0.30%) steels (formerly N.E. type)

EXAMPLE 25.6.3

A steel rod of 5-in. diameter is made of 4340 steel. This has a higher hardenability than 4140 steel. It is desired that this rod have a yield strength of at least 220,000 psi at the surface and at least 150,000 psi at the center. Will the 4340 material suffice?

Answer. Both of these steels have the same carbon content and hence the same maximum hardness. From Figure 25.6.6(b) we see that the surface hardness will exceed $52R_c$, while the inside hardness will exceed $40R_c$. From Figure 24.4.1 we note that the surface yield strength should exceed 220,000 psi and that the yield strength at the center should exceed 150,000 psi. Hence 4340 steel should suffice (as far as the strength requirements are concerned).

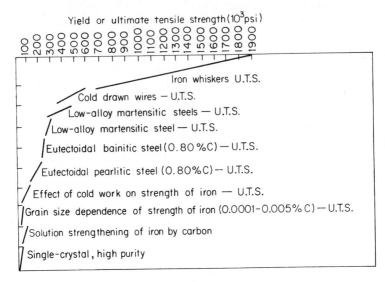

FIGURE 25.6.7. Ranges of strengths of various steels.

that the transformation is a homogeneous deformation (a homogeneous deformation is one in which straight lines remain straight lines, although their length may change and the angles between pairs of these lines may change). Also, in martensitic transformations there is a definite orientation relationship between the martensite crystal and the parent crystal. Finally, the plane separating the two phases (martensite and parent) is a particular crystallographic plane of the parent called the **habit plane**.

The martensitic transformation in steel is just one of many martensitic transformations. One of the simplest is the transformation in cobalt from hexagonal closest packed to face centered cubic. In this case, the motion of partial dislocations on successive planes will change the packing of the closest packed layers from the . . . ABAB . . . sequences of the hcp crystal to the . . . ABCABC sequence of the fcc crystal. The partial dislocation can move from plane to plane (normal to the . . . ABAB . . . sequence) if a screw dislocation lies in the normal direction. This is called a **pole mechanism**. It is thought that the pole mechanism is the likely mechanism in martensitic transformations [see C. M. Wayman, *Introduction to the Crystallography of Martensitic Transformations*, Macmillan, New York (1964)].

Martensitic transformations are not restricted to metals. Thus, the transformations of RbCl, RbBr and RbI at high pressure from the NaCl-type to the CsCl-type crystal structure show the features characteristic of a martensitic transformation [see P. J. Reddy and A. L. Ruoff, in *Physics of Solids at High*

Pressure, edited by C. T. Tomizuka and R. M. Emrick, Academic Press, New York (1965) p. 510].

A partial list of materials which undergo martensitic transformations is shown in Table 25.6.2. Many of these are discussed in detail in the book by Wayman.

Table 25.6.2. SYSTEMS SHOWING MARTENSITIC TRANSFORMATIONS

Cu—(10–15) at. % Al
Co
Ti
Carbon Steel
Stainless Steel
Au—47.5 at. % Cd
Rubidium Halides

EXAMPLE 25.6.4

A single crystal of martensite is more stable than the fcc steel below M_s. Why does a sample of the steel only partially transform to martensite at a given temperature?

Answer. Martensite platelets originate at many points. When numerous platelets are formed, two additional energy terms arise: interfacial energy and elastic strain energy. Both terms change the energetics so that additional lowering of the temperature would be required before additional martensite could be formed.

There is currently considerable research underway involving not only how martensitic transformations occur, but why and when. For example, many materials exist in the metastable state without transforming. We can understand why diffusion controlled transformations do not occur; the kinetics are too slow. But why do these materials not exhibit a martensitic transformation? The author believes that the nucleation of the latter involves an elastic instability in the bad material (such as dislocations) where the stress state is considerably different than the bulk (see Section 22.2 where it is noted that the core of a dislocation is, to a rough approximation, under a high hydrostatic pressure). Thus a condition for martensitic transformation of a cubic crystal [for which C_{11}, C_{44}, $(C_{11} - C_{12})/2$ and K are positive in the good material] is that the pressure derivative of one or more of these quantities be sufficiently negative, so that at the pressure at the core of the dislocation, the elastic constant itself becomes negative. For the rubidium halides mentioned earlier, $dC_{44}/dp < 0$. Further research may show that this is true for other systems, such as Fe-Ni, Li, Na, etc.

25.7   STRENGTHENING AT HIGH TEMPERATURES

One engineering criterion of high-temperature strength ($T > 1000°$K) is the maximum temperature at which a material will support a tensile stress of 20,000 psi without failure in 100 hr. Strengthened metals usually become soft and weak at $T_M/3$ to $T_M/2$. A first thought is that we should forget ordinary metals and use either intermetallics or oxides because of their very high melting points. HfC melts at 4160°C and MgO at 2800°C, while iron melts at 1540°C. The problem here is one of fabrication difficulties and lack of ductility at low temperatures.

One way of strengthening metals is by dispersion hardening. Austenitic stainless steels also have greater creep resistance. Iron-base alloys with up to 15% Cr (to provide corrosion resistance) and up to 25% Ni (to retain the austenitic structure) are often used. They are often hardened by the precipitation of Ni_3Al or Ni_3Ti. Research is underway on cobalt-based, nickel-based, and chromium-based alloys as well, and a number of these are used in commercial practice. Currently superalloys can support 20,000 psi for 100 hr at about 1000°C.

An important possibility for high-temperature use involves composites, which are discussed later.

25.8   STRENGTHENING MECHANISMS IN POLYMERS

Several hundred million tons of iron and steel are produced throughout the world every year. The *volume* of plastics exceeds the volume of all metals produced. The applications of plastics are diverse but certainly in some cases we are concerned with high strength (nylon rope, for example).

Polymeric materials are usually used because of their mechanical properties although their use as electrical insulators (dielectrics) also is important.

EXAMPLE 25.8.1

Name some applications of polymers.

Answer.
1. Fibers for textiles.
2. Plastic for chairs.
3. Elastomers for conveyor belts.
4. Polyurethane foams for insulation.
5. Foams for synthetic sponges.
6. Electric light switches.
7. Paints and coatings.
8. Adhesives for bonding metals, wood, plastics, etc.

9. Insulation of electrical wire and cable.
10. Bottles, tanks, and other containers.
11. Filament for 6-in. diameter anchor cables.
12. Cases for radios, phonographs, television sets, etc.

Often plastics are used because of their ease of production by casting, molding, extruding, blowing, etc. [see A. K. Schmidt and C. A. Marlies, *Principles of High Polymer Theory and Practice*, McGraw-Hill, New York (1948)].

We have already noted in Table 2.1.2 that nylon fibers are about ten times as strong as bulk nylon. The difference is one of chain alignment. Nylon fibers are permanently stretched several hundred percent as they come out of the extrusion hole. In this process the linear polymer chains become aligned, the hydrogen bonding between parallel chains (see Section 14.7) is maximized, and there is some increase in crystallinity. Thus one important strengthening mechanism is to obtain chain alignment. Another mechanism is to attain maximum crystallinity; to achieve this, say, for polyethylene it would be necessary that the molecules have essentially no side chains.

Another important polymer strengthening process is to increase the number of primary bonds between chains. Ordinary rubber may have a small shear modulus ($G \simeq 10$ psi), while the addition of 30% sulfur which crosslinks the isoprene chain may increase the modulus to $G \simeq 10^5$ psi, which is characteristic of bowling balls, automobile battery cases, etc. In the process, there is also a very large increase in strength. (Diamond and graphite can be considered the extreme limit of such a crosslinked polymer, and diamond and graphite are indeed very strong.) The phenolic resins are an example of such materials. They are thermosetting plastics which do not melt and hence hold their strength to high temperatures (for plastics).

Current research is being carried out on epoxy resins and imido compounds. Certain of these materials have tensile strengths of 50,000 psi at room temperature and maintain reasonable strengths at higher temperatures.

One of the important ways which plastics are strengthened is by adding fillers, e.g., asbestos to phenolics. The resultant material is a composite. Composites as a group are discussed in Section 25.9.

25.9   COMPOSITES

Composites were previously mentioned in Section 15.5. It was noted there that the tremendous strengths of whiskers, filaments, and wires (see Table 25.9.1) could be utilized by placing them in a ductile matrix.

There is usually a size effect associated with strong filaments, etc., such that their strength decreases as their diameter increases. This is illustrated for the case of sapphire whiskers in Figure 25.9.1. A similar relationship applies to filaments and wires.

Table 25.9.1. STRENGTHS OF FILAMENTS, WIRES, AND WHISKERS

	psi	kbars	
Asbestos	750,000	50	(Short fibers)
Glass	500,000	33	Filaments
Boron	400,000	27	Filaments
Graphite	300,000	20	Filaments
Be	180,000	14	Wire
Patented steel	600,000	40	Wire
Tungsten	550,000	37	Wire
Iron	1,900,000*	140	Whiskers
Graphite	2,800,000*	200	Whiskers
Sapphire	7,500,000*	60–500	Whiskers

* Maximum values are quoted. Thus the stronger sapphire whiskers *usually* have strengths of about 1,000,000 psi.

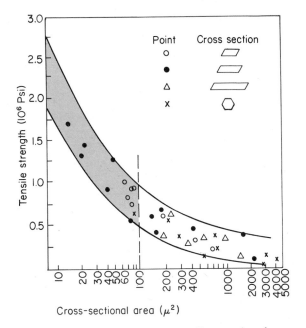

FIGURE 25.9.1. Relation between tensile strength and cross-sectional area of α-Al_2O_3 whiskers. [From E. Scala in *Fiber Composite Materials*, American Society for Metals, Cleveland (1964).]

Thus the very high-strength materials have a diameter of about 1 μ and are therefore not easy materials to handle. Figure 25.9.2 shows whisker growth from solids. Materials scientists and engineers are actively engaged in research to grow whiskers more cheaply, to grow them continuously, etc.

FIGURE 25.9.2. Whiskers of B₄C. (*Courtesy of Iqbab Ahmed, Watervliet Arsenal.*)

ELASTIC MATRIX. The role of the matrix is to distribute the stress from one fiber to another, and in so doing it must support shear stresses. We shall not delve into the mechanism of stress distribution in a composite at this time. There are a few simple rules which composites often obey. Figure 25.9.3 illustrates the **volume fraction rule**. Matrices are commonly ductile metals or plastics. However, graphite filaments have also been embedded in graphite (by first embedding in rayon and then heating to decompose the rayon).

The strength of such composites will vary drastically with direction. Thus silica fibers in a "pure" aluminum matrix result in a composite which has a tensile strength of 110,000 psi along the fibers but has a tensile strength of only 14,000 psi at 45 to 90 deg to the fiber axis. There are several mechanisms of failure:

1. Fracture of the fibers.
2. Shear failure of the matrix along the fibers.
3. Fracture of the matrix in tension normal to the fibers or failure of the fiber-matrix interface.

The mechanism which will be responsible for failure may be any one of these depending on the angle ϕ between the fibers and the specimen axis, as shown in Figure 25.9.4. It is left as an exercise to the student to develop these fracture equations. A micrograph of a fractured composite is shown in Figure 25.9.5.

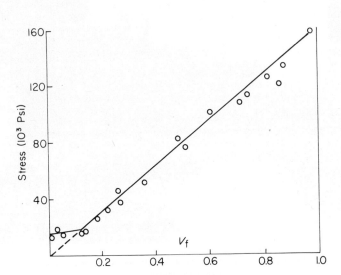

FIGURE 25.9.3. Measured ultimate tensile strength of copper reinforced with continuous brittle tungsten wires of 0.5-mm diameter, as a function of volume fraction of tungsten. [After A. Kelly, *Strong Solids*, Oxford, New York (1966).]

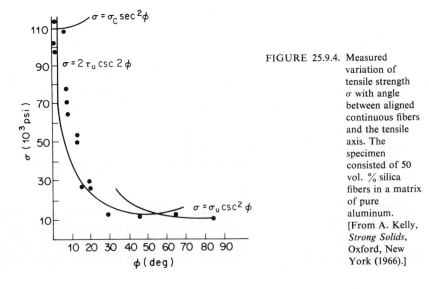

FIGURE 25.9.4. Measured variation of tensile strength σ with angle between aligned continuous fibers and the tensile axis. The specimen consisted of 50 vol. % silica fibers in a matrix of pure aluminum. [From A. Kelly, *Strong Solids*, Oxford, New York (1966).]

The scanning electron microscope, because of its depth of focus, is one of the important tools for studying fracture surfaces.

The composites considered so far consist of aligned continuous filaments. Many composites are made in which the filament windings of one layer are at an angle (say 90 deg) to the windings of the next layer. Such composites are strong in two dimensions, but not as strong in any one direction as the uniaxial composites. Other composites are made of short fibers (although long with respect to their thickness). These fibers may be nearly aligned, they may have their axes randomly oriented in a plane, or they may even be randomly oriented in three directions. The volume fraction of fiber would be expected to decrease in the latter cases. The strongest composites are made from continuous filaments (other things being equal).

FIGURE 25.9.5. Scanning electron micrograph of fracture surface of composite. (*Courtesy of R. K. Matta, U. Pittsburgh*).

VISCOUS MATRIX. In many applications the matrix may be a highly viscous material such as asphalt which is reinforced with gravel and sand. Likewise, when fiber composites are used at high temperatures, the matrix material creeps readily and thus also behaves in a viscous manner, that is, as a highly viscous fluid.

Figure 25.9.6 shows the variations in viscosity of a fluid to which spheres are added. This could be considered to be a model for sand or gravel in asphalt and hence is important in highway behavior. It could also be considered to be a model for wet concrete. Usually we add much more water to the concrete mix than is needed for hydration because the viscosity of the mix has to be reduced

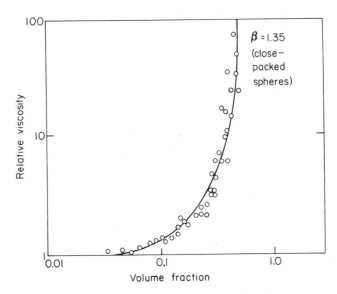

FIGURE 25.9.6. Variation of relative viscosity with volume fraction. [Experimental data after J. G. Brodnyan, *Transactions of the Soceity of Rheology*, **3**, 61 (1959).]

to a point where the mix can readily be handled. In so doing we obtain a weaker cured concrete which is an elastic matrix composite similar (except for particle size) to grinding wheels, which are also important composites.

The solid line which closely fits the points follows **Mooney's equation**:

$$\ln \frac{\eta}{\eta_0} = \frac{2.5 V_r}{1 - \beta V_r}, \tag{25.9.1}$$

where V_r is the volume fraction of spheres (which can be as large as 0.741 for closest packing of spheres) and $\beta = 1.35$ for the closest-packing case. It should be noted that there is a very rapid increase in viscosity as the available solid volume space is nearly occupied. The viscosity coefficient, however, does not go to infinity as (25.9.1) predicts. It is clear, however, that adding a strong second phase to a viscous matrix will enormously increase its resistance to creep flow.

If very fine fibers are placed in a metal matrix at high temperatures, it is possible (likely) that the fiber will dissolve in the metal or undergo some chemical reaction which will weaken it. Protecting the fiber from its environment is a problem undergoing considerable research.

The area of composites is a rapidly growing one in which there is active research. It is likely that there will be interesting developments in this field in the near future.

25.10 MATERIALS SELECTION FOR MECHANICAL STRENGTHS

Suppose that you have carefully analyzed a device and know the force-time-temperature conditions under which it must operate. How do you proceed to select a suitable material from the multitude of available materials, with due account for the economic factors, which necessarily also include the cost of producing the component? (There may be properties, other than strength, which are also important—such as corrosion—or the material may have to be non-magnetic.) Perhaps the best thing to do first is to become familiar with the monthly magazine *Materials Engineering*—perhaps to subscribe to it and in particular to learn to use the annual *Materials Selector Issue* of the magazine.

In addition there are a large number of handbooks concerned with specific materials such as metals or ceramics or plastics, etc.

Finally the producers of steel, aluminum, nylon, phenolics, alumina, etc., provide various brochures, catalogs, etc., on their materials.

EXAMPLE 25.10.1

Your research and development group is given the job of attempting to find a better and cheaper material for the front wheel spindle part for Ford cars. This part is a complicated shape containing the spindle (which supports the wheel on bearings) which is supported by a base; three arms project from the back of the base. The base, spindle and arms form a single part which is formed by forging. The base has four holes drilled in it to hold the brake drum. Your group is told that the part should show *no* failures for 100,000 miles rough usage of each of 10,000,000 spindles. Your group is also told that the existing manufacturing facilities must be used since these involve a tremendous capital investment. The same material has been used since 1929. The forged part is rough machined, heat treated, and finished ground. What factors would you use in materials selection or development?

Answer. The important factors here are how durable the material is, its initial cost, and how readily it can be produced. The finished part must satisfy the design criteria at the lowest possible cost. Here the single important factor in durability is fatigue. The loading history is complex. It is modified in a major way because of the advent of polyglass tires, which show much less roll in cornering than an ordinary tire, so that more severe stresses are placed on the spindle. You must know the loading history and you must be able to reliably predict the fatigue lifetime.

Production involves formability, machineability, and hardenability. Formability involves the ease of forging in a die, and the most important factor is die life and die cost (a different die material may be better for different steel). The most important factors in machineability are tool life followed by cutting speed; moreover, the material must have sufficient hardenability so that the part is hardened sufficiently throughout its thickness.

As an illustration: If the new material is available at 2.5¢ less per pound, the cost of a 40 pound spindle would be decreased by $1.00. Suppose a similar savings would accrue because a lighter component could be used due to improved material properties. This $2.00 saving per spindle, if repeated on a weight basis throughout the car, would reduce the cost of the car by $150–200.

25.11   A REVIEW OF MECHANICAL BEHAVIOR

When the author first came to Cornell, he established a weekly seminar on dislocations. One unexpected result was the poem 'A Crystals Lament' by H. D. Block which was first published in *Metals Progress* **77**, 100 (1960). It is reproduced here with the kind permission of Professor Block.

A Crystal's Lament

I am a little crystal
In a polycrystal sea,
There are many other crystals
And they're pretty much like me.
I don't have much to do with most
Except quite distantly
But some of us get together
At the old grain bound-ary
At the old grain bound-aree!

That's where the dislocations meet
When the tension gets too high
Warming their backs, against the stacks
Of amorphous nuclei.
Bawdy songs and ribald laughter
Fill interatomic space
As they chide a sessile companion
Who has been just put in his place.

Then over potential hills they leap
The cry is "Up and at 'em!"
To free a friend who is caught by his end
On an oversized solute atom;
Or with voices hushed and with mournful shudders

They bemoan the fate of two lost lovers,
Beyond the help of resurrectors;
They embraced—having opposite Burger's vectors.

At the old grain boundary—
That's where the dislocations go
When the tension gets too high;
On a summer's day I've seen it show
A million p.s.i.!
How can I stand it?
It baffles old Mott,
He thinks that I'm perfect
But I know that I'm not.

I've got my faults; in fact quite a few
I've got Frank-Read Sources;
And dislocations, edge and screw.
But when stresses race madly amid my cross sections
And my lattice is twisting in several directions
What keeps me from yielding in plastic deflections,
Even in spite of my own predilections?
It's not the same force that binds Isoldes to Trist'ems—
It's the fact that I've got only four slip systems.

H. D. BLOCK

REFERENCES

Kelly, A., *Strong Solids*, Oxford, New York (1966).

Felbeck, D. K., *Introduction to Strengthening Mechanisms*, Prentice-Hall, Englewood Cliffs, N.J. (1968).

Zachay, V. F., *High Strength Materials*, Wiley, New York (1965).

Parker, E. R., and Zachay, V. F., "Strong and Ductile Steels," *Scientific American* (Nov. 1968) p. 36.

Kingery, W. O., *Introduction to Ceramics*, Wiley, New York (1960).

Hove, J. E., and Riley, W. C., *Modern Ceramics*, Wiley, New York (1965).

Burke, J. J., Reed, N. L., and Weiss, V., *Strengthening Mechanisms: Metals and Ceramics*, Syracuse University Press, Syracuse, N.Y. (1966).

Broutman, L. J., and Krock, R. H., *Modern Composite Materials*, Addison-Wesley, Reading, Mass. (1967).

Rauch, H. W., Sr., Sutton, W. H., and McCreight, L. R., *Ceramic Fibers and Fibrous Composite Materials*, Academic Press, New York (1966). Excellent discussion of high-temperature composites.

Holiday, L., ed., *Composite Materials*, Elsevier, New York (1966). A wide range of materials are described.

Frazer, A. H., "High Temperature Plastics," *Scientific American* (July 1969) p. 3.

Frazer, A. H., *High Temperature Resistant Polymers*, Wiley, New York (1968).

Winters, R. F., ed., *Newer Engineering Materials*, Macmillan, New York (1969).

Strengthening Mechanisms in Solids, American Society for Metals, Cleveland (1962).

Parker, E., *Materials Data for Engineers and Scientists*, McGraw-Hill, New York (1967).

PROBLEMS

25.1. Show that substitutional solute additions follow the corollary to the fundamental theorem of strengthening. Give the functional dependence on mean free path.

25.2. Show that cold working follows the corollary to the fundamental theorem of strengthening. Find the functional dependence on mean free path.

25.3. In what way does the strength of a polycrystalline aggregate depend on grain size?

25.4. What grain size would be needed in order that the yield stress of the low carbon steel be increased to 2,000,000 psi? Criticize your extrapolation assumptions.

25.5. There appears to be a rapid increase in ductility of the low carbon steel (of Figure 25.2.1) when tested in tension when the grain size is about 0.08 mm. Explain in macroscopic terms.

25.6. Explain why precipitation hardened alloys are not useful for
a. High-temperature applications.
b. Unlimited cyclic loading applications.

25.7. OFHC 99.95% annealed copper has a yield strength of 10,000 psi. A properly heat-treated alloy of copper with 1.7 wt. % Be can have a yield strength of 200,000 psi. The phase diagram can be found in the *Metals Handbook*. Explain what the heat treatment does.

25.8. Name four commercial age precipitation-hardenable alloys.

25.9. Describe the sequence of heat treatments used to form aged alloys.

25.10. Describe the heat treatments used to form pearlite.

25.11. Assuming it was possible to make very fine pearlite, what mean ferrite path would be needed to obtain a strength of 10^6 psi in the steel of Figure 25.5.1.

25.12. A strong steel rod is sawed off. The ends are carefully ground and then lapped together. The rod is then joined with a *very* thin layer of indium. Indium has a melting point of 156°C and can readily be indented by a thumbnail at room temperature. The new rod is tested in tension and the indium joint does not break until a stress of 150,000 psi is reached. This is about what is expected of a perfect indium crystal. Explain.

25.13. Give several reasons why medium to high carbon martensite is so strong.

25.14. Define hardness and hardenability.

25.15. Why is the hardenability of a steel such an important commercial property?

25.16. It is desired to make a shaft of 4-in. diameter having 0.40% carbon and at least 50% martensite throughout. Will SAE 1040 suffice or must we use the more expensive 4140 steel?

25.17. Describe briefly how each of the materials in Figure 25.6.7 derives its strength.

25.18. Austenitic stainless steels have the fcc crystal structure. Explain why such materials might have higher creep resistance than bcc materials.

25.19. Give two mechanisms of strengthening of polymeric materials.

25.20. Estimate from binding theory the ratio of the tensile strength of nylon fibers parallel to the chains and normal to the chains.

25.21. Give some reasons for the size effect or tensile strength shown by whiskers, metal wires, and glass filaments.

25.22. Derive the volume fraction rule for composites.

25.23. Discuss some of the problems one might have in making composites of
a. Continuous filaments.
b. Short fibers.

25.24. Make a list of ten different composites of which at least two are natural and at least two are synthetics known to the Romans.

More Involved Problems

25.25. For small concentrations C of the solute in a liquid (or solid) the concentration of solute in the vapor is found to be proportional to the concentration C. This is known as **Henry's law.** Hence show that the solute concentration near a small solute sphere is greater than that near a very large sphere. A good starting point is Kelvin's equation (see Section 18.14).

25.26. Discuss the structure changes which occur during forming of patented steel wire. See J. D. Embury and R. M. Fisher, *Acta Metallurgica*, **14**, 147 (1966).

Prologue

The engineer is concerned with three electrochemical processes: generation of electricity by batteries and fuel cells using direct chemical conversion; electrolytic processes such as electrorefining, electroplating, and electrolysis; and corrosion. To understand these processes it is necessary to understand both the *equilibrium* and the *kinetic behavior* of the *electrochemical cell*. Electrochemical cells include *dissimilar electrode cells*, *differential concentration cells*, and *differential temperature cells*. Corrosion always involves an electrochemical cell. To prevent corrosion this cell must be made inoperative. Techniques for doing this are discussed; these techniques are based on an understanding of how the cell operates.

26

ELECTROCHEMICAL PROPERTIES

26.1 INTRODUCTION

There are three important areas where electrochemical processes are directly involved:

1. Direct chemical generation of electricity by batteries and fuel cells. Batteries have an important place in our lives in flashlights, automobiles, submarines, etc. The development of a battery or fuel cell which could produce sufficient power for our automobiles would help solve the air pollution problem.
2. Electroplating (which can include electrorefining) and electrolysis. Aluminum and many other metals are produced by electrorefining processes. Electroplating is an important commercial process. For example, memory drums for computers are produced by plating the base metal with copper and then with a nickel-cobalt layer which serves as the memory core. The reverse process, namely electromachining, is also important commercially. In this case material is removed from a component.
3. Corrosion. Corrosion is an important economic factor. It is estimated that it costs $10 billion in the United States annually.

To understand these various phenomena we must understand the electrolytic cell.

Figure 26.1.1 shows one example of an electrolytic cell. The electrodes are of different materials so we call this a **dissimilar electrode cell.** If there is a very high external resistance R, there will be a very small current flow. Under these conditions there will be a potential difference E^0 between the two electrodes. The chemical reaction taking place is

$$Zn + Cu^{2+} \longrightarrow Zn^{2+} + Cu, \qquad E^0 = 1.1 \text{ V}. \qquad (26.1.1)$$

Thus the Zn electrode dissolves, adding Zn^{2+} ions to the solution, which then move

691

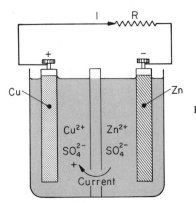

FIGURE 26.1.1. Dissimilar electrode cell. The aqueous $CuSO_4$ solution is separated from the $ZnSO_4$ solution by a porous barrier.

to the left. Cu^{2+} ions also move to the left and plate out on the Cu electrode. Thus positive ions move from the Zn electrode to the Cu electrode through the electrolyte liquid. Hence the motion of *positive* charge is as shown. (*Note:* In the external circuit, the current is actually due to *negative* electrons moving in the *opposite* direction, which results in the same current, I.) The sign of the electrodes refers to the motion of positive charge in the external circuit, such motion being toward the negative electrode. If the cell is not operated reversibly but is operated at a high current flow, the potential will be less than E^0; this is extremely important from the practical viewpoint. For the present we restrict our discussion to reversible processes (so that the current flow is negligible). When E^0 is actually measured, it is measured potentiometrically, so there is an applied external voltage "exactly" opposite in sign to the cell voltage. Hence there is "no" current flow and the cell behaves reversibly. The value of E^0 given refers to the situation in which the activities of the Cu^{2+} ions and Zn^{2+} ions in solution are 1. (The activities are quantities which are functions of concentration. For very low concentration the activity equals the concentration. We shall study the quantity called activity later in this section.)

Suppose the cell operates reversibly and in so doing eventually plates 1 mole of Cu onto the Cu electrode. Let us make $\mathscr{F}$ (the **Faraday equivalent**) equal to 1 mole of charge. (One Faraday equivalent is 96,500 coulombs.) When 1 mole of Cu^{2+} is transferred, two Faraday equivalents are transferred because of the valence of 2.

Let n be the charge of the species being transferred ($n = 2$ for Cu^{2+}). Then $n\mathscr{F}$ is the charge transferred when 1 mole of Cu is transferred. The change in energy of the chemical system under conditions of constant pressure and temperature is the Gibbs free energy per mole:

$$\Delta \bar{G}^0 = -n\mathscr{F}E^0. \tag{26.1.2}$$

Note that the cell reaction takes place if the potential, E^0, of the reaction is positive [see Equation (26.1.1)] and if the free energy of the reaction is negative. The work done *by* the chemical system is $n\mathscr{F}E^0$. (If the reaction is not carried out reversibly, less work will be done when 1 mole of Cu is plated out.) Suppose we had written the reaction with the products and reactants reversed:

$$Cu + Zn^{2+} \longrightarrow Cu^{2+} + Zn. \tag{26.1.3}$$

This would imply that the Zn is the positive electrode and the Cu the negative electrode (of the external circuit). Measurement would show that just the opposite is true so that we would say the potential is -1.1 V for this reaction, which implies that the reaction actually proceeds to the left in Equation (26.1.3). It should be noted that the cell potential represented by the reaction of (26.1.1) is 1.1 V only if the activity of each of the ions is one. For extremely dilute electrolytes (10^{-4} moles/liter or less) the activity equals the concentration, but this is not true for higher concentrations. The **activity** a is defined so that the molar Gibbs free energy is given by

$$\bar{G} = \bar{G}^0 + RT \ln a \tag{26.1.4}$$

even for nonideal solutions in analogy to the case of the ideal solution where

$$\bar{G} = \bar{G}^0 + RT \ln C. \tag{26.1.5}$$

[See Equation (17.9.13) where this is given in terms of the chemical potential (Gibbs free energy per ion instead of per mole).] If we consider a two component system, the activity of the *solvent* always approaches one as the concentration (mole fraction) of the solvent approaches 1 (Raoult's law). The standard state is pure solvent; then the molar Gibbs free energy of the solvent approaches the standard state value of $\bar{G}^0$ as the mole fraction approaches 1. However, we are concerned here with the activity of the solute at dilute concentrations. It is always true that at sufficiently low concentrations the solute obeys Henry's law. Consequently electrochemists usually define the standard state for the *solute* to be such that Equation (26.1.4) applies and reduces to (26.1.5) when the concentration of the solute approaches zero. The activity is then called a **rational activity**. In general it is related to molarity, M, by

$$a = \gamma^M M \tag{26.1.6}$$

or molality, m, by

$$a = \gamma m, \tag{26.1.7}$$

such that $\gamma \to 1$ as $m \to 0$. The activity coefficient γ (which is used in this chapter) is defined in terms of Henry's law and is called a **rational activity coefficient**.

Table 26.1.1 shows activity versus concentration for the ions in HCl. (Concentration is measured in molality here.) Data are often tabulated in terms of the **activity coefficient** $\gamma = a/m$ vs. m, the molality.

Table 26.1.1. RATIONAL ACTIVITY OF HCl—
CONCENTRATION FOR HCl IN H_2O
AT 25°C

m	a	$a/m = \gamma$
0.001	0.000966	0.966
0.002	0.001904	0.952
0.005	0.00464	0.928
0.01	0.00904	0.904
0.05	0.0415	0.830
0.1	0.0796	0.796
0.5	0.379	0.758
1.0	0.809	0.809
2.0	2.02	1.01
4.0	7.04	1.76

There are various kinds of electrolytic cells:

1. **Dissimilar electrode cells.**
2. **Differential concentration cells.** Both electrodes are copper, say, but they are immersed in solutions of different concentrations, say 0.01 m $CuSO_4$ on one side and 0.001 m $CuSO_4$ on the other.
3. **Differential temperature cells.**

26.2   ELECTRODE POTENTIALS

There are numerous electrode combinations possible. Suppose there are 100 different half-cells which could be combined with 100 other half-cells (the Cu and $CuSO_4$ solution of Figure 26.1.1 is a **half-cell**). We could then make a total of 10,000 different cells. Would we have to measure 10,000 cell potentials or would a smaller number of measurements suffice to define all 10,000 cell potentials? The answer is that 200 measured quantities would suffice. These are called **electrode potentials**. We would write each **electrode reaction** as an **oxidation process.** Thus

$$Cu \longrightarrow Cu^{2+} + 2e.$$

We would compare each half-cell with a standard half-cell (which we shall discuss later). The potential of the resultant cell is then arbitrarily defined to be the *electrode* potential of our test electrode. Thus for the above electrode we have

$$E^0 = -0.337 \text{ V.}$$

Likewise

$$Zn \longrightarrow Zn^{2+} + 2e$$

would give $E^0 = 0.763$ V against the standard reference half-cell. Consequently we have

$$Cu^{2+} + 2e \longrightarrow Cu \qquad 0.337 \text{ V}$$
$$Zn \longrightarrow Zn^{2+} + 2e \qquad 0.763 \text{ V}$$
$$\overline{Cu^{2+} + Zn \longrightarrow Cu + Zn^{2+} \qquad 1.100 \text{ V.}}$$

In other words by combining the oxidation electrode reactions in an appropriate fashion we can obtain the overall cell reaction and the resultant cell potential.

What is the **standard reference half-cell**? This is a cell in which the reactive species is hydrogen: The electrode may be some (relatively) inert material such as platinum surrounded by a solution containing H^+ ions in HCl solution with an activity of one in the presence of hydrogen gas at a pressure of 1 atm at 25°C. Thus the electrode reaction is

$$H_2 \longrightarrow 2H^+ + 2e, \qquad E^0 = 0.000 \text{ V}$$

and the electrode potential is *arbitrarily* set equal to zero.

Table 26.2.1 gives a few standard electrode oxidation potentials. Recall that these are defined for fixed activities (or pressures if a gas is involved) and temperature. Thus if both electrodes were copper, but the $CuSO_4$ solution had different concentrations (activities) in the two half-cells, then the resultant cell would have a nonzero potential. Likewise if each half-cell were Cu in a $CuSO_4$ solution but the half-cells were at different temperatures, the resultant cell would show a nonzero potential.

Table 26.2.1. STANDARD OXIDATION POTENTIALS*

Electrode Reaction	E^0	
$Li = Li^+ + e$	3.045	Anodic
$Mg = Mg^{2+} + 2e$	2.37	
$Al = Al^{3+} + 3e$	1.66	
$Zn = Zn^{2+} + 2e$	0.763	
$Fe = Fe^{2+} + 2e$	0.440	
$Cr^{2+} = Cr^{3+} + e$	0.41	
$Cd = Cd^{2+} + 2e$	0.403	
$Sn = Sn^{2+} + 2e$	0.136	
$Pb = Pb^{2+} + 2e$	0.126	
$Fe = Fe^{3+} + 3e$	0.036	
$\frac{1}{2}H_2 = H^+ + e$	0	Neutral
$Cu = Cu^{2+} + 2e$	-0.337	
$Ag = Ag^+ + e$	-0.7991	
$Hg = Hg^{2+} + 2e$	-0.854	
$Cl^- = \frac{1}{2}Cl_2 + e$	-1.359	
$Au = Au^{3+} + 3e$	-1.50	
$F^- = \frac{1}{2}F_2 + e$	-2.65	Cathodic

* W. M. Latimer, *Oxidation Potentials*, Prentice-Hall, Englewood Cliffs, N.J. (1952).

EXAMPLE 26.2.1

What is the cell potential if the electrodes are a standard Ag electrode and a standard Zn electrode?

Answer. $2Ag^+ + Zn \longrightarrow 2Ag + Zn^{2+}$ is obtained from

$$Zn \longrightarrow Zn^{2+} + 2e \qquad 0.763 \text{ V}$$
$$Ag^+ + e \longrightarrow Ag \qquad \underline{0.800 \text{ V}}$$
$$1.563 \text{ V.}$$

Hence the cell has a potential of 1.563 V and Ag is the positive electrode.

The electrode at which reduction occurs is called the **cathode** and the electrode at which oxidation occurs is called the **anode**. Reduction is the process whereby a positive ion such as Ag^+ is reduced to the metallic state. (As an example, when CuS ore is processed to produce copper metal we say the copper is reduced.) A reaction in which electrons are consumed is called **reduction**. Thus $Ag^+ + e \longrightarrow Ag$ is a reduction process.

A reaction in which electrons are produced is called an **oxidation** reaction. Thus $Ag \longrightarrow Ag^+ + e$ is an oxidation reaction.

The electrode potential for copper is supposedly -0.337 V. But what is copper? Is it a perfect single crystal? Is it a pure polycrystalline material? Is it a cold-rolled sheet of copper? Or does it not matter? The answer is that the electrode potential can be changed by

1. Cold working.
2. Impurities.

Thus the potential of a grain boundary is slightly different than the potential of a grain center. These small differences can be very important in corrosion applications. For example, an ordinary iron nail is highly cold-worked at the tip and head. The tip and head will have a different potential than the shank of the nail. Similarly a piece of Monel may have a uniform composition or it may be cored as in Figure 16.4.7, in which case the electrode potential varies from position to position. Such differences are important in corrosion.

EXAMPLE 26.2.2

Copper is cold-worked and the stored elastic energy is 5×10^{10} ergs/mole. Such copper would tend to dissolve and plate out on annealed copper. Why? What would be the cell potential if the cold-worked copper is the anode?

Answer.

$$E = \frac{\Delta \bar{G}}{n\mathscr{F}} = \frac{5 \times 10^3 \text{ J/mole}}{2 \times 96,000 \text{ C/mole}} = 0.026 \text{ V.}$$

In some textbooks and in some of the corrosion literature the electrode reactions are written as reduction potentials instead of oxidation potentials. This means that all the E^0 signs are reversed. However, when the total cell reaction is written down, such as in (26.1.1), the calculated cell voltage would be the same.

It is useful to have a **galvanic series** for real alloys in a real environment as shown in Table 26.2.2. It is not surprising that Mg wishes to return to the sea

---

Table 26.2.2. GALVANIC SERIES OF METALS
AND ALLOYS IN SEA WATER*

Anodic	Magnesium
↑	Zinc
	Alclad 3S
	Aluminum 3S
	Aluminum 61S
	Aluminum 63S
	Aluminum 52
	Low steel
	Alloy steel
	Cast iron
	Type 410 stainless steel (active**)
	Type 430 stainless steel (active)
	Type 304 stainless steel (active)
	Type 316 stainless steel (active)
	Ni-Resist
	Muntz metal
	Yellow brass
	Admiralty brass
	Aluminum brass
	Red brass
	Copper
	Aluminum bronze
	Composition G bronze
	90:10 copper-nickel
	70:30 copper-nickel—low iron
	70:30 copper-nickel—high iron
	Nickel
	Inconel
	Silver
	Type 410 (passive)
	Type 430 (passive)
	Type 304 (passive)
	Type 316 (passive)
	Monel
↓	Hastelloy C
Cathodic	Titanium

---

* From *Corrosion in Action*, International Nickel Co., New York (1955).
** The words active and passive as used here are explained in Section 26.7.

from where it came. In this table the metal near the top would be an anode in a cell with a lower metal.

26.3 CONCENTRATION CELLS

In this section we consider how to calculate the cell potential of a concentration cell. **Concentration cells** can be due to concentration differences in the electrolytes or concentration differences in the electrodes.

HYDROGEN ELECTRODE. The **hydrogen electrode reference** is shown in Figure 26.3.1. The reaction of this electrode is

$$\tfrac{1}{2}H_2 \longrightarrow H^+ + e.$$

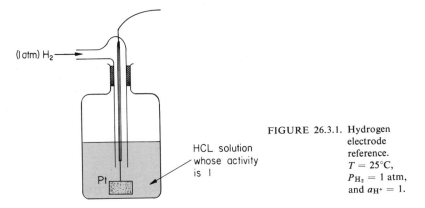

(1 atm) H₂

HCL solution
whose activity
is 1

Pt

FIGURE 26.3.1. Hydrogen electrode reference. $T = 25°C$, $P_{H_2} = 1$ atm, and $a_{H^+} = 1$.

CONCENTRATION CELL. Suppose we now have two hydrogen electrodes with a barrier through which only H^+ can pass. The HCl concentration is the same everywhere. Suppose one electrode has a hydrogen pressure of p_1 atm, while the other electrode has a hydrogen pressure of p_2 atm (where $p_2 > p_1$). At the high-pressure p_2 electrode we will have

$$\tfrac{1}{2}H_2 \longrightarrow H^+ + e,$$

while the converse will occur at p_1. The p_1 electrode will be the positive electrode or the cathode.

The free energy change when $\frac{1}{2}$ mole of H_2 gas is taken from an initial pressure, p_2, to a final pressure, p_1, is

$$\Delta G = \frac{RT}{2} \ln \frac{p_1}{p_2},$$

and since

$$\Delta G = -\mathscr{F}E$$

for the flow of 1 mole of charge, we have

$$E = \frac{RT}{2\mathscr{F}} \ln \frac{p_2}{p_1}. \tag{26.3.1}$$

Another example of an electrode concentration cell consists of two amalgam electrodes of Cd–Hg, a solid solution, which dip in a solution containing Cd^{2+} (but not Hg^{2+}). The cell is

$$Cd-Hg(a_1) \,|\, CdSO_4 \,|\, Cd-Hg(a_2).$$

If the activity $a_2 > a_1$, then the a_1 electrode will be the positive electrode. The change in free energy per mole of Cd transferred from a_2 to a_1 is

$$\Delta \bar{G} = RT \ln \frac{a_1}{a_2}$$

and using

$$\Delta \bar{G} = -2\mathscr{F}E$$

(two electrons are transferred per mole of Cd) we have

$$E = \frac{RT}{2\mathscr{F}} \ln \frac{a_2}{a_1}. \tag{26.3.2}$$

For dilute solutions

$$\frac{a_2}{a_1} = \frac{x_2}{x_1}, \tag{26.3.3}$$

where x_1 and x_2 are mole fractions. It is found for this system using (26.3.2) and (26.3.3) that calculated E's are within 0.5% of the experimental values [see G. Hulett, *Journal of the American Chemical Society*, **30**, 1805 (1908)].

Electrode concentration cells are often used to study the activity of metallic alloys [see J. Chipman, *Discussions of the Faraday Society*, **4**, 23 (1948)].

26.4 ELECTRODE POTENTIAL AND NERNST EQUATION

Suppose the Cu in Figure 26.1.1 is immersed in a $CuSO_4$ solution whose activity is different from 1. What is the electrode potential then? Suppose the Cu^{2+} ion is being created in the presence of a solution whose Cu^{2+} ion activity is $a_{Cu^{2+}} > 1$. The free energy will exceed the free energy for the case where $a_{Cu^{2+}} = 1$ by

$$\Delta \bar{G} = RT \ln \frac{a_{Cu^{2+}}}{1}.$$

Hence

$$\Delta \bar{G}_{total} = \Delta \bar{G}^0 + \Delta \bar{G}$$

and

$$-n\mathscr{F}E = -n\mathscr{F}E^0 + RT \ln a_{Cu^{2+}}$$

so

$$E = E^0 - \frac{RT}{n\mathscr{F}} \ln a_{Cu^{2+}}. \tag{26.4.1}$$

EXAMPLE 26.4.1

Calculate the cell potential of the cell shown in Figure 26.1.1 if the activity of the Cu^{2+} in $CuSO_4$ is $a_{Cu^{2+}}$ and of the Zn in $ZnSO_4$ is $a_{Zn^{2+}}$.

Answer. We have

$$
\begin{aligned}
E &= E_{Zn} - E_{Cu} \\
&= \left(E_{Zn}^0 - \frac{RT}{n\mathscr{F}} \ln a_{Zn^{2+}} \right) - \left(E_{Cu}^0 - \frac{RT}{n\mathscr{F}} \ln a_{Cu^{2+}} \right) \\
&= E_{Zn}^0 - E_{Cu}^0 - \frac{RT}{n\mathscr{F}} \ln \frac{a_{Zn^{2+}}}{a_{Cu^{2+}}}. \tag{26.4.2}
\end{aligned}
$$

Since $R = 8.314$ J/mole °K and $\mathscr{F} = 96{,}500$ C, we have at 25°C (298.13°K)

$$E(\text{volts}) = E_0(\text{volts}) - \frac{0.059}{n} \log_{10} K, \tag{26.4.3}$$

where K is the ratio of activities for the reaction [which is Equation (26.1.1) in the present case]:

$$K = \frac{a_{Zn^{2+}}}{a_{Cu^{2+}}}. \tag{26.4.4}$$

Equation (26.4.3) is called the **Nernst equation**.

EXAMPLE 26.4.2

Calculate the cell potential for the cell

$$Ag \,|\, AgCl \,|\, HCl(a_2) \,|\, H_2 \,|\, HCl(a_1) AgCl \,|\, Ag.$$

The AgCl is deposited on a silver screen. The hydrogen pressure is a constant in both electrodes but the HCl molality is

$$m_2 = 2.0 \quad \text{and} \quad m_1 = 1.0.$$

Answer. Equal numbers of H^+ and Cl^- are effectively transferred from one side to the other. Actually Cl^- is created by

$$AgCl + e \longrightarrow Ag + Cl^-$$

and destroyed by the reverse process (the Cl^- is not directly transferred from one side to the other). A similar conclusion applies to the hydrogen ion where

$$H^+ \longrightarrow \tfrac{1}{2}H_2 + e.$$

Thus the reaction

$$H^+ + Cl^- + Ag \longrightarrow AgCl + \tfrac{1}{2}H_2$$

occurs at the left electrode and the reverse at the right.

HCl is taken from a_2 to a_1. Hence the free energy change when 1 mole of HCl is transferred from a solution where the mean activity of each ion is a_2 to one where it is a_1 is

$$\Delta \bar{G} = RT \ln \frac{(a_{HCl})_1^2}{(a_{HCl})_2^2}$$

and since one mole equivalent of electrons is transferred

$$\Delta \bar{G} = -\mathscr{F} E$$

and

$$E = -\frac{RT}{\mathscr{F}} \ln K = -0.059 \log_{10} K,$$

where

$$K = \frac{(a_{HCl})_1^2}{(a_{HCl})_2^2};$$

[Note that we could readily have obtained the same result from the Nernst equation (26.4.3) since $E_0 = 0$ and $n = 1$.]

From Table 26.1.1 we see that $a_2 = 2.02$ and $a_1 = 0.809$, so that

$$E = 0.059 \log_{10} \frac{(2.02)^2}{(0.809)^2}$$

$$= 0.022 \, V.$$

We note for future reference that we could write for this case

$$E = 2 \frac{RT}{\mathscr{F}} \ln \frac{a_2}{a_1}. \tag{26.4.5}$$

26.5   TRANSFERENCE

Thus far we have been concerned with cells in which there is *no direct* ion transference from one side to the other. Let us now consider cells in which there is such flow; i.e., there is a liquid junction between the half-cells. As an example, consider the cell

$$Pt \,|\, H_2(1 \text{ atm}) \,|\, HCl(a_2) \,\vdots\, HCl(a_1) \,|\, H_2(1 \text{ atm}) \,|\, Pt.$$

The dashed line means a liquid junction. Both H^+ ions and Cl^- ions can flow through here. The velocity of the ions in terms of the driving force is given by

$$v = BF, \tag{26.5.1}$$

where F is the force acting on the ions and B is the mobility (which is directly proportional to the diffusion coefficient; see Section 18.13). We define **transference numbers** as follows:

$$t_+ = \frac{B_+}{B_+ + B_-} \tag{26.5.2}$$

and

$$t_- = \frac{B_-}{B_+ + B_-}, \tag{26.5.3}$$

where B_+ and B_- are the mobilities of the positive and negative ions, respectively.

Let $1\mathscr{F}$ of charge pass at the electrodes: At the left electrode $H^+ + e \rightarrow \frac{1}{2}H_2$, while the reverse occurs at the right electrode. The net result is $H^+(a_2) \rightarrow H^+(a_1)$.

At the liquid junction t_- equivalents of Cl^- move from left to right, i.e., from a_2 to a_1, while t_+ equivalents of H^+ move from a_1 to a_2.

Since Cl^- is not created or destroyed at the electrodes, we have an overall result: t_- moles of HCl going from a_2 to a_1. (The solution at a_1 is the product and at a_2 is the reactant.) The free energy change is therefore

$$\Delta\bar{G} = t_- RT \ln \frac{a_1}{a_2}.$$

Then since $\Delta\bar{G} = -n\mathscr{F}E_t$ and $n = 1$ here,

$$E_t = t_- \frac{RT}{\mathscr{F}} \ln \frac{a_2}{a_1}. \qquad (26.5.4)$$

(*Note:* t_- is the effective average transference number, which may be a function of activity or concentration.) An HCl concentration cell without transference gives the potential of Equation (26.4.5). These two cells together provide a method of measuring t_-.

EXAMPLE 26.5.1

What is the positive electrode in the cell just considered if $a_2 > a_1$?

Answer. The left electrode, i.e., the electrode with the highest concentration where reduction $H^+ + e \rightarrow \frac{1}{2}H_2$ occurs.

26.6   CORROSION

DIFFERENTIAL OXYGEN CELL. An **oxygen concentration cell** is shown in Figure 26.6.1. The oxygen concentration is high near the air-water interface where the reduction reaction

$$H_2O + \tfrac{1}{2}O_2 + 2e \longrightarrow 2OH^- \qquad (26.6.1)$$

takes place.

EXAMPLE 26.6.1

The reaction represented by (26.6.1) might take place near the surface (high O_2 region) or below the surface (low O_2 region). Where does it preferentially occur?

Answer. We are considering a concentration cell. Thus Le Chatelier's principle (or the law of mass action) applies. The tendency for

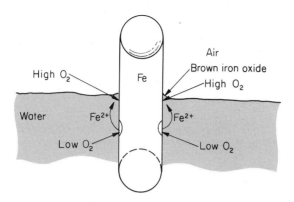

FIGURE 26.6.1. Corrosion of a steel (iron) pier or of a steel fence post.

(26.6.1) to occur is greatest where the O_2 concentration is the greatest— hence near the surface.

Since there is a tendency for a reduction reaction to occur near the surface, there must be a tendency for an oxidation reaction to occur in the low oxygen region; this is the reaction

$$Fe \longrightarrow Fe^{2+} + 2e. \qquad (26.6.2)$$

Thus the iron would corrode away and form a pit, as shown in Figure 26.6.1, at the point where the concentration of oxidizing agent (oxygen) is the *least*. Thus **oxygen starvation** causes corrosion.

In general, **corrosion** requires an electrolyte, a cathode, an anode, and an electrical connection. Note that all of these are present in the cell of Figure 26.6.1. The corrosion of the metal between the chrome strip on an automobile is an example of oxygen starvation. Something as trivial as a speck of dirt on a metal bumper can be responsible for the formation of an **aeration cell** and a corrosion pit.

DISSIMILAR ELECTRODES. Suppose a ship has a bronze propeller and a steel hull. Clearly an electrolytic cell is present [the oxidation potential of the bronze is about -0.25 V and of the steel (approximately iron) is about 0.44 V]. The iron hull would tend to dissolve, i.e., to behave as the anode.

Intergranular corrosion is another example of a dissimilar electrode cell. The atoms in the grain boundary have a higher free energy and hence tend to dissolve by the oxidation process, i.e., to behave as the anode. There may also be segregation at the grain boundaries so the composition is different from the interior. In addition grain boundaries often are the site of heterogeneous nucleation and hence of preferential precipitation. These second phase particles may also act as anodes, and intergranular corrosion may be particularly bad.

DEZINCIFICATION.   In certain brasses (Cu–Zn alloys) which are high in Zn content (Muntz metal), the β solid solution (bcc solution) is subject to **dezincification**. Here the zinc atoms are preferentially leached away from the β phase and the porous copper is left behind. This reaction takes place in water with high chlorine content.

The preferential removal of one of the constituents of an alloy can take place in many systems. Figure 26.6.2 shows an example of the removal of iron from gray cast iron (α-iron plus graphite flakes; the alloy contains 4.5 wt. % carbon) in which the porous graphite is left behind.

FIGURE 26.6.2. Porous graphite residue of a corroded cast iron pipe elbow.

EFFECT OF STRESS ON CORROSION.   The presence of stress enhances corrosion. Two types of processes are considered.

CORROSION FATIGUE AND STRESS CORROSION CRACKING.   When steel is subjected to cyclic loading in the absence of a corrosive environment it exhibits an endurance limit. However, if the same steel is subjected to a corrosive environment, the stress for failure decreases with an increasing number of cycles; i.e., the endurance limit does not exist. This is called **corrosion fatigue**.

The strength of a material may be drastically weakened by a crack formed by a combination of stress and corrosion. An example of such a crack is shown in Figure 26.6.3, which illustrates **stress corrosion cracking**.

Recall that stress increases the free energy per unit volume owing to the presence of strain energy. This increase can be particularly profound at the tip of a crack, which is therefore the anode.

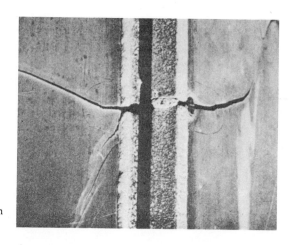

FIGURE 26.6.3. Stress corrosion cracking.

EXAMPLE 26.6.2

Why is the crack tip the anode?

Answer. Because the highly strained metal M at the crack tip will tend to dissolve, $M \longrightarrow M^{2+} + 2e$, and redeposit, $M^{2+} + 2e \longrightarrow M$, as unstrained metal on the material which is nearly free from strain. The reaction $M \longrightarrow M^{2+} + 2e$ is an oxidation reaction which occurs at the anode.

Removal of material from the crack tip deepens the crack and increases the chance of fracture. In brasses such corrosion is called **season cracking**, while in steels and irons it is called **caustic embrittlement**.

UNIFORM CORROSION. Uniform corrosion includes ordinary tarnishing, high-temperature oxidation (but not always), and rusting. Uniform corrosion is usually measured in inches per year (ipy) or in milligrams per square decimeter per day (mdd).

26.7 PROTECTING AGAINST CORROSION

There are several ways of avoiding corrosion difficulties:

1. Change materials.
2. Use special heat treatments.
3. Coat the material.

4. Use sacrificial anodes.
5. Use impressed potential.
6. Add inhibitors to the "electrolyte."

HEAT TREATMENTS.   We have already noted that the presence of a second phase or a composition variation such as in a cored Cu–Ni alloy gives rise to dissimilar electrode cells. In the case of a cored alloy a simple homogenization treatment removes the difficulty. In the case of second phase precipitates, one may form larger precipitates. For example, in a tempered martensitic steel the maximum rate of corrosion occurs for the finest form of ϵ-carbide.

NOBLE COATINGS.   If the component is completely coated with another more noble metal (such as gold), a nonporous paint, etc., then the metal is isolated from the environment and cannot corrode. However, there is one serious drawback as illustrated in Figure 26.7.1. The material below a scratch through

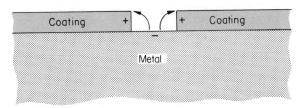

FIGURE 26.7.1. Pitting at a scratch in a noble coating.

the paint on an automobile fender will corrode very rapidly because this material is an anode in the cell in which the noble coating is the cathode. Another important mechanism which usually operates in this situation is oxygen starvation (see Section 26.6).

OXIDE COATINGS.   One look at the oxidation potential of aluminum tells us that this is a material which is extremely reactive to an oxidation environment. Aluminum readily forms an oxide layer. In fact aluminum is often systematically oxidized in such a way that a adherent oxide film forms on the surface. This is known as **anodizing**. Further oxidation can proceed only by diffusion of the aluminum (as an ion) through the ionic aluminum oxide layer. The motion through this solid electrolyte occurs by solid state diffusion and is, of course, slow. A solution which dissolves the oxide layer would wreak havoc with this mechanism of protection.

In stainless steels (containing, say, 18 % Cr and 8 % Ni) which are placed in an oxidizing environment, e.g., sulfuric or nitric acid, the reaction

$$Cr + 2O_2 + 2e \longrightarrow (CrO_4)^{2-}$$

occurs. The chromate ions are adsorbed on the anode surface and prevent further reaction. We say then that stainless steels are **passive** in the presence of oxidizing acids. However, they are **active** in the presence of other acids such as hydrochloric acid.

SACRIFICIAL COATING.   Iron is often coated with zinc by dipping the iron in molten zinc (**galvanizing**). Zinc is less noble than iron (Table 26.2.1). Hence a scratch through the zinc will not have the same consequence as a scratch through a coating more noble than iron such as chromium. In fact the zinc will *tend* to dissolve. This is known as a **sacrificial coating**.

SACRIFICIAL ANODES.   Sir Humphrey Davy was the first to use **sacrificial anodes** to control corrosion of ships' hulls in 1824 [see I. A. Denison, *Corrosion*, **3**, 295 (1947)].

Here chunks of zinc were connected to the ship's hull by electrical conductors and, as with the galvanized coating, the zinc rather than the iron of the ship's hull dissolved. The zinc anodes could be easily and periodically replaced.

IMPRESSED POTENTIALS.   A steel pipe line buried in the ground will ordinarily act as an anode. We can apply an *external* voltage (**impressed potential**) so that the steel pipe becomes the cathode (see Figure 26.7.2).

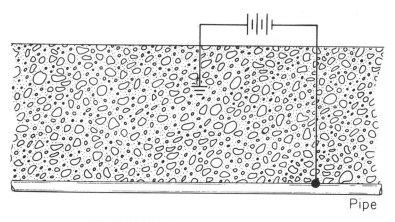

Pipe

FIGURE 26.7.2. Impressed potential protection.

RUST INHIBITORS.   Your friendly service station attendant will be glad to furnish you with a can of rust inhibitor for your radiator which probably contains chromates, phosphates, and tungstates (high oxide-containing complexes) which are adsorbed on the metal surface and isolate it from the electrolyte.

This behavior is identical to the passivation of stainless steel. The most effective inhibitor of this sort is TcO^{4-} (the oxide complex of the rare earth metal technetium). It is not known why this is so.

EXAMPLE 26.7.1

Suggest methods for preventing corrosion of an offshore oil platform. The structure is made of steel and there are three critical areas with corrosion rates in ipy (inches per year):

The immersed zone,	0.025 ipy
The splash zone,	0.055 ipy
The atmospheric zone,	0.005–0.010 ipy.

Answer. Because an electrolyte is always present in the immersed zone, cathodic protection can be provided by using an impressed potential.

In the splash zone oxygen starvation is critical and the electrolyte is moving; hence, we would use isolation techniques or sacrificial coatings.

In the atmospheric zone paint could be used as an isolation technique.

26.8   POLARIZATION AND OVERVOLTAGE

When electrochemical cells are operated at finite rates (rather than infinitesimal rates) they behave irreversibly and their cell potentials differ from the equilibrium values (reversible). When the cell operates as a battery, the potential falls below the equilibrium value. If we are carrying out electrolysis, the reverse is true; i.e., we must supply a voltage greater than the equilibrium value. There are three sources of this difference:

1. Internal resistance (joule heating, i.e., i^2R losses).
2. Concentration polarization.
3. Overvoltage.

Consider a cell of Cu and Pt dipped in $CuSO_4$ solution. When current flows,

$$Cu \longrightarrow Cu^{2+} + 2e \quad \text{(copper electrode)},$$

while at the platinum electrode

$$Cu^{2+} + 2e \longrightarrow Cu \quad \text{(platinum electrode)}.$$

Because the transfer of Cu^{2+} is diffusion-controlled, the solution around the Pt electrode will be depleted of Cu^{2+}. Hence under dynamic conditions we have a concentration cell. If we ignore the dissimilar electrode potential, then we have

$$Pt \,|\, \text{low } Cu^{2+} \text{ region} \quad \text{and} \quad \text{high } Cu^{2+} \text{ region} \,|\, Cu$$

and the reactions which would tend to occur because of the **concentration polarization** cell alone are, respectively,

$$Pt \,|\, Cu \longrightarrow Cu^{2+} + 2e \quad \text{and} \quad Cu^{2+} + 2e \longrightarrow Cu \,|\, Cu.$$

That is, Cu^{2+} would be produced in the low Cu^{2+} region and eliminated in the high Cu^{2+} region, tending to form uniform concentration. The concentration cell potential *opposes* the equilibrium potential of the actual cell. In electrolysis we would want to eliminate this **back voltage**. Vigorous stirring of the solution and an increase in temperature would succeed in doing this. However, in corrosion, a high back voltage may radically slow down the corrosion process, hence stirring would have an adverse effect. This is shown in Figure 26.8.1.

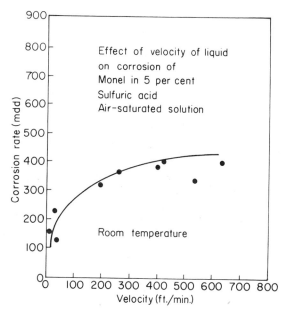

FIGURE 26.8.1. Effect of velocity of liquid on corrosion of Monel in sulfuric acid. The corrosion rate is milligrams per square decimeter (100 cm²) per day. [From *Corrosion*, The International Nickel Co., New York (1949).]

Overvoltage is due to the slowness of a chemical reaction at the electrode. Usually this is very small for metal deposition on metal. However, it is quite large for reactions involving liberation of gases. Thus if we attempt electrolysis of an HCl solution of unit activity, we would have to apply a voltage of 1.3596 V under equilibrium conditions to produce hydrogen and chlorine gases:

$$Cl^- = \tfrac{1}{2}Cl_2 + e, \qquad E_0 = -1.3595 \text{ V}$$
$$\tfrac{1}{2}H_2 = H^+ + e, \qquad E_0 = 0.$$

However, if we have a current flow of 1 A/cm² and both electrodes are ordinary platinum and the liquid is well stirred, we would have to apply a voltage at least 1 V higher than this, i.e., 2.4 V. We could measure the overvoltage at an electrode where hydrogen is liberated by preparing a cell in which the overvoltage at the other electrode is negligible. Such results are shown in Figure 26.8.2. Platinized platinum has a very fine dispersion of platinum on platinum sheet which acts as a catalyst for the hydrogen liberation process so the overvoltage is relatively small.

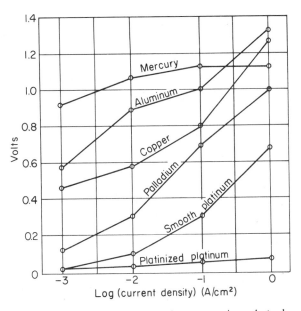

FIGURE 26.8.2. Hydrogen overvoltage on various electrode materials at various current densities.

EXAMPLE 26.8.1

Can zinc be plated from an aqueous solution?

Answer. Here the possible reactions and voltages are

$$\tfrac{1}{2}H_2 \longrightarrow H^+ + e \qquad 0 \text{ V}$$
$$Zn \longrightarrow Zn^{2+} + e \qquad 0.763 \text{ V}.$$

Under equilibrium conditions, the answer is clearly "no" since zinc would tend to dissolve and hydrogen ions would tend to be converted to gas. However, because of the sizable overvoltage, it is indeed possible to plate zinc from solution at rapid rates.

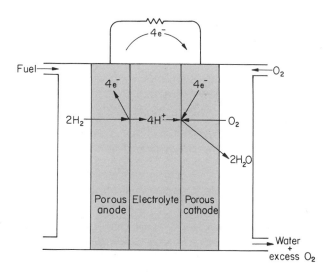

FIGURE 26.8.3. The hydrogen-oxygen fuel cell.

These kinetic problems are also of utmost importance in the operation of batteries and fuel cells. The hydrogen-oxygen fuel cell is shown in Figure 26.8.3. It has a reversible voltage of 1.22 V. However, when operated at finite power the voltage output is considerably less (see Figure 26.8.4). The initial voltage drop is primarily due to overvoltage, while the final rapid drop is due to concentration

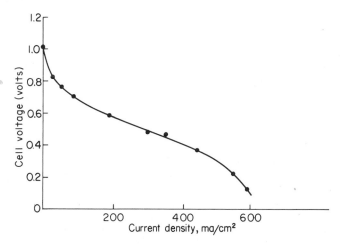

FIGURE 26.8.4. Cell voltage versus current density for a hydrogen-oxygen electrode with a 5N H_2SO_4 electrolyte.

polarization. The reader should take note of the challenging materials problems associated with the anodes and electrolyte, inasmuch as it is desirable to operate the cell at high temperatures.

REFERENCES

Corrosion in Action, International Nickel Company, New York (1955). An excellent qualitative, short (47 pages) discussion of corrosion. It also contains a description of about a dozen experiments and a good list of references divided into several areas of subject matter.

Corrosion, International Nickel Company, New York (1949).

Swann, P. R., "Stress Corrosion Failure," *Scientific American* (Feb. 1966) p. 72.

Uhlig, H. H., *Corrosion and Corrosion Control: An Introduction to Corrosion Science and Engineering*, Wiley, New York (1963).

Grubb W. T., and Niedrach, L. W., "Fuel Cells," in *Direct Energy Conversion* (G. W. Sutton, ed.), McGraw-Hill, New York (1966).

Liebhafsky, H. A., and Cairns, E. J., *Fuel Cells and Fuel Batteries*, Wiley, New York (1968).

West, J. M., *Electrodeposition and Corrosion Processes*, Van Nostrand, Princeton, N.J. (1965).

Lowenheim, F. A., *Modern Electroplating*, Wiley, New York (1963).

PROBLEMS

26.1. Phenolphthalein is an indicator which turns pink in the presence of excess hydroxyl ions. Potassium ferricyanide is an indicator which turns blue in the presence of ferrous ions. Design a simple laboratory experiment to show that in Figure 26.6.1, the high oxygen concentration region is the positive electrode, i.e., the cathode or the electrode where reduction occurs.

26.2. A cell contains copper electrodes in $CuSO_4$ solution. In one cell the concentration is 10^{-4} m. In the other cell it is 10^{-3} m. What is the potential difference? Which electrode is positive?

26.3. A car battery is in a car in Alaska on a cold night with no external heating. One student claims that when the starter is pressed, the battery will cool down and the electrolyte might even freeze. Another claims that this is nonsense and that in fact the electrolyte will be heated. Who is correct and why?

26.4. A sheet of copper metal is stressed (elastically) by a clamp such that the state of stress is given by the principal stresses $-\sigma, \sigma/3, \sigma/3$, for Poisson's ratio $v = \frac{1}{4}$. What is the cell potential between copper in this stressed state and unstrained copper?

Hint: The stored elastic energy can be equated to the change in Gibbs free energy.

26.5. What is the cell potential between a large-angle grain boundary and a grain interior in copper? Assume that the copper is ultrapure so there is no possibility of segregation of impurities at the grain boundaries. Assume the grain boundary is 5×10^{-8} cm thick. What would the anode be?

26.6. If in Problem 26.5 there were impurity atoms segregated at the grain boundary, is it possible that the sign of the cell potential could be reversed? Assuming that ultrapure copper can be made and that impurities can be systematically added, could the cell potential between the grain boundary and grain interior be used to measure segregation?

26.7. A pearlitic steel corrodes rapidly.
a. Why?
b. How may this be reduced?

26.8. List several types of electrochemical processes.

26.9. List the various types of electrochemical cells.

26.10. Explain why the head and tip of a nail would behave as anodes relative to the shank of the nail.

26.11. List the various processes of preventing corrosion.

26.12. There are three reasons the potential of an operating battery is less than the equilibrium potential. Discuss them.

26.13. Unused railroad tracks in a humid environment appear to corrode more rapidly than regularly used tracks. Why?

26.14. Season cracking is an example of stress corrosion in brasses. It was found that brass cartridges stored near cavalry stables showed such behavior. Give some possible reasons.

26.15. Tin plating is often used to protect iron. In fact, it is regularly used in making tin cans. A scratch in the tin is usually not a cause for concern, although a look at the oxidation potentials suggests otherwise. Explain.

26.16. A replaceable Mg rod is sometimes inserted into hot water heaters. Describe its purpose.

26.17. Pb should not react with H_2SO_4 in our typical car batteries and produce H_2 and $PbSO_4$; i.e., the lead electrode should completely dissolve according to equilibrium considerations. It does not. Why?

More Involved Problems

26.18. A chloroplatinic acid solution is used to plate platinum on platinum. The platinum present in solution is in the form of $PtCl_6^{2-}$ ions. The solution surrounding the cathode loses 5.85 g of Pt and the solution around the anode gains 1.95 g of Pt and 2.13 g of Cl⁻. Find the transference number of the complex ion.

26.19. Evaluate Sir Humphrey Davy's contribution to electrochemistry and corrosion.

26.20. Discuss how cathodic charging of steel introduces hydrogen and causes crack nucleation [see A. S. Tetelman and A. J. McEvily, Jr., *Fracture of Structural Materials*, Wiley, New York (1967)].

Sophisticated Problem

26.21. Gold is obtained from ores by leaching with cyanide solution (this was a revolutionary development in gold mining which made many low concentration mines operational). The gold dissolves in such a solution because it forms a complex chemical ion with several CN⁻ units attached. Describe how you would set up a transference cell to obtain the formula of the complex gold ion.

Prologue

In this chapter the nature of the valence electrons in crystals is discussed. A simple model for a metal involves the assumption that the valence electrons form an electron gas which behaves in a classical fashion. This model has certain weaknesses, many of which are removed by the *quantized electron gas* (whose behavior is based on a quantum model discussed in Chapter 8, namely, the particle in a box model). However, to understand many of the properties of solids (why is copper a metal, germanium a semiconductor, and diamond an insulator) it is necessary to take into account the fact that electrons move in a *periodic potential*. This leads to the concept of *forbidden energy bands* and *allowed energy bands*. The manner in which the allowed levels are occupied determines whether a solid is a metal, semiconductor, or insulator. The terms *valence band* and *conduction band* are introduced. The *Fermi surface* is introduced. The origin of resistivity in metals is described. Various processes causing *electron emission* from solids are described. The important quantum mechanical concept of *electron tunneling* is introduced.

27

ELECTRONS IN
CONDENSED PHASES

27.1 THE ELECTRON GAS

In 1905, P. Drude proposed that a metal contained free electrons whose movement was similar to the motion of atoms in an ideal gas [see P. Drude, *Ann. Physik* **1**, 566 (1900)]. This is called the **classical free electron gas model** or **Drude's model**. We recall that the ionization potential of a metal atom is quite small (see Table 8.4.1). Thus in a metal such as copper (atomic number 29) one electron per atom might be ionized (copper is usually monovalent chemically) and this electron would be free to move throughout the metal, which can be considered as a box. (We shall see later why this reduces the overall energy of the system.) In Drude's model, the ion cores consisting of the nucleus and 28 electrons remain at the lattice site.

The root-mean-squared velocity of an "ideal gas" particle is

$$\bar{v} = \sqrt{\overline{v^2}} = \sqrt{\frac{3kT}{m}} \qquad (27.1.1)$$

(from $\frac{1}{2}mv^2 = \frac{3}{2}kT$). Ordinarily the gas particles (electrons) would be moving about randomly with a mean free path λ. In the presence of an electrical field ξ acting in the x direction the electron would obey the equation of classical mechanics:

$$m\ddot{x} + \dot{x}/B = -e\xi. \qquad (27.1.2)$$

We have assumed the presence of a viscous force term $\dot{x}/B$, where B is called the **mobility coefficient** (see Section 18.13). (The quantity $\dot{x}$ is measured relative to the normal random motion of the electron.) The general solution is

$$x = C + De^{-t/mB} - Be\xi t,$$

717

and for $x = 0$ and $\dot{x} = 0$ at $t = 0$ we have

$$x = B^2 em\xi[1 - e^{-t/mB}] - Be\xi t. \tag{27.1.3}$$

EXAMPLE 27.1.1

Obtain an expression for the average **drift velocity** $\bar{\dot{x}} = \bar{v}$ in terms of the **time between collisions** τ.

Answer. We have

$$\bar{\dot{x}} = \frac{\int_0^\tau \dot{x}\, dt}{\int_0^\tau dt} = \frac{B^2 em\xi[1 - e^{-\tau/mB}] - Be\xi\tau}{\tau}.$$

If $\tau/mB \ll 1$, we can represent the exponential by

$$e^{-\tau/mB} = 1 - \frac{\tau}{mB} + \frac{(\tau/mB)^2}{2}$$

and we have

$$\bar{\dot{x}} = -\frac{e\tau}{2m}\xi + 0(\tau^2).$$

We drop the higher-order terms.

The magnitude of the electrical current density is given by

$$J = Nq\bar{\dot{x}},$$

where N is the number of current carriers per unit volume and q is the charge $(q = -e$ in the present case). Hence we have

$$J = \frac{Ne^2\tau}{2m}\xi,$$

and since the electrical conductivity is defined by $J = \sigma\xi$, we have

$$\sigma = \frac{Ne^2\tau}{2m}. \tag{27.1.4}$$

EXAMPLE 27.1.2

Show that the terminal velocity is $B(-e\xi)$.

Answer.

$$\lim_{t\to\infty} \dot{x} = \lim_{t\to\infty} (-e\xi)B[1 - e^{-t/mB}]$$
$$= (-e\xi)B.$$

The **charge mobility** μ, or the drift velocity per unit field (rather than per unit force), is often used and is usually referred to simply as the mobility. Then the conductivity is given by

$$\sigma = Ne\mu, \tag{27.1.5}$$

where μ is given by

$$\mu = e\tau/2m. \tag{27.1.6}$$

The relaxation time τ can be expressed as

$$\tau = \lambda/\bar{v},$$

where $\bar{v}$ is given by (27.1.1) in Drude's model. Hence we have

$$\sigma = \frac{Ne^2\lambda}{2m\bar{v}}. \tag{27.1.7}$$

Drude suggested that λ was the distance between the ions.

EXAMPLE 27.1.3

Calculate the value of λ for copper which gives, according to (27.1.7), the correct experimental value of σ at 20°C.

Answer. From Table 3.1.1 and Equation (3.1.4) we have $\sigma = 6.0 \times 10^7 \ \Omega^{-1} \, m^{-1}$. From (27.1.1) we have $\bar{v} = 6.7 \times 10^4$ m/sec since $m = 9.11 \times 10^{-31}$ kg.

Copper is fcc and contains four atoms per cubic unit cell whose lattice parameter is 3.61 Å (see Table 10.6.1). Hence the number of electrons per unit volume is

$$\frac{4}{(3.61)^3} \times 10^{30} \, m^{-3} \quad \text{or} \quad N = 8.5 \times 10^{28} \, m^{-3}.$$

Since $e = 1.6 \times 10^{-19}$ C, we have

$$\lambda = 34 \times 10^{-10} \, m = 34 \, \text{Å}.$$

We note that the classical theory gives a mean free path which is of the order of 10–100 Å (see Example 27.1.3). However, λ cannot be directly predicted from theory by the classical model; it is not equal to the atomic spacing as Drude proposed. This is one of the shortcomings of the classical model.

EXAMPLE 27.1.4

Suggest an experiment by which we can directly measure the mean free path λ.

Answer. We noted in Section 3.2 that there is a size effect for resistivity (or conductivity) when a thin enough foil or wire is studied (see also Problem 3.20). Thus if the resistivity in the plane of the film is measured for various thicknesses of thin films, λ can be equated to the thickness of the film at which there is a rapid increase in resistivity as the film thickness decreases. The electrons in the box are colliding with the boundary of the box and this then determines the mean free path between collisions. It is found experimentally by this technique that λ is on the order of a few hundred Ångstroms (a hundred or so atomic diameters).

DIFFICULTIES WITH DRUDE'S THEORY.   There are three obvious short-comings to the classical free electron theory:

1. λ is an ill-defined quantity which cannot be calculated *a priori*.
2. There is a **specific heat paradox**.
3. Many metals show a **Hall coefficient anomaly** which suggests that the charge carriers are positive (see Table 9.4.2).

In gases the mean free path is readily calculable in terms of the number of particles per unit volume and the molecular diameter. There is no similar classical theory for the calculation of λ for electrons in metals; quantum mechanics is required.

The specific heat of crystalline copper owing to the ion vibrations is $3R$ (5.96 cal/mole-°K) at high temperatures [see Equation (17.11.8)]. The classical electrons in the Drude metal would have the same specific heat as an ideal monatomic gas, i.e., $3R/2$ [from (17.2.5) and (17.11.6)]. Hence the total specific heat for copper should be

$$C_v = C_v^L + C_v^E = 3R + \tfrac{3}{2}R.$$

In fact, the total specific heat is very close to C_v^L so the electronic contribution has been vastly overestimated (by a factor of about 50 at room temperature). This paradox can readily be explained by a model which assumes that the electrons obey quantum mechanical laws (rather than classical laws) but in which the other assumptions of Drude's model are kept: The electrons move in a constant potential ($V = 0$) unaware of the existence of ion cores or other electrons (at least from a potential viewpoint). This will be discussed in Section 27.3.

The Hall effect was described in Section 3.3. There it was shown that in a material for which one charge carrier dominates, the Hall coefficient is given by

$$R_H = \frac{1}{Nq}.$$

Thus a Drude metal would have a negative Hall coefficient. It was shown in Chapter 9 (see Table 9.4.2) that the calculated Hall coefficient for a Drude metal is in fair agreement with the measured value for alkali metals and copper, silver, and gold. However, for a metal such as cadmium (which is ordinarily divalent) the experimental Hall coefficient is positive (even though cadmium is a fairly good conductor). In later sections we shall see why this is the case. To understand this behavior we shall have to make a second change in Drude's model. Not only must we assume that quantum mechanics describes the electron behavior, but we must also account for the fact that the electron moves in the periodic potential of the ion cores. As we shall see, such a model will also account for the presence of energy bands and will lead directly to a characterization of solids as conductors, semiconductors, or insulators.

27.2   THE QUANTIZED ELECTRON GAS

In classical theory the total energy of a particle is given by the sum of the kinetic and potential energy,

$$\frac{p^2}{2m} + V = E, \tag{27.2.1}$$

where $p^2 = p_1^2 + p_2^2 + p_3^2$ is the momentum squared. In quantum mechanics these various energy terms behave as operators on the function ψ and the operator for p_1 is $(h/2\pi i)(\partial/\partial x_1)$, etc. The one-dimensional wave equation (which we consider for simplicity) is

$$\frac{h^2}{8\pi^2 m}\frac{d^2\psi}{dx_1^2} + (E - V)\psi = 0. \tag{27.2.2}$$

We assume that the particle is restricted to the region $0 < x_1 < L_1$, that $V = 0$ on this range, and that $V \to \infty$ as $x_1 \to 0$ and $V \to \infty$ as $x_1 \to L_1$. This is called the one-dimensional problem of a **particle in a box with infinite walls**. Since the particle cannot exist in the region when $V = \infty$, ψ must be zero there; i.e., the probability $\psi^2\, dx_1$ of finding the particle in the element of length dx_1 (where $V = \infty$) must be zero.

The solution to this problem is discussed in Equations (8.1.6) through (8.1.13). A solution exists only when E takes on one of the discrete values

$$E = \frac{h^2 n_1^2}{8mL_1^2}, \qquad n_1 = 1, 2, 3, \ldots. \tag{27.2.3}$$

The solutions are standing waves:

$$\psi = \sqrt{\frac{2}{L_1}}\sin\frac{n_1\pi}{L_1}x_1. \tag{27.2.4}$$

EXAMPLE 27.2.1

Starting with $E = p_1^2/2m$ and $p_1 = h/\lambda_1$, where h is Planck's constant and λ_1 is the wavelength (λ is also used for mean free path but there should be no confusion), give an alternative derivation of (27.2.3) based on the assumption that the solutions are standing waves.

Answer. We have $E = h^2/2m\lambda_1^2$. What are the allowed λ_1's? We must always have nodes at $x_1 = 0$ and $x_1 = L_1$. The longest wave which can be used is $\lambda_1 = 2L_1$. In general, $n_1\lambda_1 = 2L_1$. Hence

$$E = \frac{h^2 n_1^2}{8mL_1^2}.$$

The student should plot several of the wave functions in (27.2.4) at this time.

The two-dimensional problem (which is discussed in detail in Problems 8.15–8.18) has solutions if and only if

$$E = \frac{h^2}{8m}\left[\frac{n_1^2}{L_1^2} + \frac{n_2^2}{L_2^2}\right]. \tag{27.2.5}$$

Similarly, a three-dimensional particle in a box with infinite walls and with $V = 0$ within the box obeys Equation (8.1.5). It has nontrivial solutions if and only if

$$E = \frac{h^2}{8m}\left[\frac{n_1^2}{L_1^2} + \frac{n_2^2}{L_2^2} + \frac{n_3^2}{L_3^2}\right], \tag{27.2.6}$$

where the n's are integers and the L's are the dimensions of the box. In addition to the quantum numbers n_1, n_2 and n_3, the electron has a fourth quantum number associated with electron spin. There are two spin states.

This model of the electrons in a metal is called the **Sommerfeld model** or the **quantized electron in a box model**. In the discussion which follows we shall use $L_1 = L_2 = L_3 = L$ as a mathematical convenience. Let us consider the *two-dimensional* case. Suppose we have 10^{22} electrons each with different quantum numbers $(n_1, n_2,$ and spin). What is the energy of the highest energy electron assuming all available lower energies are filled? Figure 27.2.1 shows the quantity $n_1^2 + n_2^2$ plotted in $n_1 n_2$ space. Note that each unit square in $n_1 n_2$ space represents two wave functions (since there are two spins) so that the area within the arc

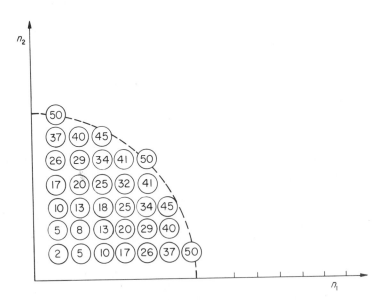

FIGURE 27.2.1. The quantity $n_1{}^2 + n_2{}^2$ is shown within the small circles. The dashed line which is the arc of a circle is a constant energy profile.

of the quarter-circle represents (to an approximation) twice the number of wave functions. The approximation gets better as the n's get larger. We note the presence of **degenerate states**: Thus $n_1 = 5, n_2 = 5$ and $n_1 = 7, n_2 = 1$ or $n_2 = 7$, and $n_1 = 1$ all give the same energy. Such degeneracy is very common in the solid state. Let

$$r^2 = n_1^2 + n_2^2 = \frac{8mL^2}{h^2}E.$$

Then the area or half the number of wave functions is given by

$$\frac{\pi r^2}{4},$$

for large numbers of wave functions.

If there are $\mathcal{N}$ allowable states altogether (we define a state by *three* quantum numbers in the two-dimensional case) then

$$\frac{\mathcal{N}}{2} = \frac{\pi r^2}{4}$$

or

$$\mathcal{N} = \frac{4\pi mL^2}{h^2}E.$$

The total area A of the two-dimensional box is L^2 so we have

$$\frac{\mathcal{N}}{A} = \frac{4\pi m}{h^2}E.$$

In three dimensions

$$r^2 = n_1^2 + n_2^2 + n_3^2 = \frac{8mL^2}{h^2}E \qquad (27.2.7)$$

and the volume in n space which contains $\mathcal{N}$ allowable states is

$$\frac{\mathcal{N}}{2} = \frac{1}{8}\frac{4}{3}\pi r^3 = \frac{\pi}{6}\left(\frac{8mL^2}{h^2}E\right)^{3/2}. \qquad (27.2.8)$$

An allowable state in three dimensions involves n_1, n_2, n_3, and spin. In this case $V = L^3$ and

$$N = \frac{\mathcal{N}}{V} = \frac{\pi}{3}\left(\frac{8mE}{h^2}\right)^{3/2}. \qquad (27.2.9)$$

The quantity N is the number of allowable states (four quantum number states) per unit volume.

If the total number of electrons per unit volume is N_F, then the highest filled allowable state will have energy E_F called the **Fermi energy**:

$$E_F = \frac{h^2}{8m}\left(\frac{3N_F}{\pi}\right)^{2/3}. \qquad (27.2.10)$$

The energy surface in n space which has $E = E_F$ is called the **Fermi surface**. For the present model it is a spherical surface. [Note the analogy between the representation of an elastic constant in x space by an elastic body (Figure 19.3.4) and the representation of the Fermi energy in n space by the Fermi surface.]

EXAMPLE 27.2.2

Compute the Fermi energy for copper on the basis of the above model.

Answer. Note first of all that E_F is independent of the size of the crystal. From Example 27.1.3, $N_F = 8.5 \times 10^{22}$ cm^{-3}. Since $h = 6.62 \times 10^{-27}$ erg/sec and $m = 9.11 \times 10^{-28}$ g, we have

$$E_F = 11.2 \times 10^{-12} \text{ erg}$$

or

$$E_F = 7.0 \text{ eV}.$$

The actual Fermi surface for copper is not quite spherical, as we shall see later.

The Fermi energy of metals is of the order of 5 eV. See Table 27.2.1.

Table 27.2.1. FERMI ENERGY

Metal	E_F, eV
Li	3.7
Na	2.5
Cu	7.0
Al	11.8

EXAMPLE 27.2.3

If E_F is the maximum energy of an electron in a metal at absolute zero, then what is the average energy (according to the present model) at absolute zero?

Answer.

$$\bar{E} = \frac{\int_0^{E_F} E \, dN(E)}{\int_0^{E_F} dN(E)}, \tag{27.2.11}$$

where $dN(E)$ is the number of states between E and $E + dE$. From (27.2.9) we have

$$dN = \frac{\pi}{2} \left(\frac{8m}{h^2}\right)^{3/2} E^{1/2} \, dE. \tag{27.2.12}$$

Then we can show that

$$\bar{E} = \frac{3E_F}{5}. \tag{27.2.13}$$

We can also write (27.2.12) in the form

$$dN = \rho(E)\, dE, \tag{27.2.14}$$

where $\rho(E)$ is the **density of states**. It is the number of allowable states per unit volume and per unit energy and is given by

$$\rho(E) = 4\pi \left(\frac{2m}{h^2}\right)^{3/2} E^{1/2}. \tag{27.2.15}$$

We note that at absolute zero the allowable states are filled as shown in Figure 27.2.2.

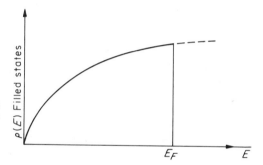

FIGURE 27.2.2. Filled allowable states at absolute zero. Dashed line shows density of allowed states, $\rho(E)$.

This is very much different than for the classical model where at absolute zero *all* the electrons would have zero energy. Here *none* of the electrons have zero energy and their average energy is $3E_F/5$. Although we represent the curve in Figure 27.2.2 as a continuous curve, it really is not. It is actually a series of energy levels with a *very tiny* separation.

EXAMPLE 27.2.4

What temperature would a classical electron gas particle in copper need to have so that its energy would equal the average (actual) quantum mechanical energy?

Answer. According to ideal gas theory $U_K = \frac{3}{2}kT$, which must equal $3E_F/5$, where $E_F = 7$ eV in the present case. Hence $T = 2E_F/5k$ or $T > 30,000°$K. Thus the free electrons in a metal are very energetic particles.

It is common to define a **Fermi temperature** T_F by

$$kT_F = E_F. \tag{27.2.16}$$

Note that the Fermi temperature is many times the melting temperature of the solid. See Table 27.2.2.

Table 27.2.2. CALCULATED FERMI SURFACE PARAMETERS FOR FREE
ELECTRONS

	Electron Concentration N, per cm³	Velocity v_F, cm/sec	Energy E_F, eV	Temperature $T_F = E_F/k$, °K
Li	4.6×10^{22}	1.3×10^8	4.7	5.5×10^4
Na	2.5	1.1	3.1	3.7
K	1.34	0.85	2.1	2.4
Rb	1.08	0.79	1.8	2.1
Cs	0.86	0.73	1.5	1.8
Cu	8.50	1.56	7.0	8.2
Ag	5.76	1.38	5.5	6.4
Au	5.90	1.39	5.5	6.4

27.3 FERMI-DIRAC DISTRIBUTION

Electrons obey Fermi-Dirac statistics in which only one particle is allowed
per wave function. It can then be shown that the **Fermi-Dirac distribution function** is

$$F(E) = \frac{1}{e^{(E-E_F)/kT} + 1}. \tag{27.3.1}$$

$F(E)$ is the probability that an energy level is filled. (For a derivation of this
expression, see J. E. Mayer and M. G. Mayer, *Statistical Mechanics,* Wiley,
New York (1940) Chapter 5.) From our previous discussion we note that $F(E) =$
1 for $E < E_F$ and $F(E) = 0$ for $E > E_F$ at 0°K. This is shown in Figure 27.3.1.

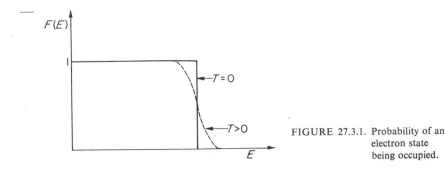

FIGURE 27.3.1. Probability of an electron state being occupied.

Note $F(E_F) \equiv \frac{1}{2}$. We are now prepared to explain the electronic specific heat
paradox [after A. Sommerfeld, *Z. Physik* **47**, 1 (1928)]. Recall that at room temperature $kT \approx \frac{1}{40}$ eV; also recall that $E_F \approx 5$ eV. Note that for $E - E_F \gg kT$,
$F(E) \approx 0$, while for $E_F - E \gg kT$, $F(E) \approx 1$. It is only for those values of E
lying close to E_F (approximately kT or less away) that the $F(E)$ curve for $T > 0$

differs from that at $T = 0$. One could say that all the other electrons (those for which $E_F - E > kT$) are unaware of the temperature T; such electrons therefore do not contribute to the specific heat. Only those electrons within kT of E_F have thermal energy which we approximate by $3kT/2$. Hence the thermal energy per unit volume is

$$U^E = E\rho(E_F)\,\Delta E = \frac{3kT}{2}\rho(E_F)kT.$$

Hence

$$C_v^E\left(\frac{\text{energy}}{\text{volume-}{}^\circ K}\right) \approx \frac{3T}{T_F}\left[\frac{3kN_F}{2}\right] \tag{27.3.2}$$

since

$$\rho(E_F) = \frac{3N_F}{2E_F}. \tag{27.3.3}$$

Inasmuch as $T_F \sim 45{,}000^\circ K$ we have at $T = 300^\circ K$

$$C_v^E \approx \frac{1}{50}\frac{3kN_F}{2}. \tag{27.3.4}$$

A more exact analysis would use the result that the energy of the system is

$$U = \int_0^{E_F} EF(E)\rho(E)\,dE, \tag{27.3.5}$$

and the electronic specific heat could then be shown to be

$$C_v^E = \frac{\pi^2}{3}\rho(E_F)k^2T. \tag{27.3.6}$$

[For a derivation of this, see J. E. Mayer and M. G. Mayer, *Statistical Mechanics*, Wiley, New York (1940).] Note that the electronic specific heat (not the energy) is proportional to T.

This can also be written

$$C_v^E = \frac{\pi^2 T}{3T_F}\left[\frac{3}{2}N_Fk\right], \tag{27.3.7}$$

a result in close agreement with the rough approximation of (27.3.2).

At low temperatures the electronic specific heat exceeds the lattice specific heat (atomic vibrations) since the latter is proportional to the cube of the temperature [see (17.11.13)] while the former varies linearly with the temperature

$$C_v^E = \gamma T, \tag{27.3.8}$$

where

$$\gamma = \frac{\pi^2}{3}\rho(E_F)k^2. \tag{27.3.9}$$

The experimental determination of C_v^E and hence of γ provides an experimental method of evaluating the density of states at the Fermi surface [$\rho(E_F)$]. An example of such data is shown in Figure 27.3.2.

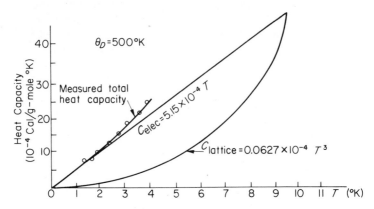

FIGURE 27.3.2. Specific heat data for a 20: 80 V–Cr alloy. [From C. Chang, C. Wei, and P. Beck, *Physical Review*, **120**, 426 (1960).]

EXAMPLE 27.3.1

Sketch the density of filled states versus energy at $T > 0$.

Answer. The density of filled states is given by $\rho(E)F(E)$, where $\rho(E)$ is given by (27.2.15) and $F(E)$ by (27.3.1).

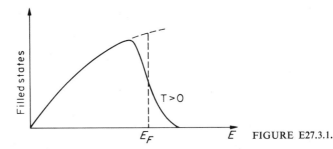

FIGURE E27.3.1.

27.4   ELECTRONS IN A PERIODIC POTENTIAL

A completely free electron has energy given by

$$E = \frac{p^2}{2m} = \frac{h^2}{2m\lambda^2} = \frac{h^2 k^2}{8\pi^2 m},$$ (27.4.1)

where we have used de Broglie's relation and where we have defined the **wave vector**

$$k = \frac{2\pi}{\lambda}s,$$ (27.4.2)

where **s** is the unit vector in the direction of propagation of the wave and λ is the wavelength (λ has also been used for the mean free path in this chapter). For mathematical simplicity we shall now assume that the wave propagates in the x_1 direction only:

$$E = \frac{p_1^2}{2m} = \frac{h^2}{2m\lambda_1^2} = \frac{h^2 k_1^2}{8\pi^2 m}. \tag{27.4.3}$$

A plot of this function is shown in Figure 27.4.1. This represents a continuum.

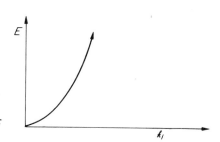

FIGURE 27.4.1. Energy versus
        wave vector for
        free electron.

The wave function which represents a free electron is

$$\psi \propto e^{ik_1 x_1} \tag{27.4.4}$$

or in the general case

$$\psi \propto e^{i\mathbf{k}\cdot\mathbf{r}}. \tag{27.4.5}$$

For the quantized electron in a one-dimensional box, only certain values of λ_1 or k_1 are possible:

$$n\lambda_1 = 2L_1, \tag{27.4.6}$$

$$k_1 = \frac{n\pi}{L_1}, \qquad n = 1, 2, 3, \ldots. \tag{27.4.7}$$

We can use the same curve in Figure 27.4.1 to represent the E vs. k_1 relation, but we must remember that it is not a continuous curve but rather a representation of *very* closely spaced levels.

An electron in a crystal moves through a periodic potential such as that shown in Figure 27.4.2 rather than in a potential $V = 0$ as assumed in Section 27.3.

This has a resultant profound effect on the energy wave vector curve as shown in Figure 27.4.3. The effect is to introduce **forbidden energy bands**

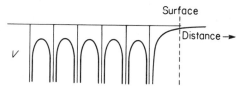

FIGURE 27.4.2. Potential energy
        in a crystal in a
        given direction
        along a given
        path.

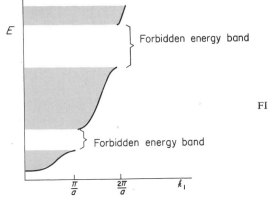

FIGURE 27.4.3. Origin of forbidden energy bands as a result of the periodic potential. Here *a* is the period of the potential in a specific direction.

(**energy gaps**). [The quantized electron in a box ($V = 0$) had a single allowed energy band which includes all energies from zero to infinity, although only very closely spaced discrete energies within the band are allowed.] The origin of the forbidden energy bands is given in the sequence of Problems 27.20–27.30. The student is urged to work through these problems (which discuss the **Kronig and Penney model**) in detail (but not necessarily now).

Note that the discontinuities in E are periodic in k with period π/a (we would expect a different E vs. k curve in another direction). Hence, if we include waves with negative k,

$$k = \pm \frac{n\pi}{a}, \qquad n = 1, 2, 3, \ldots. \tag{27.4.8}$$

In the present example we started with a free electron and showed how the introduction of a periodic potential led to energy bands. One could also start with the electron tightly bound to the atom and bring the atoms together and show that the energy levels spread out into bands as shown in Figure 27.4.4.

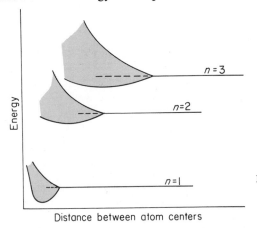

Distance between atom centers

FIGURE 27.4.4. Formation of energy bands in the tight binding approximation.

This tight binding approximation is discussed with unusual clarity in R. L. Sproull, *Modern Physics*, Wiley, New York (1956) and will not be repeated here. The energy levels for the two extreme cases, namely, free electrons and tight binding, are shown along with the actual intermediate case in crystals in Figure 27.4.5.

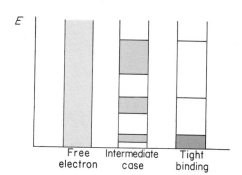

FIGURE 27.4.5. Allowed energy levels for different binding strengths.

Free electron   Intermediate case   Tight binding

27.5   BRILLOUIN ZONES

We have noted that the discontinuities in E occur when

$$k = \pm\frac{n\pi}{a}, \qquad n = 1, 2, 3, \ldots, \tag{27.5.1}$$

in the one-dimensional case. The negative k's involve waves moving in the negative direction (which we have ignored previously). The first region of k values where there is no discontinuity extends over $-\pi/a < k < \pi/a$. It is called the **first Brillouin zone**. The zones for the one-dimensional case are shown in Figure 27.5.1.

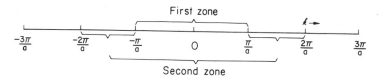

First zone

$k \rightarrow$

$-\frac{3\pi}{a}$ $-\frac{2\pi}{a}$ $-\frac{\pi}{a}$ 0 $\frac{\pi}{a}$ $\frac{2\pi}{a}$ $\frac{3\pi}{a}$

Second zone

FIGURE 27.5.1. Brillouin zones in one-dimensional lattice.

The solution of the wave equation leads to traveling waves [see Problem 27.12 except for the case where k is given by (27.5.1), which is the equation for Bragg reflection from a one-dimensional lattice].

The condition for Bragg reflection from a two-dimensional square lattice can be written

$$k_1 n_1 + k_2 n_2 = \frac{\pi}{a}(n_1^2 + n_2^2). \tag{27.5.2}$$

(For a derivation of this expression, see Charles Kittel, *Introduction to Solid State Physics*, Wiley, New York (1968) p. 52. It is necessary to read about the reciprocal lattice and the Ewald construction.) Examples of two-dimensional Brillouin zones for a simple square lattice are shown in Figure 27.5.2. The energy shows a discontinuity as the k values cross a zone boundary. We note that the total area of each zone is the same.

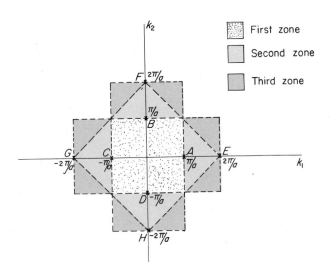

FIGURE 27.5.2. First three Brillouin zones for a two-dimensional square lattice.

The condition for Bragg reflections from a cubic lattice is

$$k_1 n_1 + k_2 n_2 + k_3 n_3 = \frac{\pi}{a}(n_1^2 + n_2^2 + n_3^2). \tag{27.5.3}$$

Here the n_1, n_2, and n_3 take on the same values as do H, K, and L for a specific crystal (see Chapter 11).

EXAMPLE 27.5.1

Construct the first Brillouin zone for a fcc crystal such as copper or silver.

Answer. The allowed *HKL* values are 111, 200, 220, etc. (see Section 11.3). The first zone boundary is the region closest to the origin in k space for which Bragg reflection occurs. For the {111} planes we have $\pm k_1 \pm k_2 \pm k_3 = 3\pi/a$.

This represents eight equations for planes in k space. These are the octahedral planes. Note that they intersect each of the k axes at $3\pi/a$. Note that the {200} planes intersect an axis at $2\pi/a$. The first Brillouin zone is thus an octahedron in which the six tips are sliced off by the planes $k_1 = \pm(2\pi/a)$, $k_2 = \pm(2\pi/a)$, and $k_3 = \pm(2\pi/a)$. This figure is shown in Figure 27.5.3.

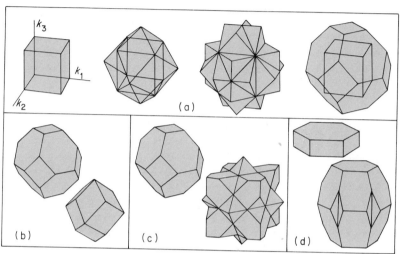

FIGURE 27.5.3. Three-dimensional Brillouin zones. (a) The first four zones for a simple cubic lattice. (b) The first two zones for a bcc lattice. (c) The first two zones for a fcc lattice. (d) The first two zones for a close-packed hexagonal lattice. [From T. S. Hutchinson and David C. Baird, *The Physics of Engineering Solids*, Wiley, New York (1963).]

27.6 CONSTANT ENERGY CURVES

A completely free electron in a two-dimensional box has energy given by

$$E = \frac{h^2}{8\pi^2 m}(k_1^2 + k_2^2), \tag{27.6.1}$$

so lines of constant energy are represented by circles in k space.

Near the center of the Brillouin zone the electrons are nearly free and so the constant energy curves are circles. These constant energy curves become perturbed as the k values approach the boundaries. One result which we do not derive here is that in the neighborhood of the zone corners, the curves of equal energy approximate circles which are centered on the corner. This behavior is illustrated in Figure 27.6.1.

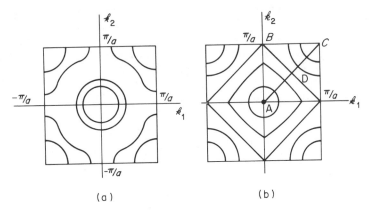

FIGURE 27.6.1. Constant energy curves in the first Brillouin zone of a simple two-dimensional lattice. (a) Weak binding. (b) Strong binding.

OVERLAP.   We note that as the energy increases from A to B in Figure 27.6.1(b) we reach an energy discontinuity at B. As we increase energy from A to D along another possible wave vector direction we reach the same energy as at B. The further increase in energy from D to C may be larger than the energy discontinuity at B. Hence the energy just past B at B' in the second zone may be less than the energy near C within the first zone. The energy bands are said to **overlap**.

THREE DIMENSIONS.   In three dimensions we shall be dealing with surfaces rather than with curves. In the nearly free electron case the constant energy surfaces are spheres, for small k values.

EXAMPLE 27.6.1

Show that electrons at the Fermi surface in copper (7 eV) have a wavelength close to the spacing between atoms, so diffraction effects should be noticeable.

Answer. We have

$$E = \frac{h^2 n^2}{8mL^2} = \frac{h^2}{2m\lambda^2}.$$

Hence using the free electron value for mass m, we have $\lambda = 4.4$ Å. The planes which are farthest apart in copper are the (111) planes. These planes have a spacing of 3.61 Å$/\sqrt{3} = 2.1$ Å.

An example of an actual Fermi surface is shown in Figure 27.6.2.

FIGURE 27.6.2. Fermi surface of copper is a distorted sphere with eight "necks" pulled out to touch the boundary of the first Brillouin zone. The faces of the Brillouin zone represent energy discontinuities in k space (momentum or velocity space).

There are several experimental techniques related to the measurement of effective mass (studied later), the Fermi surface, etc. These include

1. Cyclotron resonance.
2. de Haas-van Alphen effect.
3. Attenuation of sound wave by conduction electrons.
4. Same as technique 3 but with magnetic field present.
5. Anomalous skin effect.
6. Electronic specific heat.
7. Magnetoresistance.

These effects are discussed in solid state physics textbooks [for example, C. Kittel, *Introduction to Solid State Physics*, Wiley, New York (1966) Chapter 9].

The **de Haas-van Alphen effect** is an oscillation in magnetic susceptibility of a crystal as a function of the applied magnetic field. It can be shown that the period of the oscillations, when the magnetic moment is plotted versus $1/H$, gives the area of a maximal or minimal cross section of the Fermi surface normal to the magnetic field.

27.7 NUMBER OF STATES IN AN ENERGY BAND AND EFFECTIVE MASS

How many different wave functions are possible in an energy band?

Let us consider for mathematical simplicity a one-dimensional lattice. Let there be M atoms of spacing a so that the total length of the array is

$$d = Ma. \qquad (27.7.1)$$

The allowed λ's are, in the terms of the integer $n = 1, 2, \ldots,$ or n_{max},

$$\frac{\lambda}{2} = \frac{d}{n} = \frac{Ma}{n}. \qquad (27.7.2)$$

Now for the first zone

$$k_{max} = \left(\frac{2\pi}{\lambda}\right)_{max} = \frac{\pi}{a} \qquad (27.7.3)$$

so that by (27.7.2) and (27.7.3) we have

$$\left(\frac{2\pi}{\lambda}\right)_{max} = \frac{\pi n_{max}}{Ma} = \frac{\pi}{a} \qquad \text{so} \qquad n_{max} = M.$$

If we also include electron spin states, then the number of allowed states in the first band is $2M$, i.e., twice the number of atoms. The same result applies for simple cubic, bcc, and fcc crystals. For diamond cubic crystals the number of allowed states per allowed energy band is $4M$.

EFFECTIVE MASS.   The energy of a wave is given by

$$E = hv = \frac{h}{2\pi}\omega. \qquad (27.7.4)$$

The **group velocity** of a wave is equal to

$$v_g = \frac{d\omega}{dk} = \frac{2\pi}{h}\frac{dE}{dk}. \qquad (27.7.5)$$

[The importance of group velocity lies in the fact that the energy is propagated with this velocity. For an elementary discussion of group velocity, see C. A. Coulson, *Waves*, Wiley-Interscience, New York (1952) Chapter 8.]

Hence the acceleration is given by

$$\frac{dv_g}{dt} = \frac{2\pi}{h}\frac{d}{dt}\left(\frac{dE}{dk}\right) = \frac{2\pi}{h}\frac{d^2E}{dk^2}\frac{dk}{dt}. \qquad (27.7.6)$$

The change in energy of an electron dE when it is acted on by a field ξ is

$$dE = -e\xi \, dx = -e\xi v_g \, dt$$
$$= -e\xi\frac{2\pi}{h}\frac{dE}{dk} dt \qquad (27.7.7)$$

so that

$$\frac{dk}{dt} = \frac{-2\pi e\xi}{h}.$$ (27.7.8)

Hence the expression for acceleration becomes

$$\frac{dv_g}{dt} = \frac{-4\pi^2 e\xi}{h^2}\frac{d^2 E}{dk^2}$$

and the resultant expression for force on an electron, $F = -e\xi$, is given by

$$F = \frac{h^2}{4\pi^2(d^2 E/dk^2)}\frac{dv_g}{dt}.$$ (27.7.9)

We can define an **effective mass** by

$$m^* \equiv \frac{h^2}{4\pi^2(d^2 E/dk^2)}$$ (27.7.10)

so that the electron (wave) in the solid obeys Newton's second law.

An examination of each allowed band of the E vs. k curve of Figure 27.4.3 shows that $dE/dk = 0$ at the top and bottom of an allowed band (see Problem 27.17) and $d^2 E/dk^2$ is positive near the bottom of the band and *negative* near the top. Moreover, there is an inflection point $d^2 E/dk^2 = 0$ in between. The student should sketch v_g vs. k and m^* vs. k using E vs. k from Figure 27.4.3 for the first band.

The effective mass approaches m for a loosely bound electron (nearly free, as the $3s$ or valence electron in sodium) and may be much larger than m in the case of a tightly bound electron (for the $3d$ electron in nickel, $m^*/m \approx 28$). It is m^* at the Fermi surface which is important in all applications.

27.8   REAL CRYSTALS AND CONDUCTIVITY

Sodium has a bcc crystal structure. It is a monovalent metal and it is an excellent conductor of electricity (and heat) at room temperature. The energy bands for sodium are illustrated in Figure 27.8.1.

EXAMPLE 27.8.1

What is the first ionization potential of the isolated sodium atom?

Answer. This is the energy to excite the electron from the $3s$ state to the free electron state. As can be seen from Figure 27.8.1, this is about 5 eV. A large separation corresponds to isolated atoms.

Note that at the equilibrium separation the $2p$ state has not spread out (appreciably) into a band but appears approximately as a single state just as in the isolated atom. This is illustrated in Figure 27.8.2. We note that soft X-rays

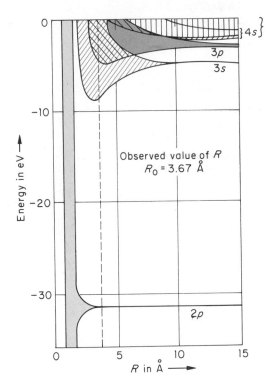

FIGURE 27.8.1. Energy bands in sodium as a function of the distance between nearest neighbors. An energy equal to zero corresponds to an unbound or free electron. Dashed line shows equilibrium distance in the solid.

[energies of the order of 30 eV or so (compare with the X-rays used in ordinary diffraction work where the energies are of the order of 20,000 eV)] can be used to obtain the *occupied* width of the 3s band, i.e., to measure the Fermi energy. In the soft X-ray experiment the 2p electron would be knocked out of the crystal by bombardment with electrons. Then an electron from the 3s band would fall into the vacated state with the emission of a soft X-ray.

In sodium metal the 3s electron is ionized and is part of the electron sea. The first band of available states for these valence electrons is called the **valence**

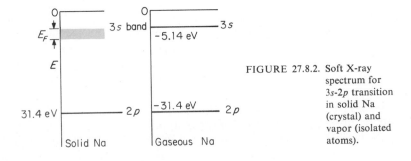

FIGURE 27.8.2. Soft X-ray spectrum for 3s-2p transition in solid Na (crystal) and vapor (isolated atoms).

band. The density of states for the first band (valence band) is sketched in Figure 27.8.3 and the occupied states (at absolute zero) are shown shaded. Note that the band is only half filled. Note also the deviations from the free electron case illustrated in Figure 27.2.2. In fact the actual density of states curve is represented by two parabolic curves except for the cusp near the middle.

FIGURE 27.8.3. Sketch of the density of occupied states (shown shaded) and allowed states for the valence band of sodium (3s band)

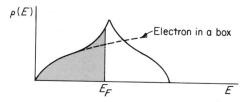

We note that the filled states follow the quantized electron in a box model (limit of weak binding) except very near to the Fermi surface. We would therefore expect the Fermi surface of sodium to be nearly spherical—a fact borne out by experiment.

CONDUCTION. When an electrical field is applied to sodium the net result is to move the Fermi "sphere" as shown in Figure 27.8.4.

FIGURE 27.8.4. Square represents view of first Brillouin zone in a simple cubic metal viewed from [100] direction. Solid sphere represents Fermi surface (plotted in $k_1k_2k_3$-space) in absence of field. Dashed sphere represents Fermi surface in the presence of the field ξ.

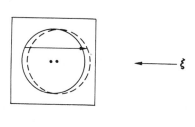

EXAMPLE 27.8.2

Estimate the energy gained by an electron owing to the presence of the applied field between collisions. Assume that there are 100 V across a 1-m length and that the mean free path $\lambda = 200$ Å.

Answer. A collision scatters the electron in random directions. Then the applied field acts on it while it moves the distance λ. The force acting has a magnitude $e\xi$ so the work done is $e\xi\lambda = (100 \text{ V/m}) \times 200 \times 10^{-10} \text{ m} = 2 \times 10^{-6} \text{ eV}$.

Thus note that the perturbation owing to the electrical field is extremely small (and hence vastly exaggerated in Figure 27.8.4) since the Fermi energy is several electron volts.

Recall that there are numerous available states within the zone. The vast majority of electrons are in states common to each sphere. A tiny fraction changes states as shown by the arrow. This tiny fraction always involves electrons *very* near (at) the Fermi surface; hence, the effective mass m^* associated with conductivity is the effective mass evaluated at the Fermi surface.

INSULATORS.   In the case of NaCl the density of states is as shown in Figure 27.8.5. In NaCl all the allowed levels in the valence band are filled. The application of ordinary electrical fields is certainly not sufficient to excite electrons across the energy gap of several electron volts (see Example 27.8.2) to the next available band called the **conduction band**. The absence of nearby allowed levels means that NaCl will be an electrical insulator insofar as electrons are concerned (we have noted earlier that at high temperature point defects are present in equilibrium concentration and their motion leads to conductivity).

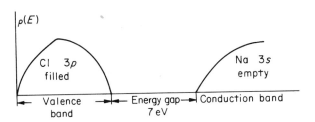

FIGURE 27.8.5. Sketch of the density of occupied and allowed states for NaCl.

The situation for diamond is quite analogous. The valence band in the diamond cubic crystal contains four allowed states per atom and the tetravalent carbon atom contributes four electrons. The valence band is filled. The conduction band is empty. Thus at room temperature and below, pure diamond is an insulator (see Table 3.1.1). However at quite high temperature pure diamond is a semiconductor.

SEMICONDUCTORS.   The energy bands for semiconductors (or insulators) are shown in Figure 27.8.6. At low temperature in the pure material the valence band is filled and the conduction band is empty. The material behaves as an insulator. However, depending on the size of the energy gap and the tempera-

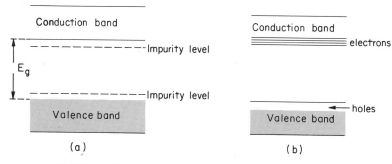

FIGURE 27.8.6. Energy bands in semiconductors. The impurity states are nonexistent in the pure materials. (a) At absolute zero. (b) At room temperature.

ture, the material may behave as a semiconductor. Thus if the energy gap is small, electrons can be thermally excited from levels near the top of the valence band to levels near the bottom of the conduction band. We call the empty state at the top of the valence band a **hole**. Electrons at the top of the band behave as if they are nearly free electrons with a negative effective mass and a negative charge. A hole is defined as having positive effective mass (equal in magnitude to the effective mass of the electrons at the top of the band) and positive charge. The conductivity which arises as a result of there being empty states near the top of the valence band can be ascribed to the motion of holes. Let us arbitrarily call the energy at the top of the valence band $E = 0$. Then the energy at the bottom of the conduction band is $E = E_g$, where E_g is the energy gap.

The probability of existence of a hole in the valence band must equal the probability of existence of an electron in the conduction band. Thus

$$F(E_g) = 1 - F(0). \tag{27.8.1}$$

From (27.3.1) we have

$$\frac{1}{e^{(E_g - E_F)/kT} + 1} = 1 - \frac{1}{e^{-E_F/kT} + 1}$$

from which we can solve for the Fermi energy (the chemical potential of the electrons):

$$E_F = \frac{E_g}{2}. \tag{27.8.2}$$

Thus in a pure (**intrinsic**) semiconductor the Fermi level lies midway between the top of the valence band and the bottom of the conduction band. Hence the number of electrons in a conduction band (or holes in the valence band) can be seen to be proportional to $e^{-E_g/2kT}$. The exact expression will be discussed in Chapter 28, where we shall see that in germanium near room temperature the number of electrons in the conduction band is about $10^{17}/cm^3$. Since the number of atoms is about $10^{23}/cm^3$, we might expect a conductivity which is about 10^{-6}

of the conductivity of a metal (assuming the mobilities are similar). See Table 3.1.1 for comparison. We are close to understanding the exponential dependence of resistivity of germanium (shown in Figure 3.3.1) which has $E_g = 0.72$ eV.

DIVALENT METALS.   In a divalent metal such as Cd (hcp) there are two available states per atom in the valence band. Since Cd is divalent, we might expect to find the first band completely filled and hence expect Cd to be an insulator. However, cadmium is a good conductor (Table 3.1.1). This is due to the fact that there is overlap of the first and second bands, as shown in Figure 27.8.7. Thus there are holes in the first band and electrons in the second band. The conductivity in the first band dominates and is due to the motion of positive charge carriers with positive mass, i.e., holes. This explains the anomalous sign of the Hall coefficient shown in Table 9.4.2 for cadmium.

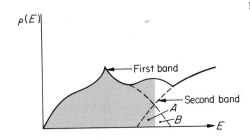

FIGURE 27.8.7. Occupation of the valence bands of a divalent metal. There are electrons at the bottom of the second band at A and holes at the top of the first band at B.

27.9   THE ORIGIN OF RESISTIVITY IN METALS

In the Kronig-Penney model discussed in Problems 27.20–27.30, once the electron is in a given state it remains in that state. Bloch has proved quite generally that in a perfect static crystal (no atom vibrations) the electron would not change states, except in the presence of external fields. Hence there would be no resistivity. Note how this differs from Drude's original idea in which it was assumed that the ions were static and that the mean free path λ was equal to the interatomic spacing. Real metals show finite resistivity owing to a finite path length λ or time τ between "collisions." The mean free path is defined by

$$\lambda = v_F \tau, \tag{27.9.1}$$

where v_F is the velocity at the Fermi surface given by

$$v_F = \frac{h}{2\pi m}(3\pi^2 N)^{1/3}. \tag{27.9.2}$$

[This is obtained from $E_F = \frac{1}{2}mv_F^2$ and Equation (27.2.10), which assumes a spherical Fermi surface.] For Cu, v_F is 1.56×10^6 m/sec. The scattering is a consequence of the departure of the lattice from perfect periodicity. These imperfections include lattice vibrations or phonons and point, line, and planar defects. Point defects may include vacancies, impurities, etc. If the crystal does not have the latter stacking imperfections (defects), it would ordinarily be called perfect. In this case the resistivity would be entirely due to phonons. The resistivity is given by

$$\rho = \frac{m^*}{Ne^2\tau} = \frac{m^* v_F}{Ne^2\lambda}.$$ (27.9.3)

[Note the factor of two variation between (27.9.3) and (27.1.4). This is due to the fact that in Drude's theory τ is the time between collisions whereas in scattering theory τ is the relaxation time for the terminal velocity, v_D, for motion of an electron through viscous media which is described by $m(\dot{v}_D + v_D/\tau) = e\xi$.] The temperature dependence is

$$\rho \propto T, \qquad T \gtrsim \theta_D,$$ (27.9.4)

$$\rho \propto T^5, \qquad T \ll \theta_D.$$ (27.9.5)

Here θ_D is the Debye temperature. The probability of scattering is directly proportional to the number of phonons. At high temperatures the number of phonons is directly related to the mean squared displacement of the atoms (this can be measured by measuring the intensity of the Bragg X-ray diffraction peak; see Section 11.4).

EXAMPLE 27.9.1

Calculate the relaxation time τ and the mean free path λ for electrons in copper at 20°C.

Answer. We obtain ρ from Table 3.1.1. Using the mks system,

$$\tau = \frac{9 \times 10^{-31} \text{ kg}}{(8.5 \times 10^{28} \text{ m}^{-3})(1.67 \times 10^{-8} \ \Omega\text{-m})(1.6 \times 10^{-19} \text{ C})^2}$$
$$= 2.48 \times 10^{-14} \text{ sec}.$$

Hence

$$\lambda = v_F\tau \approx 4 \times 10^{-8} \text{ m} \approx 400 \text{ Å}.$$

The mean free path has been measured directly (see Example 27.1.4) and a value of $\lambda = 450$ Å is obtained at room temperature [see F. W. Reynolds and G. R. Stillwell, *Physical Review*, **88**, 418 (1952)].

EXAMPLE 27.9.2

"Pure" copper has a resistivity ratio of 50,000. What is the mean free path at 4°K?

Answer. $\lambda = 450$ Å $\times$ 50,000 $= 0.2$ cm.

At low temperatures the number of phonons varies as T^3 but the scattering is inefficient; e.g., electrons are scattered only by a few degrees of angle. The **thermal resistivity** is given by

$$\rho_T = \frac{m^*}{Ne^2} \frac{1}{\tau_T}. \qquad (27.9.6)$$

There is an additional resistivity owing to stacking imperfections, often called **residual resistivity**, because it remains at absolute zero temperature

$$\rho_R = \frac{m^*}{Ne^2} \frac{1}{\tau_R}. \qquad (27.9.7)$$

Here τ_R will depend on the concentration and nature of the defects. Often τ_R will be independent of temperature unless the defect anneals out. Three possible reasons why τ_R might change are:

1. Cold-worked structures might undergo recovery.
2. A supersaturated solution may show precipitation.
3. Quenched-in vacancies may anneal out.

The total resistivity is

$$\rho = \rho_R + \rho_T = \frac{m^*}{Ne^2}\left(\frac{1}{\tau_R} + \frac{1}{\tau_T}\right) = \frac{m^*}{Ne^2} v_F \left(\frac{1}{\lambda_R} + \frac{1}{\lambda_T}\right). \qquad (27.9.8)$$

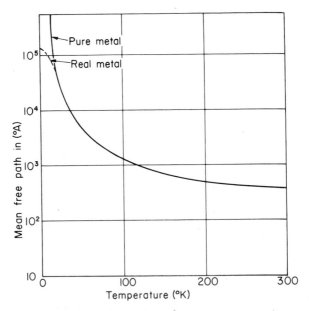

FIGURE 27.9.1. Electron mean free path in copper.

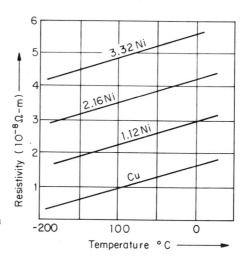

FIGURE 27.9.2. Resistivity of Cu and Cu–Ni alloys.

The additivity of resistivity is known as **Matthiessen's rule**. If the defect structure is thermally stable, then the resistivity equals a constant plus a term which increases as the temperature increases. The residual resistivity term is of overwhelming importance at very low temperatures and is essentially insignificant around room temperatures for materials which are chemically pure. See Figure 27.9.1. It appears therefore that very little can be done to increase the conductivity of a metal near room temperature, although its resistivity can be greatly increased by alloying, as shown in Figure 27.9.2. However, at low temperatures large increases in the conductivity can be made by purification and elimination of stacking defects. This tends to eliminate the residual resistivity, as shown in Figure 27.9.3. It is conceivable that there will be many applications for high-conductivity metals in the normal conducting state at low temperatures. Considerable research is going on to achieve high purities and the corresponding high conductivities.

FIGURE 27.9.3. Resistivity of Cu relative to the value of resistivity at 0°C, i.e., ρ_0. Note the residual resistivity below 20°K. Above liquid nitrogen temperature, the curve is nearly linear.

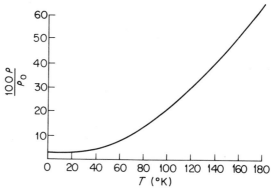

Electrical resistivity is a convenient tool for studying defects in solids. In the case of metals the resistivity change is quite small but still measurable. Thus the $\Delta\rho$ owing to the presence of defects is usually 10^{-10}–10^{-12} Ω-m or less. This is a small perturbation on the room-temperature value. Even so, the annealing of defects caused by radiation damage (neutron irradiation, etc.) can be studied by resistance measurements. In this case the defects can be destroyed by re-combination of vacancies and self-interstitials or by dislocation climb.

Metals are used in various applications as resistors. The resistance ther-mometer (Pt) is a useful laboratory tool. Precision resistors whose temperature variation is negligible at room temperature are often made of manganin (83% Cu–13% Mn–4% Ni). Manganin is also used as a resistance pressure gauge. One of the important resistor applications of metals is for heating ele-ments. A desirable property of such metals is a relatively high resistivity and a high oxidation resistance. Nichrome wire (80% Ni–20% Cr) is often used (a chromium oxide film forms). The semiconductors SiC and $MoSi_2$ are also used (here an SiO_2 film forms). For very high temperatures (2200°C), tungsten is used in a reducing (hydrogen gas) atmosphere.

27.10   ELECTRON EMISSION

Electrons can be emitted from solids by three mechanisms:

1. **Photoelectric emission** (bombardment with light whose energy is transmitted to the electron).
2. **Thermionic emission** (heating a filament causes a current to flow between it and another electrode when a small potential is applied; the filament has a negative potential relative to the other electrode).
3. **Field emission** (the application of enormous fields, on the order of 10^9 V/m, causes electron flow).

PHOTOELECTRIC EFFECT.   The energy levels of electrons in the valence band of sodium are sketched in Figure 27.8.2. The energy which must be given to the electrons having the Fermi energy in order to free the electron from the crystal is known as the **work function** ϕ. This is illustrated in Figure 27.10.1.

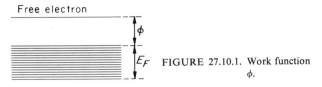

FIGURE 27.10.1. Work function ϕ.

When light below a certain **threshold frequency**, ν_0, falls on the solid, there will be no emission, i.e., no current flow in the photoelectric tube shown in Figure 27.10.2.

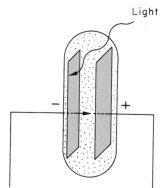

FIGURE 27.10.2. Photoelectric tube. Electrons are collected at positive electrode.

However, for light of frequency ν_0 or greater there will be electron emission and current flow. Einstein (1905) gave the explanation of this behavior, noted by R. A. Millikan, namely, that the energy of photons was quantized in units of $h\nu$ and that

$$\phi = h\nu_0. \tag{27.10.1}$$

Examples of some work functions and the corresponding wavelengths are shown in Table 27.10.1. Note that the work functions are of the same magnitude as the Fermi energies. A good cathode has a low work function and a high **quantum yield** (number of free electrons produced per quantum of incident light). The alkali metals or composite structures involving alkali metals are usually used.

Table 27.10.1. WORK FUNCTIONS

Element	ϕ, eV	Wavelength, λ_0, Å
Li	2.46	5040
Na	2.28	5430
K	2.25	5510
Cs	1.94	6390
Cu	4.48	2770
Ag	4.70	2640
Ge	4.62	2680
Si	3.59	3450
W	4.5	2750
Ta	4.1	2530

One of these composites (antimony-cesium) has a quantum yield of 20% at 4700 Å, a value about 100 times as large as that for the alkali metals. This is illustrated in Figure 27.10.3. The nature of the surface is clearly very critical. We note that in the infrared (above 7800 Å) the Ag–O–Cs electrode has a much higher yield than the Sb–Cs electrode, while in the optical range the reverse is

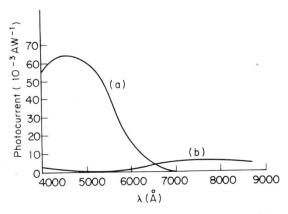

FIGURE 27.10.3. Yield versus wavelength for (a) antimony-cesium and (b) silver-oxygen-cesium.

true. (The wavelengths and energies of light in the visible spectrum are given in Figure 31.1.2.)

Figure 27.10.4 shows the current versus voltage relations for a phototube. Note that the current reaches a plateau in any specific case. This is known as **saturation**. It is due to the **space charge**; i.e., the space between the electrodes contains electrons which tend to repel an emitted electron and hence prevent it from reaching the other electrode.

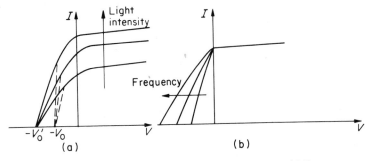

FIGURE 27.10.4. Photocurrent versus applied voltage. (a) For different light intensities at one frequency at room temperature (solid line) and near 0°K (exaggerated difference, dashed line). (b) For different frequencies at one intensity near 0°K.

THERMIONIC EMISSION. If the negative electrode in a tube (see Figure 27.10.2, but with no incident light) is heated to a high temperature, current will flow as first noted by Edison. The current density is given by the Richardson-

Dushman equation

$$J = AT^2 e^{-\phi/kT},\qquad(27.10.2)$$

where

$$A = \frac{4\pi mek^2}{h^3} = 10^6 \left[\frac{A}{m^2(°K)^2}\right].\qquad(27.10.3)$$

For the derivation of this expression see T. S. Hutchison and D. C. Baird, *The Physics of Engineering Solids*, Wiley, New York (1963) p. 208. The surfaces of emitters are usually coated with BaO (or some other oxide) which greatly lowers ϕ (thus BaO on tungsten has $\phi = 1.45$ eV, while ϕ for pure tungsten is 4.5 eV).

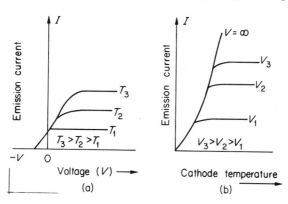

FIGURE 27.10.5. Thermionic emission current versus voltage. (a) Various cathode temperatures. (b) Various plate potentials.

  The emission current versus the voltage curve of a thermionic emitter also shows saturation (as did the photoelectric tube) owing to space charge as illustrated in Figure 27.10.5. There is a lot of current research on thermionic converters based on the hope of using these as direct energy conversion devices. Here the thermal energy of a nuclear reactor would heat the cathode and cause thermal emission.

  Thermal emission devices operate at very high temperatures. Thus tungsten operates at 2500°K ($\phi = 4.5$ eV, $A = 0.60 \times 10^6$ amps/m²°K²) while thorium-coated tungsten operates at 1900°K ($\phi = 2.6$ eV, $A = 0.03 \times 10^6$ A/m²°K²).

  FIELD EMISSION. The applied potential at the surface of a solid in the presence of an external field varies approximately as shown in Figure 27.10.6. In addition to the applied potential there is another potential which has to be taken into account. Consider an electron which is at a perpendicular distance x from the surface of a metal (conductor). It is shown in electrostatics that there is a force acting on the electron and that this is simply equal to the force that

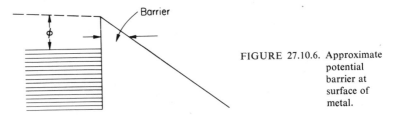

FIGURE 27.10.6. Approximate potential barrier at surface of metal.

exists between an electron at x and its mirror image at $-x$ (with the conductor absent). This is called the **image force**. The potential energy for the image force can be shown to be (in mks units)

$$eV_{image} = -\frac{1}{4\pi\epsilon_0}\frac{e^2}{4x}. \tag{27.10.4}$$

Thus if we applied a potential as at the tip of the field emission specimen, the potential actually appears as in Figure 27.10.7. The work function is reduced from ϕ to ϕ_ξ where

$$\phi_\xi = \phi - \frac{e\sqrt{e\xi}}{2\sqrt{\pi\epsilon_0}}. \tag{27.10.5}$$

This is known as the **Schottky effect**. Hence the current flow as given by the Richardson-Dushman equation is increased.

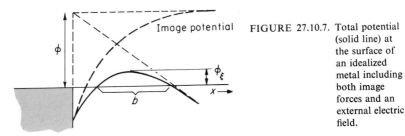

FIGURE 27.10.7. Total potential (solid line) at the surface of an idealized metal including both image forces and an external electric field.

EXAMPLE 27.10.1

Show that the work function is decreased by the term shown in Equation 27.10.5.

Answer. The potential energy of the electron is

$$eV = -\xi ex - \frac{e^2}{4\pi\epsilon_0}\frac{1}{4x}. \tag{27.10.6}$$

This has a maximum when $x = x_m$ and

$$x_m = \frac{1}{4}\sqrt{\frac{e}{\pi\epsilon_0\xi}}. \tag{27.10.7}$$

The value of the potential energy is then obtained by substitution and is

$$eV_m = -e\sqrt{\frac{e\xi}{4\pi\epsilon_0}}.$$

EXAMPLE 27.10.2

Evaluate the decrease in work function for a field of 1 volt/Å. Such fields are typically used in field emission microscopy.

Answer. To find the decrease in electron volts we must evaluate

$$\sqrt{\frac{e\xi}{4\pi\epsilon_0}}.$$

In mks units $\xi = 10^{10}$ V/m; $e = 1.6 \times 10^{-19}$ coulombs and $\epsilon_0 = 8.85 \times 10^{-12}$ farads/m. The result is 3.8 eV. Here $x_m = 1.9$Å.

EXAMPLE 27.10.3

Evaluate the width b of the potential hump in Figure 27.10.7 if the field is 1 V/Å and $\phi = 4.5$ eV.

Answer. This can be obtained by setting $eV = -\phi$ in (27.10.6) and solving for the two values of x. Their difference is b where

$$b = \sqrt{\left(\frac{\phi}{\xi e}\right)^2 - \left(\frac{e}{4\pi\epsilon_0\xi}\right)}.$$

In the present case $b = 2.4$ Å.

Before we delve further into the problem involving very high fields, we must study the concept of electron tunneling. Figure 27.10.8 shows a finite potential well. Let us suppose that the total energy E of the electron is as shown

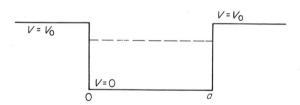

FIGURE 27.10.8. Finite potential well.

by the dashed line, i.e., $E < V_0$. This is a conservative system; on the basis of classical mechanics the electron would stay in the box. On the basis of quantum mechanics, however, there is a certain finite probability that the electron will be found outside of the box. This is shown in Figure 27.10.9. In some cases barriers of finite dimensions exist as in Figure 27.10.10. The passage of an electron through a barrier which would be forbidden by classical mechanics is called **electron tunneling**. The fact that there is a finite probability of finding the electron on the other side of the barrier with its energy unchanged is called the **tunnel effect**. The probability of tunneling can be shown to drop off exponentially with the thickness, b, of the rectangular barrier. It is important only when b is

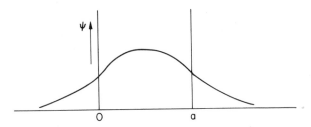

FIGURE 27.10.9. Ground state wave function for finite potential well.

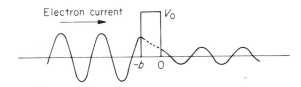

FIGURE 27.10.10. Barrier which allows tunneling.

of the order of 20 Å or less. It plays a vital role in many solid state devices. In the presence of very high fields an electron will tend to tunnel through the barrier showing in Figure 27.10.7. The current density is in this case given by the Fowler-Nordheim equation [L. Nordheim, *Proc. Roy. Soc.*, London, **A121**, 626 (1928)].

We recall that the potential energy of an isolated crystal varies along the surface, so ϕ can be expected to vary along the surface. Thus, if a tiny wire with a rounded tip has a high negative potential at the tip, and if there is a grounded fluorescent screen as shown in Figure 27.10.11, the emitted electrons will strike the screen. The intensity of the image on the screen at a given spot will be determined by the intensity with which electrons were given off in that direction and hence by the local value of the work function. The surface of the tip, which has a very tiny radius, is magnified onto the screen which has a very large radius. The image seen on the screen will show the periodicity of the work function and hence will effectively show atomic positions. This is called **field emission microscopy**. If the experiment is carried out in the presence of atoms such as helium or argon it is possible to use the ionized gas particles for image production. In this case the screen is at a negative potential relative to the tip. When atoms of helium, for instance, approach the surface of the tip (where the field is about 4.5 V/Å), electrons tunnel into the metal. The positive ion is then repelled by the tip and follows the field lines to the screen. This is called **field ion microscopy**. It has been used particularly by Müller to observe many of the atomic features of surfaces of refractory metals, particularly tungsten [see E. W. Müller, *Adv. in Electronics*

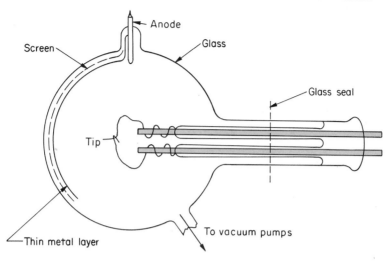

FIGURE 27.10.11. Field emission microscope.

and Electron Physics **13**, 83 (1960) Academic Press, New York]. More recently it has been used by D. N. Seidman and others to study metals with much lower melting points, such as gold [see D. G. Ast and D. N. Seidman, *Applied Physics Letters* **13**, 348 (1960)]. By the use of **field evaporation** it is possible to peal off a layer of atoms at a time and hence to observe the structure of successive remaining layers (field evaporation is really field-aided sublimation; the sublimation energy is reduced by the presence of a field). In this way vacancy concentrations, interstitial concentrations, or defect distribution in radiation-damaged regions can be studied carefully.

CONTACT POTENTIAL. Figure 27.10.12 illustrates the Fermi level in two metals prior to contact and upon contact. Since the Fermi energy is the chemical potential of the electrons, it must be the same in the two metals. Hence the

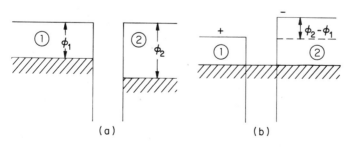

FIGURE 27.10.12. Fermi level in two metals. (a) Prior to contact. (b) Upon contact.

situation shown in (b) must develop. This means that there will be a **contact potential** between two metals whose magnitude is given by the difference of the work functions. If ϕ_1 is known and the contact potential is measured, the work function ϕ_2 can be obtained. A dynamic method for obtaining contact potentials follows: Make the two metals opposite plates of a capacitor. Connect the plates. This gives rise to charges on the plates. A rapid change of the distance between the plates gives a voltage pulse because of the change in C. In the dynamic method one of the plates is oscillated at a fixed frequency and an external dc voltage is applied and adjusted so that the voltage pulse equals zero. The applied potential then equals the contact potential. This is called the **Kelvin method** [Lord Kelvin, *Philosophical Magazine* **46**, 82 (1898)].

REFERENCES

Mackintosh, A. R., "The Fermi Surfaces of Metals," *Scientific American* (July 1963) p. 110.

Ehrenreich, H., "The Electrical Properties of Materials," *Scientific American* (Sept. 1967) p. 194.

Kittel, C., *Introduction to Solid State Physics*, Wiley, New York (1968) Chapters 7–9.

Blue, E. H., and Ingold, J. H., "Thermionic Energy Conversion," in *Direct Energy Conversion* (G. W. Sutton, ed.), McGraw-Hill, New York (1966).

Sommer, A., *Photoelectric Tubes*, Methuen, London (1951).

Nottingham, W. B., "Thermionic Emission," *Encyclopedia of Physics*, Vol. 21, Springer, Berlin (1956) p. 1.

Müller, E. W., and Tsong, T. T., *Field Ion Microscopy*, Elsevier, Amsterdam, (1969).

PROBLEMS

27.1. Using the experimental resistivity and Hall coefficients for sodium from which one deduces that the effective electron mass is $0.8m$ (sodium is bcc and the atom diameter is 3.67 Å), calculate
   a. The relaxation time τ at 300°K.
   b. The mean free path between collisions at 300°K assuming $\bar{v}$ is given by ideal gas theory.

27.2. Plot the first five wave functions for the one-dimensional particle in the box.

27.3. a. Calculate the energy difference between the ground state and first excited state of the valence electrons in solid copper.

b. Do the same for the difference between levels at the Fermi surface in copper.

c. Explain why copper would absorb rather than transmit optical radiation.

27.4. Show that for a spherical Fermi surface $N_F = \frac{2}{3}\rho(E_F)E_F$.

27.5. Compare the electronic specific heat of a 20–80 V–Cr alloy at $1000°K$ with the lattice specific heat. See Figure 27.3.2.

27.6. Discuss without detailed derivations the effect of a periodic potential on the electrons in a metal. Include a discussion of the Kronig-Penney model.

27.7. a. Prove that the first three Brillouin zones of the simple square crystal are those shown in Figure 27.5.2.

b. What is the significance of the zone boundaries?

27.8. Sketch for the valence band three graphs with a common k axis:

a. E vs. k.

b. v_g vs. k.

c. m^* vs. k.

27.9. a. Give a qualitative discussion based on Equation (27.7.10) and Figure 27.4.3 which suggests why a material with a filled valence band (and no overlap) is an insulator.

b. Explain the difference between an insulator and a semiconductor.

c. Why is magnesium a metal instead of an insulator?

27.10. Explain with reference to Figure 27.8.4 why only those electrons with velocities very near to v_F are scattered, i.e., why we use v_F in the equation $\lambda = v_F\tau$ and not $\bar{v}$.

27.11. a. Explain why it is possible to increase but not to decrease appreciably the resistivity of 99.9% Cu at $300°K$.

b. Is this true at $4°K$?

c. Discuss the potential of perfect crystalline copper in the electrical industry including various problems in using it.

27.12. Explain the origin of the phototube current-voltage relationship.

27.13. It can be shown theoretically that the probability that an electron will be scattered from a displaced atom is proportional to the square of the displacement. Why is this a plausible result?

27.14. Explain the origin of different values of $-V_0$ and $-V_0'$ in Figure 27.10.4(a).

27.15. Explain the different values of $-V_0$ in Figure 27.10.4(b).

27.16. Show that $\phi - \phi_\xi \sim 0.12$ eV for $\xi = 10^5$ V/m.

27.17. The work function of a metal can be drastically altered by the addition of a *monolayer* of electropositive atoms (e.g., sodium on tungsten). Such electropositive ions give up an electron to the metal and are in turn positive charges adsorbed on the surface. This results in a **double charge layer**. Assume the negative layer is separated from the positive layer by a distance d and that the layers are separated by vacuum. Show that the change in the work function is $\Delta\phi = (ne^2 d/\epsilon_0)$, where n is the number of ions per unit area. Show that $\Delta\phi \sim 1$ eV.

27.18. Using the classical expressions for conductivity by an electron gas, show that the ratio $k_T/\sigma T = L$ (the Lorentz constant), where k_T is the thermal conductivity [Equation (17.14.3)] and σ is the electrical conductivity. Use the quantum mechanical expression for the specific heat. Show that the Lorentz number is

$$L = \frac{\pi^2}{3}\left(\frac{k}{e}\right)^2.$$

Compare with the values for metals in Tables 3.1.1 and 6.2.1 (1 cal = 4.186 J). We have assumed that the mean free path for the thermal and electrical behavior is the same; i.e., all types of collision affect the two equally. Advanced analysis shows that this is nearly true at high temperatures but not at low temperatures.

27.19. Discuss the potential engineering applications, including economic factors, of the concepts discussed in this chapter.

More Involved Problems

Problems 27.20–27.30 involve a learning sequence whose purpose is to illustrate the origin of the forbidden energy bands. It is based on the Kronig-Penney model.

27.20. It is known from Floquet's theorem that the solution of a differential equation

$$\frac{d^2\psi}{dx^2} + \frac{8\pi^2 m}{h^2}[E - V(x)]\psi = 0,$$

where the potential $V(x)$ is a periodic function of the form $\psi = u(x)e^{ikx}$, where $u(x)$ is a periodic function with the same period as $V(x)$. In the present case we consider a periodic potential as shown:

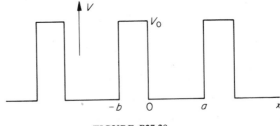

FIGURE P27.20.

Here the spacing is $a + b$ and the depth of the wells is greater than the electron energy. Show that Schrödinger's equation takes the form

$$\frac{d^2\psi_{\text{I}}}{dx^2} + \alpha^2\psi_{\text{I}} = 0, \qquad 0 < x < a,$$

and

$$\frac{d^2\psi_{\text{II}}}{dx^2} - \beta^2\psi_{\text{II}} = 0, \qquad a < x < b,$$

where

$$\alpha^2 = \frac{8\pi^2 mE}{h^2}$$

and

$$\beta^2 = \frac{8\pi^2 m}{h^2}(V_0 - E).$$

27.21. Assuming a solution in region I,

$$\psi_{\text{I}} = u_{\text{I}}(x)e^{ikx},$$

show that

$$\frac{d^2u_{\text{I}}}{dx^2} + 2ik\frac{du_{\text{I}}}{dx} - (k^2 - \alpha^2)u_{\text{I}} = 0.$$

It can also be shown that in region II

$$\frac{d^2u_{\text{II}}}{dx^2} + 2ik\frac{du_{\text{II}}}{dx} - (k^2 + \beta^2)u_{\text{II}} = 0.$$

27.22. Show that a solution of the differential equation for u_{I} is

$$u_{\text{I}} = Ae^{-(ik+i\alpha)x} + Be^{-(ik-i\alpha)x}.$$

It can also be shown that

$$u_{\text{II}} = Ce^{(-ik+\beta)x} + De^{(-ik-\beta)x}.$$

27.23. The functions in Problem 27.22 are subject to the conditions that

$$u_{\text{I}} = u_{\text{II}} \qquad \text{at } x = 0,$$

$$\frac{du_{\text{I}}}{dx} = \frac{du_{\text{II}}}{dx} \qquad \text{at } x = 0,$$

$$u_{\text{I}}(a) = u_{\text{II}}(-b),$$

$$\left(\frac{du_{\text{I}}}{dx}\right)_a = \left(\frac{du_{\text{II}}}{dx}\right)_{-b}.$$

a. Show that the first of these conditions leads to $A + B = C + D$.

b. Obtain three other relations involving A, B, C, and D.

27.24. From the four relations obtained in Problem 27.23 it can be shown that $[(\beta^2 - \alpha^2)/2\alpha\beta]\sinh \beta b + \cosh \beta b \cos \alpha a = \cos k(a + b)$. Show this.

27.25. The equation in Problem 27.24 is very complicated. We can make a convenient mathematical simplification. We let b decrease while V_0 increases in such a way that the product $V_0 b$ remains a constant, even when $b \longrightarrow 0$. We call the product the **barrier strength**. In the limit as $b \longrightarrow 0$ we have $P(\sin \alpha a/\alpha a) + \cos \alpha a = \cos ka$, where $P = [4\pi^2 ma(V_0 b)]/h^2$. Show that this is so.

27.26. Note that the period of this potential is now a since $b = 0$. Note that $-1 \leq \cos ka \leq 1$.
   a. Why?
      Note that the function $y(\alpha a) = P(\sin \alpha a/\alpha a) + \cos \alpha a$ is really a function of the energy only (assuming P and a are fixed).
   b. Why?
   c. Sketch $y(\alpha a)$ vs. (αa) for $P = 3\pi/2$ for values of αa from -4π to 4π.
   d. What can you say about a range of αa for which $y(\alpha a) > 1$?

27.27. Suppose that in Problem 27.25, $P = \pi/2$. Find a general expression for $d\alpha/dk$ and show that for values of $k = 0, \pm(\pi/a), \pm(2\pi/a), \dots$, $d\alpha/dk = 0$ and hence $dE/dk = 0$.

27.28. a. Show that if $P \longrightarrow 0$ in Problem 27.25 the resultant energy takes the form of the free electron equation (27.4.3).
   b. Show that if $P \longrightarrow \infty$, then the solution takes the form of the quantized particle in the box for which $L_1 = a$.

27.29. Note that the discontinuities in energy occur where $\cos ka = \pm 1$. What can one say about k?

27.30. a. Show for the case where k is close to zero that the equation in Problem 27.25, $P(\sin \alpha a/\alpha a) + \cos \alpha a = \cos ka$, reduces to $E = c + dk^2$, where c and d are constants. Thus for small k the parabolic E vs. k behavior which is a characteristic of the free electron is found.
   b. Show that when k is near to π/a (but less than π/a) that $E = c' - d'k^2$; i.e., we have near the top of a band an inverted parabola:

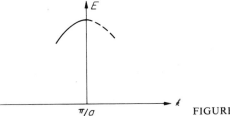

π/a FIGURE P27.30.

27.31. Discuss the state of the purity of metals. What is the highest conductivity ever obtained in normal conductors?

27.32. Discuss photoelectric behavior (beyond what is discussed in the text).

27.33. Extend the discussion of thermionic behavior given in the text. Include an up-to-date discussion of the effect of adsorbed layers.

27.34. Give an up-to-date review of the theory and capabilities of field ion microscopy.

27.35. The theory of resistivity discussed here is legitimate for either dc fields or for ac fields ($\xi = \xi_0 \cos \omega t$) in which the frequency is not too high; i.e., the period T is not too small (the period should be very long compared to the relaxation time τ). We noted in Example 27.9.1 that τ was of the order of 4×10^{-14} sec in copper at room temperature. If $T \gtrsim \tau$, then the electron would not be scattered during a cycle and we could write for the equation of motion $m\ddot{x} = -e\xi_0 \cos \omega t$. Show that this leads to an instantaneous dipole moment and a polarization $P = (Ne^2\xi_0 \cos \omega t)/m\omega^2$. Since the dielectric constant is $\kappa = 1 + (P/\epsilon_0\xi)$ and since $\kappa = n^2$, where n is the refractive index, show that n may be either imaginary or real, depending on ω. Calculate for sodium the value of ω_0 and wavelength λ_0 where n just becomes real. Compare with the experimental value, $\lambda_0 = 2100$ Å, where sodium becomes transparent.

Sophisticated Problems

27.36. a. Derive in vector form von Laue's equations for diffraction.
  b. Discuss the Ewald construction for X-ray diffraction.
  c. Show that Equation (27.5.3) is simply a condition for Bragg reflection.

27.37. Give a clear quantitative discussion of phase velocity and group velocity of a wave.

27.38. Derive Equation (27.3.6).

27.39. Derive the Richardson-Dushman equation.

27.40. Describe in detail a modern field ion microscope apparatus.

27.41. Describe in detail a modern apparatus for measuring contact potential by the Kelvin method [see T. Delchar et al., *Journal of Scientific Instruments*, **40**, 105 (1963)].

27.42. Describe how high pressures may be used to obtain information about energy bands in solids.

Prologue

Developments in the understanding of the behavior of semiconductors and junctions (fundamental theory) along with developments in making microelectronic circuitry by *large-scale integration*, LSI (structural processing), have brought about a revolution in information storage, processing, retrieval, and transmission. The concepts needed to understand the basis of operation of some important semiconductor devices are developed in this chapter. They are based on elementary quantum mechanical and statistical mechanical concepts.

We study pure and *doped* semiconductors. We shall study *p-type* and *n-type* semiconductors and *recombination processes*. The *p-n* junction is then described and we explain how it acts as a *rectifier*. We study the *Zener diode* and the *solar cell*. The basis of the *n-p-n junction transistor* (which acts as an amplifier) is then explained. Other devices are discussed.

28

SEMICONDUCTORS

28.1   INTRODUCTION

Materials which have a full valence band and an empty conduction band at the absolute zero of temperature, with an energy gap in the neighborhood of 1 eV, behave as semiconductors at higher temperature. Electrons can be promoted from the top of the valence band to the bottom of the conduction band by thermal energy, leaving *holes* in the valence band. Both the electrons in the conduction band and the holes in the valence band serve as charge carriers. If the only possible energy states present are the states in the conduction band and the valence band, the material is called an **intrinsic semiconductor** or it is said to be **intrinsic**. (It is possible to add impurity atoms to the crystal such that energy states within the band gap appear; such materials are not intrinsic. Thus the addition of phosphorus leads to a state in silicon which is only 0.045 eV below the conduction band. This means that an energy of only 0.045 eV is needed to ionize one electron from the pentavalent phosphorus atom. Room-temperature thermal energy can readily accomplish this. This introduces an electron into the conduction band.)

EXAMPLE 28.1.1

What is the room-temperature thermal energy of an electron?

Answer. We wish to know the value of kT at room temperature. Since $k = 8.62 \times 10^{-5}$ eV/°K and $T = 300$°K, we have $kT = 0.0257$ eV ($\frac{1}{40}$ eV).

Let us suppose that we have intrinsic silicon available (this means truly *pure* and defect free silicon). The number of holes and the corresponding

761

number of electrons generated by thermal excitation across the energy gap are quite small at room temperature. One would not have to add much phosphorus to make the number of electrons introduced into the conduction band by ionization greatly exceed the thermal electron concentration. Assuming that the mobilities of the holes and the electrons are roughly equal (more about that later), the conductivity in the phosphorus-doped material will therefore be due entirely (nearly) to electrons in the conduction band (ionized from the phosphorus). This is called an **n-type extrinsic semiconductor** or simply **n-type**. The *n* refers to the fact that a negative charge carrier is responsible for conduction. As we shall see later the addition of a trivalent impurity such as boron to an intrinsic material such as silicon would lead to a **p-type extrinsic semiconductor** (positive charge carriers).

EXAMPLE 28.1.2

What experimental technique can be used to measure the sign of the charge carrier?

Answer. In Equation (3.3.3) we noted that the measurement of the Hall coefficient gave the result

$$R_H = \frac{1}{Nq}, \qquad (28.1.1)$$

where q is the sign of the charge carrier and N is the number of charges per unit volume. The equation in this form is correct only when one charge carrier clearly dominates.

With this brief background we note some of the things we must consider:

1. What are the carrier concentrations in an intrinsic semiconductor?
2. What are the mobilities of the charge carriers and what determines these mobilities?
3. What is the nature of the impurity levels in semiconductors?

There are two other things which we must consider:

4. Recombination of carriers present in excess of the equilibrium concentration.
5. Contact potentials and junctions.

With this background we shall be prepared to understand solid state rectifiers and the transistor as well as other semiconductor devices.

28.2   INTRINSIC SEMICONDUCTORS

We have already noted in Chapter 27 that the Fermi surface lies at the center of the (first) energy gap in intrinsic semiconductors. We must now complete the calculation of the number of electrons per unit volume in the conduction band, N_n, and the corresponding number of holes, N_p, in the valence band. (In the literature the symbols n_e and n_h are widely used.)

The actual number of electrons in the conduction band in an energy range dE is the product of the density of allowed states $\rho(E)$ times the Fermi-Dirac probability that a state is occupied $F(E)$. Thus the total number of electrons in the conduction band is

$$N_n = \int_{E_g}^{\infty} \rho(E)F(E)\, dE, \qquad (28.2.1)$$

where we have chosen $E = 0$ at the top of the valence band as shown in Figure 28.2.1. Note that there are no allowable energy states in the energy gap. The

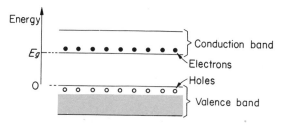

FIGURE 28.2.1. Energy diagram for intrinsic semiconductors.

valence band is full except for a small fraction of the electrons which were thermally excited from near the top of the valence band (leaving holes behind) up to the bottom portion of the conduction band. Near the bottom of the conduction band the density of states will be given by the parabolic relation [see (27.2.15)]

$$\rho(E) = 4\pi\left(\frac{2m}{h^2}\right)^{3/2}(E - E_g)^{1/2}; \qquad (28.2.2)$$

i.e., the electron behaves effectively as a particle in a box with $V = 0$. In this case the $E^{1/2}$ term is replaced by $(E - E_g)^{1/2}$ because the energy at the bottom of the conduction band is $E = E_g$, not $E = 0$. Likewise $F(E)$ is

$$F(E) = \frac{1}{e^{(E-E_F)/kT} + 1}. \qquad (28.2.3)$$

Because we are concerned with $E \gg E_g$ and since $(E_g - E_F)/kT \gg 1$, we can write

$$F(E) \doteq e^{-(E-E_F)/kT}. \qquad (28.2.4)$$

We next must substitute (28.2.2) and (28.2.4) into (28.2.1) and carry out the integration. (Inasmuch as students learn by doing, this is left to the student who may want to use the fact that

$$\int_0^{\infty} y^2 e^{-y^2}\, dy = \frac{\sqrt{\pi}}{4}.)$$

The result is

$$N_n = 2\left(\frac{2\pi mkT}{h^2}\right)^{3/2} e^{-(E_g-E_F)/kT}. \tag{28.2.5}$$

Since $E_g - E_F = E_g/2$ for the intrinsic material [see (27.8.2)], we have

$$N_n = 4.83 \times 10^{21} T^{3/2} e^{-E_g/2kT}, \tag{28.2.6}$$

where N_n has units of electrons per cubic meter and T is given in degrees Kelvin.

Table 28.2.1 shows some values of the energy gaps for semiconductors. Strictly speaking the mass m in (28.2.5) is the geometric average of the effective electron mass and the effective hole mass. In obtaining (28.2.6) we used the free electron mass.

Table 28.2.1. ENERGY GAPS IN SEMICONDUCTORS

Element	E_g, eV	Compound	E_g, eV
Diamond	5.3	AlP	3.0
Silicon	1.1	GaP	2.25
Germanium	0.72	AlSb	1.52
Gray tin*	0.08	GaAs	1.34
		GaSb	0.70
		InAs	0.33
		InSb	0.18

* Gray tin is the diamond cubic polymorph of tin. Tin is usually found in the metallic phase which has a tetragonal crystal structure.

EXAMPLE 28.2.1

Discuss the possibility of making germanium intrinsic at room temperature, 300°K.

Answer. Germanium is diamond cubic (Figure 10.4.3) and hence has 8 atoms/unit cell. The lattice parameter is (from Table 10.6.1) 5.65 Å. The number of atoms per cubic meter is therefore

$$N_a = \frac{8}{(5.65)^3 \times 10^{-30}} = 4.4 \times 10^{28} \text{ atoms/m}^3.$$

From Equation (28.2.6), using $E_g = 0.72$ eV and $k = 8.62 \times 10^{-5}$ eV/°K,

$$N_n = 1.1 \times 10^{23} \text{ electrons/m}^3.$$

Since a pentavalent impurity atom such as phosphorus would eventually contribute one electron per impurity atom, we would require that the impurity concentration N_{imp} be less than N_n. This means that

$$N_{imp}/N_a < N_n/N_a = 2.5 \times 10^{-6}.$$

From our discussions in Chapter 17, we came to the conclusion that such purities were difficult to obtain (because of the large entropy of mixing per atom) but presently possible through the use of zone refining.

EXAMPLE 28.2.2

Discuss the possibility of making silicon which is intrinsic at room temperature.

Answer. The electron concentration would be approximately the value found for germanium times the factor $e^{-(1.1-0.72)/2kT}$, which is about 10^{-4}. Hence the impurity concentration would have to be lower by a factor of 10,000. This is not presently possible.

28.3   MOBILITIES

The conductivity of a semiconductor is a result of the motion of both charge carriers so that

$$\sigma = N_n e\mu_n + N_p e\mu_p, \qquad (28.3.1)$$

where μ_n is the (charge) mobility of the negative carrier (electrons) and μ_p is the mobility of the holes. Equation (28.3.1) is a generalization of (27.1.5). Recall from Section 27.1 that the mobility is defined as the drift velocity per unit electrical field.

The Hall coefficient can be shown with some effort to be

$$R_H = -\frac{1}{e}\frac{N_n c^2 - N_p}{(N_n c + N_p)^2}, \qquad (28.3.2)$$

where

$$c = \frac{\mu_n}{\mu_p}. \qquad (28.3.3)$$

For an intrinsic semiconductor, inasmuch as $N_n = N_p$,

$$R_H = -\frac{1}{N_n e}\frac{(c-1)}{(c+1)}. \qquad (28.3.4)$$

Some values of mobilities are given in Table 28.3.1.

Table 28.3.1. MOBILITIES IN UNITS OF SQUARE METERS PER VOLT-SECOND

Element	μ_n	μ_p	Compound	μ_n	μ_p
Diamond	0.18	0.12	GaP	0.045	0.002
Silicon	0.14	0.048	GaAs	0.85	0.45
Germanium	0.39	0.19	GaSb	0.50	0.085
Gray tin	0.20	0.10	InSb	8.00	0.070

EXAMPLE 28.3.1

What is the sign of the Hall coefficient in pure germanium?

Answer. Since $\mu_n > \mu_p$ and $N_n = N_p$, R_H is negative.

Depending on the exact geometry, the Hall coefficient (28.3.2) may contain a factor of $3\pi/8$ (see References).

At high temperatures, the resistivity of a semiconductor is due to collisions between the conduction electrons with the vibrating atoms. This is called an electron-phonon interaction. It can be shown that the mobility μ_L based on these lattice interactions is given by

$$\mu_L = BT^{-3/2}, \tag{28.3.5}$$

where B is a parameter depending on the material.

EXAMPLE 28.3.2

What is the temperature dependence of the conductivity of pure germanium?

Answer. Combining (28.2.6), (28.3.5), and (28.3.1), we have

$$\sigma \propto e^{-E_g/2kT}, \tag{28.3.6}$$

i.e., a conductivity which *increases* exponentially with temperature. Note that this is what is found experimentally at high temperatures (see Figure 3.3.1). Since the resistivity $\rho = 1/\sigma$ (for cubic crystals), the resistivity decreases exponentially with temperature. This rapid variation of resistance with temperature forms the basis of the **thermistor**, a resistance-based temperature measuring device.

As with metals, there are a number of scattering mechanisms which become important at low temperatures. However, because of the much lower charge carrier concentration in semiconductors, low temperatures may be in the vicinity of room temperature rather than at several degrees Kelvin as in metals. Charge carriers are scattered by (1) ionized impurities, (2) neutral impurities, (3) grain boundaries, and (4) dislocations. These mechanisms are discussed by K. Lark-Horowitz and V. A. Johnson, in *The Science of Engineering Materials* (J. E. Goldman, ed.), Wiley, New York (1957) p. 336.

The mobilities of carriers in semiconductors may be 10^2 or 10^3 times that in metals at room temperature.

We recall that the mobility is inversely proportional to the effective mass m^*. For germanium, $m^* \approx 0.1m$ and for silicon $m^* \approx 0.2m$. The value of m^* can be obtained by cyclotron resonance (see Problem 28.46), as was first suggested by Shockley. m^* varies with direction; i.e., m^* is an anisotropic quantity.

28.4   EXTRINSIC SEMICONDUCTORS

The semiconducting compounds listed in Table 28.2.1 all have the diamond cubic structure illustrated in Figure 10.4.3. Thus in silicon, each silicon atom is tetrahedrally bonded to four other silicon atoms. In the pure crystal the outer electrons are all in the localized pair bonds and hence do not contribute to conduction. From the free electron approach the valence band is filled and the conduction band is empty and there is no conductivity. The III-V compounds listed in Table 28.2.1 have the zinc blende structure (see Section 12.7). Thus in the compound GaAs each gallium atom is bonded tetrahedrally to four arsenic atoms and each arsenic atom is bonded tetrahedrally to four gallium atoms as with the elements that have four valence electrons associated with each atom.

Let us consider what happens when a pentavalent impurity atom such as phosphorus is purposely added to a semiconducting material such as germanium (this is called doping) as a substitutional impurity. After the four covalent bonds are taken care of there is still an excess of one positive charge on the phosphorus ion and an excess electron. This electron is very weakly bound to the ion, as we shall now show. A simple model assumes that the electron and ion form a hydrogen-like atom in a dielectric medium whose dielectric constant is κ as shown in Figure 28.4.1. This would be described by Schrödinger's equation

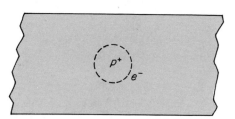

FIGURE 28.4.1. Hydrogen-like atom in $\kappa \gg 1$ medium.

(8.2.1) but with the potential given by [in cgs units]

$$V = -\frac{e^2}{\kappa r}. \tag{28.4.1}$$

For germanium $\kappa = 16$, while for a vacuum $\kappa = 1$. The solution for $\kappa = 1$ is given by Equation (8.2.8) and it obviously follows for the present case that

$$E = -\frac{2\pi^2 m e^4}{\kappa^2 h^2}\frac{1}{n^2}, \tag{28.4.2}$$

where $n = 1, 2, 3, \ldots$. Hence

$$E = -\frac{13.6}{\kappa^2}(\text{eV}) \qquad \text{for } n = 1. \tag{28.4.3}$$

Thus for germanium the ground state energy is $E = -0.053$ eV. This means that an energy of 0.053 eV is needed to excite the electron from the bound state on phosphorus to the fully ionized state ($n = \infty$) in which the electron is now at the bottom of the conduction band. Thermal energy kT equals about 0.025 eV at room temperature so that thermal energy is sufficient to strip the fifth electron from phosphorus and to place it in the conduction band of germanium. Impurity elements which can easily contribute an electron to the conduction band in this way are called **donors**. Donors are usually pentavalent substitutional impurities, but they may also involve interstitial impurities with a different valence such as lithium. If a trivalent atom such as boron is added to germanium, it enters a substitutional site. It is tetrahedrally coordinated, and to complete the four paired bonds, the boron must extract an electron from the valence band, which leaves a hole there. Such an impurity is called an **acceptor**. The formation of holes in the valence band and electrons in the conduction band is illustrated in Figure 28.4.2.

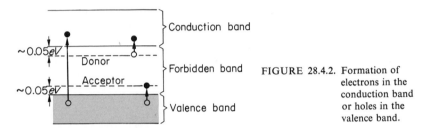

FIGURE 28.4.2. Formation of electrons in the conduction band or holes in the valence band.

The ionization energies of some solutes in germanium and silicon are shown in Table 28.4.1. Note that the values are of the same order of magnitude as predicted.

Table 28.4.1. IONIZATION ENERGIES OF SOLUTES (eV)

	Site	In Germanium	In Silicon
Donors			
Li^+	Interstitial	0.0093	0.033
P^+	Substitutional	0.012	0.045
As^+	Substitutional	0.0127	0.049
Sb^+	Substitutional	0.0096	0.039
Acceptors			
B^-	Substitutional	0.0104	0.045
Al^-	Substitutional	0.0102	0.057
In^-	Substitutional	0.0112	0.16
Cu^-	Interstitial	0.04	0.49

K. Lark-Horowitz and his associates at Purdue in 1942 showed that the trivalent impurities in the quadrivalent elements, germanium, silicon, and tin, act as acceptors, while pentavalent impurities act as donors. The lithium atom or ion is in an interstitial position.

The ionization energies shown are for very low concentrations only. At higher concentrations (where there is solubility) the ionization energy often drops to zero.

EXAMPLE 28.4.1

Explain why at low temperatures the resistivity of gallium-doped germanium (see Figure 3.3.1) becomes nearly independent of temperature. [It actually decreases slightly as T decreases ($1/T$ increases).]

Answer. At high temperatures $N_n = N_p$, but at low temperatures $N_p \gg N_n$ since gallium is an acceptor. At low temperatures the charge carrier concentration is controlled by the impurity concentration, which does not vary with temperature. The mobility varies slowly with temperature. Thus the resistivity varies slowly.

A semiconductor in which the conductivity or resistivity is due to carriers caused by donors or acceptors is called an **extrinsic** semiconductor. If the conductivity is due to equal concentration of holes and electrons owing to thermal excitation of electrons across the energy gap (as in a *pure* semiconductor), the material is called an **intrinsic** semiconductor. A doped semiconductor can be intrinsic at high temperatures and extrinsic at low temperatures (see Figure 3.3.1). If the conductivity is primarily due to negative charge carriers, the material is called *n*-type, while if the conductivity by holes (positive charge carriers) predominates, the material is called *p*-type.

THE FERMI LEVEL. The Fermi level E_F of an intrinsic semiconductor is located midway in the energy gap. The situation is considerably different for a doped semiconductor. (**Doping** is the purposeful addition of an impurity such as phosphorus to a "pure" material such as germanium.) For a donor-doped material at 0°K the Fermi energy lies below the conduction band by $E_D/2$, where E_D is the donor ionization energy. As the temperature is increased, the Fermi level decreases until at sufficiently high temperatures it reaches the midway point of the energy gap. The actual evaluation of $E_F(T)$ for a given impurity concentration, donor ionization energy, and energy gap involves the use of the Fermi-Dirac probability function $F(E)$.

EXAMPLE 28.4.2

What is the Fermi level of boron-doped silicon at low temperatures?

Answer. It lies 0.045/2 eV above the valence band. The 0.045 eV for ionization energy was obtained from Table 28.4.1. This is a *p*-type semiconductor. An electron is easily excited from the valence band (leaving a hole there) to the boron state to form a B$^-$ ion.

EXAMPLE 28.4.3

What is the Fermi energy of the germanium crystal for which resistivity data are shown in Figure 3.3.1 for

a. High temperatures?
b. Low temperatures?

Answer. a. At high temperatures the conductivity owing to electron-hole pairs which are thermally generated predominates and the Fermi level approaches the midway point in the energy gap, i.e., $E_g/2$ above the valence band.

b. At low temperatures, holes in the valence band owing to the presence of gallium impurities (gallium is trivalent and hence an acceptor according to the Lark-Horowitz criterion) predominates, and as 0°K is approached the Fermi level is $E_D/2$ above the valence band. At intermediate temperatures the Fermi level is a function of temperature and increases from $E_D/2$ to $E_g/2$ as the temperature is increased.

The exact evaluation of the Fermi level for a crystal (doped with a concentration N_D of one donor and a concentration N_A of one acceptor) follows from the equation of charge neutrality

$$N_n + n_D + N_A = N_p + n_A + N_D \qquad (28.4.4)$$

(where n_D is the concentration of electrons bound to donors and n_A is the concentration of holes bound to acceptors) and the use of the Fermi probability function for N_n, N_p, n_D, and n_A. The evaluation (numerical) of the Fermi level is described by W. Shockley, *Electrons and Holes in Semiconductors*, Van Nostrand, Princeton, N.J. (1950) pp. 465–475.

It can be shown that the product $N_n N_p$ is equal to an equilibrium constant K which is the same as for the intrinsic material (where $N_n = N_p = N_i$; here N_i is the electron or hole carrier concentration in the intrinsic material) at the same temperature. Hence

$$N_p N_n = K = N_i^2 \qquad (28.4.5)$$

for the case of a spherical Fermi surface. This is a very useful expression inasmuch as N_i is given then by Equation (28.2.6) and

$$N_p N_n \propto e^{-E_g/kT} \qquad (28.4.6)$$

regardless of whether the material is intrinsic or extrinsic. Thus if "pure" silicon is heavily doped with a donor (phosphorus, say) such that $N_D \gg N_i$, then $N_n \doteq N_D$ and N_p can be calculated from (28.4.5); N_p would be very small.

28.5   RECOMBINATION OF EXCESS CARRIERS

Consider an extrinsic semiconductor which has an equilibrium concentration of carriers at a given temperature. If this is irradiated with light of sufficient frequency,

$$hv > E_g, \qquad (28.5.1)$$

an electron and a hole will be created for each photon absorbed. When irradiation stops, the excess carriers combine and disappear. This process, called **recombination**, involves an electron dropping from the conduction band to the valence band where it combines with a hole. Recombination reduces the energy of the system. The rate of recombination is proportional to the product of the minority carrier concentration, N_m, times the majority carrier concentration, N_M:

$$\frac{dN_m}{dt} \propto -N_m N_M. \tag{28.5.2}$$

In an *n*-type semiconductor, $N_n \gg N_p$; we say the electrons are the **majority carriers** and the holes are the **minority carriers**. Because the majority carriers in the extrinsic material are such that $N_M \gg N_m$, we can assume that N_M remains a constant during the recombination process. Hence we can write

$$\frac{dN_m}{dt} = -\frac{N_m}{\tau}, \tag{28.5.3}$$

where τ is the proportionality constant, called the **minority carrier lifetime**.

If τ does not vary with N_m or N_M (as in the case for small minority carrier concentrations in silicon and germanium), then (28.5.3) can be integrated subject to the condition that at $t = 0$ when the irradiation stops,

$$N_m(0) \approx N_m^e(0), \tag{28.5.4}$$

where N_m^e is the excess minority carrier concentration, i.e., the carriers created by irradiation. We assume that in the intrinsic material the equilibrium minority carrier concentration is virtually zero; then

$$N_m^e(t) = N_m^e(0)e^{-t/\tau}; \tag{28.5.5}$$

i.e., there is an exponential decay.

What determines the decay time τ? It is, in fact, determined by the number of traps and recombination centers within the forbidden band (forbidden in the pure perfect crystals), as shown in Figure 28.5.1. The direct transition from the conduction band to the valence band has a very low probability because of the large amount of radiation energy which takes place. Instead the transition from the conduction band to a level within the forbidden band (owing to an impurity, vacancy, dislocation, etc.) called a **trap** takes place. Some traps capture electrons preferentially, while others capture holes. Certain imperfections act as traps for both electrons and holes: these are called **recombination centers**. As is clear from the figure, the electron need not descend monotonically from one trap to another but may return momentarily to the conduction band. Still the decay of the electron from the conduction band back to the valence band is much faster owing to the presence of the imperfections for small minority carrier concentration than would be the case in the perfect crystal. (At very high minority carrier concentrations, direct recombination may dominate.)

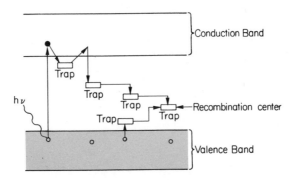

FIGURE 28.5.1. Minority carrier (electron in conduction band) produced in *p*-type crystal by external radiation followed by eventual recombination of the electron with a hole.

It is often useful to use the **diffusion length** (the distance the carrier moves before recombination) rather than the lifetime τ. Thus we have the relationship

$$L = \sqrt{D\tau} \qquad (28.5.6)$$

for the diffusion depth (compare with the answer in Example 18.5.4). Let us suppose specifically that a *small* concentration of holes is involved as minority carriers. Then

$$L_p = \sqrt{D_p\tau_p}, \qquad (28.5.7)$$

and, moreover, the diffusion coefficient for holes is related to the mobility B_p [and the charge mobility μ_p (which we call mobility in this chapter) by the Nernst relation; see (18.13.11)]:

$$\mu_p = q_p\frac{D_p}{kT} = q_pB_p. \qquad (28.5.8)$$

The Nernst relation is often called the Einstein relation. This is incorrect and unfair to Nernst, who derived and used the relation fairly extensively (1885) before Einstein studied diffusion.

Typical values of minority carrier diffusion lengths are 10^{-1}–10^{-3} in. This is an important parameter in the design of semiconductor devices such as the *n-p-n* transistor, which is discussed later. For methods of measuring minority carrier lifetimes, see the References.

28.6   THE *p-n* JUNCTION

Consider a single crystal of silicon. To the right of a certain interface it is doped with a donor and is hence *n*-type, and to the left of that interface it is doped with an acceptor and is hence *p*-type. The interface region is called a *p-n* junction. In this case it is an idealized **p-n junction** because of the sharp transition.

EXAMPLE 28.6.1

Why, in actual *p-n* junctions, is the transition from *p-* to *n*-type material not sharp?

Answer. Consider the various ways by which junctions are made. In Section 18.12, the diffusion of phosphorus into a boron-doped silicon crystal (*p*-type) is discussed. If the phosphorus concentration is much greater than the boron concentration, i.e., an excess of donor exists, the material will be *n*-type. Since the phosphorus concentration decreases to zero with distance of penetration according to Equation (18.12.7), there must be a gradual change from strongly *p*-type to strongly *n*-type.

Recall that when gradients in composition exist (as with the phosphorus and boron in Example 28.6.1) there is a *tendency* for the two to interdiffuse and to eliminate these gradients. However, for this to occur at any reasonable *rate* high temperatures must be present (see Section 18.12). At room temperature the boron and phosphorus ions do not move (for all practical purposes). However, there is also a gradient of holes and electrons and, unlike the ions, these are highly mobile at room temperature. Consequently, as the great Russian scientist Y. Frenkel was the first to note, there is diffusion of charge carriers as shown in Figure 28.6.1. This results in a charge distribution as shown in Figure 28.6.2.

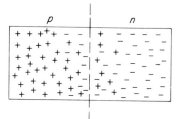

FIGURE 28.6.1. Diffusion of mobile holes across a *p-n* junction into the *n* side, and of mobile electrons into the *p* side.

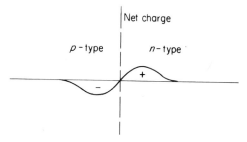

FIGURE 28.6.2. Transition layer charge barriers at *p-n* junctions.

EXAMPLE 28.6.2

Why do the holes and electrons not interdiffuse until their concentration gradients are completely destroyed, i.e., until their respective concentrations do not vary with distance?

Answer. In Section 28.5 we discussed how minority carriers could be destroyed by recombination. This means the holes can only diffuse a distance L_p into the *n*-type material before recombination occurs. A double space charge region (see Figure 28.6.2) is created.

If the two minority carriers are being destroyed by recombination processes, then we might expect, in the absence of an external field, a *continuous* flow of holes to the right (into *n*-type) and of the oppositely charged electrons to the left (into *p*-type). Opposite fluxes of oppositely charged particles lead to a net current flow. However, this does *not* happen because of the space charge in the transition layer (Figure 28.6.2), which gives rise to an electric field which opposes further diffusion by causing a compensatory flow in the opposite direction. Thus an equilibrium (no flow) rather than a steady-state flow situation exists.

Figure 28.6.3 shows an energy diagram for the *p-n* junction. Note that the Fermi level (which is near to the valence band for *p*-type material—see Section 28.4—and near to the conduction band for *n*-type material) is the same in the entire block of material (the Fermi energy is the chemical potential of the electrons and hence is the same in different interconnected parts of a system). In

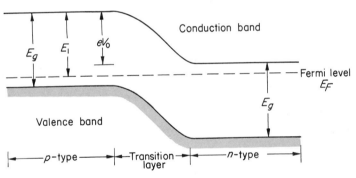

FIGURE 28.6.3. Energy diagram for electrons at a *p-n* junction in equilibrium with no applied voltage.

Figure 28.6.3, the contact potential is V_0 and the energy gap is E_g. The concentration of electrons on the left side (*p*-type) will be low, and on the right side (*n*-type) it will be high. However, if the probability of an electron on the left moving to the right is 1, the probability of the reverse process, i.e., the electron moving up the potential barrier of height eV_0, is only $e^{-eV_0/kT}$; i.e.,

$$p \propto e^{-eV_0/kT}. \tag{28.6.1}$$

EXAMPLE 28.6.3

What is the ratio of the concentration of electrons on the left side of the junction to the concentration on the right side?

Answer. This must also be $e^{-eV_0/kT}$ since the electron flow to the right, which we shall call j_n, must equal that to the left, j_f. The electron flow is given by the product of the concentration times the probability that the transfer will occur. Under no applied voltage,

$$j_f = j_n. \tag{28.6.2}$$

Suppose we now apply an external voltage V such that the p-type region is made more positive relative to the n-type region. This is called **forward bias**. This is shown in Figure 28.6.4.

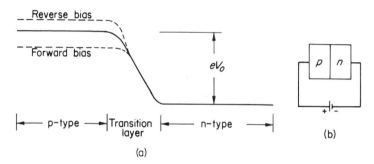

(a)

FIGURE 28.6.4. (a) Electron energy diagrams for p-n junction with applied voltages. For forward bias the p-type is made positive relative to the n-type. The curve for no applied voltage is the top curve of Figure 28.6.3. (b) Forward-biased p-n junction.

The probability of an electron moving to the right is still 1, so the current flow to the right is still j_n. The probability of the electron moving up the potential barrier to the left (which now has height $eV_0 - eV$) is now proportional to $e^{-(eV_0-eV)/kT}$ and hence the electron flow to the left is now $j_n e^{eV/kT}$. Consequently the net electron flow through the junction to the left is

$$j_n^{\text{net}} = j_n(e^{eV/kT} - 1) \tag{28.6.3}$$

and the net electrical current density to the *right*, J_n, owing to electrons, is

$$J_n = e j_n^{\text{net}}.$$

See Figure 28.6.5.

FIGURE 28.6.5. Fluxes of electron and holes, j_n^{net} and j_p^{net}, under an applied field ξ and corresponding charge flow.

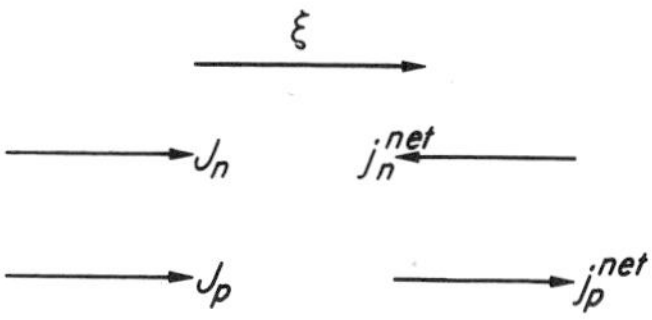

It can be shown that an expression of the same type holds for the flow of holes, so that

$$J = J_n + J_p = J_0(e^{eV/kT} - 1), \qquad (28.6.4)$$

where J_0 is a constant.

For a **reverse bias**, V is negative. As V increases negatively, J approaches $-J_0$, the **saturation current density**.

EXAMPLE 28.6.4

How large is eV/kT for a voltage of 0.118 V?

Answer. Since $kT = 0.0257$ eV (about $\frac{1}{40}$ eV), we have $eV/kT = 0.118/0.0257 = 4.6$. Hence for forward bias, $J = J_0(e^{4.6} - 1) = J_0(10^2 - 1) \approx 100J_0$, while for reverse bias $J = J_0(e^{-4.6} - 1) = J_0(10^{-2} - 1) \approx -J_0$.

The current versus voltage for a specific *p-n* junction is shown in Figure 28.6.6.

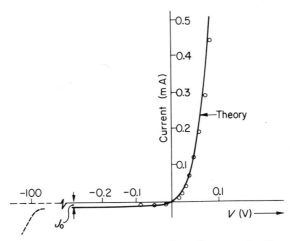

FIGURE 28.6.6. Current versus voltage for a *p-n* junction. [After W. Shockley, *Electrons and Holes in Semiconductors*, D. van Nostrand Co., Princeton (1950).]

RECTIFICATION.   As is clear from Figure 28.6.6 and Example 28.6.4, if a voltage $V = 10 \cos \omega t$ is applied across the *p-n* junction, current will readily flow during the positive portion of the cycle and will essentially not flow during the negative portion of the cycle. The input has been **rectified** so that the output is a dc with a ripple. A typical low-cost battery charger has such a ripple. Other electrical devices can be used to smooth out the ripple.

The saturation current density is given by

$$J_0 = J_n^0 + J_p^0,$$ (28.6.5)

where J_p^0 can be shown to be

$$J_p^0 = b(1 + b)^{-2}\left(\frac{\sigma_i^2}{\sigma_n L_p}\right)kT.$$ (28.6.6)

J_n^0 is given by a similar expression. Here

$b = \mu_n/\mu_p,$
$\sigma_i = $ conductivity of intrinsic material,
$\sigma_n = $ conductivity of n-type material,
$L_p = $ diffusion length of holes in n-type material.

For a derivation of (28.6.6), see W. Shockley, *Electrons and Holes in Semiconductors*, Van Nostrand, Princeton, N.J. (1950). One important parameter which affects J_p^0 is L_p, the minority carrier diffusion distance. As noted earlier, this can be increased by decreasing trapping and recombination processes.

BREAKDOWN. At high reverse voltages a **breakdown** occurs and large currents begin to flow. Current increases of several orders of magnitude occur at nearly constant voltage. Recall that carriers in semiconductors have a large mobility and hence a large mean free path. Under an electrical field ξ, such carriers can achieve between collisions an energy $e\xi\lambda$, which is correspondingly large. Such high-energy carriers are capable of producing by collision electron-hole pairs, which in turn are accelerated and produce additional carriers. An avalanche on a mountainside is often started by one region of snow falling and knocking others loose. This is known as the **avalanche effect.** When the semiconductor is flooded with carriers, high currents can pass. The current flow can increase by several orders of magnitude with only small voltage changes. Does this suggest a constant voltage control? Yes; it is called the **Zener diode**.

OTHER RECTIFIERS. Metal-semiconductor rectifiers, of which the most common are the copper oxide and the selenium rectifier, are important devices in low-frequency power engineering (high amperages). They also depend upon junction effects for rectification.

SOLAR CELLS. Figure 28.6.7 illustrates a solar cell. In this device solar energy is used to create electron-hole pairs. The additional *minority* carriers on each side of the junction diffuse across it in opposite directions. Since these are also oppositely charged, the electrical current is additive. There is a critical length involved in the design of solar batteries. Only those minority carriers created by radiation within a diffusion length of the junction will contribute to the current; the others will recombine and hence decrease the efficiency of the

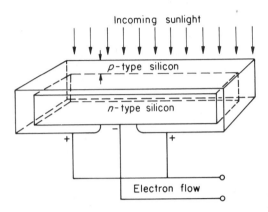

FIGURE 28.6.7. Schematic of a solar cell.

device. This is the reason for the thinness of the *p*-type layer. Using the absorption spectrum of the silicon, we can show (see Problem 28.41) that most of the solar radiation will be absorbed in a distance of about 10^{-4} cm from the surface. Note that the current in the *p* region is parallel to the junction, and hence if this region is made too thin, its resistance and hence joule heating losses will increase and the efficiency of the device will decrease. For a discussion of solar cells, see J. P. Elliott, "Photovoltaic Energy Conversion," in *Direct Energy Conversion* (George W. Sutton, ed.), McGraw-Hill, New York (1966). The theoretical efficiency of a silicon solar cell is 20%, and operating cells with efficiencies of 15% have been made. A cell such as that shown in Figure 28.6.7 might have an operating voltage of 0.4 V and a power output of 12 mW/cm². The reader should calculate the number of square miles of solar cells in sunny Arizona which would be needed to produce the annual U.S. electrical energy output of 1 trillion kWh.

28.7 THE JUNCTION TRANSISTOR

In 1947 J. Bardeen and W. H. Brattain invented the point contact transistor. This was rapidly followed by the invention of the *n-p-n* junction transistor by Shockley in 1948. This was an invention based on fairly sophisticated scientific knowledge. Shockley had synthesized from the knowledge base built up by F. Bloch, R. Kronig, W. G. Penney, A. H. Wilson, J. Frenkel, W. Schottky, Lark-Horowitz, and Bardeen and Brattain and in the process had created a device for amplifying power which was to cause a *revolution* in the communications and information transfer and processing industry. This revolution will profoundly affect all of us. Between 1948 and 1970 the increase in density of electronic circuits was a factor of 100,000. This involves the use of **large-scale**

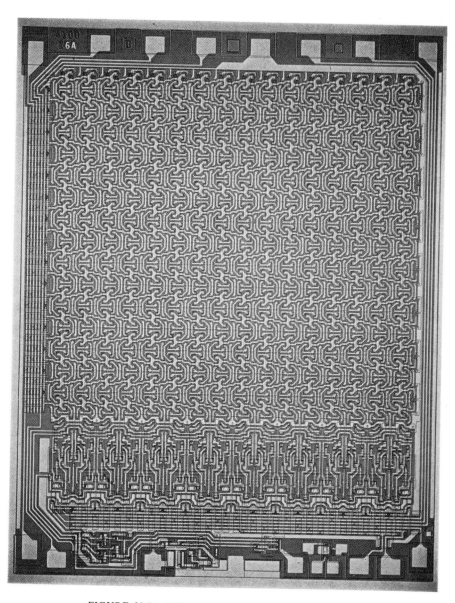

FIGURE 28.7.1. LSI computer memory circuit with 1244
transistors, 1170 resistors, and 71 diodes.
Actual size is 0.11×0.14 in. (*Courtesy of
Fairchild Semiconductor.*)

integration, LSI, i.e., building into and on a single crystal disc of silicon various *p-n* junctions, *n-p-n* junctions, electrical connectors, etc., with up to 10^5 components/in.[2]. Figure 28.7.1 shows an enlargement of an LSI circuit. It was briefly discussed in Section 18.12 how this might be made. There is also an excellent article on this: F. C. Heath, "Large-Scale Integration in Electronics," *Scientific American* (Feb. 1970) p. 22. A *n-p-n* junction transistor is shown in Figure 28.7.2.

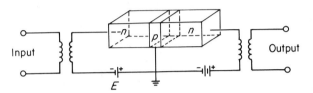

FIGURE 28.7.2. *n-p-n* junction transistor.

In the absence of the applied potentials (owing to the two power supplies and the input voltage) the energy levels appear as in Figure 28.7.3(a). This is exactly what we would expect from placing two *p-n* junctions back to back. In Figure 28.7.3(b), the potential curve is shown for the forward-biased left junction and

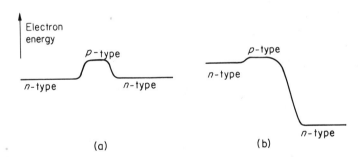

FIGURE 28.7.3. Electron energy versus distance in *n-p-n* junction. (a) Unbiased. (b) Biased as an amplifier as in Figure 28.7.2.

reverse-biased right junctions of Figure 28.7.2. Since the left junction is now forward-biased, a large number of electrons are injected from the left into the *p* region (**electron injection**). The *n* region at the left is called the **emitter**. The *p* region is very thin so that electrons are not destroyed by recombination [this means the thickness is less than the recombination distance for the minority carriers (the electrons) in the *p* region]. The *p* region, called the **base**, is therefore flooded with electrons (compared to the usual electron concentration there, which is quite small). This flow of electrons into the *p* region from the *n* region on the left depends very strongly (exponentially) on the voltage of the emitter battery, V_E, plus the oscillating input voltage, V_i:

$$J \propto e^{e(V_E+V_i)/kT} = e^{eV_E/kT} e^{eV_i/kT}. \tag{28.7.1}$$

Because of the thinness of the p region, the injected electrons do not recombine there but can pass into the n region on the right, which is called the **collector** (think of these electrons as coasting down the potential barrier). The donor concentration in the n region is made much higher than the acceptor concentration in the base. Therefore almost all the current is due to electrons, and the holes make very little contribution.

The current flow in the collector loop is nearly the same as in the emitter loop. The input impedance Z_i is high compared to the impedance in the emitter. Hence

$$V_i = Z_i i_i. \tag{28.7.2}$$

Similarly, for the output,

$$V_0 = Z_0 i_0. \tag{28.7.3}$$

Since

$$i_i \doteq i_0, \tag{28.7.4}$$

we have for the voltage gain

$$\frac{V_0}{V_i} \doteq \frac{Z_0}{Z_i}. \tag{28.7.5}$$

Gains of the order of 10^5 can be achieved.

EXAMPLE 28.7.1

What sweeps the excess electrons from the p region through the collector and the output impedance?

Answer. The voltage of the collector battery, V_C. The power output relative to the power input is, of course, similarly amplified. The energy for this amplification comes from the collector battery.

EXAMPLE 28.7.2

Suppose the base of the n-p-n junctions was 0.1 in. thick instead of very thin. What would happen?

Answer. The electrons injected into the p region from the emitter would flow around the emitter circuit only. The collector circuit would not "know about" the existence of this current.

The **injection efficiency** is the ratio of i_0/i_i and can be of the order of 0.99 for a good design or nearly zero for a poor design, as in Example 28.7.2.

28.8   SEMICONDUCTOR DEVICES

Semiconductor devices include rectifiers, *n-p-n* transistors (or *p-n-p*), field effect transistors, modulators, thermistors, varistors, detectors, electristors, photocells, tunnel diodes, Hall effect devices and thermoelectric generators. Semiconductor crystals include Ge, Si, Se, Cu_2O, Bi_2Te_3, PbTe, PbS, SiC, GaAs, and InSb.

The **field effect transistor** (see Problem 28.37) has certain advantages over the *n-p-n* transistor: It can operate with lower input current (input current of 10^{-9} amps), has a higher frequency response, and is relatively unaffected by extraneous rf fields.

When operated at low potentials, the thermistor (see Section 3.3) is ohmic. However, at high potentials there is considerable joule heating which causes the resistivity to drop rapidly. This results in a negative resistance, i.e., at high voltages the slope of the voltage-current curve is negative. The thermistor, operated at high voltage in the negative resistance region, is called a **varistor**. The nonohmic varistor can be used as an active element in a circuit and can serve as an amplifier, oscillator, and modulator at low frequencies through the audio-frequency range (see Problem 28.44).

The tunnel diode is particularly important because of its rapid switching time [less than 1 nsec (10^{-9}sec)]. See Problem 28.40. Thermoelectric heating and cooling can be explained in terms of the energy diagram in Figure 28.6.3. The electrons which move to the left over the barrier are high-kinetic energy electrons (hot electrons). In the *p*-type material, these electrons have high potential energy but low kinetic energy (cold electrons). Thus, hot electrons have been removed from the *n*-type material and cold electrons introduced into the *p*-type material: the temperature in both regions and hence at the junction is lowered. If electrons were to flow in the opposite direction, the junction temperature would be raised. (The flow of holes would result in temperature changes of the same sign as the electron flow.) Thus, a device as shown in Figure 6.8.2 would be heated at one junction and cooled at the other. The Peltier coefficient in semiconductors is about 100 times larger than in metals.

REFERENCES

Heath, F. C., "Large Scale Integration in Electronics", *Scientific American* (Feb. 1970) p. 22.

Sproull, R. L., *Modern Physics*, Wiley, New York (1963).

Shockley, W., *Electrons and Holes in Semiconductors*, Van Nostrand, Princeton, N.J. (1950).

Moll, J. L., *Physics of Semiconductors*, McGraw-Hill, New York (1964).

Hunter, L. P., *Introduction to Semiconductor Phenomena and Devices*, Addison-Wesley, Reading, Mass. (1966).

Holmes, R. A., *Physical Principles of Solid State Devices*, Holt, Rinehart and Winston, New York (1970).

Henisch, H. K., "Amorphous Semiconductor Switching," *Scientific American* (Nov. 1969) p. 30.

Runyan, W. R., *Silicon Semiconductor Technology*, McGraw-Hill, New York (1965).

PROBLEMS

28.1. Discuss the process of floating zone refining and the important role it plays in the semiconductor industry.

28.2. Give typical values for the electrical conductivity coefficient of metals, semiconductors, and insulators.

28.3. a. Give the electron configuration of carbon.
b. Describe the bonding in diamond.
c. Why is diamond an insulator at absolute zero?
d. Is diamond an insulator at 300°K? 1500°K?

28.4. Where does the Fermi surface lie for an intrinsic semiconductor?

28.5. Of what use are Hall effect studies in the science of semiconductors?

28.6. If the silicon described in Table 3.1.1 contains only one pentavalent impurity, what is the atom fraction of this impurity?

28.7. From the viewpoint of building a temperature-sensing device, why would a thermistor be preferred to a metal resistor?

28.8. The "radius" of the ground state electron in hydrogen is given by (8.2.14). What is the radius for the ground state electron bound to P^+ in germanium?

28.9. Sketch $\ln N_n$ (or $\ln N_p$) vs. $1/T$ from the melting point to very low T for a germanium crystal which becomes intrinsic at 300°K.

28.10. Show for energies such that $(E - E_F) \gg kT$ that the Fermi-Dirac statistics and Boltzmann statistics give similar results.

28.11. Could the energy gap of a semiconductor be measured by absorption of radiation? If so, describe in detail how you would do this.

28.12. Criticize Holden's comments on the *p-n* junction as a rectifier: "A voltage applied across the junction changes the height of the energy hill. When it is applied in the 'forward' direction it lowers the hill, permitting many more majority carriers to *climb* it. When the voltage is applied in the 'backward' direction, raising the hill still further, it turns back the majority carriers even more effectively, and the current is limited by the small numbers of the minority carriers normally present (which can *coast down* the hill)."

28.13. a. Discuss why a solar cell should have a large energy gap.

b. Why should this energy gap be restricted, however, to the range 1–3 eV?

28.14. Estimate the power output of a square mile of solar cells on a bright (typical) Arizona day. How many square miles would be needed to produce the annual power output in the United States of 1 trillion kWh?

28.15. A semiconductor of unknown purity shows a Hall coefficient of zero at room temperature. How can this be?

28.16. Sketch ln p vs. $1/T$ curves as in Figure 3.3.1 for specimens A, B, C, and D, which are progressively doped with more gallium.

28.17. Discuss quantitatively the analogy between extrinsic semiconductors and extrinsic conduction in ionic crystals such as NaCl doped with Ca.

28.18. For some devices it is desirable to have the semiconductor intrinsic, while for other devices this would be disastrous. Discuss.

28.19. The junction transistor is not a thermodynamically stable device. Its behavior deteriorates reversibly at moderate operating temperatures, while at still higher temperatures the deterioration is irreversible.

a. Explain each of these phenomena.

b. Give an energy which is associated with each.

28.20. Show how the Fermi level varies as an intrinsic semiconductor is doped with

a. Donors.

b. Acceptors.

28.21. Show how the Fermi level varies with temperature in truly pure silicon to which some phosphorus has been added.

28.22. Is rectification possible for the following junctions:

a. Metal—p-type.

b. Metal—n-type.

c. Metal—intrinsic?

28.23. Impurities in silicon such as Fe, Co, and Mn have large ionization potentials (roughly $E_g/2$ in silicon). Of what possible use are such impurities?

28.24. Discuss the critical design features of a *n-p-n* junction.

28.25. Discuss how a *n-p-n* junction transistor works as a power amplifier. Try to do this with the same conciseness that Holden achieves in Problem 28.12.

28.26. In a *n-p-n* junction, the electrons injected into the base move across by diffusion. If an alternating high-frequency signal is applied, many of these electrons will not have time to reach the collector. Show that there is a cut-off frequency which varies inversely with the square of the thickness of the base.

28.27. If you were told to learn three fairly simple equations in semiconductor physics which you could use to form the basis for further discussion, list the equations which you would choose and why.

28.28. Sketch the total donor and total acceptor impurity concentrations versus distance in a *p-n* junction.

28.29. Compare the mobility of electrons in sodium or copper with the mobility of carriers in germanium.

28.30. a. What is the electrical conductivity of pure germanium at 300°K?
   b. What is the thermal conductivity of germanium?
   c. What is its heat capacity at 300°K?
   d. If 20 V are applied across a slab 0.010 cm thick, what is the current density?
   e. Would the specimen tend to be heated?
   f. Sketch the voltage-current behavior for this thermistor. Is it ever possible that dV/di is negative?

More Involved Problems

28.31. Derive Equation (28.2.5).

28.32. Show that for an intrinsic semiconductor the Fermi energy is midway between the top of the valence band and the bottom of the conduction band.

28.33. Derive the general expression for the Hall coefficient.

28.34. Show how the Hall effect can be used as the basis for designing a watt-meter.

28.35. Derive Equation (28.4.4).

28.36. Check a chemistry book such as A. F. Wells, *Structural Inorganic Chemistry*, Oxford, New York (1962) and study the types of compounds formed by silicon. Based on your studies invent a way to obtain ultrapure silicon.

28.37. A **field effect transistor** is shown in Figure P28.37 (after H. Johnson,

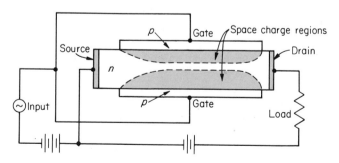

FIGURE P28.37.

"Basic Transistor Device Concepts," in *Transistors I*, RCA Laboratories, Princeton, N.J. (1956). Discuss how this operates.

28.38. Discuss the manufacture of the selenium rectifier. Use outside references. A starter is T. S. Hutchison and David C. Baird, *The Physics of Engineering Solids*, Wiley, New York (1963) p. 276.

28.39. A CdS photodetector receives radiation of 4000-Å wavelength over an area of 2×10^{-6} m^2 and with an intensity of 40 W/m^2. The energy gap is 2.4 eV.

    a. Calculate the number of electron-hole pairs generated per second if each quantum generates a pair.

    b. Calculate the increase in conductivity if the electron lifetime is 10^{-3} sec and $\mu_n = 10^{-2}$ m^2/V-sec.

28.40. The **Esaki** or **tunnel diode** involves tunneling in heavily doped semiconductors. Describe its operation. See, e.g., C. A. Wert and R. M. Thomson, *Physics of Solids*, McGraw-Hill, New York (1964), p. 263.

28.41. Using experimentally available values of the absorption coefficient versus wavelength for GaAs [see G. W. Sutton, *Direct Energy Conversion*, McGraw-Hill, New York (1966) p. 11], explain why a photon from solar radiation is either absorbed in a layer of 10^{-4}–10^{-6} cm from the surface or not absorbed at all. The **absorption coefficient**, α, is defined by $dI/dL = -\alpha I$, where I is the intensity and L is the distance of penetration.

28.42. Selenium is a **high dark-resistance material** with a resistivity greater than 10^9 Ω-m (see Table 3.1.1 for comparison). Illumination profoundly lowers the resistivity: This is the basis of **xerography**. Discuss xerography. See, e.g., R. M. Schaffert and C. D. Oughton, *Journal of the Optical Society of America*, **38**, 991 (1948).

28.43. Sketch the design of an experimental apparatus used to dope an *n*-type silicon with phosphorus to create a thin *p* region for solar cell applications.

28.44. Suppose a varistor consists of a slab of germanium 0.020 cm thick. It has a thermal conductivity of 0.8 W/cm-deg (1 W = 0.239 cal/sec), a density of 5.32 g/cm^3, a specific heat of 0.077 cal/g, and a melting point of 937.4°C. Estimate the upper limit on the frequency for this device.

28.45. Discuss how thermistors can be used for

    a. Manometers.

    b. Vacuum gauges.

    c. Flow meters.

    d. Thermal conductivity meters.

    e. Liquid level gauges.

28.46. Charge carriers in a semiconductor in a magnetic field tend to move in paths having a circular projection in the plane perpendicular to the magnetic field. Show that the angular frequency ω (**cyclotron frequency**) is given by $\omega = eB/m^*$.

28.47. Describe how recombination lifetimes are measured by the light probe method. See W. C. Dunlap, Jr., *An Introduction to Semiconductors*, Wiley, New York (1957).

Sophisticated Problems

28.48. Discuss in quantitative detail the scattering mechanisms for electrons in semiconductors.

28.49. Derive the expression for the saturation current density, J_0.

28.50. Derive the form of the space charge distribution shown in Figure 28.6.2.

Prologue

The occurrence of superconductivity in elements and compounds is described. We then show, using thermodynamics, that the entropy of the material in the superconducting state is less than the entropy of the normal state. The superconducting state is an *ordered* state. It is also shown that there is a discontinuity in the specific heat between the two states at T_c. The specific heat of the material in the superconductive state depends exponentially on the negative reciprocal of temperature. This is consistent with the *BCS theory* which postulates an *energy gap* at the Fermi surface. In earlier discussions we noted that bulk Type I superconductors are perfectly diamagnetic ($B = 0$). Actually, there is flux penetration near the surface, where the magnetic induction decays approximately exponentially. The characteristic depth of penetration is the *London penetration depth*. Flux penetrates very thin films. Because of the presence of a *negative interfacial energy* between the normal and superconducting phases in Type II materials, there is *phase separation*. This gives rise to *fluxoids.* The structure-sensitive behavior of these materials is due to *pinning of fluxoids*. The current-carrying capacity of hard superconductors is determined by the effectiveness of this pinning. Materials scientists are actively involved in studies of these pinning mechanisms as well as in the search for materials with higher transition temperatures and higher critical fields.

29

SUPERCONDUCTIVITY

29.1 INTRODUCTION

In 1911 Kamerlingh Onnes discovered a new state of matter, the **super-conducting phase**, when he found that the ohmic resistance of mercury suddenly vanished near 4.15°K. This is to be compared with the **normal phase** of a conductor which shows continually decreasing resistance as the temperature is lowered.

EXAMPLE 29.1.1

If a material remains in the normal conductive phase at all attainable temperatures, what is its resistivity as it approaches absolute zero?

Answer. It approaches the residual resistivity as was discussed in Chapter 27. Recall that pure crystals (one isotope only) with no imperfections would have a bulk electrical resistivity approaching zero as the absolute zero of temperature is approached.

EXAMPLE 29.1.2

List some of the phenomena associated with superconductivity which you studied in Chapter 5.

Answer.

1. Existence of T_c. See Table 5.2.1.
2. Existence of $H_c(0)$. See Table 5.2.1.
3. Parabolic form of $H_c(T)$. See Equation (5.2.1).
4. Meissner effect. See Section 5.2.

5. Silsbee effect. See Section 5.2.
6. Type I versus Type II *M-H* curves.
7. Existence of H_{c_1} and H_{c_2}. See Figure 5.2.3.
8. Structure-sensitive nature of J_c-*H* curves in Type II materials. See Figure 5.3.2.

It is interesting to note that Onnes' discovery of superconductivity followed only three years after he achieved, for the first time, the liquefaction of helium. Figure 29.1.1 shows the boiling temperature versus vapor pressure curve for liquid helium. The normal boiling point (at 1 atm) is 4.2°K.

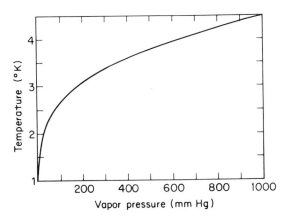

FIGURE 29.1.1. Boiling temperature versus vapor pressure for liquid helium.

EXAMPLE 29.1.3

How would you proceed to achieve a temperature of 2°K?

Answer. Vacuum pumps could be used to pump the vapor from the boiling liquid helium down to a pressure of about 30 mm Hg. The temperature of the liquid would fall to 2°K when this pressure is held constant.

BCS THEORY.   Fundamental understanding as well as extensive practical applications of superconductivity were not achieved until recently. In 1957 J. Bardeen, L. N. Cooper, and J. R. Schrieffer, *Physical Review* **106**, 162 (1957), developed a model in which *pairs* of conduction electrons were condensed into a *lower energy phase*. The electrons in a **Cooper pair** are bound to each other over a distance called the **coherence length**. The electrons in the pair have no net momentum. The difference of energy between the bound pair states and the normal state is called the **energy gap**. We shall discuss evidence for this energy gap in Section 29.2. This theory, known as the **BCS theory**, accounts for most of the properties exhibited by superconducting phases. However, not all superconductors exhibit the energy gap and hence the BCS theory may not be universal. The normal electron has a high velocity and hence momentum and by the de Broglie relation a small wavelength of the order of the atom diameter. It is therefore readily scattered by variations in potential energy due to atom vibrations. However, the coupled Cooper pair has a zero net velocity in the absence of an electric field and only a small velocity when the superconductor is carrying a large current. Moreover, all the Cooper pairs have the same small velocity, so that they can be described by a single macroscopic de Broglie wave whose wavelength is huge compared to the atom diameter. Hence there is no scattering.

EXAMPLE 29.1.4

Of what order of magnitude must the energy gap between supercon-ductive electrons and normal electrons be?

Answer. Since the phenomenon disappears at a critical temperature T_c, this energy must be of the order of kT_c. In the BCS theory the energy gap, E_g, varies between $3.52kT_c$ at $T = 0$ to zero at $T = T_c$.

OCCURRENCE. There is an excellent recent discussion which relates superconductivity and the periodic table: B. T. Matthias, "Superconductivity and the Periodic System," *American Scientist* (Jan.–Feb. 1970), p. 80. See also: B. T. Mathias, *Physics Today*, **24**, 23 (1971). It appears that insulators, ferro-magnetics, and antiferromagnetic materials do not change directly into the superconducting state; however, an insulator may undergo a change in crystal structure at which time it becomes a semiconductor or metal and this new struc-ture may on cooling become a superconductor.

Matthias notes that the transition temperature is a function of the number of valence electrons, as shown in Figure 29.1.2. Note that for electron ratios somewhat below 5 (and also 7) electrons/atom, T_c approaches a maximum.

EXAMPLE 29.1.5

Would Nb_3Sn be expected to be a high-temperature superconductor?

Answer. It has an electron/atom ratio as follows:

$$\frac{\text{electron}}{\text{atom}} = \frac{3 \times 5 + 1 \times 4}{4} = 4.75.$$

This ratio is around the maximum in Figure 29.1.2. Experimentally $T_c = 18.05°K$ for Nb_3Sn, which is high, as expected.

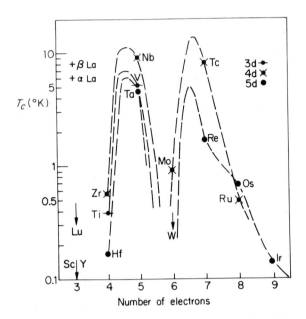

FIGURE 29.1.2. Transition temperatures as a function of the number of valence electrons. [From B. T. Matthias, *American Scientist* (Jan.–Feb. 1970) p. 81.]

Table 29.1.1 shows transition temperatures for some high field superconductors [a high field superconductor has a relatively high $H_c(0)$ value, say, $H_c(0) \sim 10^7$ A/m].

Table 29.1.1. T_c FOR COMPOUNDS

Compound	T_c, °K	Compound	T_c, °K
$Nb_3Al_{0.8}Ge_{0.2}$*	20.98	V_3Si	17.1
$Nb_3Al_{0.8}Ge_{0.2}$	20.05	V_3Ga	16.5
Nb_3Sn	18.05		
Nb_3Al	17.5		

* After special annealing treatments.

We note that the high transition temperature compounds have a cubic crystal structure, called the β-tungsten structure, as shown in Figure 29.1.3.

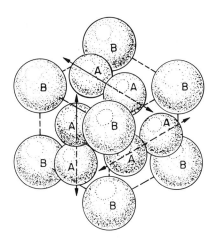

FIGURE 29.1.3. β-tungsten structure. The B atoms form a bcc structure, while the A atoms are in the tetrahedral void sites. Hence for Nb_3Sn, the B atoms are Sn and the A atoms represent Nb chains.

The reason why these β-tungsten structure materials favor high T_c's is a challenging problem.

The concern with T_c is a practical engineering matter, as shown in Table 29.1.2.

Table 29.1.2. BOILING POINTS (AT 1 ATM)
OF CRYOGENIC LIQUIDS

Liquid	T_B, °K	Dollars/liter of liquid
Nitrogen	77.32	0.20
Hydrogen	20.37	3.00*
Helium 4	4.216	4.00
Helium 3	3.195	~100.

* However, the heat of vaporization is about 20 times as high as that of helium.

The cost of keeping a superconductor cold enough changes sharply in steps as shown. Although T_c for one alloy is slightly above the hydrogen boiling point, it would have to be several degrees higher so that high currents could be carried at liquid hydrogen temperatures. The hope is that it will be possible to produce a high field superconductor whose transition temperature is above room temperature.

ISOTOPE EFFECT. In 1950, experiments showed that T_c for various isotopes of lead was inversely proportional to the square root of the masses [as is also the Debye temperature defined in Equation (17.11.10); recall that the frequency of an harmonic oscillator varies inversely with the square root of the mass as in Equation (2.4.2)]. This suggests that T_c should be proportional to the Debye temperature (although this isotope effect is not shown by all superconductors). Theory suggests that hydrogen should become metallic at very high pressures and that this metallic phase, which has a Debye temperature of nearly 4000°K, will be a superconductor with a value of T_c possibly approaching room temperature.

29.2   SUPERCONDUCTING PHASE AND THE NORMAL PHASE

What is a phase? Clearly, fcc copper and fcc nickel are different phases. So are bcc iron and fcc iron. In one case the materials have a different chemical composition. In the other they have a different crystal structure. In both cases the different phases have different physical properties and exhibit different thermodynamic behavior. This is also true of lead in the superconducting state and lead in the normal state.

THERMODYNAMIC DIFFERENCES.   We shall now proceed to consider the difference in Gibbs free energy between the normal and superconducting states for a *reversible* system. For the present we shall focus our attention on a Type I superconductor. We have already noted in Chapter 5 that such materials are perfectly diamagnetic materials (we shall modify this slightly later) and hence exclude lines of induction completely, as shown in Figure 29.2.1. Recall the *M-H* curve shown in Figure 5.2.2. (The reader should make a similar sketch for a ferromagnetic material and for an ordinary diamagnetic or paramagnetic material.)

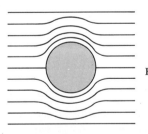

FIGURE 29.2.1. Exclusion of lines of flux by a perfect diamagnetic material.

It can be shown that (see the learning sequence of Problems 29.18–29.22) the difference in Gibbs free energy between the superconducting and normal phase at zero magnetic field is given at a fixed temperature and pressure by

$$G_S^m(0) = G_N^m(0) + \mu_0 \int_0^{H_c} M \, dH \tag{29.2.1}$$

$$= G_N^m(0) - \frac{\mu_0 H_c^2}{2}. \tag{29.2.2}$$

Recall that the equilibrium phase under conditions of constant temperature, pressure, and field will be the phase with the lowest free energy. Here H_c refers to *any* point along the $H_c(T)$ curve in Figure 5.2.1.

EXAMPLE 29.2.1

Which phase has the lowest free energy in the absence of a magnetic field below T_c?

Answer. Clearly the superconducting phase has since $H_c > 0$; see Figure 5.2.1.

It can also be shown that the entropy of the *superconducting* phase is less than that of the *normal* phase since

$$S_S = S_N + \frac{d}{dT}\left(\frac{\mu_0 H_c^2}{2}\right) \tag{29.2.3}$$

and since dH_c/dT is negative [see Figure 5.2.1 or Equation (5.2.1)].

For the derivation of (29.2.3), see Problem 29.24. Inasmuch as the superconducting state represents a state of lower entropy, it also represents a state of higher order.

EXAMPLE 29.2.2

What is the difference in entropy between the superconducting phase and the normal phase at $T = T_c$?

Answer. Since

$$S_S(T) = S_N(T) + \mu_0 H_c(T)\frac{dH_c(T)}{dT}$$

and since $H_c(T_c) = 0$ while $dH_c(T_c)/dT$ is finite,

$$S_S(T_c) = S_N(T_c).$$

It can also be shown (see Problem 29.25) that the difference in specific heat is

$$C_S = C_N + \mu_0 T\left[H_c\frac{d^2 H_c}{dT^2} + \left(\frac{dH_c}{dT}\right)^2\right]. \tag{29.2.4}$$

Note that there is no discontinuity of entropy at T_c but that there is a discontinuity in the specific heat.

EXAMPLE 29.2.3

What is the difference in specific heat between the superconducting phase and the normal phase *assuming* that Equation (5.2.1) is true?

Answer.

$$C_S(T_c) = C_N(T_c) + \frac{4\mu_0 H_c^2(0)}{T_c}.$$

In general,

$$C_S(T_c) > C_N(T_c);$$

i.e., there is a discontinuity in C at T_c. An actual example in shown in Figure 29.2.2.

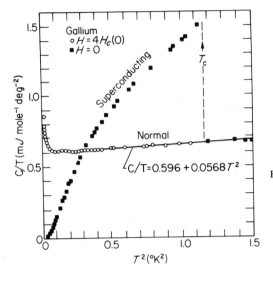

FIGURE 29.2.2. C/T vs. T^2 curves for gallium in the normal and superconducting states. [From N. E. Phillips, *Physical Review*, **134**, 385 (1964).]

EXAMPLE 29.2.4

How does the phase transition at T_c differ from a phase transition such as melting?

Answer. In melting, thermodynamic quantities such as enthalpy, entropy, and volume show discontinuous changes at T_M. We call such a process a **first-order phase transition**. In the superconducting transition at zero magnetic field, the entropy change is zero. However, thermodynamic quantities such as specific heat $[C_p = (\partial H/\partial T)_p]$ are discontinuous. We call such a process a **second-order phase transition**. Except at $T = T_c$, the superconducting transition is first order.

EXAMPLE 29.2.5

Assuming $H_c(T)$ is given by Equation (5.2.1), is $C_S(T) > C_N(T)$ for all T?

Answer. It can be shown that for $T^2 < T_c^2/3$, $C_S < C_N$, while for $T^2 > T_c^2/3$, $C_S > C_N$ for this assumed function. Thus the two specific heat curves cross. For gallium, $T_c = 1.09$, so $T_c^2/3 \approx 0.40$, while the crossover occurs for $T^2 \approx 0.33$.

At low temperature $C_S(T)$ depends *exponentially* on temperature. The argument of the exponential is $E_g/2kT$, where E_g is the gap energy at absolute zero [see N. E. Phillips, *Physical Review*, **134**, 385 (1964)]. Figure 29.2.3 shows the energy gap in the superconducting state.

Recall that the energy levels in a metal are extremely close together (perhaps 10^{-10} eV; $kT \sim 10^{-3}$ eV when $T = 12°K$). The specific heat of the

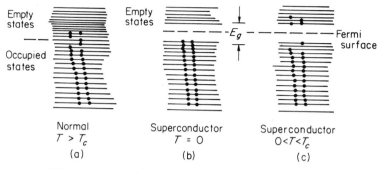

FIGURE 29.2.3. Electron levels in conductors. (a) Normal state. (b) Superconductor at $T = 0$. (c) Superconductor at $0 < T < T_c$.

superconducting electrons is *zero* (since there is no nearby energy level to which they can be excited). The specific heat of the superconducting state is due to the presence of a small fraction of normal electrons, which are those electrons which have been excited across the energy gap (above the energy gap there are numerous states lying close together).

The situation of forming some normal electrons in the superconducting state is somewhat analogous to the formation of an electron plus a hole in an intrinsic semiconductor. In that case also the Fermi level lies midway in the energy gap. Recall that the concentration of electrons (or holes) varied as $\exp(-E_g/2kT)$. A similar situation applies in the present case. The electronic specific heat of a normal conductor is

$$C_{eN} = \gamma T. \tag{29.2.5}$$

Thus the electronic specific heat of the superconductor is

$$C_{eS} = \gamma T_c e^{-E_g/2kT}. \tag{29.2.6}$$

I. Giaever has performed a tunneling experiment which directly confirms the existence of the energy gap [*Superconductivity*, Wiley-Interscience, New York (1962).]

29.3   THE PENETRATION DEPTH

The Meissner effect ($B = 0$) holds for a bulk Type I superconductor in which the external magnetic field is parallel to a cylindrical conductor for $H < H_c$. In Chapter 5 and in previous discussions in this chapter we interpreted this to mean that $H = -M$ *everywhere* within the specimen.

The actual situation is that *within* a bulk specimen $B = 0$ except near to the surface, where B varies approximately as

$$B = B_0 e^{-x/\lambda_L}. \tag{29.3.1}$$

Here λ_L is called the **London penetration distance**. A typical value of λ_L at $0°K$ is 500 Å.

EXAMPLE 29.3.1

What is the average value of B within a wide, flat specimen for which $B = B_0$ at the surfaces if the specimen is 1 cm thick?

Answer. Let $x = 0$ represent the surface and $x = 0.5$ at the plane midway between the flat surfaces. Then

$$\bar{B} = \frac{\int_0^{0.5} B\,dx}{0.5} = \frac{\lambda}{0.5} B_0 = 10^{-5} B_0.$$

EXAMPLE 29.3.2

Design an experiment which illustrates the penetration depth concept.

Answer. If we consider a very thin film (300 Å) of lead, say, then the field should penetrate readily in the superconducting state and the magnetic induction within the sample should not be zero. For example, one might find $M = -0.04H$; i.e., $B = 0.96\mu_0 H$.

EXAMPLE 29.3.3

Will the critical field for a thin film be greater than the critical field for the bulk material?

Answer. Let us illustrate for the case in which $M = -0.04H$, as in Example 29.3.2. Then from (29.2.1) we have

$$G_S^m(0) = G_N^m(0) - \frac{0.04\mu_0^2 H_c^2}{2},$$

where ^tH_c represents the critical field for the thin films. The energy difference between the superconducting and normal phases must be the same for the bulk and the thin films. Comparison with (29.2.2) shows that

$$H_c^2 = 0.04^t H_c^2$$

or

$${}^tH_c = 5H_c.$$

In Chapter 5 we mentioned synthetic high field superconductors. Do you see now why these can exist?

London has shown that the penetration depth is given by

$$\lambda_L = \left(\frac{m}{\mu_0 n_s e^2}\right)^{1/2}, \tag{29.3.2}$$

where n_s is the number of electrons per unit volume in the superconducting state, m is the electron mass, and e is the electronic charge. Since the latter decreases (according to BCS theory) as T increases toward T_c, λ_L increases as $T \rightarrow T_c$.

Values of λ_L at $T = 0$ are given in Table 29.3.1 for several "pure" elements.

Table 29.3.1. VALUES OF λ_L AS $T \rightarrow 0$

Material	λ_L, Å
Al	500
In	640
Pb	390
Nb	470

Maxwell's electromagnetic equations and the conditions that $B = 0$ (Meissner effect) and $\xi = 0$ (no resistivity, hence no voltage drop) lead in the one-dimensional case to the expression

$$\frac{d^2 J_s}{dx^2} = \frac{J_s}{\lambda_L^2}, \tag{29.3.3}$$

where J_s is the current density along the surface, x is the normal distance from the surface, and λ_L is given by (29.3.2). It can then be shown that (29.3.1) follows with λ_L given by (29.3.2).

29.4   PHASE SEPARATION

The fact that a thin film of lead might have $M = -0.05H$ while the bulk has $M = -H$ suggests that we consider an alternating array of normal and superconducting phases as shown in Figure 29.4.1.

This postulate is based on thermodynamic grounds. It is shown (see Problem 29.23) that

$$G_S^m(H) = G_N^m(H) - \frac{\mu_0 H_c^2}{2} - \mu_0 \int_0^H M \, dH. \tag{29.4.1}$$

FIGURE 29.4.1. Postulated layer-like structure of superconducting phase and a thin normal phase.

As an example consider the situation for $H = H_c$. Then for the bulk, since $M = -H$,

$$G_S^m(H_c)_{\text{bulk}} = G_N^m(H_c),$$

while for the thin film, since $M = -0.05H$ in the example chosen here,

$$G_S^m(H_c)_{\text{film}} = G_N^m(H_c) - \frac{0.95\mu_0 H_c^2}{2}.$$

Note: H_c refers to the critical field for the bulk. Clearly, $G_S^m(H_c)_{\text{film}} < G_S^m(H_c)_{\text{bulk}}$.
This is true in general. Consequently, assuming that the normal regions in Figure 29.4.1 occupy a tiny fraction of the volume and that the superconducting layers have thickness of the order of λ_L, the overall energy of the phase-separated material should be reduced, *unless (a positive) interfacial energy between the phases exists*. It is therefore necessary to postulate that Type I superconductors have positive surface energy between the phases which is sufficiently large to offset the reduction in the magnetic energy.

On the other hand, a negative surface energy would *cause* phase separation. Under these conditions, penetration would occur ($B \neq 0$) and the energy of the system would be lowered. A. A. Abrikosov (1957) predicted, on the basis of such an assumption, M-H curves characteristic of Type II materials (see Figure 5.2.3). H_c for the Type II material is defined in a thermodynamic fashion; namely, $H_c^2/2 \equiv -\int_0^{H_{c_2}} M\,dH$. This means that we find the actual area under the M-H curve, and if the material were Type I, this area would be $H_c^2/2$.

According to Abrikosov's theory, the magnetic induction penetrates the specimen along lines rather than layers as in Figure 29.4.1. See Figure 29.4.2

FIGURE 29.4.2. Example of penetration as calculated by Abrikosov. Dark areas are normal phases where flux penetrates completely. The lines with arrows represent the vortices. [From A. A. Abrikosov, *Journal of Physics and Chemistry of Solids*, **2**, 199 (1957).]

where a section normal to these lines is shown. [The square lattice predicted by Abrikosov is not the thermodynamically stable form because his calculation was in error. See: W. M. Kleiner, L. M. Roth, S. H. Autler, *Phys. Rev.* **133A**, 1226 (1964). A triangular lattice is more stable.]

In this model, a normal phase region (surrounding a flux line) is surrounded by a current vortex. The combination of a flux line and a current vortex is known as a **fluxoid**. There is a quantum of energy associated with each fluxoid. (An important feature of the superconducting state is the fact that the flux is quantized.) In a Type II superconductor there is no flux penetration until $H = H_{c_1}$ (see Figure 5.2.3). Fluxoids then appear and increase in number as H is increased further. The size of the normal cores usually (but not always) remains constant when H increases and overlap occurs. Figure 29.4.3 illustrates an experimentally observed **fluxoid lattice**.

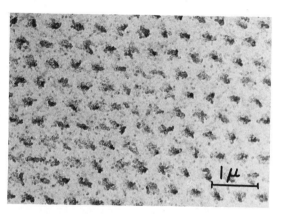

FIGURE 29.4.3. Fluxoids in Pb–6.3 at. % In. [*Courtesy of Uwe Essmann, Max Planck Institute, Stuttgart, Germany.* See also H. Trauble and U. Essmann, *Journal of Applied Physics*, **39**, 4052 (1968).]

If we were to plot B vs. x and also the number of superconducting electrons vs. x (here x is measured along a line connecting fluxoid centers), we would find results as shown in Figure 29.4.4. In general at high fields, B does not decay all the way to zero; similarly at high fields, the maximum value of n_s is small relative to the maximum value of n_s for small fields.

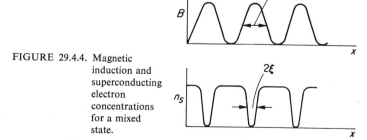

FIGURE 29.4.4. Magnetic induction and superconducting electron concentrations for a mixed state.

The width of the $B(x)$ peaks is $2\lambda_L$ and of the normal electron concentration peaks is 2ξ. ξ is called the **coherency length**. The ratio

$$\kappa = \frac{\lambda_L}{\xi} \qquad (29.4.2)$$

is called the **Ginsburg-Landau parameter**.

It can be shown that if $\kappa > 1/\sqrt{2}$, the superconductor is Type II, while if $\kappa < 1/\sqrt{2}$, it is Type I.

Theoretical expressions for κ based on the quantum theory of superconductivity are given below. We shall not attempt to derive these here.

For a pure material (low residual resistivity),

$$\kappa = \kappa_0 = 2.7 \times 10^9 \left| \frac{dH_c}{dT} \right|_{T=T_c} \lambda_L^2(0). \qquad (29.4.3)$$

Here H_c has units of amperes per meter and λ_L has units of meters.

For a material with high residual resistivity,

$$\kappa = \kappa_0 + 2.4 \times 10^5 \gamma^{1/2} \rho_0. \qquad (29.4.4)$$

Here γ is the temperature coefficient of the electronic specific heat and has units of joules per cubic meter-degrees Kelvin squared, and ρ_0, the residual resistivity, has units of ohm-meters.

The derivation of the Ginsburg-Landau relation and the theoretical expressions for γ are given in texts on superconductivity.

EXAMPLE 29.4.1

Assuming a parabolic temperature dependence as in Equation (5.2.1), estimate the Ginsburg-Landau parameter for lead.

Answer. We have

$$T_c \left| \frac{dH_c}{dT} \right|_{T=T_c} = 2H_c(0).$$

Since $H_c(0) = 7 \times 10^4$ A/m (from Table 5.2.1) and $\lambda_L(0) = 6.4 \times 10^{-8}$ m from Table 29.3.1, we estimate $\kappa = 0.57$. Hence we expect pure lead to be Type I since $\kappa < 0.7$.

By adding impurities to Pb we can greatly increase its residual resistivity, as was already noted in Chapter 27. Hence, as is clear from (29.4.4), κ can be increased and in fact the alloy is a Type II superconductor. We consider types of M-H curves exhibited by lead as shown in Figure 29.4.5. In (a) the bulk lead exhibits the Meissner effect ($B = 0$), while the thin film of lead in (b) does not. The addition of the alloying element to lead in (c) increases κ so that the alloy is Type II instead of Type I.

Note that H_c for this Type II material is defined so that the area under the dashed lines equals the area under the actual M-H curves.

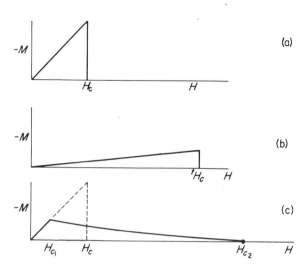

FIGURE 29.4.5. *M-H* curves for three forms of lead. (a) Pure bulk Pb. (b) Pure thin film Pb. (c) Lead with 20 at. % In.

29.5   HARD SUPERCONDUCTORS

The Type II superconductors have higher critical fields than Type I, but the ideal Type II superconductor cannot carry large currents at large fields. This is because there is a force on the fluxoids equal to the Lorentz force,

$$\mathbf{F} = \mathbf{J} \times \mathbf{B}. \tag{29.5.1}$$

(The student may previously have noted that the force exerted on a particle of charge q moving with a velocity $\mathbf{v}$ by a magnetic induction $\mathbf{B}$ is $\mathbf{F} = q\mathbf{v} \times \mathbf{B}$.) Thus the fluxoids in a soft Type II superconductor are readily swept out of the material if $\mathbf{J}$ and $\mathbf{B}$ are large. What is needed is a mechanism for pinning the fluxoids. One mechanism is illustrated in Figure 29.5.1. Here the impurity

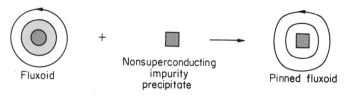

FIGURE 29.5.1. Pinning of a fluxoid.

precipitate and the normal region allow penetration. The magnetic energy of the system is lowered when the two combine. Thus it will take a force to separate them (just as it took a force to move a dislocation away from a precipitate atom). Fluxoids interact with magnetic precipitates, normal (non-magnetic, nonsuperconductive) precipitates, voids (which are really non-magnetic regions), dislocation arrays, dislocations, etc. All of these imperfections impede the motion of fluxoids and hence increase the current-carrying capacity at a given field. The presence of these imperfections produces a **hard super-conductor**, i.e., a superconductor which can carry large currents. Figure 29.5.2 illustrates the structure-sensitive nature of the current-carrying capacity of a superconductor.

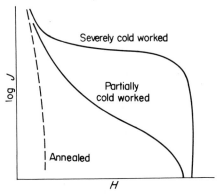

FIGURE 29.5.2. Sketch of the structure-sensitive nature of J vs. H for Nb–Ti alloy.

FUNDAMENTAL THEOREM FOR HARDENING SUPERCONDUCTORS.   To pro-duce high current-carrying superconductors, disrupt the continuum.

In the case of the Nb–Ti alloy, which is a ductile material which can be cold-drawn into wire, cold working produces an extremely fine dislocation cell structure which is responsible for the pinning of fluxoids. The relative ease of manufacturing explains why Nb–60% Ti alloys are used even though H_c at 4.2°K is only about 0.8×10^7 A/m compared to a value twice that high for Nb$_3$Sn. Nb$_3$Sn is often produced by passing niobium strips through a region of tin vapor which diffuses into the heated niobium. The compound, Nb$_3$Sn, is brittle.

The study of the interaction of imperfections with fluxoids is an active area of research. So is the study of the motion of fluxoids which is a dissipative process. The J-H curves of Figure 29.5.2 refer to dc currents. When ac currents are used the fluxoids move and the material does not have high current-carrying capacity. Fluxoid lattices such as shown in Figure 29.4.3 often show "disloca-tions." One possible way by which fluxoids move is by motion of fluxoid lattice dislocations.

REFERENCES

Parks, R. D., "Quantum Effects in Superconductors," *Scientific American* (Oct. 1965) p. 57.

Kittel, C., *Introduction to Solid State Physics*, Wiley, New York (1968) Chapter 11.

Lytton, E. A., *Superconductivity*, Wiley, New York (1962). An excellent brief introduction.

Giaever, I., *Superconductivity*, Wiley-Interscience, New York (1962).

Livingston, J. D., and Schadler, H. W., "The Effect of Metallurgical Variables on Superconducting Properties", *Progress in Materials Science*, **12**, 183 (1964).

deGennes, P. G., *Superconductivity of Metals and Alloys*, Benjamin, New York (1966). A good, sophisticated book on Type II superconductors.

Newhouse, V. L., *Applied Superconductivity*, Wiley, New York (1964).

PROBLEMS

29.1. Create an experimental method of showing that $B = 0$ for a Type I superconductor.

29.2. Explain why the superconducting state is not just a perfect conduction state and nothing more [if necessary, use Newhouse (see the References), p. 6].

29.3. List, with a short discussion for each, some of the phenomena associated with superconductors.

29.4. The armature of a generator is very heavy owing to the presence of iron cores. Why would superconducting armatures be of interest to the utility industry?

29.5. The electron/atom ratio of Nb_3Sn falls close to the first peak of Figure 29.1.2. Suggest some possible reasons for the ratios needed for high T_c.

29.6. Invent a switch which is based on the superconducting phase transition.

29.7. Nb–Ti alloys are often used for magnet materials, although their *J-H* curves fall considerably below those of Nb_3Sn compounds. Why?

29.8. Comment on the following: "The superconducting state is a more ordered state than the normal state."

29.9. A metal in the normal state exhibits two specific heat terms: One is proportional to temperature, the other to temperature cubed. What is the origin of each term?

29.10. a. Define the London penetration depth, the coherence length, and the Ginsburg-Landau parameter.

b. What is the order of magnitude of the London penetration depth?

29.11. Why do Type II superconductors have H_{c_1}, while Type I does not?

29.12. Explain the large value of H_c exhibited by synthetic high field superconductors (see Section 5.4).

29.13. Explain with equations why a void would pin a fluxoid.

29.14. Sketch the M vs. H curves for Pb, Pb–10 at. % In, and Pb–20 at. % In. Assume the two alloys are both Type II and that H_{c_1} is less for the alloy with 20% In.

More Involved Problems

29.15. A question often asked is, "Is the dc resistivity of a superconductor actually zero?" Comment on this. See, e.g., D. J. Quinn and W. B. Ittner, *Journal of Applied Physics*, **33**, 748 (1962).

29.16. Describe the operation of the cryotron.

29.17. Describe how T_c can be measured (see L. C. Jackson, *Low Temperature Physics*, Wiley, New York (1962).

Note: Problems 29.18–29.22 are a learning sequence whose purpose is to derive Equation (29.2.1).

29.18. The internal energy change for a reversible process is given by $dU = T\,dS + dW$. What is the expression for the work done in magnetizing a solid? *Note:* See Table 7.1.2.

29.19. For a system in which both $-P\,dV$ and $H\,dB$ work are done we have for a reversible process $dU = T\,dS - P\,dV + H\,dB$. The function $U(S, V, B)$ has a minimum under conditions of constant S, V, and B. However, experiments are rarely, if ever, carried out while S, V, and B are held constant. Consider a function G^m, called the Gibbs magnetic function, where $G^m = U - TS + PV - HB$. Show that equilibrium under conditions of constant $T, P,$ and H corresponds to a minimum of G^m.

29.20. In arriving at the answer to Problem 29.19 we obtained $dG^m = -S\,dT + V\,dP - B\,dH$. Under conditions of constant T and P we have $dG^m = -B\,dH = -\mu_0 H\,dH - \mu_0 M\,dH$, where we have used Equation (4.2.3). Why does the term $-\mu_0 \int_0^H M\,dH$ make a negligible contribution for a normal material? (This is, however, usually not true of high H_{c_2} superconductors; for example, the H_{c_2} of Nb-Ti alloys is limited by the paramagnetism of the normal state and because of this the fluxoid cores expand rapidly near H_{c_2}.)

29.21. Show that $G_S^m(H) - G_S^m(0) = -(\mu_0 H^2/2) - \mu_0 \int_0^H M\,dH$ for a superconductor.

29.22. a. Why is it true that $G_S^m(H_c) = G_N^m(H_c)$ for a Type I superconductor?

b. Use this result to show that $G_S^m(0) = G_N^m(0) + \mu_0 \int_0^{H_c} M\, dH$.

29.23. Use the result of Problem 29.22 to show that $G_S^m(H) = G_N^m(H) + \mu_0 \int_H^{H_c} M\, dH$.

29.24. Using the fact that $(\partial G_N/\partial T)_P = -S_N$ and $[\partial G_S(0)/\partial T]_P = -S_S$, derive (29.2.3).

29.25. Find $(dS_S/dT) - (dS_N/dT)$ from (29.2.3). Use this to derive (29.2.4).

29.26. Discuss the current status of superconducting magnet construction.

29.27. Discuss the current status of the use of superconducting computer elements.

Sophisticated Problem

29.28. Discuss the phenomenological London equation (which can be considered an adjunct to the Maxwell equations) and show how this leads to the penetration depth. (Use outside references.)

Prologue

The dielectric coefficient of a material is determined by the *polarizability*. All materials exhibit *electronic polarization*. Materials may also exhibit *ionic polarization*, *space charge polarization*, and *orientation polarization*. The origin of these different forms of polarization is discussed first qualitatively and then quantitatively. We show that the orientation polarization is described by the *Langevin function* and that this polarization is inversely proportional to the absolute temperature, while electronic and ionic polarization are virtually independent of temperature. We discuss the *molar polarizability* and the famous *Clausius-Mossotti equation*. We show that dielectric constants vary with frequency and study why they do, first qualitatively and then quantitatively. We note that there is a mathematical analog between damping in a mechanical material and damping in a dielectric material and that the same mechanism sometimes is responsible for both effects. There is a brief discussion of electrostriction, piezoelectricity, and ferroelectricity.

30

DIELECTRICS

30.1 INTRODUCTION

Dielectric materials (electrical insulators) have a wide variety of applications. Perhaps the most common is the use of plastics or ceramics for electrical insulation and in capacitors. However, piezoelectric, ferroelectric, and ferromagnetic materials are also special dielectric materials with special applications.

EXAMPLE 30.1.1

Name three of the most important electrical properties of ordinary dielectric materials.

Answer. These were discussed in Chapter 3. They are dielectric constant (relative permeability), Section 3.4; dielectric strength, Section 3.4; and loss angle, Sections 3.4 and 2.4.

30.2 MACROSCOPIC AND MICROSCOPIC MEANING OF POLARIZATION

The polarization **P** (see Section 3.4) is related to the electric field ξ by

$$\mathbf{P} = \epsilon_0 \chi \xi, \tag{30.2.1}$$

where ϵ_0 is the permittivity of vacuum and χ is the electrical susceptibility. We shall concern ourselves only with those materials for which χ is isotropic (this includes single cubic crystals but not other single crystals). The dielectric constant is then

$$\kappa = \chi + 1. \tag{30.2.2}$$

EXAMPLE 30.2.1

A parallel plate condenser for which the distance between the plates is very much less than the diameter of the plates has a capacitance C_0. What is its capacitance if the gap is filled with a slab of dielectric material with dielectric constant κ?

Answer.

$$C = \kappa C_0.$$

Note that $\kappa = 1$ for vacuum. It may have a value up to 12 for ordinary dielectrics and may have a value of the order of 10^3 for special dielectrics, such as ferroelectrics.

EXAMPLE 30.2.2

The charge on the plates of the capacitor (positive Q on one plate, negative on the other) is related to the applied voltage V by

$$Q = CV. \tag{30.2.3}$$

What is the fractional increase in charge on the plates when the vacuum is replaced by the dielectric (the voltage is the same)?

Answer.

$$\frac{Q - Q_0}{Q_0} = \frac{C - C_0}{C_0} = \kappa - 1 = \chi. \tag{30.2.4}$$

Note that this is a macroscopic meaning of susceptibility.

The capacitance of the parallel plate condenser under consideration here is

$$C = \frac{\kappa \epsilon_0 A}{d}, \qquad d \ll \sqrt{A}, \tag{30.2.5}$$

where A is the area of a plate and d is the distance between the plates. The charge Q_0 on a plate in a vacuum is

$$Q_0 = \frac{\epsilon_0 A}{d} V. \tag{30.2.6}$$

Note that if you have a device capable of measuring charge and another device capable of measuring voltage, you can readily verify this relationship. We choose to write (30.2.6) in terms of the charge per unit area

$$\sigma_0 = \frac{Q_0}{A} = \frac{\epsilon_0 V}{d}. \tag{30.2.7}$$

With the dielectric present we have

$$\sigma = \frac{Q}{A} = \frac{\kappa \epsilon_0 V}{d}. \tag{30.2.8}$$

The additional charge per unit area because of the presence of the dielectric is

$$\sigma' = \sigma - \sigma_0 = (\kappa - 1)\epsilon_0 \xi = \epsilon_0 \chi \xi. \tag{30.2.9}$$

However, we can also write

$$\sigma' = \frac{Q - Q_0}{A} = \frac{Q'}{A} = \frac{Q'd}{Ad} = \frac{\text{dipole moment}}{\text{volume}} = P. \qquad (30.2.10)$$

Recall the definition of dipole moment given in Section 9.5. Note that opposite pairs of charges q and $-q$ on a parallel plate condenser where these charges are separated by d leads to a dipole moment $p = qd$. Summing over all such pairs gives a dipole moment $p' = Q'd$. Dividing by the volume gives the dipole moment per unit volume, P. Combining (30.2.9) and (30.2.10) gives

$$P = \epsilon_0 \chi \xi.$$

Both P and ξ are vectors and what we really have is

$$\mathbf{P} = \epsilon_0 \chi \mathbf{\xi}. \qquad (30.2.11)$$

We can write

$$\mathbf{P} = N\bar{\mathbf{p}}, \qquad (30.2.12)$$

where N is the number of dipole moments per unit volume and $\bar{\mathbf{p}}$ is the average moment of an individual dipole. We can write that

$$\bar{\mathbf{p}} = \alpha \mathbf{\xi}', \qquad (30.2.13)$$

where ξ' is the **internal field** at the dipole and α is a proportionality constant called the **polarizability**.

ξ' is not as a rule the same as the *external field* ξ. For a dilute gas there is a negligible difference at low fields. Later on we shall discuss how this internal field is related to the external field.

Thus to calculate χ or κ, we must calculate two quantities. First, we must know how $\bar{\mathbf{p}}$ is related to ξ'; i.e., we must calculate α. Second, we must know how the internal field ξ' is related to ξ.

EXAMPLE 30.2.3

Why is ξ' often different than ξ?

Answer. Because the internal field at a point is due to ξ and due to all the other dipoles surrounding that point. This will be discussed further in Section 30.5.

Equations (30.2.11) and (30.2.12) provide the connection between the macroscopic and microscopic worlds.

30.3   MOLECULAR MECHANISMS OF POLARIZATION

As is clear from the previous discussion, to calculate $\mathbf{P}$ we need to find the net dipole moment per unit volume. Let us consider the molecular origin of these dipole moments.

ELECTRONIC POLARIZATION.   Consider an argon atom. The electron cloud has a spherically symmetric distribution in the ground state. In the presence of an electric field the positive ion is displaced in the direction of the electric field, while the electron cloud is displaced in the opposite direction, as shown in Figure 30.3.1(a). This is called **electronic polarization**.

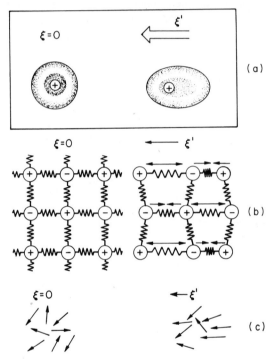

FIGURE 30.3.1. Polarization   mechanism.   (a)   Electronic. (b) Ionic. (c) Orientation.

IONIC POLARIZATION.   Consider an NaCl crystal (see Figure 9.2.2). In the presence of an electrical field the positive sodium ions are displaced in the direction of the field, while the negative chlorine atoms are displaced in the opposite direction, as shown in Figure 30.3.1(b). This is called **ionic polarization**.

EXAMPLE 30.3.1

What can you say about the frequency dependence of ionic polarization?

Answer. This depends on the motion of the atoms. Thus at very high frequencies, the atoms would not have time to undergo much displacement.

EXAMPLE 30.3.2

Is ionic polarization the only mechanism possible in the NaCl crystal?

Answer. Clearly, the ions (e.g., the positive core of the chlorine ion and the negative electron cloud around it) can be electronically polarized. Since this involves the displacement of electrons, it can occur at a much higher frequency than atomic polarization. Thus at low frequencies one could expect both ionic and electronic polarization in NaCl, while at very high frequencies only electronic polarization would occur.

EXAMPLE 30.3.3

Would NaCl have a higher dielectric constant under a static electric field or under a very high frequency alternating field?

Answer. The static field, because the polarization then would be due to two mechanisms which are essentially additive.

ORIENTATION POLARIZATION. Molecules such as HCl and H_2O contain permanent dipoles. Thus the hydrogens are positively charged, while the chlorine or oxygen atoms are negatively charged (in all cases with a fraction of an electron charge). At high temperatures in the gas phase such molecules are randomly oriented, and hence the *net* dipole moment of the system is zero. However, when an electric field is applied the dipoles have a tendency to align with the field, as shown in Figure 30.3.1(c).

SPACE CHARGE POLARIZABILITY. It is possible in multiphase materials for a charge to accumulate at the interface between the phases. This is particularly likely to happen when the resistivity of one phase is much higher than the others. This is called **space charge polarizability**. Depending on the frequency and temperature and the material, all four mechanisms may contribute to the polarization and hence to the dielectric constant or the permittivity.

30.4 POLARIZABILITY

INDUCED POLARIZABILITY. The classical mechanics model for induced polarization assumes that the positive nucleus is a point charge Ze and that the electron cloud is a uniformly smeared-out spherical charge $-Ze$ within a sphere of radius R. When the field is applied the positive charge is displaced by d relative to the center of the negatively charged sphere (as a first approximation it is assumed that the electron cloud remains spherical and of uniform density). See Figure 30.4.1. It can be shown that the dipole moment per atom (see Prob-

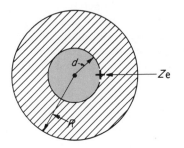

FIGURE 30.4.1. Uniform negative charge cloud of radius R. Hatched section shows electronic charge surrounding the positive charge. Shaded section shows electronic charge not surrounding the positive charge at d. The charge density in the two regions is the same.

lems 30.15–30.17) is

$$\bar{p} = 4\pi\epsilon_0 R^3 \xi'. \tag{30.4.1}$$

Inasmuch as

$$\bar{p} = \frac{P}{N}, \tag{30.4.2}$$

where N is the number of molecules per unit volume, and

$$p = \alpha\xi', \tag{30.4.3}$$

where α is called the polarizability, we have in the present case

$$\alpha_e = 4\pi\epsilon_0 R^3, \tag{30.4.4}$$

where the subscript denotes the *electronic* polarizability.

EXAMPLE 30.4.1

Estimate α_e from (30.4.4) for argon.

Answer. The radius of the atom can be estimated from van der Waals' b (see Problem 17.36) or from viscosity data. In each case $R = 1.43$ Å. Hence

$$\alpha_e = 4\pi(8.85 \times 10^{-12})(1.43 \times 10^{-10})^3$$
$$= 3.2 \times 10^{-40} \text{ F-m}^2.$$

Some measured values are given in Table 30.4.1. These quantities are essentially *independent* of temperature. The measured value is not far from our calculated value. The exact calculation is a difficult quantum mechanical problem.

IONIC POLARIZABILITY. Consider an ionic molecule in a gas for which, in the absence of field, the singly charged ions have a separation r_0 but in the presence of an electric field $r = r_0 + d$, as shown in Figure 30.4.2.

Table 30.4.1. MEASURED ELECTRONIC
POLARIZABILITIES*

Gas	10^{-40} F-m^2
He	0.23
Ne	0.4
Ar	1.83
Kr	2.3
Xe	3.7

* After C. A. Wert and R. M. Thomson, *Physics of Solids*,
McGraw-Hill, New York (1964).

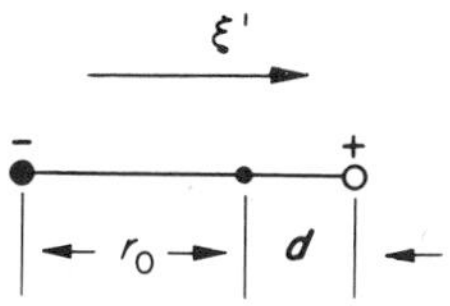

FIGURE 30.4.2. Ionic
polarization of
an ionic
molecule such
as a single CsCl
molecule.

We assume there is coulombic attraction between the two ions and a Born repulsive term (see Section 9.2). We have for the interaction potential between the ions,

$$U = -\frac{e^2}{4\pi\epsilon_0 r} + \frac{b}{r^n}. \tag{30.4.5}$$

(We use the rationalized mks system.)

Since $\partial U/\partial r = 0$ for $r = r_0$, we can eliminate b. Then recall that force is related to potential by

$$F = -\frac{\partial U}{\partial r}.$$

This force is positive for a repulsive force and negative for an attractive force. The *magnitude* of the attractive restoring force is therefore

$$F = \frac{e^2}{4\pi\epsilon_0 r_0^2}\left[\frac{1}{(r/r_0)^2} - \frac{1}{(r/r_0)^{n+1}}\right]. \tag{30.4.6}$$

If we write

$$\frac{r}{r_0} = 1 + x,$$

where

$$x = \frac{d}{r_0},$$

and if

$$d \ll r_0,$$

we can write

$$\frac{1}{(1+x)^2} \approx \frac{1}{1+2x} \approx 1 - 2x$$

and

$$\frac{1}{(1 + x)^{n+1}} \approx 1 - (n + 1)x.$$

Hence

$$F = \frac{e^2}{4\pi\epsilon_0 r_0^3}(n - 1)d.$$

Inasmuch as the force owing to the electric field is $e\xi'$ we have

$$\frac{e^2(n - 1)}{4\pi\epsilon_0 r_0^3}\mathbf{d} = e\xi'. \tag{30.4.7}$$

The induced ionic dipole is therefore

$$\mathbf{p} = e\mathbf{d} = 4\pi\epsilon_0 r_0^3\left(\frac{1}{n - 1}\right)\xi'. \tag{30.4.8}$$

Recall from Chapter 9 that $n \approx 10$.

EXAMPLE 30.4.2

Calculate the **ionic polarizability** α_i for a NaCl molecule if $n = 10$ and the bond distance is 2.51 Å.

Answer.

$$\alpha_i = \frac{4\pi \times 8.85 \times 10^{-12} \times (2.51 \times 10^{-10})^3}{9}$$

$$= 1.9 \times 10^{-40} \text{ F-m}^2.$$

For molecules such as HCl, the value of α_i is much less than this. First, as we note later in this section, the net charge on the hydrogen is only $\frac{1}{6}e$ and on the chlorine $-\frac{1}{6}e$. Furthermore, the molecules are randomly oriented in the absence of a field and slightly oriented with a field; this causes a further reduction in α_i. We would expect α_i to be independent of temperature.

For crystalline materials such as NaCl, a good calculation could be made on the basis of the Born theory. We would then have to sum over many interactions, which makes the problem more involved.

ORIENTATION POLARIZABILITY. Suppose we have many permanent dipoles in a gas such as water vapor or hydrogen chloride. Ordinarily, because of thermal energy, these are randomly oriented, so the average dipole moment is zero. Figure 30.4.3 shows a permanent dipole $\mathbf{p}$ of magnitude $p = qa$ in the presence of a field ξ'.

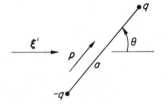

FIGURE 30.4.3. Permanent electric dipole $\mathbf{p}$ in electric field.

EXAMPLE 30.4.3

What is the magnitude of the force tending to stretch the dipole in Figure 30.4.3?

Answer. It is $F_p = q\xi' \cos \theta$, since ξ', resolved along the dipole, is $\xi' \cos \theta$.

In Figure 30.4.4, the force F_p tends to stretch the dipole and cause ionic polarization. We are not concerned with this at the present time. The force F_N

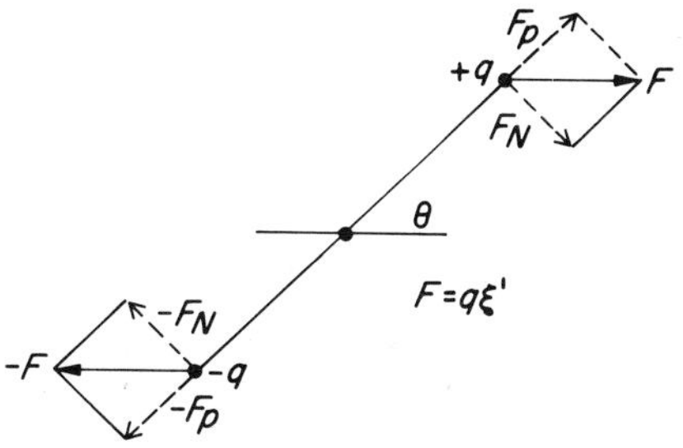

FIGURE 30.4.4. Forces on the charges in Figure 30.4.3. The force F is resolved into components shown by dotted lines.

tends to rotate (orient) the dipole. When $\theta = 0$, $F_N = 0$. In general,

$$F_N = q\xi' \sin \theta. \qquad (30.4.9)$$

The torque on the dipole is

$$\mathbf{T} = \mathbf{p} \times \boldsymbol{\xi}' \qquad (30.4.10)$$

or

$$T = F_N a = p\xi' \sin \theta. \qquad (30.4.11)$$

EXAMPLE 30.4.4

At what orientation does the dipole have the maximum potential energy?

Answer. When the dipole is antiparallel to the field, i.e., when $\theta = 180$ deg. It has the lowest potential energy when $\theta = 0$. More generally the potential energy of orientation is

$$u = -p\xi' \cos \theta. \qquad (30.4.12)$$

Because of thermal agitation, the dipoles are not all at minimum potential energy, i.e., aligned with $\theta = 0$. Rather, if the electrical field is low and the temperature is high, the distribution is nearly (not quite) random.

EXAMPLE 30.4.5

Approximately how large must the electric field be for the orientation energy to be comparable to the thermal energy at room temperature?

Answer. We must have

$$p\xi' \approx kT.$$

If $a = 2$ Å and $q = e$, then if we measure ξ' in volts per centimeter we have $p = 2 \times 10^{-8}$ eV. Since $kT \approx 0.02$ eV at room temperature, we have

$$\xi' \approx 10^6 \text{ V/cm} = 10^8 \text{ v/m}$$

(although we use mks units in this chapter, electron volts seemed to be convenient to use here.)

In the absence of a field the number of dipoles between θ and $\theta + d\theta$ (see Figure 30.4.5) is

$$dN = C\,dA = C\,2\pi \sin\theta\,d\theta, \tag{30.4.13}$$

where C can be evaluated by noting that integration over all θ gives the total number of dipoles, N. Figure 30.4.5 shows the octant of a sphere. Here one fourth of the area element dA is shown by the shaded element. In the presence of an electric field the fraction dN is weighted by the Boltzmann factor so that

$$dN = C2\pi \sin\theta e^{-u(\theta)/kT}\,d\theta. \tag{30.4.14}$$

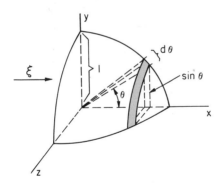

FIGURE 30.4.5. Octant of a unit (radius) sphere.

EXAMPLE 30.4.6

Write the expression for the average value of dipole moment (averaged over all orientations) in the direction of ξ'.

Answer.

$$\bar{p} = \frac{\int (p\cos\theta)\,dN(\theta)}{\int dN(\theta)}. \tag{30.4.15}$$

From (30.4.15) we can write

$$\bar{p} = \frac{\int_0^\pi p\cos\theta \sin\theta e^{p\xi'\cos\theta/kT}\,d\theta}{\int_0^\pi \sin\theta e^{p\xi'\cos\theta/kT}\,d\theta}. \tag{30.4.16}$$

This gives (see Problem 30.19)

$$\bar{p} = p\left[\coth x - \frac{1}{x}\right] = pL(x), \tag{30.4.17}$$

where

$$x = \frac{p\xi'}{kT}. \tag{30.4.18}$$

The function in brackets in (30.4.17) is known as the **Langevin function**, $L(x)$. $L(x)$ has an initial slope of $1/3$ so that $L(x) = x/3$ for $x \ll 1$. See Figure 30.4.6.

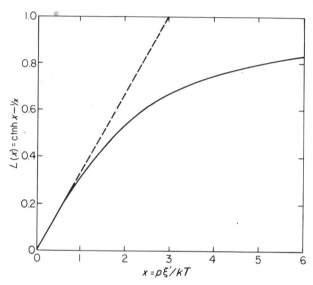

FIGURE 30.4.6. Langevin function.

The Langevin function is also of importance in magnetism and in the theory of rubberlike elasticity in the large strain region where the random chain approximation fails. We have from (30.4.17)

$$\bar{p} = \frac{px}{3} = \frac{p^2\xi'}{3kT}, \qquad x \ll 1.$$

Thus the orientation polarizability is

$$\alpha_0 = \frac{p^2}{3kT} \qquad \text{for } \frac{p\xi'}{kT} \ll 1. \tag{30.4.19}$$

For fields $\xi' < 10^7$ V/m, $p\xi'/kT \ll 1$ at room temperature.

Note that α_0 is *dependent* on temperature.

For a dilute gas, $\xi' = \xi$ and

$$\mathbf{P} = \mathbf{P}_e + \mathbf{P}_i + \mathbf{P}_0 = N\left(\alpha_e + \alpha_i + \frac{p^2}{3kT}\right)\xi. \tag{30.4.20}$$

Thus the measurement of the polarizability (or of the dielectric constant) at different temperatures enables one to measure p (as well as $\alpha_e + \alpha_i$). Examples of dipole moments are shown in Table 30.4.2. The **debye** is a commonly used unit for dipole moment. It equals 3.33×10^{-30} C-m.

Table 30.4.2. ELECTRIC DIPOLE MOMENTS OF SIMPLE
MOLECULES

Molecules	10^{-30} C-m	Debyes
HCl	3.5	1.05
CsCl	35	10.5
H_2O	6.2	1.87
NH_3	4.9	1.47
CO_2	0	0

If the distances a between the H and Cl are known (from electron diffraction, say) and p is known, then the charge q on H and $-q$ on Cl can be calculated. This is how we deduce that $q \approx e/6$ for HCl.

EXAMPLE 30.4.7

What is the shape of the CO_2 molecule? See Table 30.4.2. Does CO_2 have a dipole moment?

Answer. We would ordinarily expect C to have a net positive charge and O to have a negative charge. With this in mind a reasonable expectation is that the molecule is linear, O=C=O. Such a symmetric linear molecule would not have a dipole moment.

30.5   APPROXIMATE EVALUATION OF INTERNAL FIELDS

The internal field ξ' within a material is due to the external field ξ and to all the dipoles in the media surrounding the point. Suppose that we place an imaginary spherical surface as in Figure 30.5.1 at a reasonable distance (several molecule diameters from the point A). Beyond this sphere the dielectric can be treated as a continuum. Suppose we remove the molecules (atoms, ions) from within this sphere while the polarization outside it remains fixed. There would then be charges on the internal surface of the sphere equal to the discontinuity in the normal component of P at that point. It can then be shown (see Problem 30.18) that

$$\xi' = \xi + \frac{P}{3\epsilon_0}. \tag{30.5.1}$$

This is the **Mossotti field** (1850), although it is often called the Lorentz field

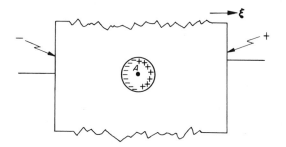

FIGURE 30.5.1. Spherical cavity in dielectric.

(1880). In this derivation the effect of the molecules within the sphere, the so-called local field, ξ_{loc}, was ignored; i.e.,

$$\xi_{loc} = 0. \tag{30.5.2}$$

It can be shown that this is reasonable for simple cubic crystals, bcc crystals, fcc crystals, and NaCl-type crystals. It is also reasonable when the atoms or molecules are neutral and have no permanent dipole or when they are arranged randomly, such as for a gas or (often) for a liquid [see A. J. Dekker, *Solid State Physics*, Prentice-Hall, Englewood Cliffs, N.J. (1960) p. 143]. It is only materials which behave isotropically in an external field that we are concerned with in this chapter.

Combining (30.5.1) and (30.2.1), (30.2.2), (30.2.12), and (30.2.13), we have

$$\frac{N\alpha}{3\epsilon_0} = \left(\frac{\kappa - 1}{\kappa + 2}\right) \tag{30.5.3}$$

for the **polarizability per unit volume.**

If N_0 is Avogadro's number (the number per mole) we have

$$N = \frac{N_0 \rho}{M}, \tag{30.5.4}$$

where ρ is the density and M the molecular weight. This gives for the **molar polarizability,** α_M

$$\alpha_M = \frac{N_0 \alpha}{3\epsilon_0} = \frac{M}{\rho}\left(\frac{\kappa - 1}{\kappa + 2}\right). \tag{30.5.5}$$

This equation, called the **Clausius-Mossotti equation**, provides the connection between atomic polarizability α and dielectric constant. α_M has the units of volume.

The density of liquid argon is of the order of 10^3 times as large as the density of the gas at 1 atm pressure. Yet the measured atomic polarizabilities are 1.83×10^{-40} F-m^2 for the atoms in the gas and 1.86×10^{-40} F-m^2 for the atoms in the liquid. In more strongly bonded liquids, fairly large differences in α might be expected between the liquid and the gas states.

EXAMPLE 30.5.1

The measured dielectric constant of argon gas at 20°C is $\kappa = 1.000517$. What is an approximate form of (30.5.5)?

Answer. Clearly, $\kappa + 2 \approx 3$, so

$$\alpha_M = \frac{M(\kappa - 1)}{\rho} \frac{1}{3} = \frac{M}{\rho}(0.000172).$$

Figure 30.5.2 shows molar polarizabilities for methane and its chlorine derivatives. Recall that from Equation (30.4.20) we expect a straight line, the slope of which is determined by the presence of permanent dipoles, while the intercept is determined by α_e and α_i.

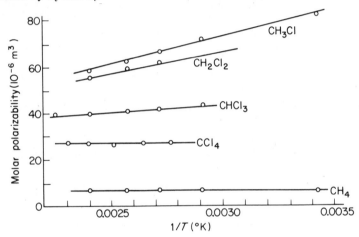

FIGURE 30.5.2. Molar polarizability vs. $1/T$.

30.6   BEHAVIOR OF DIELECTRICS IN VARYING FIELDS

The previous discussion was concerned with the calculation of the dielectric constants in static fields. We have already noted in Chapter 3 that the dielectric constants (of certain materials) vary with frequency. This is also illustrated in Table 30.6.1. At optical frequencies (approximately 10^{15} Hz, see Figure 31.1.1)

Table 30.6.1. DIELECTRIC CONSTANTS AT ZERO (STATIC)
AND "INFINITE" (OPTICAL) FREQUENCIES

Material	κ_0	κ_∞
LiF	9.27	1.90
LiCl	11.05	2.68
NaCl	5.62	2.32
KCl	4.64	2.17

the dielectric constant is related to the refractive index

$$\kappa = n^2. \tag{30.6.1}$$

Here κ_∞ (the dielectric constant at very high frequencies) is due to electronic polarization, which can occur rapidly, while κ_0 (the value in dc fields) is due to the electronic polarization plus the ionic polarization. The ionic polarization involves the movement of ions. We would therefore expect that these ions could not respond at frequencies above about 10^{12} Hz, so that only electronic polarization would contribute at high frequencies.

    In other materials, such as plastics, the slow process can involve the orientation of dipoles. Since the rotation of dipoles takes place in a viscous medium the orientation polarization can depend strongly on frequency and temperature. We can represent the behavior of the material by the model shown in Figure 30.6.1. This is a series analog of the mechanical system of Figure 20.4.2.

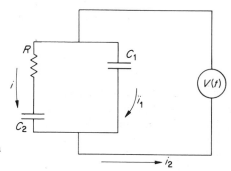

FIGURE 30.6.1. Model of a
        dielectric
        material with an
        applied voltage.

We have the following analogies:

$$\mu \rightarrow R,$$

$$k \rightarrow \frac{1}{C},$$

$$F(t) \rightarrow V(t),$$

and

$$\dot{x}(t) = i(t) = \dot{Q}(t).$$

We have

$$\frac{Q_1}{C_1} = V(t), \tag{30.6.2}$$

where Q_1 is the charge on capacitor C_1,

$$Ri + \frac{Q}{C_2} = V(t),$$

and Q is the charge on capacitor C_2. We can write

$$i = i_2 - i_1$$

and

$$Q = Q_2 - Q_1$$

so we have

$$R(\dot{Q}_2 - \dot{Q}_1) + \frac{1}{C_2}(Q_2 - Q_1) = V(t). \tag{30.6.3}$$

We can eliminate Q_1 and $\dot{Q}_1$ from (30.6.3) by using (30.6.2). We then have

$$R\dot{Q}_2 + \frac{1}{C_2}Q_2 = RC_1\dot{V} + \left(\frac{C_1}{C_2} + 1\right)V. \tag{30.6.4}$$

This model represents a **standard linear dielectric**. If a voltage $V = V_0 \cos \omega t$ is applied, then we could seek a solution of the form

$$Q_2 = A \cos \omega t + B \sin \omega t, \tag{30.6.5}$$

which can also be placed in the form

$$Q_2 = \sqrt{A^2 + B^2} \cos (\omega t - \delta), \tag{30.6.6}$$

where

$$\tan \delta = \frac{B}{A}. \tag{30.6.7}$$

The $A \cos \omega t$ term is in phase with the voltage so the capacitance will be related to this term, while the $B \sin \omega t$ term is out of phase and the loss angle δ will be related to this term. The power loss will be the product of the cycles per second ν times the energy loss per cycle: Power loss $= \nu \oint V \, dQ_2$.
    It can be shown that

$$\text{power loss} \propto \kappa \tan \delta. \tag{30.6.8}$$

Here the **loss angle** is given by

$$\delta(\omega) \doteq \tan \delta = A_0 \frac{\omega\tau}{1 + (\omega\tau)^2}. \tag{30.6.9}$$

Here

$$A_0 = \frac{C_0 - C_\infty}{\sqrt{C_\infty C_0}}, \tag{30.6.10}$$

where $C_0 = C_1$ and $C_\infty = C_1 + C_2$ and τ, the **relaxation time**, is

$$\tau = \frac{R}{\sqrt{\dfrac{(C_1 + C_2)}{C_1 C_2^2}}}. \tag{30.6.11}$$

We also have

$$C(\omega) = C_\infty - \frac{(C_\infty - C_0)}{1 + (\omega\tau)^2}. \tag{30.6.12}$$

It is of more interest to consider the material property than the behavior of the component. We then have

$$\kappa(\omega) = \kappa_\infty - \frac{\kappa_\infty - \kappa_0}{1 + (\omega\tau)^2}. \tag{30.6.13}$$

Note that δ has a maximum when $\omega\tau = 1$.

Equations (30.6.9) and (30.6.13) are often referred to as **Debye equations**. Sketches of $\kappa(\omega)$ and $\delta(\omega)$ are shown in Figure 30.6.2.

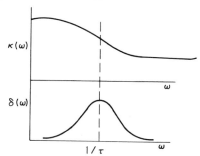

FIGURE 30.6.2. Debye curves.

It is a simple matter to show that A is related to the capacitance (or if we considered material properties, the dielectric constant) and B to the loss angle. Often a **complex dielectric constant** is defined such that

$$\kappa^* = \kappa e^{-i\delta}$$

which for small δ has the form

$$\kappa^* = \kappa - i\kappa\delta. \tag{30.6.14}$$

In such a case a complex displacement D^* and field ξ^* are also used such that

$$D^* = \kappa^* \epsilon_0 \xi^*, \tag{30.6.15}$$

where $\xi^* = \xi_0 e^{i\omega t}$. This leads to the same results as before.

Only in certain simple cases is the process characterized by a single relaxation time, τ.

EXAMPLE 30.6.1

Assume that the dipoles rotate between only two possible orientations and that this orientation involves motion over an activation barrier of height Q. What is the jump frequency?

Answer. From Chapter 18, we have

$$\Gamma = \nu e^{-Q/kT}$$

and

$$\tau = \frac{1}{\Gamma} = \frac{e^{Q/kT}}{\nu}.$$

Sketches of the $\kappa(T)$ and $\delta(T)$ curves are shown in Figure 30.6.3. Note that κ increases as T increases. Clearly, at low T the dipoles will not be able to

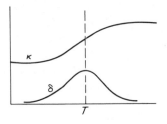

FIGURE 30.6.3. Temperature dependence of the dielectric constant and the loss angle at a given frequency.

respond so κ is entirely due to electronic polarizability. As the temperature is increased, the dipoles will orient. The $\delta(T)$ curve for dielectric loss is identical in nature to that for mechanical loss (see Figure 20.5.2).

Consider a water molecule, which has a dipole moment, to be a sphere which rotates in a viscous medium whose viscosity coefficient is η. A student of hydrodynamics could show that the torque $\mathbf{T}$ necessary to keep the sphere of radius a rotating at an angular velocity $\dot{\theta}$ is

$$T = K_v \dot{\theta},$$

where

$$K_v = 8\pi\eta a^3.$$

In the present case the torque is provided by the electric field acting on the dipole. Debye has shown that

$$\tau = \frac{K_v}{2kT} = \Omega \frac{3\eta}{kT}, \tag{30.6.16}$$

where Ω is the volume per molecule.

EXAMPLE 30.6.2

Estimate τ for water at 20°C.

Answer. From Table 2.8.1 we note that $\eta = 10^{-2}$ P. Water has a molar volume of 18 cm³/mole and Avogadro's number is 6×10^{23} so $\Omega = 3 \times 10^{-23}$ cm³. So

$$\tau = \frac{9 \times 10^{-23} \times 10^{-2}}{1.38 \times 10^{-16} \times 293}$$
$$= 2.2 \times 10^{-11} \text{ sec.}$$

(Here cgs units were used.)

This estimate is not far from the experimental value of 2×10^{-10} sec.

The structural changes which accompany melting of materials of single polar compounds or the change from the glasslike state to the rubberlike state in polymers are associated with drastic changes in the dielectric constant and the

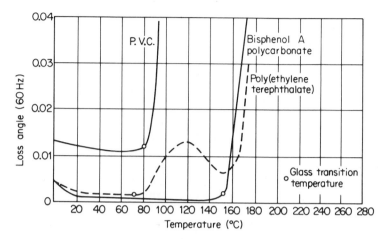

FIGURE 30.6.4. Loss angle vs. temperature for different polymers [After J. A. Brydson, *Plastic Materials*, D. Van Nostrand Co., Inc. Princeton, N. J. (1966)].

loss angle. We have noted analogous behavior for the mechanical loss angle in Section 20.7. The electrical behavior is illustrated in Figure 30.6.4.

30.7 THE CLASSICAL THEORY OF ELECTRONIC POLARIZATION AND RESONANCE ABSORPTION

Consider a charged particle moving in a harmonic potential well. (This means the potential energy is of the form $kx^2/2$ and the force is $-kx$, where x is the displacement from equilibrium.) Its equation of motion is

$$m\ddot{x} + \mu\dot{x} + kx = e\xi_0 \cos \omega t. \tag{30.7.1}$$

The $\mu\dot{x}$ term is a damping term which results from the fact that according to classical mechanics the particle emits radiation as a result of its acceleration. The solution of an analogous problem is given in Problem 2.43. If we define $mr = \mu$ and $m\omega_0^2 = k$ and recall that the dipole moment is ex, then we can solve (30.7.1) by analogy with Problem 2.43. Then it is easy to show that for the case where N is the number of dipoles per unit volume,

$$\kappa = 1 + \frac{Ne^2}{\epsilon_0 m} \frac{\omega_0^2 - \omega^2}{(\omega_0^2 - \omega^2)^2 + (r\omega)^2} \tag{30.7.2}$$

and

$$\delta = \frac{Ne^2}{\epsilon_0 m} \frac{r\omega}{(\omega_0^2 - \omega^2)^2 + (r\omega)^2}. \tag{30.7.3}$$

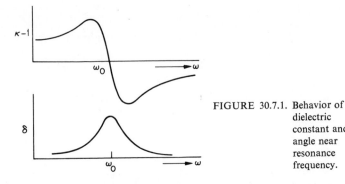

FIGURE 30.7.1. Behavior of dielectric constant and loss angle near resonance frequency.

These expressions are illustrated in Figure 30.7.1. Note that near to the resonance frequency, ω_0, κ varies strongly with ω; this is called **dispersion**; absorption of radiation is occurring.

Figure 30.7.2 summarizes the effect of frequency on polarization.

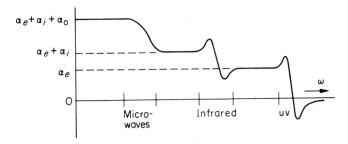

FIGURE 30.7.2. Polarizability as a function of frequency for a dipolar material with one resonance frequency. Electronic, ionic and orientation polarizabilities are involved.

30.8   DIELECTRIC STRENGTH

Dielectric breakdown in a gas can be visualized as follows. Suppose an atom becomes ionized. Both the ion and the electron are subject to the electric fields and are accelerated. These can reach sufficient energies such that upon collision with atoms, more ionization occurs. Eventually an avalanche of charged particles forms. Current then easily passes through the material. A detailed discussion of the breakdown process is given in A. von Hippel, *Dielectric Materials and Applications*, Wiley, New York (1954) Chapter III.

Breakdown in solid insulators is similar in form but different in detail. Here, breakdown begins with the appearance of a number of electrons in the conduction band. Note that because of impurity levels, even the best of insulators could be expected to have some electrons in the conduction band at room temperature. These conduction electrons attain high velocities at high fields and, by collisions, can transfer part of their energy to valence electrons which are excited to the conduction band. Above a certain field (which is often very sensitive to impurities in the material), an avalanche of electrons is produced in the conduction band. Because of the increased conductivity, due to the increased number of carriers, the current increases rapidly and the dielectric may melt, burn or vaporize locally due to joule heating. A plastic, such as polyethylene, may have a line of material decomposed to carbon which acts as a conducting path even when the field is decreased below the previous breakdown field.

EXAMPLE 30.8.1

Sketch the P vs. ξ curve for a dielectric. *Hint:* An analog would be the stress-strain curve for glass.

Answer.

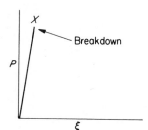

FIGURE E30.8.1.

30.9   OTHER DIELECTRIC PROPERTIES

ELECTROSTRICTION.   An electrical field produces rotation of permanent dipoles or changes in electron configurations. These changes can be associated with dimensional changes in the solid. This is called **electrostriction**. It is a small but measurable effect.

EXAMPLE 30.9.1

A solid exhibits isotropic electrostriction; i.e., the strain is a function of the field, $\epsilon = \epsilon(\xi)$. Why must this be an even function?

Answer. Replacing ξ by $-\xi$ (or instead rotating the sample by 180 deg) does not change the magnitude of the property. It must therefore be an even function of the field.

Experimentally, it is found that the strain is given by

$$\epsilon = A\xi^2. \tag{30.9.1}$$

For glass $A \approx 10^{-17}$ m²/V². The effect is very small. There is no inverse electrostrictive effect.

PIEZOELECTRICITY.   In Section 10.7 we noted how the application of stress to a material not having a center of symmetry would cause polarization. This is piezoelectricity, which was also discussed in Section 3.6; it exhibits an inverse effect. This effect, while small, is much larger than the electrostrictive effect.

FERROELECTRICITY.   In Section 12.9 we noted that when $BaTiO_3$ was cooled below a certain temperature (120°C), it underwent a phase transition from a cubic to a tetragonal form, which was permanently polarized. The behavior of such ferroelectric materials was discussed in Section 3.5. Note that each

No field applied   +2000 volts/cm

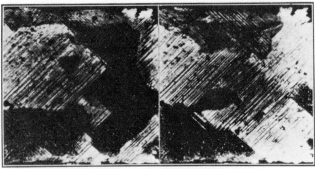

Field removed   −2000 volts/cm

FIGURE 30.9.1. Domains in $BaTiO_3$. [After P.W. Forsbergh, Jr., *Phys. Rev.* **76**, 1187 (1949)].

tetragonal region would have its own characteristic polarization in the absence of a field; such regions are called **domains**. The c axis of each domain corresponds to a cube edge of the original crystal(s). The index of refraction of a tetragonal crystal along the c axis is different than that along other directions (this is discussed further in Chapter 31). Hence, it is possible to send polarized light through a specimen and to view the emerging light through an analyzer. The domains then show up light or dark as in Figure 30.9.1. The transformed crystal might have no net polarization because of the random orientation of the domains. If the transformation is carried out in the presence of a field, then a net polarization exists when the field is removed.

Figure 30.9.2 shows the spontaneous saturation polarization for $BaTiO_3$.

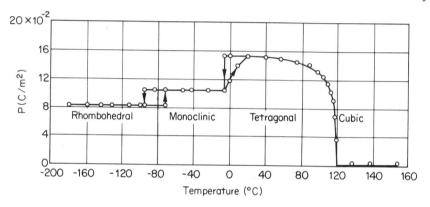

FIGURE 30.9.2. Spontaneous polarization referred to the cube edge of cubic BaTiO. [After J. W. Merz, *Physical Review*, **76**, 221 (1949).] The specimens for which this data applies did not transform completely as shown by more recent results for the cubic to tetragonal transformation; see J. W. Merz, *Physical Review*, **91**, 513 (1953) which indicates that the slope near 120°C is infinite and that the other P values are greater by about 1.6.

The direction of spontaneous polarization is along the body diagonal below $-80°C$, the face diagonal below $0°C$, and the edge below $120°C$.

Compare the $P_S(T)$ curve of Figure 30.9.2 with the $\mathscr{L}(T)$ curve of Figure 17.9.2, which is characteristic of long-range order. Such ordering phenomena are characterized in the present case by a spike in the dielectric constant at T_C. More generally, we can write

$$\kappa = \frac{\text{constant}}{T - T_c}. \tag{30.9.2}$$

This type of behavior is shown in Figure 30.9.3.

The theory of ferroelectricity is discussed by Kittel (see the References).

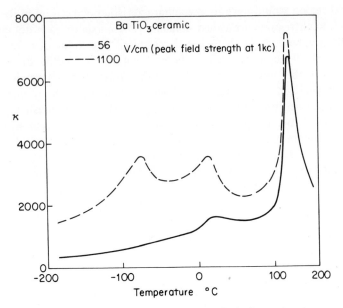

FIGURE 30.9.3. Dielectric constant of barium titanate ceramic as function of temperature. (Measurements of W. B. Westphal, Laboratory for Insulation Research, M.I.T.)

Note that if the adjacent dipoles were antiparallel there would be no resultant polarization, although there may be changes in dielectric constant. Such materials exist and are known as **antiferroelectrics**.

REFERENCES

von Hippel, A. R., ed., *Dielectric Materials and Applications*, Wiley, New York (1954).

von Hippel, A. R., *Dielectrics and Waves*, Wiley, New York (1954).

Smyth, C. P., *Dielectric Behavior and Structure*, McGraw-Hill, New York (1955).

Kittel, C., *Introduction to Solid State Physics*, Wiley, New York (1968). Chapter 12 is concerned with dielectric properties and Chapter 13 with ferroelectricity.

Mason, W. P., *Piezoelectric Crystals*, Van Nostrand, Princeton, N.J. (1950).

Jona, F., and Shirane, G., *Ferroelectric Crystals*, Pergamon Press, New York (1962).

PROBLEMS

30.1. The static dielectric constant of nitrobenzene jumps from a value of 3 to 35 at the melting point. Explain.

30.2. Sketch $\kappa(T)$ for nitrobenzene. See Problem 30.1.

30.3. H_2S freezes at 187.7°K. However, there is jump in the static dielectric constant from 3 to 20 as the temperature is increased through 103.5°K. Explain.

30.4. NaCl has a dielectric constant which varies with the frequency. Explain.

30.5. Sketch the electric field lines around a dipole.

30.6. Sketch Figure 30.7.2. Sketch directly below this $\delta(\omega)$.

30.7. Given for SO_2 gas at 1 atm:

κ	$T,\ °K$
1.009918	267.6
1.008120	297.2
1.005477	336.9
1.003911	443.8

Calculate **p** for an SO_2 molecule.

30.8. Using a lattice constant of 4 Å for $BaTiO_3$, calculate P_S (coulombs per square meter) based on the data given in Section 12.9. The experimental value is 0.26 C/m².

30.9. The resonant frequency of a certain cut of quartz crystals is

$$v = \frac{1}{2t}\sqrt{\frac{c}{\rho}}.$$

Here t is the thickness, c is an elastic stiffness, and ρ is the density. Use $c = 6.7 \times 10^{11}$ dynes/cm² and $\rho = 2.65$ g/cm³. If $v = 10$ MHz, what is t?

30.10. An electron is placed 10 Å from an argon atom.
a. Calculate p_e for the argon atom.
b. Calculate the force exerted on the atom. Is it attractive or repulsive?

30.11. Quartz crystals are often used for frequency control. What properties of a dielectric material are important for precise frequency control?

30.12. Give plausible arguments for the existence of the relationship shown in Figure 3.4.4.

30.13. Fused quartz has a low electrical conductivity and a low loss angle. How do these properties change if Na_2O is added (to lower the melting point and increase the formability)?

30.14. Show that the field ξ at a point $\mathbf{r}$ from the center of a dipole $\mathbf{p}$ is

$$\xi = \frac{3(\mathbf{p}\cdot\mathbf{r})\mathbf{r} - r^2\mathbf{p}}{r^5}.$$

30.15. A sphere contains a uniformly distributed negative charge in the space $r_i \leq r \leq R$. If a positive charge is placed in the region $r < r_i$, what is the net force exerted on it?

30.16. In Problem 30.15 the answer is zero. Thus consider a positive point charge Ze and a negatively charged electron cloud of radius R consisting of a uniform charge distribution of total charge $-Ze$. Suppose the application of a field ξ' moves the positive charge a distance d relative to the center of the electron charge which still remains spherical. Show that the fraction of the negative charge with which the nucleus reacts is $-Ze(d^3/R^3)$.

30.17. In Problem 30.16 the force that ξ' exerts on the positive charge is $Ze\xi'$. The negative charge exerts an equal and opposite force on the positive charge.

a. Why?

b. Hence show that

$$\frac{(Ze)^2(d/R)^3}{4\pi\epsilon_0 d^2} = e\xi'.$$

c. Use this result to derive (30.4.4).

30.18. Assume the sphere in Figure 30.5.1 is a unit sphere. Then use Figure 30.4.5 as a reference. The surface charge on a complete ring (one quarter of which is shown shaded in Figure 30.4.5) is the area $2\pi \sin\theta\, d\theta$ times the charge density, which is $P\cos\theta$. The resultant field also involves $(\cos\theta)/(4\pi\epsilon_0 r^2) = (\cos\theta)/(4\pi\epsilon_0.)$ Show that the Mossotti field (owing to the charge on the cavity) is $P/3\epsilon_0$.

30.19. Derive the Langevin function by substituting $y = p\xi\cos\theta/kT$ and $x = p\xi/kT$ in the integrals for $\bar{p}$.

30.20. a. Show that as the voltage is slowly increased across a capacitor and charge dQ is carried at voltage V from one plate to the other that the work done is $V\, dQ$.

b. Hence show that the energy stored is $\int V\, dQ$.

c. Show that the energy stored per unit volume is $\int \xi\, dD$. For a parallel plate capacitor, $C = \kappa\epsilon_0 A/d$.

d. Show that $\int \xi\, dD = (\epsilon_0\xi^2/2) + \int \xi\, dP$. The latter term on the right side is the **energy of polarization**. It is the energy stored in unit volume of the dielectric in excess of that which would be stored in unit volume of vacuum.

30.21. Show that a function suitable for describing equilibrium in experiments carried out at constant pressure, temperature, and electric field is

$$G^{\text{elect}} = U - TS + PV - \xi D.$$

Sophisticated Problems

30.22. A composite dielectric consists of a sheet of dielectric constant κ, zero electrical conductivity, and thickness d placed on top of a sheet of dielectric constant 0 (for simplicity), conductivity σ, and thickness bd. Show, if this fills the space between a parallel plate condenser, that it behaves as if the space were filled with a single material with dielectric constant κ^*:

$$\kappa^* = \frac{(1 + b)}{1 - i\kappa\omega b/\sigma}.$$

The capacitor is subjected to a varying voltage $V = V_0 e^{i\omega t}$. Note that very high κ values are possible (10^5) but not without high loss factors. This is an extreme case of space charge polarization.

30.23. Discuss in a quantitative fashion the theory of ordering in ferroelectrics.

Prologue

The engineering application of the optical behavior of materials involves such diverse subjects as the proper design of house paint and the design of lasers. This chapter begins with a brief review of the *electromagnetic spectrum*. The origin of the *refractive index* of solids is discussed on the basis of the theory of polarization described in Chapter 30. The *anisotropy* of the refractive index is described. The origin of *birefringence* and *photoelastic* behavior is given. The *dichroism of Polaroid* sheet is discussed. The basis of *absorption* of light is described in terms of the energy levels present in the material. We show why metals are opaque while single crystals of insulators are transparent in the visible portion of the spectra. We also see why silicon is transparent in the infrared but not the visible; i.e., we study how the *absorption coefficient* varies with frequency. We discuss the origin of *color centers* in solids. We describe the processes involved in *latent image* formation in photography. Various electro-optical phenomena are discussed including the *electro-optic effect*, the *photochromic effect*, and field effects on *nematic crystals*, along with potential applications. *Exciton* energy levels and exciton *fission* and *fusion* are studied. Some of the most important aspects of *luminescence* are explained. We study *spontaneous emission* and *stimulated emission* of light. Finally, we consider the possibility of *population inversion*, which forms the basis of *l*ight *a*mplification by *s*timulated *e*mission of *r*adiation (the *laser*).

31

OPTICAL PROPERTIES

31.1   INTRODUCTION

The optical properties of matter fall into two areas: those describable by the classical wave theory of light and those in which the photon behaves as a quantum of energy. Many of the properties of light such as reflection and refraction can be described by the classical wave theory, while other phenomena such as the atomic *origin* of the refractive index, absorption, color, photographic processes, luminescence, and laser behavior depend on the quantum states of electrons.

For the discussion which follows it is useful to have a knowledge of the spectrum of electromagnetic waves in general (Figure 31.1.1) and the visible spectrum in particular (Figure 31.1.2).

31.2   REFRACTION

The **index of refraction** n of a material is

$$n = \frac{c}{v},$$ (31.2.1)

where c is the velocity of light in vacuum and v is the velocity of light in the material. In an isotropic media (for mks units)

$$v = (\epsilon\mu)^{-1/2},$$ (31.2.2)

where ϵ is the permittivity and μ is the permeability. For the purposes of the present discussion we shall consider only nonmagnetic materials (diamagnetic and paramagnetic materials are for the present purposes nonmagnetic); hence

837

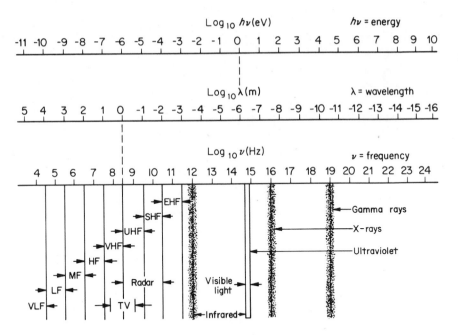

FIGURE 31.1.1. Spectrum of electromagnetic waves. The abbreviations VLF, LF, MF, etc., mean, respectively, very low frequency, low frequency, medium frequency, high frequency, very high frequency, ultrahigh frequency, superhigh frequency, and extremely high frequency. The limits indicated by the shaded region are approximate. The quality $h\nu$ represents the photon energy (h is Planck's constant and ν the frequency). Here λ is the wavelength.

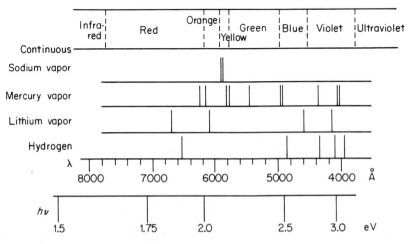

FIGURE 31.1.2. Visible spectrum.

$\mu \doteq \mu_0$. Therefore

$$n^2 \doteq \frac{\epsilon}{\epsilon_0} = \kappa, \tag{31.2.3}$$

where ϵ_0 is the permittivity of empty space and κ is the relative permittivity, or dielectric constant.

In Section 30.6 we noted that κ was a function of frequency (or wavelength). The refractive index and the velocity of light are therefore functions of frequency. This phenomenon is known as **dispersion**. The separation of the rays of the sun into a colored spectrum by passage through a glass prism is an example of dispersion. This phenomenon was first noted by Newton. Newton's prism arrangement was the first spectroscope.

EXAMPLE 31.2.1

Suppose that you had available various elementary data such as melting points and densities but did not have refractive indices or dielectric constants. How would you proceed to choose those dielectrics which had high refractive indices?

Answer. First, we would not concern ourselves with permanent dipoles. We learned in Chapter 30 that these could not respond to very high-frequency fields. Visible radiation is an electromagnetic wave with a frequency near 10^{15} Hz.

Second, an inspection of Table 30.4.1 might be helpful. We note that the electronic polarizabilities of the inert atoms are about equal to $Z \times 10^{-41}$ F-m^2, where Z is the atomic number (2, 10,18, 36, and 54 for the inert gases). Elementary theory [Equation (30.4.4)] suggests that $\alpha_\epsilon \approx Z V_a$, where V_a is the volume of the atom. Hence the polarizability per unit volume of an inert gas atom (or, we might guess, an ion with the inert gas configuration) is proportional to the number of electrons in the atom or ion. We would therefore expect that compounds formed from atoms of high atomic number would have a high dielectric constant or refractive index.

Table 31.2.1 illustrates the fact that compounds with high atomic numbers and hence, roughly, compounds of high density, have a high refractive index.

Table 31.2.1. DEPENDENCE OF REFRACTIVE INDEX
ON TOTAL ELECTRON DENSITY

Compound	n	n^2	Electrons Molecule	Molar Volume, cm^3
LiF	1.39	1.9	12	9.84
PbS	3.91	15.2	98	31.5

EXAMPLE 31.2.2

The index of refraction of silica glasses can be modified by the addition of other oxides to the SiO_2. What would be the effect of adding BaO and PbO?

Answer. This would involve the substitution of Ba ($Z = 56$) or Pb ($Z = 82$) for Si ($Z = 14$). It would result in a higher refractive index.

Table 31.2.2 gives some other values of refractive index.

Table 31.2.2. REFRACTIVE INDEX

Material	n
Silica glass	1.46
Flint glass	1.7
NaF	1.33
CaF_2	1.43
MgF_2	1.38
Al_2O_3	1.76
MgO	1.74
PbO	2.61

ANISOTROPY.   The permittivity (dielectric constant or electron susceptibility) in a strain-free carefully annealed glass is independent of direction. The same applies to cubic crystals. In general dielectric behavior involves a relationship between a displacement vector which has components D_i, $i = 1$, 2, 3, and electric field ξ_j, $j = 1, 2, 3$, of the form

$$D_i = \sum_{j=i}^{3} \epsilon_{ij}\xi_j \tag{31.2.4}$$

or in the Einstein notation (Chapter 7)

$$D_i = \epsilon_{ij}\xi_j. \tag{31.2.5}$$

The array of nine components is a second rank Cartesian tensor. It is a simple exercise in thermodynamics to show that the tensor is symmetric; i.e., $\epsilon_{ij} = \epsilon_{ji}$.

Inasmuch as ϵ_{ij} is a symmetric second rank tensor (similar to the state of stress without body couples, discussed in Section 7.3) it transforms under rotation of axes in the same manner as σ_{ij}. Thermal expansion is also a symmetric second rank matter tensor like ϵ_{ij}. It is shown in Section 10.8 that α_{ij} for a tetragonal crystal has the form of (10.8.2). Hence ϵ_{ij} has the form

$$\begin{matrix} \epsilon_a & 0 & 0 \\ 0 & \epsilon_a & 0 \\ 0 & 0 & \epsilon_c. \end{matrix} \tag{31.2.6}$$

This means that the permittivity parallel to the c axis is ϵ_c, while the permittivity parallel to either the *a* axis or *b* axis is ϵ_a (see Table 10.3.2). Also by analogy with Equation (10.8.3), the permittivity in a direction inclined at θ to the *c* axis is

$$\epsilon = \epsilon_a \sin^2 \theta + \epsilon_c \cos^2 \theta. \tag{31.2.7}$$

A similar relationship applies for the dielectric tensor and the polarizability or susceptibility tensor.

It is clear that for the tetragonal crystal the velocity of light parallel to the *c* axis is different than perpendicular to the *c* axis (it is the same in all directions perpendicular to the *c* axis). It should be emphasized that we arrived at this solution not by any detailed examination of the electronic structure of the ions or atoms or molecules making up the crystal. Rather, we had only to know that the crystal exhibited tetragonal symmetry to know that in general $\epsilon(\theta)$ is given by (31.2.7). However, this pure mathematics approach does not tell us the magnitudes of ϵ_a and ϵ_c; that requires a detailed study of the electronic structure. Crystals belonging to the hexagonal and trigonal systems also exhibit the behavior shown by (31.2.6) and (31.2.7). Cubic crystals have $\epsilon_a = \epsilon_c$.

EXAMPLE 31.2.3

How does the velocity of light vary with direction in a cubic crystal?

Answer. From (31.2.7) we have for $\epsilon_c = \epsilon_a$

$$\epsilon(\theta) = \epsilon_a(\sin^2 \theta + \cos^2 \theta) = \epsilon_a.$$

Hence $\epsilon(\theta)$ does not vary with θ. Hence the velocity of light does not vary with direction. Cubic crystals exhibit isotropic behavior with respect to second rank tensor properties.

Orthorhombic crystals have (ϵ_{ij}) given by

$$\begin{matrix} \epsilon_a & 0 & 0 \\ 0 & \epsilon_b & 0 \\ 0 & 0 & \epsilon_c. \end{matrix} \tag{31.2.8}$$

Light has a different velocity along each of the crystal axes **a**, **b**, and **c**.

For triclinic crystals which possess only onefold symmetry axes, the permittivity tensor does not simplify. However, there is a set of orthogonal axes x_1^P, x_2^P, x_3^P for which we have

$$\begin{matrix} \epsilon_{11}^P & 0 & 0 \\ 0 & \epsilon_{22}^P & 0 \\ 0 & 0 & \epsilon_{33}^P; \end{matrix} \tag{31.2.9}$$

i.e., all the off diagonal terms vanish. These axes are called the **principal axes**. The orientation of these principal axes relative to the crystal axes is different for each triclinic crystal. This is, of course, not the case for the other crystals discussed here. The three components of the diagonalized array are the principal permittivity coefficients. Crystals in which these differ (all except cubic) are called **birefringent**.

EXAMPLE 31.2.4

The crystal structure of $CaCO_3$ can be visualized as follows: Start with a NaCl-type crystal with Ca^{2+} instead of Na^+ and CO_3^{2-} instead of Cl^- ions. The latter ions are planar with their planes on the (111) plane. The crystal is then "compressed" along the [111] direction. To which crystal system does a calcite crystal formed in this way belong? Is it birefringent?

Answer. The crystal structure is rhombohedral. Inasmuch as it is noncubic it is birefringent.

EXAMPLE 31.2.5

Suppose that a perfect single crystal of NaCl was squeezed elastically along the [100] direction. Would it be birefringent in the elastically deformed state?

Answer. The deformed crystal is tetragonal (temporarily), not cubic. Hence it would be birefringent. Stress induces birefringence in other optically isotropic media such as glass and polymers. This is called the **photoelastic effect**. Stress-induced birefringence is the basis of **photoelasticity**. The state of stress in a complicated shape under complicated loading (for example, a steel drive-chain link) can be readily studied by making a transparent plastic model of the link. This is then loaded and the birefringence (which is proportional to the stress) is measured as a function of position in the specimen. Actually the birefringence gives rise to fringes under certain conditions [see A. W. Hendry, *Photoelastic Analysis*, Pergamon, New York (1966)]. Birefringent crystals are **doubly refracting**; i.e., a beam of light is broken into two beams, as shown in Figure 31.2.1, which emerge polarized at right angles to each other. Certain crystals, called **dichroic crystals**, absorb one polarized component much more than the other so the emergent beam (the other is absorbed) is polarized. Polaroid is an example of such a material. There have been several examples of Polaroid sheet invented by Land and his

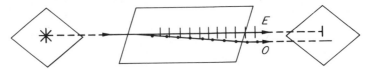

FIGURE 31.2.1. Polarization by a calcite crystal. *O* represents the ordinary ray, and *E* the extraordinary ray. The electrical vectors of these rays are normal.

coworkers. The first, called J-sheet, was invented by Land (1928) while an undergraduate at Harvard. The common Polaroid of today is H-sheet, invented by Land in 1938. It is made from the polymer polyvinyl alcohol. The polymer molecules are aligned by stretching the plastic to very large extension ratios (factor of 10). During the plastic flow most of the initially random chains become aligned. This sheet is cemented to a rigid cellulose acetate sheet [so the polyvinyl alcohol sheet does not contract]. This composite is dipped in iodine. The I_2 molecules are absorbed *along* the chains, so there are "chains" of I_2 molecules. The iodine chains are strongly absorbing along their axis but transmit freely normal to their axis. This is the basis of the dichroism.

An excellent elementary discussion of polarized light and its applications is found in W. A. Shurcliff and S. S. Ballard, *Polarized Light*, Van Nostrand, Princeton, N.J. (1964).

The refractive indices of crystals are relatively easy quantities to measure. One can obtain considerable information about the symmetry of a crystal from such measurements. Crystallographers use such measurements in the characterization of crystals (see E. E. Wahlstrom, *Optical Crystallography*, Wiley, New York (1943).

31.3   ABSORPTION

In Chapter 8 we noted that the hydrogen atoms possessed specific energy states and so could absorb or, if the electron was in an excited state, emit light in certain quantums of energy (recall that $\Delta E = h\nu$). For hydrogen, the spectrum is fairly simple and the emission or absorption lines are sharp, narrow absorption lines. [The so-called line is not a line but really a peak of intensity versus frequency (or wavelength) which has a small half-width $\Delta\nu$; for a typical line $\Delta\nu/\nu \approx 10^{-6}$.] Although the spectrum for isolated atoms or molecules (such as might occur in a gas at low pressures) may be extremely complicated, the absorption lines are nonetheless still sharp and narrow. However, as the pressure of the gas is increased so that the interactions between the atoms (or molecules) increase, these lines broaden. In liquids and solids, when the ion, atom, or molecules are condensed into a dense arrangement, absorption lines become very broad **(absorption band)**. We have already noted in Chapter 27 the effect of condensation on energy levels. This is clearly illustrated for sodium in Figure 27.8.1. Here we see that $3s$ and $2p$ electrons have specific energy levels in the isolated atom, but when N atoms are brought together to form the condensed state (the bcc crystal, in this case) the N $3s$ levels spread into a band of levels. Within the band the levels are distinct but very close together; thus absorption to adjacent energy levels would involve overlapping lines.

EXAMPLE 31.3.1

Metals such as sodium are not transparent to visible radiation. Why?

Answer. Recall that the 3*s* band is half-full of electrons (which essentially fill the lower half of the band). The unfilled band of energies (from the Fermi surface to the top of the band) provides an enormous array of possible energy states to which electrons from the filled portion of the band can be excited. The energy of visible radiation is about 1.6–3.2 eV. Since the 3*s* band is several electron volts wide, all visible radiation can be absorbed.

EXAMPLE 31.3.2

Silicon is not transparent in the visible but is transparent in the infrared. Why?

Answer. Silicon is a semiconductor with a filled valence band and an empty conduction band at absolute zero. Thus an electron at the top of the valence band must absorb an energy at least equal to the energy gap since there are no intermediate states. The energy gap is 1.1 eV (see Table 28.2.1). Light with higher energy will be absorbed, while light with lower energy (infrared) will be transmitted. Could you suggest on the basis of this example a method of measuring the energy gaps of crystals?

Figure 31.3.1 shows the absorption curve for a red glass. Recall that visible radiation extends from about 3800 to 7800 Å (0.38–0.78 μ). The glass under discussion shows strong absorption of blue, particularly green, and also

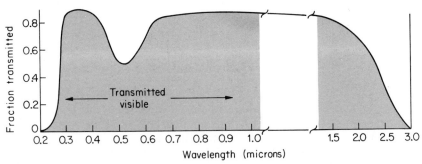

FIGURE 31.3.1. Transmitted light for a red glass.

orange, with only slight absorption of red. Consequently, the red readily passes through. This is the light which we observe.

Absorption of a given wavelength can be described quantitatively by the expression

$$\frac{dI}{I} = -\alpha\,dx; \tag{31.3.1}$$

i.e., the infinitesimal fractional decrease in intensity is proportional to the infinitesimal path traveled. The quantity α is called the **absorption coefficient**. Integration of (31.3.1) gives

$$\frac{I(x)}{I_0} = e^{-\alpha x}, \qquad\qquad (31.3.2)$$

where I_0 is the intensity of the incident (at normal incidence) radiation, x is the length of the path traveled, and $I(x)$ is the intensity at x. The ratio I/I_0 is called the **transmittance**.

EXAMPLE 31.3.3

Why does the glass show such strong absorption in the ultraviolet?

Answer. When the energies get large enough (wavelengths small enough) there is a continuum of energy levels available. Note, e.g., the nature of the overlapping higher energy bands in sodium (Figure 27.8.1).

EXAMPLE 31.3.4

Why does the glass of Figure 31.3.1 show such strong absorption in the far infrared?

Answer. The answer, of course, has to be that there are numerous energy levels available with energy differences (for allowable transitions) in the infrared. Recall that atoms vibrate with fundamental frequency up to about the Debye frequency $\nu_D \approx 10^{13}$ Hz (Table 17.11.1). The energies associated with vibrations are

$$E = (n + \tfrac{1}{2})h\nu. \qquad\qquad (31.3.3)$$

The quantum change of energy is

$$\Delta E = h\nu. \qquad\qquad (31.3.4)$$

For $\nu = 10^{13}$ Hz, this energy is about 0.05 eV. It clearly falls in the infrared.

ALKALI HALIDES.   In the visible range pure NaCl crystals are transparent. However, NaCl has an absorption peak at 61.1 μ ($1\mu = 10^{-4}$ cm) associated with the motion of ions of opposite charge toward each other. The absorption coefficient at this wavelength is 3.5×10^4 cm^{-1}.

EXAMPLE 31.3.5

Estimate the thickness of the sample to use in an experiment designed to measure α of NaCl at 61.1 μ.

Answer. It would be reasonable to have a transmittance of about 0.50. Hence from (31.3.2)

$$L = \frac{-\ln 0.50}{3.5 \times 10^4} \approx 0.20 \times 10^{-4} \text{ cm}$$
$$= 0.20 \, \mu.$$

A very thin sample is required. Note that a sample of 1 cm thickness would have essentially a zero transmittance:

$$\frac{I}{I_0} = e^{-3.5 \times 10^4} = 10^{-(3.5/2.3) \times 10^4} \doteq 0.$$

Figure 31.3.2 shows the transmittance in the infrared for NaCl. If the fraction absorbed were plotted against frequencies, an absorption peak similar to that of Figure 30.7.1 would be obtained.

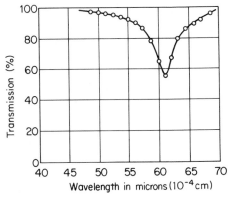

FIGURE 31.3.2. Transmittance versus wavelength for NaCl. [After R. B. Barnes, *Zeitschrift für Physik*, **75**, 723(1932).]

COLOR CENTERS. Alkali halide crystals can be colored in the visible range by the introduction of suitable chemical impurities or vacancy concentrations. For example, it is possible to introduce transition elements which have excited energy levels separated from the ground state by a quantum of energy in the visible range. This technique is also used to color glass as well as various crystalline compounds. The red color of ruby is due to the presence of chromium ions in ordinary transparent aluminum oxide crystals. This point will be discussed later in connection with lasers. When sodium chloride is heated at high temperature in the presence of sodium vapor and then quenched to room temperature the crystal has a yellow color.

When the crystal is heated in the presence of sodium vapor, sodium atoms are deposited on the surface. Chlorine ions migrate from the interior to combine with the sodium:

$$Na + Cl^- \longrightarrow NaCl + e^-.$$

This leaves an excess of chlorine ion vacancies (net positive charge) within the crystal and electrons at the surface. The latter move to the vacancies.

The yellow color is due to the energy levels of the electrons attached to excess chlorine ion vacancies. When crystals are colored by either of the above methods we say that **color centers** are present.

EXAMPLE 31.3.6

What is the charge associated with a chlorine ion vacancy?

Answer. It is a positive charge. The negatively charged chlorine ion has been removed from the lattice site.

The electrons generated at the surface combine with positive-charged vacant sites to form a hydrogen-like "atom." This is a simple model for the *F*-center. (This color center is called an *F*-center after the German word for color, *Farbe*. Recall that a similar model was used to account for the acceptor and donor impurity levels in germanium.) The energy of such hydrogen-like states when the electron moves in a media of dielectric constant κ is

$$E = \frac{-13.6}{\kappa^2 n^2} \text{ (eV)},$$

where $n = 1, 2, 3, \ldots$ [See Section 8.3. Note that this case differs from the hydrogen atom in that the potential is $-e^2/\kappa r$ (in cgs units) instead of $-e^2/r$.]

EXAMPLE 31.3.7

Estimate the *F* center energy for NaCl.

Answer. The dielectric constant is, from Table 3.4.1, 2.25. Hence for total ionization

$$\Delta E = \frac{13.6}{(2.25)^2}\left(\frac{1}{1^2} - \frac{1}{\infty^2}\right) = 2.7 \text{ eV}.$$

The experimental value is 2.7 eV. The agreement should be regarded as partially fortuitous.

The energies at the center of the absorption peak for some *F* centers are shown in Table 31.3.1. *F* centers can also be produced by X-ray and γ-ray

Table 31.3.1. *F*-CENTER ENERGIES

Alkali Halide	Energy, eV
NaF	3.6
NaCl	2.7
NaBr	2.3
KF	2.7
KCl	2.2
KBr	2.0

irradiation or neutron and electron bombardment. Illumination with light absorbed in the *F* band ionizes the *F* centers. When they are all fully ionized, light is no longer absorbed; the crystals are then said to be **bleached** and are then transparent. The bleaching process is accompanied by photoconductivity. There are a number of other point-defect-based centers in the alkali halides which we shall not discuss here.

31.4   REFLECTION

The amount of light reflected from a transparent substance depends on the angle of incidence and the refractive index of the substance. The French physicist A. J. Fresnel (1788–1827) showed that for natural light perpendicularly incident from a medium of refractive index n_1 upon a transparent substance of refractive index n_2, the ratio of the reflected intensity I to the incident intensity I_0 is

$$\frac{I}{I_0} = \left(\frac{n_2 - n_1}{n_2 + n_1}\right)^2 = R. \tag{31.4.1}$$

The fraction R is called the **reflectance**. For light in air incident against glass of refractive index 1.5, this gives $I/I_0 = 0.04$. In general R depends on the refractive indices and the angle of incidence. Translucency or total opacity (no transmission) of dielectric materials is due to extensive multiple internal reflection and refraction. This is illustrated in Figure 31.4.1.

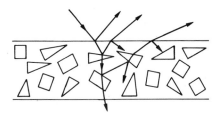

FIGURE 31.4.1. Multiple reflection and refraction.

White house paint, e.g., consists of transparent linseed oil with transparent crystals of TiO_2 (or lead oxide) dispersed in the liquid. (The white "color" is not produced by suspending white particles.) To produce colored paints, colored transparent particles can be used or added to the white paint. Then the light which returns from the interior to the air passes through these tiny filters. Its color is determined by the wavelength which passes through the filters.

Light passing through a fine dispersion obeys the relation

$$I = I_0 e^{-sx}, \tag{31.4.2}$$

where s is the **scattering coefficient**.

31.5 THE PHOTOGRAPHIC PROCESS

Tiny crystals of AgBr are suspended in gelatin which is coated on glass plates or cellulose acetate sheets. The absorption of light by the AgBr causes partial decomposition of the AgBr. Further decomposition of the light-sensitized region and a resultant growth of silver metal crystals take place during development of the film in an alkaline solution of hydroquinone. Following this, the remaining undissolved AgBr is removed by the bath of sodium thiosulfate (fixing bath). No further reaction is then possible.

The development of photography depended on the unique combination of properties of silver bromide. (Not one suitable substitute offering both high speed and high resolution has yet been found. Because of the high demand and scarcity of silver, the development of a cheap substitute is badly needed.)

According to the model of N. F. Mott and R. W. Gurney for the development of images in AgBr, a photon (of energy greater than the energy gap) is absorbed, creating an electron-hole pair. The hole is in the valence band, and the electron is in the conduction band. The hole combines with a bromine ion at the surface to form a neutral atom on the surface,

$$Br^- + hole \longrightarrow Br.$$

The electron is trapped at the surface by Ag_2S or some other impurity. An interstitial Ag^+ ion is attracted to the trapped electrons at the surface. (Note that this involves diffusion at room temperature!) Then the Ag^+ interstitial and the electron combine at the surface to form a silver atom adsorbed at the surface. The process is repeated. Then there is a combination of bromine atoms to yield bromine gas, Br_2, which desorbs. Similarly, the silver atoms form a tiny nucleus of silver. This nucleus grows into a tiny colloidal precipitate called a **latent image**.

The growth of the silver particle continues in the presence of the developer until the particular sensitized AgBr grain in question is completely converted to Ag (and Br_2). Note that grains which were not sensitized by the light remain as AgBr at this point. The fixer removes this AgBr. The high sensitivity of AgBr photographs is due to the fact that only a small fraction of the AgBr needs to be decomposed to form the latent image.

The details of image formation are discussed in C. E. K. Mees, *The Theory of the Photographic Process* (ed. by T. H. James), Macmillan, New York (1966).

31.6 LIQUID CRYSTALS

Some organic crystalline solids change at a certain temperature to liquid crystals and only melt totally to the isotropic liquid state at a higher temperature.

The liquid crystal range may be a few or several hundred degrees Kelvin. Until recently, the range was above room temperature. A **liquid crystal** is a fluid, yet it exhibits some of the optical anisotropy characteristics of noncubic crystalline solids. The fluid viscosity in the liquid crystal state is often comparable to glycerine. When nematic liquid crystal fluid is placed in an electric field, it changes from transparent to opaque in a fraction of a second (often 10^{-3} sec). This is called **nematic behavior**. The reason for such behavior is shown in Figure 31.6.1.

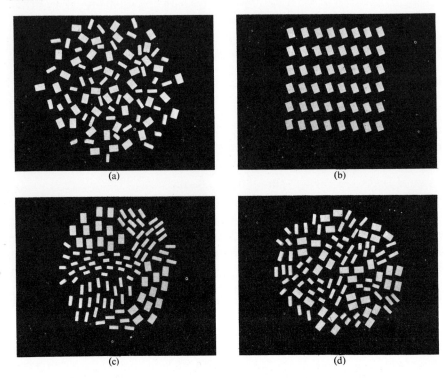

FIGURE 31.6.1. (a) Molecular arrangement in true liquid. (b) Molecular arrangement in true crystal. (c) Molecular arrangement in liquid crystal; well-defined domains appear upon microscopic examination at × 100. (d) Smaller domains present in electric field.

Liquid crystals have considerable potential for graphical displays, including flat television screens and telephone screens for visual communication. Their study is an active area of research.

31.7   SOME ELECTRO-OPTIC PHENOMENA

ELECTRO-OPTIC EFFECT.   The Scottish physicist John Kerr noted that a transparent dielectric such as glass or turpentine becomes doubly refracting when located in a strong electric field. This is the **electrooptic effect**.

Such a specimen, placed in a beam of polarized light between a polarizer and analyzer, will cut off the light when the field is established. The field may be alternated at 10^8 Hz. This effect has applications as a shutter for high-speed photography (e.g., in studying explosive forming of materials); the **Kerr cell** was used by W. C. Andersen in his measurements of the velocity of light [*J. Optical Society of America* **31**, 187 (1941)].

PHOTOCHROMIC EFFECT.   Many materials become colored or darkened under radiation by high-energy light or by electrons (or other particles). If this change can be reversed by heat or radiation, the materials are said to be photochromic. Specific images can be created by bombardment of specific areas. The resultant images can be viewed directly or can be projected with light which does not cause bleaching. A sheet of the material is therefore an electronic image storage device.

The alkali halide crystals were examples of materials which show this behavior, although there are many examples. Photochromic materials are less sensitive by a factor of 10^{-7} than silver bromide films, which, of course, are not reversible. At low temperatures and in the dark the photochromic effect is nearly permanent. At room temperature, the darkening decreases with time. The effect is not completely reversible, and repeated darkening and bleaching leads to deterioration.

31.8   EXCITONS

Suppose you have a pure stoichiometric crystal of cuprous oxide. Ordinarily one might expect that there would be no absorption of light whose energy is less than the gap energy. However, a series of absorption bands are found, as shown in Figure 31.8.1, with energies below the energy gap. We recall that a photon with energy equal to the gap energy can, when absorbed by the crystal, create an electron-hole pair which is free to move through the crystal. There is, however, a coulomb attraction between the particles so it is possible to form a bound state between the electron and the hole called the **exciton**.

A simple model is the Wannier model, which assumes that the pair forms a "hydrogen-like molecule." The binding energy of this pair is (cgs units)

$$-\frac{2\pi^2 \mu e^4}{h^2 \kappa^2 n^2}, \qquad n = 1, 2, \ldots,$$

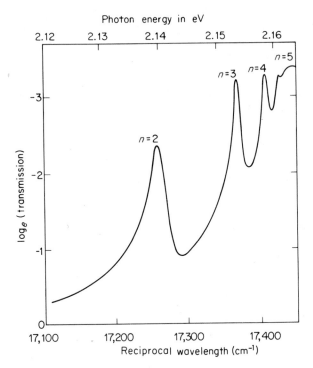

FIGURE 31.8.1. Exciton absorption peak in Cu_2O. [After P. W. Baumeister, *Physical Review*, **121**, 359 (1961).]

or if the top of the valence band is used as a reference point,

$$E = E_g - \frac{2\pi^2 \mu e^4}{h^2 \kappa^2 n^2}.$$ (31.8.1)

Here κ is the dielectric constant and μ is the reduced mass, where

$$\frac{1}{\mu} = \frac{1}{m_n} + \frac{1}{m_p},$$

and m_n is the effective electron mass and m_p the effective hole mass. When one mass is enormous compared to the other, e.g., electron around a proton, then the reduced mass is very nearly equal to the proton mass. However, in the present case if both m_n and m_p equal the free electron mass m,

$$\mu = 0.5 \, m.$$

Expected exciton levels are shown in Figure 31.8.2.

Note that the exciton can move through a crystal. In so doing it transports energy but *no net* charge. The exciton in Cu_2O is often called the **Wannier exciton** or the weakly bound exciton. There is another exciton, called the **Frenkel exciton**

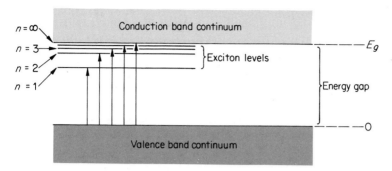

FIGURE 31.8.2. Exciton levels.

or tightly bound exciton, in which the pair is more closely associated, say, on an individual atom. In the alkali halides the excitons are on the halogen ion. Frenkel excitons are also associated with molecules in molecular crystals such as anthracene. In anthracene there is the possibility of generating radiation of 3.14 eV by initially exciting the crystal with 1.8-eV radiation. This is due to the presence of an exciton state 1.8 eV above the ground state and another exciton state 3.14 eV above the ground state. Two of the first excitons combine to yield the latter exciton with the excess energy forming a phonon. This is called **exciton fusion**. The resultant 3.14-eV exciton then may undergo a transition which results in the transmission of a 3.14-eV photon.

In other systems **exciton fission** is possible [see N. Geacintov et al., *Physical Review Letters*, **22**, 110 (1969)]. It is conceivable that exciton fission plays a vital role in basic life processes such as photosynthesis. Exciton systems have a number of interesting potential applications in information storage and transmission as well as in optical devices.

The detailed study of excitons involves a thorough study of quantum mechanics and of the electron states available in atoms, ions, molecules, and condensed phases.

31.9   LUMINESCENCE

Insulating and semiconducting materials are often capable of emitting radiation in excess of their thermal radiation. These are called **luminescent** materials. Before luminescent emission can occur, the excitation must first be created by one of the following ways:

1. By photons—**photoluminescence**.
2. By electron bombardment—**cathodoluminescence**.
3. By an alternating electric field—**electroluminescence**.
4. By certain chemical reactions—**chemiluminescence**.

Luminescent materials are divided into two classes based on the time delay between absorption of the excitation energy and luminescent emission: rapid emission (less than 10^{-8} sec) is called **fluorescence**. Slower emission (taking longer than 10^{-8} sec) is called **phosphorescence**. Most commercial **phosphors** have a delay time of several milliseconds, although materials exist with the delay times of hours. An important area of application of these phenomena is in electronic image storage [see B. Kazan and M. Knoll, *Electronic Image Storage*, Academic Press, New York (1968)].

PHOTOLUMINESCENCE.   Zinc sulfide has an energy gap of 3.7 eV. Various impurities can be introduced [in tiny concentrations (10^{-6})] substitutionally for the zinc atoms. These introduce energy levels just above the valence band called **activator levels** (Cu, Ag, Au) or just below the conduction band called **coactivator levels** (In, Ga). The former can act as acceptors, the latter as donors. All the delay times in luminescent materials are strongly influenced by the nature

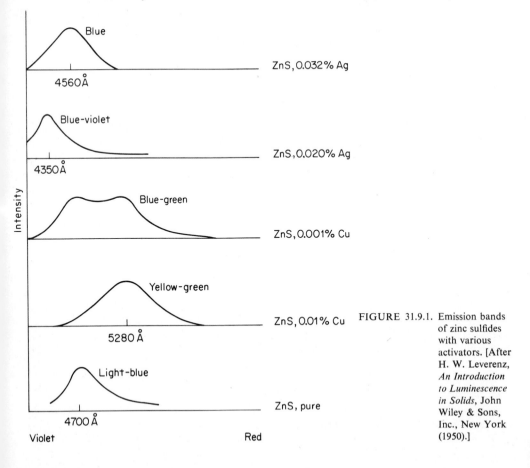

FIGURE 31.9.1. Emission bands of zinc sulfides with various activators. [After H. W. Leverenz, *An Introduction to Luminescence in Solids*, John Wiley & Sons, Inc., New York (1950).]

of the coactivator, while the activator primarily determines the spectrum; the latter is illustrated in Figure 31.9.1.

Figure 31.9.2 shows a coactivator level acting as a trap, with the fraction of time that the electron stays in the conduction band being proportional to $\exp(-E_T/kT)$, where E_T is the trapping energy. A detailed study of quantum mechanics and transition probabilities, which are briefly discussed in the next section, are needed to understand why the conduction band to activator level transition is the most probable in the system shown in Figure 31.9.2 relative to transitions such as conduction band to valence band, coactivator level to activator level, or coactivator level to valence band. The addition of 0.01 % Cu to ZnS gives behavior consistent with that shown in Figure 31.9.2.

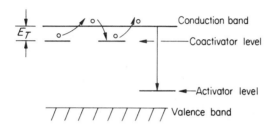

FIGURE 31.9.2. Energy band in a phosphor showing how the coactivator levels determine delay times, while activator levels determine the energy of emission or the wavelength.

ELECTROLUMINESCENCE. Consider a composite made of copper-doped ZnS containing precipitates of ZnO, which is a very good conductor relative to the ZnS matrix. When fields of 10^6 V/m (which might involve a voltage of 100 V across a film 0.01 cm thick) are applied, then local fields a few orders of magnitude higher can be present near the precipitates. Electrons may be liberated by very high fields and in turn may ionize the activators. These then act as luminescent centers. Electroluminescence increases with applied field and frequency to a saturation frequency of 10^5 Hz. Note that the voltage can be applied to a tiny area of the sheet which is luminescent independently of the rest of the area.

31.10   LASERS

The emission of radiation from an excited atom discussed thus far has been **spontaneous emission**, i.e., independent of the incident radiation. There is also the possibility of stimulated emission. Absorption is clearly stimulated. The probability of such absorption taking place in unit time is $\rho_\nu B_{12}$, where B_{12} is the probability of transition from a state of energy E_1 to a state of energy E_2, where $E_2 > E_1$, and ρ_ν is the energy density of the incident radiation of frequency ν, where $h\nu = E_2 - E_1$. There is also the reverse stimulated process $\rho_\nu B_{21}$, called **stimulated emission**. Here, too, the emitted radiation is *phase coherent* with the incident radiation. Einstein first showed that $B_{12} = B_{21}$. This says that the probability of external radiation inducing an upward transition from an atom in state 1 to 2 is exactly equal to the probability of the same radiation inducing a downward transition from an atom in state 2 to 1. The calculation of B_{12} is under certain conditions a straightforward though tedious quantum mechanical problem [see H. Eyring, J. Walter, and G. E. Kimball, *Quantum Chemistry*, Wiley, New York (1949) p. 108].

Einstein showed that the probability coefficient for spontaneous emission from a system in state 2 to 1 could be obtained from a thermodynamic argument as follows. If the number of systems in the state with energy E_2 is N_2 and the number in the state with energy E_1 is N_1, then the Boltzmann distribution states that, for equilibrium,

$$\frac{N_2}{N_1} = \frac{e^{-E_2/kT}}{e^{-E_1/kT}} = e^{-(E_2-E_1)/kT}. \tag{31.10.1}$$

Similarly, for equilibrium the number of upward transitions must equal the number of downward transitions or

$$N_1\rho_\nu B_{12} = N_2\rho_\nu B_{21} + N_2 A_{21}, \tag{31.10.2}$$

where A_{21} represents the probability of spontaneous emission occurring in unit time. A_{21} can be obtained from these two equations.

EXAMPLE 31.10.1

Would photoluminescent radiation be phase coherent or incoherent?

Answer. This represents spontaneous emission from atoms. The emission from one atom is independent from another so the emission from different atoms takes place at different times. The total radiation is phase incoherent.

INVERTED POPULATIONS.

EXAMPLE 31.10.2

Consider the case where $E_2 > E_1$ and $E_2 - E_1 = 2.3$ eV. The ground state of the atom (or ion) is E_1. There are no other nearby energy levels except for E_2. What is the fraction of atoms in the excited state at room temperature?

Answer. Recall that $kT \approx \frac{1}{40}$ eV at room temperature. From (31.10.1) we have

$$\frac{N_2}{N_1} = e^{-2.3\times40} = 10^{-40}.$$

From the previous problem it is clear that the number of excited states in thermal equilibrium at room temperature in a system capable of emitting visible radiation is very small. To get any net emission it is necessary to increase the number of electrons in excited states above the equilibrium concentration. To get high-intensity emission it is necessary to have a large number of electrons in the excited state (leaving few in the ground state). Such a condition is called **population inversion**.

LASERS.   As a result of work by J. Weber (1952) and C. H. Townes (1955), the **laser** was developed. Laser is an acronym for *light amplification by stimulated emission of radiation*. The radiation which lasers emit is phase coherent, and it has high intensity, an extremely sharp emission line (small frequency spread), and narrow beam divergence. (Radio waves emitted from an antenna are also phase coherent but with a much different wavelength.) These four unique characteristics of the laser in the range of visible and near-visible radiation lead to a number of interesting applications. Because of the phase coherence of the radiation they emit, lasers can be used to carry signals by amplitude modulation. The possibilities for communication are enormous. A laser source is an ideal source for interferometry measurements, spectroscopy, holography, high-speed photography, etc. Because of the narrow beam, it can provide localized heating for surgical purposes. Because of the high intensity (particularly of the CO_2 laser) and the high focusing, lasers can be used in such materials processes as drilling of holes in diamonds to be used for wire drawing, cutting of inch-thick titanium plate, welding of different metals, etc.

There are four major types of lasers: the chemical laser, the gas-discharge laser, the semiconductor diode laser, and the homogeneous optically pumped liquid or solid state laser. We shall briefly consider only the latter and specifically the ruby laser. A crystal of corundum, Al_2O_3, is doped with 0.05% Cr. The chromium is present as a trivalent substitutional ion, Cr^{3+}, for Al^{3+}. Pure corundum is an insulator which is transparent and colorless. The Cr^{3+} ion is a color center which gives the material its characteristic ruby color (by providing

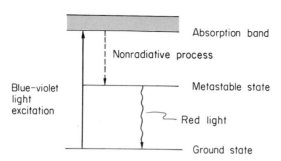

FIGURE 31.10.1. Simplified picture of the energy state of the ruby laser.

an absorption band). Figure 31.10.1 is a simplified picture of the energy level diagram.

The Cr^{3+} ion can be excited by radiation of about 4100 Å wavelength (blue-violet). Such light can readily be provided by a fluorescent xenon tube. The electron then falls from the excited state of the absorption band to a metastable state by a nonradiative process; instead of a photon being created, a phonon is created; i.e., a lattice vibration is excited to a higher energy. This nonradiative process is extremely rapid. The metastable state has, relatively speaking, a very long lifetime (for spontaneous decay) of about 5×10^{-3} sec.

EXAMPLE 31.10.3

How far can light travel in a media of refractive index 1.76 in 5×10^{-3} sec?

Answer. The speed of light in a vacuum is 3×10^8 m/sec and in the above media, which could be Al_2O_3, $3 \times 10^8/1.75 = 1.7 \times 10^8$ m/sec. Hence the distance is 8.55×10^5 m. If light is traveling back and forth in a crystal 4.27 cm long it could make 10^7 passes through the crystal after a specific ion was excited before that ion would spontaneously decay.

Because of the long lifetime of the metastable state, the ions can essentially all be "pumped" into the metastable state quickly by a xenon flashtube. Hence an inverted population is created. If now a single emission occurs and this in turn induces (stimulates) a second emission, the latter will be phase coherent with the former. If these in turn stimulate other emissions, a situation can rapidly be reached wherein we have a high-density coherent radiation supply in the crystal. A schematic of the system is shown in Figure 31.10.2.

The coherent radiation is reflected back and forth from the silver mirrors (the silvered surfaces at the ends of the specimen). Finally there is essentially a population explosion of emission, and a high-intensity pulse of short duration then is emitted from the lightly silvered face.

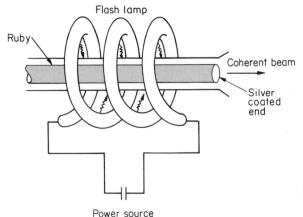

FIGURE 31.10.2. Ruby laser.

REFERENCES

Wood, E. A., *Crystals and Light*, Van Nostrand, Princeton, N.J. (1964).

Sears, F. W., *Optics*, Addison-Wesley, Reading, Mass. (1949).

Javan, A., "The Optical Properties of Materials," *Scientific American* (Sept. 1967) p. 239.

Patel, C. K. N., "High-Power CO_2 Lasers," *Scientific American* (Aug. 1968) p. 23.

Leverenz, H. W., *An Introduction to Luminescence in Solids*, Wiley, New York (1950).

Weber, S., ed., *Optoelectronic Devices and Circuits*, McGraw-Hill, New York (1964). This is a superb collection of articles by different authors which bridges the gap between pure theory and application.

Kazan, B., and Knoll, M., *Electronic Image Storage*, Academic Press, New York (1968).

Goldberg, P., editor, *Luminescence of Inorganic Solids*, Academic Press, New York (1966).

Gray, G. W., *Molecular Structure and Properties of Liquid Crystals*, Academic Press, New York (1962).

Thornton, P. R., *The Physics of Electroluminescent Devices*, Barnes and Noble, Inc., New York (1967).

PROBLEMS

31.1. a. Explain the origin of the strong absorption of glass in the ultraviolet.
      b. The infrared.

31.2. What kind of ions should be added to silica glass to increase its refractive index? Why?

31.3. a. Why does calcite, $CaCO_3$, split radiation (if not at normal incidence) into two beams?
      b. Are there certain directions of propagation for which this does not happen?

31.4. Ground (but not polished) glass is not transparent. Why?

31.5. Thallium iodide has a hexagonal crystal structure.
      a. Will it be optically isotropic?
      b. Suppose tiny crystals are compacted by pressure into a randomly oriented polycrystalline slab having the theoretical density. The slab is annealed to relieve stress. The single crystals are transparent in the visible range. Will the slab be?

31.6. Sodium chloride crystals are compacted into a thick lens having essentially the theoretical density. Will the lens be transparent? Discuss your answer.

31.7. In the manufacture of a glass lens there is a great concern about eliminating residual interior stresses, residual surface stresses, inclusions, bubbles, flaws, etc. Explain why in each case.

31.8. Ceramists are able to make certain glasses which separate into two interconnected phases. Suppose the "radius" of a given phase region on the average is 3×10^4 Å. Would you in general expect the glass to be transparent? Why?

31.9. LiF dissolves a considerable quantity of MgF_2 at the eutectic temperature. However, this precipitates out at low temperatures. Single crystals of LiF and of MgF_2 are transparent.
a. Would the alloy be transparent at room temperature?
b. Does your answer to (a) depend in any way on the thermal history?

31.10. To reduce the reflectance at a surface, lenses are often coated with a thin layer of MgF_2 (the latter is transparent and optically isotropic). How thick should the coating be and on what principle is it based?

31.11. Light passes through a glass filter which is blue and one which is yellow. The transmittance for these filters is shown in Figure P31.11. Suppose the two filters are placed in contact and inserted in a beam of white light. Plot the transmittance of the pair. What color transmits?

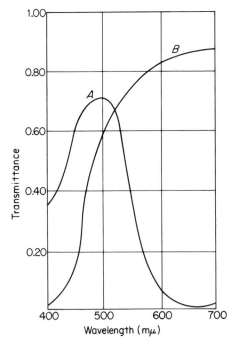

FIGURE P31.11.

31.12. Glass is often colored by adding certain oxides. Which class of oxides is used and why does this lead to coloration?

31.13. A KCl crystal is irradiated with γ-rays. It is colored purple. The crystal is placed in a glass dessicator at room temperature. In time the color fades. Explain.

31.14. Discuss ten possible applications of luminescent materials.

31.15. Explain the ruby laser.

More Involved Problems

31.16. Explain the activity of the "lightning bug."

31.17. Discuss the high-energy CO_2 laser. (See the References.)

31.18. Write a paper in which the quantitative details of the photographic processes are discussed.

31.19. Discuss the operation of an actual Kerr cell. (See Weber in the References.)

31.20. Discuss the current state of development of photochromic behavior.

31.21. Discuss why organic semiconductors and exciton behavior in them is of interest to biologists.

31.22. Discuss the behavior of the gallium-arsenide laser (outside reading).

Sophisticated Problem

31.23. Discuss the current state of developments in the area of exciton fission. Discuss some possible applications.

Prologue

All materials have a *diamagnetic* contribution to their magnetic behavior. The origin of this is discussed briefly. Many materials exhibit paramagnetism owing to the presence of *permanent magnetic moments* associated with atoms. In some materials (primarily in the transition and rare earth metals) there is a strong *cooperative* interaction between the magnetic dipoles. This leads to *spontaneous alignment* and the formation of *domains* (regions in which the dipoles are similarly aligned). These materials are *ferromagnetic*. In *ferrimagnetic* materials, a fixed fraction of dipoles is aligned in one direction; the remainder are antiparallel, but there is a net magnetic dipole moment in one direction. A number of oxides, in particular the *ferrites*, exhibit this behavior. Domains are also present in these materials. They are strongly magnetic but have high electrical resistance. Hence they exhibit small eddy losses. There is an upper limit on the size of a domain owing to energetic considerations. Tiny elongated particles about 1000 Å in diameter exist as *elongated single domains*. These can be used to make hard (permanent) magnets. However, a large block of a single-phase material such as iron consists of many domains. The reasons for this are studied in this chapter. A ferromagnetic material is soft if the domain walls can move easily and is hard if the domain walls are pinned. Mechanisms for pinning domain walls are discussed. Methods for producing very hard magnets (oriented elongated single domains), by both mechanical means and metallurgical means, are discussed.

32

MAGNETIC MATERIALS

32.1   INTRODUCTION

Diamagnetic materials repel flux lines slightly. Their permeability is only slightly less than μ_0 (the permeability of vacuum); their susceptibility, which is the fractional change in permeability relative to vacuum, is of the order of -10^{-5} to -10^{-6}. We shall study diamagnetic materials briefly because of their scientific interest. Diamagnetism is not (now) directly important in engineering.

Paramagnetic materials attract flux lines weakly. Their permeability is only slightly greater than μ_0; their susceptibility is of the order of 10^{-4}–10^{-5}. Paramagnetism has some direct applications, but at present paramagnetic materials do not form a class of important engineering materials. We shall study paramagnetism primarily for the insight it provides to ferromagnetism.

Ferromagnetism is by far the most important magnetic behavior from an engineering viewpoint and one of the most interesting phenomena from the theorists' viewpoint. However, ferrimagnetism, as exhibited by the ferrites, is also very important.

EXAMPLE 32.1.1

List some of the important physical properties associated with ferromagnetic materials.

Answer.
1. Permeability, $\mu \gg 1$.
2. Hysteresis of *B-H* (or *M-H*) curves. See Section 4.2.
3. Saturation induction, B_S. See Section 4.2.
4. Saturation magnetization, M_S. See Section 4.2.
5. Remanent induction, B_R. See Section 4.2.

6. Coercive force, H_c. See Section 4.2.
7. Energy loss per cycle, $\oint H \, dB$. See Section 4.2.

With the exception of the saturation induction B_S (or M_S) all of the physical properties are strongly structure-sensitive. This means that the materials scientist, when he understands the origin of the behavior, can design and with sufficient ingenuity construct a specific microstructure to achieve specific behavior.

Ferromagnetism will be studied in detail. We have already noted in Chapter 4 the unique variation of M_S vs. T (see Figure 4.2.5). Note that the degree of ferromagnetism decreases gradually as temperature is increased up to T_c. This is a general characteristic of all ferromagnetic materials. Note that this type of behavior is also characteristic of superconductivity (see Figure 5.2.1) and long-range order (see Figure 17.9.2).

There are two additional types of magnetic behavior: antiferromagnetism and ferrimagnetism. Ferrimagnetic materials, like ferromagnetic materials, have $\mu \gg \mu_0$ below T_c and exhibit a temperature dependence below T_c which is qualitatively similar to the ferromagnetic materials. The ferrites are important commercial ferrimagnets because along with their strong magnetism they have a high resistivity which reduces eddy current losses.

UNITS.   In the rationalized mks system, $\mathbf{B} = \mu\mathbf{H}$, where $\mu_0 = 4\pi \times 10^{-7}$ Henry/m (or kg-m/C²). Also $\mathbf{B} = \mu_0(\mathbf{H} + \mathbf{M})$ and $\mathbf{M} = \chi_m H$.

The scientific literature often uses the cgs electromagnetic units (emu). In these units (we use primes here to distinguish between the two systems)

$$\mathbf{B} = \mu'\mathbf{H},$$

and the units are such that $\mu_0' = 1$. B and H are related by

$$\mathbf{B} = \mathbf{H} + 4\pi\mathbf{I}$$

and

$$\mathbf{I} = \chi_m'\mathbf{H}.$$

Table 32.1.1. UNITS FOR MAGNETIC FIELDS AND PROPERTIES

Factor	mks rationalized	cgs emu	Conversion
B	Wb/m²	Gauss	1 Wb/m² = 10⁴ Gauss
H	A-turns/m	Oersteds	1 A/m = $4\pi/10^3$ Oe
Magnetization	M, A-turns/m	I, Mx/cm²	1 A/m = $1/10^3$ Mx/cm²
Permeability of vacuum	$4\pi \times 10^{-7}$ Henry/m	1*	
Relative permeability	μ_r	μ'	$\mu_r = \mu'$
Susceptibility	χ_m	χ_m'	$\chi_m = 4\pi\chi_m'$

* This is a dimensionless quantity.

Hence

$$\chi'_m = \frac{\mu' - 1}{4\pi}.$$

Table 32.1.1 shows the correspondence of the units.

32.2   DIAMAGNETISM

A current i circulating in a planar loop of area A leads to a **magnetic dipole $\mathbf{p}_m$**:

$$\mathbf{p}_m = i\mathbf{A}. \qquad (32.2.1)$$

(Recall that the representation of area by a vector enables one to describe both the magnitude of the area and the unit normal to the area.)

Consider a single electron moving in a circular orbit about an ion core. The current is then

$$i = -\frac{ev}{2\pi r}. \qquad (32.2.2)$$

Note that $v/2\pi r$ is simply the number of revolutions per second. Therefore the electron of charge $-e$ passes a given point $v/2\pi r$ times/sec.

The magnetic moment is

$$p_m = -\frac{ev}{2\pi r}\pi r^2 = -\frac{evr}{2}. \qquad (32.2.3)$$

The magnitude of angular momentum is mvr or, more generally,

$$\mathbf{L} = m(\mathbf{r} \times \mathbf{v}) = \mathbf{r} \times \mathbf{p}, \qquad (32.2.4)$$

where $\mathbf{p}$ is the linear momentum (this should not be confused with the dipole moment $\mathbf{p}_m$, which always has the subscript m).

Thus

$$\mathbf{p}_m = -\frac{e}{2m}\mathbf{L}. \qquad (32.2.5)$$

The quantity $e/2m$ is the **gyromagnetic ratio, γ**.

Let us now consider the motion of an electron around a positive ion core in the presence of a field of magnetic induction B. The inward force on the electron whose velocity is v is evB owing to the magnetic field and $e^2/4\pi\epsilon_0 r^2$ owing to coulomb attraction:

$$F_{\text{inward}} = evB + \frac{e^2}{4\pi\epsilon_0 r^2}. \qquad (32.2.6)$$

The inward acceleration of a particle moving in a circular orbit with constant angular velocity ω is v^2/r (where $v = \omega r$). Hence by Newton's second law,

$$evB + \frac{e^2}{4\pi\epsilon_0 r^2} = \frac{mv^2}{r}. \qquad (32.2.7)$$

We want to calculate the *small* change in the angular momentum, mvr, owing to the presence of the field. To do this we first calculate v from (32.2.7). We have

$$v = \frac{eBr/m + (e/\sqrt{\pi\epsilon_0 rm})\sqrt{1 + (eBr/m)^2(\pi\epsilon_0 rm/e^2)}}{2}. \qquad (32.2.8)$$

Note that for $B = 0$, we have

$$v = v_0 = \frac{e}{\sqrt{4\pi\epsilon_0 rm}}. \qquad (32.2.9)$$

We can also expand v in a Taylor series in B about $B = 0$. We have

$$v = \frac{eBr}{2m} + \frac{e}{\sqrt{4\pi\epsilon_0 rm}} + 0(B^2) \qquad (32.2.10)$$

so that the change in v is given by

$$\Delta v \doteq \frac{eBr}{2m}. \qquad (32.2.11)$$

(Note that we have ignored any possible changes in r, as well as higher-order terms.) The change in angular momentum owing to the field is $m\,\Delta vr$ and combining (32.2.11) and (32.2.5) we have for the change in dipole moment

$$p_m = -\frac{e^2 r^2}{4m} B \doteq -\frac{e^2 r^2}{4m} \mu_0 H. \qquad (32.2.12)$$

The susceptibility per atom is $-e^2 r^2 \mu_0/4m \approx -10^{-34}$ m³ since $r \approx 1$ Å.

   There are about 10^{28} atoms/m³ so the susceptibility per unit volume is $\chi_m \approx -10^{-6}$; this is called the **volume susceptibility**; it is a dimensionless quantity.

   While this classical model leads to a reasonable estimate of χ_m for diamagnetic solids, it should be emphasized that a direct calculation of χ_m is a quantum mechanical problem. For a spherical electron distribution in an atom or ion with Z electrons with N atoms per unit volume, the result is

$$\chi_m = \frac{-NZe^2 \mu_0 \overline{r^2}}{6m}, \qquad (32.2.13)$$

where $\overline{r^2}$ is the mean-squared distance of the electrons from the nucleus. This can be evaluated from

$$\overline{r^2} = \int_0^\infty \psi^* r^2 \psi 4\pi r^2 \, dr. \qquad (32.2.14)$$

Note that the diamagnetic susceptibility is essentially *independent* of temperature.

   Table 32.2.1 shows some values of the susceptibilities. All materials have a diamagnetic component of susceptibility. However, χ_m will be positive for many of these materials because the paramagnetic or ferromagnetic component is much larger.

Table 32.2.1. VOLUME SUSCEPTIBILITIES

Material	χ_m (dimensionless)
Cu	-0.7×10^{-6}
Au	-2.8×10^{-6}
He (STP)	-0.8×10^{-10}
Ne (STP)	-3.0×10^{-10}
Ar (STP)	-9.0×10^{-10}
Kr (STP)	-13.0×10^{-10}
Xe (STP)	-19.6×10^{-10}

32.3   PARAMAGNETISM

ORIGIN OF MAGNETIC DIPOLES.   Paramagnetism owes its origin to the permanent dipoles of atoms or ions. Before the advent of modern quantum mechanics N. Bohr postulated a model for the hydrogen atom as follows: The electron moves in circular orbits about the proton at radius r with velocity v. It is attracted to the nucleus by the coulombic attraction $e^2/4\pi\epsilon_0 r^2$. The inward acceleration is v^2/r so a force balance gives

$$\frac{e^2}{4\pi\epsilon_0 r^2} = \frac{mv^2}{r}.$$ (32.3.1)

(*Note:* In the cgs system we would drop the $4\pi\epsilon_0$ term.)

The energy of the system is the kinetic energy plus the potential energy so

$$E = \frac{mv^2}{2} - \frac{e^2}{4\pi\epsilon_0 r}.$$ (32.3.2)

It is a result of electromagnetism and classical mechanics that an accelerating charge emits radiation and hence loses energy and slows down. Bohr (1912) postulated that there were certain stable orbits for which this did not happen. He postulated that the angular momentum (mvr) must exist in units of $h/2\pi$. From this postulate he obtained

$$mvr = \frac{nh}{2\pi}, \quad n = 1, 2, 3, \ldots.$$ (32.3.3)

By combining these equations one gets an expression for the possible energy states of the hydrogen atom in mks units. This, of course, is the same result as obtained from the wave equation in Section 8.2 (in cgs units).

The magnitude of the magnetic moment of a circulating electron with angular momentum $h/2\pi$ is, from (32.2.5),

$$\beta = \frac{e}{2m}\frac{h}{2\pi} = 9.27 \times 10^{-24}\text{A-m}^2.$$ (32.3.4)

This is known as the **Bohr magneton**. It is the unit dipole moment. It is of interest that Langevin did not know of the origin of permanent dipole moments when he proposed (1905) his theory of paramagnetism, which will be discussed later in this section.

We now know that the Bohr theory was inadequate in many respects and was replaced by modern quantum mechanics. In terms of modern quantum mechanics, the angular momentum L is given in terms of the quantum number l, where $l = n - 1, n - 2, \ldots, 0$. (The quantum number l arises in a natural way in the solution of Schrödinger's equation for the hydrogen atom by the separation of variables technique.) It can be shown (see Problems 32.23–32.26) that the total orbital angular momentum is

$$L = \sqrt{l(l+1)}\frac{h}{2\pi} \qquad (32.3.5)$$

and that the z component of the angular momentum is

$$L_z = m_l\frac{h}{2\pi}, \qquad m_l = 0, \pm 1, \ldots, \pm l. \qquad (32.3.6)$$

This leads to magnetic dipoles

$$p_m(\text{orbital}) = \frac{e}{2m}\sqrt{l(l+1)}\frac{h}{2\pi} \qquad (32.3.7)$$

and

$$p_{m_z}(\text{orbital}) = \frac{e}{2m}m_l\frac{h}{2\pi}. \qquad (32.3.8)$$

Recall that spherically symmetric eigenfunctions, such as s-state electrons, have $l = 0$.

Electron spin also contributes to the magnetic moment:

$$p_{m_z}(\text{spin}) = \frac{e}{m}\frac{h}{2\pi}s, \qquad s = \pm\tfrac{1}{2}. \qquad (32.3.9)$$

Note that

$$p_{m_z}(\text{spin}) = \pm\beta. \qquad (32.3.10)$$

To obtain the magnetic moment of a more complex atom it is necessary to sum the magnetic moments of all the electrons. We would find that the inert gases have a net magnetic moment of zero. For example, helium has two $1s$ electrons ($l = 0$) so their orbital moment is zero and their spin moments are of opposite sign and cancel. Neon has two $1s$ electrons (no net moment), two $2s$ electrons (no net moment), and six $2p$ electrons (no net moment). The six $2p$ electrons together have a spherical distribution (see Example 8.3.2) and hence no net orbital moment. [From the quantum mechanical viewpoint the orbital momentum is calculated by operating on the wave function ψ with the operator L^2 and equating this to the magnitude squared of the angular momentum (the unknown) times the wave function ψ. It can be shown (see the sequence of Problems 32.23–32.25) that the operator is

$$L^2 = -\frac{h^2}{4\pi^2}\left[\frac{1}{\sin\theta}\frac{\partial}{\partial\theta}\left(\sin\theta\frac{\partial}{\partial\theta}\right) + \frac{1}{\sin^2\theta}\frac{\partial^2}{\partial\phi^2}\right]. \qquad (32.3.11)$$

Clearly, if the ψ for all six $2p$ electrons together is a function of r only, then $L^2\psi = 0$ and the angular momentum of this sextet is zero.]

EXAMPLE 32.3.1

What is the magnetic moment of a fluorine ion, F^-?

Answer. This ion has the neon electron configuration $(1s)^2(2s)^2(2p)^6$. Electron spins are all paired and hence there is no net spin moment. Likewise specific electron shells are filled and hence spherically symmetric. Thus there is no net orbital moment. The F^- ion has no permanent magnetic dipole owing to the electrons.

In general full shells have a net angular momentum of zero.

Many molecules such as H_2 have no permanent dipole. However, many atoms have permanent magnetic dipoles owing to unpaired spins or a net orbital momentum. **Hund's rule** states that the electron spins are arranged in such a way that they make a maximum contribution to angular momentum.

EXAMPLE 32.3.2

What is the maximum spin magnetic moment of a d shell? An f shell?

Answer. There are ten possible electrons in a d shell. Five of these can have plus spins and five can have minus spins. A shell with *only* five d electrons would, according to Hund's rule, have five unpaired electrons, i.e., either $5+$ or $5-$. A d shell with more or less than five electrons would have a net spin magnetic moment less than 5 Bohr magnetons. For the f shell, the maximum spin moment would be 7β (Bohr magnetons).

EXAMPLE 32.3.3

Consider the electron configuration of the silver atom (see Appendix 8A). Except for the $5s$ valence electron, the shells present are all filled. Recall that individual s states have spherically symmetric wave functions. Therefore the total wave function for the electrons is also spherically symmetric. If the silver atom has a magnetic moment, to what must it be attributed?

Answer. The total orbital angular momentum is zero [see Equation (32.3.11)] and hence the orbital magnetic moment is zero. Hence the magnetic moment must be attributed to the intrinsic spin angular momentum of the single valence electron. This spin was measured by O. Stern and W. Gerlach in a classic experiment [*Zeitschrift für Physik*, **8**,

110 (1927) and **9**, 349 (1927)] in which a beam of silver atoms was passed through a strong inhomogeneous magnetic field. The beam separated sharply into two beams. One beam was deflected as though atoms in it had a magnetic dipole β and the other $-\beta$.

The exact manner in which the orbital and spin moments combine in solids will not be given here.

ORIENTATION OF MAGNETIC DIPOLES.   The calculation of the average magnetic moment of an assembly of independent magnetic dipoles, which tend to have a random arrangement owing to the thermal field and tend to be aligned owing to the magnetic field, is *directly* analogous to the calculation for electric dipoles carried out in Section 30.4. See Figures 30.4.4 and 30.4.5 and the associated discussion. In fact, the analysis given there was first given for the magnetic situation by Langevin in 1905.

The **P** vs. ξ result of Equation (30.4.20) for electric dipole orientation is

$$\mathbf{P} = \frac{N p_{\text{elect}}^2}{3kT} \xi'. \tag{32.3.12}$$

For the present case we obtain

$$\mathbf{M} = \frac{N p_m^2}{3kT} \mathbf{B}' = \frac{N p_m^2 \mu}{3kT} \mathbf{H}'. \tag{32.3.13}$$

(In general ξ', $\mathbf{B}'$, and $\mathbf{H}'$ in these two equations represent the local fields within the material. If the system is a dilute gas, these local fields are nearly equal to the applied fields.) We have in such cases for the paramagnetic susceptibility

$$\chi_m = \frac{N p_m^2 \mu_0}{3kT}. \tag{32.3.14}$$

Materials which exhibit this behavior are said to obey the **Curie law**. Note that as with dielectric materials the induced susceptibility (diamagnetism) is always present, while the effect owing to permanent dipoles (paramagnetism) is also sometimes present. Usually when paramagnetism is present it is an order of magnitude larger than the diamagnetism of the material.

When one studies the experimental data of solids carefully, one finds the **Curie-Weiss relation**

$$\chi_m \doteq \frac{C}{T - T_c}. \tag{32.3.15}$$

Langevin's theory does not account for T_c. Pierre Weiss (1912) assumed that this deviation was due to the interaction of the dipoles. This interaction could be represented as an effective internal molecular field (whose magnitude was proportional to the degree of magnetization) which had to be added to the external field. This additional field gives rise to the additional constant in the denominator of (32.3.15).

Our derivation of (32.3.13) assumed that all angles of orientation of the atomic dipoles in the magnetic field are possible. This is not the case. Quantum mechanics allows only certain discrete angles of orientation. Thus rather than integrating over all angles as in Equation (30.4.15), it is necessary to sum over the allowable angles. This was first carried out by Brillouin. For our present purpose, there are only quantitative differences between the Brillouin and the Langevin function. The Brillouin theory is discussed by C. Kittel, *Introduction to Solid State Physics*, Wiley, New York (1968) p. 434.

EXAMPLE 32.3.4

At very low temperatures or very high fields, paramagnetic materials saturate. What is the saturation magnetization?

Answer. By analogy with (30.4.17),

$$\bar{p}_m = p_m L(x),$$

where $L(x)$ is the Langevin function and $x = p_m B/kT$. $L(x) \longrightarrow 1$ for large x so $\bar{p}_m \longrightarrow p_m$ and $M \longrightarrow Np_m = M_S$.

EXAMPLE 32.3.5

Estimate the field needed at room temperature to approach saturation for a paramagnetic material for which $p_m = 5\beta$, where β is the Bohr magneton.

Answer. We shall use $x = 5$ in the Langevin function $[L(x) \approx 0.8]$ as our criterion for saturation (see Figure 30.4.6). Then

$$B \approx \frac{5kT}{p_m} = \frac{kT}{\beta}.$$

Since $k = 1.38 \times 10^{-23}$ J/°K and β is given by (32.3.4), we have $B \approx 500$ Wb/m². This is a very high field. Superconducting magnets can generate fields of magnetic induction of about 15 Wb/m².

OCCURRENCE OF PARAMAGNETISM.   Paramagnetism is exhibited by

1. Atoms, molecules, and lattice defects which have an odd number of electrons. Examples: alkali metal atoms in gas state; F centers in NaCl.
2. Atoms or ions with only partially filled shells. Example: Mn^{2+}, which has the configuration $Ar(3d)^5$. The shells of argon are full.
3. Certain compounds with an even number of electrons. Example: oxygen molecule.
4. Metals.

CONDUCTION ELECTRONS.   The free electron gas in a metal might be expected to have a susceptibility of the Curie form of (32.3.14), where $p_m = \beta$ and N is the number of free electrons per unit volume. Recall, however, that this electron gas obeys Fermi-Dirac statistics. Hence only those unpaired electrons near the top of the distribution can flip over in a field. [The situation is analogous to the electronic specific heat (see Section 27.3).] Roughly, only a fraction T/T_F, where T_F is the Fermi energy, is available. Thus N should be replaced by NT/T_F, resulting in a paramagnetic susceptibility *independent* of temperature.

ADIABATIC DEMAGNETIZATION. Paramagnetism, like diamagnetism, is primarily of scientific interest because of its relationship to atomic structure. However, there are a few interesting applications. One of these is the refrigeration principle, proposed independently in 1926 by W. F. Giauque and P. Debye, called **adiabatic demagnetization**. Certain rare earth salts have a large paramagnetic susceptibility. Suppose such a salt (gadolinium sulfate is an example) is cooled to a low temperature, say 1.5°K. In the absence of a field the magnetic dipoles would be random. The application of a large magnetic field, say a magnetic induction of 1 Wb/m², would align these magnets. Suppose that such a system is now present at 1.5°K at 1 Wb/m² and that it is thermally isolated. The field is then turned off. The dipoles can now randomize. In so doing the spin entropy increases. However, because the process is adiabatic, $dQ = 0$; but $dQ = T\,dS$, so the *net* entropy change must be zero. Consequently the entropy of the lattice vibrations must decrease. Hence the temperature of the salt must fall. In Giauque's initial experiments the temperature fell to 0.25°K. Much lower temperatures have since been obtained by this technique.

32.4 SPONTANEOUS MAGNETIZATION

CURIE-WEISS TEMPERATURE. Weiss postulated that the effective field B' is equal to the external field B plus an internal field proportional to the magnetization:

$$B' = B + aM. \tag{32.4.1}$$

Then x in Langevin's theory [see (30.4.18)] is

$$x = \frac{p_m B'}{kT} = \frac{p_m(B + aM)}{kT}. \tag{32.4.2}$$

Since $L = M/Np_m = \frac{1}{3}x$ [see (30.4.17) and note that $M = N\bar{p}_m$],

$$M = \frac{Np_m^2(B + aM)}{3kT}. \tag{32.4.3}$$

Solving for M we have

$$M = \frac{(Np_m^2/3k)\mu_0}{T - Np_m^2 a/3k} H. \tag{32.4.4}$$

Hence χ_m has the form

$$\chi_m = \frac{C}{T - T_c}, \tag{32.4.5}$$

where

$$T_c = \frac{Np_m^2 a}{3k} = \frac{Ca}{\mu_0}. \tag{32.4.6}$$

SPONTANEOUS MAGNETIZATION.   We now examine, as Weiss did, the situation when the external field is zero. We then have $B' = aM$ so $x = p_m Ma/kT$ and $M = kTx/p_m a$. Hence

$$\frac{M}{M_S} = \frac{kT}{p_m M_S a} x. \tag{32.4.7}$$

We also have from Langevin's theory

$$\frac{M}{M_S} = L(x). \tag{32.4.8}$$

We recall that the initial slope of the Langevin function is $\frac{1}{3}$. The reader should now sketch $L(x)$ vs. x [see Equation (30.4.17) or Figure 30.4.6] and should draw a dashed line which is tangent to $L(x)$ at $x = 0$. The slope of M/M_S from Equation (32.4.7) is $kT/p_m M_S a$. If

$$\frac{kT}{p_m M_S a} > \frac{1}{3} \tag{32.4.9}$$

(draw the line), the line intersects $L(x)$ only at $M = 0$, $x = 0$. However, if

$$\frac{kT}{p_m M_S a} < \frac{1}{3} \tag{32.4.10}$$

(draw the line), the line intersects the curve $L(x)$ at some finite value of M and x as well as at the origin. This means that the material can be magnetized when $B = 0$. Weiss called this **spontaneous magnetization**.

EXAMPLE 32.4.1

   If $kT/p_m M_S a = \frac{1}{6}$, what is the spontaneous magnetization in terms of M_S?

   Answer. From the figure for $L(x)$, intersection of the line of (32.4.7) with $L(x)$ occurs at about 0.8; hence the spontaneous magnetization is about $0.8 M_S = 0.8 N p_m$.

It can be shown that the temperature dependence of the saturation magnetization (see Figure 4.2.5) follows from the Weiss theory.

Iron can be found either magnetized or demagnetized. To explain this, Weiss postulated the presence of **domains** which are regions which are small compared to the ordinary sample size and within which the material is magnetized in a given direction (see Figure 32.8.4). The formation of domain structures is described in Section 32.8. The magnetization behavior is related to the ease or difficulty of domain boundary motion and domain rotation. This is studied in Section 32.9. The random orientation of domains leads to no net macroscopic magnetization.

32.5   THE ORIGIN OF FERROMAGNETISM

If one associates the critical temperature at which spontaneous magnetization disappears with the Curie-Weiss temperature T_c, then

$$aM_S = \frac{3kT_c}{p_m}.$$

Then using the fact that $p_m = 2.2\beta$ and $T_c \approx 1000°K$ for iron and $k = 1.38 \times 10^{-23}$ J/°K, we have

$$aM_S \approx 2000 \text{ Wb/m}^2.$$

This is a huge field compared to the external fields ordinarily in use. The presence of this huge field cannot be explained on the basis of classical magnetic interaction forces.

EXAMPLE 32.5.1

What is the order of magnitude of kT_c in electron volts for iron if $T_c = 1043°K$?

Answer. Recall that $kT \approx \frac{1}{40}$ eV at room temperature. Hence $kT_c \approx 0.1$ eV. Hence the energy which causes spontaneous alignment must be of this order.

EXAMPLE 32.5.2

Estimate the field at one iron ion owing to another iron ion in the metal. The field owing to a dipole is

$$B \approx \mu_0 \frac{p_m}{r^3}.$$

(The field varies with direction. We have assumed that the dipoles are widely separated.)

Answer. We assume $p_m \approx \beta$. Hence

$$B \approx \frac{10^{-6} \times 10^{-23}}{10^{-29}} = 1 \text{ Wb/m}^2.$$

(Note how small this is compared to the internal field $aM_S \approx 2000$ Wb/m².)

Since the interaction energy of a dipole with a field is given by

$$U = -\mathbf{p}_m \cdot \mathbf{B},$$

the energy of this interaction is only about 10^{-4} eV. When all the interactions in the crystal are summed the energy is still of this order.

The situation is somewhat analogous to the bonding of two hydrogen atoms to form a hydrogen molecule. Classical interaction forces could not explain the magnitude of this interaction either. However, quantum mechanical concepts could. A most elementary model for the hydrogen bond is given in Example 9.3.1. The essence of this model is that each electron has become delocalized and may be found in any part of the molecule. The electrons have opposite spin. A similar interaction called an **exchange interaction** leads to ferromagnetism. However, in the latter case the energy is lowered when the spins are aligned or parallel. The *d* band electrons of iron can be considered as quasi-free. They therefore obey Fermi-Dirac statistics. We would therefore expect the spins to be paired (the electron configuration of iron is $Ar3d^64s^2$). If the *d* electrons are unpaired, the Fermi energy of the *d* band is increased. However, there is an energy decrease owing to the exchange interaction which occurs when the unpaired spins are aligned. The latter term tends to dominate in those elements in which the *d* band is narrow. This also corresponds to a high density of states at the Fermi level. In such a case a number of electrons can be excited to a higher state without a large increase in energy. See Figure 32.5.1. This situation exists only for a few transition elements and rare earth 4*f*-band elements.

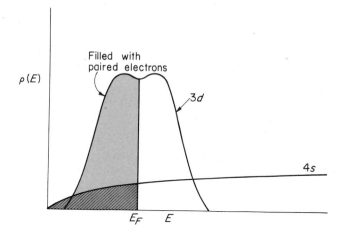

FIGURE 32.5.1. Sketch of density of states versus energy of typical 3*d* band and 4*s* band for typical ferromagnetic material. Note that when a fixed number of the electrons are unpaired the energy increase for the 3*d* band is not nearly as large as for the *s* band.

Table 32.5.1 shows the principal ferromagnetic elements. Note that in the isolated iron atom we would expect according to Hund's rule five spins up and one down or a net electronic moment of four magnetons (for reasons which we shall not give here, the orbital moment of the *d* electrons is zero). In the condensed

Table 32.5.1. PRINCIPAL FERROMAGNETIC ELEMENTS

Element	Electron Configuration	Atomic Radius Å	M_S (at 0°K), 10^6 A/m	Number of Bohr Magnetons Per Atom	T_c, °K	T_M, °K
Fe	$3d^64s^2$	1.24	1.69	2.2	1043	1808
Co	$3d^74s^2$	1.25	1.36	1.7	1404	1753
Ni	$3d^84s^2$	1.25	0.47	0.6	631	1728
Gd	$4f^75d^16s^2$	1.78	5.66	7.12	289	1585

metal the number of Bohr magnetons per atom is experimentally found to be 2.2 instead of 4, which would be expected on the basis of Hund's rule.

32.6   ANTIFERROMAGNETISM

In certain compounds of the transition metals the exchange interaction is such that the energy is minimized when neighboring dipoles are antiparallel. Such materials, below a certain critical temperature called the Néel temperature, are antiferromagnetic, i.e., exhibit spontaneous demagnetization. An example is MnO, which from the chemical viewpoint has the NaCl-type structure. However, from the magnetic viewpoint the unit cell has eight times the volume. See Figure 11.5.2. Recall that neutron diffraction can be used to study this behavior.

EXAMPLE 32.6.1

Why can neutron diffraction be used to show that MnO is antiferromagnetic?

Answer. The neutron has a permanent magnetic moment which interacts with the permanent magnetic moment of the Mn^{2+} ions.

32.7   FERRIMAGNETISM

The **ferrites** are a class of ionic crystals of composition $MeFe_2O_4$, where Me is a metal ion; these crystals have the **inverted spinel structure**. The compound Fe_3O_4, i.e., $Fe^{2+}O^{2-}Fe_2^{3+}O_3^{2-}$, is an example. So is $NiFe_2O_4$. The cubic unit cell of Fe_3O_4 contains 56 ions: 32 O^{2-}, 8 Fe^{2+}, and 16 Fe^{3+}. One fourth of the unit cell is shown in Figure 32.7.1.

The iron ions occupy two types of sites, eight which are tetrahedrally coordinated and sixteen which are octahedrally coordinated. Eight of the Fe^{3+} ions occupy the tetrahedral sites and eight occupy eight of the octahedral sites. The

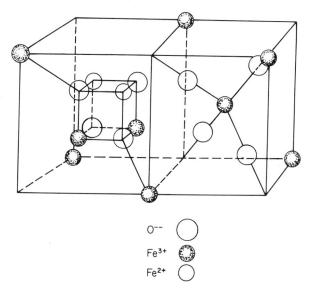

FIGURE 32.7.1. One fourth of the unit cell of Fe_3O_4, lodestone.

eight Fe^{2+} ions occupy the remaining eight octahedral sites. The spin moments of the two groups of Fe^{3+} ions are antiparallel so the Fe^{3+} ions contribute no net moment. The eight Fe^{2+} ions have their moments aligned so the solid has a net moment. Solids which have a resultant moment because of incomplete cancellation are called **ferrimagnetic materials**.

EXAMPLE 32.7.1

Calculate the magnetization at 0°K for Fe_3O_4. The lattice parameter is 8.37 Å.

Answer. The oxygen atoms shield the Fe^{2+} ions so they have a magnetic moment characteristic of the free ions. The electron configuration is $Ar(3d)^6$. From Hund's rule we would expect a spin configuration of five up and one down or a net spin of 4β (Bohr magnetons).

Hence the magnetic dipole per unit volume expected is

$$M = \frac{4 \times 8(9.27 \times 10^{-24} \text{A-m}^2)}{(8.37 \times 10^{-10} \text{ m})^3}$$
$$= 0.5 \times 10^6 \text{ A/m}.$$

The experimental value is 0.48×10^6 A/m.

The fact that we can essentially predict the magnetization saturation of ferrites in general is shown in Figure 32.7.2.

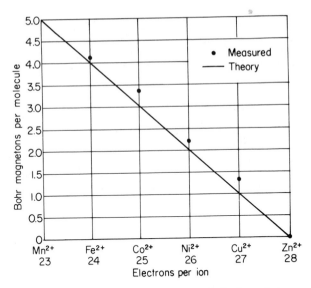

FIGURE 32.7.2. Molecular moments of ferrites of iron-group elements.

Ferrites, because of their high resistivity (10^6–10^9 higher than common magnetic metal alloys), show much smaller losses in high-frequency applications because the eddy current loses are less. The ferrites can be used for information storage with switching speeds in the microsecond range. The market for ferrites is illustrated in Table 32.7.1.

Table 32.7.1. SOFT FERRITE, MARKET-ESTIMATED, 1968*

Television receivers	20×10^6
Communications and radio components, recording heads, magnetostrictive transducers	20
Telephone communication	12
Microwave ferrite components	3
Computer memories	55
Total	110×10^6

* After I. S. Jacobs, *Journal of Applied Physics*, **40**, 917 (1969).

32.8   DOMAINS

Very tiny elongated pieces of iron exist as single domains. Large crystals of iron consist of many domains, as illustrated in Figure 32.8.1. The external

FIGURE 32.8.1. (a) Single domain established by exchange energy. (b) Magnetic energy, roughly proportional to the spatial extension of the field, about halved. (c) Magnetic energy reduced virtually to zero by closure domain.

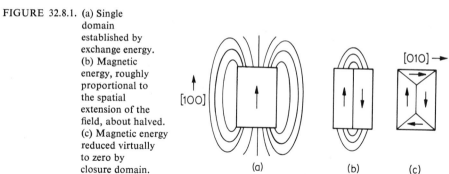

(a)   (b)   (c)

fields and hence the external energy are eliminated because of the domain formation.

There are several energy terms which have to be considered:

1. Exchange energy.
2. Magnetocrystalline anisotropic energy.
3. Magnetostatic energy.
4. Domain boundary energy.
5. Magnetostriction strain energy.

The exchange energy is of the order of 10^9 J/m³.

When a pure single crystal of a ferromagnetic material with no net moment is placed in a magnetic field, it magnetizes at a relatively small field. The magnitude of the field needed to achieve full magnetization depends on crystallographic direction, as shown in Figure 32.8.2. We say that the [100] direction is the easy direction in iron.

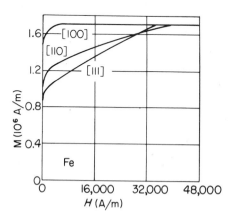

FIGURE 32.8.2. Magnetization curves for single crystals of iron.

For nickel the easy direction is [111], while [100] is the hard direction. For cobalt (which has the hcp crystal structure) the easy direction is [0001] and the hard direction is [10$\bar{1}$0].

The magnetic energy is given by $\int H \, dB$. We can write

$$\int H \, dB = \mu_0 \int H \, dH + \mu_0 \int H \, dM.$$

The first term would be present in a vacuum. The second term is due to the magnetization and is called the **magnetization energy**. Thus for iron, the work needed to magnetize the crystal in the [100] direction is nearly negligible. The extra work required to magnetize it in the [111] direction, say, is the **magneto-crystalline anisotropic energy**. It is of the order of 10^4 J/m³ (compare with the exchange energy). Thus the preferred direction of dipole alignment in iron is the [100] direction, and we might expect the domains in Figure 32.8.1 to have [100] magnetization.

The **magnetostatic energy** is the energy which results from the macroscopic dipole moment of the material. How the formation of domains lowers this energy is shown in Figure 32.8.1.

When a domain structure is formed, domain boundaries are formed. There is an energy per unit area associated with such boundaries analogous to grain boundary energy. However, there is an important difference. Grain boundaries are relatively narrow (a few atoms), as has been clearly shown by the field ion microscope (see Chapter 13). Domain boundaries are wide (300 atoms). Let us examine the reason for this. The exchange interaction between neighboring spins S_1 and S_2 is

$$U = -2J\mathbf{S}_1 \cdot \mathbf{S}_2, \tag{32.8.1}$$

where J, the exchange integral, is a constant of proportionality. For small angles

$$U = -2JS^2\left(1 - \frac{\theta^2}{2}\right),$$

so the difference in energy owing to rotation is $JS^2\theta^2$. Clearly, the spins must rotate as we change from one domain to another. It is clear that from the view-

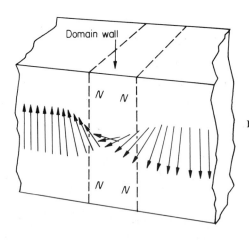

FIGURE 32.8.3. Domain wall. All magnetic moments lie in the plane of the wall. The N's represent poles formed on the surface of the material.

point of the exchange energy it is desirable to have the rotation between neighbors as small as possible and hence to have a very wide boundary. However, in passing from one [100] domain to a [$\bar{1}$00] domain in iron, the direction of spin rotates relative to the crystal, passing through various harder magnetization directions. Thus from the viewpoint of magnetocrystalline energy a narrow boundary is preferred. The balance of these two energy terms determines the width of the boundary. For iron, this boundary is about 1000 Å thick and has an energy of 2×10^{-3} J/m² (for comparison purposes, grain boundary energies are of the order of 1 J/m²). A **domain wall** (also called a **Bloch wall**) is illustrated in Figure 32.8.3. The poles formed on the surface of the material (shown by N's in Figure 32.8.3) are delineated by bright lines in **Bitter patterns**. Such a Bitter pattern is formed by placing a drop of colloidal suspension of magnetic material on the surface of a crystal, as shown in Figure 32.8.4. Domains can also be made visible by using the reflection of polarized light from a surface; such light is rotated by a magnetized surface in direct proportion to the magnetization.

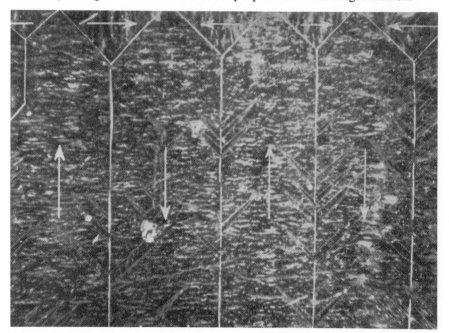

FIGURE 32.8.4. Bitter pattern (with arrows added) of (100) plane of iron.

MAGNETOSTRICTION.   When a crystal becomes magnetized it changes dimensions. This is called **magnetostriction**. (It is an analog of electrostriction.) The longitudinal strain versus field for three crystallographic directions in iron is shown in Figure 32.8.5. There is also a linear fractional increase in volume of

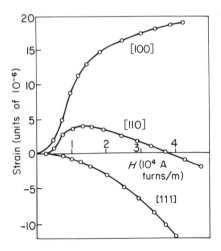

FIGURE 32.8.5. Longitudinal magnetostrictive strains versus field.

about 4×10^{-11} m/A for fields up to 2.5×10^4 A/m. Thus a crystal magnetized in the [100] direction will shrink in the [010] or [001] direction.

Consider now Figure 32.8.1(b). Both slabs will expand along the [100] direction and *contract* normal to it in the [010] direction. There are no constraints and hence no strain energy. However, in Figure 32.8.2(c), the closure domain would tend to *expand* in the [010] direction. It is constrained from doing so by the vertical domains. Hence there is elastic strain energy present because of magnetostriction. The strain energy can be decreased by reducing the volume of

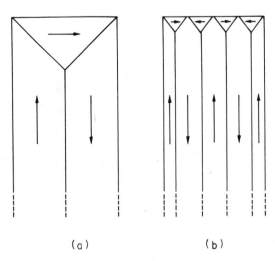

(a)   (b)

FIGURE 32.8.6. Elastic strain energy can be decreased by decreasing the volume of closure domain.

the closure domain, as shown in Figure 32.8.6. This results in an increase in domain boundary energy. The total balance between these various energy effects (along with the presence of externally induced strains, imperfections, etc.) determines the domain structure of a crystal.

32.9   MAGNETIZATION PROCESSES ACCORDING TO DOMAIN THEORY

Ferromagnetic materials are soft or hard depending on the ease or difficulty of moving domain boundaries and of rotating domains. This is illustrated in Figure 32.9.1. It should now be clear why the saturation magnetization M_s is

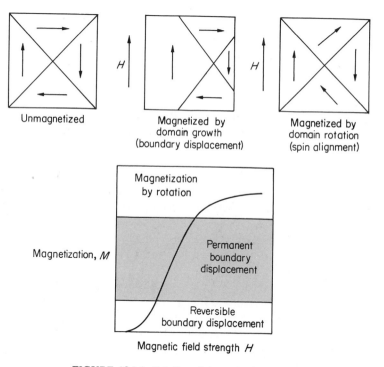

FIGURE 32.9.1. Relation of magnetization behavior.

essentially a structure-insensitive property, while other properties such as the relative permeability are strongly structure-sensitive. The magnitude of structure-sensitive properties depends on the ease of domain boundary movement and domain rotation. Note from Figure 32.9.1 that domain rotation is always difficult.

EXAMPLE 32.9.1

List several of the structure-sensitive properties of ferromagnetic materials.

Answer.
1. Permeability, μ. See Section 4.2.
2. Remanent induction, B_R. See Section 4.2.
3. Coercive force, H_c. See Section 4.2.
4. Hysteresis loss, $\oint H \, dB$. See Section 4.2.
5. Maximum energy product. See Section 4.4.

EXAMPLE 32.9.2

If a ferromagnetic material is divided into very fine elongated particles, these exist as single domains, as first suggested by Néel. Discuss how such **elongated single domains** (ESD) could be used to make a permanent magnet.

Answer. These particles could be placed in a liquid matrix, such as Pb, or epoxy resin prior to curing. This mixture could be placed in a magnetic field so that the particle domains were all oriented in the same direction. The matrix would then be solidified. The resultant permanent magnet is known as an **elongated single domain permanent magnet**. Theory suggests that the energy product could be about five times as large as for Alnico V, i.e., about 2×10^5 J/m³.

Alnico V is one of the hardest permanent magnet materials readily available. It is a cast alloy of iron, cobalt, nickel, aluminum, and copper. The first Alnico alloy was roughly Fe_2NiAl and was invented by T. Mishima in Japan. This was improved by F. W. Tetley of England, who added cobalt and copper. When properly heat-treated this material is actually a fine powder magnet since it consists of two phases, one of which is a rod-like precipitate as was shown by transmission electron microscopy by E. A. Nesbitt and R. Heidenreich. Figure 32.9.2 illustrates this structure, which is obtained by pouring the melt into a cylindrical mold which has hot walls and a cold bottom. It therefore cools uniaxially, separating in the process into a magnetic component which is rich in iron and cobalt and a nonmagnetic component rich in nickel and aluminum. In this respect it is analogous to the uniaxial formation of eutectic composites discussed in Section 16.4.

To make really hard magnets, we should eliminate the possibility of domain boundary movement. The best way to do this is to eliminate the domain boundary altogether. The two techniques for doing this have been discussed above. Active research is being carried out in each area.

The General Electric Company produces an ESD magnet material called Lodex which may be stamped, cold-pressed, etc. An alloy of 65% Fe–35% Co is electrodeposited in mercury as ESD. The particles are removed from the liquid

(a)   (b)   (c)

FIGURE 32.9.2. Microstructure of Alnico V cooled at 2°C/sec
from 1300°K. (a) (100) plane, no field. (b)
(010) plane, with field along [100] during cooling. (c) (100) plane, with field along [100]
during cooling. [*Courtesy of R. Heidenreich
and E. A. Nesbitt.* For further information
see *J. Applied Physics* **23**, 352 (1952).]

and coated with antimony. They are then placed in a lead matrix. The resultant
material is ground to powder, each grain of which is a single magnet. Because
lead is a soft malleable material, this material can be used to make magnets of
complicated configuration.

The Westinghouse Electrical Corporation makes a product named Westro
by sintering ESD ferrite. The ferrite used is strontium ferrite, which has a very
large crystalline anisotropy. Recall that sintering is the powder metallurgy
process which depends on the elimination of surface energy as the driving force
and on diffusion to accomplish the mass transport.

EXAMPLE 32.9.3

A refrigerator door needs a rubber gasket for stopping air flow and a
magnetic latch. Develop a technique for "killing both birds with one stone."

Answer. Strontium ferrite particles are dispersed in the rubber
prior to curing. The domains are aligned in a magnetic field and the
rubber is cured. This composite material is both the seal and latch.

The various mechanisms of pinning domain boundaries are very important in determining the behavior of soft magnetic materials. In this particular
application we are interested in eliminating such pinning so that boundary displacement occurs readily and reversibly.

R. Becker noted that inasmuch as most materials show magnetostriction, domain configuration must be related to local strains owing to foreign atoms, grain boundaries, or other metallurgical inhomogeneities. These interactions are eliminated if the magnetostriction is zero. This is illustrated by the very high permeability of permalloy materials at compositions at which the magnetostriction is zero. Nickel and iron exhibit opposite magnetostriction so it is not too surprising that an alloy can be made which has zero magnetostriction.

Another pinning mechanism was suggested by Kersten. A large nonmagnetic inclusion in a crystal would, if located at a domain boundary, decrease the total boundary energy since it decreases the domain boundary area, as shown in Figure 32.9.3. When a field is applied a certain critical energy has to be supplied by the field to move the boundary because of the need to create the additional domain boundary. The coercive force is not large for this mechanism. For example, cementite particles of the optimum size in a 1.5 wt. % C steel cause a coercive force of only 300 A/m.

FIGURE 32.9.3. Nonmagnetic precipitate at domain boundary.

Néel suggested another pinning mechanism at large nonmagnetic inclusions. There would be a pole density on the surface of such an inclusion within a domain. Magnetostatic energy is associated with such poles. Néel noted that this energy could be decreased by the creation of V-shaped spike-like domains of reverse or perpendicular magnetization near the inclusion. As a domain wall

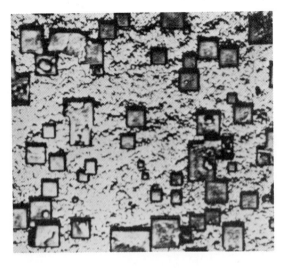

FIGURE 32.9.4. Cube-textured steel × 500. (*Courtesy of Karl Foster, Westinghouse Electric Corporation*).

moves past it is attracted to the spike; a coercive force must be applied to tear it away.

Another important factor, in addition to eliminating imperfections, in obtaining soft magnetic materials is eliminating crystalline anisotropy. The addition of molybdenum to permalloy results in zero anisotropy and an increased softness. The field needed to magnetize iron to $M = 1.5 \times 10^6$ A/m in the [111] direction is about 20,000 amps/m while in the [100] direction only about 20 amps/m are required. This suggests that transformer sheet should be manufactured with the plane of the sheet parallel to the cube faces of the crystal. N. P. Goss [*Transactions American Society for Metals* **23**, 515 (1935)] has developed techniques for rolling and heat treating polycrystalline iron in such a way that most of the crystals have a direction of easy magnetization [100] within several degrees of the rolling direction.

The hysteresis losses in such grain oriented sheet are about half that of nonoriented sheet. An example of such a cube-textured material is shown in Figure 32.9.4.

REFERENCES

G. E. Permanent Magnet Manual, General Electric Co., Edmore, Mich. Contains design data on various hard magnetic materials plus some worked-out design problems.

Keffer, F., "The Magnetic Properties of Materials," *Scientific American* (Sept. 1967) p. 222. Clear, very elementary discussion.

Bozorth, R. M., "The Physics of Magnetic Materials," in *The Science of Engineering Materials* (J. E. Goldman, ed.), Wiley, New York (1957). Excellent non-mathematical discussion.

Jacobs, I. S., "Role of Magnetism in Technology," *Journal of Applied Physics*, **40**, 917 (1969). A thorough discussion of the applications of magnetic phenomena. Includes magnetic information storage.

Nesbitt, E. A., *Ferromagnetic Domains*, Bell Telephone Laboratories, Murray Hill, N.J. (1962).

Bozorth, R. M., *Ferromagnetism*, Van Nostrand, Princeton, N.J. (1951). A classic book concerned with the structure-sensitive aspects of ferromagnetism.

Chikazumi, S., and Charap, S. H., *Physics of Magnetism*, Wiley, New York (1964). An excellent more advanced book.

PROBLEMS

32.1. The molar susceptibilities (10^{-9} m^3/kg-mole) of the inert gases are for He, Ne, Ar, Kr, and Xe, respectively, -1.88, -6.74, -19.6, -28.8, and -43.9. The atomic numbers are 2, 10, 18, 36, and 54, nearly equal in magnitude to the respective susceptibilities. Give the reasons for this.

32.2. Discuss why you would expect the Na^+ ion to be diamagnetic and the Na atom to be paramagnetic.

32.3. Would $|\chi_m|$ for the Na^+ ion be greater or smaller than $|\chi_m|$ for the neon atom?

32.4. Discuss why the paramagnetic susceptibility of conduction electrons is independent of temperature.

32.5. Discuss how adiabatic demagnetization works.

32.6. Calculate the saturation magnetization of $NiOFe_2O_3$ ferrite assuming there is no orbital moment. Compare with experiment. The lattice parameter is 8.34 Å.

32.7. In the $3d$ transition metals, the $3d$ electrons are quasi-free. Assuming this to be the case in copper and nickel, what atom concentration of copper can be added to nickel before the alloy loses its ferromagnetism?

32.8. What is the origin of ferromagnetism?

32.9. a. Why does a ferromagnetic crystal form domains?
  b. Closure domains?
  c. Why does it prefer many closure domains to a few?
  d. What limits the size of domains?

32.10. Describe the processes which occur during various portions of an *M-H* curve in ferromagnetic materials.

32.11. Name two techniques used to form very hard magnets.

32.12. List all the structural factors which must be controlled to obtain a soft magnetic material.

32.13. Explain why the squareness ratio is an important property of a memory core material.

32.14. Eddy current losses do not appear in static *B-H* curves. Comment.

32.15. Give two reasons why iron cores in transformers are laminated.

32.16. Power transformers often "hum." Why?

More Involved Problems

32.17. Show that the torque on a magnetic dipole is $\mathbf{T} = \mathbf{p}_m \times \mathbf{B}$.

32.18. Show that the potential energy of a magnetic dipole is $U = -\mathbf{p}_m \cdot \mathbf{B}$.

32.19. Calculate the molar diamagnetic susceptibility of the hydrogen atom $(-2.4 \times 10^{-12} \text{ m}^3/\text{g-mole})$ using (32.2.13).

32.20. Give at least two techniques for measuring *M-H* curves.

32.21. One of the important techniques of measuring magnetic fields is by studying nuclear magnetic resonance. Describe. For an elementary discussion, see C. A. Wert and R. M. Thomson, *Physics of Solids*, McGraw-Hill, New York (1964), Chapter 21.

32.22. Write an essay on magnetic memory devices.

32.23. For a single particle the angular momentum about the origin is $\mathbf{L} = \mathbf{r} \times \mathbf{p}$, where $\mathbf{r}$ is the position vector and $\mathbf{p}$ is the linear momentum. Evaluate L_x, L_y, and L_z.

32.24. In Problem 32.23, $L_x = yp_z - zp_y$. In quantum mechanics the x's and p's are replaced by operators. The x's are x's as operators, while the p's as operators are represented by

$$p_z = \frac{h}{2\pi i} \frac{\partial}{\partial z}, \quad \text{etc.}$$

What are the quantum mechanical expressions for the operators L_x, L_y, and L_z?

32.25. In Problem 32.24,

$$L_x = \frac{h}{2\pi i}\left(y\frac{\partial}{\partial z} - z\frac{\partial}{\partial y}\right).$$

Express L_x, L_y, and L_z in terms of the spherical coordinates r, θ, and ϕ. Here $x = r\cos\theta\sin\phi$, $y = r\sin\theta\sin\phi$, and $z = r\cos\theta$. This is not a simple problem. Help can be found in H. Eyring, J. Walter, and G. E. Kimball, *Quantum Chemistry*, Wiley, New York (1944) p. 40. Then evaluate the operator $L^2 = L_x^2 + L_y^2 + L_z^2$. See (32.3.11).

32.26. The wave functions for the hydrogen atom are given on p. 84 of H. Eyring et al. (see Problem 32.25). Use this to show that $L^2\psi = l(l+1)(h^2/4\pi^2)\psi$ and hence that $L = \sqrt{l(l+1)}(h/2\pi)$, and that $L_z\psi = m_l(h/2\pi)\psi$ and hence that $L_z = m_l(h/2\pi)$.

Sophisticated Problems

32.27. Discuss the Brillouin analysis of the Langevin model.

32.28. The OH^- ion has an electrical dipole moment of 4 D. Suppose OH^- ions are substituted for Cl^- ions so there are 2.9×10^{24} OH^- ions/m^3.
   a. Discuss how this system could be used to achieve adiabatic cooling (**paraelectric cooling**).
   b. Carry out the thermodynamics to predict the temperature drop from $1.27°K$ in a field of 75 kV/cm when the field is shut off. See I. Shepherd and G. Feher, *Physical Review Letters*, **15**, 194 (1965).

APPENDIX

It has been noted in the References to Chapter 2 that the September 1967 issue of *Scientific American* was completely devoted to materials. Upon reading that issue, John Updike composed the following poem which was published in the book *Midpoint and Other Poems*. It is reproduced here by permission of the publisher, Alfred A. Knopf, Inc.

THE DANCE OF THE SOLIDS

by John Updike

ARGUMENT: *In stanzas associated with allegory the actual atomic structure of solids unfolds. Metals, Ceramics and Polymers. The conduction of heat, electricity and light through solids. Solidity emerges as being intricate, giddy, playful.*

All things are Atoms: Earth and Water, Air
And Fire, all, *Democritus* foretold.

Swiss *Paracelsus*, in's alchemic lair,
    Saw Sulfur, Salt, and Mercury unfold
    Amid Millennial hopes of faking Gold.
    Lavoisier dethroned Phlogiston; then
    Molecular Analysis made bold
    Forays into the gases: Hydrogen
Stood naked in the dazzled sight of Learned Men.

The Solid State, however, kept its grains
    Of Microstructure coarsely veiled until
    X-ray diffraction pierced the Crystal Planes
    That roofed the giddy Dance, the taut Quadrille
    Where Silicon and Carbon Atoms will
    Link Valencies, four-figured, hand in hand
    With common Ions and Rare Earths to fill
    The lattices of Matter, Glass or Sand,
With tiny Excitations, quantitively grand.

The *Metals*, lustrous Monarchs of the Cave,
    Are ductile and conductive and opaque
    Because each Atom generously gave
    Its own Electrons to a mutual Stake,
    A Pool that acts as Bond. The Ions take
    The stacking shape of Spheres, and slip and flow
    When pressed or dented; thusly *Metals* make
    A better Paper Clip than a Window,
Are vulnerable to Shear, and, heated, brightly glow.

Ceramic, muddy Queen of human Arts,
    First served as simple Stone. Feldspar supplied
    Crude Clay; and Rubies, Porcelain, and Quartz
    Came each to light. Aluminum Oxide
    Is typical—a Metal is allied
    With Oxygen ionically; no free
    Electrons form a lubricating tide,
    Hence, Empresslike, *Ceramics* tend to be
Resistant, porous, brittle, and refractory.

Prince *Glass*, *Ceramic's* son though crystal-clear,
    Is no wise crystalline. The fond Voyeur

And Narcissist alike devoutly peer
Into Disorder, the Disorderer
Being Covalent Bondings that prefer
Prolonged Viscosity and spread loose nets
Photons slip through. The average *Polymer*
Enjoys a Glassy state, but cools, forgets
To slump, and clouds in closely patterned Minuets.

The *Polymers*, those giant Molecules,
Like Starch and Polyoxymethylene,
Flesh out, as protein serfs and plastic fools,
The Kingdom with Life's Stuff. Our time has seen
The synthesis of Polyisoprene
And many cross-linked Helixes unknown
To *Robert Hooke*; but each primordial Bean
Knew Cellulose by heart: *Nature* alone
Of Collagen and Apatite compounded Bone.

What happens in these Lattices when *Heat*
Transports Vibrations through a solid mass?
$C_p = 3Nk$ is much too neat;
A rigid Crystal's not a fluid Gas.
Debye in 1912 proposed Elas-
Tic Waves called *phonons* which obey *Max Planck's*
Great Quantum Law. Although amorphous Glass,
Umklapp Switchbacks, and Isotopes play pranks
Upon his Formulae, *Debye* deserves warm Thanks.

Electroconductivity depends
On Free Electrons: in Germanium
A touch of Arsenic liberates; in blends
Like Nickel Oxide, *Ohms* thwart Current. From
Pure Copper threads to wads of Chewing Gum
Resistance varies hugely. Cold and Light
As well as "doping" modify the sum
Of *Fermi* Levels, Ion scatter, site
Proximity, and other factors recondite.

Textbooks and Heaven only are Ideal;
Solidity is an imperfect state.

Within the cracked and dislocated Real
Nonstoichiometric crystals dominate.
Stray Atoms sully and precipitate;
Strange holes, *excitons,* wander loose; because
Of Dangling Bonds, a chemical Substrate
Corrodes and catalyzes—surface Flaws
Help Epitaxial Growth to fix adsorptive claws.

White Sunlight, *Newton* saw, is not so pure;
A Spectrum bared the Rainbow to his view.
Each Element absorbs its signature:
Go add a negative Electron to
Potassium Chloride; it turns deep blue,
As Chromium incarnadines Sapphire.
Wavelengths, absorbed, are reëmitted through
Fluorescence, Phosphorescence, and the higher
Intensities that deadly *Laser Beams* require.

Magnetic Atoms, such as Iron, keep
Unpaired Electrons in their middle shell,
Each one a spinning Magnet that would leap
The *Bloch* Walls whereat antiparallel
Domains converge. Diffuse Material
Becomes *Magnetic* when another Field
Aligns domains like Seaweed in a swell.
How nicely microscopic forces yield,
In Units growing Visible, the World we wield!

INDEX *

A

α-amino acids, **316**
α-iron, 363
Absorption band, **843**
Absorption coefficient, **786, 845**
Acceptors, **768**
Acids, organic, 309
Acrylonitrile, 305
Activated state, **443**
Activation energy, **442**
  creep, **46**
  diffusion, **448**
  diffusion, data, 462
Activator levels, **854**
Active (corrosion), **707**
Activity:
  coefficient, **693**
  HCl as a function of molality, data, 694
  rational, **693**
Addition, 1–4, **307**
Addition polymerization, **299**
Adherent oxide film, 472, 706
Adhesives, 53
Adiabatic demagnetization, **872**
Adiabatic process, **397**

Adipic acid, 309
Aeration cell, **703**
Age precipitation-hardenable alloys, examples, 363
Age precipitation-hardened alloy, **362**
Age precipitation hardening, theory, 662
Aggregate, 334
Aging, **130, 361**
  strain, **606**
Aircraft, 49
Airplanes, 1
AISI code for steel, 673, 675
Al-Cu phase diagram, 360
Allowed bands, 731
Alloy, 16
  eutectic, **352**
Alloying elements, **16**
Alloys, age precipitation-hardened, **362**
Alnico V:
  microstructure, 885
  properties, 100
Altimeter, 437
Aluminum-copper alloys:
  Al-Cu phase diagram, 360
  crystal structure, 204
  GP zone formation, 663
  heat treatment, 130, 360

* The boldface page numbers refer to the page on which the entry is defined and set in boldface type.

895

Aluminum-copper alloys (*Contd*)
  precipitation strengthening, 661
  substitutional solutions, 360
Amines, 309
Amonton's laws, **55**
Amounts of phases, 349
Amplitude, 29, **526**
Anelastic behavior, **525**
Angle:
  diffraction, **224**
  loss, **31, 526, 824**
Angular momentum, 865
Anharmonic solid, **433**
Anisotropic, **14**
  properties, data, 212
Anisotropy, 66
  diffusion, 476
  in graphite, 244
  relation to crystal structure, 211
Annealing, 609
Annihilation, dislocation, **591**
Anode, **696**
  sacrificial, **707**
Anodizing, **706**
Antiferroelectrics, **832**
Antiferromagnetic, **232**
Antiferromagnetism, 876
Arrest, thermal, **376**
Arrhenius rate equation, **445**
Asbestos, 256
Asphalt pavement, 683
ASTM grain size, **325**
Atactic, **306**
Atmospheres, dislocation, **606**
Atomic:
  coordination, 240
  percent, 247
  polarizability, **191**
  radius, data, 261
  scale, **321**
  scattering factor, **228**
  structure, **2**
Austenite, **246**
Austenitic steel, 368
Austenitizing, **130, 368**
Autofrettaging, **645**
Avalanche effect, **777**

Average molecular speeds, data, 394
Average molecular weight, **303**
Axes, principal, **841**
Axis, principal, **144**

B

Back voltage, **709**
Bainite, **365**
Bakelite, **313**
Band:
  absorption, **843**
  conduction, **740**
  forbidden energy, **729**
  gap energy, **73**
  Lüders, **604**
  valence, **738**
Band structure, electron:
  divalent metals, **742**
  insulators, 740
  monovalent metals, 739
  semiconductors, 740
Bar, **13**
Barium titanate, 83, 253
Barometric formula, **423**
Barrier:
  potential energy, **442**
  strength, **758**
Basal plane, **208, 238**
Base (n-p-n junction), **780**
Basis, **198**
$BaTiO_3$, 253
Batteries, 691
Bauschinger effect, **647**
BCS theory, **791**
Behavior, anelastic, **525**
B-H data for silicon iron, 104
Bias:
  forward, **775**
  reverse, **776**
Biaxial stress, **627**
Binary systems, 345
Binding energies:
  ionic, crystals, data, 176
Bitter patterns, **881**
Bleached, **848**
Bloch wall, **881**
Block polymer, **308**

Body-centered cubic, **203**
Bohr magneton, **867**
Bohr model, 867
Bohr radius, **158**
Boiling points of cryogenic liquids, data, 793
Boltzmann constant, **73, 387**
Boltzmann distribution, **422**
Bond:
    angles, data, 178
    covalent, **177**
    electron pair, **176**
    energies, **416**
    energies, covalent, 180
    hydrogen, **184**
    ionic, **174**
    lengths, data, 180
    metallic, **179**
Bonding, hydrogen, **184**
Bonds:
    hybridized, **178**
    secondary, **182**
    sp^3, **178**
    tetrahedral, **178**
    van der Waals, **183**
Born-Haber cycle, **191**
Born repulsive potential, **174**
Boron fiber composite, 333
Boron filaments, **334**
Boundary:
    large-angle grain, **286**
    twist, **286**
Bowling balls, 308
Boyle's law, **386**
Bragg equation, **224**
Bragg's law, **224**
Branched polymers, **300**
Brass, 704
Bravias lattices, **201**
Breakdown, **777**
Breakdown strength, **78**
Bridges, 49
Bridgman technique, **329**
Brillouin zone, **731**
Brinell hardness number, 653
Brittle behavior, **37**
Brittle material, **41**

Bubble raft model of crystal, 288
Bubbles, magnetic, **101**
Buckling, 12
Bulk moduli:
    alkali metals, 495
    data, 27
Bulk modulus, **27**
    origin, 494
Burgers vectors, **276**

C

C, phase diagram, 344
Cable, electric power, 78
Calorie, 125
Cantilever beam, 7
Capacitance, **75**
Carbide, sintered tungsten, **335**
Carbon-diamond phase diagram, 344
Carbon in iron, 246
Carburization depth, **457**
Carnot cycle, **405**
Carriers:
    majority, **771**
    minority, **771**
Case:
    carburized, 456
    hardened, 457
    nitrided, 457
Castings, 326
Cast iron, **369**
Cathode, **696**
Cathodoluminescence, **853**
Caustic embrittlement, **705**
Cell:
    aeration, **703**
    concentration, **698**
    differential concentration, **694**
    differential temperature, **694**
    dissimilar electrode, **691, 694**
    formation, **603**
    Kerr, **851**
    oxygen concentration, **702**
    parameters for some elements, data, 209
    primitive, **196**
    unit, **196**
    walls, **603**

Celluloid, 316
Cellulose, **315**
Cement, Portland, 334
Cementite, **364**
Centers:
  color, **847**
  recombination, **771**
Ceramics, **251**
  asbestos, 256
  barium titanate, 83
  $BaTiO_3$, 253, 830
  clay, 257
  concrete, 334
  corundum, 857
  diffusion in, 464
  Ferroxcube, 98
  fire clay refractories, 377
  forsterite, Mg_2SiO_4, 256
  lodestone, 877
  MgO, 175, 346
  MnO, 231
  NaCl, 174
  pyrex, 260
  quartz, 83, 120
  ruby laser, 857
  silica glass, 258
  silicates, 254
  soft magnet, 97
  talc, 257
  yttrium iron garnet, 99
  zeolites, 258
Cesium chloride crystal structure, **249**
Chain:
  helical, **306**
  polypeptide, **316**
Chain structures, **254**
Charge:
  dipole interaction energy, 191
  induced dipole interaction energy, 191
  mobilities in semiconductors, data, 765
  mobility, **718**
  space, **748**
Charpy V-notch impact test, **47**
Chemical force, **475**
Chemical potential, **410**

Chemiluminescence, **853**
Cis-configuration, **308**
Classical free electron gas model, **717**
Clausius-Mossotti equation, **821**
Clay, 257
Cleavage, **556**
Climb, **572, 582**
  dislocation, **459**
  force, **580**
Closest-packed layer, **237**
Closest packing, **237**
Closure domains, **879**
C-Ni phase diagram, 345
Coactivator levels, **854**
Coating, sacrificial, **707**
Coefficient:
  absorption, **786, 845**
  activity, **693**
  distribution, **370**
  fluidity, **52**
  friction, **55**
  friction, data, 55
  frictional viscous, **525**
  linear thermal expansion, **118**
  mobility, **474, 717**
  permeability, **54**
  rational activity, **693**
  relative permeability, **90**
  scattering, light, **848**
  segregation, **370**
  self-diffusion, **460**
  thermal conductivity, **120**
  tracer, **460**
  viscosity, **51**
  volume thermal expansion, **118**
Coefficients:
  elastic compliance, **498**
  Peltier, **132**
Coercive field, **82**
Coercive force, **92**
Coherence length, **790, 801**
Coherent boundary, **283, 663**
Coherent precipitate, 662
Coherent radiation, 857
Coherent twin, **283**
Coil, inductance of, 89

Collector, **781**
Collision, time, **718**
Color, dye, 190
Color centers, **847**
Column, 12
Columnar grains, **328**
Comet disasters, **45**
Complex dielectric constant, **825**
Complex silicates, **257**
Component, 347
Composites, 14, 16, **331**
  asphalt pavement, 683
  boron fiber, 333
  concrete, 334
  filamentary, 682
  grinding wheels, 335
  GRP, 331
  Lodex, 884
  microelectronic circuits, 336, 471
  plywood, 335
  sintered tungsten carbide, **335**
  strengthening, 681
  synthetic hard superconductors, 336
  Westro, 885
  whisker-based, 334
Composition, **347**
Compounds:
  interstitial, **246, 252**
  nonstoichiometric, **274**
  vinyl, **304**
Compressibility, linear, **211**
Compressive strength, 628
Concentration cell, **698**
Concentration polarization, **709**
Concept of local equilibrium, **369**
Concrete, **334**
Condensation polymerization, **370, 583**
Conduction, electrical by:
  electrons, 739
  holes, 741, 742
  ions, 461
Conduction band, **740**
Conductivity, electrical, **67**
  of ionic materials, 461, 475
  of metals, 740, 742
    data, 67
  Nernst equation, **475**

Conductivity (*Contd*)
  Ohm's law, **67**
    derivation, 718
  of semiconductors, 740, 765
    data, 67
Conductivity, thermal, **120**
  coefficient, **120**
    data, 122, 123, 124
  electrical insulators, 444
  Fourier heat conduction law, **120**
  gases, 433
  metals, 445
Congruent melting point, **354**
Conservation principles for dislocation,
    567
Conservative behavior, examples, 138
Conservative dislocation motion, **592**
Conservative process, **137**
Constant:
  complex dielectric, **825**
  dielectric, **76**
  equilibrium, **413**
  gas, **404**
  Hall, **73**
  Madelung, **188**
Constants (*see frontispiece*)
Contact potential, **754**
Continuity, equation of, **126**
Controlled eutectics, 359
Conversion factors (*see frontispiece*)
Cooling, paraelectric, **889**
Cooling curves, **376**
Cooperative phenomena, **422**
Cooper pair, **790**
Coordinates, lattice, **203**
Coordination:
  number, **180, 240**
  polyhedron, **240**
Copolymer, **310**
Copper:
  atomic radius, 261
  crystal structure, 209
  electrical resistivity, 67, 745
  electron mean free path, 744
  Fermi energy, 726
  ion radius, 262
  melting temperature, 129

Copper (*Contd*)
  Poisson's ratio, 27
  shear modulus, 20
  solution with nickel, 359, 669, 745
  thermal conductivity, 121, 123
  with beryllium, 363
  Young's modulus, 13
Copper-beryllium alloys, 363
Copper-gold phase diagram, 351
Copper-nickel phase diagram, 359
Core, dislocation, **570**
Core effects, **583**
Coring, **359**
Correlation factor, **460**
Corrosion:
  effect of liquid velocity, 709
  fatigue, **704**
  intergranular, **703**
  protection, 705
  requirements for, **703**
Corrosion, types of:
  caustic embrittlement, **705**
  dezincification, **704**
  fatigue, 704
  intergranular, **703**
  oxygen starvation, **703**
  season cracking, **705**
  stress corrosion cracking, 705
Corundum, 857
Coulomb attraction, 174
Covalent bond, **177**
Covalent radii, data, 262
Cracking:
  season, **705**
  stress corrosion, **704**
Cracks, 47
Creep, **45**
  activation energy, **46**
  diffusional, **518**
  high-temperature, theory of, **621**
Cristobalite, 242, 257
Critical:
  crack size, 626
  field, superconducting, **108**
  fields at 0°K, data, 108
  nucleus, **480**
  radius, **480**

Critical (*Contd*)
  resolved shear stress, **545**
  stress for crack propagation, 626
  temperature, superconducting, **108**
Crosslinking, **307**
Cross slip, **578, 588**
Cryogenic temperatures, **119**
Cryotron, 112
Crystal, faceting, **289**
Crystal, liquid, **850**
Crystal cell dimensions, data, 209
Crystal growth, 329
Crystal structure, **198**
  body centered cubic, **203**
  CsCl type, **249**
  diamond cubic, **204**
  face centered cubic, **204**
  hexagonal closest packed, **204**
  ice, 185
  ideal hcp, **238**
  inverted spinel, **876**
  metals, data, 243
  perovskite, **253**
  polymers, 314
  sodium chloride, 248
  tridymite, 258
  wurzite, **250**
  zinc blende, **249**
Crystal systems, **200**
Crystallization:
  in castings, 327
  in polymers, strain induced, 508
Crystallographic:
  direction, **206**
  plane, **206**
  point groups, **216**
Crystals:
  dichroic, **842**
  noncentrosymmetric, **209**
CsCl crystal structure, **249**
Cu-Au phase diagram, 351
Cu_3Au ordered solution, 247, 351
Cu-Be alloys, 363
Cu-Ni phase diagram, 359
Cube-texture, 887
Cubic:
  body centered, **203**

Cubic (*Contd*)
 diamond, **204**
 face-centered, **204**
 layer, 238
Cup and cone fracture, **633**
Curie:
 law, **870**
 temperature, **94**
Curie point:
 ferroelectric, **83**
  data, 83, 94
 ferromagnetic, **94**
  data, 94
Curie-Weiss relation, **870**
 theory, 872
Current density, **67**
 saturation, **776**
Curve:
 cooling, **376**
 hysteresis, **92**
 S-N, 44, 634
 time-temperature-transformation, **669**
 TTT, **669**
Cycle, Carnot, **405**
Cyclic:
 fatigue, **44**
 stress softening, 634
Cyclotron:
 frequency, **787**
 resonance, 766
Czochralski technique, **329**

D

Damped natural frequency, **526**
Damping:
 magneto-mechanical, **537**
 mechanical, 28
 thermoelastic, **589**
Darcy's equation, **54**
 derivation, 475
Darken equation, 467
Dark resistance material, **786**
Dashpot, **525**
Data:
 activation energies for diffusion, 462
 activity, HCl, 694
 angle of loss:

Data (*Contd*)
 electrical, 81
 mechanical, 31
 anisotropic properties, 212
 ASTM grain size, 327
 atomic radii, 261
 average molecular speeds, 394
 B-H data for silicon-iron, 104
 boiling points of cryogenic liquids, 793
 bond angles, 178
 bond energies:
  covalent, 180
  ionic, 176
  metals, 181
  organic compounds, 180
 bond lengths, 180
 bulk moduli, 27
  alkali metals, 495
 cell parameters of some elements, 209
 charge mobilities in semiconductors, 765
 Charpy V-notch energy, dependence on temperature, 48
 covalent radii, 262
 critical fields at 0°K, 108
 crystal structure of metals, 243
 Curie points, ferromagnetic, 94
 Debye frequencies, 429
 Debye temperatures, 429
 density, 49
 diamagnetic susceptibilities, 867
 dielectric constants, 76
  alkali halides, 822
 dielectric loss angle, 81
 dielectric strength, 79
 diffusion:
  gases in polymers, 469
  interstitial impurity, 454
  ionic crystals, 464
  metals, 462
  NaCl, 464
  vacancies in metals, 458
 dipole moment, electric, 820
 elastic constants:
  bulk modulus, 27
  Poisson's ratio, 27

Data (*Contd*)
   shear modulus, 20
   Young's modulus, 13
electrical resistivities, 67
electrical resistivity:
   of metals versus temperature, 71
   of semiconductors versus temperature, 72
electric loss angle, 81
electron concentrations, 726
electron configurations, 164
electronic polarizabilities, 815
energies of vacancy formation, 414
energies of vacancy motion, 458
energy gaps in semiconductors, 764
entropies (standard state), 432
F-center energies, 847
Fermi energies, 724
Fermi energy, calculated, 726
Fermi temperature, calculated, 726
Fermi velocity, calculated, 726
ferroelectric crystal data, 83
ferroelectric Curie points, 83
ferromagnetic data, 94
ferromagnetic elements, data, 876
ferromagnetics, soft, data, 96
galvanic series of metals in sea water, 697
glass transition temperatures of elastomers, 502
glass transition temperatures of polymers, 538
grain size, ASTM, 327
Hall coefficients, 182
heat capacity, molar, temperature dependence for solids, 124
heats of melting and vaporization, 127
heats of sublimation of molecular crystals, 184
heats of vaporization, metals, 181
hydrogen overvoltage, 710
hysteresis losses, magnetic, 96
ionic radii, 262
ionization energies of solutes in semiconductors, 768
ionization potentials, 163
lattice energy, ionic crystals, 176

Data (*Contd*)
   lattice parameters for some elements, 209
lattice parameters of martensite, 366
loss angle, dielectric, 81
   temperature dependence, 827
loss angle, mechanical, 31
magnetization saturation, 94
maximum energy product, 100
melting points, interstitial compounds, 252
melting temperatures, 129
metallic radii, 261
minimum radius ratios, 250
mobilities in semiconductors, 765
molecular diameters of gas molecules, 510
molecular speeds, 394
moments, electric dipole, 820
oxidation potentials, 695
penetration depths at 0°K, 798
permeability of gases in polymers, 469
photoelectric yield, 748
planar defect energies, 282
Poisson's ratio, 27
polarizabilities, electronic, 815
radii:
   covalent, 262
   ionic, 262
   metallic, 261
refractive index, 840
relative permeability, 96
resistivities, electrical, 67
resistivity:
   dependence of metals on temperature, 71
   dependence on temperature in semiconductors, 72
saturation:
   induction, 96
   magnetization, 94
   polarization, 83
self-diffusion, 462
shear moduli, 20
slip planes and directions, 546
specific stiffness, 49
specific tensile strength, 49, 333

Data (*Contd*)
  standard oxidation potentials, 695
  strength, tensile yield, 15
  strength of filaments, wires and whisk-
    ers, 680
  superconducting critical field, 108
  superconducting critical temperature,
    108
  superconducting transition tempera-
    tures, 791
  superconducting transition tempera-
    tures for compounds, 792
  surface energies, 282
  susceptibilities, magnetic, volume, 867
  tensile strength:
    of filaments, wires and whiskers,
      680
    of sapphire whiskers, size effect,
      680
  tensile yield strength, 15
  thermal conductivity, 123
    of copper, 121
    of quartz, 122
  thermal emf's of junctions, 131
  thermal expansion coefficient, linear,
    118
  transition temperature, superconduc-
    tivity, 791
  twin planes and directions, 559
  vacancy formation energies, 414
  vacancy motion energies, 458
  vibrational frequencies of atoms, 429
  viscosity coefficient, 52
    dependence on pressure, 59
    dependence on temperature, 59
    soda glass, 514
  work functions, 747
  yield strength, 15
  Young's modulus, 13
de Broglie relation, **155, 230**
Debye, **820**
  equations, **825**
  frequencies, data, 429
  frequency, 444, **429**
  model of specific heat, 428
  temperature, **429**
  temperatures, data, 429

Debye-Scherrer method, **227**
Decomposition, spinodal, **484**
Decrement, logarithmic, **526**
Deep sea submergence, 17
Defect:
  Frenkel, **269, 271**
  impurity, **271**
  modulus, **533**
  Schottky, **271**
  self-interstitial, **269**
Deformation:
  elastic, 22
  plastic, **35**
  twinning, **559**
  viscous, 51
Degenerate states, **154, 723**
Degree of crystallinity, **260**
Degree of polymerization, **303**
Degrees of freedom, **341**
de Haas–van Alphen effect, **735**
δ-iron, 364
Demagnetization, adiabatic, **872**
Dendrites, **328**
Density:
  crystals, from lattice parameter, 206
  current, **67**
  of material, **49**
  of states, **725**
  strain energy, **26**
Depth, carburization, **457**
Desalinization, 469
Deviation, standard, **439**
Devitrification, **260, 626**
Dezincification, **704**
Diamagnetic, **91**
Diamagnetism, theory of, 865
Diameter, molecular, **439**
Diameters, molecular collision, **453**
Diamond, 244
  band gap energy, 764
  charge mobility, 765
  crystal structure, **179**
  cubic, **204**
  electrical resistivity, 67
  metastability, origin of, 444
  origin of electrical resistance, 740

Diamond (*Contd*)
  synthesis, 344, 346
  Young's modulus, 13
Dichroic crystals, **842**
Dielectric, standard linear, **824**
Dielectric constant, **76**
  alkali halides, data, 822
  frequency dependence, 825, 827
  frequency dependence data, 76
Dielectric loss angle, **80**
Dielectric loss factor, **80**
Dielectric material, **76**
Dielectric strength, **78**
  data, 79
  origin of, 828
Dielectric susceptibility, **82**
  owing to induced dipoles, 814
  owing to permanent dipoles, 816
Differential concentration cells, **694**
Differential temperature cells, **694**
Diffraction:
  angle, **224**
  electrons, 230
  line, **227**
  low-energy electron, **230**
  neutrons, 230
  peak, 224, **226**
  X-rays, 223
Diffractometer, 225
Diffusion:
  activation energy for, **448**
  applications, 471
  atomic mechanism, 451
  coefficient, relation to mobility, 475
  coefficient, relation to random walk, 451
  controlled processes, list, 471
  distance, one dimensional, estimate, 449
  gases, 452
  grain boundaries, 465
  interstitial impurity, data, 454
  ionic crystals, 461
  ionic crystals, data, 464
  length, **772**
  liquids, 452
  measurement by tracers, 447

Diffusion (*Contd*)
  in microelectronic circuit manufacture, 47
  in NaCl, data, 464
  pipe, 465
  in polymers, 467
  self-, data, 462
  steady-state, **446**
  stress-assisted, **534**
  surface, 465
  time-dependent, 446
  tracer, 447
  vacancies, 457
Diffusional creep, **518**
Diffusionless transformation, **367**
Diffusivity, thermal, **119**
Diode:
  tunnel, **786**
  Zener, **777**
Dipole:
  dislocation, **585**
  electric:
    field created by, 833
    torque due to electric field, 817
  magnetic, **865**
    torque due to magnetic induction, 888
  moment, **82, 183**
  moments, electric data, 820
  permanent, **183**
  relation of electric dipoles and polarization, 811
  relation of magnetic dipoles to magnetization, 870
Direction:
  crystallographic, **206**
  slip, **546**
Dislocation, **274**
  annihilation, **591**
  atmospheres, **606**
  in cell walls, 603, 612
  climb, **459, 572**
  conservation principle, **567**
  core, **570**
  cross slip, **578**
  density, 567
  dipole, **585**

Dislocation (*Contd*)
edge, **276**
edge, stress field, 571
energy, 573
etch pits, 278, 285
fluxoid lattice, 804
force on, 580
formation energy, 573
Frank-Read source, **586**
Frank sessile, **579**
glide, 567
in grain boundaries, 285, 505
history, 565
impurity atmospheres, 596, 606
jogs, 585
kinks, 585
mixed, 585
motion, conservative, **592**
motion, nonconservative, **592**
multiple, **575**
multiplication, **586**
partial, **576**
Peierls-Nabarro stress, **583**
pileup, **593**
pinning, 596
plastic deformation, 567
poem, 686
reactions, 576
rearrangement, 611
screw, **281**
screw stress field, 569
sessile, **576**
total, **575**
velocity, 589
viewed by electron microscopy, 278
width, **583**
Disordering temperature, **421**
Dispersion, **828, 839**
Dispersion forces, **183**
Dispersion hardening, 666
Displacement, electric, **77**
Dissimilar electrode cell, **691, 694**
Dissipative behavior, examples, 138
Dissipative process, **137**
Distance, London penetration, **797**
Distance between planes in simple cubic
crystals, 207

Distinct configurations, **402**
Distribution coefficient, **370**
Distribution formula for Boltzmann's
statistics, **422**
Divacancies, **272**
Divalent metal, 742
DNA, 317
Domain:
basis for occurrence of, 878
boundary pinning, 883
closure, 879
elongated single, **884**
ferroelectric, **254, 830**
magnetic, **873**
magnetic, micrograph, 881
magnetic bubble, **101**
origin, 879
rotation, 883
wall, ferroelectric, **254, 881**
wall motion, 883
wall pinning, 885
Donors, **768**
Doping, **769**
Dorn relationship, **621**
Double chain silicates, **254**
Double charge layer, **756**
Double cross slip, **588**
Doubly refracting, **842**
Drift velocity, **718**
Drude's model, **717**
Duality of matter, 151
Ductile-brittle transition, effect of pres-
sure, 650
Ductile-to-brittle transition temperature,
632
Ductility, **36, 40**
pressure effects on, 650
Dulong and Petit, specific heat, **124,**
427
Dye molecules, color, 190

E

Easy direction, 879
Eddy currents, 95
Edge dislocation, **276**
Effect:
avalanche, **777**

Effect (*Contd*)
  Bauschinger, **647**
  de Haas-van Alphen, **735**
  electro-optic, **851**
  Hall, **73**
  inverse piezoelectric, **83**
  isotope, 793
  Meissner, **109**
  Peltier, **132**
  photoelastic, **842**
  piezoelectric, **83, 210**
  pyroelectric, **131**
  Schottky, **750**
  Seebeck, **132**
  Silsbee, **109**
  skin, **71**
  tunnel, **752**
Effective mass, **737**
Efficiency, electron injection, **781**
Efficiency, thermal, 45, 407
Eigenvalue problems, 153
Einstein convention, **141**
Einstein model of specific heat, 426
Einstein relation, **156**
Elastic:
  anisotropy, 500
  bodies, 501
  compliance coefficients, **498**
  constant matrix, 498
  constant measurements, 28
  energy per unit volume, **26**
  instability, **12**
  stiffnesses, **496**
Elastically:
  anisotropic, **26**
  isotropic, **26**
  unstable, **499**
Elastic constants:
  bulk modulus, **27**
    data, 27
  Poisson's ratio, **26**
    data, 27
  shear modulus, **25**
    data, 20
  Young's modulus, **22**
    data, 13

Elastomers, **308, 501**
  examples, 502
Electrical:
  conductivity, **67**
  ionic crystals, 462
  field, **67**
  force, **475**
  resistivity, data, 67
  resistivity, origin, 742
Electric current density, **67**
Electric dipole, **183**
Electric dipole moments of molecules,
    data, 820
Electric displacement, **77**
Electrochemical cell, 691
Electrode, 691
  potentials, **694**
  reaction, **694**
Electrodialysis, **470**
Electroluminescence, **853**
Electrolysis, 709
Electron:
  concentrations, data, 726
  configurations, table, 164
  delocalization, 177
  diffraction, **230**
  gas, 182, 717
  injection, **780**
  microscope, 279
  pair bond, **176**
  in a periodic potential, 729
  spin, **154**
  tunneling, **751**
Electronic polarizabilities, data, 815
Electronic polarization, **812**
Electronic specific heat, 727
  of superconductors, 796
Electroplating, 691
Electrooptic effect, **851**
Electrostriction, **829**
Element, symmetry, **199**
Elongated single domain, **884**
  permanent magnet, **884**
Embrittlement, caustic, **705**
Emission:
  field, **746**

Emission (*Contd*)
  photoelectric, **746**
  spectra, 156
  spontaneous, **855**
  stimulated, **856**
  thermionic, **746**
Emitter, **780**
Endurance limit, **45, 634**
Energy:
  activation, **442**
  band gap, **73**
  bands, origin of, 730
  bond, **416**
  Fermi, **723**
  formation of vacancy, **269**
  gaps, **730**
    semiconductors, data, 764
    superconductors, **790**
  Gibbs free, **409**
  Helmholtz free, **411**
  internal, **123, 395**
  jog, **584**
  magnetization, **880**
  magnetocrystalline anisotropic, **880**
  magnetostatic, **880**
  vacancy formation, data, 414
  vacancy motion, data, 458
  vibrational, of atoms, 425
Engine efficiency, 45
Engineering tensile strain, **21**
Enthalpy, **123, 409**
Entropy, **404, 408**
  measurement, 431
  mixing, 409
  standard state, data, 432
  vibrational, 430
Environment, **385**
Epitaxy, **291**
Equation:
  Arrhenius rate, **445**
  Bragg, **224**
  Clausius-Mosotti, **821**
  continuity, **126**
  Darcy's, **54**
  Darken, 467
  Debye, **825**

Equation (*Contd*)
  Eyring, **512**
  Fick's first, **445**
  Fick's second, **447**
  Gibbs-Curie, **295**
  Kelvin, **479**
  Mooney's, **684**
  Nernst, **475, 700**
  Newton's viscosity, **509**
  parabolic diffusion, **447**
  Schrödinger, **152**
  of state, **387**
  van der Waals, **438**
Equiaxed, **325**
Equilibrium:
  constant, **413**
  local, **583**
  state, **126**
Error function, **456**
  table, 456
ESD magnet, 885
Etchants, **323**
Etch pits, 278
Ethane molecule, 179
Ethylene, 305
Eutectic:
  alloy, **352**
  composition, 352
  controlled, 359
  lamellar, 357
  point, **352**
  rod-like, 358
  structure, **356**
  temperature, **352**
  transformation, **352**
  uniaxially solidified, 358
Eutectoid, **355**
  structure, **365**
Evaporation, field, **753**
Exact differential, **399**
Exchange interaction, **875**
Exciton, **851**
  fission, **853**
  Frenkel, **852**
  fusion, **853**
  Wannier, **852**

Exclusion principle, Pauli, **154**
Expansion, thermal:
   linear, **118**
   volume, **118**
Extension ratio, 507
External surfaces, **286**
Extrinsic region, **464**
Extrinsic semiconductor, **769**
Eyring equation, viscosity, **512**

F

Face-centered cubic, **204**
Factor, correlation, **460**
Failure:
   dielectric, 79
   mechanical, 8, 45, 47
Faraday equivalent, **692**
Fatigue:
   Comet disasters, 45
   corrosion, **704**
   cyclic, **44**
   endurance limit, **45**
   static, **45**
Fault, stacking, **284, 576**
F-center, **847**
   energies, data, 847
Fe-C phase diagram, 364
$Fe-Fe_3C$ phase diagram, 364
Fermi-Dirac distribution function, **726**
Fermi energy, **723**
   data, 724
   extrinsic semiconductors, 769
   intrinsic semiconductors, 763
Fermi surface, **723**
   parameters, data, 726
Fermi temperature, **725**
   data, calculated, 726
Fermi velocity, 726
   data, calculated, 726
Ferrimagnetic materials, **877**
Ferrite, **246**
Ferrites, **876**
Ferroelectric, **82**
   Curie point, **83**
   data, 83
   domains, **254**
Ferromagnetic, **91, 232**

Curie point, 94
   domains, 880
   elements, data, 876
   transition temperature, **94**
Ferromagnetism, **91**
   occurrence, 876
   origin, 874
Fick's first equation, **445**
Fick's law, **445**
   derivation of, 453
Fick's second equation, **447**
Field:
   coercive, **82**
   effect transistor, **782, 785**
   electrical, **67**
   emission, **746,** 749
      microscopy, **752**
   evaporation, **753**
   intensity, magnetic, **90**
   internal, **811**
   ion microscopy, **752**
   Mossotti, **820**
   tensor, **144**
Filaments, boron, **334**
Films, thin magnetic, **101**
Fine particle magnets, **101**
First-order phase transition, **795**
First-rank Cartesian tensor, **141**
Fission, exciton, **853**
Flow:
   plastic, 35
   viscous, 50
Fluid flow, 51
Fluidity coefficient, **52**
Fluids, Newtonian, **51,** 509, 512
Fluorescence, **854**
Flux exclusion, 109
Fluxoid, **800**
   lattice, **800**
      dislocations, 804
   pinning, 803
Folded-chain structure, **314**
Forbidden energy bands, **729**
Force:
   chemical, **475**
   climb, **580**
   dispersion, **183**

Force (*Contd*)
  electrical, **475**
  glide, **581**
  image, **750**
  interatomic, 174
  magnetic, 87
  Mott-Nabarro, **580**
  repulsive, **174**
  van der Waals, 183
Formaldehyde, 312
Forming of materials, 43
Forsterite, Mg_2SiO_4, 256
Forward bias, **775**
Fourier heat conduction law, **120**
  derivation of, 434
Fractional decrease in:
  amplitude per cycle, 31
  stored energy per cycle, 31
Fracture:
  brittle materials, 37, 41, 624, 625
  conchoidal, **633**
  cup and cone, **633**
  environmental effects, 630
  shear, **556**
  theory of, 624
Framework silicates, **257**
Framework structures, **257**
Frank-Read multiplication mechanism, **586**
Frank sessile dislocation, **579**
Freedom, degrees of, **341**
Free electron theory of metals, 181
Free energy:
  formation, standard, **412**
  partial molar Gibbs, **411**
  of reaction, **412**
Frenkel defect, **269, 271**
Frenkel exciton, **852**
Frenkel's model of ultimate strength, **552**
Frequency:
  cyclotron, **787**
  damped natural, **526**
  Debye, **429**
  natural, **29**
  threshold, **746**
Friction, **55**

Frictional coefficient, **525**
Fringed micelle structure, **314**
Fuel cells, 711
Fundamental theories, **2**
Fused silica glass, 258
Fusion, exciton, **853**

G

GaAs:
  band gap, 764
  charge mobility, 765
  crystal structure, 767
Gain, junction transistors, 781
Galloping Gertie, 35
Galvanic series, **697**
  of metals in sea water, data, 697
Galvanizing, **707**
γ-iron, 363
Gaps, energy, **730**
  superconductors, **790**
Gas, ideal, **386**
Gas constant, **404**
Gases:
  diffusion coefficient, 452
  diffusion in polymers, 469
  ideal, 386
  mean free path, **433,** 452
  mixing, 408
  molecular diameters, **439**
  thermal conductivity, 433
  viscosity, 509
Gauss's law, **77**
Gel, **334**
Generalized flow stress, **642**
Generalized form of Hooke's law, **496**
Generalized strain, **647**
Geometric scattering factor, **228**
Gibbs-Curie equation, **295**
Gibbs free energy, **409**
Gibbs phase rule, **341**
Ginsburg-Landau parameter, **801**
Glass, **259**
  blowing, 513
  reinforced plastics, **332**
  silica, **258**
  transition temperature, **259**

Glass (*Contd*)
  of elastomers, data, 502
    natural rubber, dependence on sulfur content, 503
    polymers, data, 538
Glide, **546**
  force, **581**
  mirror operation, **217**
  packets, **546, 547**
  planes, **546**
GP zone, 663
Gradient, 139
Graft polymer, **308**
Grain, **324**
  boundary, 282
    field ion micrograph of, 288
    small angle, 595
    small angle tilt, **285**
  columnar, **328**
  equiaxed, 328
  growth, 614
  -oriented transformer steel, 887
  preferred orientation, **326**
  size:
    ASTM, **325**
    effect on strength, 660
Graphite, 344
Griffith theory, **626**
Grooving, thermal, **324**
Ground state, **156**
Group, **214**
  multiplication table, 215
  velocity, **736**
Growth:
  parabolic, **472**
  spiral, 281
GRP, **332**
Guinier-Preston zone, 663
Gutta percha, **308**
Gyromagnetic ratio, **865**

H

Habit plane, **676**
Hair, 317, 501
Half-cell, **694**
  standard reference, **695**
Hall coefficient, **73**
  anomaly, **720**

Hall coefficient (*Contd*)
  data, 182
  for free carrier, 74
Hall constant, **73**
Hall effect, **73**
Hard direction, 879
Hardenability, **672**
Hardened materials:
  magnetic, 884
  mechanical, 659
  superconductive, 803
Hardening, mechanical:
  age, 661
  dispersion, 666
  grain boundary, 660
  martensite, 670
  polymers, 678
  second phase, 667
  spinodal decomposition, **669**
  strain, 607
Hard magnetic materials, **99**
Hardness, **651**
  Brinell, 653
  maximum, 674
  Rockwell, 653
  Vickers, 653
Hard superconductor, **107, 111, 804**
Harmonic oscillator:
  classical mechanical, **396**
  quantum mechanical, 426
Hcp crystal structure, **204**
Heat capacity, **123**
  molar, temperature dependence for solids, data, 124
  solids, 427
Heat flow, 125
  steady-state, **126**
  time-dependent, **126**
Heating element, 65
Heats of:
  melting, **127**
  melting and vaporization, data, 127
  sublimation, **127**
    molecular crystals, data, 184
  vaporization, **127, 343**
    metals, data, 181
Heat treatment, **16**, 130
  age precipitation hardening, 130

Heat treatment (*Contd*)
 solution, **130**
 steel, 130
Helical chain, **306**
Helical spring, 19
Helmholtz free energy, **411**
Henry's law, **689**
Heterogeneous nucleation, **481**
Hexamethylene diamine, 309
High-current carrying superconductors, 803
High-temperature creep, **621**
Hindered rotation, **300**
Hindrance, steric, **306**
Hole, **741**
Home permanent, 317
Homogeneous, macroscopically, **25**
Homogeneous nucleation, **481**
Homogenization, 359
Hooke's law, **22, 496**
H_2O phase diagram, 343
Hume-Rothery rules, 245
Hume-Rothery solubility rule, **247**
Hund's rule, **869**
Hybridized bonds, **178**
Hydrogen atom:
 Bohr radius, **158**
 energy levels, 156
 quantum numbers, 156
Hydrogen bond, **184**
Hydrogen bonding, **184**
Hydrogen bridge, 311
Hydrogen electrode, standard reference, **698**
Hydrogen-like atom, **158**
Hydrogen-like wave functions, **161**
 examples, 159
Hydrogen molecule, 176
Hydrogen overvoltage, data, 710
Hydrostatic pressure, 494
Hysteresis curves, **92**
Hysteresis loop, square, **98**
Hysteresis loss, magnetic, data, 96

I

Ice, crystal structure, 184
Ideal gas, **386**
Ideal gas temperature scale, **387**

Ideal hcp crystal structure, **238**
Idealized plastic material, **644**
Ideal solution, **418**
Identity period, **239**
Image, latent, **849**
Image force, **750**
Impact toughness, 47
Imperfections, **267**
Impressed potential, **707**
Impurity defect, **271**
Index of refraction, **837**
Indices, Miller, **206**
Indices, Miller-Bravais, **208**
Indifferent point, **350**
Inductance:
 mutual, 92
 self, **89**
Induction:
 magnetic, **90**
 remanent, **92**
 saturation, **92**
Induction effect, **183**
Ingot, 327
Initial permeability, **92**
Injection, electron, **780**
Injection efficiency, **781**
Instability:
 elastic, **40**
 plastic, **40**
Insulators, **66,** 75, 740
Integrating factor, **400**
Integration, large-scale, **778**
Interaction:
 exchange, **875**
 short range, **176**
Interface, Matano, **466**
Interfacial angles, 195
Interfacial energy, 282
Intergranular corrosion, **703**
Internal energy, **123, 395**
Internal field, 811, 820
Internal friction:
 brass, 535
 iron, 534
 irradiated copper, 536
 polymers, 537
 rubber, 529
Interplanar spacings, 207

Interstitial compounds, **246, 252**
   examples, 253
Interstitial impurity diffusion:
   application, 456
   data, 454
   mechanism, 454
Interstitial solution, **245**
Intrinsic, **741**
Intrinsic region, **463**
Intrinsic semiconductor, **761, 769**
Intrusions, 634
Invariant point, 354
Inverse piezoelectric effect, **83**
Inversion, population, **857**
Inversion operation, **201**
Ionic:
   bond, **174**
   conductivity, 462
   crystals, diffusion, 462
   polarizability, **816**
   polarization, **812**
   radii, **175**
      data, 262
Ionization energy of impurity atoms, es-
      timation of, 767
Ionization energy of solutes in semicon-
      ductors, data, 768
Ionization potential, **156**
   data, 163
Iron:
   carbide, 246, 364
   carbon phase diagram, 365
   cast, **369**
   crystal structure, 204, 209, 243
   elastic properties, 13, 20, 27
   ferromagnetic domains, 881
   ferromagnetic properties, 94, 96, 100,
      101, 876
   mechanical strength properties of
      various alloys, 15, 652, 676
   polymorphism, 243
   solution with carbon, 245, 363
   in steel, 365, 368
Island silicates, **254**
Isotactic, **306**
Isothermal process, **397**

Isotope effect, 793
Isotropic, **14**

J

Jog, **581, 584**
   energy, **584**
Jominy test, **672**
Junction:
   p-n, **75, 772**
   transistors, **75**

K

Kelvin equation, **479**
Kelvin method, **754**
Kerr cell, **851**
Kinetics, **441**
Kink, **458, 584**
Kirkendall effect, **467**
Kronig and Penny model, **730**

L

Lamellar spacing, pearlite, 365
Lamellar structure, **331, 356**
   effect of spacing on strength, 668
Lamé parameter, **498**
Langevin function, **819**
Lang technique, **279**
Large-angle grain boundary, **286**
Large-scale integration, **778**
Larson-Miller parameter, **622**
Laser, **857**
Latent image, **849**
Lattice, **196**
   Bravais, **201**
   coordinates, **203**
   fluxoid, **800**
   parameter, **197**
   parameters for some elements, data,
      209
   plane, **196**
   space, **196**
   vector, **197**
Laue method, **227**
Law:
   Amonton's, **55**
   Boyle's, **386**

Law (*Contd*)
Curie, **870**
Dulong and Petit, **124**
Fick's, **445**
Fourier heat conduction, **120**
Gauss's, **77**
generalized form of Hooke's, **496**
Hooke's, **22**
mass action, **413**
of mixtures, **375**
Ohm's, **67**
Raoult's, **418**
Schmid's, **549**
Sohncke's, **556**
Vegard's, **245**
Layer:
closest-packed, **237**
cubic, 238
double charge, **756**
Lead-tin phase diagram, 352
LEED, **230**
Length:
coherence, **790**
coherency, **801**
diffusion, **772**
Lennard-Jones potential, **495**
Leveling, zone, **372**
Levels:
activator, **854**
coactivator, **854**
Lever rule, **350**
LiBr-LiC1 phase diagram, 350
Lifetime, minority carrier, **771**
Light waves, 157
Limit, endurance, **634**
Linear:
behavior, examples, 137, 138
compressibility, **211**
elastic range, **22**
polymers, **301**
properties,
conservative, 138
dissipative, 138
response, **19**
solid, standard, **532**
thermal expansion coefficient, **118**
Line defects, **268**

Line tension, 661
Liquid crystal, **850**
Liquidus, **347**
Local equilibrium, **583**
concept of, **370**
Lodestone, Fe_3O_4, 877
Lodex, 884
Logarithmic decrement, **526**
London penetration distance, **797**
Long-range order parameter, **421**
Loop, hysteresis, 93
Loss angle, **31, 526, 824**
dielectric, **80**
dielectric, temperature dependence,
data, 827
electric, data, 81
mechanical, data, 31
relation to:
fractional decrease in amplitude
per cycle, 31
fractional decrease in energy per
cycle, 31
Q-factor, 33
relaxation time, 31
resonance amplification, 33
resonance peak width, 33
rubber, 529
Losses:
dielectric, 80
magnetic hysteresis, 94
mechanical, 31
ohmic, 66
Loss factor, dielectric, **80**
Low-energy electron diffraction, **230**
Lower yield point, **604**
Lubricants, 55
Lüders band, **604**
Luminescent, **853**

M

Machining, role of plastic deformation,
43
Macroscopically homogeneous, **25**
Macroscopic de Broglie wave, 791
Macrostructure, **321**
Madelung constant, **188**

Magnesium oxide:
  crystal structure, 249
  dislocation in, 665
  melting temperature, 129
  with nickel oxide, 346
  refractory, 129
Magnet, 68
Magnetic:
  bubbles, **101**
  dipole, **865**
    orientation, 870
  field intensity, **90**
  induction, **90**
  oxides, 97, 101, 876, 885
  scattering, **231**
  susceptibility, **91, 93**
    diamagnetic, data, 867
    paramagnetic, 870
  tape, 101
Magnetism:
  antiferro-, 876
  dia-, 91, 865
  ferri-, 877
  ferro-, 91, 874
  para-, 91, 867
Magnetization, **92**
  energy, **880**
  saturation, **93**
  spontaneous, **873**
Magnetocrystalline anisotropic energy,
    880
Magneto-mechanical damping, **537**
Magneton, Bohr, **867**
Magnetostatic energy, **880**
Magnetostriction, **101, 881**
Magnets, fine particle, **101**
Majority carriers, **771**
Martensite, **366**
  effect on strength of steel, 670
  finishes temperature, **367**
  lattice parameters, 367
  starts temperature, **366**
  transformation, **675**
    examples, 677
Masking, oxide, **471**
Mass, effective, **737**
Matano interface, **466**

Materials balance, 348
Materials selection, 685
Matrix, rotation, **143**
Matter tensors, **144**
Matthiessen's rule, **744**
Maximum energy product, **99**
  data, 100
Maximum hardness, steel, 674
Maximum permeability, **92**
Maximum real strain, **40**
Maximum shear stress criteria, **643**
Mean free path, **433, 509**
  relation to molecular diameter, 452
Mechanical damping, 28
Mechanical twinning, **283**
Meissner effect, **109**
Melting, heat of, **127**
Melting point:
  congruent, **354**
  data, 129
  interstitial compounds, data, 252
Membrane, semipermeable, 469
Mer, **299**
Metallic bond, **179**
Metallic radii, data, 261
Metallic radius, **181**
Metallography, **324**
Metals, **66**
  Al-Cu alloys, 360
  alloy activity, 699
  Alnico V, 885
  aluminum, 16
  brass, 704
  cast iron, **369**
  copper, 14
  Cu-Be alloys, 363
  iron, 96
  manganin, 746
  Mg-Al alloys, 363
  Nb_3Sn, 110, 804
  Nb-Ti alloy, 804
  nichrome, 746
  origin of electrical conductivity, 737
  precipitation in, 481
  samarium-cobalt magnets, 100
  SAP, **666**
  sintered tungsten carbide, 252

Metals (*Contd*)
  solder, 353
  stainless steel, 48
  steel, 16, 331, 368
  TD nickel, **667**
Metastable, **324**
Metastable polycrystalline configuration, **324**
Metastable structures:
  age precipitation hardened alloys, 362
  cold worked solids, 41, 609
  cored structures, 359
  diamonds, 344, 444
  eutectics, 355
  eutectoids, 365
  grain boundaries, 324
  iron carbide, Fe_3C, 364
  martensite, 368
  pearlite, **365**
  p-n junction, 474
  supersaturated solids, 361
Methane molecule, 179
Mg-Al alloys, 363
MgO-NiO phase diagram, 346
Microduplex structure, **623**
Microelectronics, **778**
Micrograph, **323**
Microscopy:
  field emission, **752**
  field ion, **752**
  scanning electron, **289**
  quantitative, **336**
  transmission electron, **230**
Microstrain, **634**
Microstructure, **321**
  effects on:
    current carrying capacity of super-conductors, 803
    hardness of magnetic materials, 885
    mechanical strength, 607, 659
Microstructures:
  Alnico-V, 885
  cored, 360
  dislocation tangles, 604
  domains in $BaTiO_3$, 830
  domains in iron, 881

Microstructures (*Contd*)
  filamentary composite, 683
  filamentary eutectic, 358
  fluxoids, 801
  lamellar, 357, 358
  martensite, 367
  microduplex, **623**
  pinned dislocation, 665
  polycrystalline aluminum, 322
  precipitate, 362, 662, 665
  spherulite, 315
  surface, 289
Miller-Bravais indices, **208**
Miller indices, **206**
Minimum radius ratio, **250**
  data, 250
Minority carriers, **771**
  lifetime, **771**
Mixed state, 800
Mixtures, law of, **375**
Mobilities in semiconductors, data, 765
Mobility:
  charge, **718**
  coefficient, **474, 717**
Modulus:
  bulk, **27**
  defect, **533**
  origin, 494, 499, 501
  shear, **25**
  tensile, **22**
  Young's, **22**
Molality, 693
Molar heat capacity at constant pressure, **123**
Molar heat capacity at constant volume, **123**
Molar polarizability, **821**
Molecular:
  beam experiment, 393
  collision diameters, **453**
  diameter, **439**
  diameters of gases, data, 510
  solids, **182**
  speeds, average, 394
  speeds, distribution, 392
Moment:
  dipole, **82, 183**

Moment (*Contd*)
  electric dipole, data, 820
  magnetic dipole, **865**
Mooney's equation, **684**
Mossotti field, **820**
Mott-Nabarro force, **580**
Multiphase materials, 330, 341
Multiple dislocation, **575**
Muscles, 501

N

NaF-MgF$_2$ phase diagram, 354
Natural frequency, **29**
Natural rubber, **308**
Nb$_3$Sn, 110, 804
Necking, **40**
Nematic behavior, **850**
Nernst equation, **475, 700**
Network-forming ions, **260**
Network-modifying ions, **260**
Network polymers, 312
Neutron diffraction, 230
Newtonian fluids, **51**
Newton's equation of viscosity, **509**
  derivation of, 509
N-fold rotation, **199**
Nitriding, 457
Noble coatings, 706
Nominal tensile strain, **21**
Nominal tensile stress, **21**
Noncentrosymmetric crystals, **209**
Nonconservative dislocation motion, **592**
Noncrystalline materials, 258, 303
Nonequilibrium structures (*see* metastable structures)
Nonmagnetic-magnetic phase diagram, 94
Nonstoichiometric compounds, **274**
Normal phase, **789**
Normal stress, **22**
Normal-superconductive phase diagram, 109
Notch fracture, 47
N-type, **74, 762**
  extrinsic semiconductor, **762**
Nuclear scattering, **231**

Nucleation, 476
  crack, 629
  folded-chain polymer crystals, 483
  Gibbs free energy of, 479
  hetergeneous, **481**
  homogeneous, **481**
  in perfect crystal growth, 483
  rate, 480
Nucleus, critical, **480**
Number, coordination, **180, 240**
Nylon 66, **310**

O

Octahedral:
  coordination, 240
  planes, **238, 548**
  void, **240**
Offshore oil platform, 708
Ohm's law, **67**
  derivation, 718
One-dimensional particle-in-box, 153
One-dimensional random walk, **387**
1-4 addition, **307**
Operation, translation, **197**
Operations, symmetry, **198**
Optical microscope, 323
Optical pumping, 858
Ordered substitutional solutions, **247**
Order of the reflection, **224**
Organic acids, 309
Orientation:
  effect, **183**
  electric dipole, 816
  grain, 887
  magnetic dipole, 870
  preferred, **326, 653**
Orlon, 304
Orthoferrites, **101**
Oscillator, harmonic, **396**
Osmosis, reverse, **469**
Osmotic pressure, **303**
Overaging, **362, 665**
Overlap, **734**
Overvoltage, **709**
  hydrogen, data, 710
Oxidation, 471, **696**
  potentials, data, 695

Oxidation (*Contd*)
  process, **694**
Oxide, adherent film, 472
Oxide masking, **471**
Oxygen concentration cell, **702**
Oxygen starvation, **703**

P

Packets, glide, **546, 547**
Packing, closest, **237**
Packing imperfections, **268**
Pair, Cooper, **790**
Parabolic diffusion equation, **447**
Parabolic growth, **472**
Parabolic heat equation, **126**
Paradox, specific heat, **720**
Paraelectric cooling, **889**
Paramagnetic, **91, 232**
Paramagnetism, theory of, 867, 868
Parameter:
  Ginsburg-Landau, **801**
  Lamé, **498**
  Larson-Miller, **622**
  lattice, **197**
    data, 209
Partial dislocations, **576**
Partial molar Gibbs free energy, **411**
Particle-in-a-box problem, **153**
Particle-in-a-box with infinite walls, **721**
Particle strengthening, 661, 666
Passivation, 706
Passive, **707**
Path:
  mean free, **433, 509**
  possible reaction, **441**
Patterns, Bitter, **881**
Pauli exclusion principle, **154**
Pauling's rules, **251**
Pb-Sn phase diagram, 352
Peach-Koehler equation, **581**
Peak, diffraction, **226**
Peak-width, **33**
Pearlite, **365**
  formation, 365
  strength, 667
Peierls-Nabarro stress, **583**

Peltier coefficients, **132**
Peltier effect, **132**
Penetration depth at $0°K$, data, 798
Peptide linkage, **316**
Percent:
  atomic, 351
  weight, 351
Percent reduction in area, **41**
Perfect diamagnetic material, 109
Period, indentity, **239**
Periodic potential, 728
Peritectic, **355**
Peritectoid, **355**
Permanent dipole, **183**
Permeability coefficient, **54**
  flow, 469
  initial, **92**
  magnetic, vacuum, **89**
  maximum, **92**
Permeability of gases in polymers, data,
  469
Permittivity, **76**
Perovskite structure, **253**
Petit and Dulong, "law" of, **124**
Phase, **330**
Phase, normal, **789**
Phase, separation, superconductors, 799
Phase, superconducting, **789**
Phase compositions, 347
Phase diagram, 340
  Al-Cu, 360
  applications, 374
  C, 344
  $CaO\text{-}Al_2O_3\text{-}SiO_2$, 379
  C-Ni, 345
  Cu-Au, 351
  Cu-Ni, 359
  determination, 376
  Fe-C, 364
  Fe-Cr-Ni, 379
  $Fe\text{-}Fe_3C$, 364
  H_2O, 343
  LiBr-LiCl, 350
  magnetic-nonmagnetic, 94
  MgO-NiO, 346
  $NaF\text{-}MgF_2$, 354
  Ni-C, 345

Phase diagram (*Contd*)
  nonmagnetic-magnetic, 94
  normal-superconductive, 109
  Pb-Sn, 352
  Pb-Sn-Bi, 378
Phenol, 312
Phenol-formaldehyde plastic, **312**
Phenolic, **313**
Phenomena, yield point, **604**
Phenomenology, **1**
Phonon, **428**
Phosphorescence, **854**
Phosphors, **854**
Photochromic, 851
Photoelastic effect, **842**
Photoelasticity, **842**
Photoelectric emission, **746**
Photoelectric yield, data, 749
Photograph, 849
Photoluminescence, **853**
Photon scattering, by atoms, 224
Piezoelectric effect, **83, 210**
Piezoelectric frequency reference, 83,
      833
Piezoelectric motor, 84
Pileup, **593**
  90-deg double, **624**
  120-deg double, **624**
Pinning:
  domain walls, 885
  fluxoids, 803
Pitting structures, **289**
Planar defect, **268**
  energies, data, 282
Planck's constant, **152**
Plane:
  basal, **208, 238**
  crystallographic, **206**
  glide, **546**
  habit, **676**
  lattice, **196**
  octahedral, **238, 548**
  reflection, **199**
  slip, **546**
  twinning, **559**
Plastic, glass reinforced, **332**
Plastic, phenol-formaldehyde, **312**

Plastic, silicone, **311**
Plastic deformation, **35**
Plastic instability, **40**
Plasticizers, **539**
Plastic work term, **628**
P-n junction, **75, 772**
  origin of rectification, 774
Point:
  defects, **268**
  eutectic, **352**
  groups, crystallographic, **216**
  indifferent, **350**
  lower yield, **604**
  upper yield, **604**
Poise, **52**
Poiseuille's equation, 51
Poisson ratio, **26**
  data, 27
Polarizability, **811**
  ionic, **816**
  molar, **821**
  space charge, **813**
  total, 819
  volume, **821**
Polarization, **82,** 809
  concentration, **709**
  electronic, **812**
  energy, **834**
  ionic, **812**
  remanent, **82**
  saturation, **82**
    data, 83
Polaroid, 842
Pole mechanism, **676**
Polyacrylonitrile, 304
Polycrystalline, **321**
Polyethylene, **299**
Polygonization, **610**
Polyhedron, coordination, **240**
Polyisoprene, **307**
  stereisomerism, 308
Polymerization:
  addition, **299**
  condensation, **310**
  degree of, **303**
Polymers:
  Bakelite, 313

Polymers (*Contd*)
  block, **308**
  branched, **300**
  cellulose, 315
  crystals, 314
  glass transition temperatures, 538
  graft, **308**
  linear, **301**
  nylon, 14
  Nylon 66, 310
  Orlon, 305
  phenolics, 312
  polyacrylonitrile, 305
  polyethylene, 304
  polyisoprene, 307
  polypropylene, 306
  polystyrene, 305
  polytetrafluoroethylene, 307
  protein, **316**
  silicones, 311
  strengthening, 678
  Teflon, **307**
  vinyl compounds, 305
Polymer structures:
  folded chain crystals, 314
  fringed micelle, 313
  glassy, 468
  random chain, 301
  spherulitic, 315
Polymorphism, **243**
Polymorphs, **244**
Polypeptide chain, **316**
Polyphase materials, 330
Polypropylene, 305
  stereoisomerism, 306
Polystyrene, 304
Polytetrafluoroethylene, 307
Polyvinylacetate, 304
Population inversion, **857**
Porous media, 54
Portland cement, 334
Possible reaction paths, **441**
Potential:
  chemical, **410**
  contact, **754**
  electrode, **694**
  energy barrier, **442**

Potential (*Contd*)
  impressed, **707**
  ionization, **156**
  Lennard-Jones, **495**
  oxidation, 695
  6-12, **495**
Powder metallurgy, 474
Powder X-ray method, 227
Power, thermoelectric, **132**
Power generation, 95
Precipitates, effect on strength, 661
Preferred orientation, **326, 653**
Pressure:
  effect on ductility, 44, 650
  effect on viscosity coefficient, 59, 510
  effect on volume, 494
Pressure gradient, 54
Primitive cell, **196**
Principal axes, **144, 841**
Principal stresses, **144**
Probability density, **152**
Process:
  adiabatic, **397**
  conservative, **137**
  dissipative, **137**
  isothermal, **397**
  rate, **441**
  reversible, **398**
Product, **214**
Property:
  structure-insensitive, **16**, 268
  structure-sensitive, **16**, 268
Protective films, 472
Protein, **316**
PTFE, **307**
P-type, **74**
  extrinsic semiconductor, **762**
Pulling technique, **329**
Pyrex, 260
Pyroelectric effect, **131**

Q

Quality factor, **33**
  quartz, 84
Quantitative microscopy, **336**
Quantized electron gas, **722**
Quantum number, **154**

Quantum yield, **747**
Quartz, 83, 196, 257, 342
  crystals, resonance, frequency, 833
Quasichemical model, **416**
Quenched, **130**
Quenching, **361**

R

Radial distribution function, **391**
Radial probability density, **158**
Radii:
  covalent, data, 262
  ionic, **175**
    data, 262
  metallic, data, 261
Radius:
  Bohr, **158**
  critical, **480**
  metallic, **181**
  ratio, **241**
    for voids in bcc crystal, 242
    for voids in closest packing, data, 241
    minimum, **250**
Random:
  chain configuration, **303**
  distribution function, **390**
  solid solutions, **246**
  walk:
    in diffusion, 451
    relation to polymer chain length, 301, 451
    velocity distribution in gases, 392
Raoult's law, **418**
Rate, temperature-reduced strain, **621**
Rate equation, Arrhenius, **445**
Rate processes, **441**
Ratio:
  gyromagnetic, **865**
  Poisson, **26**
  radius, **241**
  resistivity, **70**
  squareness, **99**
  supersaturation, **481**
  Wiedemann-Franz, **120**
Rational activity, **693**
  coefficient, **693**

Reaction, electrode, **694**
Real strain, **38**
  at fracture, 41
Real stress, **38**
Recombination, **771**
Recombination centers, **771**
Recovery, **610**
Recrystallization, **484,** 612
Rectified, **776**
Rectifiers, 777
Reduction in area, percent, **41**
Reduction reaction, **696**
Refining, zone, **372**
Reflectance, **848**
Reflection:
  microscope, 321
  order of the, **224**
  planes, **199**
Refracting, doubly, **842**
Refractive index:
  anisotropy, 840
  data, 840
Refractory materials, **128**
Regular solution, **419**
Reinforced:
  concrete, 335
  metals, 682
  plastics, 331
Relation, the de Broglie, **230**
Relative permeability:
  coefficient, **90**
  data, 96
Relaxation:
  strain, **532**
  stress, **532**
  time, **31, 83, 526, 824**
    dielectric, origin of, 824
  Zener, **535**
Remanent induction, **92**
Remanent polarization, **82**
Repulsive potential energy, 174
Residual resistivity, **744**
Resistivity, **66**
  effects of impurities, 70
  electrical, data, 67
  metals versus temperature, data, 71
  ratio, **70**

Resistivity (*Contd*)
  residual, **744**
  semiconductors, versus temperature, data, 72
  size effect, 71, 87
  thermal, **743**
Resonance amplification, **33,** 34
Resonance frequency, **33**
Reverse bias, **776**
Reverse osmosis, **469**
Reversible processes, **398**
Rockwell C hardness, 653
Rotation:
  hindered, **300**
  matrix, **143**
  n-fold, **199**
Rotoinversion, **201**
Rubber:
  natural, **308**
  silicone, **311**
  stress-strain curve, 508
  synthetic polyisoprene, **307**
Rubberlike elasticity, origin, 505
Rubberlike materials, examples, 502
Rule:
  Gibbs phase, **341**
  Hume-Rothery solubility, **247**
  Hund's, **869**
  lever, **350**
  Pauling's, **251**
  selection, **229**
  Trouton's, **128**
Rust inhibitors, 707

S

Sacrificial anodes, **707**
Sacrificial coating, **707**
Safety factor, 9
St. Venant's hypothesis, **646**
SAP, **666**
Saturation, phototube, **748**
Saturation current density, **776**
Saturation induction, **92**
  data, 96
Saturation magnetization, **93**
Saturation polarization, **82**

Scale:
  atomic, **321**
  ideal gas temperature, **387**
Scanning electron microscope, **289**
Scattering, magnetic, **231**
Scattering, nuclear, **231**
Scattering coefficient, light, **848**
Scattering factor:
  atomic, **228**
  geometric, **228**
Schmid's law, **549**
Schottky defects, **271**
Schottky effect, **750**
Schrödinger's wave equation, **152**
Screw axis, 217
Screw dislocation, **281**
Season cracking, **705**
Secondary bonds, **182**
Second-order phase transition, **795**
Second phase strengthening, 661, 666, 667
Second-rank Cartesian tensor, **141**
Seebeck effect, **132**
Segregation, **371**
Seismology, 28
Selection of materials, 685
Selection rules, **229**
Self-diffusion, **448**
  coefficient, **460**
  data, 462
Self-inductance, **89**
Self-interstitial, **269**
  viewed by field ion microscopy, 270
Semiconductor, **66,** 740
  intrinsic, **741, 761**
  n-type extrinsic, **762**
  p-type extrinsic, **762**
Semipermeable membrane, **303**
Sequence, stacking, **238**
Sessile dislocation, **576**
Shakedown pressure, **645**
Shear:
  fracture, **556**
  modulus, **25**
    data, 20
  stiffness constant, **25**
  strain, **23**

Shear (*Contd*)
  stress, **23**
  yield strength, 19
  yield stress, ultimate, 552
Sheet silicates, **256**
Sheet structures, **256**
Shock, thermal, **117**
Short-range interaction, **176**
Side chains, **300**
Silica glass, **258**
Silica polymorphs, 257
Silicate structural units, examples, 254
Silicate structures:
  double chain, **254**
  framework, **257**
  island, **254**
  sheet, **256**
  silica glass, **258**
  single chain, **254**
Silicon:
  band gap energy, 764
  charge mobilities, 765
  crystal structure, 204, 209
  diffusion of phosphorous in, 473
  electrical resistivity, 67
  microelectronic technology, 471, 779
  oxidation, 472
Silicone plastics, **311**
Silicone rubber, **311**
Silk, 317
Silsbee effect, **109**
Silver bromide, 849
Simple cubic crystal, 207
Simple harmonic motion, 29
Single chain silicates, **254**
Sintered aluminum powder, **666**
Sintered tungsten carbide, 252, **335**
Sintering, 290
SiO_2, 258
SiO_4^{4-} tetrahedron, 254
6-12 potential, **495**
Skin effect, **71**
Slicing method, diffusion, **448**
Slip, **546**
  cross, **578, 588**
  direction, **546**
  double cross, **588**

Slip (*Contd*)
  planes, **546**
  planes and directions, data, 546
  system, **548**
Small angle tilt grain boundary, **285**
S-N curve, 634
Snowflake, 200
Soda glass, **260**
Sodium chloride structure, **174**
Soft ferromagnetics, data, 96
Soft magnetic material, **95**
Soft superconductors, **108**
Sohncke's law, **556**
Solar cells, 777
Solder, 353
Solenoid, 68, 89
Solid, anharmonic, **433**
Solidification, uniaxial, 356, 884
Solids, molecular, **182**
Solid solution:
  interstitial, **245**
  ordered, **247**
  random, **246**
  substitutional, **245**
  supersaturated, 361, 366
Solidus, **347**
Solubility, basis of, 416
Solution:
  ideal, **418**
  interstitial, **245**
  ordered substitutional, **247**
  regular, **419**
  substitutional solid, **245**
Solution heat treatment, **130**
Solutionizing, **361**
Sommerfeld model, **722**
Sonar, 83
Sound velocities, 28
Sp^3 bonds, **178**
Space charge, **748**
Space charge polarizability, **813**
Space lattice, **196**
Spacings, interplanar, 207
Spatial defects, **268**
Specific heat:
  electrons, paradox, **720**
  gases, 124

Specific heat (*Contd*)
  solids, 427
Specific stiffness, **49**
  data, 49
Specific tensile strength, **49**
Specific tensile strength, data, 49
Spectroscopy, notation, 157
Spinel structure, inverted, **876**
Spinodal decomposition, **484**
  hardening, **669**
Sphalerite, **249**
Spherulites, 315
Spontaneous emission, **855**
Spontaneous magnetization, **873**
Spring, helical, 19
Square hysteresis loop, **98**
Squareness ratio, **99**
Stacking fault, **284, 576**
Stacking sequence, **238**
Stainless steel:
  austenitic, 368
  corrosion properties, 706
  low temperature ductility, 48
Standard deviation, **439**
Standard linear dielectric, **824**
Standard linear solid, **532**
Standard oxidation potentials, data, 695
Standard reference half-cell, **695**
Starch, **316**
Starvation, oxygen, **703**
State, activated, **443**
State, equation of, **387**
State function, **387**
States, degenerate, **723**
Static fatigue, **45**
Statistics:
  Boltzmann's, **422**
  Fermi-Dirac, **726**
Steady-state, **126**
  diffusion, **446**
  heat flow, **126**
Steel, **368**
  AISI code, 675
  austenitic, 368
  bainite formation, 365
  designations, AISI, 675
  ductile-to-brittle transition, 47

Steel (*Contd*)
  elastic constants, 13, 20, 27
  Fe-Fe$_3$C phase diagram, 364
  heat treatment, 365, 669
  hypoeutectoid, 365
  martensite formation, 367, 675
  patented steel wire, 667
  pearlite formation, 365, 669
  plastic properties, 15, 676
Steps, **458**
Stereoisomers, **306**
Steric hindrance, **306**
Stern-Gerlach experiment, 869
Stiffness:
  cantilever beam, 8
  constant, shear, **25**
  constant of helical spring, 20
  constant of shaft in torsion, 18
  specific, **49**
Stiffnesses, elastic, **496**
Stimulated emission, **856**
Stirling's approximation, 389
Strain:
  aging, **606**
  component, 145
  energy density, **26**
  engineering tensile, **21**
  generalized, **647**
  hardening, **36,** 554
  hardening exponent, **39**
  infinitesimal, 523
  matrix, 496
  maximum real, **40**
  nominal tensile, **21**
  plastic, relation to dislocation motion, 567
  rate, plastic, relation to dislocation motion, 567
  real, **38**
  shear, **23**
Strength:
  breakdown, **78**
  creep, 45, 618
  dielectric, **78**
  fatigue, 44, 634
  filaments, wires and whiskers, 680
  fracture, 47, 624

Strength (*Contd*)
  high temperature, 45, 618
  specific tensile, **49**
  tensile fracture, **14**
  tensile yield, **1, 22**
Strengthening concepts, 659
Strengthening mechanisms:
  composites, 679
  dispersed particles, 666
  eliminate imperfections, 659
  grain boundary, 660
  at high temperatures, 616, 618, 678
  lamellar structures, 667, 669
  martensite formation, 669
  of piano wire (patented steel wire),
    667
  of polymers, 679
  precipitate, 661
  second phase, 666, 667
  solute, 553, 596, 604
  spinodal decomposition, **669**
  strain hardening, **36,** 554, 607
  of viscous matrices, 683
Stress:
  assisted diffusion, **534**
  biaxial, **627**
  concentration factor, **42**
  corrosion cracking, **704**
  critical resolved shear, **545**
  general form, 139
  generalized flow, **642**
  matrix notation, 496
  nominal tensile, **21**
  normal, **22**
  Peierls-Nabarro, **583**
  real, **38**
  relaxation, **532**
  relaxation time, **532**
  shear, **23**
  tensor, **141**
  units, 13
Stresses, principal, **144**
Structure:
  atomic, **2**
  crystal, **198**
  CsCl crystal, **249**
  diamond, **179**

Structure (*Contd*)
  euctectic, **356**
  eutectoid, **365**
  folded-chain, **314**
  fringed micelle, **314**
  -insensitive property, **16**
    table of, 268
  lamellar, **331, 356**
  microduplex, **623**
  sensitive property, **16**
    table of, 268
  sodium chloride, **174**
Styrene, 305
Subcells, 603
Sublimation, heat of, **127**
Substitutional solution:
  ordered, **247**
  random, **246**
  solids, **245**
Superconducting critical field, **108**
  data, 108
Superconducting critical temperature,
    108
  data, 108
Superconducting phase, **789**
Superconductive phase, specific heat,
    796
Superconductivity, occurrence, 791
Superconductor, **66**
  applications, 112
  hard, **107, 111, 804**
  soft, **108**
  synthetic hard, 112
  type I, **109**
  type II, **109**
Supercooling, 481
Superlattices, **247**
Superplasticity, **41, 622**
Supersaturated solid solution, 361
Supersaturation ratio, **481**
Surface:
  energies, data, 282
  Fermi, **723**
  Gibbs-Curie equation, **295**
  steps, **289**
    electron micrograph replica, 289
Surrounding, **382**

Susceptibility:
  dielectric, **82**
  magnetic, **91, 93**
    volume, data, 867
  volume, **866**
Symmetry, **198,** 214
  element, **199**
  operations, **198**
Syndiotactic, **306**
Synergism, 331
System, **385**
  crystal, **200**
  slip, **548**

T

Talc, 257
TD nickel, **667**
Teflon, **307**
Temperature:
  change, 118
  compensated time, creep, **622**
  cryogenic, **119**
  Curie, **94**
  Debye, **429**
  disordering, **421**
  ductile-to-brittle transition, **632**
  effects destroyed by, 128
  effects on properties, 128
  eutectic, **352**
  exponential effects, 129
  Fermi, **725**
  ferromagnetic transition, **94**
  gradient, 120
  martensite finishes, **367**
  martensite starts, **366**
  reduced strain rate, **621**
Temperature effects on:
  chemical reaction rates, 443
  conductivity:
    electrical, ionic, 463
    electrical, metals, 71
    electrical, semiconductor, 72
    thermal, 121, 122
  creep, 46, 620
  diffusion, 448, 460
  electron concentration, semiconductors, 764

Temperature effects on (*Contd*)
  electron emission rate, 748
  engine efficiency, 45, 407
  impact toughness, 48
  strain hardened materials, 41, 609
  vacancy:
    concentrations, 414
    motion, 458
  vapor pressure, 416
  viscosity coefficients:
    gases, 510
    liquids, 53
  yield strength, 617
Tempered, **130**
Tempering, **368**
Tensile fracture strength, **14, 20**
Tensile modulus, **22**
Tensile strength, specific, **49**
Tensile yield strength, **1, 22**
  data, 15
Tension test, 649
Tensor, **141**
  field, **144**
  matter, **144**
  stress, **141**
  symmetric, **144**
Ternary systems, 377
Tetrafluoroethylene, 307
Tetragonal distortion, 597
Tetrahedral bonds, **178**
Tetrahedral void, **240**
Thermal arrest, **376**
Thermal conductivity:
  coefficient, **120**
    data, 123
    units, 120
  of copper, temperature dependence data, 121
  metals, origin, 435
  origin, 433
  temperature dependence of quartz, 122
Thermal diffusion:
  distance, 119
  time, 119
Thermal diffusivity, **119**
Thermal emf's, data, 131

Thermal expansion, theoretical basis, 432

Thermal expansion coefficient:
  linear, **118**
    data, 118
    volume, **118**

Thermal grooving, **324**
Thermal resistivity, **744**
Thermal response time, **126**
Thermal shock, **117**
Thermal stresses, 117
Thermionic emission, **746,** 748
Thermistor, **766**
Thermocouple, **132**
Thermodynamics, **385**
  classical, function of, 413
  rubber, 503
  statistical, function of, 414
Thermoelastic damping, **589**
Thermoelectric:
  effect, origin, 782
  generator, 133
  heat pump, 132
  power, **132**
  refrigerator, 133
Thermoplastic, **313**
Thermosetting, **313**
Thin magnetic films, **101**
Three-element model, 531
Threshold frequency, **746**
Tie line, **348**
Time:
  between collisions, **718**
  dependent heat flow, **126**
  independent heat flow, **126**
  relaxation, **83, 526, 824**
  stress relaxation, **532**
  temperature-compensated, **622**
  temperature transformation curve, **669**
  thermal response, **126**
Titanate, 83, 253, 831
Total dislocation, **575**
Toughness, **36**
  impact, 47
Tracer coefficient, **460**

Trans-configuration, **308**
Transducer, 83
  magnetostrictive, 101
  piezoelectric, 83
Transference numbers, **701**
Transformation kinetics:
  condensation, 479
  eutectic, 356, 450
  martensitic, 670
  nucleation, 479
  oxidation, 472
  pearlitic, 669
  precipitation, 361
  recovery, 609
  solidification, 481
Transformation of axes, 141
Transformations:
  diffusionless, **367**
  eutectic, **352**
  eutectoid, **355**
  ferroelectric, 83
  ferromagnetic, 94
  indifferent, 350
  inversion, 355
  martensitic, **675**
  melting, 354
  peritectic, **355**
  peritectoid, **355**
  spinodal, **484**
  superconductive, 109
Transformer sheet, 887
Transistor, 779
  field effect, **782, 785**
  junction, **75**
Transition:
  brittle-to-ductile, 650
  ductile-to-brittle, 48
  ferromagnetic, 94
  first-order phase, **795**
  second-order phase, **795**
  temperature, glass, **259**
Translation operation, **197**
Transmission electron microscopy, **230**
Transmittance, **845**
Trap, **771**
Tresca yield criteria, **643**

Tridymite, 257, 342
  crystal structure, 258
Trouton's rule, **128**
TTT-curve, **669**
T-2 tanker, 47
Tungsten carbide, sintered, 252
Tunnel diode, **786**
Tunnel effect, **752**
Tunneling, electron, **751**
Twin, coherent, **283**
Twinning:
  deformation, **559**
  direction, **283**
  plane, **283, 559**
    examples, 283
Twin planes and directions, data, 559
Twist boundary, **286**
Type I superconductor, **109**
Type II superconductor, **109**

U

Ultimate shear strength, theory, 550
Ultimate tensile strength, theory, 557, 562
Uniaxial solidification, 358
Unit cell, **196**
Unit tangent vector to the dislocation line, **573**
Units, magnetic, 864
Universal constants (*see frontispiece*)
Unstable, elastically, **499**
Upper yield point, **604**

V

Vacancies, **268**
Vacancy:
  concentrations in metals in equilibrium, 414
  diffusion, 457
  energy of formation, **269**
    data, 414
  energy of motion, 441
    data, 458
  sources and sinks, 459
Vacuum, permeability coefficient of a, **89**

Valence band, **738**
Van der Waals:
  b, **438**
  bonds, **183**
  equation of state, **438**
Vaporization, heat of, **127, 343**
Variance, **341**
Varistor, **782**
Vector:
  Burger's, **276**
  lattice, **197**
Vegard's law, **245**
Velocity:
  distribution in gases, 392
  drift, **718**
  group, **736**
  sound, 28
Vibrational:
  energy of atoms, 425
  entropy, 430
  frequency of atoms, 429
Vibration fatigue, 44
Vibrations:
  damped, 31
  forced, 32
  free, 29
Vinyl acetate, 305
Vinyl chloride, 305
Vinyl compounds, **304**
Viscoelastic behavior, 525
Viscosity:
  gases, 510
  liquids, theory, 512
  Newton's equation of, **509**
  solids, 517
Viscosity coefficient, **51**
  crystalline solids, derivation, 517
  data, 52
  dependence on pressure, data, 59
  dependence on temperature, data, 59
  gases, derivation of, 510
  soda glass, data, 514
  units, 52
Void:
  octahedral, **240**
  radius ratio, **241**

Void (*Contd*)
 tetrahedral, **240**
 to atom ratio, 241
Voigt model, **525**
Voltage, back, **709**
Volume:
 fraction rule, **681**
 of formation of a vacancy, 414
 polarizability, **821**
 susceptibility, **866**
 thermal expansion coefficient, **118**
Von Mises criteria for yielding, **642**
V_3Si, 499
Vulcanization, **307**

W

Wall:
 Bloch, **881**
 cell, **603**
 domain, **881**
Wannier exciton, **852**
Water of hydration, 335
Waterproofing, 312
Wave function, **152**
Wave functions, hydrogen-like, **161**
Wave vector, **728**
Westro, 885
Whiskers, **280, 334,** 550
 discovery, 566
 size effect on strength, 680
Wiedemann-Franz ratio, **120**
Work function, **746**
 data, 747
Wurtzite, **250**
Wustite, **274**

X

Xerography, **786**
X-rays:
 diffraction by crystals, 223
 origin of, 162
 small angle scattering, 362

Y

Yield criteria, 642
 maximum shear stress, **643**
 pressure vessel example, 643
 Von Mises, **642**
Yielding, Von Mises criteria for, **642**
Yield point phenomena, **604**
Yield quantum, **747**
Yield strength, dependence on temperature, data, 617
Yield stress:
 relation to dislocation density, 607
 in shear, 19
 in tension, 1, 15
Young's modulus, 8, **22**
 data, 13

Z

Zener diode, **777**
Zener relaxation, **535**
Zeolite, 258
Zinc blende structure, **249**
Zone:
 Brillouin, **731**
 leveling, **372**
 refining, **372**